THE SELECTED PAPERS OF
DENIS NOBLE CBE FRS
A Journey in Physiology Towards Enlightenment

ICP Selected Papers

Published

THE SELECTED PAPERS OF

DENIS NOBLE CBE FRS

A Journey in Physiology Towards Enlightenment

Editors

Denis Noble
University of Oxford, UK

Zhu Chen
Rui Jin Hospital, Shanghai Jiao Tong University School of Medicine, China

Charles Auffray
CNRS, Claude Bernard University, European Institute for Systems Biology &
Medicine, Lyon, France

Eric Werner
University of Oxford, UK

Imperial College Press

ICP

Published by

Imperial College Press
57 Shelton Street
Covent Garden
London WC2H 9HE

Distributed by

World Scientific Publishing Co. Pte. Ltd.
5 Toh Tuck Link, Singapore 596224
USA office: 27 Warren Street, Suite 401-402, Hackensack, NJ 07601
UK office: 57 Shelton Street, Covent Garden, London WC2H 9HE

British Library Cataloguing-in-Publication Data
A catalogue record for this book is available from the British Library.

ICP Selected Papers — Vol. 2
THE SELECTED PAPERS OF DENIS NOBLE CBE FRS
A Journey in Physiology Towards Enlightenment

For photocopying of material in this volume, please pay a copying fee through the Copyright Clearance Center, Inc., 222 Rosewood Drive, Danvers, MA 01923, USA. In this case permission to photocopy is not required from the publisher.

ISBN-13 978-1-84816-842-8
ISBN-10 1-84816-842-X

Printed in Singapore.

Foreword

The Musical Life of a Troubadour of Modern Science and Culture

During most of the past century, biological and medical sciences have made impressive progress in describing the basic processes operating at the molecular and cellular levels. Analytical reductionism has provided the framework for the main stream of research and engineering applications, driving them further away from humanities and culture. As we enter the second decade of the twenty-first century, it is becoming clear that our ability to translate this knowledge into understanding complex biological systems and diseases is much more limited than we would need to develop efficient treatments for diseases. There is growing awareness that the solution will come from a renewed encounter of the biological and medical sciences with both the formal sciences and humanities. Fortunately, during the last half century when the central dogma of molecular biology was dominating the scene, overshadowing the more traditional biological sciences altogether, the spirit of physiological sciences as defined by its founding fathers of the past centuries was kept alive by people such as Denis Noble. We are thus in the privileged position of rejuvenating physiology through systems biology and medicine, leveraging advances made in genomics and bioengineering.

But this is easier said than done. The recipe for doing this successfully is becoming somewhat clearer as one follows the fascinating path taken by Denis Noble as described in this book. First of all, it is a testimony that once a relevant and challenging problem has been identified, such as the normal and pathological functioning of the heart, nothing should distract us from developing the proper tools to resolve it, knowing that it may take a long time — spanning up to several decades. Second, the early achievements of Denis Noble in modelling cardiac impulses have required his acquisition of advanced mathematical and computational skills, which were all too often absent in the biological science curriculum; this calls for reintegration of formal and biological sciences at an early stage of training. Third, Denis Noble has taken clear advantages in learning as much from his failures as from his successes in the development of successive generations of cardiac models. These are now finding practical applications, for example in assessment of drug safety.

This is certainly necessary to open new avenues, but is it sufficient to explore them efficiently, to advance knowledge and practical applications? Chapter after chapter, as we progress in his footsteps, we learn from Denis Noble that what is at stake is no less than the foundation of a new philosophy of nature, and that there is much more required to tackle this grand challenge, including a deeper understanding of the foundations of the scientific method, its philosophical and epistemological bases, and a revisit of basic notions such as causation. To deconstruct the current dominating paradigm of molecular biology and construct the novel framework to sustain systems approaches in biology and medicine, Denis Noble invites us to navigate through metaphorical spaces. We are led to move away from the purely mechanistic views of biological systems that have been prominent in the recent years, and to engage again with the more integrative views which are the signatures of physiology and medicine since their inception. In the process, basic principles are provided for the foundation of a theory of life that will certainly inspire experienced and young researchers alike.

The selected papers of Denis Noble and his magnificent Introduction represent a treasure that many will now navigate to initiate new intellectual, scientific and human adventures. They provide an opportunity to bring together again views of Nature that have long traditions in Western and Eastern cultures, but developed separately to a large extent in the recent period. For this to be possible it has required the steady pursuit of one consistent line of inquiry for over half a century, resisting distraction by dominating streams of research; learning the languages of mathematics, computer science and engineering, challenging established practices in biological education; promoting the results of failures in model predictions, when prominent journals are striving to publish only success stories based on positive findings. Many should be thankful for his mentorship at Balliol College and University of Oxford, and his leadership in the launch of the Physiome Project of the International Union of Physiological Sciences. But it would not have been a successful 'journey to enlightenment' if it had not been illuminated through deep exposures to a rich variety of cultures and languages, shared with many around the world through lectures and musical performances, sometimes wonderfully combined together. Hopefully many more will have a chance to listen to him introducing a lecture on systems biology principles in English, French or Japanese by playing the guitar and singing a poem in Occitan.

Charles Auffray,[1] Zhu Chen[2] and Eric Werner[3]

Afterthought: A Very Personal View of Day-to-Day Conversations with Denis Noble

When I first approached Denis Noble about having a seminar where we look at the foundations of systems biology, I found a very willing partner. We both thought that the area of systems biology was gaining more and more popularity and intensity without much

[1]Principle Investigator, Genexpress Team, CNRS — Institute of Biological Sciences.
[2]Vice-President of Chinese Academy of Sciences (CAS) and current Health Minister, People's Republic of China.
[3]Balliol Graduate Centre, The Computing Laboratory, and Dept. of Physiology, Anatomy and Genetics, University of Oxford, and Cellnomica, Inc. and Oxford Advanced Research Foundation, Inc.

thought given to the basic assumptions, presuppositions and concepts underlying the methods used to model and simulate biological systems. It was and is a Wild West atmosphere fuelled by massive governmental spending. Suddenly everyone was a systems biologist. Physicists, computer scientists, mathematicians of all flavours (OEDs, logic, numerical methods, algebraic methods, Petri nets and many more), linguists and philosophers jumped into this exciting new challenge. Everyone with an intellectual method, formal or informal, wanted to apply it to biology. Biologists were to supply the data and the interlopers would crunch and analyse the data, building a model that could then be used to simulate the biological process in question, leaving the biologist with little more to do than observe if the model did capture biological reality.

With the sequencing of the human genome as well as many other organisms, data was growing at an exponential rate. Bioinformatics was the field that analysed data and used database theory to construct huge distributed databases to contain the newly acquired and exploding information. It soon became clear that data and its analysis was not enough to gain understanding of how biological systems function. Since Watson and Crick's discovery of the structure of DNA and its role in protein production, it was initially thought or hoped that once we have the genes and their proteins we would understand biological systems. This hope was disappointed by the very first step required, namely, to understand proteins. The amino acid sequence, generated by the process of RNA translation, has to fold into a protein. Modelling and simulating this folding process turned out to be extremely difficult and computationally intensive. Denis Noble and I were both at an IBM Blue Gene Symposium in Edinburgh where the atmosphere was palpably pessimistic. The times involved to model protein folding accurately in a liquid environment would take the fastest computers (yet to be built) years of computational processing!

It was there that I first heard Denis Noble make some fundamental, deeper points that rang true and offered hope. Because of his insightful comments, he stood out above all the rest. It was obvious that the bottom up effort to model and understand biological systems by way of DNA to RNA to proteins to cells was doomed to failure. We needed to combine bottom up data with higher level system information to constrain the vast complex search space confronting any molecular-based, bottom up methodology. At the same Blue Gene Symposium, I presented a higher level view of multicellular systems biology where a society of cells is modelled and simulated and where regulatory networks become as important as the protein parts generated by individual genes. Unfortunately, Denis had already left when I gave my talk, else we might have started our discussions even earlier.

As the Balliol College seminar progressed and enough trust was gained between the participants, Denis and I began to express disagreements about basic issues such as: Are there programs in the genome? Which is more important — the cell or the genome? Where does the complexity and information needed to develop an organism lie? We have been debating these and other questions for over three years now and while we have come closer on many issues we still get into interesting debates.

My point here is that while many a professor I have known cannot stand to be contradicted, Denis Noble thrives on differences! This I think is one of the traits that make him such a successful scientist. He is able to integrate opposing views with ease. He consistently treats his opponents with gentleness, politeness and humour, much like a Zen master. This shows the greatness of the man and not just the scientist. Because of these qualities, it has been a privilege and an honour to be able to work and indeed to fight with Denis. And the enjoyment continues; we are still debating in the Balliol Interdisciplinary Institute, which we founded as a result of our seminar.

Those readers who have a difficult time following the technical details of the early papers will find the later papers fascinating, opening wide new vistas. You will get a flavour of the debates we had and are having in our seminar at Oxford on the Conceptual Foundations of Systems Biology.

Eric Werner

Contents

PART 2

Papers Reproduced
The chapters correspond to those in Part 1

Introduction

Journey Towards Enlightenment

With a selection of the papers of
Professor Denis Noble CBE FRS

When Imperial College Press proposed to publish a collection of my scientific papers, I was not at all sure what to think. It was easy enough to give permission, though I admit I was sceptical. But then I learned that some commentaries would be required to set the scene for each set of publications. That was when I suddenly warmed to the idea. The warmth became hot enthusiasm once I started writing. The book, for that is what it has become, just wrote itself in a period of about eight weeks.

The reasons are easy to explain. My field of research is not usually thought to be readily accessible to a general reader. Electrophysiology is the kind of area that most biology and medical students steer clear of. It requires considerable knowledge of the relevant physics and mathematics as well as biology. Moreover, in the case of the heart, it has been highly controversial. My first reaction was therefore to doubt whether I could write the story clearly enough for a general audience.

Controversy: that is what changed my mind. I realised that if, in an autobiographical account, I could bring the controversies out into the open and let people see how scientists deal with them, there would be a story to tell. Possibly even an exciting one. Moreover, I have been involved in and have been the centre of much of that controversy for a period of 50 years. There must be few scientists who have pioneered a field and who are still highly active in that same field half a century later. I have seen heresies become orthodoxies in my own scientific lifetime. That is a rare privilege.

But why did the book almost write itself? That is easy to explain. It is not just the story of controversies and their successful resolutions — it is a journey. As a scientist and a somewhat amateur philosopher, I have been profoundly affected by this journey. The person who started the journey 50 years ago is so foreign to me that, in Chapter 8, I even talk about him in the third person.

So, this book is also the story of a personal transformation. Particularly since the publication of *The Music of Life* in 2006 and its subsequent rapid translation into seven foreign languages, I am sometimes described as one of the pioneers of systems biology. That is correct in the sense that what I was doing 50 years ago in discovering potassium channels in the heart and then building the first mathematical model of cardiac rhythm certainly was a form of systems biology. I was putting together a system of interacting proteins within the context of a cell. Rhythm was an integrative property of that system. But, as I explain in the first chapter of this book, this is not how I would have explained it at the time. Indeed, I nearly lost the opportunity to use the early mainframe computer I needed precisely because I could not explain it, as I would be able to today. The guardians of the machine, on which time was so precious, were highly sceptical both of my ability to do what I was proposing and of whether the proposal itself was coherent.

I was lucky. Not only to have the experimental material to make such a proposal to model heart activity, but also in many other respects. University College London, where I had studied and where I was working 50 years ago, accommodated an oddball with open arms. I was allowed to roam around the maths lectures, the philosophy seminars and much else, while pursuing my PhD in physiology. I was then lucky enough to obtain a post at Oxford where I could not only set up my own laboratory, but also interact with some of the best professional philosophers and mathematicians in the world.

The result is what I call a journey towards enlightenment. I chose that title very deliberately, knowing that it could easily be misunderstood. There are Buddhist overtones, and a certain degree of presumption. How does anyone know that they are enlightened? Notice though that the word is 'towards' not 'to'. I can say that, at least, since I know that I have moved a long way from my starting point — not in terms of my research subject, that has stayed the same — in how I think about biological science in general. I call the state I have reached 'enlightened' because that is precisely how many others have described it. I lecture frequently all over the world on my view of systems biology (as expressed in *The Music of Life*). These lectures have been given to an astonishing variety of audiences, not all of them scientific. Many have been in the humanities, the social sciences, the performing arts and even religious communities. The reaction has been uniformly exciting, as though a cloud has been lifted. Of course, I would think that! Who would not think of themselves as enlightened? Perhaps the first point at which I started to accept the description, at least as an approach towards a goal, was after a debate in Paris where I had lectured to a congress of psychoanalysts. The philosopher in that debate, Clotilde Leguil, wrote a quite remarkable critique both of the book and my lecture. It is on the French translation page of my website, http://musicoflife.co.uk/pdfs/Clotilde%20Leguil.pdf.

Buddhist overtones? Those are welcome and also deliberate. As I explain in Chapter 9, I do not think of myself as particularly religious, though many of the ideas I have expressed in my recent work seem to have resonated with religious thinkers, even to the extent of drawing me into a debate in St Paul's Cathedral, and with HH the Dalai Lama during a visit to Oxford. Anyway, I don't think of the tradition I feel closest to, which is the Buddhist tradition, as being, itself, a religion, at least not in the Western sense. It is more a philosophy of man. The acknowledgement of Buddhist ideas simply came naturally as I was writing the last two chapters of *The Music of Life*. They flow directly from the relativistic interpretation that I give to systems biology. Readers who wish to know more about this aspect of my thinking will find the relevant material in Chapter 9, where I explain how *The Music of Life* came to be written.

One of the reviewers of my work on Amazon wrote that I should be 'chained to a desk … and forced to write more books' (Hayward, 2009). Well, I suppose I was 'forced' to write this one. Sometimes, the prisoner can be willing.

References

Leguil, C. (2008) http://musicoflife.co.uk/pdfs/Clotilde%20Leguil.pdf.

Hayward, Matthew (2009) http://musicoflife.co.uk/reviews.html.

Acknowledgements

I would like first to thank all those who have worked with me over a period of fifty years, starting with my supervisor, Otto Hutter, who did so much to transmit his enthusiasm for physiology, and for the heart in particular, when he welcomed me as a graduate student at UCL in 1958. He, and many others referred to in this book, also kindly checked my memory of the events in the story. Their names are in the relevant chapters and as co-authors on the papers reproduced in Part 2. Many of them kindly lent photos, some of which are reproduced here. Some also checked my memory on various chapters: Stephen Bergman, Dick Tsien, Hilary Brown, Susan Noble, Penny Noble, Dario DiFrancesco, Don Hilgemann, Raimond Winslow, Peter Hunter, Junko Kimura, Wayne Giles, Yung Earm, SungHee Kim, Anthony Spindler, Alan Montefiore, Peter Kohl and Christoph Denoth.

The idea for this book came from the editors, Charles Auffray, Zhu Chen and Eric Werner. I think they were keen on the idea that one of the examples of systems biology in action should be recorded in print. But I am not sure that they anticipated that a story would flow from my writing quite as quickly or extensively as it did. I would also like to thank Laurent Chaminade and Jacqueline Downs at Imperial College Press for excellent collaboration, particularly in the aftermath of the events described in the postscript. They kept the show on the road while I recovered.

PART 1

Chapter 1

Discovery of Potassium Channels and the First Heart Cell Model

1.1. Introduction

University College London (UCL) was established in the Bloomsbury area of London in 1826. The impressive central campus boasts a great classical portico resembling that of the National Gallery. That is not surprising, since it is by the same architect, William Wilkins. The campus is surrounded by elegant Georgian squares, each with gardens in the centre. The town houses around each square are all built to the same successful design, with three main floors where the spacious living rooms of the original occupants would have been. Above these, there are small attic rooms; below them, a basement, half underground and usually reached by a separate small set of steps to the side of the main entrance. A century earlier, these would have been used by the servants. By the 1950s, when I studied at UCL, most of these houses had been converted to other uses. Some had become residential halls for students; others were used by university departments overflowing from the crowded central campus.

In the early 1960s, as a graduate student and then a young lecturer, my life revolved around two of these Georgian houses. In one of them, I had a flat in the attic. I had become the vice-warden of a student residence, Connaught Hall. In the other, I helped to wear down the already-worn stone steps to the basement. That basement housed a newly created twentieth-century god. It was heavily guarded by its own set of high priests. And, just as prayers are submitted on paper messages in many temples around the world, it was fed with supplications, often very long ones, written on paper. Only the high priests and a select band of supplicants could read these messages. After some time, often hours, of thought the god would reply with more paper messages. Einstein would have been pleased to know that the 'mind of god' spoke in the language of mathematics.

1.2. The Mercury Computer

This god was an early electronic valve computer, programmed in gibberish code punched out in holes on rolls of paper tape. When these machines were first built in the 1940s, they were extremely expensive, and very rare. So rare that the then chairman of IBM, Thomas Watson, estimated the world market to be around five! They not only ate paper tape, they

Figure 1.1   The Ferranti Mercury Computer, circa 1960.

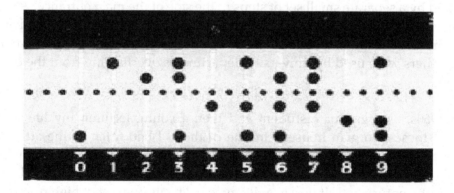

Figure 1.2   The paper-tape code used by Mercury. Each row of up to five punched holes corresponded to a number.

were also power hungry. The 'priests' were of course a new generation of mathematicians and engineers called computer scientists.

Now that the world market in computers runs to billions it is hard to imagine the days when a computer was such a rarity, and so lacking in function. There was no screen, no graphics, no windows, very little memory, and not even Fortran to help the user — the programming was done in a mixture of machine code and a primitive structure called autocode. The keyboards that we used were those of teleprinters that converted type into patterns of holes on rolls of paper tape to be fed to the computer via a tape reader. The machines were, in effect, dedicated calculators, used only for heavy numerical work. It took hours to produce a result that, today, can be achieved in the blinking of an eye on a simple laptop computer. The speed was around 10^4 flops[1] per second. Today we can reach 10^{16} on the fastest computers. A thousand billion operations can now be done in the time it took to do just one.

Time on the computer was precious. There was severe competition between the numerical analysts, the crystallographers and particle physicists, all beseeching the guardians of the machine to grant them time. The primitive coding and functionality also created a serious barrier to entry. A user not only had to understand the mathematics he was using or developing, he also had to master the difficult code. User-friendliness was an unknown quality in those days.

I believe I was the only biologist in the whole university to have dared to ask for time on Mercury. Yet, the idea that mathematics could be used in biology was an old one. William Harvey used careful calculations in his demonstration of the circulation of the blood (see Auffray and Noble, 2009). In 1865, Claude Bernard insisted that 'the application of mathematics to natural phenomena is the aim of all science, because the expression of the laws of phenomena should always be mathematical' (see Noble, 2008: p. 17). But exceedingly few biologists had followed his vision in the 100 years since his book was published.

My credentials to do so looked hopeless. I had stopped studying mathematics at the age of 16, when the English school system forced an absurd choice between maths and biology. Not only did I not have a degree in maths, I didn't even have the school qualification, the A level, that would have taught me differential equations and the maths of integration. Little wonder therefore that my request was declined. They clearly feared that I would be wasting scarce computer time. Yet I knew that I needed to use the machine.

1.3. Experimental Discoveries

My need to use Mercury began with a major experimental discovery, first described in the *Nature* paper (Hutter and Noble, 1960)[2] with my supervisor, Otto Hutter. Following the lead of Hodgkin and Huxley's (1952) work on the nerve impulse, we decided to look in the heart muscle for the potassium channel current responsible for repolarisation at the end of each electrical impulse. As an undergraduate student in 1958 I had been enormously impressed with the Hodgkin–Huxley equations. Here, at last, was physiological analysis that could rival the use of mathematics in physics: precise experimental measurements of the relevant nerve electrical parameters, accurate fitting of mechanistic equations to the data, followed by integration of those equations to produce a complete explanation of the nerve impulse and its conduction. I was far from alone in being impressed. Just five years later, in 1963, Hodgkin and Huxley won a Nobel Prize for their achievement.

---

[1]flop = floating point operation. The number of flops per second is often used to measure computer performance.
[2]Reprinted in Part 2, page 151.

Figure 1.3   Denis Noble and Otto Hutter at the IUPS Congress in Glasgow 1993.

I naturally wondered how their work could be applied to the heart. There was a big puzzle to be solved. In the case of nerve, the permeability of the membrane[3] increases during the whole duration of the impulse. This is what we would expect, since it is generated by the opening, first of sodium ion channels, and then of potassium ion channels. But, in heart cells, the reverse happens. After an initial increase in permeability, the permeability rapidly falls. Even worse, in the experiments of Silvio Weidmann (1951), who measured the membrane resistance (the inverse of permeability) throughout the action potential, the permeability apparently continues to *decrease* throughout the long, slow repolarisation. How could this be if it required the opening of potassium channels to bring that repolarisation about?

My experiments with Otto Hutter explained the low permeability. We measured the electrical current flowing through the cell membranes in conditions where the great majority of the current is carried by potassium ions. The result was a surprise. Contrary to the situation in nerve, the potassium ion channels that are open in the resting state immediately close when the membrane is depolarised. The graphical convention for current–voltage relations was different in 1960, so in Figure 1.4 I have re-plotted those experiments using the modern convention of making the voltage the abscissa, and the current the ordinate. Our results were interpreted to show the presence of two types of potassium channel. The first channel, naturally called i_{K1}, closes when the membrane is depolarised. In fact, its permeability becomes almost zero. This would explain the low permeability during the long

---

[3]'Permeability' is a measure of the speed with which substances (such as ions) cross the cell membrane. When these are electrically charged, the movement creates an electric current. The permeability can then be measured as an electrical conductance. The inverse measurement is an electrical resistance. 'Polarization' refers to the electrical potential across the cell membrane created by these movements of ions. At rest, the inside of the cell is negative. 'Depolarisation' refers to the change towards a positive potential. 'Repolarisation' is the recovery of the resting negative potential. Many of the technical terms used in this book are explained in the Glossary.

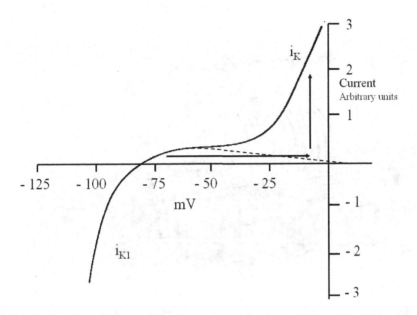

Figure 1.4   Re-drawing of the Hutter–Noble (1960) results using the modern convention that the voltage is displayed on the abscissa. The curve has also been corrected numerically for the cable properties of the fibres on which the results were obtained. The horizontal arrow indicates the first effect of membrane depolarisation, which is to rapidly reduce the potassium permeability towards zero (extrapolated dotted line). The vertical arrow indicates the slow onset of i_K.

cardiac action potential. But how, then, does repolarisation occur? Our results explained the repolarisation process since the second kind of potassium channel i_K (we called it i_{K2}), slowly activates with time, eventually overcoming the depolarising currents.

1.4.   Building the Computer Model

So far, so good. These potassium channel characteristics can explain a lot of the properties of the cardiac impulse. But how can they possibly explain the slow decrease in permeability measured in Weidmann's experiments? I could see the glimmering of an explanation in the properties of the sodium channel in Hodgkin and Huxley's work. The sodium channels have two kinds of gates: one that opens the channel and one that closes it. There is a range of membrane potentials over which a small fraction of both gates are open even after a long period of time. Could the potassium channel behaviour be taking the membrane potential slowly into this region and so producing an apparent increase in membrane resistance when, in fact, the net membrane permeability is increasing? This is the kind of counterintuitive explanation of a phenomenon that cries out for a calculation to demonstrate that it works quantitatively.

Frustrated by the refusal to be allocated time on Mercury, I started calculating by hand using a Brunsviga calculator that I found in the physiology laboratory. I got as far as computing the rapid upstroke of the action potential (Noble, 1962a: Figure 7). But then I did

Figure 1.5   Brunsviga hand calculator, model 20. Numbers were entered by moving levers and an arithmetical operation was implemented by turning the main handle. This is the machine on which Andrew Huxley performed the computations of the Hodgkin–Huxley 1952 paper. It took six months. He developed a very strong right hand. I think that is why he once beat me easily in a game of table tennis.

quite a different calculation. How long would it take me to go all the way through the cardiac action potential using a hand calculator? In 1952 it took Andrew Huxley 8 hours on this machine to compute 8 msec of a nerve impulse. The cardiac action potential lasts about 500 msec. So it would require several months to do a single calculation. Clearly, although working on the hand machine taught me a lot about the process of integration (a fact that subsequently stood me in good stead when it came to programming my own fast integration routines), it was completely impractical for my project.

But, ringing in my ears were the biting comments of the guardians of Mercury: 'You don't know enough mathematics and you don't even know how to program!' In their position I would have made exactly the same judgement. I did the only thing possible, which was to sign myself up for a course that the mathematicians were giving to the engineering students, and I bought a book (I still have it) on how to program Mercury. The maths lectures were nearly a disaster. The lecturer, Dr Few, was excellent, and he kindly agreed to mark my assignments. But I was so far behind in mathematical knowledge that I couldn't follow anything he said during the first three or four lectures. Matrices, Bessel functions, and much else — they were just slavishly copied down as I panicked and wondered whether to give up. What came to my rescue was an innate talent. Although I had been forced to abandon maths at the age of 16, up till then I was nearly always top of the class. I must have had some natural gift for it. Slowly, just like relearning a language after years of non-use, those instincts came back. I still have the assignments marked by Dr Few and his astonished comments that, after such a faltering start, I had moved on to getting full marks.

Armed with this knowledge I then returned to the guardians of Mercury, carrying my book of programming and some examples that I hoped would convince them that I was ready.

They were more sympathetic this time, so the discussion moved on to what I actually wanted to do. What, they asked, was my numerical problem, and how did I propose to solve it? Naïvely, I sketched out on a piece of paper the cyclical variations in electrical potential recorded experimentally, showed them the equations I had fitted to my experimental results and said that I was hoping that they would generate what I had recorded experimentally. A single question stopped me in my tracks. 'Where, Mr Noble, is the oscillator in your equations?'

Here, a word of explanation is required on the vexed question of scientific strategy. Twentieth-century biological science was largely ruled by a naïve reductionist strategy. I know because I was part of it. All we needed to do was to characterise the components of the system we were studying. The rest would follow since, after all, it is 'just a bunch of molecules anyway'. That was my mindset at the time, too. My reply should have been: 'The oscillator is an inherent property of the system, not of any of the individual components, so it doesn't make any sense for the equations to include explicitly the oscillation it is seeking to explain. That would be an empty hypothesis, not even open to the critical criterion of scientific sense, the ability to be falsified.' Instead, I continued to sketch on the paper, as well as use some hand waving to try and indicate the cycle of activity that I thought could happen. It was all just qualitative speculation. Even so, they were convinced to give me the two hours per day that they thought would be necessary for my project to be developed. Did I say 'per day'? I should have said 'per night', for I was offered the worst time slot: 2–4 a.m.

I now think that was the best time they could have offered. I could work the night on Mercury and then go to the slaughterhouse at 5 a.m. to gather the fresh sheep hearts that would form the basis of the experiments during the day. Often enough those experiments did not finish until midnight, by which time it was necessary to have a cup of coffee and write the next modifications of the paper tapes to be fed into Mercury by 2 a.m. That cycle would continue for two or three days before I crashed out to catch up on long overdue sleep. As I explain in *The Music of Life* (2006: Chapter 6), that experience deeply disturbed my circadian rhythm which can wander almost freely away from the norm, with the convenient result that I can travel, east or west, with minimal jet lag.

The absence of a screen, graphics, windows, and all the features that make modern computers so user-friendly, forced me to consider how to monitor the progress of my computations. To wait for two hours just to find that a simple error or a bad choice of a parameter had made the result useless was clearly a waste of time. But I found that this god could speak! Mercury had a loudspeaker and there was an instruction that sent a pulse to the loudspeaker. It didn't take long to realise that by putting this instruction into a repetitive loop it was possible to generate notes of any duration or pitch. So I introduced various

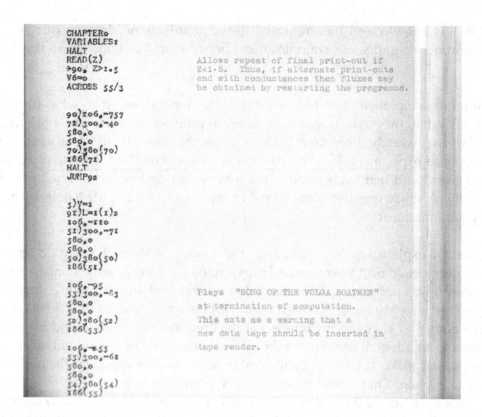

Figure 1.6   Section of code for Mercury playing a melody. The lines, playing 'Song of the Volga Boatmen', were executed at the termination of each computation and so gave a warning to feed more data to the computer. Source: Noble, 1961.

bells and whistles to inform me of the progress of the computation. There was even an error routine that played the melody 'Oh dear, what can the matter be?' I was initially told that this was wasting computer time until I explained that this was precisely what enabled me to save computer time. So, I eventually made my god hum and whistle its way through the night. Once the paper tape had been fed into the computer I could turn to my maths assignments confident that Mercury would soon tell me if something had not worked. When all was well there would be a warbling of hunting sounds and chants, breaking out into the slow 'Song of the Volga Boatmen' to warn me to be ready with the next tape of data.

1.5.   The Model Works

The crazy schedule, including the maths and the experiments during the day, worked. A few months later, the two *Nature* publications appeared (Hutter and Noble, 1960; Noble, 1960), to be followed two years later with the full paper in the *Journal of Physiology* (Noble, 1962a)[4] Rhythm did indeed emerge as an integrative property of the interactions of the components of the system, a fundamental goal of any form of systems biology (Figure 1.7).

---

[4]These papers are reprinted in Part 2, pages 151 and 154.

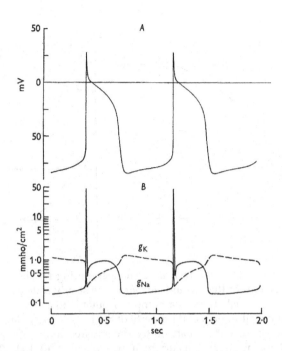

Figure 1.7 Electrical potential changes (A) and sodium and potassium conductance changes (B) computed from the first biophysically detailed model of cardiac cells. Two cycles of activity are shown. The conductances are plotted on a logarithmic scale to accommodate the large changes in sodium conductance. Note the persistent level of sodium conductance during the plateau of the action potential, which is about 2% of the peak conductance. Note also the rapid fall in potassium conductance at the beginning of the action potential. This is attributable to the properties of the inward rectifier i_{K1}, and it helps to maintain the long duration of the action potential, and in energy conservation, by greatly reducing the ionic exchanges involved.

The model also correctly explained Weidmann's experimental result on the conductance changes occurring during each cycle of activity (Noble, 1962a: Figure 10). The delayed K current, i_K, does take the membrane potential into a region where the sodium current system displays a rapidly changing and even a negative slope resistance (Noble, 1962a: Figure 9), so giving the impression of a decrease in permeability when, in fact, the permeability is increasing. The mechanism is revealed in Figure 1.8.[5]

In fact the reconstruction was remarkably good, quantitatively (Figure 1.9). The computed and experimental results could almost be superimposed on each other. The only difference was that Weidmann used square current pulses. For my computations it was easier to program a sinusoidal current.

The prediction that there is a range of membrane potentials over which a 'hump' or 'window' of steady-state sodium current should flow was eventually demonstrated experimentally in the Oxford laboratory (Attwell *et al.*, 1979).

---

[5]Some general readers might wish to skip this figure and its explanation. It is sufficient to know that the experimental results could be explained by the theory (Figure 1.9).

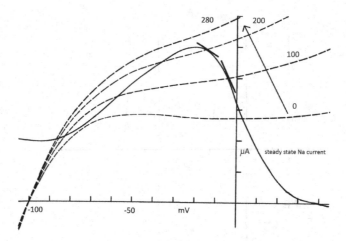

Figure 1.8   Mechanism of the counterintuitive resistance changes during repolarisation in the heart. The continuous curve shows the behaviour of the sodium current in the steady state as a function of membrane potential. Over a wide range of potentials there is a hump attributable to a small fraction of channels that remain open. This is a general and well-understood property of the Hodgkin–Huxley equations. The sodium current is plotted here as a positive current to enable it to be compared to the opposite-flowing potassium current. The interrupted lines show the potassium current at different times (0, 100, 200 and 280 msec) as the i_K channels open (indicated by the arrow). The two thick lines are tangents to the sodium current curve. At 100 msec, the tangent is steep, meaning that a large current change would occur for a given voltage change. This produces a low 'slope resistance'. At 200 msec, the tangent is shallow, so producing a much larger slope resistance. Yet, the net membrane current (conductance) is increasing. Source: Figure adapted from Figure 22 of Noble, 1961.

1.6. Conclusions

In each of these chapters, I will draw conclusions with a balance sheet of pluses and minuses as each stage of the story develops. What can we say on the 1962 model and its experimental basis?

1.6.1. *Pluses*

The model was surprisingly successful in explaining many experimental facts:

1. The counterintuitive resistance measurements I have already referred to.
2. The existence of an oscillator in the interactions.
3. The existence of thresholds for the termination of the action potential (Noble, 1962a: Figure 11) as well as thresholds for its initiation.
4. The classification of the potassium current channels into two types, i_{K1} and i_K. This classification remains correct today and I will elaborate on it and how it has developed in Chapter 2.
5. The paradoxical effect of potassium on action potential duration (Noble, 1965), which has turned out to be important in explaining some of the changes during ischemia (reduced blood flow).

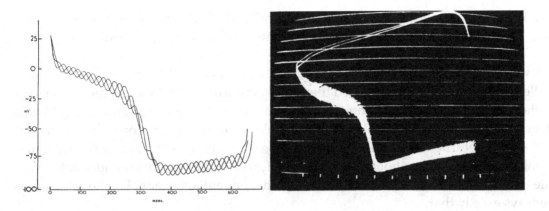

Figure 1.9 Left: computed result using successive sinusoidal current changes to reconstruct the increase in apparent resistance during the slow phase of repolarisation, shown as a progressive increase in the width of the deflections. Right: the experimental result obtained by Silvio Weidmann using square current pulses. These illustrations were taken directly as scans from Noble's 1961 PhD thesis. The experimental result was displayed on an oscilloscope, which explains why the traces are white on a black background — this is a photo of the oscilloscope screen. The early part of the recording was also displayed as a rapid return recording on the oscilloscope. The oscilloscope photo has been scaled to use almost the same scales as the computed result.

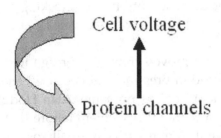

Figure 1.10 The Hodgkin cycle.

The existence of thresholds for repolarisation (item 3 in the list above) was the focus of an argument that raged soon after I published the *Nature* article in 1960.[6] Johnson and Tille (1961), working in Australia, had performed an experiment that seemed to be a knock-down disproof of its existence experimentally in the ventricle, even though Weidmann (1951; 1956) had, equally unambiguously, shown its existence in Purkinje fibres. It looked as though the model was going to be disproved almost as soon as it was published. The existence of these thresholds was critical to the application of any form of Hodgkin–Huxley equations to cardiac muscle. Those equations involve highly nonlinear feedback between the electrical potential and the gating of the ion channels (Figure 1.10). To show that such feedback does not exist would clearly disprove applications of equations of this kind.

Ted Johnson was fully aware of that deep implication. If they were correct, no amount of fine-tuning of the equations would avoid this inconvenient fact. Their paper was there in *Nature* just months after my own. To say that I was worried would be an understate-

---

[6]General readers might wish to skip these paragraphs and jump to Section 1.6.2.

ment. I walked around the UCL quadrangle, and back and forth to my flat in a state of alarm. The ball was firmly in my court. It was either a disaster or a godsend. Which?

What was Johnson and Tille's experiment? It was clever. They had made a double-barrelled microelectrode[7] that enabled them to inject current into the same cell from which the other barrel was recording. They plunged this apparatus into the ventricle and found that, no matter how far they repolarised the membrane during the action potential, the potential simply returned to its original time course once the current pulse was terminated. Not only was there no threshold, there was no effect whatsoever. This is impossible to reconcile with a mechanism dependent on voltage-dependent ion channels. The title of their paper says exactly that.

Somewhere during those anxious walkabouts, I must have approached the building where Dr Few's lectures were held. Ah, Bessel functions![8] I went back to my room to find the slavish scribbles I had made when I didn't understand what he was talking about. There it was in my handwriting. If you perturb a two- or three-dimensional network the disturbance does not decay exponentially from the source, it decays much faster. Could it decay so fast that the effect would be nullified? I coded it all up, ran to my appointment with Mercury and tested the idea. The output was shocking. All I got were unwanted oscillations!

By that time Andrew Huxley had moved from Cambridge to take the chair of physiology at UCL. There is a widespread impression that because I used their equations I must have been trained by one of them. This is not true. Alan Hodgkin examined my thesis in 1961, but I had no interaction with either of them before that. Huxley's reputation in maths was phenomenal so I went to him with my oscillatory output from Mercury and asked for his advice. It was swift. 'You need Bessel functions of an imaginary argument.' *Imaginary* argument? I had no idea what an imaginary argument was, and couldn't even imagine what it *might* be. Yet, it is fundamental to the very nature of mathematics that it does not have to deal only with what we imagine to be reality. Many important mathematical developments use constructs that we find difficult to imagine as objects in the real world. It is easy enough to imagine the numbers from 0 to ∞ arranged on an infinite line. We can even extend this line backwards to −∞ to include negative numbers. But suppose we expand the line of numbers to become a two-dimensional sheet. This can then include a line at 90° to that of our 'real' numbers. To what could this possibly correspond? The answer is fascinating. It can be represented as being the set of 'real' numbers multiplied by an 'imaginary' number which is $\sqrt{-1}$. The sense of 'imaginary' is now obvious. Any real negative number multiplied by itself would always produce a positive number, yet $\sqrt{-1}$ needs to be a number which, if multiplied by itself, would produce a negative number. Not only is this fascinating mathematically, it turns out to produce mathematics that has important practical applications. This is true for Bessel functions.

---

[7]A glass micropipette with a tip diameter less than 1 μm, but with 2 such pipettes stuck together so that they can be inserted into the same cell.

[8]The definition of Bessel functions and other technical terms can be found in the Glossary.

So it was back once again to the notes of Dr Few's lectures and a quick climb up the learning curve on $\sqrt{-1}$ and all its ramifications. Then back to coding Mercury. That night I must have emerged from the basement singing the praises of its god. Using Bessel functions of an imaginary argument for the spread of current flow from the source, and my equations for the membrane channels, the result was exactly what Johnson and Tille had found. I wrote the paper in a frenzy. It was published in the newly established *Biophysical Journal* (Noble, 1962b) and it shows unambiguously that the Bessel function decay is sufficiently rapid to completely linearise the current–voltage relations even in a highly nonlinear system (Noble, 1962b: Figure 3).

That experience was not only absolutely critical to the development of the theory of cardiac electrophysiology, it also introduced me to the theory of electric current flow in excitable cells in general. That became the title of a major book (see Chapter 3). Meanwhile, the 1962 model had survived the most fundamental challenge to its basis. It was fundamental because that challenge could not have been met simply by further developments of the model. It required a mathematical proof that the result was a function of the way in which the current had been applied.

Perhaps the 1962 model was too successful. It found its way rapidly into an edition of Hugh Davson's famous Textbook of General Physiology. By 1970, when that edition of Davson's book was published, the cracks were already appearing. I was somewhat shamefaced when I saw it all in Davson's book. He never asked me about it. I could have told him that calcium channels had already been discovered (by Harald Reuter) and that the model was ready for a major update at least (I will describe that update in Chapter 2). Textbooks have a hard time keeping pace with a rapidly developing field, and this was a field developing at breakneck speed.

1.6.2. *Minuses*

The cracks mentioned above were large:

1. The model made the sodium current equations perform all the roles that were soon to be revealed to be also attributable to calcium channels (Reuter, 1967). In retrospect, this failure can also be seen to anticipate the findings that sodium channels are in fact different in nerve and cardiac cells.
2. There was no representation of changes in ionic concentrations. These subsequently turned out to be critical to one of the next stages of development (see Chapter 5).
3. There was nothing in the model corresponding to the time-dependent pacemaker channels to be discovered later, first under the guise of a new form of i_{K2}, then as the i_f channel. So, pacemaker rhythm in Purkinje tissue (for that was the part of the heart I was working on) was attributed to the wrong mechanism. It was, however, one of the mechanisms that eventually turned out to be correct for the real pacemaker tissue of the heart.

In relation to the failures, it is important to note that, at each stage in this story, we will find that the failures often revealed more than the successes. It should be remembered that it is the function of a scientific hypothesis to be falsifiable. If the model was developed well and on the basis of the best interpretation possible at the time, we should be celebrating the failures as much as the successes. To anticipate a later discussion in this book, there was no circularity in the thinking. Each model was a stage in the continuous iteration between theory and experiment that is also a fundamental characteristic of systems biology.

1.6.3. *Contributions to systems biology*

In each chapter in the current work I will highlight the insights that were gained for a systems approach to biology.

1.6.3.1. *Downward causation*

As Alan Hodgkin had already pointed out in the case of nerve, electrical excitation involving voltage-dependent ion channels can be seen as an example of a cycle (originally called the Hodgkin cycle) in which ion channels contribute charge to determine the cell potential, which in turn influences the channel kinetics themselves. It is characteristic of such cycles that their behaviour requires quantitative analysis since intuition often fails to see the important properties of the whole system.

The feedback from the cell voltage to the ion channels is an example of what, in *The Music of Life*, I call 'downward causation' since it involves an influence of higher levels (in this case the cell as a whole) on lower-level processes, in this case the ion channels.

1.6.3.2. *Energy conservation*

The main integrative insight that the i_{K1} discovery led to was an energy-saving mechanism. The flow of ions down their electrochemical gradients must eventually be reversed by processes that consume energy. A long action potential can be generated without the i_{K1} mechanism, as shown by Fitzhugh (1960), who was working on the Hodgkin–Huxley equations at the same time as I was working on my first models. But the price is substantial in terms of energy demand. By greatly reducing the permeability during the long action potential, only relatively small ionic currents flow during most of the time, so conserving energy. It is easy to see why channels like i_{K1} were selected during the evolutionary process.

1.6.3.3. *Genetic program*

This chapter cannot be concluded without attention being drawn to the way in which the feeding of paper (or magnetic) digital tape into a computer influenced thinking about the role of the genome in biological systems. Monod and Jacob introduced their famous description of the genome as a genetic program using precisely this analogy. Jacob was quite specific about it: 'The program is a model borrowed from electronic computers. It equates the genetic material with the magnetic tape of a computer.' (Jacob, 1982). I think that this analogy is misleading. I will return to this question in Chapter 9.

References

Papers reprinted in Part 2:

Hutter, O.F. and Noble, D. (1960). Rectifying properties of heart muscle, *Nature*, **188**, 495.

Noble, D. (1960). Cardiac action and pacemaker potentials based on the Hodgkin–Huxley equations, *Nature*, **188**, 495–497.

Noble, D. (1962a). A modification of the Hodgkin–Huxley equations applicable to Purkinje fibre action and pacemaker potentials, *Journal of Physiology*, **160**, 317–352.

Others:

Attwell, D., Cohen, I., Eisner, D. *et al.* (1979). The steady state TTX sensitive ('window') sodium current in cardiac Purkinje fibres, *Pflügers Archiv: European Journal of Physiology*, **379**, 137–142.

Auffray, C. and Noble, D. (2009). Conceptual and experimental origins of integrative systems biology in William Harvey's masterpiece on the movement of the heart and the blood in animals, *International Journal of Molecular Sciences*, **10**, 1658–1669.

Davson, H. (1970). *A Textbook of General Physiology*, 6th edition, volume 2, London, Churchill.

Fitzhugh, R. (1960). Thresholds and plateaus in the Hodgkin–Huxley nerve equations, *Journal of General Physiology*, **43**, 867–896.

Hodgkin, A.L. and Huxley, A.F. (1952). A quantitative description of membrane current and its application to conduction and excitation in nerve, *Journal of Physiology*, **117**, 500–544.

Jacob, F. (1982). *The Possible and the Actual*, Pantheon Books, New York.

Johnson, E.A. and Tille, J. (1961). Evidence for independence of voltage of the membrane conductance of rabbit ventricular fibres, *Nature*, **192**, 663.

Noble, D. (1961). *Ion Conductance of Cardiac Muscle*. PhD thesis, University College London.

Noble, D. (1962b). The voltage dependence of the cardiac membrane conductance, *Biophysical Journal*, **2**, 381–393.

Noble, D. (1965). Electrical properties of cardiac muscle attributable to inward-going (anomalous) rectification, *Journal of Cellular and Comparative Physiology*, **66 (Suppl 2)**, 127–136.

Noble, D. (2006). *The Music of Life*, OUP, Oxford.

Noble, D. (2008). Claude Bernard, the first systems biologist, and the future of physiology, *Experimental Physiology*, **93**, 16–26.

Reuter, H. (1967). The dependence of slow inward current in Purkinje fibres on the extracellular calcium concentration, *Journal of Physiology*, **192**, 479–492.

Weidmann, S. (1951). Effect of current flow on the membrane potential of cardiac muscle, *Journal of Physiology*, **115**, 227–236.

Weidmann, S. (1956). *Elektrophysiologie der Herzmuskelfaser*, Huber, Bern.

Chapter 2

Discovery of Multiple Slow Channels

2.1. Introduction

The Rhodes Scholarships bring brilliant graduate students to Oxford University from various parts of the world, including some of the Commonwealth countries. Many of them also come from the United States.[1] In the 1960s the North American students made the journey on the *Queen Elizabeth*, sailing across the Atlantic Ocean to Southampton before the bus journey to Oxford where they were taken to their colleges. The 1966 intake included two men who came to work in my small and already-crowded laboratory: Stephen Bergman from Harvard and Dick Tsien from MIT.

Three years earlier I had taken up the award of a Tutorial Fellowship at Balliol College. Leaving UCL was a heart-rending decision. The department there was stacked full of Nobel Prize winners, including AV Hill, Bernard Katz and Andrew Huxley. UCL was a clear world leader in cell biophysics. Those were the last glory days of cell biophysics, as the focus of the Nobel Prize in Physiology or Medicine shifted firmly towards the spectacular growth of molecular biology.

I had also spent some of my time interacting with philosophers like AJ Ayer (who had already moved from UCL to Oxford) and Stuart Hampshire (still at UCL at that time). My intellectual wings were rapidly spreading way beyond the laboratory. I think that once people have tasted the academic joys of multidisciplinary work, the temptation to go further afield is irresistible.

Oxford had two attractions for me. First, there was the exciting challenge to create a laboratory of cell biophysics and cardiac physiology, neither of which existed at Oxford, despite the fact that the first professor of physiology in Oxford had been John Burdon-Sanderson, the discoverer of the long duration of the cardiac action potential (Burdon-Sanderson and Page, 1883). After his work, the heart and cell biophysics had been neglected at Oxford, which had been dominated by neurophysiology (Sherrington, Liddell, Phillips) and respiration (Douglas, Lloyd, Cunningham).

---

[1]This is not a nostalgic throwback to when the American states were British colonies. Germany is also allocated Rhodes Scholarships.

The second attraction of Oxford was its phenomenal reputation in philosophy. I had already read books and papers by AJ Ayer, Gilbert Ryle, JL Austin, Peter Strawson, Richard Hare and Bernard Williams – some of the seminal Oxford philosophers. Not long after moving to Oxford I was already engaging some of them in debate and publications (Noble, 1966, 1967a, 1967b), including interactions with Anthony Kenny, Charles Taylor and Alan Montefiore. The significance of these interactions and the grounding in professional philosophy that this established will become clear in Chapters 8 and 9 of this book.

2.2. The Nerve Group in the Noble Lab

Let's return to the Rhodes Scholars of the 1960s. Stephen Bergman came to work on the neurophysiology of learning and memory, using insect models. Dick Stein (from MIT, as a Marshall Scholar) had rapidly established work with me in nerve physiology, and we had published substantial papers on nerve excitation (Noble and Stein, 1966) that owed a lot to my experience of the argument on excitation and repolarisation thresholds described in Chapter 1. In fact, Dick Stein also built a computer in the laboratory, designed for neurophysiological work, a phenomenal achievement in those days. Keir Pearson was also working on control of movement. Both of these men eventually took tenured posts in Canada. Stephen Bergman was the last in this line in my lab. He later became an MD and wrote the best-selling novel *The House of God*, on which, 30 years later, I wrote a commentary (Noble, 2008). It has become a classic book and almost mandatory reading for medical students.

2.3. Establishing the Oxford Cardiac Lab

The reason that the neural part of the laboratory I had established with Dick Stein died out after that has a lot to do with what I will call the phenomenon of Dick Tsien. There simply wasn't room for both areas when the cardiac area was expanding to tackle some very difficult and contentious problems. Anyway, the tradition in nerve biophysics was soon taken over in Oxford by Julian Jack (see Chapter 3). The reputation that my laboratory has enjoyed in cardiac electrophysiology ever since was firmly established by the work that Dick Tsien did during his thesis research. Out of that work we published around ten papers together and an enormous book (see Chapter 3) on the mathematics of excitable cells. Not surprisingly, he went on to become a member of the National Academy of Sciences after greatly extending his reputation both in cardiac and neural science following his return to the US.

The field of cardiac electrophysiology was, in any case, being revolutionised by the introduction of the voltage clamp technique. The main criticism of my experimental work leading up to the 1962 model was that it had all been done without control of the membrane potential. Since this was also what controlled the ion channels (the 'downward causation' described in Chapter 1) it was seriously limiting to control the current but not the voltage.

Figure 2.1   The author with Stephen Bergman at Balliol College in the early 1970s before a dinner speech by the former Prime Minister Harold Macmillan. Asked by Stephen how to give a speech, Macmillan replied, 'First, remember that a speech is about just one thing. Second, start with a surprise – don't tell them what they think you will tell them at the beginning.' He reduced everyone to tears as his speech recalled the devastation of the First World War. My father, George, was wounded in that war, in the battles of the Somme. The dinner suit I am wearing on this occasion was made in 1958 by my mother, Ethel, to the last pattern cut by George before he died in 1957 (see Chapter 8). I still have that suit and sometimes wear it – Ethel replaced the silk facing before she died in 2001.

The first successful voltage clamp technique in the heart was achieved in Trautwein's laboratory in Heidelberg (Deck and Trautwein, 1964). My supervisor, Otto Hutter, was also involved in developing the method at the same time (Hecht *et al.*, 1964). These developments occurred while I was busy building the apparatus for my Oxford laboratory (we built virtually all our own electronic equipment in those days), so I was two years behind when the technique was first used in my own lab (McAllister and Noble, 1966, 1967). By then it was already clear that there was a central and challenging puzzle to be resolved. Despite the experimental evidence described in Chapter 1, neither the early Trautwein work (Dudel *et al.*, 1967) nor the Hecht *et al.* experiments showed the delayed potassium current, i_K! This was a serious, even fatal, challenge to the 1962 model, or indeed any model dependent on such mechanisms for repolarisation. Today, the story is so completely accepted

Figure 2.2   Denis Noble with Dick Tsien in Montpellier, 2004.

that it is hard now to realise how difficult it was to establish it. Therefore, I will try here to explain the problems.

Hecht *et al.*'s work can now be seen in retrospect as revealing a completely different ionic current, now called the transient outward current, i_{to}. I suspect that this channel simply masked the onset of i_K. As to why the Trautwein group had difficulty in recording it initially (they eventually did do so), the best explanation I can offer is that, as the work in this chapter shows, there are many different potassium-carrying channels.[2] Dissecting them out was difficult. And, as the McAllister and Noble paper acknowledges (1966: summary item 4), 'in some fibres, no delayed rectification is observed even when the membrane potential is made positive' (see Noble and Tsien, 1968: p. 212). Even so, the early voltage clamp experiments with Eric McAllister (McAllister and Noble, 1966) fully confirmed its existence, and that sodium ions were critical for observing the current changes at very negative potentials. Chapter 5 will reveal the explanation of this surprising but significant finding.

The key to my own lab's research in working out how to detect slow potassium channel currents in voltage clamp experiments came from work with Eric McAllister (also a Rhodes Scholar). We argued that if i_K was responsible for the pacemaker depolarisation in Purkinje fibres, then why not apply voltage control in the relevant voltage range, negative to about −60 mV? This had the additional advantage that we could avoid the technical difficulties of controlling the voltage once the threshold for the fast sodium current had

---

[2]There are over 80 mammalian genes forming templates for subunits of potassium channels. Potassium channels form the largest family of ion channel proteins. Not all of these are expressed in the heart.

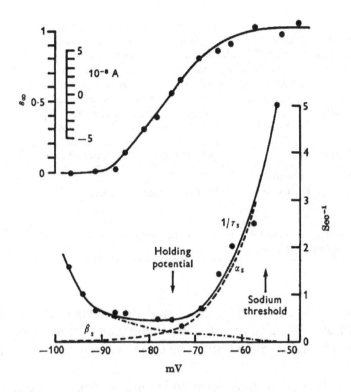

Figure 2.3 Kinetics of the slow ion channel current activated in the pacemaker range of potentials. This was the first complete analysis of the gating kinetics of a cardiac ion channel. In Noble's 1968 paper it was interpreted as a potassium channel activated by depolarisation. Subsequent work (see Chapter 5) showed that it is a combined sodium and potassium channel activated by hyperpolarisation. The activation curve (top figure) should therefore be the other way round. Source: Noble and Tsien, 1968: Figure 4.

been passed. We succeeded in recording the expected slow current changes, but we did not know then that it was a completely different current, and that it was critical that we had kept sodium ions in the bathing solution! The full explanation for this will come in Chapter 5.

2.4. The Arrival of Dick Tsien

This work opened the way for the first major contribution of Dick Tsien. Mine and Tsien's 1968 paper (reproduced in Part 2, page 192) was a very ambitious one since it aimed to produce precisely the same degree of rigour in the experimental measurements of the channel kinetics as Hodgkin and Huxley had achieved in the squid axon. Yet the cardiac fibres on which we were working were very small compared to the size of the squid giant axon (roughly 50 μm compared to up to 1000 μm). Instead of inserting a wire along the inside of the fibre, which was possible for Hodgkin and Huxley, we had to access it with fine glass pipettes called microelectrodes. These differences seriously limited the speed with which we could control the membrane potential. Nevertheless, we succeeded in analysing the slow current changes in the pacemaker range in great detail, including the kinetics (see

Table 2.1. Summary of the channels carrying potassium ions discovered in the early work, and their current designations. For reasons that will be elaborated in Chapter 5, what was originally called i_{K2} became two channel systems activated in completely different voltage ranges (plateau range and pacemaker range). The mechanisms in the plateau range became further subdivided. The existence of the two distinct voltage ranges was fully established in Figure 7 of Noble and Tsien, 1969b.

Channel	original name	current name
Resting K channel	i_{K1}	i_{K1}
Delayed K channels	i_{K2} (plateau range)	i_K
fast	i_{x1}	i_{Kr}
slow	i_{x2}	i_{Ks}
'Pacemaker' channel	i_{K2} (pacemaker range)	i_f
Transient K channel	i_{to}	i_{to}

Figure 2.2; Noble and Tsien, 1968: Figure 4), voltage dependence, the transfer function – conceived as the ion flux in the absence of gating (Noble and Tsien, 1968: Figure 8; Tsien and Noble, 1969) – and the reversal potential. We found several completely new results.

First, the application of transition state theory to Hodgkin–Huxley type channels in Tsien and Noble (1969) in the newly-established *Journal of Membrane Biology*, clarified the nonlinear behaviour of ion channels. It was already clear that the gating of ion channels produces highly nonlinear behaviour. We highlighted the possibility that an ion channel might display nonlinear behaviour even in the absence of the main kinetic gating. This idea later turned out to be correct. One of the major components of i_K shows strong nonlinearity similar to what I originally found for i_{K1}. This is clear in Figure 9 of the Noble and Tsien (1969a) paper. The fast component (what we called i_{x1}, but is now called i_{Kr}) is nonlinear, while the slow component (i_{x2}, later called i_{Ks}) is linear.

The terminology here becomes very confusing because it changed as the story developed. Thus, Table 2.1 summarises the various components and how their names changed in subsequent work.

The second result found showed that the energies involved in the gating of the channel in the pacemaker range were unusual: instead of the temperature dependence (expressed as a Q_{10} – the change in speed for a 10°C change in temperature) falling in the expected range of 2–3 it was around 6 (Noble and Tsien, 1968: Figure 12), one of the highest values known. Expressed as an activation energy, this means that the reaction would never proceed in a biologically relevant time scale without something to help it do so. In transition state theory, that would require negative entropy of activation. The origins of this are still unknown.

Most importantly though, we applied the critical test for a potassium channel current: does the reversal potential (the potential at which the current is zero because the electrical and chemical gradients cancel each other) change with potassium ion concentrations following the Nernst equation? Apparently, the results passed this test very well indeed (Noble and Tsien, 1968: pp. 195–196). Figure 5 of that paper shows how 'clean' this reversal often appeared in the results. We had no idea at that time that Nature had set a fabulous trap for us, and that the Nernst equation test might be faulty.

This work led to the first reinterpretation of the mechanism of the pacemaker depolarisation. Dick Tsien did the computation shown as Figure 13 of the 1968 paper. Together with the papers with McAllister, this clearly established that the pacemaker depolarisation was attributable largely to a channel whose voltage-dependent gating occurred *within the pacemaker range of potentials* (see Figure 1) and not at strongly depolarised levels as in the 1962 model. This was the first stage in revising that model.

Clearly then, there are two very different voltage ranges in which slow, potassium-dependent current changes occur. The 1968 paper took the analysis of the new mechanism in the pacemaker range as far as possible at that time. We also knew why it was not recorded in the 1960 experiments. Its sodium dependence meant that in the sodium-deficient solutions used in the 1960 experiments it was completely absent. I will return to further analysis of this mechanism in Chapter 5.

2.5. Repolarisation Mechanisms

The scene was now set for tackling the channels activated in the depolarised range of potentials, i.e. the equivalent of the original i_K in the 1962 model. As can be seen by the problems encountered by the Trautwein group (Dudel *et al.*, 1967) and by Hecht (Hecht *et al.*, 1964) this was not an easy project.

Before I describe the results of the 1969 paper with Dick Tsien,[3] I should explain some technical problems. Figure 3 in that paper will serve as an example. Part of this figure is reproduced below in Figure 2.4. Each current trace lasted about 20 seconds, and to achieve a steady state before recording the next one, we had to wait for a minute or more (McAllister and Noble, 1966: Figure 9). To collect all the traces (around 60) shown in that figure – which were, in any case, laboriously pasted together from continuous pen recorder traces – took around 4 hours.[4] Keeping two electrodes inside the heart cells for that length of time, and keeping the cells themselves in a stable state for such a long period of time, was extremely difficult. In the conditions used today with patch clamp recording from single cells (rather than multicellular tissue) and with automated computer recording, it is difficult to realise what an achievement this was. I freely acknowledge that this was attributable to Dick Tsien's skill and persistence. I could hardly believe the value of the treasure he brought to be analysed when this experiment was completed. Great rolls of pen-recorder paper contained the data. It took weeks to analyse it properly. This is

---

[3]Reprinted in Part 2, page 222.
[4]Dick Tsien tells me, 40 years later, that he had to miss an important meeting to complete this extraordinary experiment.

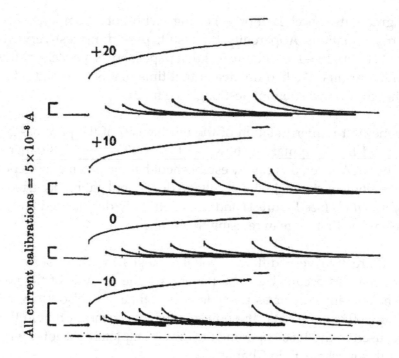

Figure 2.4   Part of the extensive set of recordings of slow potassium channels in the plateau range of potentials. The membrane was held at −30 mV and then depolarised to the potentials indicated for varying periods of time to record both the onset and decay of ionic current as the protein channels opened and closed. The time scale at the bottom is seconds, so each of these recordings required many tens of seconds to occur and even longer periods of rest to allow full recovery. Source: Noble and Tsien, 1969: Figure 3.

also a suitable point to remind readers that, in those days, many journals, like the *Journal of Physiology*, insisted on the alphabetical listing of authors. Dick Tsien was the major author of that and the 1968 paper. In today's conventions, he would have been the first author.

The analysis was lengthy for another reason also. Even a cursory glance at the recordings shows that the changes do not follow a single exponential time course. Figure 2.5 shows this clearly by plotting some of the data as bi-exponential components (another form of analysis that had to be carried out entirely by hand). The results of the analysis were clear, with some minor exceptions at strong and long depolarisation, all the results were consistent with the existence of two types of channel underlying i_K changes, one faster than the other. So, the two main components of i_K were identified. We called them i_{x1} and i_{x2} for reasons that need not concern us here, but they are clearly what later workers (for example, Sanguinetti and Jurkiewicz, 1990) called i_{Kr} and i_{Ks}. The work with Dick Tsien preceded those studies by more than 20 years. The only significant difference between the results is that ours were obtained on multicellular tissue, whereas later recordings used patch clamp on single cells.

Armed with this detailed analysis of the channel kinetics and other properties, we could revisit the calculations of the repolarisation phase. Tsien performed the numerical integra-

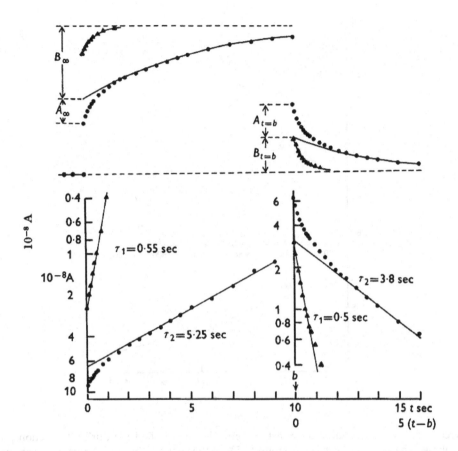

Figure 2.5   Analysis of one of the recordings of slow current changes in Figure 2.4, showing the presence of two components (A and B) each with exponential time courses. These were referred to as x_1 (A) and x_2 (B) in Noble and Tsien (1969) and are clearly what we now call i_{Kr} and i_{Ks}. The lower half of this figure shows the changes in the upper half plotted semilogarithmically to reveal the time constants of the two exponentials. Source: Noble and Tsien, 1969: Figure 4.

tion following a procedure similar to that used in the 1968 paper for the pacemaker depolarisation. The result was impressive, as shown in Figure 2.6. In particular, it explained why the repolarisation threshold exists only during the first part of the repolarisation process. After about 200 ms it disappears.

This analysis was important because it revealed the time courses of two key parameters in repolarisation. The first, V_{SP}, is the potential that would form a stable state were the slow time-dependent changes to cease. This is the potential towards which the cell is tending. During the period for which this point exists there is a kind of stability. We call it a quasi-stable state since it is also, itself, slowly moving. The second, V_{TH}, is the threshold point for initiating all-or-nothing repolarisation. These two points slowly approach each other and they disappear together. After that time, about halfway through the action potential, the repolarisation process becomes less stable. The voltage time course is then best described as a 'free-fall', or to use another terminology favoured by some cardiac electrophysiologists, there is no 'repolarisation reserve' left. The system can no longer resist perturbations

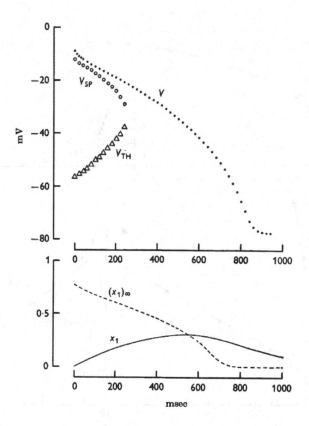

Figure 2.6 Reconstruction of the variation in activation of i_{Kr} (here called i_{x1}) during the action potential repolarisation phase. The results showed that around 35% of the channels become activated during repolarisation. They also reproduced the changes in the (quasi-) stable plateau potential (V_{SP}) and the threshold for all-or-nothing repolarisation (V_{TH}). Source: Noble and Tsien, 1969b: Figure 3.

in either direction. This is an insight that is fundamental to the understanding of cardiac arrhythmias that occur through failure of the repolarisation process. I will return to this issue in Chapters 7 and 8.

2.6. Development of the McAllister–Noble–Tsien Model

The success in reproducing both the pacemaker depolarisation and the plateau phase of the action potential naturally led to the question of whether a complete revision/replacement of the 1962 model could be attempted. But it took some time to achieve this, partly because it was necessary to incorporate equations for several other channel mechanisms that had been discovered elsewhere, such as the calcium channels and the transient outward current. The McAllister–Noble–Tsien model eventually appeared in 1975 (McAllister *et al.*, 1975). As we will see in Chapter 5, it contained a major fault in attributing what was then called i_{K2} to a pure potassium channel, but the work that went into creating it was meticulous and highly accurate. As an indication of the accuracy, Figure 14 from that paper is included here as Figure 2.7. This is a reconstruction of the beautiful experiment of Weidmann (1951) that applied short current pulses during the pacemaker depolarisation.

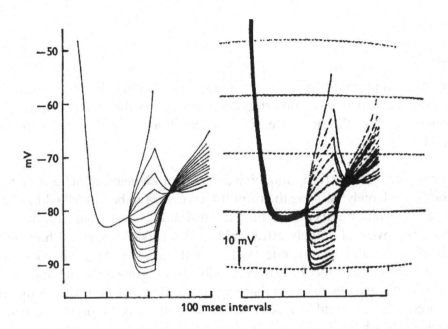

Figure 2.7   Chronotropic effects of short current pulses applied in the pacemaker range of potentials. Left: computed effects of depolarising and hyperpolarising current pulses applied during the pacemaker potential. Note that subthreshold depolarising currents slow the subsequent approach to threshold, whereas hyperpolarising pulses speed the subsequent depolarisation (computed from the MNT model by Hauswirth (1971)). Right: experimental records obtained by Weidmann (1951).

This figure was calculated by Otto Hauswirth, an Austrian pharmacologist working as a postdoctoral fellow in my laboratory. The result is extraordinarily accurate, and it is counterintuitive since one would expect depolarising pulses to shorten the interval to firing, and hyperpolarising pulses to lengthen it. Exactly the reverse applies since the voltage deflections produced by the pulses directly influence the channel gating in this range of potentials. This is the kind of counterintuitive result in cardiac electrophysiology that has frequently required computation to reveal what is happening. The accuracy of the reconstruction also shows that the measurement of the gating characteristics was good. How could this be so accurate when the model itself was not correct on the identification of the ions carrying current through this channel? I will defer the answer to that question to Chapter 5.

I will finish the general description of the articles for this chapter with reference to the paper of Hauswirth *et al.* (1968).[5] This has been chosen for inclusion in Part 2 of the current work because it shows a novel action of a hormone/transmitter. Adrenaline (epinephrine) has been shown to activate ion channels – both i_{Ca} and i_K are upregulated by adrenaline. What this paper shows is that it can also shift the voltage dependence of the gating mechanisms. Tsien subsequently followed this discovery up by showing that the regulation occurs via intracellular cAMP.

---

[5]Reprinted in Part 2, page 190.

2.7. Conclusions

2.7.1. *Pluses*

The discovery of multiple components of the potassium channels involved in repolarisation has major significance for work on drugs and other agents that cause arrhythmia. This work therefore opened up the later extensive collaborations with the pharmaceutical industry (Chapter 4).

The McAllister–Noble–Tsien (1975) model was a major development, which has languished in the shade largely because the tumultuous events to be described in Chapter 5 occurred only a few years after its publication. But it should be recognised for the accuracy and detail that was involved, largely attributable to Dick Tsien's work. I have included Figure 2.7 showing the chronotropic effects of current pulses in the pacemaker range to illustrate that fact. A beautiful theory, however, only requires one ugly fact to destroy it. It was painful to see this one destroyed in that way, even though, as I acknowledge repeatedly in these accounts, we should be celebrating the failures of theories as much as the successes, particularly when those failures reveal major insights.

2.7.2. *Minuses*

Clearly, the big minus in this work is that we missed the warning signs that the channel identified in the pacemaker range of potentials was not a pure potassium ion channel. The warning signs were its sodium dependence (but recall that many channels are controlled by ions other than those they transport), as well as that the fit to the Nernst equation for the reversal potential was both highly accurate (the slope of the line against external potassium concentration was virtually exactly 60 mV) yet always a little too negative. This story will be continued in Chapter 5.

2.8. Contributions to Systems Biology

An essential feature of successful systems biology involving computational models is the iteration between theory and experiment. The McAllister–Noble–Tsien model was an essential stage in that iteration and so laid the groundwork for the next stages in the cycle. In fact, one of those followed quite rapidly since the first model of ventricular muscle cells was that of Beeler and Reuter (1977), which was essentially a modification of the McAllister–Noble–Tsien model. The Beeler–Reuter model was later to be the starting point for the Luo–Rudy models of ventricular cells (Luo and Rudy 1991, 1994a, 1994b). Many of the insights of both of these models have been carried forward into subsequent developments. That is particularly true of the later generations of models developed in my laboratory.

This period of work laid the foundations for the classification of cardiac potassium channels. All the main types of channel were revealed. This advance has tended to be obscured by later work, partly through changes in nomenclature, but also through identification of the genes involved, of which there is now a bewildering array. The functionality, however, lies at the level of the channels themselves and their interactions in the cell as a whole. All the main cardiac functions of these channels were revealed in the work described in this chapter and in Chapters 1 and 5.

References

Papers reprinted in Part 2:

Hauswirth, O., Noble, D. and Tsien, R.W. (1968). Adrenaline: mechanism of action on the pacemaker potential in cardiac Purkinje fibres, *Science*, **162**, 916–917.
Noble, D. and Tsien, R.W. (1968). The kinetics and rectifier properties of the slow potassium current in cardiac Purkinje fibres, *Journal of Physiology*, **195**, 185–214.
Noble, D. and Tsien, R.W. (1969). Outward membrane currents activated in the plateau range of potentials in cardiac Purkinje fibres, *Journal of Physiology*, **200**, 205–231.

Others:

Beeler, G.W. and Reuter, H. (1977). Reconstruction of the action potential of ventricular myocardial fibres, *Journal of Physiology*, **268**, 177–210.
Burdon-Sanderson, J. and Page, F.J.M. (1883). On the electrical phenomena of the excitatory process in the heart of the frog and of the tortoise, as investigated photographically, *Journal of Physiology*, **4**, 327–338.
Deck, K.A. and Trautwein, W. (1964). Ionic currents in cardiac excitation, *Pflügers Archiv.*, **280**, 65–80.
Dudel, J., Peper, K., Rudel, R. *et al.* (1967). The potassium component of membrane current in Purkinje fibres. *Pflügers Archiv: European Journal of Physiology*, **296**, 308–327.
Hecht, H.H., Hutter, O.F. and Lywood. (1964). Voltage current relations of short Purkinje fibres in sodium-deficient solution, *Journal of Physiology*, **138**, 5–7P.
Luo, C.-H. and Rudy, Y. (1991). A model of the ventricular cardiac action potential – depolarisation, repolarisation and their interaction, *Circulation Research*, **68**, 1501–1526.
Luo, C.-H. and Rudy, Y. (1994a). A dynamic model of the cardiac ventricular action potential – simulations of ionic currents and concentration changes, *Circulation Research*, **74**, 1071–1097.
Luo, C.-H. and Rudy, Y. (1994b). A dynamic model of the cardiac ventricular action potential: II. After depolarisations, triggered activity and potentiation, *Circulation Research*, **74**, 1097–1113.
McAllister, R.E. and Noble, D. (1966). The time and voltage dependence of the slow outward current in cardiac Purkinje fibres, *Journal of Physiology*, **186**, 632–662.
McAllister, R.E. and Noble, D. (1967). The effect of subthreshold potentials on the membrane current on cardiac Purkinje fibres, *Journal of Physiology*, **190**, 381–387.
McAllister, R.E., Noble, D. and Tsien, R.W. (1975). Reconstruction of the electrical activity of cardiac Purkinje fibres, *Journal of Physiology*, **251**, 1–59.
Noble, D. (1966). The biological origins of self, *Common Factor*, **4**, 24–31.
Noble, D. (1967a). Charles Taylor on teleological explanation, *Analysis*, **27**, 96–103.
Noble, D. (1967b). The conceptualist view of teleology, *Analysis*, **28**, 62–63.

Noble, D. (2008). 'The birth of the house of God', in Kohn, M. and Donley, C. (eds), *Return to the House of God: Medical Resident Education, 1978–2008*, Kent State University Press, Kent, OH, pp. 1–8.

Noble, D. and Stein, R.B. (1966). The threshold conditions for initiation of action potentials by excitable cells, *Journal of Physiology*, **187**, 129–162.

Noble, D. and Tsien, R.W. (1969a). Outward membrane currents activated in the plateau range of potentials in cardiac Purkinje fibres, *Journal of Physiology*, **200**, 205–231.

Noble, D. and Tsien, R.W. (1969b). Reconstruction of the repolarisation process in cardiac Purkinje fibres based on voltage clamp measurements of the membrane current, *Journal of Physiology*, **200**, 233–254.

Sanguinetti, M.C. and Jurkiewicz, N.K. (1990). Two components of cardiac delayed rectifier K+ current. Differential sensitivity to block by class III antiarrhythmic agents, *Journal of General Physiology*, 96, 195–215.

Tsien, R.W. and Noble, D. (1969). A transition state theory approach to the kinetics of conductance changes in excitable membranes, *Journal of Membrane Biology*, **1**, 248–273.

Weidmann, S. (1951). Effect of current flow on the membrane potential of cardiac muscle, *Journal of Physiology*, **115**, 227–236.

Chapter 3

Analytical Mathematics of Excitable Cells[1]

3.1. Introduction

The mathematical lectures of Dr Few at UCL (Chapter 1) must have left a lasting impression on me. My first reason for attending them was the entirely practical necessity of convincing people to let me have time on the Mercury computer. That strategy was successful. But his lectures also rekindled a deep respect for mathematics itself. There is something immensely satisfactory in finding an analytical solution to a problem. Computational models are necessary, of course, but from a mathematical viewpoint they can be seen as just high-level number crunching. They can create excitement, certainly, for example, when they are used to clarify a counterintuitive result, as in the apparent permeability changes in Chapter 1 and the pacemaker pulse comparison in Chapter 2 (Figure 2.5). But the satisfaction that comes from deriving closed form analytical solutions is much deeper, for a very simple reason: *such solutions are general.* The story of the Bessel function solutions that solved the challenge to the existence of thresholds for repolarisation in Chapter 1 illustrates this point very well indeed. It was not just a particular computation that resolved the issue. That might have been a special case. Rather, it was the general conclusion resulting from the maths itself. Any point excitation in a two- or three-dimensional network will decay so rapidly that even gross nonlinearities in membrane properties get hidden. Point polarisation of a tissue was never used again in current or voltage control of the heart. An elegant method had been destroyed by a single ugly mathematical fact. (Experimentalists sometimes forget that it can be that way round too.)

This kind of generality is also what is required for the future of the systems approach to biology. I will return to that question at the end of this chapter, and again in Chapter 9.

3.2. Current Flow in Excitable Cells

One of the reasons why most biological models are based on computation using differential equations is that this approach can deal with almost any degree of nonlinearity in the system being studied. By contrast, analytical solutions are easiest to obtain in linear

[1]Readers without mathematical skills may choose to skip this chapter.

systems. A good example of this distinction in excitable cell theory is that of cable theory, which is required to study the conduction of the electrical impulse. For voltage ranges in which the system is reasonably linear, which is usually the subthreshold region, analytical solutions in the form of error functions can be obtained for the spread of electrical current (Hodgkin and Rushton, 1946). By contrast, once the excitation threshold is reached the linear solutions are no longer valid. The conduction of the impulse requires nonlinear differential equations to be solved. But must this always be by numerical analysis? Or are there ways in which closed form solutions can be obtained?

Frustrated by the limits of the relatively slow computers of those days, this was the question that I was tackling with Peter McNaughton in the early 1970s. Peter is now a professor and head of the department of pharmacology at Cambridge. The background to this particular project was a collaboration with Julian Jack and Dick Tsien. Since the arrival of Julian Jack in the department in 1964, he and I had organised a Saturday-morning seminar on excitation theory. It was restricted to Saturday mornings since most faculty members were not at all sure that this degree of mathematics was required in a department of physiology, though there were exceptions, notably Brian Lloyd, a respiratory physiologist.[1] We started with work done with Dick Tsien on a book on the mathematics of excitation theory (Jack *et al.*, 1975), planned in the late 1960s, but which was taking us far longer to write than we expected. My interest in the project had reached Chapter 12, the last chapter apart from a mathematical appendix, entitled 'Nonlinear cable theory: Analytical approaches using polynomial models'.

As this chapter title says, the idea was to represent the nonlinearity of the ion channel current by polynomial functions. Any continuous function can be fitted to varying degrees of accuracy with polynomials, the accuracy being determined by the order. For functions of interest in electrophysiology, these start with third-order polynomials, which can generate the kind of current–voltage relation seen in excitable cells, with three crossing points (see the continuous curve in Figure 3.1). The lower (V_A) represents the resting state. The upper one (V_D) represents the voltage towards which excitation carries the potential. The middle one (V_B) is a threshold beyond which excitation produced by a uniform current applied to the membrane is all-or-nothing.

Examples of fitting of such polynomials to experimental data are to be found in Figure 5 of the reprinted paper. A cubic polynomial can be a very close fit to data from squid nerve, investigated in Hodgkin and Huxley's work. A quintic (fifth order) works well for a cardiac Purkinje fibre.

My reaction to the use of polynomial functions in this context was to realise that we had a tool for analytical investigation of thresholds in excitable cells. The derivation of the threshold V_C in Figure 3.1 is an example of this approach. The curve leading up to it is the current that would have to be applied in point excitation of a uniform cable. This

---

[1]Brian Lloyd was the father of mathematical respiratory physiology. His equations for control of respiration by O_2 and CO_2 are still used today.

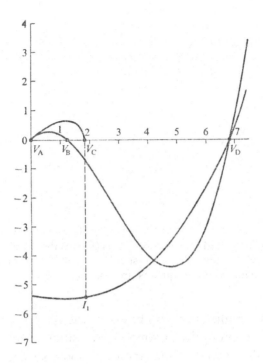

Figure 3.1   Third-order polynomial representation of the idealised current–voltage relation of an excitable cell (see Figure 12.1 and Equation 12.12 in Jack *et al.*, 1975).

requires a larger voltage change since a longer section of the cell needs to be generating inward current. This is called the liminal length (see Noble, 1972, for an analysis of this parameter). There is a simple relationship between the threshold V_C and the form of the current–voltage relation, which is that the integrals of the positive and negative current areas should be equal, as in Figure 3.2, which is from a review of the Hodgkin–Huxley equations that I wrote some years earlier (Noble, 1966), reproduced as Figure 9.1 in Jack *et al.* (1975).

I immediately used this approach to revisit the repolarisation threshold problem discussed in Chapter 1. The results for uniform excitation, point excitation of a cable and point excitation of a sheet are shown in Figure 3.3 below (from Figure 12.2 of Jack *et al.*, 1975). The predictions are clear. Uniform excitation can reveal the true membrane ionic current thresholds, and point excitation of a cable will lead to very negative thresholds (as is the case experimentally), while point excitation of a sheet will find no thresholds at all, as in Johnson and Tille's work (1961: Chapter 1).

The significance of these results on analytical models is that natural forms of excitation of nerves and muscles in the body are not uniform. The geometry of the cells and tissues involved, and of the origins of the excitatory stimuli, are therefore important. For example, one of the mechanisms of arrhythmia to be discussed in Chapters 5 and 7 reflects the existence of a focus where a region of tissue is ischemic or shows a similar pathology. The

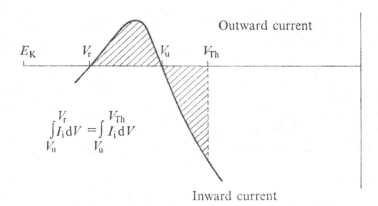

Figure 3.2   Relation of the voltage threshold for cable excitation (V_{Th}) to the membrane current–voltage relation. V_r is the resting potential and V_u is the voltage threshold for excitation by uniform polarisation. V_{Th} is given by the point at which the two integrals are equal. Source: Noble, 1966.

question of how large such a region must be to excite the heart as a whole is precisely a question of geometry and the strengths of sources and sinks of ionic current. So also is the study of how the natural pacemaker, the sinoatrial node, succeeds in reliably exciting the whole heart. Just a few cells would not do the job. Hundreds of thousands of cells are required.

The idea of using polynomials to represent membrane currents was not new. As long ago as 1928, Van der Pol and Van der Mark used a cubic equation (see Equation 11.21 in Jack *et al.*, 1975) to create a model of heart rhythm (Van der Pol and Van der Mark, 1928). Their approach involved allowing the cubic i(V) relation to relax upwards or downwards according to a kinetic equation determined by the membrane potential just as real ionic channels have kinetics that are voltage-dependent. Fitzhugh (1961) developed extensions of this kind of model, notably the Fitzhugh–Nagumo model, which has been widely used in neuroscience. As mentioned in Chapter 1, Fitzhugh was also investigating the ability of the Hodgkin–Huxley equations to generate long action potentials at the same time as I was developing the 1962 model.

Introducing kinetics into models using polynomial functions was therefore already established. Our interest was in whether these kinetics could be made to model more closely the actual Hodgkin–Huxley kinetics, including delayed activation.

Peter McNaughton and I were pouring over some pages of mathematical scribble when Peter Hunter joined us to look at what we were doing. He saw one of the solutions immediately, and that was what led us to invite him to join us on the paper.[2] I did not realise then that this would be the first of many interactions with Peter Hunter, leading eventually to the Physiome Project.

---

[2]Reprinted in Part 2, page 249.

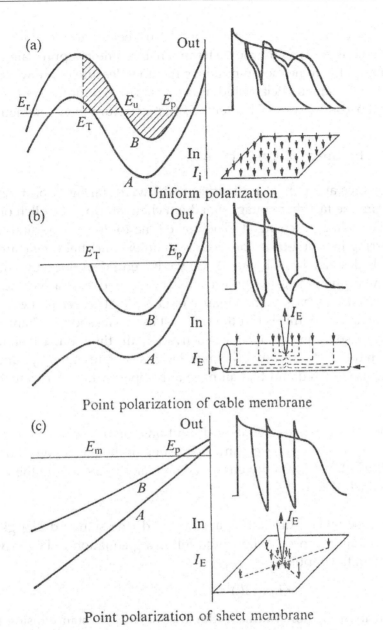

(a) Uniform polarization

(b) Point polarization of cable membrane

(c) Point polarization of sheet membrane

Figure 3.3   Effect of tissue and stimulus geometry on the repolarisation threshold in cardiac muscle. In each case, the early and late i(V) relations are shown on the left, and the expected voltage changes with time on the right. (a) Uniform polarisation: The thresholds always lie between the resting and plateau levels and approach the plateau as repolarisation occurs. (b) Point polarisation of a cable: The early threshold then disappears. The late threshold is shifted strongly in a negative direction. (c) Point polarisation of a two-dimensional sheet: The current–voltage relations become linearised and thresholds disappear. Source: Jack *et al.*, 1975.

As will be clear from the articles reproduced in this book, and many others that are not, I have interacted with many mathematicians, and mathematically inclined engineers and physicists. I find they belong to (at least) two classes: those who have to work hard through the developments and proofs (a class to which I certainly belong) and those to whom maths is a natural language of thought, who see solutions before they even begin to prove them.

The same is true of the language of music, where also I belong to the world of the slow plodders. But belonging even to the slow stream enables one to appreciate the work of those who swim naturally in the medium. I have had the pleasure of knowing and working with some brilliant mathematicians and musicians. The significance of the interaction with musicians will become clear in Chapter 9. Let's now return to the mathematicians.

3.3.   Equations for the Speed of Conduction

The problem Peter McNaughton, Peter Hunter and I were staring at on those sheets of paper stemmed from the fact that voltage-gated ion channels do not switch on or off immediately when the voltage is changed. They take time to do so. The simplest kinetics leads to this occurring in a first-order process, with an exponential time course, like the potassium channels described in Chapter 2. So we have to deal not only with the non-linearity but also with the gating reaction itself. The exponential case could be treated analytically, which in itself was a major advance (an insight also seen at the same time by Alan Hodgkin). But, as I said, this is just the simplest case. The sodium channels in nerve and muscles usually display an initial delay in activation; the time course is sigmoid rather than exponential. In the Hodgkin–Huxley equations this is represented by supposing that more than one gate is involved and that all have to be opened for the conduction of ions to occur.

The question therefore was whether analytical solutions could be obtained even for these more usual cases. The answer was yes, and that is one of the major contributions of the Hunter *et al.* paper of 1975. It was only improved on many years later in the work of Rob Hinch in my lab (Hinch, 2002).

The result for the case where three gates are involved (the standard Hodgkin–Huxley sodium channel activation process) allows the following equation to be derived for the conduction velocity of the impulse:

$$\theta = k_1(a/2R)^{1/2}.(k_2g)^{1/8}.C^{(-5/8)}$$

where θ is the velocity, a the fibre radius, R the longitudinal resistance inside the fibre, g the maximal channel conductance, C the membrane capacitance, and the ks are constants. Some parts of this result had been known for a long time. That the speed of conduction increases as the square root of fibre size and inversely with internal resistance (first term) was well known. So also was the dependence on membrane capacitance (third term). The novelty lies in the extraordinarily small dependence on the ion channel conductance, g. For delayed activation involving three gates (m particles in the Hodgkin–Huxley equations), the velocity increases only as the eighth root of the conductance.

These findings were fully incorporated into Jack *et al.* (1975). In fact, so far as nonlinear cable theory is concerned, this was one of the major contributions of that book, which grew out of an initial desire simply to teach the mathematics of nerve and muscle excitation to students, but then developed into a set of major original contributions to the field.

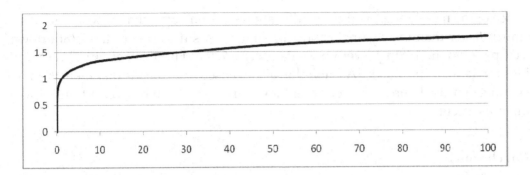

Figure 3.4   Graph showing how an eighth root increases with its argument. Most of the increase occurs at very small values. Between 10 and 100, in increase of an order of magnitude, the eighth root increases by only 33%. This gain in speed could be achieved by just an 80% increase in fibre size.

It remains the standard text, and is now freely available as a PDF file on the website of *The Music of Life*.

The equation for the conduction velocity has many consequences for the biology of the transmission of impulses. Consider, first, the need for an organism to transmit information rapidly. The faster it knows about a challenge in the environment, and the faster it can react to that in its own motion, the more likely it is to survive. One might think that the best solution to this problem is to tell the genome to make more sodium channels and pack them into the cell membrane as much as possible. This would increase g in the equation. But, after an initial gain in speed as channels are added, the gain becomes very small (Figure 3.4). To double the speed would require 256 (2^8) times as many channels, an increase of more than two orders of magnitude! It would be much more effective to simply increase the fibre size, where a doubling of speed can be achieved with just a fourfold increase in radius. This is precisely what has happened in evolution. The most spectacular example in invertebrates is the squid giant axon, responsible for the rapid escape that the squid achieves when it activates its huge mantle muscle to jet-propel itself away from a predator. It also ejects a cloud of ink to confuse the predator. This was the nerve fibre discovered in the squid by JZ Young in 1937 and which formed the basis of the Nobel-prizewinning work of Hodgkin and Huxley in 1952. I have often reflected that the prize might reasonably have been shared with JZ Young whose discovery and understanding of the function of large nerve fibres in invertebrates was so important.

The solution for high-conduction velocity found in vertebrates is to enclose the majority of the length of the nerve in an insulating myelin sheath with only occasional nodes acting as transmission stations with dense sodium channels. Some of the mathematics of conduction in myelinated nerves is dealt with in Chapter 10 of Jack *et al.* (1975).

Not only would packing more sodium channels along the entire fibre length produce only a very modest gain in speed, eventually it actually *reduces* the speed. The reason is that the gating process in channels involves transient movement of charge, which adds to the capacitance of the membrane. As the equation shows, adding capacitance would slow the speed much faster than the channel conductance increases it (compare the 5/8th power for C, compared to the 1/8th power for *g*). Pages 128–129 of Hunter *et al.* (1975) deal with equations for gating capacitance. As Hodgkin also realised, this means that there is an optimal channel density. In order for cells to achieve this, some form of electro-transcription coupling must occur.

3.4. Conclusions

3.4.1. *Pluses*

I believe that the work in Jack *et al.* (1975), together with the developments described in the paper with Hunter *et al.* (1975), form a high watermark in mathematical physiology. Given the difficulty and density of the mathematics, it was reasonable to expect only very modest sales. It surprised me therefore that the book was sufficiently appreciated to have a second edition (in paperback) in 1983. A fact of which the authors are immensely proud is that the corrections amounted to just one bracket in one equation, a tribute to the careful proofreading and the help of some of our students and colleagues. The book remains the standard text 35 years later and I am delighted that OUP gave permission for the PDF to be freely available (http://musicoflife.co.uk/pdfs/Jack_ElectricCurrentFlowInExcitableCells_1975.pdf).

3.4.2. *Minuses*

Like a beached whale, though, the book and the articles appeared during the period when the tsunami of molecular biology was sweeping away almost everything in its path. Classical physiology took a back seat during those years. Moreover, despite the paperback publication, the number of physiologists who could understand the book and the papers on which it was based was small. The contribution of mathematics to physiology is still a vexed question.

3.4.3. *Contributions to systems biology*

It may be, however, that the recent growth of systems biology will help to change that situation. And systems biology itself could benefit from the lessons of our work on the mathematics of excitation. Not everything of value in the quantitative analysis of biological function has to be cranked out as differential equations solved numerically on fast computers. On the contrary, and as the general nature of the solutions to nonlinear problems in excitation theory shows, greater understanding can be achieved when closed form analytical results are obtained. That is possible sometimes even in highly nonlinear systems.

The insight into the existence of an optimal density of ion channels is important. The electro-transcription coupling that must occur to ensure this is also a form of 'downward causation' to which I will return in Chapter 9.

References

Paper reprinted in Part 2:

Hunter, P.J., McNaughton, P.A. and Noble, D. (1975). Analytical models of propagation in excitable cells, *Progress in Biophysics and Molecular Biology*, **30**, 99–144.

Others:

Fitzhugh, R. (1961). Impulses and physiological states in theoretical models of nerve membrane, *Biophysical Journal*, **1**, 445–466.

Hinch, R. (2002). An analytical study of the physiology and pathology of the propagation of cardiac action potentials, *Progress in Biophysics and Molecular Biology*, **78**, 45–81.

Hodgkin, A.L. and Huxley, A.F. (1952). A quantitative description of membrane current and its application to conduction and excitation in nerve, *Journal of Physiology*, **117**, 500–544.

Hodgkin, A.L. and Rushton, W.A.H. (1946). The electrical constants of a crustacean nerve fibre, *Proceedings of the Royal Society B*, **133**, 444–479.

Hunter, P.J., McNaughton, P.A. and Noble, D. (1975). Analytical models of propagation in excitable cells, *Progress in Biophysics and Molecular Biology*, **30**, 99–144.

Jack, J.J.B., Noble, D. and Tsien, R.W. (1975). Electric Current Flow in Excitable Cells, OUP, Oxford.

Johnson, E.A. and Tille, J. (1961). Evidence for independence of voltage of the membrane conductance of rabbit ventricular fibres, *Nature*, **192**, 663.

Noble, D. (1966). Applications of the Hodgkin–Huxley equations to excitable tissues, *Physiological Reviews*, **46**, 1–50.

Noble, D. (1972). The relation of Rushton's 'liminal length' for excitation to the resting and active conductances of excitable cells, *Journal of Physiology*, **226**, 573–591.

Van der Pol, B. and Van der Mark, J. (1928). The heartbeat considered as a relaxation oscillation and an electrical model of the heart, *The Philosophical Magazine Supplement*, **6**, 763–775.

Chapter 4

Insight into the T Wave of the Electrocardiogram

4.1. Historical Background

John Burdon-Sanderson was the first professor of physiology at Oxford University. In moving from UCL to Oxford I was following in his footsteps since he was also at UCL before Oxford. There seems to be a tradition of UCL handing heart physiologists on to Oxford. But the similarities do not stop there. Sanderson (I will come to his double-barrelled name later) was the great grandfather of the cardiac action potential. He was the first to show its very long duration.

To assess the significance and technical achievement of this work, we need some imagination. In the 1880s there were no computers of course (other than Charles Babbage's pioneering work on his 'difference engine' – but that never worked in his lifetime). Moreover, there were no electronic recording systems of any kind. The only instrument available was the capillary electrometer, which uses a mercury column that can move because of changes in surface tension when an electrical signal is applied to it. Sanderson showed great ingenuity. He shone a light beam at the column and the resulting shadow of the column was projected onto photographic apparatus. There were no microelectrodes that could be employed to access the interior of a cell. The connection to the intracellular potential of the heart muscle was achieved by placing one of the electrodes on a damaged region of the heart.

The result was a remarkably accurate recording of the shape and duration of the cardiac action potential. Moreover he demonstrated quite clearly that one of the waves, called the T wave, of the electrocardiogram corresponds to the repolarisation phase of the action potential. The classical paper in which all this is described is that of Burdon-Sanderson and Page (1883). Notice the double-barrelled name.

I naturally wondered what caused him to change his surname. Could the change have been influenced by the alphabetical order of names insisted on in those days by the *Journal of Physiology*? I checked in the early minute books of the Physiological Society. All the early references to him, including his own signature, are to John Sanderson or to John B Sanderson. I suspect therefore that he was also faced with the problem of the alphabetical

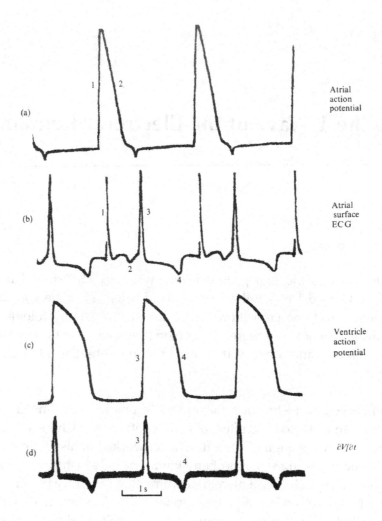

Figure 4.1   Simultaneous recordings of atrial (a) and ventricular (c) action potentials, the electrocardiogram (b) and the first derivative of the ventricular action potential (d) made in a tortoise with the injury method originally used by Burdon-Sanderson and Page (1883), but using a modern pen recorder instead of a capillary electrometer. Source: Figures 1–5 in Noble, 1975.

order. Dick Tsien (see Chapter 2) should have been called Dick Chien, one of the alternative transliterations of his family name, used by another Californian scientist with the same Chinese surname, Shu Chien.

4.2.   Repeat of the Classic Experiment

Burdon-Sanderson and Page's 1883 work was on the tortoise and the frog. In both cases the relationship between the action potential and the surface electrocardiogram was very simple, as shown in Figure 4.1, which was a repeat of his experiment (done in my laboratory), also using injury to gain access to the intracellular potentials. The figure was also used to illustrate my book *The Initiation of the Heartbeat* (Noble, 1975: Figure 1.5).

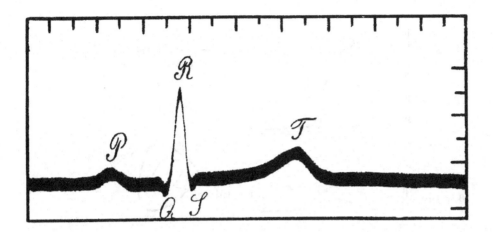

Figure 4.2 Electrocardiogram recorded between the surfaces of the two arms in man using a string galvanometer. Time in 50 ms intervals, potential in 0.5 mV intervals. The labelling of the waves, P, Q, R, S and T is in Einthoven's own hand. Source: Einthoven, 1913.

Notice that each deflection in the electrocardiogram (labelled 1, 2, 3 and 4) is related to a large change in the corresponding action potential. Each action potential generates a rapid positive wave at its beginning and a slower negative wave at its end. This is what we would expect since the current that flows externally should, to a first approximation, be proportional to the first derivative of the action potential.[1] The first derivative of the ventricular action potential is shown in the lowest trace; its resemblance to the ventricular waves (3 and 4) in the electrocardiogram is obvious.

4.3. The Mammalian T Wave is Positive

So far, the story seems simple. One might even imagine that what doctors can do when they see an electrocardiogram is to infer directly what is happening in the cells of the heart. In mammals, that is simply not true. The T wave of the electrocardiogram (wave 4 in Figure 4.1, T in Figure 4.2), is not negative. It is usually positive! The first recordings of electrocardiograms in humans by Waller (1887) showed this fact, and so did the later, more accurate recordings by Einthoven (1913) using the faster string galvanometer that he introduced (Figure 4.2).

Why that is the case has occupied electrophysiologists of the heart for many years. An important clue came from Burdon-Sanderson's own work. He showed that simply heating or cooling one region of the heart, which has the effect of shortening or lengthening the action potentials in that region, can invert the T wave or change its shape. It will now be obvious why in 1984, when the British Heart Foundation offered me one of its prestigious chairs, I chose for it to be named after Burdon-Sanderson, as the most natural tribute to a great predecessor.

---

[1]Strictly speaking, this should be the first derivative of the *spatial* variation. In a uniformly propagating impulse, the time derivative is proportional to the spatial derivative.

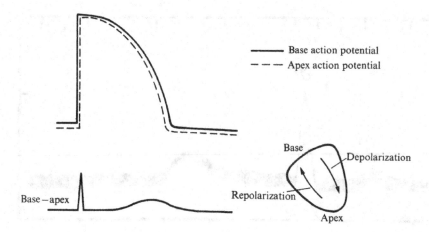

Figure 4.3   Mechanism by which action potential duration differences can give rise to a positive T wave. This diagram shows the difference between the base and apex. The electrocardiogram can then be represented, to a first approximation, as the difference between these, which is always positive. Source: Noble, 1975.

What the heating experiments showed is that differences in action potential duration, however they might be caused, could have profound effects on the T wave. This is why the T wave is perturbed during cardiac ischemia (reduced blood flow). As I showed in the work reviewed in Chapter 1, hyperkalemia (excess extracellular potassium – one of the effects of ischemia) alters the action potential duration, and these changes can explain the T-wave effects during ischemic heart attacks.

But why is the normal T wave positive in mammals? The most natural explanation would be that, even in the normal heart, there are natural variations in action potential duration between different regions. This is the case. There are variations across the ventricular wall and between the base and apex of the ventricle that follow a rule: the regions of the heart tissue that are excited first repolarise last. This is illustrated diagrammatically in Figure 4.3.

4.4.   The 1976 *Nature* Paper[2]

The question that remains to be answered is: What generates the action potential differences that are responsible for the direction and shape of the T wave? The paper reprinted for this chapter (Cohen *et al.*, 1976) reveals that, surprisingly, these differences are not present in quiescent tissue. They develop rapidly during repetitive activity. This was shown by cutting isolated tissue from different regions of the ventricle, allowing the tissue to rest for a period of time (30–60 seconds), and then stimulating it repetitively. The first action potentials in the train are not significantly different. The second and subsequent action potentials all show the difference.

---

[2]Reprinted in Part 2, page 295.

To unravel what was happening we used two interventions, both expected to reduce the activity of the sodium-potassium exchange pump: cooling and application of the drug ouabain, which specifically inhibits the sodium-potassium exchanger. In both cases, the duration differences were reduced or abolished.

What might cause these effects? In the article (Cohen *et al.*, 1976) we speculate that activity-dependent changes in ion concentrations may be involved, but we were not able to test this idea.

How do these findings correlate with what is known to happen to the whole heart *in situ*? There is a clear and simple prediction. If the heart is allowed to rest quiescent for a period, the normal positive T wave should change in form and even reverse. In their exhaustive book on clinical observations, Scherf and Schott (1973) devote a whole section to changes in the shape of the T wave following pauses created by extrasystolic (additional) beats of the human heart. The predicted T-wave inversion is seen in many cases (see Figures 3–22, 3–23 and 3–24 of their book, and the example reproduced as Figure 19 in Noble and Cohen (1978)). It is worth quoting from that paper:

> We can also make one other important prediction from our hypothesis. This is that if the duration gradients are activity-dependent they should be reduced or abolished by periods of quiescence. We have, therefore, looked carefully at the electrocardiographic literature for examples of this. One of the most important papers on this subject was published by Scherf (1944). He investigated the electrocardiogram during beats following the pause that usually occurs following a premature extrasystole. He found that in some, though not in all, cases the T wave is greatly changed in the post-extrasystolic beat. Scherf correctly noted that this phenomenon, which is seen in about 30% of cases, is directly related to the long diastole. Ashman *et al.* (1945) also reached this conclusion. It does not depend on the abnormality of the previous ventricular extrasystole. This is shown by the fact that it also occurs following atrial extrasystoles and following long diastolic intervals in patients with atrial fibrillation and A-V block. (Scherf and Schott, 1973: p. 26)

4.5. Subsequent Developments

This venture into the interpretation of the T wave of the electrocardiogram has had far-reaching consequences. Failures in the process of repolarisation and changes in the T wave are critically important in the pharmaceutical industry. Failure of repolarisation is one of the causes of cardiac arrhythmias, sometimes even fatal. The T wave is one of the easiest measurements that can be made non-invasively to monitor the repolarisation process. Many failures of drugs through side-effects on the heart have shown T-wave changes as the first warning sign. It is therefore used as a biomarker for potential adverse effects on the heart.

These were the reasons why I found myself, in 1997, lodged at the Watergate Hotel in Washington waiting to give evidence at a hearing of the Food and Drug Administration (FDA). My host was the pharmaceutical giant, Hoffmann-La Roche.

A few months earlier, I had been contacted by Jean-Paul Clozel who was leading the preclinical assessment of a new compound, mibefradil (posicor). To understand the significance of this compound, I need to explain that just as potassium channels have been found to come in various types, so also have calcium channels. In the case of the heart, there are two main calcium channels. The L-type channels (L for long-lasting since they switch off slowly) form the largest group and they are responsible for triggering contraction. The T-type channels (T for transient) are activated at lower voltages and so play a role in pacemaker activity (Lei *et al.*, 1998). In the ventricle, though, they play a very minor role. Blocking these channels does little either to the action potential or to contraction. By contrast, they play a major role in arterial smooth muscle, so blocking them can relax these muscles and lower blood pressure. A specific T-type calcium channel blocker could therefore be a useful drug, helping to lower blood pressure without adversely affecting the mechanical power of the heart. Mibefradil had the distinction of being the first compound to have such an effect. Naturally, there was considerable excitement about this development and whole symposia were devoted to T-type channels and the actions of mibefradil (Tsien *et al.*, 1998).

Clozel's problem was that mibefradil had been found to modify the shape and duration of the T wave (Figure 4.4). For the FDA, this was a bad sign. The T wave was spread out over a longer period of time and even broken up into several waves. Yet the action potential was not prolonged by the usual criteria. How could this be? The situation was confusing because the usual criteria for measuring the timing of repolarisation from the T wave could not be applied. How could one measure the peak time of the T wave if it sometimes had more than one peak? There was no way in which any experiment could be conceived at that time that could even begin to answer the question.

Could modelling help? That was the question Jean-Paul put to me. With Rai Winslow and Peter Hunter (see Chapter 7) we were already incorporating cellular models of the heart into tissue and organ models using large parallel computers. We could even reproduce a version of the electrocardiogram. Could we therefore demonstrate that, at the least, the action of mibefradil on the T wave of the electrocardiogram could be understood and that it was not necessarily a bad sign? Rai and I set to work. In fact, as I remember it, Rai set aside around six weeks of time and work on the Johns Hopkins parallel supercomputer to work flat-out over the Christmas and New Year period to do the calculations. They worked. It was possible to show that, given the effects that mibefradil was known to have at the molecular and cellular level, the T-wave changes were, in principle, understandable. In particular, a broadening of the T wave could be consistent with a shortening of the action potential.

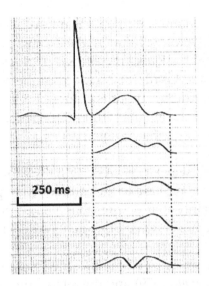

Figure 4.4   Examples of some of the strange T wave patterns produced by the T-channel blocker, mibefradil.

I need to be very clear here about what the modelling was showing. Understanding was the key, not prediction. And that understanding meant that the T-wave changes did not necessarily mean that a hidden electrophysiological effect was lurking somewhere that could itself be arrhythmogenic. It could not be interpreted to mean that mibefradil was entirely safe since we were modelling only the heart, and only one aspect of heart activity. As it turned out, mibefradil did subsequently have to be withdrawn because of metabolic interactions in the liver. One would need to have good models of that organ also to deal with those problems computationally. I will return to this issue in Chapter 7.

Some time before the FDA hearing, Roche organised a meeting in New York to assess the evidence. I was asked to attend at very short notice, but had to decline because of a prior commitment that meant I could not travel to the US in time. That problem was solved very quickly. Roche agreed to fly me out via Concorde, the only time I have had the opportunity to fly in that remarkable aircraft.[3] I left after breakfast in London to arrive before breakfast in New York. I still have a Concorde cut-glass paperweight on my desk as a reminder of that extraordinary experience. People said that you could see the curvature of the earth because Concorde flew at such a high altitude. Certainly, I saw curvature of the horizon, but I am not sure that can be interpreted as sufficient evidence that the earth is a sphere. Even a flat earth would have a horizon and, from any given vantage point, it would seem to meet the sky at the same distance from the observer in all directions, so creating a curved horizon. I think one has to be much higher up to see that the earth really is a sphere.

The FDA hearing was also an extraordinary experience. Unlike in Europe, these hearings in the US are held in full public view. There is almost an element of theatre in the proceedings. Big issues are at stake. The investment in a single drug can represent around

---

[3]Remarkable for speed and cruising height, of course. The fuel consumption was unacceptably high.

$1 billion. That experience has been put to good use in my subsequent interactions with the pharmaceutical industry to try to find a solution to the T-wave problem. The reason is that I was deeply impressed by one of the FDA officials, Ray Lipicky. He was the official who directly interrogated me before the Committee was given the chance (Noble, 1997). But, more relevantly, he had spent enormous amounts of time checking the electrocardiograms himself. I think he came to the conclusion that, from those alone, you could conclude very little. I agree. The T wave is a seriously flawed biomarker.

The reason goes deeper than the T wave itself. As originally shown by Burdon-Sanderson, the T wave reflects the repolarisation phase of the cardiac action potential – also an unreliable biomarker on its own. Certainly, many drugs that delay repolarisation cause arrhythmia, but some do not. It all depends on what other actions they have.

Many of my subsequent publications have been concerned with this question (Noble, 1984; Noble and Cohen, 1978; Taggart *et al.*, 1979) and the related one of developing and assessing drugs that may avoid the problem (Fink *et al.*, 2009; Fink and Noble, 2010; Noble, 2008; Noble and Noble, 2006; Noble and Noble, 2000; Noble and Varghese, 1998; Rodriguez *et al.*, 2006). This project has grown into a European-funded consortium called PreDiCT (http://www.vph-predict.eu/) in collaboration with many pharmaceutical companies to help in developing better biomarkers. Ray Lipicky is one of its scientific advisors, while Roche (the developer of the first T-channel blocker – the focus of that 1997 hearing at the FDA) is one of the members of the consortium.

4.6. Conclusions

4.6.1. *Pluses*

So far as I am aware, the main conclusions of the *Nature* paper on the T wave (Cohen *et al.*, 1976) and the subsequent 1978 article in *Cardiovascular Research* with Ira Cohen (Noble and Cohen, 1978) are both correct and have not been improved on.

4.6.2. *Minuses*

Yet no one seems to have followed up the lead showing the activity-dependence of the T wave.

4.6.3. *Contributions to systems biology*

The subsequent developments leading to the PreDiCT project clearly form an example of application of systems biology to clinical problems and in the pharmaceutical industry. In fact, the systems approach has attracted considerable interest from clinical and pharmaceutical research, as shown by two very recent articles published in a special issue of *Clinical Pharmacology and Therapeutics* (Kohl *et al.*, 2010; Rodriguez *et al.*, 2010).

References

Paper reprinted in Part 2:

Cohen, I., Giles, W.R. and Noble, D. (1976). A cellular basis for the T wave of the electrocardiogram, *Nature*, **262**, 657–661.

Others:

Ashman, R., Ferguson, F.P., Gremillion, A.L. *et al.* (1945). The effect of cycle length changes upon the form and amplitude of the T deflection of the electrocardiogram, *American Journal of Physiology*, **143**, 453.

Burdon-Sanderson, J. and Page, F.J.M. (1883). On the electrical phenomena of the excitatory process in the heart of the frog and of the tortoise, as investigated photographically, *Journal of Physiology*, **4**, 327–338.

Einthoven, W. (1913). Uber die Deutung des Elektrokardiograms, *Pflügers Archiv: European Journal of Physiology*, **149**, 65–86.

Fink, M. and Noble, D. (2010). Pharmacodynamic effects in the cardiovascular system: The modeller's view, *Basic and Clinical Pharmacology and Toxicology*, **106**, 243–249.

Kohl, P., Crampin, E., Quinn, T.A. *et al.* (2010). Systems biology: An approach, *Clinical Pharmacology and Therapeutics*, **88**, 25–33.

Lei, M., Brown, H.F. and Noble, D. (1998). 'What role do T-type calcium channels play in cardiac pacemaker activity?' in Tsien, R.W., Clozel, J.-P. and Nargeot, J. (eds), *Low-voltage-activated T-type Calcium Channels*, Adis International, Chester, pp. 103–109.

Noble, D. (1975). *The Initiation of the Heartbeat*, OUP, Oxford.

Noble, D. (1984). The T wave of the electrocardiogram in relation to intracellular action potentials, *Proceedings of 3rd Einthoven Symposium on Past and Present Cardiology*, Spruyt, Leiden, pp. 34–42.

Noble, D. (1997). Transcript NDA 20-689, *80th Meeting of Cardiovascular and Renal Drugs Advisory Committee*. FDA, Washington.

Noble, D. (2008). Computational models of the heart and their use in assessing the actions of drugs, *Journal of Pharmacological Sciences*, **107**, 107–117.

Noble, D. and Cohen, I. (1978). The interpretation of the T wave of the electrocardiogram, *Cardiovascular Research*, **12**, 13–27.

Noble, D. and Noble, P.J. (2006). Late sodium current in the pathophysiology of cardiovascular disease: Consequences of sodium-calcium overload, *Heart*, **92**, iv1–iv5.

Noble, D. and Varghese, A. (1998). Modeling of sodium-calcium overload arrhythmias and their suppression, *Canadian Journal of Cardiology*, **14**, 97–100.

Noble, P.J. and Noble, D. (2000). Reconstruction of the cellular mechanisms of cardiac arrhythmias triggered by early after-depolarizations, *Japanese Journal of Electrocardiology*, **20 (Suppl 3)**, 15–19.

Rodriguez, B., Burrage, K., Gavaghan, D. *et al.* (2010). Cardiac applications of the systems biology approach to drug development, *Clinical Pharmacology and Therapeutics*, **88**, 130–134.

Rodriguez, B., Trayanova, N. and Noble, D. (2006). Modeling cardiac ischemia, *Annals of the New York Academy of Sciences*, **1080**, 395–414.

Scherf, D. (1944). Alterations in the form of the T waves with changes in the heart rate, *American Heart Journal*, **28**, 332.

Scherf, D. and Schott, A. (1973). *Extrasystoles and Allied Arrhythmias*, Heinemann, London.

Taggart, P., Carruthers, M., Joseph, S. *et al.* (1979). Electrocardiographic changes resembling myocardial ischaemia in asymptomatic men with normal coronary arteriograms, *British Heart Journal*, **41**, 214–225.

Tsien, R.W., Clozel, J.-P. and Nargeot, J. (eds), (1998). *Low-voltage-activated T-type Calcium Channels*, Adis International, Chester, UK.

Waller, A.D. (1887). A demonstration on man of electromotive changes accompanying the heart's beat, *Journal of Physiology*, **8**, 229–234.

Chapter 5

The Surprising Heart[1]

5.1. A Telephone Call from Milan

> He telephoned me from Milano in January 1980 to tell me this result and the
> same night I was able to use a computer program he and I had developed
> together to show that his new interpretation of i_{K2} as a non-specific inward
> current i_f could give a full and accurate theoretical account of the i_{K2} results.
> (Noble, 1984: p. 10)

'He' was Dario DiFrancesco; 'this result' was to cause turmoil for several years. And it was
itself the outcome of turmoil.

5.2. Ion Concentration Changes

The turmoil began with the discovery that some components of electrical current change
recorded from multicellular heart tissue arise from changes in the concentrations of
sodium, potassium and calcium ions rather than just from the opening and closing of protein channels.

In Chapter 2, I referred to the fact that the analysis of the slow potassium ion channels
that we now call i_{Kr} and i_{Ks} had also revealed some 'minor exceptions' in the ionic current traces. These exceptions were even slower components that Dick Tsien and I were
reluctant to attribute directly to ion channel activity. We preferred the idea that they may
have resulted from slow changes in potassium ion concentrations, perhaps in the spaces
between the heart cells. My wife, Susan Noble, worked on these changes in frog atrial
muscle, described first in her 1972 doctoral thesis (Noble, 1972) and in later papers (Brown
et al., 1976a; Brown *et al.*, 1976b; Noble, 1976).

These developments revealed a difficult dilemma. The analysis of the kinetics and voltage-
dependence of ionic channels depends on knowing where the electrical and chemical energy gradients balance each other at what is called the reversal potential. If the relevant ion
concentrations are changing, so also is this potential. How could one possibly separate out

---

[1]This is the longest chapter. Take a deep breath or have a cup of coffee!

changes attributable to that process from the real kinetics of the channels? Some critics of our work said it was impossible and that the results obtained from voltage clamp work on multicellular tissue of the heart were, quite simply, an inextricable mess. In fact, Johnson and Lieberman (1971) wrote a long review saying precisely that, even describing the frog atrial muscle preparation voltage clamped by the double sucrose gap method as an 'insanitary preparation'. The cardiac electrophysiology sessions at the 1977 World Congress of the International Union of Physiological Sciences, in Paris, and a satellite meeting in Poitiers, were dominated by this controversy. It didn't help that I gave my plenary lecture at that Congress in French (this was the last IUPS Congress to be multilingual).[2]

Later, we worked on the equations for this kind of problem. Using the maths of perturbation theory, DiFrancesco and I were able to show that it is possible to dissect out the gating kinetics from other components (DiFrancesco and Noble, 1980b), while Susan Noble and Wayne Giles (Brown, Giles and Noble, 1977) used Provencher's (1976) DISCRETE program to confirm the accuracy of their hand analysis of the multi-exponential changes seen experimentally. These two mathematical analyses formed the background to the work that ensured that I had a computer program ready for that fateful telephone call.

5.3. Discovery of the 'Funny' Current

But before we come to what that telephone call revealed there is another important discovery to note. In the 1970s, we developed a method to study induced pacemaker rhythm in strips of frog atrial muscle (Brown *et al.*, 1976a; Brown *et al.*, 1976b). When Wayne Giles came to Oxford for two years we extended this work to spontaneously beating strips of frog sinus venosus (the natural pacemaker of the heart). When the electrical potential was made very negative we recorded a slowly developing inward current (also observed by Seyama (1976) in the rabbit at about the same time). We called it the 'additional current', without at that time realising its importance.

When Dario DiFrancesco came from Cambridge to join our group in 1977, he was keen to investigate pacemaking in mammalian tissue and to use the so-called 'small preparation' of rabbit sinoatrial node tissue recently pioneered by Aki Noma and Hiroshi Irisawa (Noma and Irisawa, 1976). It was technically challenging to use this preparation so that the tiny 200 micron diameter ball of tissue continued its spontaneous beating and extremely difficult to impale it with two microelectrodes and obtain a uniform control of voltage. Dario was the lead experimenter in this work and his persistence and skill was rewarded by recordings of a remarkable (as it seemed to us at the time) inward current which appeared in the potential range of the pacemaker, precisely the same range as i_{K2} in the

---

[2]An American wag said I must have done that to confuse my critics, who didn't think to get the earphones to listen to the simultaneous translation. It was, of course, done, as I nearly always do in Francophone countries, as a courtesy. Nor did it help that the slide projector light bulb blew in the middle of the lecture so that I had to continue for about 15 minutes without notes or slides. It was the first occasion on which I had lectured in French in free conversational form. A Russian lecturer following me said (in English) that he could now believe that pigs could fly!

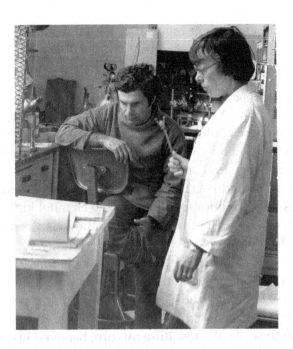

Figure 5.1 'What a funny current!' Dario DiFrancesco and Hilary Brown contemplate an experimental problem in the lab. It might have been the first recording of the 'funny' current.

Purkinje fibres that I had worked on with Dick Tsien. But, unlike i_{k2}, it did not show reversal at the potassium equilibrium potential. Instead it continued to increase even beyond that potential. 'There's that funny current,' we would say. So, i_f, it became (Brown and DiFrancesco, 1980). A paper we published in *Nature* showed that it was reversibly increased by adrenaline (Brown *et al.*, 1979), contributing to the acceleration of heart rhythm by adrenaline.

5.4. The Critical Experiment

On his return to Milan, Dario carried out a critical experiment. Reluctant to accept that i_f in the sinus node and i_{K2} in Purkinje tissue really were two different channels that happened to share a lot of characteristics, he wondered whether ion concentration changes could account for the difference. The Purkinje cells have many i_{K1} channels; the sinus node cells have few or none. Suppose one blocked i_{K1} in the Purkinje tissue? He used barium ions, which were known to do this. The result was dramatic. The reversal potential that had identified the channel as a pure potassium channel simply disappeared! The Purkinje tissue then resembled that of the sinus node.

After he told me this result, I immediately turned to the computer program that we had been using to analyse the effects of ion concentration changes. With a few tweaks it was ready to address an audacious question. The reversal potential results in Purkinje fibres looked clean, and the dependence on potassium ion concentration followed the Nernst equation faithfully. Similar results had also been obtained by Shrier and Clay (1982) using embryonic chick hearts. In fact, their reversal potential recordings were even cleaner than ours.

5.5. Mapping the Two Theories

The audacious question was how could this possibly arise if the channels were really not pure potassium ion channels? Could Nature have set a cruel trap for electrophysiologists? This was an even greater challenge for computational modelling than anything I had tackled so far. The question to be settled was this: Could the current variations attributable to ion concentration changes have kinetics so similar to the channel kinetics that they could cancel each other out cleanly? And not only do that, they also had to do so in a way that created an accurate illusion of a Nernstian reversal. Twenty years on from 1960, the computers were much faster. I didn't need weeks for calculations. By the next morning I was able to tell Dario that it had all worked out like a dream. If one replaced i_{K2} in the McAllister–Noble–Tsien (1975) model with a mixed (sodium and potassium) current, i_f, like that in the sinus node, and included the accumulation and depletion of potassium ions in the spaces between the cells, the resulting mixture behaved just like i_{K2} (see Noble, 1984: Figure 5). Not only did this explain the Nernstian behaviour of the false reversal potential, it also explained why the current disappeared when one removed sodium ions (Chapter 2) and why the apparent reversal was always a few mV negative to the expected reversal. We set to work to analyse this, initially very strange, result mathematically and published it the same year (DiFrancesco and Noble, 1980a). The full details were published two years later (DiFrancesco and Noble, 1982).[3]

If you are not an ion channel electrophysiologist, it is hard to appreciate the full nature of the shock this result produced. The Nernst equation was, after all, the gold standard for identifying the ionic composition of a channel current. Yet we had shown that it could 'lie'. So unbelievable was the result that I had many rounds of correspondence with those who had also identified the i_{K2} mechanism in other species and tissues of the heart. It took some time for the significance to sink in. Was it just a coincidence? If so, why should it occur so widely? In fact, the mathematics we did showed that it was far from a coincidence, and it only required very moderate (10%) changes in ion concentration to produce the effects. Yet again in the work of my laboratory, not only was mathematical modelling necessary to reveal the relevant insights, mathematical *analysis* was also required; the apparently (and initially) obscure work on perturbation theory had borne fruit, just as the use of Bessel functions had done (Chapter 1). Once again, analytical mathematics had complemented computation to produce results of complete generality, which could provide powerful explanations of counterintuitive experimental results.

When Dario and I wrote the 1982 article, we looked for an appropriate piece of literature that could reflect the painful, yet joyful, nature of this journey of discovery, and which would also reflect his and my interests. I had already been exploring the medieval

---

[3]Reprinted in Part 2, page 300.

Troubadour poets, and had found that Dante Alighieri had praised one of them, Arnaud Daniel (circa 1180) in the *Purgatorio* of *La Divina Commedia* (circa 1308–1321). Not only did he praise Arnaud as the best craftsman (*il miglior fabbro*) of poetry in the language of the people, in his case Occitan, he also wrote these verses of his great work in Occitan rather than Italian as his tribute. The verse fully expresses the pain of discovery, yet how easy it is for others to follow where the discoverer has led:

> Ara vos prec, per aquella valor
> que vos guida al som de l'escalina,
> sovenha vos a temps de ma dolor.
> (Purg., XXVI, 140–147)

The lines can be translated as:

> Therefore do I implore you, by that power
> Which guides you to the summit of the stairs,
> Remember my suffering, in the right time.

The stairs could, of course, be the stairs of mount Purgatory in the *Purgatorio*. I rather like to think of them as the difficulty the researcher experiences in blazing a path up Mount Discovery. Others coming later can climb readily what he found difficult. Thirty years later, no one today finds the i_f story difficult at all. But its transformation from the apparently secure i_{K2} story was far from easy.

Figure 5.2 The author (right) with Dario DiFrancesco (centre) after he won the Grand Prix Lefoulon-Delalande in Paris in 2008. On the left is Professor Alain Carpentier, president of the committee that chooses the laureate. The prize was awarded for DiFrancesco's discovery of the i_f channel.

5.6.   The DiFrancesco–Noble (1985) Model

The obvious next step was to develop the McAllister–Noble–Tsien model of 1975 to replace i_{K2} by i_f. But that was much easier said than done. It took a full five years of development. This was because it was not just a matter of replacing one ionic channel mechanism by another. It also involved modelling global ion concentration changes for the first time[4] in an electrophysiological model of the heart, including the intracellular calcium signalling. Dario and I did that because it was necessary to explore fully what we had discovered. We did not know then that we would be creating the seminal model from which virtually all subsequent cardiac cell models would be developed. There are now over a hundred such models for various parts of the heart and many different species (downloadable from www.cellml.org).

Extending biological models is often like tumbling a row of dominoes. Once one has fallen, many others do too. The reason is that all models are necessarily partial representations of reality. The influence of the parts that are not modelled must either be assumed to be negligible or to be represented, invisibly as it were, in the assumed boundary conditions and other fixed parameters of the model. Once one of those boundaries is removed, by extending out to a different boundary, other boundaries become deformed too. In this case, modelling external potassium changes required modelling of the influence of those changes not only on the ion channels already in the model, but also on exchange mechanisms, like Na-K-ATPase (sodium pump) and the Na-Ca exchanger. That, in turn, required the model to extend to modelling internal sodium concentration changes, which in turn required modelling of intracellular calcium changes, which then required modelling of the sarcoplasmic reticulum uptake and release mechanisms. For a year or two it was hard to know where to stop and where to stake out the new boundaries.

It was this process that uncovered a major insight. To simplify the story I will focus on just one of the new mechanisms: the sodium-calcium exchanger. This had been discovered in the heart by Harald Reuter (Reuter and Seitz, 1968), who also discovered cardiac calcium currents (Reuter, 1967) (see Chapter 2). Harald and I complemented each other experimentally, with his focus being on calcium ion transport, while mine was on potassium ion transport. We first met on a course in Homburg, Germany, in 1966 where I was invited to lecture on electrophysiological theory (the basis of what became *Electric Current Flow in Excitable Cells*, Chapter 3). Harald was a student on the course. As such, he certainly benefited from it much more than I did. To go on to discover both the calcium current and the sodium-calcium exchanger was phenomenal.

Until the work with Dario DiFrancesco, Harald and I had kept off each other's territory. But now Dario and I were forced to enter his area. But in doing so, we were also forced to change it. Harald's work strongly suggested that the sodium-calcium exchanger was electrically neutral. To extrude one calcium ion carrying two positive charges, the exchanger

---

[4]Beeler and Reuter (1977) had incorporated changes in calcium concentration in a subspace, but they did not incorporate the ion exchangers and pumps that were required in our work.

would transport two sodium ions into the cell. It was using the sodium gradient to maintain the calcium gradient, and doing so in an assumed 2:1 ratio. Sodium ions carry one charge, so the charge balance would be neutral.

We put this ratio into the model. It didn't work. Instead of driving the intracellular calcium down to around 100 nM, a level at which the mechanics of the cell would be quiescent, it barely achieved 10 times that level, i.e. around 1 μM, at which level the cell would be in a permanent state of contraction. Clearly, during each heartbeat the levels of intracellular calcium must oscillate between these extremes, but not be permanently at the high end of the range. This discovery forced us to abandon the assumption of neutrality. But that was to fly in the face of the best experimental evidence at that time.

Finding some equations to do so was not difficult. Lorin Mullins (1981) had already explored this possibility mathematically. So we adopted his equations and tried out a 4:1 ratio (his favoured one). That also did not work. It drove the calcium levels far too low. Obviously, 3:1 was the right choice. That is what we went for, and subsequent experimental work has fully confirmed that choice.

It is hard to tell a complex story like this one without initially telling a lie to simplify it. I will now correct the lie. It is not true to say that we had *no* experimental leads to favour an electrogenic activity of the exchanger. We had the experimental evidence in our own laboratory in the form of electrical currents recorded by a Japanese student, Junko Kimura, working with Hilary Brown, Susan Noble and Anne Taupignon. They were finding an extra component of inward current in the sinus node of the heart (Brown *et al.*, 1983). Per Arlock in Sweden had found a similar component (Arlock and Noble, 1985). The problem was that these results were controversial and regarded with suspicion. Perhaps they were yet another example of Johnson and Lieberman's (1971) strictures against our work: were they just a set of artefacts? If these results were correct then not all the slow components of inward current in the heart were attributable to calcium channels carrying calcium into the cells to activate contraction, some were also attributable to sodium-calcium exchange driving calcium out of the heart as part of the process of relaxation.

I think there is a general rule for revolutionaries in experimental physiology. You are allowed one revolution at a time. If the substitution of i_f for i_{K2} had been the only major change in the DiFrancesco–Noble model then it might have passed the critics. The evidence for it rapidly became uncontroversial and strong. Since we were forced by the logic of the new modelling to ask people to accept a second revolution at the same time, for which the evidence was much weaker, the critics thought we were going too far out on a limb with our speculations. A set of papers, including the DiFrancesco–Noble model, but also including all the experimental work of Hilary Brown, Junko Kimura, Susan Noble and Anne Taupignon (Brown *et al.*, 1984a; Brown *et al.*, 1984b; Brown *et al.*, 1984c; Noble and Noble, 1984) was sent off to the *Journal of Physiology*. They were rejected as a set, with the modelling paper singled out as unsuitable. This is the reason why that set of papers,

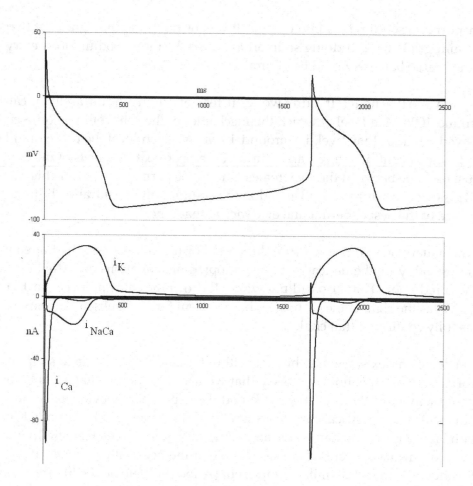

Figure 5.3   The DiFrancesco–Noble 1985 Model. This figure was made using COR and the CellML coding of the model. It shows the temporal relationship between activation of the L-type calcium current and the almost immediately following activation of the sodium-calcium exchange current.

and the DiFrancesco–Noble model itself,[5] were published in journals of the Royal Society instead. We were very reluctant to split up the experimental and theoretical work.

There was, however, a curious twist to the story. Coincidentally, I had been asked to give the Physiological Society's Annual Review Lecture that year. This is the lecture that became *The Surprising Heart* (Noble, 1984),[6] which incidentally not only contains all the new heresies, but fully explains the reasons for them. So, the major results were published in the *Journal of Physiology*, after all, albeit in review form.

Heresies, you may ask? Twenty-five years later it is very hard to see what all the fuss was about. Almost everything in the DiFrancesco–Noble model is now part of the standard story, and it forms the canonical model on which all subsequent ones in many laboratories around the world have been based.

---

[5]Reprinted in Part 2, page 336.
[6]Reprinted in Part 2, page 439.

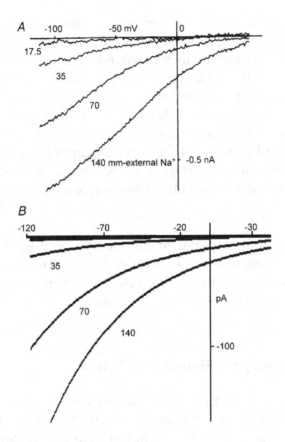

Figure 5.4 Comparison between the experimental results on the sodium-calcium exchange current (Kimura *et al.*, 1987) with those given by the equations used in the DiFrancesco–Noble 1985 model. The curves were obtained at different levels of external sodium ions between 17 and 140 mM.

In fact, the current generated by sodium-calcium exchange during and following the action potential became used as an indicator of intracellular calcium changes both in our laboratory, for example, in work done with Trevor Powell, Terry Egan, Jean-Yves LeGuennec, David Fedida, Yakhin Shimoni and many others (Egan *et al.*, 1989; Fedida *et al.*, 1987; Noble *et al.*, 1991; Noble and Varghese, 1998), and in other laboratories around the world. Here in the UK, David Eisner designed an accurate method for estimating the calcium content of the sarcoplasmic reticulum, from which he has developed an elegantly complete account of calcium cycling in cardiac cells (Eisner and Sipido, 2004; Eisner *et al.*, 2000; Venetucci *et al.*, 2007). This method is now widely used.

5.7. Sodium-Calcium Exchange

As for the second heresy – that of attributing components of inward current to the sodium-calcium exchanger – that rapidly became confirmed in a spectacular way by experiments performed by Junko Kimura after she returned to Japan.

She returned to work in the National Institute of Physiological Sciences in Okazaki. I think she must have reported on the work she did with us in Oxford, to be met by a considerable degree of scepticism from her mentor, Akinori Noma. But they did exactly the right thing, which was to put the whole story to a rigorous experimental test. They eliminated virtually all other currents than the exchanger and measured its voltage and ion dependence (Kimura *et al.*, 1986; Kimura *et al.*, 1987). The test was simple. If the exchanger was electrogenic there had to be a detectable current. I could hardly believe the stroke of luck we had. Not only did they record the expected current, the comparison between the theoretical and experimental results (Figure 5.5) was remarkable and speaks for itself. Rarely does theoretical speculation benefit from this degree of good fortune in subsequent experimental work.

The main difference between the experiment and theory is that the increase in negative current at very negative potentials is not as great in the experiments. Something clearly limits the speed of the exchanger at these extremes. This difference hardly affects the results obtained in the normal range of electrical activity of the heart, which lies between -90 and $+30$ mV.

5.8. Sodium-Calcium Exchange in Sinus Node Rhythm

Rhythm in the natural pacemaker of the heart, the sinus node, differs considerably from that in Purkinje tissue. The main inward current involved in the upstroke of the action potential is not the sodium current. It is the calcium current. Hilary Brown and Susan Noble, in the Oxford laboratory, were exploring this mechanism and the role of the i_f channel, together with their collaborators, Dario DiFrancesco, Junko Kimura and Anne Taupignon. Some of the results found their way, discreetly, into the bottom drawer, thought initially to be unpublishable, even artefactual, as I noted earlier in this chapter. These were the strange slow transient components of inward current that followed the activation of the calcium channels.

The results eventually came out of the bottom drawer to see the light of day in publication because they became the focus of another fruitful interaction between the experimental and computational work. The theoretical work predicted not only that a slow transient inward current *should* follow the more rapid calcium current, as in Figure 5.2, but also that, under the right conditions, it could even *precede* it. This was precisely what those bottom-drawer experimental results showed!

The critical experiment was quite clever. Allow the pacemaker depolarisation in the nodal cells to approach, but not go beyond, the threshold for the L-type calcium current. Then hold the voltage at that level. In many experiments in which this protocol was used during the last third of the pacemaker depolarisation a slow transient inward current could be recorded that looked just like the one that usually follows activation of the calcium channel (Figure 5.4).

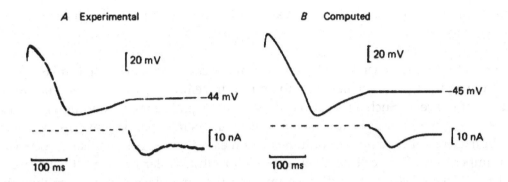

Figure 5.5   Left: Experimental result. The membrane potential in the sinus node is allowed to follow the natural time course of the pacemaker depolarisation until −44 mV is reached. The voltage is then clamped at this point. A very slow inward current develops. The activation and turn-off take almost equal periods of time, which is characteristic of the slow inward currents attributable to sodium-calcium exchange. Right: Reconstruction of this type of current record using the sinus node version (Noble and Noble, 1984) of the DiFrancesco–Noble 1984 equations. Sources: Brown *et al.*, 1983, 1984a.

By that time, Susan Noble and I had modified the DiFrancesco–Noble Purkinje model to create the first model of a sinus node cell (Noble and Noble, 1984). So, we applied the same protocol to the derived model. The results showed the same behaviour as in the experiments. This is shown in Figure 11 of Noble (1984), which was taken from the Noble–Noble 1984 paper. What is happening here? The slow transient current is attributable to calcium release from the sarcoplasmic reticulum. The calcium signal then activates current flow through the sodium-calcium exchanger.

This insight raises the possibility that there are at least two oscillator mechanisms in the natural pacemaker of the heart, one driven primarily by membrane ion channels, the other by an internal calcium oscillator. This is the idea that was later followed up in great detail by Ed Lakatta's group in the US. A recent issue of *Circulation Research* focuses on this question and on how the two oscillators entrain each other (Christoffels *et al.*, 2010; DiFrancesco, 2010; Efimov *et al.*, 2010; Lakatta *et al.*, 2010; Noble *et al.*, 2010; O'Rourke, 2010). As we will see in Chapter 6, the sinus node pacemaker is robustly backed up by several mechanisms that can substitute for each other.

5.9. Sodium-Calcium Exchange in Cardiac Arrhythmias

After leaving my laboratory, Dick Tsien moved to Yale. One of his first experiments in his new laboratory followed up some uninterpreted squiggles of ionic current that he had found in the Oxford results. With Jon Lederer (Lederer and Tsien, 1976) he demonstrated that these transients could be regularly evoked by treating cardiac cells with ouabain, digitalis and similar steroids that are known to block the sodium pump. It didn't take long for us to realise that the transient inward currents we were observing in 1984 were of the same kind. In fact, Lederer and Tsien had already proposed sodium-calcium exchange as one of the possible mechanisms. The idea was simply waiting for proof that the exchanger could be electrogenic.

This work has formed the basis of unravelling a major cause of heart arrhythmias. What a block of the sodium pump achieves is a rise in intracellular sodium as less sodium is pumped out of the cells. Such a rise also occurs in many pathological conditions, including ischemia. The mechanism is now well established. As intracellular sodium increases, so the sodium gradient used by the sodium-calcium exchanger becomes weaker. Less calcium is pumped out of the cell, leading to a rise in intracellular calcium. If this rise is sufficient to trigger release of further calcium from the sarcoplasmic reticulum then the calcium oscillator can begin cycling, either transiently or indefinitely. Each calcium release generates inward current through activation of sodium-calcium exchange, which can then trigger full excitation. This is one mechanism of what are called ectopic beats in the heart (see Chapter 7).

Some of the models developed from the DiFrancesco–Noble model reproduce this mechanism. I will return to this question after relating the next story, which was also started by a telephone call.

5.10.   A Telephone Call from Los Angeles

This call was from Don Hilgemann who was working in Los Angeles early in 1985. Before I explain what Don was excited about and how I reacted, I need to describe the political situation in the UK at that time. We were in the middle of a ferociously strong and determined Thatcher government. With virtually no opposition (she had completely crushed the opposition in the 1983 general election), her government was pursuing a severe financial policy that, amongst many other things, seriously cut the budgets for research. I was a member of a Research Council Grants Committee at that time and experienced directly the chaos this provoked. Senior scientists in the UK were driven to despair by the situation in which we found ourselves. We sometimes did not know how pitifully small the allocation would be until the grants committees met. By then it was too late to adjust our marking and thinking to a situation in which we could probably save only 1 out of maybe 20 grant applications. To say we were shocked is an understatement. The heart was being eaten out of the glory that was British science in its heyday. The period of international recognition in the form of Nobel and similar prizes was threatened. And, indeed, the frequency of those honours did fall quite dramatically during the next two decades.[7]

These were the circumstances in which the council of my university had put forward a proposal, normally uncontroversial, to award the prime minister an honorary degree.

---

[7]The average number of UK Nobel prizes in science per decade over the period of the 1950s, 1960s and 1970s was 11. The average over the 1980s and 1990s dropped to 5.5. During the first decade of the twenty-first century it is back up to 11. It is hard to interpret the figures since there must be delays between funding and results, and many other factors are involved. Over the hundred years of the Nobel prizes, the UK has performed very well when the number is expressed per capita of population.

Figure 5.6   The author (centre) with Junko Kimura (left) and Don Hilgemann (right) at the international meeting on sodium-calcium exchange in Baltimore, USA, in April 1991.

Whether wisely or foolishly, I, and many others, said 'no'. In Oxford, such a proposal requires ratification by a full meeting of academics known as a 'Congregation'. I made one of the major speeches opposing the degree, which was rejected by an overwhelming majority, perhaps the only setback Thatcher experienced during that period.

There was joy throughout the academic community in the UK. But whether it was a wise decision was another matter. There was also, and naturally, fury and anger from her supporters and from the government. By the time that the DiFrancesco–Noble 1985 paper was published in the *Philosophical Transactions of The Royal Society*, I was a national television personality, appearing regularly on behalf of the science and academic communities. This initiative eventually became the pressure group, *Save British Science*, in January 1986 (now a highly effective organisation known as the *Campaign for Science and Engineering*). It was launched by the majority of Britain's Nobel laureates, over a hundred Fellows of the Royal Society, and over 1500 fellow scientists from all over the UK.

This is not the place to go into the pros and cons of that instinctive reaction to Oxford University's Council's proposal. One of my former colleagues from UCL simply commented: 'What you did was just; whether it was wise is another question.' I agreed. But that was the decision, difficult though it was.

The telephone call from Don Hilgemann came in the middle of the hectic round of media appearances. I simply told him I was too busy to talk. His call must have been squeezed in somewhere between the BBC and *The Times*.

Fortunately, he persisted. He telephoned back later and said 'you have got to listen to me'. He was greatly excited by the DiFrancesco–Noble paper since he had experimental results that it could explain. He had been measuring the changes in extracellular calcium in the heart during each beat (Hilgemann, 1986a, 1986b). His results showed that the calcium level between the cells does indeed fall quickly (though not by much – these were difficult experiments) as calcium enters the cells through the calcium channels. But within only 20 milliseconds it was being pumped out again! The thinking up to that time was that restoration of the ionic gradients may occur over a much longer period, perhaps after relaxation of the heart. Don recently wrote to me to give his version of the thinking:

> It seemed abundantly clear when I finished my PhD in 1981 that calcium was leaving the myocyte during the contraction cycle, not between beats. The question was how and with what electrical implications. I was able to measure both influx and extrusion of Ca during single contraction cycles and relate Ca extrusion to late depolarisations in the action potential. From several suggested mechanisms, only electrogenic Na/Ca exchange could account for the data. This work was completed and published (Hilgemann, 1986b) almost simultaneously with the first measurements of exchanger currents by Junko Kimura, Akinori Noma and Hiroshi Irisawa (Kimura *et al.*, 1986).

He could therefore see that, in the DiFrancesco–Noble model, early extrusion is exactly what is predicted since the sodium–calcium exchange current is activated as soon as the calcium signal occurs, which is also responsible for activating the contractile machinery. It is obvious when one thinks about it. The calcium system is constructed to activate the contractile proteins as quickly as possible. There is no reason why it should not also activate the sodium-calcium exchange proteins.

Don's proposal was direct and to the point. He wanted to come to Oxford as soon as possible so that we could work together to follow up what was obviously a novel and important insight. He had the experiments. We had the explanation. Don came, and so my life became a fine balance between maintaining an active and sometimes controversial laboratory while also carrying the even more controversial national media role on behalf of science and universities in the UK. The political side of what I was doing had also become important. After a year or so of standoff the government eventually took us seriously. Not only did I interact effectively with the science minister (Robert Jackson), I also interacted with the prime minister's personal science advisor, George Guise. The stakes were high. Meanwhile, Julian Jack (see Chapter 3) became a trustee of the Wellcome Trust, which transformed itself into the giant funding charity that it is today, and which also greatly influenced later government policy. In fact, those initiatives eventually led to the sustained recovery of science funding that occurred during the Blair governments, particularly while David Sainsbury was minister of science. So, maybe it was 'wise' after all. But it took all of two decades to find that out.

5.11. The Hilgemann–Noble Model

In between the politics, Don and I set to work. The first problem was that he was not working on Purkinje tissue. He was working on the rabbit atrium. So, we decided to develop the DiFrancesco–Noble model to be applicable to the very different electrical waveforms found in the atrium. Don is a man of many parts. In addition to his consummate experimental ability (he went on later to invent the giant patch technique, see Collins *et al.* (1992) and Hilgemann (1989, 1990)) he has great computational skill. I convinced IBM to donate an early version of its PC to us since my lab was too poor to buy one – we had also suffered from the funding cutbacks. Don bought his own PC and wrote the programs in TurboPascal, which eventually convinced me to switch from Algol, which I had been using since the development of more structured programming languages. The days of writing machine code were over me, for me at least.

Painfully slowly (the graphs were just dots on the screen that appeared one after the other after the long integration steps) the model was adjusted. This was the first time we had used a PC. Up until that time all the computational work had been done on big mainframe computers. Often enough, the early PC (no windows, just MS-DOS) was left to work through the night. I also used it to develop the first publicly distributed program for heart modelling, OxSoft HEART. I had formed a company in 1984 to market it. I would have preferred to distribute it for free, but storage discs in those days were far too expensive at around £100 each.

It didn't take long for Don to develop an atrial model from the Purkinje one[8] and to confirm the fit to his experimental findings. But, fortunately for the future work on cardiac cell modelling, he did not stop there. He moved the boundaries of the model deep into further protein systems in the cardiac cell. First, we incorporated the calcium buffer proteins so that the total quantities of calcium being cycled were closer to reality. Then we completely overhauled the sarcoplasmic reticulum system using what we could of the experimental results of Fabiato (1983), who had performed some heroic experiments in which he removed the cell membrane so that he could study directly the process (calcium-induced calcium release) by which the full calcium signal is formed.

5.12. Suppers and Wine at Holywell Manor[9]

Much of this work, and that with Dario DiFrancesco, was carried out over impromptu suppers and convivial wines at Holywell Manor in Oxford, where Balliol College has its

---

[8]The paper is reprinted in Part 2, page 396.

[9]My home and kitchen have been the focus of such lab conviviality ever since the days of Dick Tsien, who introduced Chinese cooking. The rolls of the i_{Kr} and i_{Ks} recordings described in Chapter 2 were brought direct

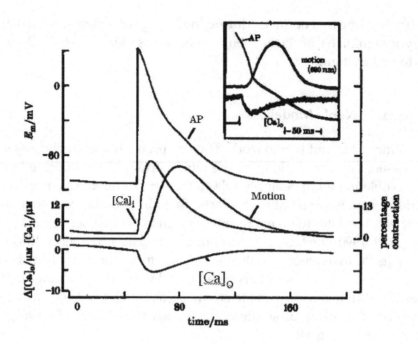

Figure 5.7   The Hilgemann–Noble 1987 model of the atrial action potential (AP), showing its reconstruction of the intracellular calcium transient, $[Ca]_i$, contraction (Motion) and the extracellular calcium transient, $[Ca]_o$. The inset shows one of Hilgemann's experimental recordings showing how close the correspondence between the experiment and the model is for the extracellular calcium transient.

Graduate Centre. At the time I was the head (Praefectus) of the Centre, living in its beautiful lodgings, a manor house dating from 1516. At the completion of our work and its submission to the journal, Don and I celebrated, not only with a good bottle of Bordeaux, but with what I believe was a cognac with a vintage around the time that the Allies defeated Hitler in the Second World War. I must have saved the bottle for some occasion like this one.

In retrospect, we had every reason to celebrate. Those improvements in the modelling have turned out to be just as seminal as the DiFrancesco–Noble model. In fact, I see the different models, all the way from 1960 to 1990, as forming a continuum of development, sometimes occurring in a revolutionary way, but often also as slow evolving of detail in the continuous interaction between experiment and theory. The experimental work in my lab, particularly on sodium calcium exchange (Ch'en *et al.*, 1998; Chapman and Noble, 1989; Egan *et al.*, 1989; Milberg *et al.*, 2008; Noble, 1996; Noble, 2002a; Noble, 2002b; Noble and Blaustein, 2007; Noble *et al.*, 1991; Noble *et al.*, 1996; Noble *et al.*, 2007; Noble and Powell, 1991; Noble and Varghese, 1998; Sher *et al.*, 2007), continued to benefit from this interaction over all the subsequent years.

---

from the laboratory to be laid out on the floor of my home. Julian Jack, during the writing of *Electric Current Flow in Excitable Cells*, introduced us to the joys of Barsac served with a kiwi fruit pavlova. Otto Hauswirth cooked Szegediner Goulash, sent as a recipe from his mother in Vienna. Carlos Ojeda produced amazing couscous, while I cooked the endless Indian curries.

Figure 5.8   Susan Noble, Junko Kimura, Denis Noble and Yung Earm at a meeting in Japan in 1985.

5.13.   The First Single Cell Model

All the models described so far were developed from experiments performed on intact tissue. By the time that Don Hilgemann was working with us, Trevor Powell had come to Oxford as a British Heart Foundation Reader. He had pioneered the technique of isolating ventricular cardiac cells so that experiments could be performed on single cells using microelectrodes or patch electrodes. We published a book together on the exploitation of his discovery (Noble and Powell, 1987). The problems associated with multicellular preparations were a thing of the past and the powerful technique of patch clamping allowed recordings from single cells to be made either in whole cell mode or from single channels within the patch itself. Parallel with the isolation of ventricular cells, the technique of obtaining single pacemaker cells from the SA node was developed and our group made progress using these also (Denyer and Brown, 1990a, 1990b). Quite quickly, all the work in our laboratory and around the world shifted to the single cell technique. This created a need for the models to also apply to single cells. One of my collaborators at that time was Yung Earm from Seoul National University in Korea. He had previously worked with me between 1979 and 1981, so he already knew the Oxford lab well. His second visit was on a British Heart Foundation Visiting Fellowship and it was during this time that we worked on developing the first single cell model (Earm and Noble, 1990).[10]

He and his colleagues (Earm *et al.*, 1990) subsequently used this in a lovely experiment that would shed more light on the role of sodium-calcium exchange. They performed an

---

[10]Reprinted in Part 2, page 382.

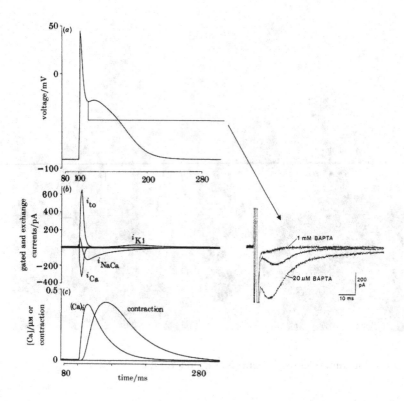

Figure 5.9   The single atrial cell model (Earm and Noble, 1990) and the experimental results (Earm *et al.*, 1990) showing the time course of the sodium-calcium exchange current during a voltage clamp at the time of the late low plateau phase of the action potential.

experiment in which they allowed the early phases of the atrial action potential to run as usual. But then, at the beginning of the low plateau phase, they clamped the voltage at a level (−40 mV) that enables the slow development and decay of the inward exchange current to be recorded. They then infused a calcium chelator into the cell through the electrode to show that, once the calcium transient was removed, the current was too.

5.14.   Women in Physiology

It will be obvious even to a casual reader that, in the work described in this central chapter, women physiologists played a major role. In the earliest years of my Oxford laboratory, women (both students and faculty) were confined to the 'women's colleges'. Somerville College had the good fortune to have Jean Banister as the medical and physiology fellow and tutor. Her main field was cardiovascular so it was natural to link up with her. We famously did live demonstrations together for the students. Jean performed the experiment while I tried to time the pivotal lecture points to arrive just in time for the experiment to illustrate it. Most of the time the synchronisation worked but there was much hilarity amongst the students when we got it wrong so that I was convincingly announcing that x had happened, while Jean hit back sharply 'No, it's y!' The tortoise heart experiment illustrated in Chapter 4 (Figure 4.1) was from one of those demonstrations in 1969.

Figure 5.10 Hilary Brown (left), Junko Kimura (middle front), David Eisner (middle back) and Anthony Spindler (right).

Jean was tutor to two of the scientists who worked in my group during these seminal years. One was my wife, Susan. The other was Hilary Brown, who had worked for her thesis on the heart of a crustacean, *Squilla mantis*, in the marine laboratory in Naples, where my former anatomy professor, JZ Young, did most of his work on the octopus and the squid. Hilary joined up with my fledgling team and was then joined by Susan when she was ready to do her thesis. This is how it came about that the seminal *Nature* paper on i_f in 1979 was by Brown, DiFrancesco and Noble (Susan).

As various parts of this book show, the contributions of Hilary and Susan and their colleagues have been a continuous and flourishing base of the work. They branched out into the more difficult parts of the heart, including the sinus node. With Tony Spindler, easily the longest-serving member of the team and the mainstay of all our developments of electrophysiological technology, they developed our lab's version of a technique known as the sucrose gap for doing voltage clamp experiments on tissues where microelectrode work was too difficult. Susan taught this technique to Wayne Giles, one of the authors of the *Nature* paper on the T wave described in Chapter 4. Wayne and Susan worked extremely well together and their collaboration was crowned with a key paper showing the inhibitory influence of acetylcholine on the calcium current in the heart (Giles and Noble, 1976).

Tony Spindler was renowned for developing one of the best sucrose gap rigs in the world. The Perspex halves of the compartments were machined and polished to produce optically flat surfaces so that, when they were brought together, a tight fit ensured a very high resistance forcing the electric current to canalise itself through the cells of the tissue rather than leak through the compartments. People from our lab took these beautiful pieces of apparatus to their own labs around the world when they left. Although Tony originally joined me as a technician, he proudly converted himself into a fully-fledged scientist and gained his doctorate in 2000, a year or two before he retired. The focus of his thesis was on i_{bNa} whose role in pacemaker activity is described in Chapter 6. *Doctor* Spindler, indeed. (Spindler *et al.*, 1998).

In addition to the science, Hilary and Susan were also responsible, directly and indirectly, for a statistic we are all proud of, which is that during almost the whole period, the proportion of women scientists working in the group became and remained substantial. This was not achieved as any kind of campaign. It happened naturally. Their presence as leading researchers in the team encouraged others to come.

There are two very appropriate endings to this part of the story. The first is that, as the newly elected President of IUPS at its Kyoto Congress in 2009, I attended a symposium on women in physiology organised by our former colleague, Junko Kimura (who also had Jean Banister as her tutor at Somerville College), whose work on sodium-calcium exchange has already been described in this chapter. That meeting produced a published report (Kimura *et al.*, 2010) to be considered by the Council of the Union. It highlights the wide variations in gender inequality in various parts of the world.

The second is that one of the last of the women to work in the group is my own daughter, Penny, who has been a mainstay of the programming and curating work all the way from taking over the developments of OxSoft HEART to curating the models on the CellML website and working on their use in problems of cardiac repolarisation (Garny *et al.*, 2002; Liu *et al.*, 2008; Noble and Noble, 2000; Noble and Noble, 2006; Noble and Noble, 2011; Noble *et al.*, 2007; Noble *et al.*, 2010; Roux *et al.*, 2001; Roux *et al.*, 2006; Sears *et al.*, 1999; Sher *et al.*, 2007; Sher *et al.*, 2008; Ten Tusscher *et al.*, 2004; Volk *et al.*, 2005).

5.15. Conclusions

5.15.1. *Pluses*

This chapter forms the central core of the work, with a flowering of interaction between experiment and theory that generated the whole family of cardiac cell models developed since that period. Given the controversy that occurred both in the development of the work, and subsequently, it is remarkable how much of it has completely stood the test of time. Without this work, the Cardiac Physiome Project (Chapter 7) could not have developed as it has, on a firm experimental and modelling basis at the cellular level.

5.15.2. *Minuses*

We could have achieved much more if funding had allowed. Streams of good people were clamouring to work in our labs. In fact, without the secure and unwavering support of the British Heart Foundation (BHF), which gave me a chair and associated funds from 1984 to

2004, the work would have collapsed. Government funding would have been inadequate. I shall always be grateful for the BHF support. BHF is the reason why cardiac science still flourishes in the UK, while the Wellcome Trust supports medical science as a whole.

5.15.3. *Contributions to systems biology*

There is much argument today about how systems biology should be conducted. The work described in this chapter shows how rich the rewards can be when experimental and theoretical work interacts in close proximity, with some people doing both. At some stages, theory was ahead. At other stages, the experimental work triggered new developments in theory. In Chapter 9, I will elaborate on the middle-out approach to multi-level analysis in biology. This chapter shows that the cell modelling clearly forms my middle, from which we were able subsequently to reach out to both lower and higher levels in what I believe to be one of the major paradigm examples of systems biology in practice.

The work of this period also illustrates the role of analytical mathematics to complement brute-force heavy computing. The one regret I have is that so few of my physiological colleagues understand what we were doing, that the role of the maths has easily been forgotten.

References

Papers reprinted in Part 2:

DiFrancesco, D. and Noble, D. (1982). 'Implications of the re-interpretation of i_{K2} for the modelling of the electrical activity of pacemaker tissues in the heart', in Bouman, L.N. and Jongsma, H.J. (eds), *Cardiac Rate and Rhythm*, Martinus Nijhoff, The Hague/Boston/London, pp. 93–128.

DiFrancesco, D. and Noble, D. (1985). A model of cardiac electrical activity incorporating ionic pumps and concentration changes, *Philosophical Transactions of the Royal Society B*, **307**, 353–398.

Earm, Y.E. and Noble, D. (1990). A model of the single atrial cell: Relation between calcium current and calcium release, *Proceedings of the Royal Society B*, **240**, 83–96.

Hilgemann, D.W. and Noble, D. (1987). Excitation-contraction coupling and extracellular calcium transients in rabbit atrium: Reconstruction of basic cellular mechanisms, *Proceedings of the Royal Society B*, **230**, 163–205.

Noble, D. (1984). The surprising heart: A review of recent progress in cardiac electrophysiology, *Journal of Physiology*, **353**, 1–50.

Others:

Arlock, P. and Noble, D. (1985). Two components of 'second inward current' in ferret papillary muscle, *Journal of Physiology*, **369**, 88P.

Beeler, G.W. and Reuter, H. (1977). Reconstruction of the action potential of ventricular myocardial fibres, *Journal of Physiology*, **268**, 177–210.

Brown, H.F., Clark, A. and Noble, S.J. (1976a). Analysis of pacemaker and repolarisation currents in frog atrial muscle, *Journal of Physiology*, **258**, 547–577.

Brown, H.F., Clark, A. and Noble, S.J. (1976b). Identification of the pacemaker current in frog atrium, *Journal of Physiology*, **258**, 521–545.

Brown, H.F. and DiFrancesco, D. (1980). Voltage-clamp investigations of membrane currents underlying pacemaker activity in rabbit sinoatrial node, *Journal of Physiology*, **308**, 331–351.

Brown, H.F., DiFrancesco, D. and Noble, S.J. (1979). How does adrenaline accelerate the heart? *Nature*, **280**, 235–236.

Brown, H.F., Giles, W. & Noble, S.J. (1977). Membrane currents underlying activity in frog sinus venosus, *Journal of Physiology*, **271**, 783–816.

Brown, H.F., Kimura, J., Noble, D. *et al.* (1983). Two components of 'second inward current' in the rabbit S.A. node, *Journal of Physiology*, **334**, 56–57P.

Brown, H.F., Kimura, J., Noble, D. *et al.* (1984a). The ionic currents underlying pacemaker activity in rabbit sinoatrial node: Experimental results and computer simulations, *Proceedings of the Royal Society B*, **222**, 329–347.

Brown, H.F., Kimura, J., Noble, D. *et al.* (1984b). Mechanisms underlying the slow inward current, isi, in the rabbit sinoatrial node investigated by voltage clamp and computer simulation, *Proceedings of the Royal Society B*, **222**, 305–328.

Brown, H.F., Noble, D., Noble, S.J. *et al.* (1984c). Transient inward current and its relation to the very slow inward current in the rabbit S.A. node, *Journal of Physiology*, **349**, 47P.

Ch'en, F.C., Vaughan-Jones, R.D., Clarke, K. *et al.* (1998). Modelling myocardial ischaemia and reperfusion, *Progress in Biophysics and Molecular Biology*, **69**, 515–537.

Chapman, R.A. and Noble, D. (1989). 'Sodium-calcium exchange in the heart', in Allen, J., Noble, D. and Reuter, H. (eds), *Sodium-Calcium Exchange*, OUP, Oxford, pp. 102–125.

Christoffels, V.M., Smits, G.J., Kispert, A. *et al.* (2010). Development of pacemaker tissues of the heart, *Circulation Research*, **106**, 240–254.

Collins, A., Somlyo, A.V. and Hilgemann, D.W. (1992). The giant cardiac membrane patch method: Stimulation of outward Na(+)-Ca2+ exchange current by MgATP, *Journal of Physiology*, **454**, 27–57.

Denyer, J. C. and Brown, H. F. (1990a). Pacemaking in rabbit isolated sion-atrial node cells during Cs^+ block of the hyperpolarisation-activated current i_f. *Journal of Physiology*, **429**, 401–409.

Denyer, J.C. and Brown, H.F. (1990b). Rabbit sinoatrial node cells: Isolation and electro-physiological properties, *Journal of Physiology*, **428**, 405–424.

DiFrancesco, D. (2010). The role of the funny current in pacemaker activity, *Circulation Research*, **106**, 434–446.

DiFrancesco, D. and Noble, D. (1980a). If i_{K2} is an inward current, how does it display potassium specificity? *Journal of Physiology*, **305**, 14P.

DiFrancesco, D. and Noble, D. (1980b). The time course of potassium current following potassium accumulation in frog atrium: Analytical solutions using a linear approximation, *Journal of Physiology*, **306**, 151–173.

DiFrancesco, D. and Noble, D. (1982). 'Implications of the re-interpretation of i_{K2} for the modelling of the electrical activity of pacemaker tissues in the heart', in Bouman, L.N. and Jongsma, H.J. (eds), *Cardiac Rate and Rhythm* (ed.), Martinus Nijhoff, The Hague/Boston/London, pp. 93–128.

Earm, Y.E., Ho, W.K. and So, I.S. (1990). Inward current generated by Na-Ca exchange during the action potential in single atrial cells of the rabbit, *Proceedings of the Royal Society B*, **240**, 61–81.

Earm, Y.E. and Noble, D. (1990). A model of the single atrial cell: Relation between calcium current and calcium release, *Proceedings of the Royal Society B*, **240**, 83–96.

Efimov, I.R., Federov, V.V., Joung, B. *et al.* (2010). Mapping cardiac pacemaker circuits: Methodological puzzles of the sinoatrial node optical mapping, *Circulation Research*, **106**, 255–271.

Egan, T., Noble, D., Noble, S.J. *et al.* (1989). Sodium-calcium exchange during the action potential in guinea-pig ventricular cells, *Journal of Physiology*, **411**, 639–661.

Eisner, D. and Sipido, K.R. (2004). Sodium calcium exchange in the heart. Necessity or luxury? *Circulation Research*, **95**, 549–551.

Eisner, D.A., Choi, H.S., Diaz, M.E. *et al.* (2000). Integrative analysis of calcium cycling in cardiac muscle, *Circulation Research*, **87**, 1087–1094.

Fabiato, A. (1983). Calcium induced release of calcium from the sarcoplasmic reticulum, *American Journal of Physiology*, **245**, C1–C14.

Fedida, D., Noble, D., Shimoni, Y. *et al.* (1987). Inward currents related to contraction in guinea-pig ventricular myocytes, *Journal of Physiology*, **385**, 565–589.

Garny, A., Noble, P.J., Kohl, P. *et al.* (2002). Comparative study of sinoatrial node cell models, *Chaos, Solitons and Fractals*, **13**, 1623–1630.

Giles, W. and Noble, S.J. (1976). Changes in membrane currents in bullfrog atrium produced by acetylcholine, *Journal of Physiology*, **261**, 103–123.

Hilgemann, D.W. (1986a). Extracellular calcium transients and action potential configuration changes related to post-stimulatory potentiation in rabbit atrium, *Journal of General Physiology*, **87**, 675–706.

Hilgemann, D.W. (1986b). Extracellular calcium transients at single excitations in rabbit atrium measured with tetramethylmurexide, *Journal of General Physiology*, **87**, 707–735.

Hilgemann, D.W. (1989). Giant excised cardiac sarcolemmal membrane patches: Sodium and sodium-calcium exchange currents, *Pflugers Archiv*, **415**, 247–249.

Hilgemann, D.W. (1990). Regulation and deregulation of cardiac Na-Ca exchange in giant excised sarcolemmal membrane patches, *Nature*, **344**, 242–245.

Johnson, E.A. and Lieberman, M. (1971). Heart: Excitation and contraction, *Annual Reviews of Physiology*, **33**, 479–530.

Kimura, J., Miyamae, S. and Noma, A. (1987). Identification of sodium-calcium exchange current in single ventricular cells of guinea-pig, *Journal of Physiology*, **384**, 199–222.

Kimura, J., Noma, A. and Irisawa, H. (1986). Na-Ca exchange current in mammalian heart cells, *Nature*, **319**, 596–597.

Kimura, J., Suzuki, Y., Mizumura, K. *et al.* (2010). Report on the Women in Physiology Symposium in IUPS 2009, *Journal of Physiological Sciences*, **60**, 227–234.

Lakatta, E.G., Maltsev, V.A. and Vinogradova, T.M. (2010). A coupled system of intracellular Ca2+ clocks and surface membrane ion clocks controls the timekeeping of the heart's pacemaker, *Circulation Research*, **106**, 659–673.

Lederer, W.J. and Tsien, R.W. (1976). Transient inward current underlying arrhythmogenic effects of cardiotonic steroids in Purkinje fibres, *Journal of Physiology*, **263**, 73–100.

Liu, J., Noble, P.J., Xiao, G. *et al.* (2008). Role of pacemaking current in cardiac nodes: insights from a comparative study of sinoatrial node and atrioventricular node, *Progress in Biophysics and Molecular Biology*, **96**, 294–304.

McAllister, R.E., Noble, D. and Tsien, R.W. (1975). Reconstruction of the electrical activity of cardiac Purkinje fibres, *Journal of Physiology*, **251**, 1–59.

Milberg, P., Pott, C., Fink, M. *et al.* (2008). Inhibition of the Na+/Ca2+ exchanger suppresses Torsade de pointes in an intact heart model of LQT2 and LQT3, *Heart Rhythm*, **5**, 1444–1452.

Mullins, L. (1981). *Ion Transport in the Heart*, Raven Press, New York.

Noble, D. (1984). The surprising heart: A review of recent progress in cardiac electrophysiology, *Journal of Physiology*, **353**, 1–50.

Noble, D. (1996). 'The functional significance of sodium-calcium exchange in the heart', in Morad, S. *et al.* (eds), *Molecular Physiology and Pharmacology of Cardiac Ion Channels and Transporters*, Kluwer, Dordrecht, London, pp. 457–467.

Noble, D. (2002a). Influence of Na-Ca exchange stoichiometry on model cardiac action potentials, *Annals of the New York Academy of Sciences*, **976**, 133–136.

Noble, D. (2002b). Simulation of Na-Ca exchange activity during ischaemia, *Annals of the New York Academy of Sciences*, **976**, 431–437.

Noble, D. and Blaustein, M.P. (2007). Directionality in drug action on sodium-calcium exchange, *Annals of the New York Academy of Sciences*, **1099**, 540–543.

Noble, D., LeGuennec, J.Y. and Winslow, R. (1996). Functional roles of sodium-calcium exchange in normal and abnormal cardiac rhythm, *Annals of the New York Academy of Sciences*, **779**, 480–488.

Noble, D. and Noble, P.J. (2006). Late sodium current in the pathophysiology of cardiovascular disease: consequences of sodium-calcium overload, *Heart*, **92**, iv1–iv5.

Noble, D., Noble, P.J. and Fink, M. (2010). Competing oscillators in cardiac pacemaking: historical background, *Circulation Research*, **106**, 1791–1797.

Noble, D. and Noble, S.J. (1984). A model of S.A. node electrical activity using a modification of the DiFrancesco–Noble (1984) equations, *Proceedings of the Royal Society B*, **222**, 295–304.

Noble, D., Noble, S.J., Bett, G.C.L. *et al.* (1991). The role of sodium-calcium exchange during the cardiac action potential, *Annals of the New York Academy of Sciences*, **639**, 334–353.

Noble, D. and Powell, T. (1987). *Electrophysiology of single cardiac cells*, Academic Press, London.

Noble, D. and Powell, T. (1991). The slowing of calcium signals by calcium indicators in cardiac muscle, *Proceedings of the Royal Society B*, **246**, 167–172.

Noble, D., Sarai, N., Noble, P.J. *et al.* (2007). Resistance of cardiac cells to NCX knockout: A model study, *Annals of the New York Academy of Sciences*, **1099**, 306–309.

Noble, D. and Varghese, A. (1998). Modeling of sodium-calcium overload arrhythmias and their suppression, *Canadian Journal of Cardiology*, **14**, 97–100.

Noble, P.J. and Noble, D. (2000). Reconstruction of the cellular mechanisms of cardiac arrhythmias triggered by early after-depolarizations, *Japanese Journal of Electrocardiology*, **20 (Suppl 3)**, 15–19.

Noble, P.J. and Noble, D. (2011). 'A historical perspective on the development of models of rhythm in the heart', in Tripathi, O.N. *et al.* (eds), *Heart Rate and Rhythm: Molecular Basis, Pharmacological Modulation and Clinical Implications*, Springer, Heidelberg, pp. 155–174.

Noble, S.J. (1972). Membrane currents in atrial muscle. DPhil thesis, Oxford University.

Noble, S.J. (1976). Potassium accumulation and depletion in frog atrial muscle, *Journal of Physiology*, **258**, 579–613.

Noma, A. and Irisawa, H. (1976). Membrane currents in the rabbit sinoatrial node cell as studied by the double microelectrode method, *Pflugers Archiv*, **364**, 45–52.

O'Rourke, B. (2010). Be still my beating heart – Never! *Circulation Research*, **106**, 238–239.

Provencher, S.W. (1976). An eigenfunction expansion method for the analysis of exponential decay curves, *Journal of Chemical Physics*, **64**, 2772.

Reuter, H. (1967). The dependence of slow inward current in Purkinje fibres on the extracellular calcium concentration, *Journal of Physiology*, **192**, 479–492.

Reuter, H. and Seitz, N. (1968). The dependence of calcium efflux from cardiac muscle on temperature and external ion composition, *Journal of Physiology*, **195**, 451–470.

Roux, E., Noble, P.J., Hyvelin, J.-M. *et al.* (2001). Modelling of Ca2+-activated chloride current in tracheal smooth muscle cells, *Acta Biotheoretica*, **49**, 291–300.

Roux, E., Noble, P.J., Noble, D. *et al.* (2006). Modelling of calcium handling in airway myocytes, *Progress in Biophysics and Molecular Biology*, **90**, 64–87.

Sears, C.E., Noble, D., Noble, P.J. *et al.* (1999). Vagal control of heart rate is modulated by extracellular potassium, *Journal of the Autonomic Nervous System*, **77**, 164–171.

Seyama, I. (1976). Characteristics of the rectifying properties of the sinoatrial node cell of the rabbit, *Journal of Physiology*, **255**, 379–397.

Sher, A., Hinch, R., Noble, P.J. *et al.* (2007). Functional significance of Na+/Ca2+ exchangers co-localisation with ryanodine receptors, *Annals of the New York Academy of Sciences*, **1099**, 215–220.

Sher, A., Noble, P.J., Hinch, R. *et al.* (2008). The role of Na+/Ca2+ exchangers in Ca2+ dynamics in ventricular myocytes, *Progress in Biophysics and Molecular Biology*, **96**, 377–398.

Shrier, A. and Clay, J.R. (1982). Comparison of the pacemaker properties of chick embryonic atrial and ventricular heart cells, *Journal of Membrane Biology*, **69**, 49–56.

Spindler, A.J., Noble, S.J., Noble, D. *et al.* (1998). The effects of sodium substitution on currents determining the resting potential in guinea-pig ventricular cells, *Experimental Physiology*, **83**, 121–136.

Ten Tusscher, K.H.W.J., Noble, D., Noble, P.J. *et al.* (2004). A model of the human ventricular myocyte, *American Journal of Physiology*, **286**, H1573–1589.

Venetucci, L.A., Trafford, A.W. and Eisner, D.A. (2007). Increasing ryanodine receptor open probability alone does not produce arrhythmogenic calcium waves: threshold sarcoplasmic reticulum calcium content is required, *Circulation Research*, **100**, 105–111.

Volk, T., Noble, P.J., Wagner, M. *et al.* (2005). Ascending aortic stenosis selectively increases action potential-induced Ca2+ influx in epicardial myocytes of the rat left ventricle, *Experimental Physiology*, **90**, 111–121.

Chapter 6

Understanding Robustness in Biological Systems

6.1. Background: The Genome Ten Years On

As I write this chapter, the June 2010 edition of *Prospect* carries an article on the genome, headed 'Too much information'. It begins

> Ten years ago, the first draft of the sequence of the human genome was heralded as the dawn of a new era of genetic medicine [...] You might have noticed that it hasn't [happened]. The medical impact of the human genome project (HGP) has so far been negligible.

The explanation given by the author, no doubt following a lead from an editorial in *Nature* (Editorial, 2010), is

> The activity of genes is affected by many things not explicitly encoded in the genome, such as how the chromosomal material is packaged up and how it is labelled with chemical markers. Even for diseases like diabetes, which have a clear inherited component, the known genes involved seem to account for only a small proportion of the inheritance [...] the failure to anticipate such complexity in the genome must be blamed partly on the cosy fallacies of genetic research. After Francis Crick and James Watson cracked the riddle of DNA's molecular structure in 1953, geneticists could not resist assuming it was all over bar the shouting. They began to see DNA as the "book of life", which could be read like an instruction manual. It now seems that the genome might be less like a list of parts and more like the weather system, full of complicated feedbacks and interdependencies.

In 2002 I wrote an article for the magazine of the Physiological Society, *Physiology News*, explaining why the genome is not the 'book of life' (Noble, 2002). A reader was so enthused by it that he approached the Editor of *Prospect* to insist that it deserved a much wider readership and thought that *Prospect* was the ideal medium. It would have been, but the offer was turned down without question. Probably it didn't fit the mindset of that time, full of confidence that molecular biology was going, finally, to deliver the goods through exploitation of the genome data.

What has changed in the subsequent eight years, so that even a staff writer for *Prospect* now expresses the main message of the 2002 article? The answer is that 2010 was the ten-year watermark after the sequencing of the genome, the year the leaders of the Human Genome Project promised that the benefits for health care would arrive. Diabetes, hypertension and mental illness were amongst the targets. Science journalists are therefore becoming uneasy that they bought into a promise that has not and, I would argue, could not have been delivered. And they are not alone. The drug industry also bought in, literally, and start-up genomics companies were bought up for hundreds of millions of dollars. The sequencing of genomes has been of great value for basic science, particularly in studies on comparison of genomes for the purposes of evolutionary biology, but the interpretation of the genome data in terms of biological functions (phenotypes) has proved vastly more difficult than anticipated.

6.2. Limitations of the Differential View of Genetics

Physiologists have been aware of many of the relevant forms of 'complicated feedbacks and interdependencies' for many years. We were largely ignored as molecular biology took centre stage during the second half of the twentieth century. One of the reasons is an unnecessary and incorrect limitation on the relationship between genes and phenotypes that developed in twentieth-century biological thought. The standard story in the neo-Darwinian view of evolution is that differences in genes (different alleles) are responsible for differences in phenotype so that selection of the phenotype equates with selection of the successful allele. Evolution is seen as occurring primarily through incremental change as new alleles (mutations) arise. I call this the *differential* view of genetics (Noble, 2011a). It is the basis of the selfish gene concept (Dawkins, 1976, 2006). In a later book, *The Extended Phenotype*, however, Dawkins acknowledges that no experiment could distinguish between the gene-centred view of evolution and its alternatives (Dawkins, 1982: p. 1).

The differential view of genetics is limiting in a way somewhat analogous to where we would be in mathematics if we were limited to differential equations and if the integral sign had never been invented. The analogy is quite good. To integrate differential equations we need the initial and boundary conditions. These are just as much a 'cause' of the particular solutions we obtain as the differential equations themselves. Likewise, differences in genomes operate in the context of the system as a whole, which form the initial (egg cell) and boundary (environmental) conditions for the development of an organism. Those conditions are just as much a cause of the phenotype as are the genes (Noble, 2008; Kohl *et al.*, 2010).

Moreover, nature does not recognise the limitation of the differential view. A gene, defined as a particular sequence of DNA, has many effects. These are not limited to the differences one can observe between that gene and one of its alleles. For a full physiological understanding of the relationship between genomes and phenotypes we therefore need an *integral* view of genetics (Noble, 2011a), in which we look for the many phenotypic effects it may have, even when those phenotypic effects are hidden by the system.

When I first read Richard Dawkins' acknowledgement in *The Extended Phenotype* ('I doubt that there is any experiment that could be done to prove my claim') I was strongly inclined to agree with it and, indeed, if you compare the selfish gene metaphor with an opposing metaphor, such as genes as prisoners, it is impossible to think of an experiment that would distinguish between the two views, as I show in my book *The Music of Life* (Noble, 2006: Chapter 1). For any given case, I still think that must be true. But I have slowly changed my view on whether this must be true if we consider *many* cases, looking at the functioning of the organism as a whole. There are different ways in which empirical discovery can impact on our theoretical understanding. Not all of these are in the form of the straight falsification of a hypothesis. Sometimes it is the slow accumulation of the weight of evidence that eventually triggers a change of viewpoint. This is the case with insights that are expressed in metaphorical form (like 'selfish' and 'prisoners'), which is not intended to be taken literally.

Consider the following thought experiment. Take an organism for which we know the complete genome. Check on the proteins for which each of the genes forms a template. Then carry out knockout experiments in turn on every single one of the genes in the organism. Suppose that we find that in the great majority of cases the knockouts have no effect whatsoever! Would we conclude that these were not, after all, genes since changing them did not have a phenotypic effect? Clearly not. Each and every one of them forms templates for the production of proteins that function in particular networks. That is what we now mean by a 'gene'. At the level of proteins, therefore, all the genes can be expressed and have functionality in that sense. But at the level of the phenotype, they appear not to be functional. I think that the sheer weight of such experimental evidence would force us to reconsider the relationship between genes and phenotypes. We would conclude that there is no sense in talking just about differences between alleles of particular genes as though that were the sole determinant of a phenotype difference. That would be to mistake the tip for the whole iceberg.

6.3. Robustness in Yeast

Actually, this is not a thought experiment at all. It is a real one! It was carried out on the 6,000 or so genes in the yeast genome (Hillenmeyer *et al.*, 2008). In the experiments, as many as 80% of knockouts revealed no phenotypic effect in the sense of affecting growth and reproduction of the organism. Nevertheless, the great majority (97%) of knockouts do have an effect under metabolic stress, when the back-up mechanisms are compromised. What is happening here? Modern geneticists call this genetic buffering. The networks of interactions formed by the proteins, the genes and the other structures in the organism with which they all interact, are robust. Like aircraft control-systems that are fail-safe so that control is not lost when one of the mechanisms fails, organisms are strongly protected against most genetic failures. If one mechanism is knocked out, another one takes over. Moreover, the organism itself continually corrects mistakes in the genome as it gets copied. The robustness of the organism is therefore a property of the system, not of the individual genes, both in terms of protecting the system against failures and repairing copying mistakes.

Now, this fact about the robustness of organisms is clearly an empirical discovery. And equally clearly it favours the idea of cooperative genes, not selfish genes. Metaphors, for that is what we are dealing with here, may not be falsifiable in the simple way, in which an empirical theory is falsifiable. They are nonetheless sensitive to empirical discovery in a more general sense. It is a matter of judgement which metaphor amongst competing metaphors best describes the situation. But that doesn't mean that the judgement is completely arbitrary and independent of the empirical evidence. The relationship is just more nuanced than it would be in a case where a single observation can falsify a scientific hypothesis. This is the reason for which, at the end of Chapter 1 of *The Music of Life*, after comparing selfish and cooperative gene metaphors, I wrote: 'it does seem more natural, and certainly more meaningful, to say that the rationale for existence lies at the level at which selection occurs. This is at the level at which we can say why an organism survived or not' (Noble, 2006: p. 22).

With this background of ideas in mind, we are ready to consider a particular example and to see what conclusions we can draw from it.

6.4. The Robustness of Heart Rhythm

The article chosen for reprinting for this chapter[1] (Noble *et al.*, 1992) shows how quantitative physiological analysis can reveal the mechanisms of these forms of robustness, how we can reverse engineer a system to avoid the problems raised by the differential view, and therefore precisely why the integrative view of genetics should replace the differential view. Let's analyse this case in detail.

The paper was based on using a model of a sinus node cell derived from the work described in Chapter 5 (DiFrancesco and Noble, 1985; Noble and Noble, 1984). First, we can use the model to answer what seems at first sight to be a simple question. Is there a genetic program for cardiac rhythm? We can answer that question since all the protein mechanisms represented in the equations of the model are made from templates formed by genes. But, try as we may, there is nothing in the DNA sequences for those proteins, nor in the properties of those proteins in isolation, that could form a 'program' for cardiac rhythm.

A simple experiment on the model will demonstrate this.

In Figure 6.1, the model was run for 1300 ms, during which time 6 oscillations were generated. These correspond to six heartbeats at a frequency similar to that of the heart of a rabbit, the species on which the experimental data was obtained to construct the model. During each beat, all the protein mechanisms also oscillate in a specific sequence. To simplify the diagram, only three of those protein channels are represented here. At 1300 ms, an experiment was performed on the model. The 'downward causation' (see Chapter 1)

---

[1] In Part 2, page 489.

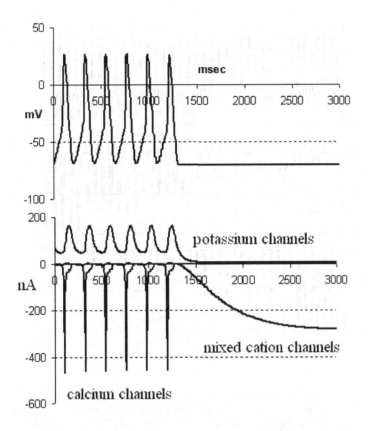

Figure 6.1   Computer model of sinus node pacemaker rhythm (using equations from Noble and Noble, 1984). Three of the protein channels are shown, a potassium channel, a calcium channel and a non-specific (sodium and potassium) channel. After six cycles, the feedback from the membrane potential onto the ionic channels was removed by holding the potential constant. All the oscillations disappear and the activity of each channel relaxes to a steady state.

between the global cell property, the membrane potential, and the voltage-dependent gating of the ion channels was interrupted. If there were a subcellular 'program' forcing the proteins to oscillate, the oscillations would continue. In fact, all oscillations cease and the activity of each protein relaxes to a steady value. In this case, therefore, the 'program' includes the cell itself and its membrane system. In fact, we don't need the concept of a program here. The sequence of events, including the feedback between the cell potential and the activity of the proteins, simply is cardiac rhythm. It is a property of the interactions between all the components of the system. It doesn't even make sense to talk of cardiac rhythm at the level of proteins and DNA.

It is important to clarify what is happening here. The demonstration that the rhythm is a property of the cell, not of individual proteins or gene-protein networks, is an empirical discovery. Subcellular oscillators can and do also exist, for example, the intracellular calcium oscillator discussed in Chapter 5. In other circumstances, for example, in ischemic conditions (sodium-calcium overload), the model used in Figure 6.1 would also show this kind of oscillation. This kind of oscillation also depends on 'complicated feedbacks

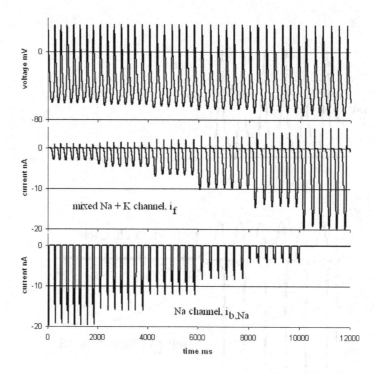

Figure 6.2   Example of the use of computational systems biology to model a genetic buffering mechanism. Top: Membrane potential variations in a model of the sinus node pacemaker of the heart. Bottom: The background sodium channel, i_{bNa}, is progressively reduced until it is eventually 'knocked out'. Middle: The mixed (sodium and potassium) cation current channel, i_f, progressively takes over the function, and so ensures that the change in frequency is minimised. (Recomputed using COR from Noble *et al.*, 1992). Coordinates: membrane potential in mV, Current in nA, time (abscissa) in ms.

and interdependencies', though the feedback involved is then dependent on the calcium signalling system, the sarcoplasmic reticulum. It can also be seen as one of the back-up systems, with each oscillator, cellular and subcellular, entraining each other (Noble *et al.*, 2010).

6.5.   Reverse Engineering the Cardiac Pacemaker

We are now ready to fully appreciate the article reproduced for this chapter. In addition to the existence of two types of oscillator, cellular and subcellular, the cellular oscillator itself is also a good example of a system with back-up.

The main experiment performed on the sinus node model in this article (reproduced as Figure 6.2) can be viewed as a progressive knockout. The background sodium current channel (Kiyosue *et al.*, 1993), which contributes around 80% of the depolarising current that generates the pacemaker depolarisation, was reduced by 20% at each stage of the computation until, finally, it was knocked out completely. Without back-up mechanisms this would be expected to have a very large effect on the frequency, even abolishing rhythm completely. Instead we find only a very small (10%) and gradual decrease in frequency.

The model also reveals the reason for such powerful buffering of this knockout experiment. As the sodium channel is reduced, another channel, i_f, progressively takes over the depolarising role. We move smoothly with no interruption of rhythm from a mechanism primarily dependent on one channel to a mechanism primarily dependent on the other. Although the computation doesn't show it, we can also carry out the reverse knockout. Progressive reduction of i_f in a model in which it is the major mechanism also leads to only a modest reduction of frequency as the background sodium channel takes over.

What conclusions can we draw from this model experiment?

First, the insights involved, all the way from the identification of the i_f channel described in Chapter 3, through to the demonstration that its inhibition could be a safe way to reduce heart frequency, led to the realisation by the drug company, Servier, that a specific blocker of the i_f channel could be a very useful drug since slowing cardiac rhythm helps patients with ischemia and other forms of heart trouble by reducing the energy demand by the heart. But, of course, the slowing must be limited in range. This is precisely what block of i_f can achieve. Ivabradine is the drug that was developed and has now achieved FDA and other regulatory approval for use in patients with ischemia. The clinical trials show effectiveness, particularly in patients with high heart rates. Figure 6.3 shows a newspaper

Heart failure pill 'could save thousands of lives'

Ivabradine 'could prevent 10,000 deaths a year'

Cardiologist calls it 'a significant breakthrough'

Helen Pidd

A pill for chest pains that costs around £10 per week could save the lives of thousands of heart failure patients. It would also save the NHS millions by reducing hospital admissions by a quarter, trial results have shown. One expert involved in the trial - the largest so far published on heart failure - described the evidence as a "significant breakthrough" and said it would compel him to change his clinical practice. At a conservative estimate, up to 10,000 deaths per year in the UK could be prevented, he said.

The drug, ivabradine, also known under the brand name Procoralan, is already available in the UK for angina and is known to be safe. The trial results, presented yesterday at the European Society of Cardiology annual meeting in Stockholm and published in the Lancet medical journal, suggest that ivabradine could be resurrected as a cost-effective treatment for many thousands of patients with moderate to severe heart failure.

Over two years, the drug cut the trial patients' risk of death from heart failure by 26%. It had a similar impact on the likelihood of their being admitted to hospital because of worsening symptoms. More than 700,000 people over the age of 45 live with the risk of heart failure, which occurs when damage to the heart leaves it too weak to pump blood efficiently round the body. An estimated 68,000 cases are diagnosed each year. Heart failure causes fatigue, breathlessness, increased heart rate, and swollen ankles. Around 40% of those affected are dead after a year.

Heart failure soaks up 1-2% of the total NHS budget; the direct medical costs alone amount to £625m a year.

The key to ivabradine is its effect on heart rate. Unlike other treatments, it lowers the number of heartbeats per minute without also reducing blood pressure.

Prof Martin Cowie, consultant cardiologist at the Royal Brompton hospital in London, who led the UK arm of the study, said: "The evidence represents a significant clinical breakthrough."

At a conservative estimate, prescribing ivabradine could save 5,000 to 10,000 lives a year, he said. Cowie added: "I rarely come back from a conference and say I've got to change my clinical practice, but these results will make me do that. There are lots of patients I'm going to have to discuss this with." He said other international experts he had spoken to were "excited" by the findings.

But one expert cautioned against rushing into treatment with ivabradine. In a comment in the Lancet, Dr John Teerlink, from the University of California in San Francisco, questioned whether all the control-group patients were being optimally treated with standard medications. Those receiving high enough doses of beta-blockers saw no significant extra benefit from ivabradine, he said.

Figure 6.3   Article from *The Guardian*, 30 August 2010.

report that appeared as I was writing this chapter. From the initial discovery of the channel in 1979 (Brown *et al.*, 1979) to the current clinical trials, there is a period of 30 years. Those who press basic scientists for more immediate impact (see Noble, 2010) should note that this is far from unusual.

Second, despite the strong buffering that almost completely hides the actual contribution of each ion channel to the net ionic current flow, the model can be used to estimate those contributions in a highly quantitative way. This demonstrates the integrative power of reverse engineering using physiological models. We do not have to be limited simply to observing differences. It is difficult to see any other way forward given what we now know of the extent of genetic buffering. Gene products act most of the time in cooperation. Working out their contributions using only the forward mode (using differences at the level of genes or proteins and their effects at the level of the phenotype) clearly won't solve the problem.

The conditions for such reverse engineering to succeed are (a) that the elementary components of the model should be individual gene products (proteins, RNAs); (b) that the models should reach beyond the level of proteins to reproduce functionality at higher levels (cells, tissues, organs, systems) and (c) that the models should be sufficiently well validated, experimentally, to have confidence in the results.

6.6. Conclusions

6.6.1. *Pluses*

This article largely resolved arguments about which channel(s) play the largest roles in pacemaker activity by showing that, depending on the conditions, they can replace each other as back-up mechanisms. The cardiac pacemaker, just like other biological oscillators such as circadian rhythm, is a multi-process fail-safe system.

6.6.2. *Minuses*

So far, there are very few models that satisfy all the conditions for reverse engineering to be successful in quantifying relative contributions of particular gene products. It may seem surprising that this should be so when so much experimental data is available on genes and proteins. The problem lies in the sizes of the networks involved. In the case of cardiac electrophysiology we have been fortunate in dealing with relatively small networks of proteins and other components that function as a module in relative isolation. By contrast, metabolic pathways and signalling networks often contain hundreds of components. Combinatorial explosion then thwarts even the cleverest analysis. Of course, the electrophysiological networks we have analysed sit 'on top' of those metabolic and signalling networks, so their 'isolation' is only relative. As we expand the models to incorporate those networks we will face the same kind of difficulty. I see this as one of the greatest challenges facing systems biology. I have not yet read a convincing strategy for meeting this challenge.

6.6.3. *Contribution to systems biology*

For me, at least, this work opened the way to appreciating the difference between the differential and integral approaches to gene-phenotype relations (Noble, 2011a, 2011b).

References

Paper reprinted in Part 2:

Noble, D., Denyer, J.C., Brown H.F. *et al.* (1992). Reciprocal role of the inward currents $i_{b,Na}$ and i_f in controlling and stabilizing pacemaker frequency of rabbit sino-atrial node cells, *Proceedings of the Royal Society B*, **250**, 199–207.

Others:

Brown, H.F, DiFrancesco, D. and Noble, S.J. (1979). How does adrenaline accelerate the heart? *Nature*, **280**, 235–236.

Dawkins, R. (1976, 2006). *The Selfish Gene*, OUP, Oxford.

Dawkins, R. (1982). *The Extended Phenotype*, Freeman, London.

DiFrancesco, D. and Noble D. (1985). A model of cardiac electrical activity incorporating ionic pumps and concentration changes, *Philosophical Transactions of the Royal Society B*, **307**, 353–398.

Editorial. (2010). The human genome at ten, *Nature*, **464**, 649–650.

Hillenmeyer, M.E., Fung, E., Wildenhain, J. *et al.* (2008). The chemical genomic portrait of yeast: Uncovering a phenotype for all genes, *Science*, **320**, 362–365.

Kiyosue, T., Spindler, A.J., Noble S.J. *et al.* (1993). Background inward current in ventricular and atrial cells of the guinea-pig, *Proceedings of the Royal Society B*, **252**, 65–74.

Kohl, P., Crampin, E., Quinn, T.A. *et al.* (2010). Systems biology: An approach, *Clinical Pharmacology and Therapeutics*, **88**, 25–33.

Noble, D. (2002). Is the genome the book of life? *Physiology News*, **46**, 18–20.

Noble, D. (2006). *The Music of Life*, OUP, Oxford.

Noble, D. (2008). Genes and causation, *Philosophical Transactions of the Royal Society A*, **366**, 3001–3015.

Noble, D. (2010). Funding the pink diamonds: A historical perspective, *Notes and Records of the Royal Society*, **64**, 97–102.

Noble, D. (2011a). Differential and integral views of genetics in computational systems biology, *Journal of the Royal Society Interface Focus*, **1**, 7–15.

Noble, D. (2011b). Neo-Darwinism, the modern synthesis, and selfish genes: Are they of use in physiology? *Journal of Physiology*, **589**, 1007–1015.

Noble, D., Noble, P.J. and Fink, M. (2010). Competing oscillators in cardiac pacemaking: Historical background, *Circulation Research*, **106**, 1791–1797.

Noble, D. and Noble, S.J. (1984). A model of S.A. node electrical activity using a modification of the DiFrancesco–Noble (1984) equations, *Proceedings of the Royal Society B*, **222**, 295–304.

Chapter 7

The Physiome Project

7.1. The First Massively Parallel Computers

In 1989 I was invited to the University of Minnesota to give a lecture at the Bacaner Research Awards ceremony. Marvin Bacaner, a physiologist at the university since the 1960s, was responsible for the development of the anti-arrhythmic drug, bretylium, which works by releasing noradrenaline and then inhibiting further release from nerve endings. He had given the University of Minnesota some of the substantial royalties from bretylium to found the prizes and lecture, and later (in 2000) gave a further $0.5 million to help found a chair at the university.

My lecture described the development of cellular models of the heart. This was shortly after the development of the DiFrancesco–Noble, Hilgemann–Noble and Earm–Noble models. Interest in using these and in extending them was already high around the world. After describing these developments at the cellular level, I ended the lecture by lamenting that we would have to wait until the twenty-first century to see the use of such models in reproducing activity at the levels of multicellular tissue and the whole organ. I was guessing that it would take that long for computational power to reach the required level. During discussion, though, I was told by Rai Winslow, then at Minnesota but now at JHU, that I was wrong. The computing power was already there in the form of the US Army High Performance computer installation coming online at the US Army High Performance Computing Research Center (AHPCRC) at the University of Minnesota. This machine was a giant in those days. Called the Connection Machine CM-2 and produced by Thinking Machines, Inc. it achieved its power by having 64,000 processors ganged together in a grid (Hillis, 1989).[1] Connection Machine was the right name for it.

Rai had already seen that this computer was ideal for some of the tissue-level simulations he had in mind. He had already used it to develop biophysically detailed models of the horizontal cell network in the retina. The structure of the machine, with each processor communicating directly with the neighbouring processors, mirrors the structure of heart tissue, with each cell communicating by electrical connections (the nexus junctions) to its

---

[1]Remarkably, the Connection Machine was Danny Hillis's PhD thesis project with Tomasso Poggio at MIT. For a while, Thinking Machines, Inc. was the leading supercomputer vendor in the world, exceeding Cray in sales.

neighbours. You could therefore map a block of tissue onto the processors themselves, in the simplest case by assigning a cell model to each processor, so generating a block of tissue corresponding to 64,000 cells.

So, I agreed to provide input on the cell models for the study he proposed. For me, this was the start of the Physiome Project, also beginning on the other side of the world in the laboratory of Peter Hunter in Auckland. I will return to Auckland later in this chapter. Three very interesting results came out of the Connection Machine work with Rai and his colleagues (Winslow *et al.*, 1991a; Winslow *et al.*, 1991b; Winslow *et al.*, 1991c).

The first was an answer to a question I had been interested in for a long time, in fact since a symposium in Holland (Bouman and Jongsma, 1982) at which Robert DeHaan (De-Haan, 1982) presented a remarkable result. Two chick ventricle heart cells synchronised their rhythms with, apparently, almost no connection between them! Actually, there were long thin projections between the cells. But, even if these touched, there wouldn't be room for more than a nexus junction or two. DeHaan's electron microscopic images showed such a nexus at the very end of the process (DeHaan, 1982: Figure 4). Could that single nexus conceivably synchronise two cells? During the talk, I did a quick back-of-envelope calculation, starting with the net current flowing in a single cell during the pacemaker depolarisation. This is very small, just a fraction of a picoamp. Then I asked the question, how much current could flow through a single nexus with a voltage difference of just a few mV, knowing the nexus conductance? The calculations matched (Noble, 1982). Although I could not prove it, the result showed that it was plausible that even a single nexus could have allowed the two cells to synchronise.

The proof had to wait for the Connection Machine simulations. We took cell models of the sinus node with different intrinsic frequencies and connected them together with different levels of conductance between the cells. Even four or five gap junctions with a conductance of 50 pS were sufficient to allow the complete network to synchronise (Winslow *et al.*, 1992).[2] At that time, we also knew that the connexin 43 (the protein that forms the nexus junctions) density in the sinus node is very low compared to the atrium and ventricle. The electrical connections between sinus node cells are indeed weak compared to those between atrial or ventricular cells. Problem solved.

The second problem we tackled involved introducing known spatial variations in cell properties within the sinus node. When isolated, cells at the centre of the node show a relatively small voltage variation and a lower frequency, whereas cells at the periphery display larger voltage changes and a somewhat higher frequency. The question was: What determines where the impulse starts in the node? In the intact heart, the impulse usually starts at or near the centre of the node, although the precise location changes with autonomic nerve and hormonal activity (Bouman *et al.*, 1968; Meek and Eyster, 1914; Shibata *et al.*, 2001).

---

[2]See related papers reprinted in Part 2, pages 498 and 516.

The computation showed precisely the wrong result. The initiation of the impulse started at the periphery and conducted towards the centre. I shall never forget the first time I presented the lovely film that Rai and his team produced of this result at a meeting in the UK. I simply showed it as an example of what could be achieved in multi-level simulation using the Connection Machine, but admitted that there was a problem with the result and that we would be investigating ways in which we could understand that, and what would be required for the initiation to begin towards the centre.

Fortunately, Mark Boyett (then working in Leeds) was in the audience. His comment was dramatic. 'You don't need to look any further. The result is correct!' I am ashamed to say that I did not know of the work he had done on this question, but his experiment was precisely what was needed. He had shown that if you carefully cut around the sinus node to isolate it from the atrial tissue, the impulse does indeed start at the periphery. The atrial cells effectively suppress the pacemaker depolarisation in the periphery by supplying hyperpolarising current. Boyett and his colleagues subsequently analysed this phenomenon in more detail than we did (Toyama *et al.*, 1995).

The third major contribution from the work on the Connection Machine was designed to answer a question that is relevant to mechanisms of arrhythmia. One of the advantages of the Hilgemann–Noble and Earm–Noble models is that they can reproduce the spontaneous depolarisations in sodium-overload conditions that are attributable to cycling of the intracellular calcium oscillator. In conditions where intracellular sodium is high, intracellular calcium also increases since the sodium gradient driving the calcium efflux through the sodium-calcium exchanger is weakened. This is the mechanism that was first studied by Dick Tsien working with Jon Lederer (see Chapter 5). An important question is how extensive must the sodium-overloaded tissue be in order to generate a propagated extra-action potential. This was an ideal question for the Connection Machine since, once again, a sheet or block of cells could be modelled in an efficient way.[3] The answer was that the region of overloaded cells does not have to be very large. In a network of about a quarter of a million cells, only about 1000 cells with sodium overload can induce a propagating ectopic beat. The explanation for this result is that the sodium-calcium exchanger is very effective as a current generator. Since the current it carries rapidly increases as the membrane is hyperpolarised (see Figure 5.3), the mechanism resists hyperpolarisation. This result could be relevant to questions concerning arrhythmic mechanisms in ischemia. The ischemic region need not be very extensive in order to generate an arrhythmia.

7.2. Auckland and the Virtual Ventricle

My collaboration with Peter Hunter goes back to the 1970s, when he was in Oxford and we worked on some of the analytical mathematics of excitation and conduction in nerve and muscle described in Chapter 3.

---

[3]See paper reprinted in Part 2, page 544.

My first visit to his laboratory in Auckland was in 1990 when I was invited to be the Butland Visiting Professor. My journey there was almost as memorable as the flight on Concorde since it is the only time I have flown right across the world in first class. I arrived at the Air New Zealand desk at Heathrow to be informed that the flight was full and that I couldn't be fitted in! Initially, that was all I heard. My mind was already on the question of how to communicate to my hosts that I would, presumably, be coming at least a day late. Then I noticed that I was being given the boarding pass saying first class, and an invitation to the first-class lounge. That was my introduction to a top New Zealand white wine as I was pampered and cosseted all the way to Los Angeles and then on to Auckland. Actually, it was far better than the Concorde experience. To have a completely flat bed on such a long journey was a miracle.

I arrived early in the morning to be ushered immediately into the laboratory where Bruce Smaill, Peter Hunter and their collaborators were developing an accurate computer model of the anatomy of the ventricle. I am not sure I paid as much attention as I should have done to what they were doing. The problem was not the jet lag (as explained in Chapter 1, I don't find that a big problem). It was rather that the Marlborough Sauvignon Blanc, from the South Island, served to the first-class passengers, had made a deep impression on me, probably more than it should have done. If you appreciate top-quality Sancerre, the nearest French equivalent, you will know what I mean. One respected wine critic has described Marlborough Sauvignon Blanc as like 'having sex for the first time' (Taber, 2005: p. 244). I am not sure I agree with that – Naomi Wolf (1998: p. 133) asks 'Is that it?', echoing many people's first-experience disappointment – of sex, not wine! Anyway, my first experience of a top New Zealand white was simply magical. What a way to begin what was to become the Physiome Project!

So, what was the Auckland team doing? They were using stereotactic instruments to computerise a whole heart by progressively shaving off very thin layers and recording the fibre orientations over the entire exposed surface, until a complete three-dimensional image was created in the form of a database. The idea was that this database could eventually be combined with equations for the mechanical, electrical and biochemical properties of the cells in each region of the heart to produce the world's first virtual organ. The critical characteristic of the database was that it included information on the fibre orientation since this determines both the direction in which force is generated when the cells contract, and the preferential route for conduction of the electrical impulse.

The Auckland team was also developing the mathematical framework for characterising this tissue structure in a way that could be used in the solution of the governing physical laws that determine organ behaviour. Most engineering structures (cars, bridges, aircraft) are designed using a numerical technique called the 'finite element method' that considers a continuous material as a collection of small regions ('finite elements') linked together through points ('nodes') on their boundaries. It is a 'divide and conquer' strategy that is well suited to dealing with complex geometries and inhomogeneous (spatially varying) material properties on digital computers. The geometry and material structure of biological organs such as the heart are of course highly complex, so it is no surprise that these techniques developed for the world of engineering should be equally relevant to the world of physiology. There is, however, a very close relationship between organ geometry and tissue structure that is not often seen in man-made engineering structures. The particular finite element methods developed by the Auckland group in the 1980s were designed to capture this close relationship and have now become the basis for much subsequent physiome modelling of the body's organ systems.

This work beautifully complemented the tissue-level work that was being done in collaboration with Rai Winslow. But the computational demands created by simulating at the level of the whole organ were even larger than for simulating multicellular blocks of tissue. The simulations described in Chapter 4 that were done by Rai Winslow and his team to reconstruct the T wave of the electrocardiogram illustrate the problem. It took around six weeks work on a supercomputer to complete that project. Even today, we are still struggling to harness enough computing resources to deal effectively with the challenges of whole organ simulation. The ten petaflop computer being constructed by Fujitsu in Japan, on which we have been offered time by the Japanese Ministry of Science, will still only just ensure that the simulations run in real time, i.e. will take no longer than the heart itself takes to complete a cycle.

It should be noted also that even the most complete of the cell models of the heart do not represent more than a small fraction of the total number of components (i.e. genes, proteins, metabolites, membrane structures) involved. Certainly, we should be using whatever computing power comes our way as the speed and capacity rapidly increase. But, I think it is also likely that the practical and theoretical limits to this approach will force us to consider alternatives to modelling the body. I will return to this question at the end of Chapter 9.

7.3. The Physiome Project of the International Union of Physiological Sciences (IUPS)

In 1993, I was responsible, as Chairman of the Organising Committee, for arranging the World Congress of Physiological Sciences held in Glasgow. Nearly 5,000 people attended the meeting, one of the largest IUPS congresses to be held. We used the opportunity to publish a book, *The Logic of Life* (Boyd and Noble, 1993), which was distributed to all members of the congress. I see that book as an early precursor of what was to lead eventually to the Physiome Project and to the physiological aspects of systems biology. It was also at this congress that Sydney Brenner gave a plenary lecture in which he challenged physiologists to respond to the need to interpret the genome.

My colleagues and I met after the congress to discuss how to react to that challenge. By then, I also knew that I was to become the next Secretary-General of the IUPS. I served in that role from 1994 to 2001, with two world congresses occurring during that period. The first was at St Petersburg in 1997. By that time, Jim Bassingthwaighte in Seattle had already proposed the idea of the Physiome Project. The ending '–ome' means 'totality', of course. Thus, genome refers to the complete set of DNA of an organism. 'Physiome' therefore means the totality of physiological function in an organism. That is precisely what physiology is, or should be. Why, therefore, invent a different word?

The main reason, I think, was to widen the scope of the project well beyond the discipline of physiology itself, to include the bio-engineers, mathematicians and computer scientists, amongst others from the physical sciences, who would need to bring their skills to bear on what is essentially a multidisciplinary project, to reconstruct the organism mathematically. We did not know then that, within a few years, similar ideas on the role of mathematics and computer modelling in biology would develop within the context of systems biology. I will discuss the relationships between systems biology, the Physiome Project and the Virtual Physiological Human (VPH) project later in this chapter.

The formal launch meeting of the Physiome Project was held just after the St Petersburg congress. About 40 of us stayed at a beautiful but rundown former palace on the coast to the west of the city, near Petrodvorets, where the famous Peterhof Palace is located. The outcome was the formation of the Physiome Committee of the IUPS, with Peter Hunter from Auckland as chair, and Sasha Popel from Johns Hopkins University as co-chair. This committee has run very successful symposia and satellite meetings at the subsequent international congresses in Christchurch (2001), San Diego (2005) and Kyoto (2009). The project has moved from small beginnings to see substantial funding, particularly in the USA, UK and Japan, while the EU has launched the related VPH project.

In the early days, however, funding was not easy to find. This was one of the reasons why Rai Winslow, Peter Hunter and I were involved for several years in the development of a venture capital company, Physiome Sciences, Inc., which attracted substantial funding both for the research and for its exploitation. In its last round of funding, it raised around $50 million of capital investment and had a substantial facility located in Princeton, New Jersey. The work done for Roche (Chapter 4) was an early success for exploitation and encouraged us to think that a viable business model was possible. The involvement of many large pharmaceutical companies in the publicly funded EU successor to the project, PreDiCT (Chapter 4) shows that may well have been true. Two factors, however, ensured that this venture capital approach did not survive. One was that we probably underestimated the time it would take to develop the work to a fully exploitable stage. The second was the catastrophic effect of the events of 9/11 on the venture capital biotech market. Quite simply, the sentiment of the market completely switched after the fall of the Twin Towers.

There are, however, some lasting gains that are attributable to Physiome Sciences, Inc. It funded important work on the Physiome Project at a time when public funding was not available on the scale required. The Johns Hopkins, Oxford and Auckland teams benefited substantially from that funding. And it raised awareness of the potential for applications of biological modelling work in the pharmaceutical world. Some of the subsequent collaborations, particularly with Novartis, owe their origins to the work of Physiome Sciences, Inc.

Possibly the most important legacy of the Physiome Sciences era was the CellML standard. Very few mathematical models of biological processes are published in the peer-reviewed scientific journals without either errors (typographical or otherwise) or missing parameter values. The more complex the model, the more difficult it is to reproduce the published results from the equations laid out in the publication. One strategy for addressing this problem is to include computer code on a website, but even that is often unsatisfactory because the equations implemented in the code may not exactly match the ones written down. The only solution is to code the mathematics in an unambiguous 'marked-up' form, rather like the HTML standard used for web pages. CellML is an XML (eXtensible Markup Language) framework for specifying the equations, together with their units and information about the model structure. It provides the syntax or 'grammar' for the model, while the biophysical and biological meaning – the semantics – of variables and parameters are specified with ontologies via the CellML metadata standard. The equations and computer code for a model can be generated automatically from the CellML syntactical encoding and the ability to combine models as components of more complex models is facilitated by the semantic metadata. The CellML standard was developed by a team from Auckland University and Physiome Sciences, primarily under the leadership of Poul Nielsen from Auckland. A related but independent effort at Caltech on SBML (Systems Biology Markup Language) began shortly after. These two efforts and a third one from Auckland around the 'FieldML' standard for encoding spatial information are now at the heart of the Physiome Project and are providing a robust foundation for biological modelling.

7.4.   The Coming of Systems Biology

As I write this chapter, there are nearly 3000 papers published with the phrase 'systems biology' in their titles or abstracts. Of these, only 2 were published before 2000, and 90% were published in the 5 years after 2005. What has happened? And how does the Physiome Project relate to this sudden development?

One explanation for the development of 'systems biology' as an idea is that the Human Genome Project, completed in draft form at the turn of the millennium, was the natural culmination of the reductionist drive to burrow right down to the smallest elements in an organism, the DNA sequences that are responsible for transmitting genetic information from one generation to its successors. Having reached this level, though, where next could the reductionist agenda take us? It might be plausible to imagine, as some do, that the whole of life is encoded in DNA (needless to say, I am not amongst those who think this way), but it is clearly implausible to suppose that it can be encoded, in anything other than an extremely general sense, below that level. I suppose that strict determinists would need to postulate that life is indeed inherent in the properties of matter even at the level of, say, string theory. But that is a very general sense of 'inherent' that is, at present at least, of no scientific value. We have no way either of coherently formulating such a hypothesis or of submitting it to any kind of experimental test. In the reductionist analysis, DNA really is the bottom level with which we have to work today. So, having reached that level, do we now understand life?

In *Science Oxford*, a public forum for science, where I recently gave a lecture, I spent some time looking at a huge book on display in the exhibition area. It consisted of the printing of the complete sequence of the DNA in a single chromosome. Occasionally in the book, as you turn the pages, you can see a highlighted section that is known to correspond to a gene. For those sections, the triplets of nucleic acids corresponding to each amino acid would tell you for which protein the sequence forms a template. Huge sections though are just, apparently, gibberish code for which we have only the glimmering of a 'meaning'. Is this book really 'the book of life'? At the least, we have to admit that from being able to scan it in the casual way I was doing in *Science Oxford*, no sudden enlightenment arises. On the contrary, one reads this much as one would read the gibberish binary data that might correspond, on my computer, to the text I am currently writing. At that level, one would need an interpretative program – called Word on my PC – which is of course the level at which the vast majority of us operate. It is part of the purpose of disciplines like bioinformatics to be able to do that with the DNA sequences of the genome.

Undoubtedly, bioinformatics of this kind has made substantial progress in comparing genomes from different species to arrive at some fundamental conclusions about evolution. We need that kind of study of biological systems and how they have developed. But there is a very simple reason why, from the sequences alone, we will not understand life. This is that those sequences do absolutely nothing until they receive signals – transcription factors and epigenetic marking of various kinds – from the rest of the organism, starting in development with the fertilised egg cell and its environment (the mother's womb in the case of mammals). If we wish to attribute meaning to something, it must surely be in those signals. They 'command' the genome to the only act of which it is capable, which is transcription. Well, even that is not strictly true – that act is performed by a dedicated set of proteins working on the DNA sequences.

That is precisely the realisation that leads to the main idea of systems biology. To understand those signals acting on the genome, you have to understand the system of which they are a part. Systems biology can therefore be seen to be a 'bouncing back' to higher levels in the organism itself after having reached the bottom in sequencing its genome.

But that is where the trouble begins. Where in the multi-level system that is an organism do you bounce back to? Some, like Sydney Brenner, would insist that you have to bounce back to the level of the cell. Others might claim that metabolic and developmental pathways will do the trick. Certainly, over 90% of what is called systems biology today is largely at the level of biochemical pathways and the analysis of huge amounts of genomic and proteomic data. Very little of what is being done under its name lies above the cell level.

One of the principles of systems biology that I formulated in an article in 2008 (Noble, 2008) is what I call the principle of biological relativity. It states that, *a priori*, there is no privileged level of causality in biological systems. In an organism in which there are multiple feedbacks and feedforwards between all the levels from the genome to the complete phenotype, and even beyond to its environment, I would say that this insight is a necessary one. The emphasis in what I have just written must be on the *'a priori'*. Before investigating a system we cannot say at which level a particular phenotype is integrated, and therefore understood. That doesn't mean to say that in investigating a particular phenotype we can't discover that, as a matter of fact, it is expressed or integrated at a particular level. Pace-maker activity of the heart, for example, can now be seen to be a function integrated at the level of a cell, though it is also greatly modulated by even higher levels.

It is in this question of level of causality that we can understand the relation between the Physiome Project and systems biology. The Physiome Project and its relative the VPH project (see below), complete the intellectual revolution in biology that systems biology represents. Without them there is no guarantee that systems biology can succeed.

7.5. The Virtual Physiological Human (VPH) Project

The systems approach to biology is being driven by many different streams of activity. One of those streams is the world of computing. As the Physiome Project has developed, it has rapidly become one of the grand challenges for supercomputers. If you want a test-bed for your nation's latest mammoth machine, biology provides that in abundance. So much so, that it has become one of the drivers of supercomputing itself. This is the origin of the European Union's flagship project in this area, the Virtual Physiological Human.

Launched under EU Framework 7, this project, or rather set of projects, is funded by the information and communications technology (ICT) section, and is therefore a good example of this area of technology acting as a major driver of work on systems biology focused at the higher levels of function.

7.6. Conclusions

7.6.1. *Pluses and contributions to systems biology*

The development of the Physiome Project[4] is essential to the success of the systems biology approach. This is appreciated by some of the major players in systems biology. I recently took part in a debate with Hans Westerhoff, who has successfully championed the systems approach worldwide in many ways. The strange situation is that, although his own focus has been on systems biology of yeast, he was arguing *for* the Virtual Human, while I was acting the part of the sceptic! That is how it should be. An approach that promises everything 'in a decade or two' will fail. To avoid the publicity over-reach of the Human Genome Project, we need the debate, and we all need to be sceptical. Science thrives on scepticism.

---

[4]See papers reprinted in Part 2, pages 531 and 535.

7.6.2. *Minuses*

I am not sure, however, that there are many like Hans Westerhoff in the systems biology camp. I would like to be proved wrong. Part of the problem is that, with the decimation of classical physiology during the decades of dominance by the molecular biological approach, we don't seem to have many world leaders left to argue the case. Or perhaps people are, naturally, more concerned with their next grant request? A major reason why I accepted the nomination to become the President of the IUPS in 2009 is that I see outreach towards our fellow scientists and the general public as a major role that an international union should be playing. This is also a suitable point at which to acknowledge the role of bioengineering in 'keeping the faith'. It is no accident that many of the major proponents of the Physiome Project – Jim Bassingthwaighte, Peter Hunter, Sasha Popel and Rai Winslow – are all bioengineers, and that many graduate students and postdoctoral fellows who are joining the work come from the physical sciences.

References

Papers reprinted in Part 2:

Bassingthwaighte, J.B., Hunter, P.J. and Noble, D. (2009). The cardiac physiome: Perspectives for the future, *Experimental Physiology*, **94**, 597–605.

Cai, D., Winslow, R.L. and Noble, D. (1994). Effects of gap junction conductance on dynamics of sinoatrial node cells: Two-cell and large-scale network models, *IEEE Transactions on Biomedical Engineering*, **41**, 217–231.

Noble, D. (2002). Modeling the heart: From genes to cells to the whole organ, *Science*, **295**, 1678–1682.

Winslow, R.L., Kimball, A., Varghese, A. *et al.* (1993). Simulating cardiac sinus and atrial network dynamics on the Connection Machine, *Physica D: Non-linear phenomena*, **64**(1–3), 281–298.

Winslow, R.L., Varghese, A., Noble, D. *et al.* (1993). Generation and propagation of ectopic beats induced by spatially localized Na-K pump inhibition in atrial network models, *Proceedings of the Royal Society B*, **254**, 55–61.

Others:

Bouman, L.N., Gerlings, E.D., Biersteker, P.A. *et al.* (1968). Pacemaker shift in the sinoatrial node during vagal stimulation, *Pflügers Archiv: European Journal of Physiology*, **302**, 255–267.

Bouman, L.N. and Jongsma, H.J. (1982). *Cardiac Rate and Rhythm*, Martinus Nijhoff, The Hague.

Boyd, C.A.R. and Noble, D. (eds), (1993). *The Logic of Life*, OUP, Oxford.

DeHaan, R.L. (1982). 'In vitro models of entrainment of cardiac cells', in Bouman, L.N. and Jongsma, H.J. (eds), *Cardiac Rate and Rhythm*, Martinus Nijhoff, The Hague, pp. 323–359.

Hillis, D. (1989). *The Connection Machine*, MIT Press, Cambridge, MA.

Meek, W.J. and Eyster, J.A.E. (1914). The effect of vagal stimulation and of cooling on the location of the pacemaker within the sino-auricular node, *American Journal of Physiology*, **34**, 368–383.

Noble, D. (1982). 'Discussion', in n Bouman, L.N. and Jongsma, H.J. (eds), *Cardiac Rate and Rhythm*, Martinus Nijhoff, The Hague, pp. 359–361.

Noble, D. (2008). Claude Bernard, the first systems biologist, and the future of physiology, *Experimental Physiology*, **93**, 16–26.

Shibata, N., Inada, S., Mitsui, K. *et al.* (2001). Pacemaker shift in the rabbit sinoatrial node in response to vagal nerve stimulation, *Experimental Physiology*, **86**, 177–184.

Taber, G.M. (2005). *Judgment of Paris: California vs France and the Historic 1976 Paris Tasting that Revolutionized Wine*, Scribner, New York.

Toyama, J., Boyett, M.R., Watanabe, E. *et al.* (1995). Computer simulation of the electrotonic modulation of pacemaker activity in the sinoatrial node by atrial muscle, *Journal of Electrocardiology*, **28 Supp 21**, 2–5.

Winslow, R., Cai, D. and Noble, D. (1992). Effects of gap junction conductance on oscillation properties of coupled sino-atrial node cells, *IEEE Computers in Cardiology*, **1992**, 579–582.

Winslow, R., Kimball, A., Noble, D. *et al.* (1991a). Computational models of the mammalian cardiac sinus node implemented on a Connection Machine CM-2, *Medical and Biological Engineering and Computing*, **29**, 832.

Winslow, R., Kimball, A., Noble, D. *et al.* (1991b). Modelling large SA node-atrial cell networks on a massively parallel computer, *Journal of Physiology*, **446**, 242P.

Winslow, R., Kimball, A., Noble, D. *et al.* (1991c). Simulation of very large sinus node and atrial cell networks on the Connection Machine CM-2 massively parallel computer, *Journal of Physiology*, **438**, 180P.

Wolf, N. (1998). *Promiscuities: A Secret History of Female Desire*, Chatto and Windus, London.

Chapter 8

Fifty Years On

8.1. Still At It

5 November 2010, the date on which I wrote this chapter, marks 50 years from the 1960 articles in *Nature*. I am still modelling cardiac cells, as the articles reprinted for this chapter show. The Ten Tusscher model in 2004 was one of the first to incorporate experimental data from work on human cells.[1] The 2007 article[2] was based on a Hodgkin–Huxley–Katz Prize Lecture given for the Physiological Society in 2004 in which I outlined how we progressed from the Hodgkin–Huxley equations of 1952 to the development of a virtual heart.

There are now around 100 cardiac cell models on the CellML website, all downloadable and ready for anyone anywhere in the world to use. They can do so on laptops vastly more powerful than the Mercury computer I used in 1960. The range of species and cell types grows every year. This range of species and cell types has proved valuable in collaborations with pharmaceutical companies who need ways in which we can more reliably extrapolate from common laboratory species like the mouse, rat and rabbit, to the human. They also need to resolve the QT problem (the arrhythmic side effects of many drugs) referred to in Chapter 4. Projects being funded by the European Union are tackling that problem (Noble, 2008a; Fink and Noble, 2009; Fink and Noble, 2010; Garny *et al.*, 2009; Niederer *et al.*, 2009; Stewart *et al.*, 2009; Rodriguez *et al.*, 2010).

Higher-level modelling, of tissue and of the whole organ, has also developed to the stage at which there are teams active around the world. The cardiac Physiome Project is in good form (Bassingthwaighte *et al.*, 2009; Hunter *et al.*, 2006). It is also being used as a paradigm for comparable work on other organs and systems of the body, including the kidney, liver and lung.

8.2. Who Was He Fifty Years Ago?

So, if I could ask the 23-year-old walking down the worn steps to the basement room where the Mercury computer lived in 1960 where he thought his research was leading,

---

[1]Reprinted in Part 2, page 551.
[2]Reprinted in Part 2, page 568.

what would he have said? Not much, probably. How many students see much beyond their thesis when they are writing it? He would have been astonished by the publication 15 years later of *Electric Current Flow in Excitable Cells* (Jack *et al.*, 1975) with 'you don't know enough mathematics!' ringing in his ears and the humbling discussion with Andrew Huxley on Bessel functions of imaginary arguments still to come. He would have been even more astonished with election to the Royal Society four years after that. He certainly would not have anticipated the congratulatory letter from JZ Young, his former anatomy professor, mixing the congratulations with the strange comment that he had 'waited rather a long time'. JZ, as he was always called, was elected very young indeed to the Royal Society, at the age of 38.

But what would astonish my 23-year-old self the most would be the philosophical journey. He was about as reductionist as you could get in 1960, so much so that I hardly recognise him as the same person. His first interactions with philosophers were motivated by the wish to explore that position. He even wrote articles on determinism for one of the UCL journals flourishing at that time. One of his professors, DR Wilkie, a muscle biophysicist, liked it and asked him when the sequel was going to appear. It never did! The reasons are now obvious. He was barking up the wrong tree, but simply didn't know it. A remark by Stuart Hampshire, then Professor of Philosophy at UCL, began the process of finding the right tree. The 23-year-old tried to give an account of action at one of Hampshire's graduate philosophy classes. The argument was based on distinguishing between causes within the organism and causes outside the organism. The response was 'you need to read Spinoza'. The relevant statement is definition VII in *The Ethics*: 'That thing is called free, which exists solely by the necessity of its own nature, and of which the action is determined by itself alone.' Needless to say, Hampshire had just published a book on Spinoza (Hampshire, 1956).

Spinoza was, of course, also the arch anti-dualist. His *Ethics* laid out the ground for opposition to Descartes' view of the relation between mind and body.[3] I read Hampshire and other material (Elwes, 1951) on Spinoza as a graduate student, but I must have squirreled it all away, for it didn't really play a major overt role in my gradual shift to a completely different philosophical position, despite the fact that Chapters 9 and 10 of *The Music of Life* are strongly non-dualist. It is impossible to exclude a subconscious influence of course. But what I can be sure of is that my 23-year-old self would not have understood the arguments well enough to make effective use of them.[4]

---

[3]Opposition to Descartes is most clearly spelled out in Spinoza's correspondence in 1661 with Oldenburg (an early secretary of the newly founded Royal Society) in which he argues that the will is not a cause. Spinoza was also, in spirit, a systems biologist. He writes: 'every part of nature agrees with the whole, and is associated with all other parts' and 'by the association of parts, then, I merely mean that the laws or nature of one part adapt themselves to the laws or nature of another part, so as to cause the least possible inconsistency'. He had an excellent understanding of the problems faced in trying to understand what we, today, would characterise as an open system.

[4]That has become clear to me in a dramatic way recently. This quotation from one of Spinoza's letters could have been, but wasn't, the basis of my use of the Silman stories in *The Music of Life*: 'Let us imagine, with your permission, a little worm, living in the blood, able to distinguish by sight the particles of blood, lymph etc, and to reflect on the manner in which each particle, on meeting with another particle, either is repulsed,

In fact, when I contemplate that person, and try to think myself back into his skin, I am surprised at his naïvety. Some of my student friends in philosophy at that time must also have wondered what hope there was for this strange mixture of cardiac experimentalist, mathematical modeller and naïve philosopher. He wasn't exactly brilliant at any one of them. He happened to like combining them, and that was unique enough to plough completely new ground. Naturally, some of those friends were women. He was greatly influenced by some of them. Notice that my philosopher in Chapter 9 of *The Music of Life* is a woman, so also is the brilliant space-travelling linguist in Chapter 10. In the heavily smoke-filled seminar classes given in the philosophy department, many of the students were women.

So that was the milieu in which the 23-year-old me began to grow up. It is hard to imagine those days now. The air of sophistication in those classes seemed to be associated with knowing how to combine making a profound argument with pointed gestures using the ritual of matches and cigarettes. Meanwhile, another physiologist elsewhere in London, Richard Doll at the Central Middlesex Hospital, was busily refining the bombshell he had published in 1950 showing the link between smoking and lung cancer. My 23-year-old self absorbed most of the influences of the seminars – and the brilliant women students – except for the cigarettes. My father had died four years earlier after many years of chain smoking. He died too soon for a young family of brothers. As the eldest, I became a substitute father. The youngest of my brothers, Ray, eventually became an academic at UCL, founding the Centre for Reproductive Ethics. Many of the stages in the transition were argued out between us, and we still have an extraordinary intellectual empathy today. The 50-year journey towards enlightenment has involved a great companion.

8.3. The Threads of Transition

But to return to the central puzzle of this book, what led a young reductionist determinist to progressively abandon the naïve position he held? The immediate background to writing *The Music of Life* will be dealt with in the next, and last, chapter. Here, I am more concerned with the earlier stages. There were two threads in the process.

8.3.1. *Scientific thread*

One was scientific. While studying ion channels in excitable membranes was near the bottom of the reductionist agenda in the 1960s, the grand sweep of molecular biology soon removed that kind of work from its apparently privileged pedestal. There was a critical period around 1980 when I had to decide whether to follow many of my colleagues down to the molecular level. Something made me hesitate. Yes, it was impressive to see the molecular structure of one type of ionic channel after another being revealed. At last, the molecular basis of, for example, the gating process could be unravelled, and it largely

---

or communicates a portion of its own motion. This little worm would live in the blood, in the same way as we live in a part of the universe, and would consider each particle of blood, not as a part, but as a whole. He would be unable to determine how all the parts are modified by the general nature of blood, and are compelled by it to adapt themselves, so as to stand in a fixed relation to one another' (letter XV to Oldenburg, 1663).

Figure 8.1   Denis and Ray Noble singing *Se Canta* – a traditional Occitan song – at a performance of the Oxford Trobadors in the Holywell Music Room, August 2010. This is the concert referred to in the postscript.

confirmed Hodgkin and Huxley's idea of charged regions of the channel that move in the electric field. In the case of the inactivation gate on a sodium channel, this is literally so: part of the intracellular amino acid chain swings just like a gate to open or close the channel. We can also visualise where the individual ions sit in a channel during the process of transport though the channel. The invention of the patch clamp method, for which Neher and Sakmann received the Nobel Prize in 1991, has enabled us to study the opening and closing of individual channel proteins. All of this has been important in establishing the molecular basis of channel protein behaviour.

The problem I could see was not whether this shift to the molecular level was important or worthwhile. It clearly was. The problem was more sociological. There is a strong herd instinct amongst scientists. There is a gold rush each time a new vista is opened up. For technical reasons, that rush tends to be towards lower and lower levels as instruments become finer and more capable of resolving smaller detail. The problem this creates for biological science is that its multi-level nature absolutely requires work at *all* levels. A reductionist would argue that, nevertheless, one starts at the bottom since that has privileged causality. It doesn't (see Chapter 9). Moreover, we often need the spectacles provided by understanding at higher levels in order to interpret the data at the lower levels.

I am instinctively someone who swims against the tide. I think it was at a meeting on ion channels in the heart at which nearly every talk concerned the same question – analysis of single channel current jumps as the channels open and close – that I came to the conclusion that, while this was important molecular biophysical work, it was not what I wanted to do and it was not going to help me much in reconstructing the heart. That judgement was largely correct. Very few models of cell function need to include the individual channels. Stochasticity and averaging out ensures that. This is a problem similar, of course, to the difference between molecular dynamics and thermodynamics. To understand the properties of a gas in large-scale work, the random motion of the individual molecules does not need to be represented.

Swimming against the tide? In Chapter 5, when describing the painful process of the reinterpretation of i_{K2} as i_f, I used the quotation from Dante's *Purgatorio*, where he makes the great Troubadour, Arnaut Daniel, speak of his pain. Those were not Arnaut's own words of course. But we do in fact have his own words for swimming against the tide in this lovely verse in which he describes his poetic acrobats:

> Ieu soi Arnaut qu'amas l'aura
> E chaç la lebre amb lo buòu
> E nadi contra suberna.

> I am Arnaut who gathers up the wind,
> And chases the hare with the ox,
> And swims against the torrent.

I first learnt this poem, and how to pronounce it, from native language speakers in the Périgord, where I have had a village farmhouse for the last 40 years. When I first arrived in the village I had no idea that a language other than French was spoken there. I simply thought that, when I couldn't understand local people, the dialect must be too broad. I had such an experience as a boy when evacuated to Yorkshire during the Second World War – broad Yorkshire dialect spoken by children in the streets was incomprehensible to a London boy. But when I put this idea to a local farmer he bridled: 'Non, Monsieur Noble, ce n'est pas notre dialecte, c'est notre langue!' It was clear that if I was to integrate well into the village I had better learn it. He introduced me to a teacher, Jean Roux (Joan Ros in Occitan), who was deep into studying the Troubadours. He also loves the above words of Arnaut Daniel. Most of what I know has been learnt from that family. I write 'family' because in a strange twist of fate, his son, Etienne (Esteve) is also not only a brilliant manipulator of the language (naturally enough) but also a physiologist working at the University of Bordeaux. We have collaborated (Marhl *et al.*, 2006; Roux *et al.*, 2001; Roux *et al.*, 2006) in extending the physiome modelling approach to the lung. All our email correspondence, even on the science, is in Occitan. When receiving an honorary doctorate from the University of Bordeaux, I was delighted to be able to use the language in my reply.

8.3.2. *Philosophical thread*

The second strand in the transition was philosophical. I wrote in Chapter 2 that a major attraction for me when I left UCL was the extraordinary reputation of Oxford philosophy. It wasn't long before the naïvety started to wear off and was replaced with a more professional... well, at least a veneer of professionalism.

My first philosophical interaction was with Dick Hare, the author of *The Language of Morals* (Hare, 1963), and *Freedom and Reason* (Hare, 1965), with whom I discussed the nature of pain. I didn't realise at first that he had been a prisoner of war in a Japanese camp from the fall of Singapore in 1942 right through to the end of the war in 1945. Not surprisingly, his interest in pain and in the way in which philosophy could address the question of the harshest conditions in which humans might live brought an immediacy and urgency to what he wrote and talked about. That, in itself, was fascinating enough to a young physiologist – recall that I still had a neural section of my laboratory in those days. But also of great interest was that, in *Freedom and Reason*, Hare approached ethics almost like a scientific experiment. A moral statement can be compared to a conjecture (hypothesis), which might be refuted if it failed various tests (in particular the test of universality – could it apply regardless of the individual to whom the moral statement was applied?). In effect, Hare introduced a version of Karl Popper's principle of falsifiability. I was sufficiently attracted to this idea that I used it once in a sermon in the Balliol College Chapel – successive chaplains in the College have been broadminded enough not to insist on a strict interpretation of what a sermon should be. Years later I even gave a sermon from a Buddhist perspective (Noble, 2008b). Those who have read Chapter 10 of *The Music of Life* will know what that was about. All this was important enough, but even more relevant was learning from Hare some of the significant distinctions, such as when a movement is an action. Moral philosophy requires that, of course, but so does analysis of the physiological basis of behaviour.

Which leads me naturally on to the next interactions, which were with Anthony Kenny, who has written far more philosophical books than I could even list here, and Alan Montefiore, a philosopher with strong connections to contemporary French philosophy. They both introduced me to a work by Charles Taylor, *Explanation of Behaviour* (1964). There were several important outcomes of those interactions. Taylor and I debated some of them in articles in *Analysis* (Noble, 1967a, 1967b). Initially, Tony Kenny and I thought that my article was a straightforward knockout of Taylor's ideas – I can't now recall whether Alan agreed with that. When Taylor replied, however, Tony Kenny immediately commented that he had not anticipated how good the reply might be. This introduced me to a fundamental concept in explanation. The knockout nature of my attack was based on what the case is when one considers a single occurrence of an event (specifically in this case a behaviour) where it was easy for me to show that for any higher level (e.g. teleological) explanation there must always be a difference at the lower level which could also count as the explanation of the event. Taylor responded with the idea that there could be order at the higher level that could form the basis of an explanation. By contrast, while there would be individual differences at the lower level in each case, they might not, as a set, conform to an explanatory order.[5] Those who are familiar with my argument between a physiolo-

---

[5]This account is necessarily brief. While I accepted Taylor's reply, it does involve a substantial shift to what I called a conceptualist's view of the problem (the title of my second paper) since it rests on whether a set of correlations at the lower level can or cannot be said to conform to some explanatory order. That is a conceptual question (what do we accept as explanatory?) not just an empirical question (does the set in fact conform to any particular order?). All that matters in the present context is that, while an explanation of a phenomenon at one level necessarily requires that there should be events at the lower level that are different depending on whether the higher level functionality exists, there is no *guarantee* that these will form the basis of an explanation.

gist and a philosopher in Chapter 9 of *The Music of Life* will recognise that this is a distant origin of one of the main ideas in that argument. I am now much closer to Taylor's position than I originally thought, though I did also acknowledge this in the second *Analysis* article (Noble, 1967b). Some of these ideas also found their way into Tony Kenny's book, *The Five Ways* (1969). These philosophical interactions are clearly the basis of the realisation that order may exist at one level of a biological system but not at lower levels, even though the higher-level states must have corresponding differences at the lower level.

Important though these interactions were, the really sustained formative interaction for me was with Alan Montefiore. As I have already indicated, Alan straddles the Anglo-Saxon and French worlds of philosophy. We were a natural pair for discussions since we share a fascination with French culture and language. We followed on from the debate with Charles Taylor by running for some years a graduate class in the philosophy of behaviour, on which we were joined by the psychology/zoology fellow of the college, David McFarland, and another philosopher, Kathy Wilkes.[6] Balliol College became a kind of multidisciplinary Centre for the years while those seminars were running. They were successful enough to lead to a book that Alan and I edited (Montefiore and Noble, 1989b) and to which I contributed several chapters (Montefiore and Noble, 1989a; Noble, 1989a, 1989b). Some of this work was also written up for another book (Noble, 1990). This was the context in which I developed the central story of the brain in Chapter 9 of *The Music of Life*. To the insight developed in the debate with Taylor, these discussions added the realisation that many higher-level 'events' are best not described as 'events'. In this sense, an intention is not an event for which we need to look for a particular corresponding neurophysiological event.[7] I will leave the development of this thread to Chapter 9 where I describe the immediate formative elements that led to writing *The Music of Life*. Here I will just note that it can take years, even decades, to allow a set of philosophical ideas to mature. While the discussions with Alan Montefiore were essential to what eventually led to *The Music of Life*, I could not have written that book at the time that *Goals, No Goals and Own Goals* was written. And, almost certainly, I could not have written *The Music of Life* without those extensive interactions with professional philosophers.

The spirit of multidisciplinarity still lives on in Balliol. The seminars, organised with Eric Werner, Tom Melham and Jonathan Bard, now centre on the conceptual foundations of systems biology, involving computer scientists, physicists, economists, biologists, philosophers and many others in exploring the origins and consequences of the systems approach. They form part of the Balliol Interdisciplinary Institute and contributed to its creation.

---

[6]It was an interesting feature of the debates in those seminars that the dividing lines rarely reflected that between science and philosophy. Often enough, Alan and I found ourselves on the same side. So, sometimes, did Kathy Wilkes and David McFarland.

[7]In those debates I used a computer analogy to express this insight. A behavioural event could correspond to the implementation at a particular time of a particular instruction. The installation of the program could correspond to the existence of an intention. An intention is then more like a capability, something that is not pinned down to a particular time, just as a computer program gives the machine its capabilities for as long as the program is installed.

Figure 8.2 Denis Noble and Peter Kohl enjoying a dessert wine at Denis's farmhouse in the Périgord. Neither can now recall what the conversation was about, but this photograph formed part of the inspiration for the story of 'pointing behaviour' in Chapter 9 of *The Music of Life*, although the context of the story was changed.

8.4. Passing on the Baton

I retired from my Oxford chair in 2004. The transition to an Emeritus Professor was relatively painless for a very good reason. Over the preceding years I had already passed the baton of experimental work onto the scientist who inherited my laboratories.

I first met Peter Kohl at a conference in Prague around 1992. He had been working in East Berlin and in Moscow and his work, at that time, was not well known in the West. His earliest publications were in Russian. Nevertheless, and with restricted facilities, he was carrying out some very ambitious experiments on a controversial area of integrative cardiac biology, and I was so impressed with what he showed me that I invited him to come to Oxford as a postdoctoral fellow. That confidence in him was fully justified. His innovative suggestions of myocyte-nonmyocyte electrical coupling, and of mechanosensitivity of the electrical activity of fibroblasts, were eventually proven by his investigations that went very much against the then established views of cardiac tissue structure and function. He brought the area of mechano-electric feedback in the heart to Oxford. There had been no previous work of this kind in the department. He was therefore an independent researcher with original projects even before joining my group in the department at Oxford. That innovative approach has continued as he has developed his own independent research team to produce some lovely results. Many people talk about the *potential* of the systems biological approach; Peter *produces* it.

By 2004, Peter was already effectively running the experimental research labs. Handing responsibility over to him was therefore simply a formality and it also enabled me to continue to interact closely with experimental research while pursuing my work on developing the intellectual base of the systems approach. The only problem we faced was how to manage without the automatic support that came from holding a British Heart Foundation chair. Those funds were critical in enabling me, over the 20-year period for which I held the chair, to finance new initiatives as pilot projects before they were mature enough to justify full grant applications. Somehow, though, we managed the transition and kept the grant money flowing in. There are now strong experimental and computational teams both in the department of physiology, anatomy and genetics and in the computing laboratory. Some of the key papers published since or just before my retirement have been in collaboration with Peter Kohl and his team (Garny *et al.*, 2005; Garny *et al.*, 2009; Iribe *et al.*, 2006; Kohl *et al.*, 2000; Kohl and Noble, 2009; Kohl *et al.*, 2010; Rodriguez *et al.*, 2010). More recently, Peter has also brought his own critical eye to help in the clarification of what systems biology is about (Kohl and Noble, 2009; Kohl *et al.*, 2010).

8.5. Conclusions

8.5.1. *Pluses*

Tracing the threads of a philosophical journey is a major challenge. In these chapters I have woven together the professional and the personal in order to reveal at least part of what makes a scientist like me 'work'. It is a story with apparently conflicting themes. On the one hand, it is a 50-year journey along a single path towards understanding a small part of the heart: the electrical signals that trigger the heartbeat. On the other hand, it is a maze rambling through the gardens of philosophy, music, languages, cuisine, and much else, yet in an integrated way that leads naturally to enable a book like *The Music of Life* to be written.

From the Oriental tradition I have encountered two poetic inspirations for these contrasting journeys. The first was a favourite of the distinguished Japanese cardiac electrophysiologist, Hiroshi Irisawa, who liked to quote the poem of Ikkyū Sōjun (1394–1481): 'To each fisherman, just one rod',[8] clearly the motto for a single-minded highly focused search.

The contrasting poem comes from a much earlier Korean monk, Won Hyo 元曉 (원효) (617–686), who can be regarded as a predecessor of Ikkyū since he was a comparable heretic, breaking almost the same rules as he did. He wrote: 'Never follow one discipline, but never neglect any discipline', which might be the motto for cross-disciplinary work.

---

[8]In Japanese, 漁父生涯竹一竿, or 漁夫生涯竹一竿 (*gyo-fu shougai take ikkan*). This poem is attributed to the great Buddhist monk, IkkyūSōjun (一休宗純) who was also the subject of my sermon to the Balliol College Chapel, 'Meditation on two heretics' (Noble, 2008b). He was one of the two 'heretics' in that sermon, having broken several of the central rules of Buddhism, yet he became the highly respected Abbot of Daitokuji, the temple of 'great virtue' in Kyoto, which he restored. The glory of what he restored can be appreciated by wandering and meditating, as I have done, through the gardens and buildings of that beautiful temple.

8.5.2. *Minuses*

The dangers of autobiography are well known. Recollected stories are riddled with gaps, and worse, because of the vicissitudes of memory. I have tried to avoid the worst errors of this kind by plaguing my colleagues with questions to check my memory. Embellishment and poetic licence also creep in. So much so as time goes by that, eventually, it is almost impossible to disentangle the story from its embellishments. If you think you can do so, remember the 1950 Japanese film *Rashomon* (羅生門) by Akira Kurosawa (黒澤明). Rashomon was a gate to the city of Kyoto, ruined after a war. A woodcutter and a priest take refuge in the gatehouse from a downpour. They and other witnesses then relate the story of a rape and murder. The stories conflict, completely. Yet, each of the witnesses saw the same event. Each account creates distance from the 'truth' by weaving in various justifications. It is one of the most powerful movies I have ever seen. If you think that only what you see with your own eyes is the truth, you need to watch this film.

8.5.3. *Contribution to systems biology*

Without this journey I would not have been ready to embrace the ideas of systems biology and to try and develop them. I will take this theme up more completely in the final chapter, Chapter 9.

References

Papers reprinted in Part 2:

Noble, D. (2007). From the Hodgkin–Huxley axon to the virtual heart, *Journal of Physiology*, **580**, 15–22.
Ten Tusscher, K.H.W.J. *et al.* (2004). A model of the human ventricular myocyte, *American Journal of Physiology*, **286**(4), H1573–1589.

Others:

Bassingthwaighte, J.B., Hunter, P.J. and Noble, D. (2009). The cardiac physiome: Perspectives for the future, *Experimental Physiology*, **94**, 597–605.
Elwes, R.H.M. (1951). *The Chief Works of Benedict de Spinoza*, Dover, New York.
Fink, M. and Noble, D. (2009). Markov models for ion channels – versatility vs. identifiability and speed, *Philosophical Transactions of the Royal Society series A*, **367**, 2161–2179.
Fink, M. and Noble, D. (2010). Pharmacodynamic effects in the cardiovascular system: The modeller's view, *Basic and Clinical Pharmacology and Toxicology*, **106**, 243–249.
Garny, A., Noble, D., Hunter, P.J. *et al.* (2009). Cellular Open Resource (COR): current status and future directions, *Philosophical Transactions of the Royal Society A*, **367**, 1885–1905.
Garny, A., Noble, D. and Kohl, P. (2005). Dimensionality in cardiac modelling, *Progress in Biophysics and Molecular Biology*, **87**, 47–66.
Hampshire, S. (1956). *Spinoza*, Faber and Faber, London.
Hare, R.M. (1963). *The Language of Morals*, OUP, Oxford.
Hare, R.M. (1965). *Freedom and Reason*, OUP, Oxford.

Hunter, P.J., Li, W., McCulloch, A. *et al.* (2006). Multi-scale modeling: Standards, tools and databases for the Physiome Project, *Proceedings IEEE*, **39**, 48–54.

Iribe, G., Kohl, P. and Noble, D. (2006). Modulatory effect of calmodulin-dependent kinase II (CaMKII) on sarcoplasmic reticulum Ca^{2+} handling and interval-force relations: a modelling study, *Philosophical Transactions of the Royal Society A*, **364**, 1107–1133.

Jack, J.J.B., Noble, D. and Tsien, R.W. (1975). *Electric Current Flow in Excitable Cells*, OUP, Oxford.

Kenny, A.J.P. (1969). *The Five Ways*, Routledge and Kegan Paul, London.

Kohl, P., Crampin, E., Quinn, T.A. *et al.* (2010). Systems biology: An approach, *Clinical Pharmacology and Therapeutics*, **88**, 25–33.

Kohl, P. and Noble, D. (2009). Systems biology and the Virtual Physiological Human, *Molecular Systems Biology*, **5**, 291–296.

Kohl, P., Noble, D., Winslow, R. *et al.* (2000). Computational modelling of biological systems: Tools and visions, *Philosophical Transactions of the Royal Society A*, **358**, 579–610.

Marhl, M., Noble, D. and Roux, E. (2006). Modeling of molecular and cellular mechanisms involved in Ca^{2+} signal encoding in airway myocytes, *Cell Biochemistry and Biophysics*, **46**, 285–302.

Montefiore, A.C.R.G. and Noble, D. (1989a). 'General introduction', in Montefiore, A.C.R.G. and Noble, D. (eds), *Goals, No Goals and Own Goals*, Unwin-Hyman, London, pp. 3–13.

Montefiore, A.C.R.G. and Noble, D. (eds), (1989b). *Goals, No Goals and Own Goals*, Unwin-Hyman, London.

Niederer, S.A., Fink, M., Noble, D. *et al.* (2009). A meta-analysis of cardiac electrophysiology computational models, *Experimental Physiology*, **94**, 486–495.

Noble, D. (1967a). Charles Taylor on teleological explanation, *Analysis*, **27**, 96–103.

Noble, D. (1967b). The conceptualist view of teleology, *Analysis*, **28**, 62–63.

Noble, D. (1989a). 'Intentional action and physiology', in Montefiore, A.C.R.G. and Noble, D. (eds), *Goals, No Goals and Own Goals*, Unwin-Hyman, London, pp. 81–100.

Noble, D. (1989b). 'What do intentions do?' in Montefiore, A.C.R.G. and Noble, D. (eds), *Goals, No Goals and Own Goals*, Unwin-Hyman, London, pp. 262–279.

Noble, D. (1990). 'Biological explanation and intentional behaviour', in Said, K.A.M., Newton-Smith, W.H., Viale, R. *et al.* (eds), *Modelling the Mind*, OUP, Oxford, pp. 97–112.

Noble, D. (2008a). Computational models of the heart and their use in assessing the actions of drugs, *Journal of Pharmacological Sciences*, **107**, 107–117.

Noble, D. (2008b). Meditation on two 'heretics', *The Balliol Record*, 28–30.

Rodriguez, B., Burrage, K., Gavaghan, D. *et al.* (2010). Cardiac applications of the systems biology approach to drug development, *Clinical Pharmacology and Therapeutics*, **88**, 130–134.

Roux, E., Noble, P.J., Hyvelin, J.-M. *et al.* (2001). Modelling of Ca^{2+}-activated chloride current in tracheal smooth muscle cells, *Acta Biotheoretica*, **49**, 291–300.

Roux, E., Noble, P.J., Noble, D. *et al.* (2006). Modelling of calcium handling in airway myocytes, *Progress in Biophysics and Molecular Biology*, **90**, 64–87.

Stewart, P., Aslanidi, O.V., Noble, D. *et al.* (2009). Mathematical models of the electrical action potential of Purkinje fibre cells, *Philosophical Transactions of the Royal Society Series A*, **367**, 2225–2255.

Taylor, C. (1964). *The Explanation of Behaviour*, Routledge and Kegan Paul, London.

Chapter 9

Systems Biology and *The Music of Life*

9.1. Basel and Music, 2003

In 2003 I was invited to lecture on systems biology of the heart at a meeting organised by Torsten Schwede at the Biozentrum of the University of Basel. The evening before the lecture a speakers' dinner was held in one of the ancient cellars of the city. A classical guitar recital preceded the dinner. I suspect that the organisers knew of my love of the guitar. They had invited a world-ranking performer, Christoph Denoth, not only to perform but also to join the dinner afterwards, with a place allocated next to mine. The performance was inspired. Christoph is to music what some of my brilliant colleagues have been to mathematics: someone who moves so easily in the medium that they can focus on interpretation and inventiveness rather than just on technicality. One of the pieces was a homage that Christoph himself had composed. He rarely plays it, but that evening the tremolo parts of the piece were magical. He made the guitar perform successively the roles of a balalaika, a multiple harmonics percussion instrument, and a sweet lute in the same piece of music. No other instrument can match the classical guitar in the hands of a world-class performer.

It is easy therefore to imagine my disappointment when, at the end of the recital, he started walking out of the room. Not knowing whether this was shyness – perhaps he did not relish dining with a bunch of academics? – I reacted instinctively, and rushed over to intercept him, asking whether I could see his guitar. He took it out of its case and, as he handed it to me, he noticed the difference in the lengths of my fingernails on the two hands. 'You play!' he exclaimed (the left-hand nails are short to enable the fingers to firmly stop the strings at the various positions on the fingerboard, while those of the right are kept at the optimal shape and length for playing the strings). 'Yes, but not like you!' I replied. Before I knew how to avoid it, he had struck a deal: 'I join the dinner if you agree to play a piece after the main course.' Was he joking?

Just as I wouldn't even presume to compete with my mathematical colleagues who 'see' solutions even before they prove them, I wouldn't dare to follow a performer of Christoph Denoth's international level with an attempt at a classical piece. Thinking quickly, I said that I would play, but I decided to accompany myself simply with a song, rather than playing a classical solo. Arpeggio accompaniment of a song is much safer than attempting to play a Villa-Lobos prelude! And, by then, I had some years' experience of publicly performing my interpretations of love songs in the language of the medieval Troubadours, Occitan (otherwise known as Provençal). So, after the main course I sang 'Arron d'Aimar' ('After Love'), a modern Gascon song in Occitan composed by one of the most successful groups in the Pyrenees, Nadau.

Someone who speaks French or Italian would follow a few of the words of a song in Occitan. It is not hard to see that *'que t'aime'* is *'que je t'aime'* ('how I love you'), even when it is pronounced quite differently (in Occitan it sounds like 'Kay tie me'). Christoph was not, however, catching just the odd phrase. He was nodding all the way through. As a remarkable coincidence he is one of the small minority (roughly 30,000) of Swiss people who speak Romansh, a language that has similar roots to Occitan. We spent the rest of the evening comparing notes on the two languages and what we perform on the classical guitar.

The next day, as I was giving my lecture, I noticed that, up there near the back of the lecture theatre, Christoph Denoth had crept in to listen. I had just reached the point at which I was explaining that it is wrong to think of genes as controlling the organism, that this is to put the cart before the horse, as it were. His presence led me to look for a way of relating this to the playing of music. Almost without thinking (I can often listen to myself giving a lecture when I am doing so in freestyle presentation) I said: 'You know, it would be just as absurd to think that the pipes in a large cathedral organ determine what the organist plays. Of course, it was Bach who did that in writing the score, and the organist himself who interprets it. The pipes are his passive instruments until he brings them to life in a pattern that he imposes on them, just as we impose patterns on our genomes to generate all the two hundred or so different types of cell in our bodies.'

And so, the title of my book, *The Music of Life*, was born and the message of Chapter 2 ('The organ of 30,000 pipes') was virtually already written. I did not know then that there are pipe organs that have as many pipes as the human genome has genes (roughly 25,000 – see footnote on page 31 of *The Music of Life*). Nor did I know that Barbara McClintock, who won the Nobel Prize late in life (in 1983 at the age of 81 for her discovery of jumping genes), had referred to the genome as an 'organ of the cell' (McClintock, 1984) responding not only to the commands of the cell but also reacting to the challenges of the environment. Yet, here we are, nearly 30 years later, still receiving new students at our universities who, at school, have absorbed the dogma that 'they [genes] created us, body and mind' (Dawkins, 1976, 2006). *The Selfish Gene* is, often enough, the only book on genetics that they have read, and we as university teachers have to undo the damage. Not only us; Richard Dawkins himself has also to distance himself from the naïve genetic determinism – a later book (Dawkins, 2003) has a chapter entitled 'Genes aren't us'.

Barbara McClintock worked on plants (mainly maize). The insight that the genome is primarily a database (another way of expressing 'organ of the cell') used by the cells, tissues, organs and systems of the body is just as true for other organisms. As Beurton *et al.* (2008) comment, 'it seems that a cell's enzymes are capable of actively manipulating DNA to do this or that. A genome consists largely of semi-stable genetic elements that may be rearranged or even moved around in the genome, thus modifying the information content of DNA.'

9.2. Oxford 2005 – Gestation of *The Music of Life*

If Basel was the conception of *The Music of Life*, Oxford was its gestation. There were several strands of intellectual activity that combined to form its womb.

9.2.1. *The scientific womb*

The first was my own and my collaborators' scientific research over many years.[1] To say that the 'selfish gene' was useless to us as a hypothesis would be understating the problem. It has no cashable value whatsoever in physiological work. Indeed, popular neo-Darwinism in general relegates the organism and its development to the category of a transient vehicle – the 'lumbering robot' destined to transmit its 'eternal' genes on to the next generation. All of this is biased metaphorical polemic, and I would argue that it has impeded the systems approach (Noble, 2011b).

Modern evolutionary biologists (see Pigliucci and Müller, 2010) now work to extend the modern synthesis (the more correct term for what is often called neo-Darwinism) and to reconnect with development in the field of evolutionary developmental biology (evo-devo). But it was not just development that was ignored: it was physiology too. Yet, it is whole organisms that live or die, which is the process essential to selection. The only way in which that could be re-interpreted to be the same as selection of genes was to maintain, as Dawkins does in *The Extended Phenotype* (Dawkins, 1982), that the gene's-eye view and the organism-eye view are essentially equivalent, just perceived in a different way (rather like the Necker Cube optical illusion used on the front cover of the first edition of *The Extended Phenotype*), so that the one can replace the other. That would be true only if one really could identify each phenotype characteristic with a difference at the level of DNA. As I explained in Chapter 6 of the present work, that is simply not true. Most of the time as physiologists, we are bedevilled by the robustness of the organism in resisting the consequences of changes at the genetic level. Moreover, it is incorrect to consider important functions, like the rhythm of the heart, as being determined by 'genetic programs'. There are no such programs. The example I gave in Chapter 6 illustrates this with the necessity of including downward causation from the complete cell as an absolutely essential part of the process of pacemaker activity, such that the attribution of cardiac rhythm simply doesn't make sense at the level of genes and proteins. I have explored this problem in the article on 'Differential and integral views of genetics' (Noble, 2011a).

---

[1]See Paton lecture reprinted in Part 2, page 576.

9.2.2. *The philosophical womb*

The second strand of intellectual activity involved in forming the womb of *The Music of Life* was academic philosophy. Many of my scientific colleagues would consider the discipline of philosophy to be a waste of time. On this view, science has progressively replaced philosophy in providing answers to important questions about the nature of the world and of us. Clearly, in unravelling facts about the world through the process of empirical discovery, and in formulating conceptual schemes to create mathematical and other formal descriptions of those facts, this is undeniably true. What used to be called 'natural philosophy' is now called 'science'. The remnants of this history can be seen in the title of one of my favourite journals, *Philosophical Transactions of the Royal Society*, the longest-running scientific journal in the world, as well as in the titles of some scientific chairs in British universities, such as the Kelvin Chair of Natural Philosophy (Glasgow).[2]

So, can we ignore philosophy, relegating it to the category of things about which we cannot speak (to quote the ending of Wittgenstein's *Tractatus*)? This is about as impossible as it would be to eat yourself to survive starvation! As soon as we open our mouths to speak, or take up our pens to write, we are guilty (if that is the right word for an action we cannot avoid) of being philosophers: all of us. There are no exceptions, not even those who may loudly proclaim that they are not philosophers; indeed, they are the most likely to err. The reason is simple but subtle. Our language contains conceptual traps for the unwary at every turn of phrase. Conceivably, science might succeed in being entirely neutral, philosophically speaking, if the only language it used was that of mathematics and logic (though even there I am not entirely sure – even in logic we have to choose the form of logic we employ – do we allow what Buddhists call four-cornered logic, for example?). But to restrict science in this way would not be to produce empirical discovery. Mathematics and logic are conceptual tools, not in themselves forms of empirical statement. So we must also use declaratory language.

But as soon as we use language, whether ordinary or technical, the conceptual frame in which our language has developed creeps in, surreptitiously, to colour our thought and expression. How else are we to understand the way in which neuroscientists like Sherrington (1940) and Eccles (1953) were 'trapped' in their dualist interpretation of the relation between the brain and the mind (see Chapter 9 of *The Music of Life*)? The reification of 'self' was a historical linguistic development in Western thought (Hacker, 2011), not a necessary 'object' for neuroscience to discover. Hardly any scientist today thinks that the dualism of Sherrington or Eccles was anything other than a philosophical prejudice. This kind of prejudice creates the greatest problems when we are unaware of the surreptitious nature of the philosophical baggage we import into our 'science'.

Here, I will take just one example of the problem. What 'causes' an organism? What, indeed, is life, to take the title of Schrödinger's famous book (Schrödinger, 1944)? I had,

---

[2]Other examples include the Chairs of Natural Philosophy in Cambridge and Edinburgh. And, of course, the basic doctorate in our universities is the PhD: Doctor of Philosophy.

briefly, thought of using the same title for *The Music of Life*. I didn't dare to do that, but I was concerned that the question has been badly misunderstood, which is why the introduction of my book starts with the question.

Take, for example, Francis Crick's statement:

> You, your joys and your sorrows, your memories and your ambitions, your sense of personal identity and free will, are in fact no more than the behaviour of a vast assembly of nerve cells and their associated molecules. (Crick, 1994)

It is difficult to know how to take such a statement. Is he joking? Crick was famous for his sense of humour and fun. But, this statement essentially *is The Astonishing Hypothesis*, the title of the book. So it can't just be dismissed as a humorous aside. It is based on a philosophical position, which is that the lower elements in a system (in this case the cells and molecules) are always to be preferred in seeking an explanation for the higher elements (memories, ambition, identity and free will – the latter two only as the 'sense of', which is already to imply that, somehow, they are just what we perceive, so not real). This is a prejudice, though a common one found in scientific writing. There is absolutely no reason, *a priori*, to favour one level over another in seeking the 'causes' of anything. To anticipate my next section, 'level', 'higher' and 'lower' are metaphors, though not always recognised as such.

Let's spell this out in terms of one of the examples I considered in Chapter 6. When we construct a model of cardiac rhythm, we do so by creating differential equations for each of the 'lower' elements in the cell that is involved, i.e. the proteins forming channels, transporters, buffers, etc. To these we add equations for the structure (cell, organelles). Then we solve the equations by integrating them. At that point we necessarily incorporate initial and boundary conditions without which the integration would be impossible. If we look for the causes of the solution, these conditions are just as much a cause as the differential equations themselves. It wouldn't make sense to say that the rhythm is 'nothing more than the activity of the molecules', precisely because the physical structure *is* something more. It constrains what the molecules do, even to the extent of rhythm being abolished if we remove that constraint. As I wrote in Chapter 1, it doesn't even make sense to ascribe cardiac rhythm to the molecules of the system.

Philosophers have analysed the concept of cause over millennia, starting at least with the work of Aristotle. It is an elementary aspect of work in philosophy to recognise the different categories of causation. I explore this kind of question in relation to genetic causation in the first of the articles reproduced here from *Philosophical Transactions of the Royal Society*. 'Genes and causation'[3] seeks answers to the following questions:

---

[3]Reprinted in Part 2, page 587.

(a) How has the concept of 'gene' changed since its invention 100 years ago, and how does that change how we view genetic causes? The answer is that the change is fundamental. From being a necessary cause (defined in that way as a hypothetical entity, postulated as an allele – a gene variant), a gene has become a particular physical, far from hypothetical, entity for which causation has to be demonstrated experimentally. Each DNA sequence that we identify as a gene can have causal consequences far outside the phenotypic domain under which it was first named.

(b) If the genome inheritance can be represented as digital information, how do we compare it with the analogue information inherited with the fertilised egg cell? I argue that the non-DNA information is at least as important. In terms of information, they are comparable (Noble, 2011a).

(c) Does it make sense to view one side of this duality of inheritance as primary? My answer is that there is no such reason. They are different, but both are completely necessary. If we wish to give primacy, though, I would favour the cell. That is why viruses are 'dead' outside the context of a cell. If you enucleate a cell, it continues to function (red cells in the blood do just that) until it needs to make more protein. But if you destroy the cell membrane you no longer have functionality.

'Genes and causation' also introduces what I call the 'genetic differential effect problem', which is produced by the neo-Darwinist insistence that what matters is differences at the genetic level that cause differences at the phenotype level. This issue is pursued at greater length in Noble (2011a).

9.2.3. *The metaphorical womb*

The third form of gestation of *The Music of Life* was the theory of metaphor. At the time that I started working on *The Music of Life* I was advising a Korean colleague, Sung Hee Kim (the wife of Yung Earm, see Chapter 5), on some of the medical terms used in body metaphors since she was working on a doctoral thesis in the Faculty of English Language and Literature of Oxford University, comparing body metaphors used in Korean and English print media (Kim, 2006). Her research identified some major differences between the two languages that could only be understood by reference to the full semantic frame of the metaphor in each language and which could be best represented by mapping the metaphorical statements onto their targets in a systematic way. A key aspect of metaphor theory used in this kind of approach is that competing metaphors illuminate different parts of the target to which they are applied. By doing so, they can both be correct. As a simple example from the work of metaphor theorists like Lakoff and Johnson (1980), love can be a journey and it can be war. The two metaphors illuminate such different aspects of the same target, love, that – as we all know only too well – they can both be true.

The trap here is that science likes to avoid competing explanations that can both be correct. We like to isolate a problem from all the complicating factors until we can see an experiment that unambiguously decides between the possible theories to arrive at a single 'true' answer. But if we have imported metaphorical ideas into our theories, how can we be sure that this is the correct approach? Competing metaphors can both be correct.

Scientists are sometimes not aware, first, that they use metaphors much more frequently than they might think and, second, that the relationship between a metaphor and reality is very different from that between a scientific theory and reality. This is the issue I explore in Chapter 1 of *The Music of Life* by comparing 'selfish' and 'prisoner' genes to demonstrate their metaphorical nature. Just like 'journey' and 'war' applies to 'love', there is no experiment that could unambiguously disprove their application. They can both be 'true'. In short, they are not empirical scientific theories. The selfish gene is an idea more in the field of metaphorical polemic than science (Noble, 2011b).

Could the selfish gene idea be rescued for science? It has been so phenomenally successful as a popularisation of genetic biology that it would be valuable if it could be interpreted as a standard scientific hypothesis. At least we would then be able to subject it to the standard scientific test of validity and see how and whether it could be falsified. I can see only one way in which that could be done. It would require a criterion of 'selfishness' that could be assessed at the level of genes (now defined as particular DNA sequences), *independently of the prediction that the theory makes*, i.e. that the frequency of that gene in the gene pool increases in subsequent generations. Without such a criterion the idea is circular, as metaphorical ideas often are.

Try as one might, though, the task of finding such a criterion is surely and inevitably a failure (Noble, 2011b). How could a particular nucleotide sequence, the only measure we have at the level of genes, tell us whether that gene is or is not selfish? The only test we have is the prediction of the statement itself, i.e. that the frequency of that gene increases in the gene pool – that is what is meant by 'selfish' here. It is a strange 'theory' that depends on its own, and only, prediction even to have meaning, let alone be testable. Moreover, that meaning depends on the context, i.e. the phenotype and the environment. A sequence that may be successful in one organism in a particular environment could be disastrous in another organism or in a different environment. And, indeed, virtually all cross-species clones fail to survive. The DNA has been put into the wrong context. The sequence by itself means nothing, just as a word of Maori might have no meaning or, worse still, the wrong meaning in English. In looking for an independent criterion to save the theory from circularity, we are therefore forced to move to the level of the phenotype after all. The concept of the selfish gene is then redundant and misleading. That, in a nutshell, is why the concept is of no use in physiological science (Noble, 2011b).

9.2.4. *The historical/religious womb*

I don't think of myself as a religious person, though I hold on to the idea of human spirituality, which is itself a 'high' level concept not capable of molecular or cellular level reduction.[4] I would also argue strongly that the idea of spirituality does not depend on

---

[4]'Spirit' comes from the Latin for breath, *spiritus*, and it is not incidental that many meditation techniques focus on the breath. 'Respiration' comes from the same root. So does 'inspire'. There are many modern definitions of 'spirit'. Here, I am using the term as almost interchangeable with, but also as an extension of, the 'self', which I think should be viewed as a process, not as an object. Processes can be causes. Viewing 'spirit' or 'self' in this way does not detract from causal efficacy. Someone who denies human spirituality would, I think, be

particular metaphysics. In fact, my stance is basically non-metaphysical, as will be clear from the last two chapters of *The Music of Life*, which is what brings me closer to the Buddhist tradition than to any other. I see questions, to which the answers are largely a matter of personal choice amongst rival metaphysics, as beyond what empirical science can tell us. But that applies both ways, to theism and atheism alike. So, I don't find the high-profile (and usually very naïve) attacks on religion by some scientists to be either convincing or helpful. They fail to address the central problem, which is that metaphysical questions are notoriously difficult to express coherently. Many of the questions are incoherent because we simply do not know what it would be for them to have an answer. Yet, as human beings, we find it difficult (theists, atheists and agnostics alike) to avoid the questions. On the one hand, it seems sensible to deal only with what we can observe, measure and understand. This is the pragmatic approach of science. Every valid scientific theory should be falsifiable, at least in principle.[5] On the other hand, it is laughably presumptuous to suppose that this resolves all questions about life. Clearly, it can't. Thus, it is relevant to our ethics in relation to human life to understand genetics, evolution, physiology – we need to know what we are made of and how we developed since those studies set limits to what is possible – but those subjects do not dictate to us what our ethics should be. Attempts to develop a purely scientific humanism as a substitute for religion don't succeed for precisely this reason.

Perhaps a major part of the problem is that, just as we have unnecessarily reified 'self' and 'spirit', we have a limited conception of what 'religion' means. I was therefore very intrigued indeed when my early attempts to learn East Asian languages (Japanese, Korean, Chinese) led me to an important discovery. The words for 'god', 'religion' and 'prayer' simply do not correspond precisely to the meanings we give them in the Western religious traditions. The 'god' word, 神 (*kami* in Japanese, *shén* in Chinese, *shin* in Korean), carries meanings closer to 'spirit' or 'essence'. This is why Christian missionaries to East Asia had to use different Chinese characters to express the concept of a Creator. The word for 'religion', 教, carries more the meaning of 'teaching'. It is used in the compound for 'professor', for example. The context in which these words have their meanings can allow for a 'religion' that, by our Western ideas, might not even count as a religion. Taoism and Buddhism seem to me to be in that category. I was therefore delighted to read *The Awakening of the West* (Batchelor, 1994), a brilliant historical account of the encounter between Western missionaries and Eastern religions. The author was a Buddhist monk for many years in

---

denying such efficacy, relegating whatever is being referred to as an epiphenomenon. I think that position is incoherent. I needed 'inspiration' in the relevant sense to write this book. That is not just a function of my own brain. It is interpersonal. Spirituality, therefore, has social connotations that take us outside the range of biology, and thus, the boundaries of spirituality cannot be precisely defined in biological terms.

[5]Here I am using 'theory' in the sense of an empirically testable proposition about the world. 'Theories' that are not testable in this way, such as mathematics and metaphysics, are better viewed as tautologies that help us in viewing and understanding the world. No experiment could ever disprove a valid geometry, for example. Nor can a 'theory' like 'force = mass x acceleration' be disproved since it has, in effect, become a definition of force or mass (take your choice!). Some biological 'theories' are also in this category. As discussed earlier in this chapter, no experiment could ever disprove (or prove) the 'selfish gene theory', which is therefore also metaphysical.

both the Tibetan and Korean traditions. Batchelor's later books (Batchelor, 1997, 2010) are also highly relevant, although I did not know about these when I wrote *The Music of Life*.[6]

This was the background to the story of Jupiterians in the last chapter. But, what led me to incorporate such ideas into a book on systems biology? The answer is another convergence between my own thinking and some aspects of Oriental philosophy. As I came to write Chapter 9 on the subject of the brain, I was reflecting on how to apply the systems approach to neuroscience. Some of my previous articles (Egan *et al.*, 1989; Noble, 1989, 1990; Noble, 2004; Noble and Vincent, 1997) had already developed the idea that the self is a construct, a useful one of course, but not one to be identified either with an immaterial substance or simply with the brain. The way I express this in Chapter 9 of *The Music of Life* is to say that it is better regarded as a process than as an object. Just as it doesn't make sense to talk about heart rhythm at the level of genes and proteins, it doesn't make sense to talk of the self at the level of neurons or hormones. At those levels, it is as though there is no self at all. The idea of no-self (*anatman* in Sanskrit, *an* = no, *atman* = self) is, of course, precisely that of Buddhism.

Or is it? It has taken me several years to try to answer that question. The original insight 2500 years ago may have been part of the general non-metaphysical stance of the historical Buddha, Siddhartha Gautama (Batchelor, 2010), but it is hard to decide precisely what this insight was. We live in such a different world from that of Gautama and it is all too easy, as Gombrich (2009) has warned, to take his words out of context. I started out thinking that it was an empirical discovery. Perhaps, during meditation, he looked for the self, the 'I', the soul, and simply didn't find it, rather as David Hume famously examined his own thoughts and perceptions two millennia later and came to the conclusion that none of them could be identified as 'the self', that in that sense such a thing did not exist.

But, to say that something doesn't exist, we do at least need to know what it would mean for it to exist, how we would recognise it if we tried to find it. And, of course, we don't know how to recognise it. I recognise you, the reader, as a person, as having a sense of self, and we know what words like 'yourself', 'myself' and 'himself' mean. To indicate these, we would point at you, me or him as the case may be. You can also point at yourself to indicate yourself. But, if we had your brain out on a dish, as it were, how could we possibly say that this is you? The brain is necessary to you, but it is not sufficient. That is the basis of my story of 'the frozen brain' in Chapter 9 of *The Music of Life*.

Looking at the question this way, we are forced to say that the concept of no-self is just that: a conceptual truth, not an empirical one. No scientific, or meditative, experiment is necessary to establish such a truth. To return to the quotation from Crick earlier in this chapter, looking for such things at the level of neurons and molecules is a conceptual mistake.

---

[6]Interested readers can view an extensive discussion between Stephen Batchelor and Denis Noble at http://www.voicesfromoxford.org/B-S-Batchelor.html.

In Chapter 10 of *The Music of Life* I used the famous Oxherder parable (Wada, 2002) from the Chinese Buddhist tradition as a way of explaining the object of meditation to, as it were, subdue the self. One of the ten pictures is just of an empty circle, as though the self (the ox in the story) has disappeared. I no longer think of it this way. It is rather a parable about how to subdue *selfishness*, not the self. Buddhist meditation has, as one of its aims, to remove selfish, greedy and angry attitudes, one of the central aims of any ethical practice.

So, there are two kinds of 'discovery' here. The first is the conceptual truth that it doesn't make sense to talk of the self as an object in the sense in which our brains are objects. The second is that, through meditative techniques we can subdue selfishness. But doing that is not equivalent to some conjuring trick of 'making the self disappear'. I am reinforced in that conviction by the idea that what the Buddha was arguing against was not so much the self, as usually conceived when we refer to 'himself' or 'myself', but rather against the idea that it was an unchanging thing (Gombrich, 2009). That idea fits well with the concept of the self as a process, as Gombrich also argues.

Does that mean that our experience, e.g. of meditation, is irrelevant? I don't think so. Experience can lead us to a conceptual truth even when it is not itself necessary to that truth. It was seeing the images of gravitational lensing produced by the Hubble Space Telescope that led me to take the idea of the bending of space by huge gravitational fields seriously. Yet the theory of general relativity does not require me to have that experience in order for it to be a valid theory of the structure of the universe.

I hesitated about writing Chapters 9 and 10 of *The Music of Life*. They were the most difficult to write. The book could have finished on evolution in Chapter 8. But that would have cut its head off. You can't ask a question as audacious as 'What is life?' and not deal with questions of the brain and the self.

9.3. *The Music of Life* – Reactions

At the time of writing, it has been four years since *The Music of Life* was published. It has already been translated into seven other languages.[7] While being carefully written (don't

---

[7]The translations have been a remarkable experience. My former collaborator Carlos Ojeda (see Chapter 5) produced the first, in French, in collaboration with a linguist, Véronique Assadas. He had been involved in many of the debates since the beginning, including a debate at Holywell Manor with Richard Dawkins in the year, 1976, when *The Selfish Gene* was published. The philosopher Anthony Kenny (see Chapter 8) challenged the concept of the selfish gene by noting that by knowing the alphabet one did not thereby know Shakespeare, to which Dawkins' reply was 'well I am just a scientist, I am only interested in truth'. From the back of the room Carlos rapidly interjected 'and what is truth?' Carlos put his heart and mind into the French translation of *The Music of Life* (*La Musique de la Vie*). Sadly he died before it appeared. It stands as a tribute to his many intellectual contributions to the debate.

More recently I have had the very different pleasure of interacting for a year with the translators Zhang Li-Fan and Lu Hong-Bing, working on the Chinese version, 生命的乐章. We have worked together through many of the problems of trying to express some of the language of *The Music of Life* in Chinese. The experience fully confirms the difficulties of translating when the semantic frames are so different, a subject I drew attention to in the book.

be misled by the easy, superficially simple style!) it does not hide its agenda, which is a radical revision of the way in which we think about and practise biological science. It turns one biological dogma after another upside down, starting with the central dogma of molecular biology and its misinterpretations. So what is missing? Quite simply, there is no reply from those who might wish to defend the standard biological story that held sway for most of the twentieth century. Perhaps that is because people have misunderstood the implications. The other articles reproduced for this chapter aim to ensure that people do understand the implications. *The Music of Life* was, deliberately, written as a short, accessible book. The background of ideas that formed its basis requires much more explanation and exploration, a project that I regard as an ongoing one as I work with colleagues in Oxford and elsewhere to explore in greater depth what the conceptual foundations of the systems approach might be.

The second of the *Philosophical Transactions* articles reproduced here, 'Biophysics and systems biology',[8] was commissioned as part of the celebrations of the 350th anniversary of the Royal Society. Each person commissioned to write was asked to provide a state-of-the-art assessment and an indication of where the field is going. The article goes well beyond the previous article, 'Genes and causation', by explicitly dealing with the status of neo-Darwinism and why it is incompatible with more recent discoveries in the physiological and related sciences.

It is, of course, hazardous, if not impossible, to foresee where any field of science is going to develop in the future. If we knew the future of discovery it would no longer be discovery. In responding to the invitation of the Royal Society, I therefore focused on where the field of systems biology is now and what issues need resolving.

First, is it a field? Is it a subject that should have a separate department in a university? The argument in my article with Peter Kohl, Edmund Crampin and Alex Quinn (Kohl *et al.*, 2010)[9] is that it is not a separate discipline. It is an *approach* that can be applied to any discipline, and to all levels in biology. That issue lies at the heart of the difficulties people have experienced in trying to define it. There is a sense of unease, which is frequently obvious when one attends conferences on systems biology. A *Nature* staff reporter clearly expressed this when, interviewing me after lecturing to such a meeting, she challenged me to say what was different about the meeting compared to many other meetings she had to report on in the fields of biochemistry, molecular biology, even physiology. Wasn't there the same almost endless genomic and proteomic data sifting? Was it really just about analysing the mountain of data?

As I show in some of the quotes from Sydney Brenner in *The Music of Life*, he is at the origin of some of the ideas that form the systems approach: 'I know one approach that will fail, which is to start with genes, make proteins from them and to try to build things bottom-up' (stated at a Novartis Foundation meeting in 2001). He first used the term 'middle-out'

---

[8]Reprinted in Part 2, page 602.
[9]Reprinted in Part 2, page 617.

to distinguish the approach from purely bottom-up or top-down methods (Brenner *et al.*, 2001). Multi-level analysis is central to the systems approach. More genomics, more information sifting, is not the systems approach.

To explain why, let me use another story of the Silmans, who appear from time to time in *The Music of Life*. In Chapter 7 I let these visitors from space be so tiny that they make the mistake of thinking that the cells in a human body are separate organisms (as indeed they are in a sense – multicellular organisms evolved by single cells coming together in symbiotic relationships). One of the reviewers on Amazon describes this chapter as a 'brilliant description of sexual intercourse'. So it is, and it is quite a different story from most such descriptions, but before you rush out to buy it for that reason, he goes on to write, 'that should utterly dispose of any simplistic ideas of "Lamarckian" inheritance of acquired characteristics as "wrong"'. Well, maybe that is even more shocking. It should be. Lamarck has not exactly had a good press in the English-speaking world for at least a hundred years. The significance of his work in introducing the concept of biology as a science and in establishing the transformation of species is largely ignored, while he is incorrectly represented as the inventor of 'Lamarckism'. In that sense, Darwin was just as much a Lamarckist (Noble, 2010). We need to reassess both Lamarck and Lamarckism. The unjustified denigration of Lamarck is one of the great historical and philosophical mistakes of neo-Darwinism. As Steven J Gould (1993) noted, this denigration began at the time of his death when Cuvier (who bitterly opposed Lamarck's arguments for evolution) wrote 'one of the most deprecatory and chillingly partisan biographies I have ever read'. Metaphorical polemics, masquerading as science, are dangerous to reputations!

But to return to the Silmans, the story of Chapter 7 was based on their being too small to appreciate the significance of a multicellular organism, and so they interpreted each cell as a separate organism and the cell types as separate species. But, through this misunderstanding they came to appreciate an important fact about how the same genome can be used to make such fundamentally different cells as bone cells and heart cells. Faced with the challenge of multi-level systems biology, we are in a similar situation, not because of our physical size, but rather because of our intellectual immaturity. It is easy to state the multi-level principle of systems biology, and to appreciate the idea of no privileged level of causality. But, at the present time, we do not possess the mathematical tools to achieve this in anything other than a piecemeal fashion. As described in Chapter 8, we use intuitive feel for the significance of events at one level to decide what details to carry up into higher levels. We are a little like the mice in Figure 9.1, each of which tries to identify what the elephant is.

Figure 9.1 represents systems biology as having an overall view that enables it to see the whole of life. But the cartoon also represents systems biology as being detached from the rest, as though it lacks the tools to implement its vision of the whole.

What should those tools be? I think we face a situation not dissimilar to the one I described in Chapter 3, when my focus turned to analytical approaches with closed form solutions.

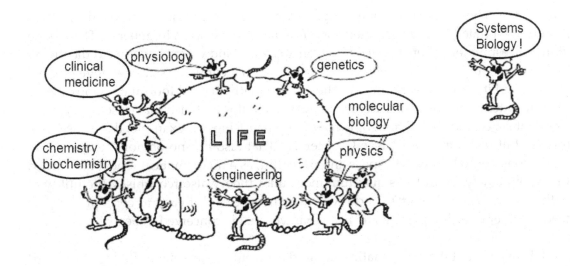

Figure 9.1 Cartoon produced by Yung Earm to introduce a lecture by Denis Noble at the IUPS World Congress in Kyoto, Japan, in July 2009. The cartoon was inspired by the ideas of the Korean Buddhist monk, Won Hyo (617–686).

The problem then was inadequate computing power. Even though now we have massive computing power (compared to 40 years ago), we still don't have enough, and if digital computing is the only form we can use, then I think we will never have enough. To illustrate this, as I write this chapter, my team is working towards using the 10-petaflop computer being constructed by Fujitsu in Kobe, Japan. But even with this machine (10 petaflops is 10^{16} flops) we will still only be able to run whole organ heart models in real time. Yet, those models currently represent only a small fraction of all the gene products that function in a heart.

Are we simply going to continue developing ever more complex differential equation models to demand ever-larger computing power? There are four reasons at least to doubt that.

First, simply building ever more complex models doesn't necessarily lead to greater understanding, since the models may be sufficiently complex to be, in themselves, in need of explanation. As I showed in earlier chapters, analytical mathematics gives greater generality and greater understanding. The kinds of mathematics I described there involved obtaining closed form solutions. But that is not the only form of mathematical analysis possible.

What kind of mathematics is required then? It has to be capable of expressing or deriving some very general principles in relation to living systems. Since these operate on multiple scales, from the molecular to the whole organism and even beyond (organisms are not Turing machines, they are open systems), the ability to relate mathematical analysis at different scales is important. This is the reason why I think it may be important to investigate the application of the principles of scale relativity (Auffray and Nottale, 2008;

Nottale, 2000). These were developed originally by the astrophysicist and relativity theorist Laurent Nottale (2000) as an extension of relativity principles in general. There is an obvious correspondence here between such an approach and what I have called the theory of biological relativity (Noble, 2008). Scale relativity is a controversial new development in physics and it is far too early to say whether the mathematical tools it develops can be applied to extend the equations we use in systems biology. It also makes some assumptions about space and time that people may find surprising, for example, that space-time is fractal. But remember $\sqrt{-1}$ (see Chapter 1). That also has no obvious 'existence' in what we consider to be 'normal' space and time. But, as a mathematical tool, it has turned out to be extremely fruitful. As quantum mechanics theory also recognises, the utility of a mathematical theory has little to do with whether we find it easy to conceptualise, or indeed whether it itself represents reality, whatever that is conceived to be.

Second, I suspect that the mathematics required for multi-scale systems biology will need to explain attractors in the development of biological systems. The oscillators underlying heart rhythm are a good example. The robustness of the system can be expressed by saying that almost wherever the system may be at any time, it tends towards the oscillator as an attractor. The existence of the attractor is far more explanatory than is any particular differential equation model since it is far more general, even if the equations for it are not capable of closed form solutions.

Third, we need more general mathematical/computational theories of evolution. The 'theory of evolution' is not so much a theory, as understood in the physical sciences. It is more a description, essentially a history, of what has happened. Biologists – except for a few fundamentalists – no longer argue about the fact of evolution. Instead, they argue fiercely about the process and the mechanisms involved. We won't be able to resolve some of those arguments without some quantitative ways of representing what we think has happened. There are reasons for thinking that the process may be predictable since mutations are far from random (Stern and Orgogozo, 2009).

Fourth, we need to consider the arguments of those who are seeking non-algorithmic forms of mathematics to describe biological systems (Simeonov, 2010).

9.4. Conclusions

9.4.1. *Pluses*

The articles reprinted for this chapter, and *The Music of Life* itself, represent a major shift in the way in which we think about biology. I am far from being the only person to think this way. Dupre (1993), Jablonka and Lamb (1995, 2005), Keller (2000, 2002), Margulis (1981) and Shapiro (2005, 2009a, 2009b) are just some of the other authors who are reinterpreting the fundamentals of biological science. And the most recent developments in evolutionary (Pigliucci and Müller, 2010) and genetic and developmental (Beurton *et al.*, 2008) theory also do so. I would like to think that *The Music of Life* contributes not only to the shift within science itself, but also to the battle for the hearts and minds of the educated public and the school students from whom the future biological scientists will be drawn. They all deserve better than the one-sided picture frequently presented to them.

9.4.2. *Minuses*

Nevertheless, we are a long way from achieving the goals. Systems biology still stands in need of good definition. It could easily lead to the same problems that have befallen the Human Genome Project, with overblown claims and promises simply not being delivered. As my imagined Silmans contemplate the elephant, I would guess that we are at least a century away from anything as audacious as the Virtual Physiological Human. That is not to say that there won't be valuable insights along the way. It is to say that the project is enormous.

With regard to *The Music of Life*, the main minus that I see is that, despite its attempt to remove or replace one metaphor after another, some central metaphors remained untouched that should have been exposed for the problems they cause. The most blatant example is the word 'code', which is used throughout the book without a single warning. I would now replace nearly all uses of phrases like 'genes code for proteins' with 'genes form templates for proteins'. A template is not an instruction. It is simply used, just as an instrument maker uses templates to cut his wood or metal to the right shape. The word 'code' encourages people, wrongly, to think of the genome as a set of instructions. There are similar problems with the word 'information'. 'Every metaphor produces its own forms of prejudice' (2006: p. 142).

9.4.3. *Contribution to systems biology*

When I first decided on how many chapters this book should be written in, I tried to make it ten. There is something satisfying about ten chapters, which is why *The Music of Life* is structured in that way. It also fitted naturally into development of the successive implementations of the musical metaphor.

So, should there be a concluding Chapter 10 to this book? Before deciding to use the postscript (*The Artist Disappears?*), I looked for something equivalent poetically to the Oxherder parable. Could I finish on such a note? In the end I decided that would be a metaphor too far. In each chapter of the book I have tried to be objective in identifying the pluses and minuses, and the contributions to systems biology. The book comes to a natural end with this chapter of reflections on the composition of *The Music of Life* and the articles that expand on its basis. Moreover, a Chapter 10 would be hazardous and likely to be out-of-date even before it was published. Many people are now engaged with the question of where the systems approach is taking us and how it will develop. Since it is an *approach*, rather than a discipline, we can't know. We don't even know what entities it will discover or have to invent in order to understand living systems. We are a little like the early Oriental medical practitioners who had to invent explanations for the efficacy, when such existed, for their methodologies. This is how concepts like *chi* (気) developed.

They are almost impossible to map onto current biological thought since they carry such a large semantic frame with them, with multiple meanings (Noble, 2009). Hsu (1999), for example, lists more than 20 meanings for *chi* alone, some of which are not even nouns as we understand them.

With regard to systems biology, we can already see some of the concepts that will feature strongly in its analysis of living systems. I would suggest that the multi-level nature of causality will be one, and that the identification of attractors will be another. But, after that, the spectacles through which we try to discern the future are cloudy. I like the comment made by Marc Kirschner (2005) when addressing an audience of young scientists about systems biology:

> You are probably running out of patience for some definition of systems biology. In any case, I do not think the explicit definition of systems biology should come from me but should await the words of the first great modern systems biologist. She or he is probably among us now.

The readership of this book may also, possibly, include the 'first great modern systems biologist'.

References

Papers reprinted in Part 2:

Kohl, P., Crampin, E., Quinn, T.A. *et al.* (2010). Systems biology: An approach, *Clinical Pharmacology and Therapeutics*, **88**, 25–33.
Noble, D. (2008). Claude Bernard, the first systems biologist, and the future of physiology, *Experimental Physiology*, **93**, 16–26.
Noble, D. (2008). Genes and causation, *Philosophical Transactions of the Royal Society A*, **366**, 3001–3015.
Noble, D. (2010). Biophysics and systems biology, *Philosophical Transactions of the Royal Society A*, **368**, 1125–1139.

Others:

Auffray, C. and Nottale, L. (2008). Scale relativity theory and integrative systems biology 1. Founding principles and scale laws, *Progress in Biophysics and Molecular Biology* **97**, 79–114.
Batchelor, S. (1994). *The Awakening of the West*, Parallax Press, Berkeley, CA.
Batchelor, S. (1997). *Buddhism Without Beliefs*, Bloomsbury Publishing, London.
Batchelor, S. (2010). *Confession of a Buddhist Atheist*, Spiegel and Grau, New York.
Beurton P.J., Falk, R. and Rheinberger, H.-J. (eds), (2008). *The Concept of the Gene in Development and Evolution: Historical and Epistemological Perspectives*, Cambridge University Press, Cambridge.
Brenner, S., Noble, D., Sejnowski, T. *et al.* (2001). 'Understanding complex systems: top-down, bottom-up or middle-out?' in *Novartis Foundation Symposium: Complexity in Biological Information Processing*, John Wiley, Chichester, pp. 150–159.

Crick, F.H.C. (1994). *The Astonishing Hypothesis: The Scientific Search for the Soul*, Simon and Schuster, London.

Dawkins, R. (1976, 2006). *The Selfish Gene*, OUP, Oxford.

Dawkins, R. (1982). *The Extended Phenotype*, Freeman, London.

Dawkins, R. (2003). *A Devil's Chaplain*, Weidenfeld and Nicolson, London.

Dupré, J. (1993). *The Disorder of Things*, Harvard University Press, Cambridge, MA.

Eccles, J.C. (1953). *The Neurophysiological Basis of the Mind. The Principles of Neurophysiology*, OUP, Oxford.

Egan, T., Noble, D., Noble, S.J. *et al.* (1989). Sodium-calcium exchange during the action potential in guinea-pig ventricular cells, *Journal of Physiology*, **411**, 639–661.

Gombrich, R. (2009). *What the Buddha Thought*, Equinox, London.

Gould, S.J. (1993). Foreword to *Georges Cuvier: an annotated bibliography of his published works*. Compiled by Jean Chandler Smith. Smithsonian, Washington, pp. iii–xx.

Hacker, P.M.S. (2011). 'The sad and sorry history of consciousness: Being among other things a challenge to the "consciousness studies community"', in Sandis, C. (ed.), *Human Nature*, Royal Institute of Philosophy, London, in press.

Hsu, E. (1999). *The Transmission of Chinese Medicine*, Cambridge University Press, Cambridge.

Jablonka, E. and Lamb, M. (1995). *Epigenetic Inheritance and Evolution. The Lamarckian Dimension*. OUP, Oxford.

Jablonka, E. and Lamb, M. (2005). *Evolution in Four Dimensions*, MIT Press, Boston, MA.

Keller, E.F. (2000). *The Century of the Gene*, Harvard University Press, Cambridge, MA.

Keller, E.F. (2002). *Making Sense of Life. Explaining Biological Development with Models, Metaphors and Machines*, Harvard University Press, Cambridge, MA.

Kim, S.-H. (2006). 'Blood and bone: A comparative study of body metaphors in Korean and British print media', in *English Language and Literature*, OUP, Oxford, p. 319.

Kirschner, M. (2005). The meaning of systems biology, *Cell*, **121**, 503–504.

Kohl, P., Crampin, E., Quinn, T.A. *et al.* (2010). Systems biology: An approach, *Clinical Pharmacology and Therapeutics*, **88**, 25–33.

Lakoff, G. and Johnson, M. (1980). *Metaphors We Live By*, University of Chicago Press, Chicago.

Margulis, L. (1981). *Symbiosis in Cell Evolution*, W.H. Freeman Co., London.

McClintock, B. (1984). The significance of responses of the genome to challenge, *Science*, **226**, 792–801.

Noble, D. (1989). 'Intentional action and physiology', in Montefiore, A.C.R.G. and Noble, D. (eds), *Goals, No Goals and Own Goals*, Unwin-Hyman, London, pp. 81–100.

Noble, D. (1990). 'Biological explanation and intentional behaviour', in ed. Said K.A.M., Newton-Smith, W.H., Viale, R. *et al.* (eds), *Modelling the Mind*, OUP, Oxford, pp. 97–112.

Noble, D. (2004). Qualia and private languages, *Physiology News*, **55**, 32–33.

Noble, D. (2006). *The Music of Life*, OUP, Oxford.

Noble, D. (2008). Claude Bernard, the first systems biologist, and the future of physiology, *Experimental Physiology*, **93**, 16–26.

Noble, D. (2009). Could there be a synthesis between Western and Oriental medicine? *Evidence-based Complementary and Alternative Medicine*, **6**, 5–10.

Noble, D. (2010). Letter from Lamarck, *Physiology News*, **78**, 31.

Noble, D. (2011a). Differential and integral views of genetics in computational systems biology, *Journal of the Royal Society Interface Focus*, **1**, 7–15.

Noble, D. (2011b). Neo-Darwinism, the modern synthesis, and selfish genes: Are they of use in physiology? *Journal of Physiology*, **589**, 1007–1015.

Noble, D. and Vincent, J.-D. (1997). *The Ethics of Life*, UNESCO, Paris.

Nottale, L. (2000). *La relativité dans tous ses états. Du mouvements aux changements d'échelle*, Hachette, Paris.

Pigliucci, M. and Müller, G.B. (eds), (2010). *Evolution – The extended synthesis*. MIT Press, Cambridge, MA.

Schrödinger, E. (1944). *What is Life?* Cambridge University Press, Cambridge.

Shapiro, J.A. (2005). A 21st-century view of evolution: Genome system architecture, repetitive DNA, and natural genetic engineering, *Gene*, **345**, 91–100.

Shapiro, J.A. (2009a). Letting E. coli teach me about genome engineering, *Genetics*, **183**, 1205–1214.

Shapiro, J.A. (2009b). Revisiting the central dogma in the 21st century, *Annals of the New York Academy of Sciences*, **1178**, 6–28.

Sherrington, C.S. (1940). *Man on his Nature*. Cambridge University Press, Cambridge.

Simeonov, P.L. (2010). Integral biomathics: A post-Newtonian view into the logos of bios, *Progress in Biophysics and Molecular Biology*, **102**, 85–121.

Stern, D. and Orgogozo, V. (2009). Is genetic evolution predictable? *Science*, **323**, 746–751.

Wada, S. (2002). *The Oxherder*, George Braziller, New York.

Postscript
The Artist Disappears?[1]

The Holywell Music Room in Oxford is the oldest (1742) purpose-built concert hall in Europe. It was used by Handel and Haydn. The acoustics are so good that it is a joy to perform in it, and particularly so to the full and enthusiastic audience on 17 August 2010.

With a standing ovation still ringing in the ears of us six musicians as we walked out, I was dragged back into the hall where a group from the audience wanted photographs. I smiled as best I could. Someone earlier in the interval had said that I looked very pale. I

---

[1]The title of Chapter 10 of *The Music of Life*.

was certainly very tired, but happy tired – the kind that comes from the joy of performing really well. In fact we had just agreed amongst the performers that we had performed our best concert ever. The percussionist, Keith Fairbairn,[2] had revealed himself as a melodic magician. Bryan Vaughan had performed with great feeling and range as the evening's anchor. Ray,[3] my brother, had excelled himself – the audience had been swinging in their seats as he thrilled them with one scintillating song after another. With an astonishing range: one moment he was a quiet troubadour singing Jaufre Rudel's (1180)[4] song about distant love (*amor de terra lonhdana*), having seamlessly transformed himself from a modern rock star, dance and all. The last song I had sung in the language of the Troubadours ends 'I know stars that will turn your head' (*coneissi las estalas qui hen virat los caps*).

As I looked up at the back of the hall to smile for the camera, my head started turning. My quick instinct was very fortunate – I sat down immediately. If I had waited any longer – the cameras were still clicking – I would have fallen with goodness knows what consequence if that spinning head had crashed hard into the polished floor of the concert hall. But even that rapid precaution was insufficient. As friends rushed to help and to hold me up I completely blacked out.

Before I knew much about what was happening I was being rushed in an ambulance with oxygen mask, beeping ECG machine . . . the complete works. The intensive care unit at the John Radcliffe Hospital appeared later as a stellar sky of blinking lights and great hush. It was obvious what people suspected had happened.

In fact, as I came round more fully and greeted my brother ('You were bloody brilliant tonight!' – he responded, 'You stole the show!'), I knew that, most likely, that could not be the case. You can't be a cardiac physiologist for half a century without some degree of clinical instinct creeping in. But for two days the medical team pursued this line to the limit, as they should have done, of course. They also thought they were treating a concert guitarist (the tell-tale fingernails) just suffering a *crise cardiaque* at the end of an exhausting performance.

Knowing that it was quite unlikely to be a heart problem, I was almost unreasonably humorous, so much so that the alarm bells kept ringing on the blinking light machines each time I laughed. The medical team must have wondered what kind of joker they had just admitted to their normally eerily quiet sanctum.

Particularly when treated by cardiologists, I am more than happy to be incognito. That they thought they were simply treating a musician suited me perfectly. But that artist's mask came off when a young doctor who had attended lectures in my department ten years ago appeared in my room: 'You must be the Denis Noble who wrote that fabulous book, *The Music of Life!*' Nice compliment of course but the basic puzzle remained. An

[2]http://www.keithfairbairn.co.uk/
[3]http://www.raynoble.com/index.html
[4]http://en.wikipedia.org/wiki/Jaufre_Rudel

older doctor summed it up: 'People don't just black out from a sitting position.' They had in fact recorded a substantial difference between my standing and sitting blood pressures. Sitting down should have restored the pressure to a safer level. Yet the home-care specialist was already working out how I could be cared for on being sent home. Fortunately, the consultant was cautious and kept me in for a further day.

The resolution came the first time I went to the bathroom myself. The swimming head returned and I was reduced to being on all fours on the floor. I was quick to pull the red alarm cord. They were even quicker with the oxygen and ECG recorders. They had the opportunity they needed: to observe the event in detail at close quarters. The next morning I was in the endoscopy unit, with a complicated apparatus down my throat, to be treated, successfully, for a duodenal ulcer from which, clearly, copious bleeding had occurred. QED.

At least I now know that my heart is in pretty good shape and, of course, that it wasn't a stroke either.

But I can't help asking some further physiological questions, because this episode which took me so close to a final disappearing act is in fact a neat example of the systems approach in a clinical context. When precisely did the trigger catastrophe actually happen? It must have happened before the concert started. My daughter had noticed that I looked tired. So did someone else who knew me well, in the interval when I took a walk in the fresh air to 'cool down' the sweating I attributed to exhaustion – I was the lead performer for most of the first half of the concert.

So, how could someone suffering an event that would produce severe hypotension continue right through a two-hour concert before succumbing to the inevitable collapse? The answer is in the small molecule we call adrenaline. Fired up as the Oxford Trobadors[5] always are before and during a performance, that little molecule, pouring out of my adrenal glands in response to my nervous signals, kept my blood pressure just above the lowest level and for just long enough to complete the performance, even including a magical encore after I had triumphantly called each of the musicians in turn to receive their deserved accolades from the audience.[6]

Some of the mechanisms of adrenaline's actions in accelerating heart rhythm and so maintaining blood pressure were worked out in my own laboratory (Brown *et al.*, 1975; Brown *et al.*, 1978; Brown *et al.*, 1980; Cohen *et al.*, 1978; Egan *et al.*, 1987; Egan *et al.*, 1988; Hauswirth *et al.*, 1968; Hauswirth *et al.*, 1969; McNaughton and Noble, 1973; Noble, 1975a; Noble, 1975b). The approach to sudden collapse of the circulatory system is also well analysed in

---

[5]http://www.musicoflife.co.uk/music.html

[6]The same mechanism must have been working for Kathleen Ferrier when she bravely performed her final role in Gluck's *Orfeo ed Euridice* at Covent Garden Opera House in 1953 despite being in great pain. She also left the performance on a stretcher. Soldiers in battle also know the phenomenon, when they can continue despite severe injury until the adrenaline surge fades and the inevitable collapse follows. Fortunately, my situation was different in all other respects. There was no pain. Just the sudden blackout.

Ray Noble (centre) leading the performance of *Los de qui cau* at the Oxford Trobadors' concert in the Holywell Music Room. This song was composed by Nadau (see Chapter 9) and is one of the most beautiful and popular of the modern Occitan songs.

systems physiology (Joyner, 2009), and the suddenness of the final event, when adrenaline can no longer maintain the system, is a subject of further investigation.

The standing ovation occurred after this encore when we sang in Korean a song, 'Mannam' (만남), that is extremely popular in Korea. It was magical because, as soon as Ray sang the first line ('Our meeting, it was not by chance'), a 'choir' of 40 people emerged gradually in a crescendo from within the audience itself. A group of teachers from Kwangju (光州 광주) in South Korea were distributed all through the audience. They started standing and swaying as the song reached its climax and Ray out-performed himself by going up a whole octave to rise above the massed singing.

The collapse afterwards was, of course, all the more dramatic. That is why sitting down was not sufficient to contain the problem.

The artist disappears? Well, no – not yet anyway.

References

Brown, H.F., McNaughton, P.A., Noble, D. *et al.* (1975). Adrenergic control of cardiac pacemaker currents, *Philosophical Transactions of the Royal Society B*, **270**, 527–537.

Brown, H.F., Noble, D. and Noble, S.J. (1978). 'The initiation of the heartbeat and its control by autonomic transmitters', in Dickinson, C.J. and Marks J. (eds), *Developments in Cardiovascular Medicine*, MTP Press, Lancaster, pp. 31–52.

Brown, H.F., Noble, D. and Noble, S.J. (1980). '*Le rythme cardiaque: les mécanismes ioniques de son contrôle sous l'influence de l'adrenaline et de l'acetylcholine*', in Akert, C. (ed.), *La transmission neuromusculaire. Les médiateurs et le 'milieu interieur'*, Masson, Paris, New York, Barcelona, Milan, pp. 207–229.

Cohen, I., Eisner, D.A. and Noble, D. (1978). The action of adrenaline on pacemaker activity in cardiac Purkinje fibres, *Journal of Physiology*, **280**, 155–168.

Egan, T., Noble, D., Noble, S.J. *et al.* (1987). An isoprenaline activated sodium-dependent inward current in ventricular myocytes, *Nature*, **328**, 634–637.

Egan, T., Noble, D., Noble, S.J. *et al.* (1988). On the mechanism of isoprenaline- and forskolin-induced depolarization of single guinea-pig ventricular myocytes, *Journal of Physiology*, **400**, 299–320.

Hauswirth, O., Noble, D. and Tsien, R.W. (1968). Adrenaline: Mechanism of action on the pacemaker potential in cardiac Purkinje fibres, *Science*, **162**, 916–917.

Hauswirth, O., Noble, D. and Tsien, R.W. (1969). Reconstruction of the actions of adrenaline and calcium on cardiac pacemaker potentials, *Journal of Physiology*, **204**, 126–128P.

Joyner, M.J. (2009). Orthostatic stress, haemorrhage and a bankrupt cardiovascular system, *Journal of Physiology*, **587**, 5015–5016.

McNaughton, P.A. and Noble, D. (1973). The role of intracellular calcium ion concentration in mediating the adrenaline-induced acceleration of the cardiac pacemaker potential, *Journal of Physiology*, **234**, 53–54P.

Noble, D. (1975a). 'Actions of catecholamines on ionic currents in cardiac muscle', in Nayler, W.G. (ed.), *Contraction and Relaxation in the Myocardium*, Academic Press, London, pp. 267–291.

Noble, D. (1975b). *The Initiation of the Heartbeat*, OUP, Oxford.

Glossary

In each entry, cross-references to other items in the glossary are indicated in bold type.

Action potential

Nerves, muscles, and other excitable cells communicate by sending electrical signals, called action potentials, along their fibres. The signal consists of a transient reversal of the normal membrane potential, first with a phase of **depolarisation** that reverses the normal negative potential to become positive, and then a phase of **repolarisation** that restores the internal negative potential.

Activation curve

Ion channels that are **gated** by voltage do not all switch on at the same voltage. There is a gradual change from all channels closed to all channels open. The curve relating the number of open channels to the membrane potential is called the activation curve.

Activation energy

A chemical reaction from state A to state B occurs at a speed that is determined by how much energy is required to make the change. When the energy is large the speed is slow. You can think of this like a mountain between the two valleys, A and B. A high mountain (high activation energy) is more difficult to cross than a low mountain. Activation energies also determine how temperature influences the speed. This is also sometimes expressed as a Q_{10}.

Allele

The precise DNA sequence for any particular **gene** can vary from individual to individual. These variants are called alleles. In a population there can be many alleles. Since most organisms have two sets of chromosomes, each individual has two copies of each gene, which may be the same or different alleles.

Atrium

The chambers into which blood flows into the heart are called atria. They then pass the blood into the **ventricles** of the heart.

Background sodium current

Not all **ion channels** that play a major role in the heart are **gated** by voltage. An important example is the channel that conducts what we called the background sodium channel, i_{bNa}. This was originally a prediction from the modelling of **sinus node** pacemaker activity. We had to postulate the existence of a channel conducting sodium ions all the time in order for the modelling to work. Later, work in Irisawa's laboratory in Japan and our own laboratory succeeded in characterising its ion selectivity and other properties.

Bessel functions

These are mathematical functions that form solutions to certain differential equations applied to cylindrical or spherical co-ordinates. There are various kinds, including Bessel functions using imaginary arguments, which is the kind referred to in Chapter 1.

Cable theory

The equations of cable theory were originally developed for the flow of electric current in metal cables, such as the first transatlantic cables. They were applied to nerve axons since these also function like cables. The most extensive treatment of cable theory applied to biological systems is *Electric Current Flow in Excitable Cells* (Jack, Noble & Tsien, 1975).

Calcium-induced calcium release

Calcium ions are stored inside muscle cells in the **sarcoplasmic reticulum**. The trigger for releasing the stored calcium to activate contraction is calcium itself. So, calcium ions, which enter the cell through calcium **ion channels**, then cause the release of a much larger quantity of calcium inside the cell. This is called calcium-induced calcium release.

Conductance

The transport of ions through **ion channels** forms an electric current. The magnitude of the current depends on the channel conductance. The higher the conductance the more current flows. An alternative measure of the speed of transport is the **permeability**, which can also apply to substances that are not charged.

Depolarisation

Cells are polarised. The interior of the cell is usually negative. During excitation, entry of cations, which are positively charged, reduces the polarisation. This process is called depolarisation. It was originally thought that this process only went as far as removing the negative potential. In fact it reverses it so that the cell becomes positive. But we still refer to this as depolarisation. The reverse process is called **repolarisation**.

Electrocardiogram

The changes in electrical potentials inside cardiac cells during each beat of the heart create electric currents that flow between different parts of the heart. A small fraction of these currents can be detected at the surface of the body. These recordings are called electrocardiograms. Doctors use them routinely to check on heart activity and to help diagnosis of disease states.

Electrogenic

This refers to any structure or mechanism that carries electric current. **Ion channels** are obvious examples since they form pores in the cell membrane through which the charged ions can flow. Proteins that exchange ions, such as sodium-potassium exchange (the **sodium pump**) or the **sodium-calcium exchanger**, are also electrogenic because the movement of charge in the two directions does not balance. The sodium pump transports 3 sodium ions in exchange for 2 potassium ions, so it carries a net positive charge out of the cell. The sodium-calcium exchanger was originally thought to be electrically neutral (2 sodium ions in exchange for one divalent calcium ion). In fact, it has been found to transport an extra sodium ion (so the ratio is 3:1) and is therefore electrogenic.

Entropy of activation

The activation energy of an **ion channel gating** reaction with high temperature dependence can be so large that, if there were nothing else to favour the reaction, it would not proceed in a biologically relevant time scale. This is the case for the **pacemaker** current, i_f, with a Q_{10} of 6 (Chapter 2). One possible explanation is that, in the formation of the activated complex, there is an entropy change favouring the reaction. A different possibility is that we are dealing with a chain of reactions, with the temperature dependence of one reaction changing the conditions for a second reaction.

Equilibrium potential

See **Reversal potential**

Error functions

Error functions are so called because they are sigmoid functions that were developed for use in probability and statistics. They have also proved very useful in solving partial differential equations, which is their application in **cable theory**.

Excitation threshold

Action potentials in excitable cells, like nerves and muscles, are 'all-or-nothing'. Once the cell has been excited to initiate the action potential it continues to generate itself using positive feedback between the membrane voltage and activation of the **ion channels**. For example, once sufficient sodium channel **gates** have been opened, sodium ions flow into the cell in sufficient quantity to **depolarise** the membrane further, which opens even more channels. This process continues until the membrane potential approaches the **reversal potential** at which the sodium chemical and electrical gradients balance each other.

Gap junction

See **Nexus**

Gating

Ion channels are formed by proteins that sit in the cell membrane and form a pore through which ions can cross the membrane. Most channels are gated. Part of the protein itself, or another subunit or chemical (sometimes, ions themselves can be gates), moves to close or open the channel. In excitable cells, the main channel proteins involved are gated by the membrane potential. The gating part or molecule moves in the electric field because it is charged. The best documented example is the sodium channel inactivation gate, which consists in a charged section of the amino acid chain that connects two of the transmembrane domains (III and IV), and which swings in the electric field to close the channel by sitting at its intracellular opening. This is the process of **inactivation**.

Gene

The modern molecular biological definition of a gene is a section of DNA sequence forming a template for the production of a protein or RNA. The original definition of a gene, introduced a century ago, was the inheritable cause of a particular phenotype. The cause(s) of any particular phenotype, however, extend way beyond any particular DNA sequence. The two definitions of a gene are not therefore equivalent. This is the cause of much confusion in the literature of genetics, both popular and scientific. For further development of this critique of the concept of a gene, see Chapter 9.

Giant patch clamp

This is a development of the **patch clamp** to enable large areas of cell membrane to be controlled to enable ionic currents to be resolved that are not necessarily generated by **ion channels**. It was developed by Don Hilgemann (see Chapter 5) and used by him to great effect in studying the **sodium-calcium exchanger**.

Hodgkin–Huxley equations

These are differential equations formulated by Alan Hodgkin and Andrew Huxley in 1952 to describe the **gating** kinetics of the sodium and potassium **ion channels** in the giant nerve fibre of the squid. The equations are based on the gating reaction being the movement of a charged particle or part of the channel in the membrane electric field.

Holding potential

In a **voltage clamp** experiment, the membrane potential is usually held constant at a level called the holding potential between each pulse. In Hodgkin and Huxley's original work the holding potential was chosen to be the resting level of the membrane potential, but this is not necessarily required. In a spontaneously beating cell, such as a cardiac pacemaker cell, there is no resting potential anyway. The holding potential is then chosen for convenience usually outside the range of the **activation curve** for the **ion channel** being studied.

Hyperpolarisation

This is the opposite of **depolarisation**. Instead of reducing the negative intracellular potential, it increases it. It was through hyperpolarising the membrane of **sinus node** cells that Brown, DiFrancesco and Noble (1979) discovered the 'funny' current, i_f. This is sometimes called a hyperpolarising-activated **ion channel** to distinguish it from channels that are activated by depolarisation.

Inactivation

In addition to displaying an **activation curve**, many **ion channels** with **gating** by voltage also display inactivation so that the ionic current flowing through the channel is transient. The first example of this process was the sodium channel investigated by Hodgkin and Huxley which inactivates quite rapidly following the activation process. The transient outward current referred to in Chapter 2 is an example of a potassium channel that displays inactivation.

Ion channels

Since ions are charged, they do not pass through the lipid bilayer forming the cell membrane. The lipids repel charged ions, just as they also repel polar molecules like water. Ion channels are pores that allow ions to cross the membrane without having to pass through the lipid layer. The pores are formed by proteins. There are ion channels that selectively pass sodium, calcium, potassium, or chloride ions, and channels that allow protons (hydrogen ions) to pass.

Matrices

A matrix in mathematics is an array of numbers arranged in rows and columns. These could correspond, for example, to the coefficients of interaction between the components of a system. Matrices are important in algebra, and they can also be of great use in solving systems of partial differential equations of the kind often used in electrophysiology.

Nernst equation

This is the equation for the **reversal potential** (equilibrium potential) of an **ion channel** selective for a single ion. In the absence of an electric field, the gradient causing flow of ions through the channel will be the chemical gradient, which exists when the ions are more concentrated on one side of the membrane than on the other. An electric field can be set up that opposes this flow. For example, potassium ions flow out of the cell under the influence of the chemical (concentration) gradient, but they can be held in the cell by the negative electrical potential of the interior. The potential at which the two forces exactly balance, and at which there is no net movement, is called the **equilibrium potential**. The Nernst equation is the equation for this balance. In the case of a single charged (monovalent) ion the equation predicts a roughly 60 mV change in the equilibrium potential for a tenfold change in concentration. This therefore became one of the tests for the ionic selectivity of a channel.

Nexus

This is a connection between two cells formed by proteins called connexins.

Nonlinearity

In mathematics we distinguish between linear and nonlinear functions. For example, the simplest model of an **ion channel** would be one in which the flow of current is proportional to the chemical and electrical gradients (see **reversal potential**). Very often, this is not true and the relationship is nonlinear. Also, a channel **gated** by voltage might be linear (or close to linear) for sudden changes but nonlinear once the gating process occurs. Linear systems are easier to solve mathematically, but as the work in Chapter 3 shows, even some highly nonlinear systems can be solved with closed form solutions.

Pacemaker

The heartbeat is a nearly regular rhythm, approximately 60 beats per minute in the human. The areas of the heart that generate rhythm are called pacemakers. The natural pacemaker is the **sinus node**, a small region of the heart situated where the veins empty blood into the **atria**. There is also a region joining the **atria** and **ventricles** which is called the atrio-ventricular node and which also generates rhythm. The **Purkinje fibres** can also display rhythm. Normally, the **sinus node** is the fastest pacemaker and so excites the heart before the other regions can do so. An ectopic pacemaker is one that arises because of a disease state (such as that described in Chapters 3 & 5) in a region that does not normally display rhythm. This can then compete with the natural pacemaker to produce arrhythmia.

Pacemaker depolarisation

Excitation of a nerve or muscle cell is achieved by depolarising it to the **voltage threshold**. In **pacemaker** regions the depolarisation occurs spontaneously as a gradual process called the pacemaker depolarisation. A major goal of cardiac electrophysiology is to work out the **ion channel** mechanisms generating this depolarisation.

Patch clamp

The earliest techniques for controlling the membrane potential (**voltage clamp**) used large electrodes that could be inserted into very large nerve fibres, like that found in the squid. Work on the small cells of the heart became possible because of the invention of glass microelectrodes with tips around 1 μm in size. These can penetrate the cell membrane, which then reseals around the glass to enable the pipette to be used to record voltage or to inject current. A major advance came when Erwin Neher and Bert Sakmann discovered that if the pipette tip was made smooth enough it was possible for the glass pipette to seal to the membrane without penetrating it. This allowed the experimenter to study the **ion channels** in the patch of membrane sealed to the electrode. Alternatively, the patch could be sucked up to break it and enable the patch electrode to record or inject current for the cell as a whole. A further development of this method involved using much larger electrodes – the **giant patch clamp**.

Permeability

This is a measure of the speed with which a substance can move through something like a membrane. **Ion channels** are what enable cell membranes to be permeable to ions. The permeability of an ion channel determines its electrical **conductance** – for the movement of charged items conductance is an alternative way of expressing permeability, though the two are not defined in the same way.

Perturbation theory

This involves mathematical methods that are used to find an approximate solution to a problem which cannot be solved exactly, by starting from the exact solution of a related problem. Perturbation theory is applied by adding a small term to the mathematical description of the exactly solvable problem. The theory then leads to an expression for the desired solution in terms of a power series in a small parameter that quantifies the deviation from the exactly solvable problem. This is the method that was used with Dario DiFrancesco in Chapter 5 to resolve problems arising from ion accumulation.

Physiome Project

This project to model the cells, organs and systems of the body is described in Chapter 7. It was launched in 1997 by the International Union of Physiological Sciences (IUPS).

Plateau phase

Depolarisation in most heart cells is very fast, just as in nerve and skeletal muscle. **Repolarisation**, however, is very slow. The cell remains depolarised for a much longer period of time. This produces a very long **action potential**. The slowest phase of repolarisation is called the plateau.

Polynomials

Polynomial functions are mathematical series of finite length consisting of terms using whole number exponents, and which can also include addition, subtraction and multiplication. Their use in electrophysiology is described in Chapter 3.

Purkinje fibres

In the hearts of large animals, the speed of activation of the **ventricle** would be too slow to ensure synchronous contraction if the conduction occurred only through the ventricular muscle itself. A fast-conducting network of fibres has developed to ensure synchrony. This network was first identified in 1839 by the Czech physiologist, Jan Purkyně. The speed is achieved by the cells being very large (see Chapter 3 for the relevant equation). This is why they were chosen for the early **voltage clamp** experiments. It was easier to penetrate large cells with two microelectrodes. All the experimental work for the Noble 1962, McAllister–Noble–Tsien 1975 and DiFrancesco–Noble 1985 models was done on Purkinje fibres.

Q_{10}

A standard way to determine the **activation energy** for a chemical reaction is to measure the way in which its speed varies with temperature. At higher temperatures more molecules have sufficient energy to reach the activation energy level. If the activation energy is very high, the temperature dependence will also be high. In biological systems, because the temperature range that can be used is fairly limited, it has become customary to measure the ratio of the speeds at two temperatures 10°C apart. This ratio is called the Q_{10}. For the **gating** reactions of most **ion channels** this ratio is in the range 2–3 which corresponds to reasonable activation energies. A surprise in the work described in Chapter 2 was that the slowly activated current, the i_f channel, in the **pacemaker** range of potentials has an unusually high Q_{10} of 6.

Repolarisation

This is the process by which the initial **depolarisation** of the **action potential** is reversed. The process is fragile in the heart, which is why many disease states and drugs cause arrhythmias.

Resistance

This is the inverse of **conductance**. A high conductance means a low resistance and vice versa.

Reversal potential

This is the potential at which an **ion channel** current reverses direction. In the case of a channel conducting just one ion species, e.g. potassium, this potential would be equal to the **equilibrium potential** given by the **Nernst equation**. When more than one ion species is carried, the reversal potential is not equal to either of the equilibrium potentials. There are equations for this situation, referred to in some of the papers, but they are not discussed in the chapters of this book.

Sarcoplasmic reticulum

This is a network (hence reticulum) in the muscle cytoplasm (hence sarcoplasmic) that stores calcium ions. The release of calcium from these stores is achieved by the **calcium-induced calcium release** process.

Sinus node

This is the normal **pacemaker** region of the heart, found in the region where the great veins empty into the **atrium**.

Slope conductance

Membrane **conductance** can be measured by applying small voltage deflections and measuring the current that flows. This might be equivalent to the net membrane conductance, but this is not necessarily true since there may be **gating** reactions occurring sufficiently rapidly to react to the voltage change before the current change is measured. The conductance that is then measured is a tangent (slope) to the relevant current-voltage relation of the cell. The slope conductance can even change in the opposite direction to the net conductance, as described in Chapter 1.

Sodium-calcium exchanger

This is the protein that uses the sodium gradient to push calcium ions across the cell membrane. Discovered in the heart by Harald Reuter, it was originally thought to be electrically neutral, transporting two univalent sodium ions in exchange for each divalent calcium ion, called a 2:1 stoichiometry. It is now known to have a stoichiometry of 3:1 (see Chapter 5) and is therefore **electrogenic**.

Sodium pump

This is a protein that forms an exchanger for sodium and potassium. Three sodium ions are pumped out of the cell in exchange for two potassium ions. The mechanism is therefore **electrogenic**. It is often called the Na^+-K^+ATPase since it transports the ions by splitting ATP.

Sodium threshold

This is the membrane potential at which the sodium channel current becomes sufficiently large to initiate an all-or-nothing **action potential**.

T wave

This is the wave of the electrocardiogram (Chapter 4) that corresponds to **repolarisation** of the **ventricles**.

Transfer function

In the Tsien and Noble (1969) analysis of **ion channel** kinetics using **transition state theory**, this corresponds to the ion transport of a channel protein in the absence of **gating**. If the channel were a simple conductance this function would be linear. In fact it often displays **nonlinearity**.

Transition state theory

This is the chemical theory of how **activation energies** determine the rate of a chemical reaction.

Ventricle

The ventricles are the thick muscles of the heart that pump the blood at high pressure into the aorta.

Voltage clamp

This is a technique for keeping the membrane potential at values chosen by the experimenter. Since voltage is what controls the **gating** of **ion channels**, it is important to control this parameter.

Voltage-dependent gating

This refers to the opening and closing of **ion channels** by gates that move in the electric field. The reactions described in the Hodgkin–Huxley nerve equations are voltage-dependent gating reactions.

Voltage threshold

This is the voltage at which an all-or-nothing action potential or repolarisation is triggered.

PART 2

Papers Reproduced
The chapters correspond to those in Part 1

NO 4749   **November 5, 1960**   N A T U R E   495

PHYSIOLOGY

Rectifying Properties of Heart Muscle

A FACTOR of likely importance in the genesis of the long-lasting action potentials of vertebrate heart muscle is the influence of membrane potential on the potassium permeability of the fibres. That a difference here exists between cardiac muscle and nerve is evident from Weidmann's[1] demonstration of a low slope conductance during the plateau of the action potential. But little information is so far available on the voltage dependence of the membrane conductance under conditions designed to avoid the generation of action potentials.

To study this point, so far as the technical limitations imposed by the structure of cardiac muscle at present allow, excised Purkinje fibres from sheep ventricles were used. While sharing the essential electrical properties of myocardial tissue, these fibres do not produce a mechanical response on depolarization; they are also insensitive to parasympathomimetic substances so that choline chloride or sucrose may be used as a substitute for extracellular sodium chloride to abolish excitability.

Fig. 1, *A* and *B*, shows how the membrane conductance of a Purkinje fibre in sodium-deficient

solution depends on the direction and magnitude of the polarizing current. An outward current of 0·31 μamp., for example, produced the same voltage deflexion as is caused by nearly twice as great an inward current. Allowing for the cable properties of the fibre, this means that the conductance to a depolarizing current may be about four times less than to a hyperpolarizing current. Since the contribution of chloride ions to the membrane conductance of cardiac muscle is small[2], a decrease in the potassium conductance presumably occurs on depolarization. The direction of the potassium conductance change in Purkinje fibres is thus opposite to the predominant change in nerve[3], but has a parallel in skeletal muscle[4].

With strong depolarizing currents a decline in the electrotonic potentials during the pulse may be observed, suggesting a gradual increase in conductance. This second effect probably accounts for the S-shaped relation between the amplitude of the electrotonic potential at the end of a long pulse and the polarizing current (Fig. 1 *B*), but over the physiological range the conductance remains below its value at the resting potential.

The slope conductance decreases appreciably over the range occupied by the slow diastolic depolarization in spontaneously beating preparations[1]. This alinearity may require consideration in interpreting the resistance changes observed during the pacemaker potential[5].

O. F. HUTTER
D. NOBLE

Department of Physiology,
University College London,
Gower Street, London, W.C.1.

[1] Weidmann, S., *J. Physiol.*, **115**, 227 (1951).
[2] Hutter, O. F., and Noble, D., *J. Physiol.*, **147**, 16, *P* (1959).
[3] Cole, K., and Curtis, A. J., *J. Gen. Physiol.*, **24**, 551 (1941). Hodgkin, A. L., and Huxley, A. F., *J. Physiol.*, **116**, 424 (1952).
[4] Katz, B., *Arch. Sci. Physiol.*, **3**, 289 (1949). Hodgkin, A. L., and Horowicz, P., *J. Physiol.*, **148**, 127 (1959). Hutter, O. F., and Noble, D., *J. Physiol.*, **151**, 89 (1960).
[5] Trautwein, W., and Dudel, J., *Pflüg. Arch. ges. Physiol.*, **266**, 653 (1958).

Cardiac Action and Pacemaker Potentials based on the Hodgkin–Huxley Equations

SINCE the equations describing the nerve action potential were formulated by Hodgkin and Huxley[1], the range of phenomena to which they have been shown to apply has been greatly extended. Huxley[2] has applied them to the influence of temperature on the propagated response and to the repetitive firing observed in low calcium concentrations. More recently, Fitzhugh[3] has shown that the long action potentials induced by tetraethylammonium ions in squid nerve may also be reproduced.

The computations described in this communication were carried out with the aim of reconstructing the long-lasting action potential and pacemaker potential of cardiac muscle. Although this work was done independently, the results agree with those of Fitzhugh in showing that action potentials of long duration may be accounted for by Hodgkin and Huxley's formulation of the membrane properties. The description of the potassium current, however, differs from that used by Fitzhugh and provides a better description of the conductance changes.

The equations I have used to describe the sodium current are very similar to Hodgkin and Huxley's,

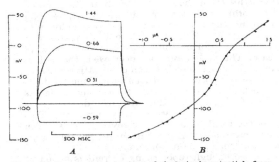

Fig. 1 *A*. Superimposed tracings of electrotonic potentials from a sheep's heart Purkinje fibre. Sodium ions in Ringer's solution replaced by choline. Depolarization is shown in an upward deflexion from a resting potential of −92 mV. Two closely spaced KCl-filled intracellular electrodes were used to record the membrane potential and to pass practically rectangular current pulses of a strength indicated by the figures on each record. *B*, Current-voltage relation for the same preparation. Ordinate, membrane potential at end of current pulse lasting 700 m.sec. Abscissa, total membrane current

496 NATURE November 5, 1960 VOL. 188

those for h (the variable describing the availability of sodium carriers) being based on Weidmann's[4] experiments on the availability of sodium carriers in Purkinje fibres.

The behaviour of the potassium battery in cardiac muscle differs from that in nerve in that the potassium conductance (G_K) falls when the membrane is depolarized[5]. A small delayed increase in conductance appears to be present when large currents are used, but this is not great enough for the conductance of the depolarized membrane to exceed the resting conductance. For the purpose of setting up equations to describe this behaviour, it is convenient to suppose that potassium ions move through two types of channel in the membrane. In one, G_K is an instantaneous function of the membrane potential and falls when the membrane is depolarized. A simple empirical equation has been used to describe the current in this channel. In the other type of channel the rectification is in the opposite direction and occurs with a delay. The conductance of this channel may be described by Hodgkin and Huxley's equations for delayed rectification with two modifications. First, the magnitude of this conductance must be small; accordingly the maximum conductance $\overline{(G_K)}$ has been made less than 1/50 of that in nerve. Secondly, the time constants have been made very much longer to take account of the slow onset of the effect.

The membrane capacity was taken as 12 μF./cm.² (ref. 4), and the absolute values of the various conductances were adjusted to give a resting membrane resistance of about 2,000 ohms/cm.² (ref. 4).

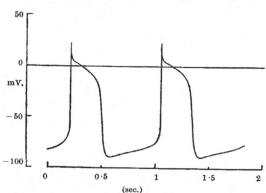

Fig. 2. Solution to equations in which the membrane potential is unstable in diastole so that pacemaker activity occurs. In this case, the potential at which the steady state sodium and potassium currents are equal and opposite is −38 mV. and an unstable state at this potential corresponds to a second solution to the equations

The equations were set up on a digital computer and were integrated by a numerical approximation using a step length of 0·05 m.sec. during the initial spike of the action potential and 0·3 m.sec. during the slower phases. In this connexion I should like to thank Dr. R. A. Buckingham, director of the University of London Computer Unit, for permission to use *Mercury* for this purpose, and Dr. M. J. M. Bernal and other members of the Computer Unit staff for advice on programming.

It was found that computed responses resembling action potentials of different durations could be obtained by inserting appropriate values for the constants in the equations for m (the variable describing the activation of sodium carriers). One of the solutions is shown in Fig. 1. It can be seen that the general shape of the action potential (A) corresponds very closely to that observed experimentally in Purkinje fibres[4]. The time-course of the computed conductance changes is shown in Fig. 1B. After an initial 'spike', the sodium conductance (G_{Na}) settles down to a value during the plateau which is about 8 times its resting value. G_K by contrast is below its resting value throughout the duration of the plateau.

The total membrane conductance ($G_K + G_{Na}$) rises during the plateau, but the slope conductance (determined by adding current to the equations at different times during the integration) was found to decrease. This is the result of a decrease in the sodium slope conductance which becomes negative towards the end of the plateau. Although the slope conductance during the plateau falls to a value which is equal to the resting conductance, the decrease was not so great as that observed experimentally[4]. This may be the result of a defect in my equations which over-estimate the potassium current produced by large depolarizations of the membrane.

If a large enough repolarizing current is added to the equations during the plateau, an all-or-nothing repolarization is initiated. The threshold for this phenomenon is about −30 mV. at the middle of the plateau, which is in fair agreement with that found experimentally by Weidmann[4].

In these equations there is always at least one potential at which the steady state sodium and potassium currents are equal and opposite. In the computations just described, this forms the resting potential and the system is stable unless excited. A

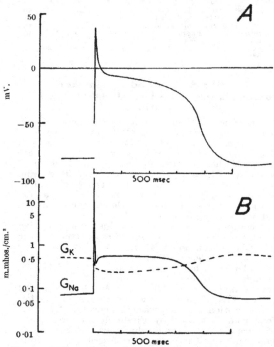

Fig. 1. A, Computed action potential. The integration was started by displacing the membrane potential to −50 mV., which is equivalent to a very short cathodal pulse of $3\cdot6 \times 10^{-7}$ coulombs/cm.². B, Time course of computed conductance changes on a logarithmic scale. G_K, potassium conductance ; G_{Na}, sodium conductance. The potassium and sodium equilibrium potentials were set at −100 mV. and +40 mV. respectively

No. 4749   **November 5, 1960**   N A T U R E   497

small change in the constants in the equations for m is sufficient to make the system unstable in diastole, and pacemaker activity then occurs. Such a solution is shown in Fig. 2. However, in order to bring about repolarization in this case a larger delayed potassium conductance had to be assumed. It is not yet known whether this is a necessary feature of the equations and further computations are being made to test this point. Nevertheless, the sensitivity of the computed pacemaker potential to changes in ionic conductance has been shown to correspond quite well with the experimental information available[6].

                                        D. NOBLE

Department of Physiology,
  University College,
    London.

1 Hodgkin, A. L., and Huxley, A. F., *J. Physiol.*, **117**, 500 (1952).
2 Huxley, A. F., *Ann. N.Y. Acad. Sci.*, **81**, 221 (1959).
3 Fitzhugh, R., *J. Gen. Physiol.*, **43**, 867 (1960).
4 Weidmann, S., *J. Physiol.*, **127**, 213 (1955); "Elektrophysiologie der Herzmuskelfaser" (Hans Huber, Berne and Stuttgart, 1956).
5 Hutter, O. F., and Noble, D. (preceding communication).
6 Noble, D., *J. Physiol.*, Proceedings of the Physiological Society, September Meeting (1960).

J. Physiol. (1962), **160**, *pp.* 317–352

With 15 *text-figures*

Printed in Great Britain

A MODIFICATION OF THE HODGKIN–HUXLEY EQUATIONS APPLICABLE TO PURKINJE FIBRE ACTION AND PACE-MAKER POTENTIALS

By D. NOBLE

From the Department of Physiology, University College London

(*Received* 19 *July* 1961)

In 1949 Hodgkin & Katz showed that the amplitude and rate of rise of the action potential of squid nerve vary with the extracellular sodium concentration in a way which suggested that the rising phase of the nerve impulse is produced by a large and specific increase in the permeability of the membrane to sodium ions. Since the sodium equilibrium potential is normally opposite in sign to that of potassium, this hypothesis readily accounted for the reversal of the membrane potential which had already been observed (Hodgkin & Huxley, 1939, 1945; Curtis & Cole, 1940, 1942).

Using the voltage-clamp technique, Hodgkin & Huxley (1952*a*, *b*, *c*, *d*) separated the membrane current into sodium and potassium components and formulated equations describing the way in which these currents vary with membrane potential and time. They showed that, when combined with the equations of cable theory, their equations could accurately reproduce many of the electrical properties of squid nerve including the shape and size of the action potential, impedance changes, velocity of conduction and the ionic exchanges. The range of phenomena to which they have been shown to apply has since been greatly extended. The original hand computations were confirmed by Cole, Antosiewicz & Rabinowitz (1955) who first set the equations up on an electronic computer. Huxley (1959*a*) has applied the equations to the influence of temperature on the propagated response and to the repetitive firing observed in low calcium concentrations, using the experimental information obtained by Frankenhaeuser & Hodgkin (1957). The prolonged action potentials produced by treating squid nerve with tetraethylammonium ions (Tasaki & Hagiwara, 1957) may be largely accounted for by greatly slowing the rise in potassium permeability (Fitzhugh, 1960; George & Johnson, 1961) and the hyperpolarizing responses obtained at high extracellular potassium concentrations (Segal, 1958; Tasaki, 1959) may be described, at least qualitatively, by introducing the appropriate change in the potassium equilibrium potential (Moore, 1959; George & Johnson, 1961).

21

318   *D. NOBLE*

The aim of the computations described in this paper is to test whether, with certain modifications, Hodgkin & Huxley's formulation of the properties of excitable membranes may also be used to describe the long-lasting action and pace-maker potentials of the Purkinje fibres of the heart. These fibres differ from squid nerve in that depolarization *decreases* the potassium permeability of the membrane (Hutter & Noble, 1960; Carmeliet, 1961). During large depolarizations part of this decrease appears to be only transient and the potassium permeability slowly increases during the passage of the depolarizing current (Hutter & Noble, 1960). The equations describing the dependence of the potassium current on potential and time have been modified to take account of this behaviour. The sodium current equations, however, are very similar to those of Hodgkin & Huxley and are in part based on Weidman's (1955) voltage-clamp experiments. The solution to these equations closely resembles the Purkinje fibre action and pace-maker potentials and it will be shown that its behaviour in response to 'applied currents' and to changes in 'ionic permeability' corresponds fairly well with that observed experimentally.

Preliminary reports of some of this work have already been published (Noble, 1960 *a*, *b*). The defect in the potassium current equations then used has now been corrected and this change accounts for the small quantitative differences between the conductance changes described in this paper and previously.

DESCRIPTION OF THE MEMBRANE CURRENT IN PURKINJE FIBRES

The basic feature of Hodgkin & Huxley's (1952 *d*) formulation of the properties of excitable membranes is that the current is carried by ions moving down their respective electrochemical potential gradients. The sodium current, for example, changes direction when the sodium electrochemical potential gradient is reversed, by changing either the membrane potential or the extracellular sodium concentration (Hodgkin & Huxley, 1952 *a*). In cardiac muscle this point has not been directly tested, since it has not yet proved possible to apply the voltage-clamp technique in its original form. The possibility that current is also produced by an electrogenic pump cannot therefore be entirely excluded. However, there is no conclusive experimental evidence for this view and in this paper it will be assumed that none of the membrane current is of direct metabolic origin. On this view the current carried by an ion species depends only on the magnitude of its electrochemical potential gradient and on the ease with which the ions may cross the cell membrane.

Hodgkin & Huxley (1952 *a*) showed that for squid nerve in sea water the permeability of the membrane to Na and K ions is best described in terms of the contributions which these ions make to the membrane conductance. The individual ionic conductances are defined by the equations

$$g_{\mathrm{Na}} = I_{\mathrm{Na}}/(E_m - E_{\mathrm{Na}}), \tag{1}$$

$$g_{\mathrm{K}} = I_{\mathrm{K}}/(E_m - E_{\mathrm{K}}), \tag{2}$$

where g_{Na} and g_{K} are the sodium and potassium conductances respectively in mmho/cm^2,

I_{Na} and I_{K} are the ionic currents in μA/cm^2,

E_{Na} and E_{K} are the equilibrium potentials in mV

and E_m is the membrane potential in mV expressed as the inside potential minus the outside potential.

In addition, a leak conductance was assumed which may be attributed, at least in part, to chloride ions. It will be convenient in this paper to refer to this as the anion conductance, g_{An}

$$g_{\mathrm{An}} = I_{\mathrm{An}}/(E_m - E_{\mathrm{An}}), \tag{3}$$

where I_{An} is the anion current and E_{An} the anion equilibrium potential. Various values for g_{An} will be inserted in order to reproduce the effects of anions of different permeabilities.

In Hodgkin & Huxley's equations the membrane potential (V) is measured with respect to a 'zero' at the resting potential and has a sign such that the action potential is a negative variation in V. The convention adopted here is different and conforms to that usually adopted in experimental work with intracellular electrodes. The potential (E_m) is the potential of the inside with respect to the outside, the resting potential is a negative quantity and the action potential is a positive variation. Positive currents are therefore outward and not inward as in Hodgkin & Huxley's equations. In comparing the equations in this paper with those of Hodgkin & Huxley the substitution $E_m = E_r - V$ should be made, where E_r is the resting potential of squid nerve (about -55 mV).

The total membrane current (I_m) is given by the sum of the ionic currents and the current flowing into the membrane capacity

$$I_m = C_m \frac{\mathrm{d}E_m}{\mathrm{d}t} + I_{\mathrm{Na}} + I_{\mathrm{K}} + I_{\mathrm{An}}, \tag{4}$$

where C_m is the membrane capacity and t is time in msec. C_m will be taken to be 12μF/cm^2 (Weidmann, 1952; Coraboeuf & Weidmann, 1954) which is 12 times larger than in squid nerve. If an action potential is initiated at all points along a fibre simultaneously, the membrane potential at each instant will be uniform. The axial current will therefore be zero, so that, in the absence of applied currents, the total membrane current will also be zero. This type of response was called a 'membrane' action potential by Hodgkin & Huxley and is given by equation (4) with $I_m = 0$. In these circumstances all the net ionic current is used in changing the charge on the local membrane capacity, so that the rate of change of potential, $\mathrm{d}E_m/\mathrm{d}t$,

320 *D. NOBLE*

is proportional to the net ionic current. In the present paper only membrane action potentials will be described and in comparing the results with experimental records of propagated action potentials it is assumed that the axial current, which must be very small during the slow phases of the action potential, may be neglected.

The equivalent electrical circuit assumed for the Purkinje fibre membrane is shown in Fig. 1. The only qualitative difference between this and the circuit for squid nerve (Hodgkin & Huxley, 1952d) is that the potassium current is assumed to flow through two non-linear resistances. The reason for making this assumption is explained below.

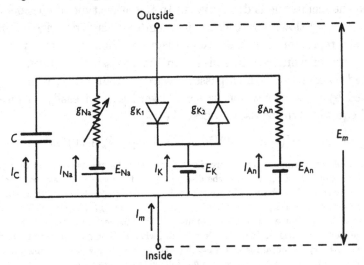

Fig. 1. Equivalent electrical circuit for Purkinje fibre membrane.
Explanation in text.

The potassium current

The equations which will be used to describe the potassium current are based on Hutter & Noble's (1960) measurements of the current–voltage relations of Purkinje fibres in sodium-deficient solutions. In contrast to the situation in squid nerve, depolarization was found to *decrease* the membrane conductance (Hutter & Noble, 1960; Carmeliet, 1961). A small and slowly developed increase in conductance occurs when large depolarizing currents are used (Hutter & Noble, 1960) but this effect is not large enough for the conductance of the depolarized membrane to exceed the resting conductance.

The chloride conductance of normal resting cardiac muscle is very small (Carmeliet, 1961; Hutter & Noble, 1961) so that the fall in conductance on depolarization must be mainly, if not entirely, attributed to a fall in g_K. For the present purpose it will be assumed that all the current

COMPUTED CARDIAC ACTION POTENTIALS   321

measured in sodium-deficient solutions is carried by potassium. So far as the action potential mechanism is concerned this assumption will not matter greatly, since, over a large range of potentials, the potassium and chloride currents flow in the same direction. It does, however, mean that the potential dependent changes in g_K given by the equations described below are likely to be rather smaller than the true changes.

For the purpose of describing the potassium current mathematically, it is convenient to suppose that K ions may move through two types of channel in the membrane. In one the potassium conductance (g_{K_1}) is assumed to be an instantaneous function of the membrane potential and falls when the membrane is depolarized. In the other type of channel the conductance (g_{K_2}) slowly rises when the membrane is depolarized. These channels are represented in the circuit diagram (Fig. 1) by two parallel rectifiers, both of which are in series with the potassium battery. g_{K_1} is represented by a rectifier which passes inward current easily, while g_{K_2} is represented by a rectifier which passes outward current easily. A purely empirical equation will be used to describe g_{K_1}

$$g_{K_1} = 1 \cdot 2 \exp\left[(-E_m - 90)/50\right] + 0 \cdot 015 \exp\left[(E_m + 90)/60\right]. \tag{5}$$

Hutter & Noble's experiments do not provide any evidence for the assumption that g_K is an *instantaneous* function of E_m, because the discharging of the membrane capacity took so long (C_m is large and, when the membrane is depolarized, r_m is also large) that it was not possible to determine the changes occurring in g_K during the first 25–50 msec of the pulses, but it seemed to be the simplest assumption to make in the absence of information obtained under voltage-clamp conditions. It will be shown later, when the computed action potentials are compared with experimental records, that this assumption may well be wrong and that there may be a small delay in the changes in g_{K_1} following changes in E_m.

The conductance of the other type of channel (g_{K_2}) will be described by Hodgkin & Huxley's potassium current equations (Hodgkin & Huxley, 1952d, equations (6), (7), (12) and (13)), with two main modifications. First, the value of $\bar{g}_{K_2}$ (the maximum value of g_{K_2}) will be made much smaller than in nerve in order that the increase in g_{K_2} produced by depolarization should not offset the decrease in g_{K_1}. Secondly, the rate constants will be divided by 100 in order to take account of the very much slower onset of this effect in Purkinje fibres (Hutter & Noble, 1960). With these modifications the equations become

$$g_{K_2} = 1 \cdot 2n^4, \tag{6}$$

$$\frac{dn}{dt} = \alpha_n(1-n) - \beta_n n, \tag{7}$$

$$\alpha_n = \frac{0 \cdot 0001(-E_m - 50)}{\exp\left[(-E_m - 50)/10\right] - 1}, \tag{8}$$

$$\beta_n = 0 \cdot 002 \exp\left[(-E_m - 90)/80\right]. \tag{9}$$

322　　　　　　　　　　　　　*D. NOBLE*

The absolute values of the conductances have been adjusted to give a resting conductance (slope conductance at $E_m = -90\,\text{mV}$) of about $1\,\text{mmho/cm}^2$ (Coraboeuf & Weidmann, 1954). The potassium equilibrium potential will be set at $-100\,\text{mV}$ so that the total potassium current is given by

$$I_K = (g_{K_1} + g_{K_2})(E_m + 100). \qquad (10)$$

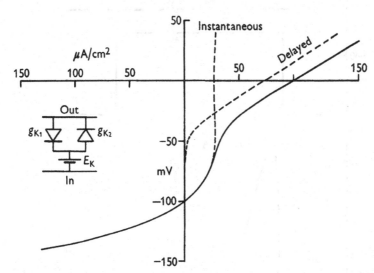

Fig. 2. Current–voltage relations described by K equations. Ordinate, membrane potential (mV); abscissa, K current in ($\mu\text{A/cm}^2$). Interrupted curves show current–voltage relations in the two types of K channel, as described in text. The continuous curve shows total steady-state current. The shape of this curve resembles that recorded experimentally, and over the voltage range of the action potential fits the curve obtained by Hutter & Noble (1960) reasonably closely when the experimental curve is corrected for the cable properties of the fibre, as described by Cole & Curtis (1941).

The current–voltage relations described by equations (5)–(10) are shown in Fig. 2. The interrupted curve labelled 'instantaneous' shows the current flowing in the first type of channel. The interrupted curve labelled 'delayed' shows the *steady-state* current flowing in the second type of channel, given by $g_{K_2}^{\infty}(E_m + 100)$, where $g_{K_2}^{\infty}$ is the value of g_{K_2} at a given potential after the potential has been held at this value for a long time. dn/dt is then zero and equations (6) and (7) give

$$g_{K_2}^{\infty} = 1 \cdot 2[\alpha_n/(\alpha_n + \beta_n)]^4. \qquad (11)$$

The continuous curve shows the sum of the currents in the two channels. The constants in the equations have been chosen so that this curve should reproduce that recorded experimentally. Over the range of potentials covered by the action potential the shape of the curve is a reasonable fit with that recorded by Hutter & Noble (1960) after correction for the cable

COMPUTED CARDIAC ACTION POTENTIALS   323

properties of the fibre (the experimental curve was obtained by polarizing the membrane at one point with a micro-electrode and the correction for curves obtained in this way is given by Cole & Curtis (1941)). Outside this range the agreement is poor, especially when the membrane is hyperpolarized, when the potassium current is underestimated by the equations.

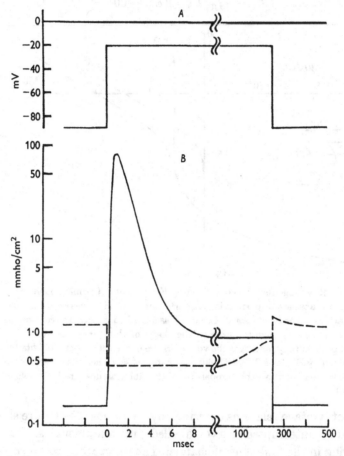

Fig. 3. Computed conductance changes occurring during depolarization of the 'membrane' from -90 to -20 mV. Ordinates: A membrane potential (mV); B ionic conductance (mmho/cm²) on a log. scale. Continuous curve, g_{Na}; interrupted curve, g_K. Abscissa: time (msec). Note change in time scale after 10 msec.

However, it did not seem worth while to try other values for the constants in the hope of obtaining a better fit because, in the computations described in this paper, the equations will not be used outside the voltage range of the action potential.

The time course of the changes in g_K given by these equations is shown in Fig. 3 (interrupted curve) in which the effect of a sudden change in E_m from

324 D. NOBLE

$-90\,\mathrm{mV}$ to $-20\,\mathrm{mV}$ is shown. g_K initially falls and then slowly rises during the period of depolarization (note the change in time scale after $10\,\mathrm{msec}$). When the potential is suddenly returned to $-90\,\mathrm{mV}$, g_K rises above its resting value, towards which it then slowly falls. Some experimental evidence for this slow fall in g_K after depolarization has been obtained (Hutter & Noble, unpublished; Noble, 1961).

The sodium current

In squid nerve changes in the membrane potential have a dual effect on the sodium conductance (Hodgkin & Huxley, 1952c). When the membrane is suddenly depolarized there is initially a very large increase in g_Na, but, even if the depolarization is maintained, g_Na soon falls again to a low value. Moreover, the magnitude of the initial increase in g_Na depends on the previous value of the membrane potential. Hodgkin & Huxley described this behaviour by supposing that g_Na is determined by two variables, m and h, which vary with the membrane potential in opposite directions and with different time constants

$$g_\mathrm{Na} = m^3 h \bar{g}_\mathrm{Na}, \tag{12}$$

where $\bar{g}_\mathrm{Na}$ is a constant and m and h obey the equations:

$$\frac{\mathrm{d}m}{\mathrm{d}t} = \alpha_m(1-m) - \beta_m m, \tag{13}$$

$$\frac{\mathrm{d}h}{\mathrm{d}t} = \alpha_h(1-h) - \beta_h h, \tag{14}$$

where α_m, β_m, α_h and β_h are functions of E_m.

The dependence of h on E_m describes the relation between the initial membrane potential and the maximum sodium current which may be produced by depolarization of the membrane. Using a modification of the voltage-clamp technique and using the maximum rate of depolarization as a measure of the sodium current, Weidmann (1955) showed that in Purkinje fibres this relation is very similar to that in squid nerve, except that the curve is shifted along the voltage axis by about $20\,\mathrm{mV}$. His method did not allow accurate measurements of α_h and β_h but he did show that these are of the same order of magnitude and vary with E_m in the same way as in squid nerve. Thus the only modifications required in the equations for h is that the functions for α_h and β_h should be shifted along the voltage axis so as to make the relation between E_m and the steady-state value of h,

$$h_\infty = \alpha_h/(\alpha_h + \beta_h), \tag{15}$$

coincide with Weidmann's experimental curve. This was done by adjusting the constants determining the position of the curve until the potential at which $h_\infty = 0.5$ became about -71 mV (Weidmann, 1955). The equations for α_h and β_h obtained in this way are:

$$\alpha_h = 0.17 \exp\left[(-E_m - 90)/20\right], \tag{16}$$

$$\beta_h = \left[\exp\left(\frac{-E_m - 42}{10}\right) + 1\right]^{-1}. \tag{17}$$

This procedure leaves the *shape* of the h_∞/E_m relation unaltered. In fact, Weidmann's curve for Purkinje fibres is slightly steeper than that for squid nerve, but, as he has pointed out, it is very likely that this is only due to differences in experimental technique.

In the voltage-clamp technique used by Weidmann, when the sodium conductance is greatly increased on depolarization of the membrane it is not possible to retain control of the membrane potential owing to the limitation on the amount of current which may be passed through a micro-electrode. He was therefore unable to obtain information on which to base equations for m. However, in view of the close similarity of the processes determining h in Purkinje fibres and in squid nerve, it seems reasonable to assume that the processes determining m are also similar. The choice of constants in the m equations must at present be a somewhat arbitrary procedure and the method used will be described in detail below (see Methods). The equations obtained are:

$$\alpha_m = \frac{0.1(-E_m - 48)}{\exp\left[(-E_m - 48)/15\right] - 1}, \tag{18}$$

$$\beta_m = \frac{0.12(E_m + 8)}{\exp\left[(E_m + 8)/5\right] - 1}. \tag{19}$$

In arriving at these equations it was assumed that a small component $(0.14\,\text{mmho/cm}^2)$ of g_{Na} is independent of E_m and t. $\bar{g}_{Na}$ was set at $400\,\text{mmho/cm}^2$ and E_{Na} at 40 mV. When these values are inserted into equation (1) I_{Na} is given by

$$I_{Na} = (400m^3h + 0.14)(E_m - 40). \tag{20}$$

The behaviour of these equations is of course very similar to that of Hodgkin & Huxley's (Hodgkin & Huxley, 1952d; Huxley, 1959a). It is, however, worth illustrating in order to note the changes which occur when a *long-lasting* depolarization is applied. This is shown in Fig. 3 (continuous curve). When E_m is suddenly changed from -90 to -20 mV it can be seen that, following the large transient increase in g_{Na}, there is a small maintained increase which persists throughout the period of the depolarization. The steady-state Na current therefore increases in spite of the decrease in the Na electrochemical potential gradient. This property allows the

326 *D. NOBLE*

equations to be extended to describe long-lasting action potentials without any serious modification to the sodium current equations (cf. Fitzhugh, 1960).

METHODS

Method of obtaining equations for α_m and β_m

In the absence of any direct experimental evidence on which to base equations for m in cardiac muscle it is not possible to describe the sodium current fully without using the action potential itself as a source of information. The problem then becomes: given the shape of the action potential and equations for h and for I_K, what equations are required to describe m? A number of different equations for α_m and β_m were tried in order to see whether functions of the same general form as Hodgkin & Huxley's could be used. This turned out to be the

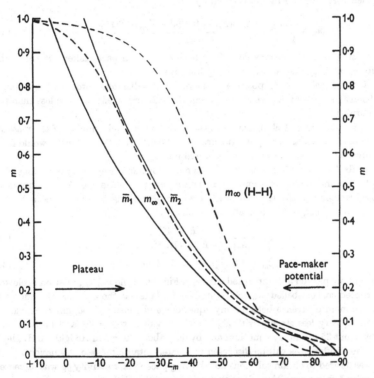

Fig. 4. Relations between m (ordinate) and membrane potential (abscissa). $\bar{m}_1$ is given by equation (22); $\bar{m}_2$ by equation (23); m_∞ by equations (18), (19) and (21); m_∞ (H–H) by equations (21), (24) and (25). Explanation in text.

case and it proved relatively easy to obtain solutions which resemble the Purkinje fibre action potential. However, it proved much more difficult to find equations for m which would also allow pace-maker activity to occur. The method of obtaining the constants by simple trial and error is excessively tedious and would have proved very costly in terms of computer time, so that it became essential to find a more direct method.

The way in which equations (18) and (19) were obtained is illustrated by Fig. 4, in which m is plotted against E_m. It was convenient to take $\bar{g}_{Na}$ as 400 mmho/cm^2 and to assume that 0·14 mmho/cm^2 is independent of E_m and t. This does not necessarily mean that g_{Na} in some channels is in fact independent of E_m and t in cardiac muscle, but it made a little easier the process of finding functions for α_m and β_m which allow pace-maker activity to occur. The continuous curve labelled $\overline{m}_1$ in Fig. 4 shows the value which m_∞, given by

$$m_\infty = \alpha_m/(\alpha_m + \beta_m), \qquad (21)$$

must have at each potential in order that the steady-state sodium current should be equal and opposite to the potassium current when $n = 0$. This is given by

$$\overline{m}_1 = \left(\frac{[g_{K_1}(E_m + 100)/(40 - E_m)] - 0\cdot14}{400 h_\infty} \right)^{\frac{1}{3}}, \qquad (22)$$

and is applicable at the end of diastole and during the spike of the action potential when n is small enough for g_{K_2} to be neglected. The continuous curve labelled $\overline{m}_2$ shows the same relation when n has the value which it would have if E_m were held constant for a long time at a potential corresponding to the termination of the plateau. $E_m = -30$ mV was chosen and n_∞ is then very nearly 0·72. $\overline{m}_2$ is given by

$$\overline{m}_2 = \left(\frac{[(g_{K_1} + 1\cdot2n^4)(E_m + 100)/(40 - E_m)] - 0\cdot14}{400\,h_\infty} \right)^{\frac{1}{3}}. \qquad (23)$$

Now in order to obtain solutions describing both action and pace-maker potentials the functions for α_m and β_m must satisfy two requirements:

(1) $m_\infty < \overline{m}_2$ for all values of E_m positive to the maximum diastolic potential (E_m approximately -90 mV), since for repolarization to occur the sodium current must be less than the potassium current.

(2) $m_\infty > \overline{m}_1$ for all values of E_m negative to the potential at the beginning of the plateau. If, for example, $m_\infty = \overline{m}_1$ at some potential around -90 mV then this potential would form a resting potential and pace-maker activity would not occur.

The curve labelled m_∞ in Fig. 4 satisfies these conditions and is given by equations (18), (19) and (21). The individual functions for α_m and β_m are plotted in Fig. 5 for comparison with those of Hodgkin & Huxley, which after adjustment along the voltage axis by the same amount as the h equations (see above) become

$$\alpha_m = \frac{0\cdot1(-E_m - 47)}{\exp{[(-E_m - 47)/10]} - 1}. \qquad (24)$$

$$\beta_m = 4 \exp{[(-E_m - 72)/18]}. \qquad (25)$$

These equations are plotted as interrupted curves in Fig. 5 and the m_∞/E_m relation is plotted as an interrupted curve labelled m_∞ (H–H) in Fig. 4. So far as the *shape* of the curves is concerned, the main difference between my equations and those of Hodgkin & Huxley is that m_∞ and β_m vary less steeply with E_m. Thus the values $m_\infty = 0\cdot5$ and $m = 0\cdot1$ are separated by about 35 mV in my equations and by only about 20 mV in Hodgkin & Huxley's equations. These differences will be discussed in greater detail later (see Discussion).

When it was desired to obtain solutions not showing pace-maker activity g_K was increased by 0·1 mmho/cm², when condition (2) above is no longer satisfied.

Numerical computation

The integration of the four simultaneous differential equations (4), (7), (13) and (14) was done on the London University digital computer 'Mercury', using a numerical approximation (the Runge–Kutta rule). Solutions for the rapid changes occurring during the spike of the action potential were obtained by using a step length of integration, Δt, of 0·1 msec. It was found that integration at shorter step lengths did not produce an appreciably different result.

By comparison with the initial spike, the plateau and pace-maker potential are very long-lasting indeed and to obtain solutions for these parts of the action potential with $\Delta t = 0\cdot1$ msec more than 60 min of computer time was required. This could not be reduced simply by increasing Δt, because computations involving differential equations with very small time

328   *D. NOBLE*

constants become unstable when $\Delta t > 2 \cdot 8\, T_{\min.}$, where $T_{\min.}$ is the smallest time constant in the system (Carr, 1958; N.P.L., 1961). This difficulty was overcome by making m an explicit function of only E_m during the plateau and pace-maker potential. This is justified, since the time constants for m are small enough compared to the duration of these phases of the action potential to allow the assumption that, at each value of E_m, m has its steady-state value, m_∞. Equation (21) may then be used to compute m. The problem now reduces to the integra-

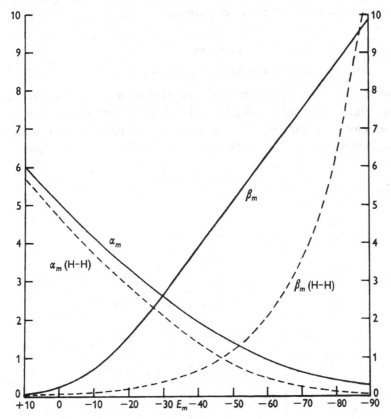

Fig. 5. Comparison of modified functions for α_m and β_m with those of Hodgkin & Huxley. Continuous curves: modified functions given by equations (18) and (19). Interrupted curves: Hodgkin & Huxley's m functions after adjustment along voltage axis by same amount as for h equations. These curves are given by equations (24) and (25). Note that β_m changes less rapidly with E_m in modified function than in original Hodgkin–Huxley function.

tion of three simultaneous equations and, since the time constants for h are about 10 times larger than those for m, Δt can be increased to 1 msec without introducing appreciable error. This was checked by comparing results obtained at $\Delta t = 0 \cdot 1$ msec with equation (13) to compute m with those obtained at $\Delta t = 1$ msec with equation (21). The computer programme was arranged so that equation (21) was used when $dE_m/dt < 0 \cdot 5$ V/sec.

In order to start the computation, the initial values of E_m, m, h and n have to be given. These were obtained by choosing an initial potential at about the middle of the pace-maker potential, when dE_m/dt is very small. m and h could therefore be given their steady-state values (m_∞ and h_∞) without introducing appreciable error. A guess was then made for the initial value of n. A small error here is not critical, however, and after one cycle the initial value chosen for n has no influence on the solution, apart from determining its position on the time scale.

Results were printed out by the machine usually after every 10 integration steps. In addition to E_m and t, the following were printed out when required: m, h, n, g_{Na}, g_K, Na efflux, Na influx, K efflux, K influx, net Na gain, net K loss. Where appropriate these are plotted in the illustrations in addition to the potential changes. Curves relating to Na are continuous, whereas curves relating to K are interrupted.

RESULTS

Computed potentials and conductances

The solution to the equations was computed over two cycles and is shown in Fig. 6. It can be seen that the shape of the potential wave (curve A) closely resembles that recorded experimentally in Purkinje fibres and shows the characteristic spike, followed by a plateau lasting

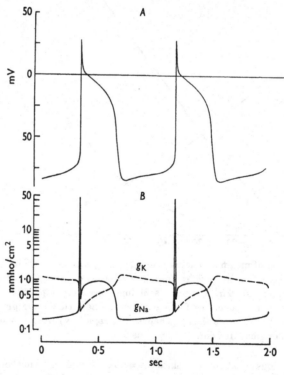

Fig. 6. A, computed action and pace-maker potentials. B, time course of conductance changes on a log. scale. Continuous curve, g_{Na}: interrupted curve, g_K.

about 300 msec which is terminated by a faster phase of repolarization. The membrane then slowly depolarizes again—the pace-maker potential—until the threshold is reached and another action potential is initiated.

Only in one major respect does the computed action potential fail to

330 *D. NOBLE*

reproduce that recorded experimentally. The maximum rate of depolarization, dE_m/dt_{max}, during the spike of the computed action potential is about 100 V/sec. This is very much less than that recorded experimentally: action potentials initiated from about -70 mV (as in the computed action potential in Fig. 6) show a maximum rate of depolarization of about 300 V/sec (Weidmann, 1955). The relation between E_m and dE_m/dt_{max} is very steep at about -70 mV (Weidmann, 1955) and a better comparison may be obtained by initiating the action potentials from a potential at

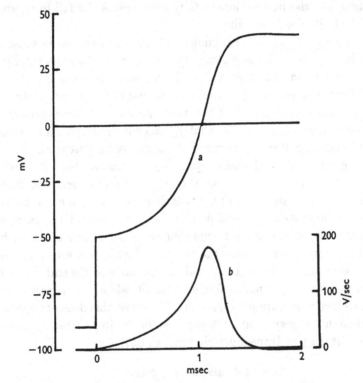

Fig. 7. *a*, Early part of computed action potential initiated by suddenly displacing E_m from -90 to -50 mV. *b*, Rate of change of membrane potential, dE_m/dt, (V/sec).

which h is very nearly 1, so that all the sodium carriers (or sites) are readily available for sodium transport. In these circumstances the value for dE_m/dt_{max} obtained experimentally is about 800 V/sec (Draper & Weidmann, 1951; Weidmann, 1956 *b*). When the computed action potential is initiated from -90 mV, dE_m/dt_{max} is only 180 V/sec (Fig. 7). The possible reasons for this discrepancy will be discussed later.

The time course of the computed conductance changes is shown in Fig. 6 *B*. A logarithmic scale was chosen in order to accommodate the large changes which occur in g_{Na}. During the spike of the action potential g_{Na}

rises to a very high value because m rises much faster than h falls. Within a few milliseconds the fall in h reduces g_{Na} again; the inactivation is not complete, however, and, after an 'undershoot', g_{Na} settles down to a fairly constant plateau value which is about 8 times larger than the lowest value reached in diastole. The relative constancy of g_{Na} in spite of changes in E_m during the plateau is due to the fact that, over this range of potentials, m and h change in opposite directions in such a way that m^3h remains almost constant. At the end of the plateau, during the final rapid phase of repolarization, the rise in h no longer fully counteracts the fall in m, and g_{Na} then falls to its diastolic value.

By contrast, g_K *falls* at the beginning of the action potential as a result of the fall in g_{K_1}. During the plateau g_{K_2} slowly rises, but the total g_K remains below the end-diastolic value throughout the duration of the plateau. A further increase in g_K occurs at the end of the action potential when g_{K_1} increases as a result of the repolarization of the membrane. g_{K_2} takes some time to fall again, so that g_K exceeds its end-diastolic value for several hundred milliseconds during the pace-maker potential.

The *total* membrane conductance, $g_{Na} + g_K$, increases about 50 times during the spike of the action potential in Fig. 6 *A* and almost 100 times during the faster spike shown in Fig. 7. This is in good agreement with the experimental observation of Weidmann (1951), who found the ratio of resting membrane conductance to maximum active conductance to be about 1:100. During the plateau the total conductance is much smaller and, at the beginning of the plateau, is about the same as the end-diastolic conductance. Since g_{Na} remains fairly constant while g_K increases, the total conductance rises during the plateau. However, this does not conflict with Weidmann's observation that the membrane impedance increases during the plateau (see Impedance changes, p. 337).

Computed currents and fluxes

The changes in g_{Na} and g_K which occur during the action potential to some extent resemble those produced by an applied 'square-wave' voltage change of similar magnitude to the plateau (Fig. 3) although, in the case of the action potential, the voltage changes themselves are in turn produced by the conductance changes. During an applied 'square-wave' voltage change current would have to be supplied to the membrane to keep the potential constant, and this is the current which would be recorded by a 'voltage-clamp' technique. The action potential, on the other hand, requires no externally applied current, all the current being supplied by the fibre itself, and in the case of a 'membrane' action potential the relation between ionic current and the potential changes is a fairly simple one, dE_m/dt being proportional to the net ionic current.

332 *D. NOBLE*

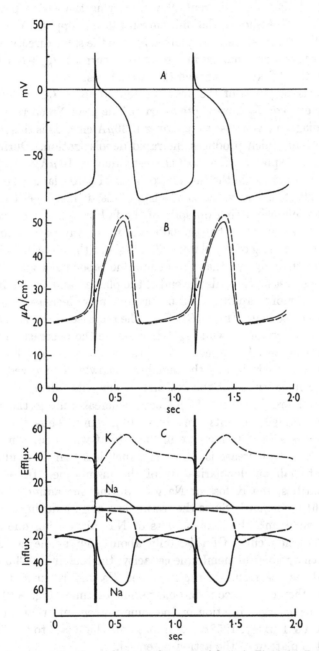

Fig. 8. *A*, computed action and pace-maker potentials (as in Fig. 6 *A*). *B*, computed ionic currents. Continuous curve, I_{Na}; interrupted curve, I_K. The peak Na current was about 1200 $\mu A/cm^2$ and is not shown on the scale used. dE_m/dt is positive when $I_{Na} > I_K$ and is negative when $I_{Na} < I_K$. *C*, Ionic fluxes given by equations (26) and (27). Effluxes are plotted upwards above the abscissa; influxes are plotted downwards below the abscissa.

COMPUTED CARDIAC ACTION POTENTIALS   333

The computed Na and K currents flowing during the action potential are shown in Fig. 8 B. Although the ionic currents flow in opposite directions (Na inwards, K outwards) they are plotted here in the same direction. This makes it easier to observe changes in the net ionic current ($I_{Na} + I_K$), which is then simply the difference between the two curves.

During the rising phase of the action potential spike I_{Na} (continuous curve) greatly exceeds I_K (interrupted curve). The peak Na current is not shown in the diagram, as it rose to just over $1200\mu A/cm^2$. It is this intense inward Na current which produces the rapid depolarization. During the falling phase of the spike I_{Na} falls to its lowest value ($\simeq 10 \ \mu A/cm^2$), partly as a result of the decrease in the Na electrochemical potential gradient and partly as a result of the low value to which g_{Na} falls at this stage (Fig. 6 B). Since I_K is continuously rising (in spite of the fall in g_K), it now exceeds I_{Na} by almost $20 \ \mu A/cm^2$, so that the potential falls fairly rapidly towards its value at the beginning of the plateau. Throughout the plateau I_K is only slightly greater than I_{Na}, so that the membrane repolarizes very slowly. This difference increases towards the end of the plateau and the membrane then repolarizes more rapidly. Both currents now decrease: I_{Na} falls because the decrease in g_{Na} (Fig. 6 B) offsets the increase in the Na electrochemical potential gradient, while I_K falls because the decrease in the K electrochemical potential gradient offsets the increase in g_K (Fig. 6 B). I_{Na} now slightly exceeds I_K and the membrane slowly depolarizes again.

The average magnitudes of the Na and K currents during the pacemaker potential are just over $20 \ \mu A/cm^2$, whereas during the action potential the average currents are about $40 \ \mu A/cm^2$. Thus the action potential involves a twofold increase in the rate at which the fibre gains Na and loses K. This increase is, however, much less than it would be if g_K were not to fall on depolarization of the membrane. In terms of chemical quantities, the K loss or Na gain during one action potential amounts to 161 pmole/cm². Since the resting loss or gain during a similar period is 83 pmole/cm², the extra K loss or Na gain during one action potential is 77 pmole/cm². Of this, only about 13·5 pmole is used in changing the charge on the membrane capacity, the remaining 63·5 pmole being 'wasted' as the result of the increased Na and K currents overlapping each other over a considerable period of time (Fig. 8 B). This 'wastage' forms a larger fraction of the ionic movements than in squid nerve (Hodgkin & Huxley, 1952 d) and represents the 'cost' to the fibre of maintaining the plateau of the action potential.

The efflux and influx components of the ionic currents were computed on the assumption that the independence principle applies to Purkinje fibres. This principle states that the influx and efflux of a given ion species are independent of one another, i.e. the influx is independent of the intra-

334   D. NOBLE

cellular concentration, while the efflux is independent of the extracellular
concentration. This assumption leads to the following equations for the
effluxes (Hodgkin & Huxley, 1952d):

$$\text{Na efflux} = I_{Na}/[\exp((E_{Na} - E_m)F/RT) - 1], \qquad (26)$$

$$\text{K efflux} = I_K/[\exp((E_K - E_m)F/RT) - 1]. \qquad (27)$$

The influxes are then given by the differences between the currents and
the effluxes. Hodgkin & Huxley showed that the independence principle
holds very well in the case of Na movement in squid nerve, but no experimental confirmation of the principle yet exists in the case of cardiac
muscle. The application of these equations to Purkinje fibres therefore
requires caution and the presence of any appreciable degree of interaction
during the movement of ions through the membrane would invalidate
them.

In Fig. 8C the effluxes are plotted above the abscissa (continuous curve,
Na; interrupted curve, K) while the influxes are plotted below the abscissa.
The most surprising feature is the very small increase in K efflux which
occurs during the action potential. When averaged out over the duration
of one cycle, this increase amounts to only about 10% of the diastolic
efflux. This is in striking contrast to the large increase in K efflux during
the squid-nerve action potential observed by Keynes (1951) and computed
by Hodgkin & Huxley (1952d). In Purkinje fibres most of the increase
in the computed K current results from a very large fall in K influx. In
the case of sodium both influx and efflux increase during the action
potential. The peak Na fluxes are not shown in the illustration. The peak
influx was nearly 1500 μA/cm^2 and the peak efflux was nearly 300 μA/cm^2.

The correlation of these computed fluxes with the experimental information at present available in cardiac muscle will be discussed later
(see Discussion).

Current–voltage relations

A simpler, though more approximate, description of the mechanism of
the action and pace-maker potentials may be obtained by plotting Na and
K current–voltage relations at different times during the cycle (cf. Hodgkin,
Huxley & Katz, 1949). This has been done in Fig. 9. The ordinate is the
membrane potential (mV) and the abscissa is the ionic current (μA/cm^2).
Although the Na and K currents flow in opposite directions, they have
been plotted here in the same direction so that intersections between the
curves represent points at which the currents are equal and opposite. The
interrupted curves show the *instantaneous* K current–voltage relations
given by equations (5), (6) and (10) with n having values appropriate to
the beginning of the action potential (0 msec, $n = 0.32$), two stages during
the plateau (100 msec, $n = 0.58$; 200 msec, $n = 0.68$) and at the end of

COMPUTED CARDIAC ACTION POTENTIALS   335

the plateau (280 msec, $n = 0.72$). These curves also apply at points during the pace-maker potential, but the temporal order is then reversed as n is then falling instead of rising. The continuous curves are Na current–voltage relations. The line intersecting the 0 msec K curve at the point A is the *instantaneous* Na current–voltage relation at the peak of the action potential

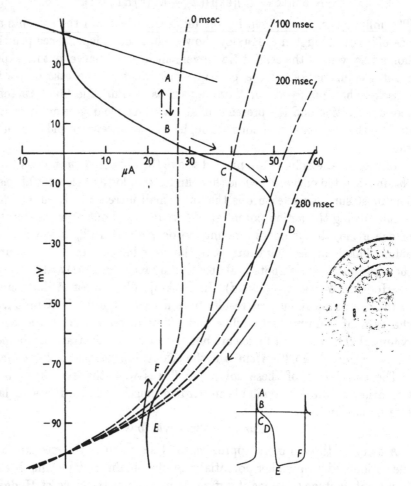

Fig. 9. Ionic current–voltage relations. Ordinate, membrane potential (mV); abscissa, ionic current (μA/cm^2). Interrupted curves are instantaneous K current–voltage relations at various stages during the action potential. The continuous curves are Na current–voltage relations. Points A–F correspond to the stages indicated on the computed action potential shown in inset. Explanation in text.

spike. This is given by equations (12) and (20) with $m = 0.9996$ and $h = 0.010$. The other continuous curve shows the *steady-state* Na current–voltage relation given by equations (12), (15), (20) and (21). Since changes in m and h follow changes in E_m fairly closely during the slower phases of

336 *D. NOBLE*

the action potential, this curve gives an approximate estimate of the Na current at different voltages during the plateau and pace-maker potential.

When dE_m/dt is small the net ionic current must be small, so that the Na and K currents are nearly equal and opposite (see Fig. 8). Some of the intersections in Fig. 9 therefore closely correspond to the values of the potentials and currents occurring at various instants during the action potential. The points labelled A–F correspond to the stages during the cycle indicated in the inset. Point A corresponds to the peak of the action potential spike, when g_{Na} is very large. As the Na-carrying system becomes inactivated g_{Na} approaches its steady-state value at the beginning of the plateau, and the potential falls to point B. I_K then increases, so that the potential slowly falls, passing through point C, until the K curve no longer intersects the Na steady-state curve in the region of the plateau (point D). The K current now exceeds the steady-state Na current at all potentials above the point E, so that the plateau can no longer be maintained and the rapid phase of repolarization commences. During this phase the steady-state Na curve does not give a good approximation to the sodium current, because the potential changes too rapidly for h to approximate closely enough to its steady-state value at each potential. Thus, at a membrane potential of -50 mV the Na current during repolarization is 27 μA/cm^2 (Fig. 8B), whereas the steady-state current is about 38 μA/cm^2 (Fig. 9). When the point of maximum repolarization, E, is reached the Na and K curves once again intersect. The K current–voltage curve now 'swings' back again towards the 0 msec curve and the potential slowly rises until the point F is reached at the end of the pace-maker potential. The Na and K curves now no longer intersect in the pace-maker region, the membrane depolarizes more rapidly and another action potential is initiated. During this phase the Na current is very much greater than the steady-state current, and the potential rapidly changes to point A. The cycle then repeats itself.

The mechanism of the pace-maker potential is thus very similar to that of the plateau. Both can be approximately represented as a point of intersection moving along the steady-state Na current–voltage curve.

In these equations there is one potential at which the steady-state Na curve intersects the *steady-state* K curve. The latter is the continuous curve in Fig. 2, but is not shown in Fig. 9. Its point of intersection with the Na curve occurs at about -32 mV and a constant potential at this point therefore corresponds to a second solution to the equations. However, this solution is unstable because it occurs at a potential at which the total membrane slope conductance is negative. A deflexion, however small, in the repolarizing direction would reduce I_{Na} more than I_K, so that the repolarization would become regenerative. Similarly, a deflexion in the depolarizing direction would increase I_{Na} more than I_K, which would lead to a regenerative depolarization.

Only small modifications to the equations are required to make such a point occur outside the region of negative slope conductance. If g_K were increased by about 0·1 mmho/cm², a point of intersection would occur in the region of the pace-maker potential. The potential would then fail to reach the point F spontaneously and the system would correspond to a normal quiescent fibre. On the other hand, if $\bar{g}_{K_2}$ were reduced by about 35 % a 'stable state' would occur in the region of the plateau at about -20 mV. This would correspond to a fibre which becomes temporarily or permanently arrested on the plateau, as is in fact observed when Purkinje fibres are cooled sufficiently (Trautwein, Gottstein & Federschmidt, 1953; Coraboeuf & Weidmann, 1954; Chang & Schmidt, 1960).

Impedance changes

It has been shown above that the increase in membrane conductance during the spike of the computed action potential agrees well with that observed experimentally by Weidmann (1951). Weidmann also recorded the changes in impedance occurring during the slower phases of the Purkinje fibre action potential. Although he used square-wave current pulses and has expressed his results in terms of membrane 'resistance' (Weidmann, 1956b, Fig. 16) it is clear from his record that, at least during the plateau, the electrotonic potentials did not reach completion, so that his experiment measured a quantity which depended on the membrane reactance as well as the membrane resistance. An equation for sinusoidal current was therefore employed in order to reproduce Weidmann's result on the computed response. The currents used in obtaining the curves shown in Fig. 10 are

$$I_m = 7 \sin\left[\frac{2\pi}{50}(t+b)\right] \mu A/cm^2 \tag{29}$$

with $b = 0, 15, 30$. This gives three currents of the same period (50 msec) but with different phase shifts (b). The integrations were started at the peak of an unmodulated action potential and the resulting potential curves have been superimposed in Fig. 10. So far as the main features are concerned the agreement between this and Weidmann's experimental record is very good. The impedance rises during the plateau, falls again at the end of repolarization and then rises slightly during the pace-maker potential. The rise in impedance during the plateau may appear surprising in view of the fact that the total membrane conductance ($g_{Na} + g_K$) increases during this time (Fig. 6B). The reason for this is that m, h, n and g_{K_1} vary periodically as a result of the periodic changes in E_m, so that g_K and more particularly g_{Na} have values which depend on the phase of the applied current. The factors determining the amplitude of the voltage wave are not therefore the ionic chord conductances (g_K, g_{Na}) but the ionic *slope* conductances ($dI_K/dE_m, dI_{Na}/dE_m$). If the period of the alternating current is long compared with the Na time constants, but short compared with the K time constants, the slope conductances will be approximately equal to the reciprocal slopes of the curves shown in Fig. 9; from which it can be seen that the Na slope conductance falls during the plateau and eventually becomes negative at about -20 mV. Although the K slope conductance rises, it does not do so sufficiently to prevent the total slope conductance from falling. The membrane impedance therefore rises during the plateau. This is a rather striking illustration of the fact that changes in membrane impedance do not necessarily closely reflect changes in ionic conductance and, as in this case, may even appear to indicate changes in the opposite direction to those which are actually occurring.

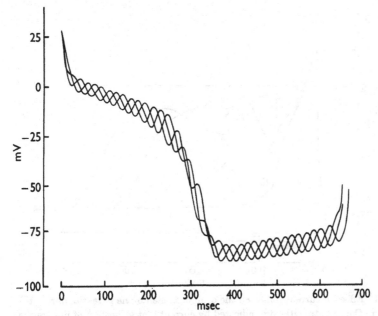

Fig. 10. Variation in membrane impedance during action and pace-maker potentials. Three superimposed potential curves obtained by setting I_m equal to sinusoidal currents given by equation (29). The impedance rises during the plateau, falls at the end of repolarization and then rises slightly during the pace-maker potential.

The rise in impedance during the pace-maker potential is due to three factors: the slow fall in g_{K_2}, the fall in g_{K_1} consequent upon depolarization and the increasing negativity of the Na slope conductance (see Fig. 9).

Vector analysis of the impedance changes in Fig. 10 into parallel resistive and reactive components gave a rather complicated result. In particular, the parallel reactance does not remain constant, as it depends not only on the constant membrane capacity assumed in the equations but also on a variable 'anomalous' reactance attributable to the periodic changes in g_{Na} and g_K described above.

COMPUTED CARDIAC ACTION POTENTIALS **339**

Regenerative repolarization

When a large enough repolarizing current is passed through the membrane during the plateau of the action potential, an all-or-nothing repolarization is initiated (Weidmann, 1951, 1956b). That this type of behaviour may be reproduced by the Hodgkin–Huxley equations, and modifications of them, is now well known (Huxley, 1959a,b; Fitzhugh, 1960; George & Johnson, 1961). The behaviour of the equations used here is illustrated in Fig. 11, which shows the potential changes produced by adding square-wave current pulses at two different times during the

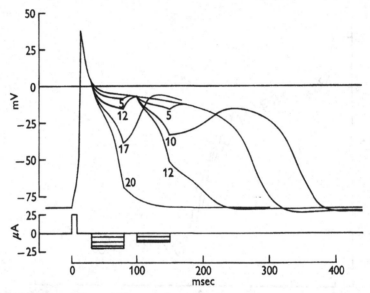

Fig. 11. Effect of current pulses on the computed membrane potential during the plateau. Current strengths are indicated by current plot at bottom of illustration and by the figures (in $\mu A/cm^2$) on the potential curves. Description in text.

plateau. When currents above a certain threshold strength are added, the potential does not return to the plateau when the current pulse is terminated. Instead, the potential returns to the resting potential. The threshold for this phenomenon is larger at the beginning of the plateau than at the end (cf. Fitzhugh, 1960) and, during the early part of the plateau, is about 40–50 mV negative to the plateau potential. This is less negative than the threshold observed by Weidmann when recording the potential changes very close to a polarizing electrode (Weidmann, 1956b, Fig. 25B) but is greater than that observed when the electrodes are inserted one or two space constants apart (Weidmann, 1956b, Fig. 25C). A difference of this kind is to be expected, since, in the first case when the applied current is switched off local circuit currents will flow in such a direction as to bring

340 *D. NOBLE*

the potential back to the plateau, so that the threshold voltage displacement will be greater than in the case where the membrane is polarized uniformly, as it is in the case of the computed action potentials. When the potential is recorded at some distance away from the polarizing electrode, the converse will apply and the local circuit current will flow through the membrane in the opposite direction. A further complication arises here because the regenerative repolarization response is probably propagated (Weidmann, 1951). This effect is not of course reproduced in computed 'membrane' action potentials.

A current which just fails to initiate repolarization (e.g. 10 μA in the middle of the plateau) *prolongs* the action potential. This effect results from the dependence of the rate constants of the 'delayed' potassium channels, α_n and β_n, on E_m so that g_{K_2} rises more slowly when the potential is altered by the repolarizing current than it does during the normal action potential. When the current is terminated, and the potential returns to the plateau, more time is required for g_{K_2} to reach the value required to bring about repolarization. Such a prolongation has been observed in Purkinje fibres (Weidmann, 1956b, Fig. 26) but the effect is not as large as in the computed action potential. Again, this difference may be due to the difference in the way in which the current is applied. In the experimentally recorded action potentials the current was not applied uniformly, so that local circuit currents occurred which are absent in the computed response.

These differences between the computed and experimentally recorded action potentials in their responses to applied currents are minor ones but are difficult to deal with satisfactorily in qualitative terms. It is clearly desirable that solutions for propagated action potentials with locally applied currents should be computed to test these points.

Repetitive stimulation

Another property of cardiac muscle which may be accounted for by these equations is the shortening of the duration of the action potential produced by an increase in the frequency of stimulation (Carmeliet, 1955a, b; Hoffmann & Suckling, 1954; Trautwein & Dudel, 1954). This results from the slow time course of the decay of g_{K_2} after the end of the action potential. If two action potentials are initiated in rapid succession, the second is shorter than the first because g_{K_2} starts at a higher value and so takes less time to rise to the value required to initiate repolarization of the membrane. A potassium current system with long time constants is also a feature of Fitzhugh's (1960) modification of the Hodgkin–Huxley equations and he has already shown how this accounts for the frequency-duration relation.

COMPUTED CARDIAC ACTION POTENTIALS 341

An interesting consequence of this relation is that an alternation in the duration of successive action potentials is observed when a resting fibre is suddenly stimulated at a high enough frequency. This effect is shown on the computed action potential in Fig. 12. The interrupted curve shows the changes which occur in n (the variable determining g_{K_2}). In the 'resting fibre' n is small, g_{K_2} is virtually zero, so that during the *first* action potential n takes a long time to rise to the value required to initiate repolarization. The *second* action potential follows very soon after the first, while n is still fairly large. Much less time is therefore required for n to rise again, so that the second action potential is very much shorter than the first. This degree of shortening represents nearly the maximum obtainable from an increase

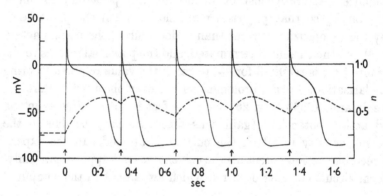

Fig. 12. Effect of repetitive stimulation on the computed action potential (continuous curve). Interrupted curve shows the changes which occur in n. 'Fibre' made 'quiescent' by adding 0·1 mmho/cm² to g_K and then suddenly stimulated at a frequency of 3/sec. Note alternation in duration of action potentials.

in frequency and is similar to that observed by Trautwein & Dudel (1954), who recorded action potentials down to durations of about 20 % of the low-frequency duration. The *third* action potential follows after a longer 'diastole' than that which preceded the second, so that it is longer than the second although shorter than the first. An alternation of this kind persists for several action potentials before the duration finally reaches a stable value. This type of behaviour is often observed in cardiac muscle during sudden increases in frequency or on stimulation of a previously quiescent fibre (e.g. Hoffmann & Suckling, 1954; Schütz, 1936).

There is, however, one feature of the effect of frequency on duration which is not accounted for by these equations. After a period of high-frequency stimulation the action potential duration may take some time to return completely to normal, even though the fibre is stimulated at a constant low frequency (e.g. Carmeliet, 1955a, b; Hoffmann & Cranefield, 1960, Fig. 7–2). It seems therefore that some other factor must be involved

342 *D. NOBLE*

in addition to the one described here. This may be the time required for the restoration of ionic concentration gradients after a period of activity, as suggested by Carmeliet (1955*b*).

The influence of permeability changes

The changes in duration produced by alterations in the frequency of stimulation mainly involve changes in the duration of the plateau, the other phases of the action potential being relatively unaffected. The same also applies when the action potential duration is altered by changes in the ionic permeabilities or in the ionic environment. In spontaneously beating fibres the frequency is also very sensitive to permeability changes.

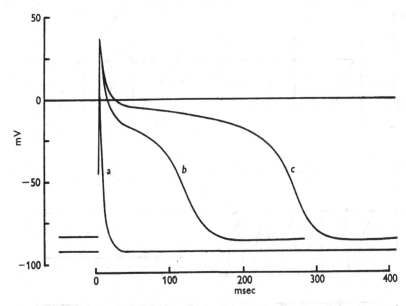

Fig. 13. Effect of additional ionic conductances on the duration of the computed action potential. *a*, No additional conductance. *b*, Additional conductance of 0·2 mmho/cm² with equilibrium potential at the resting potential. *c*, Effect of increasing g_K by 1·0 mmho/cm².

Figure 13 shows the effect of additional conductances on the shape of the computed action potential. The addition of a conductance of 0·2 mmho/cm² with an equilibrium potential at the resting potential approximately halves the duration of the plateau (curve *b*), which is similar to the effect produced by substituting Cl by NO_3 ions on the action potentials of Purkinje and ventricular fibres (Carmeliet, 1961; Hutter & Noble, 1961). An additional g_K of 1·0 mmho/cm² has a very striking effect which is shown in curve *c*. The plateau is now completely absent and the action potential becomes very short indeed. This resembles the effect produced on the sinus

COMPUTED CARDIAC ACTION POTENTIALS 343

venosus action potential when g_K is increased by stimulation of the vagus nerve (Hutter & Trautwein, 1956). The conductance increase assumed here is of the right order of magnitude. Harris & Hutter (1956) found that the rate of movement of potassium ions in the sinus venosus may be increased two to three times by acetylcholine; the conductance increase assumed in Fig. 13 is approximately twofold at the resting potential.

Like the plateau, the pace-maker potential is very sensitive to changes in ionic permeability. Thus, an increase in g_K by only 0·1 mmho/cm² is sufficient to stop pace-maker activity completely in these equations. In this respect the equations mimic the behaviour of the natural pace-maker when g_K is increased by vagal stimulation (Hutter & Trautwein, 1956).

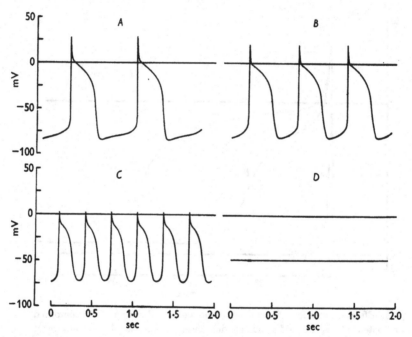

Fig. 14. Effect of various anion conductances on the computed pacemaker potential. A, $g_{An} = 0$; B, $g_{An} = 0\cdot075$ mmho/cm²; C, $g_{An} = 0\cdot18$ mmho/cm²; D, $g_{An} = 0\cdot4$ mmho/cm². $E_{An} = -60$ mV in all cases. Description in text.

Rather striking alterations in pace-maker activity are observed when different anions are present in the extracellular fluid (Hutter & Noble, 1961) and these may also be reproduced by the equations. It seems likely that in continuously active fibres the anion equilibrium potential is low (Hutter & Noble, 1961) and E_{An} was therefore set at -60 mV. The results of inserting this value for E_{An} and various values for g_{An} are shown in Fig. 14. Curve A shows the solution obtained when $g_{An} = 0$ and is the same as that in Fig. 6A. When g_{An} is increased to 0·075 mmho/cm²

344   D. NOBLE

(curve *B*) the frequency increases by about 50% while the shape and duration of the action potential are only slightly affected. This resembles the effect of Cl ions on Purkinje fibres.

When g_{An} is increased to 0·18 mmho/cm² (curve *C*) the frequency greatly increases, the maximum diastolic potential falls, the action potential is markedly shortened and the overshoot is almost completely abolished. This closely resembles the effect of prolonged exposure to NO_3 ions on spontaneously beating Purkinje fibres. The value of g_{An} assumed in Fig. 14 is almost equal to that assumed in order to reproduce the effect of NO_3 ions on a driven fibre (Fig. 13). Thus, in so far as the relative sensitivity of the plateau and pace-maker potential to changes in anion conductance is concerned the behaviour of the equations is consistent.

A further increase in g_{An} to 0·4 mmho/cm² (curve *D*) does not produce an increase in frequency but has the effect of arresting the 'fibre'. This effect is obtained in Purkinje fibres when Cl ions are replaced by I ions (Hutter & Noble, 1961).

The dual effect which a progressive increase in g_{An} has on the computed pace-maker may appear surprising, since only one factor in the equations is being varied. The explanation lies in the fact that E_{An} is assumed to be considerably less negative than E_K. A moderate increase in g_{An} accelerates the depolarization towards E_{An} during diastole and so increases the frequency. A large enough increase in g_{An}, however, stabilizes the potential at or near E_{An}, so that pace-maker activity is completely abolished.

DISCUSSION

Discrepancies between computed and recorded action potentials

So far as the potential changes are concerned the main discrepancy is that the maximum rate of depolarization, $dE_m/dt_{max.}$, during the spike is much smaller in the computed action potentials than in those recorded experimentally. This difference cannot be attributed to progressive errors in the computation, as the same result was obtained when the integration was repeated at shorter step lengths. Nor can it be attributed to the fact that the recorded action potentials were propagating, since initially, when d^2E_m/dt^2 is positive, current would be drawn away from the membrane and so reduce the current discharging the membrane capacity. This would slow the rate of depolarization and allow more time for h to fall, so that when $dE_m/dt_{max.}$ is reached, g_{Na} would be smaller in the propagating action potential than in the membrane action potential. Thus, in squid nerve the computed values for $dE_m/dt_{max.}$ are 431 V/sec for a propagated action potential and 564 V/sec for a membrane action potential (Hodgkin & Huxley, 1952*d*).

There remain two other possibilities. First, the computed maximum depolarizing current may be too small. This would appear to be a very likely error in view of the somewhat arbitrary way in which the equations for m were obtained (see Methods). It is difficult, however, to see how such an error might be rectified without radically affecting the other phases of the action potential. In order for I_{Na} to be increased at a given E_m, either $\bar{g}_{Na}$ or m or both must be increased, but this would have the effect of greatly prolonging the plateau or of preventing repolarization altogether (see Fig. 4). Furthermore, the computed increase in membrane conductance during the spike is in good agreement with that observed by Weidmann. It seems unlikely, therefore, that the peak sodium current is seriously underestimated by the equations.

The other possibility is that only a small part of the membrane capacity is discharged during the spike of the action potential. This would happen if the major part of the capacity were to be in series with a resistance which is small compared to the resting membrane resistance but fairly large compared to the membrane resistance during the spike. This part of the capacity would then be discharged mainly during the beginning of the plateau. If $10\ \mu\mathrm{F/cm^2}$ were to be discharged slowly in this way, this would leave only $2\ \mu\mathrm{F/cm^2}$ to be discharged by the sodium current during the spike and $\mathrm{d}E_m/\mathrm{d}t_{max.}$ would be increased by about the required factor. In physical terms this might mean that a large fraction of the membrane capacity is distributed along invaginated folds or tubules of membrane, the series resistance being the resistance of the 'extracellular' fluid in the folds or tubules. In the case of skeletal muscle the existence of intracellular tubules of membrane has been clearly demonstrated in observations made with the electron microscope (Bennett & Porter, 1953; Edwards & Ruska, 1955; Robertson, 1956; Porter & Palade, 1957) and Huxley & Taylor (1958) have shown that it is very likely that some process like spread of depolarization occurs along these tubules. At present, however, there is no direct evidence to support the idea that the tubules open out on the fibre surface (Huxley, 1959b), though this is not essential, provided that there is electrical continuity via a low resistance between the tubular fluid and the extracellular fluid, as suggested by Hodgkin & Horowicz (1960). Fatt (1961) has shown that the high-frequency membrane capacity of frog skeletal muscle is about $2\ \mu\mathrm{F/cm^2}$, which may be compared with the value of 6–$8\ \mu\mathrm{F/cm^2}$ obtained with 'square-wave' current analysis using intracellular micro-electrodes (Fatt & Katz, 1951). The situation in Purkinje fibres has not yet been investigated and another possibility which cannot be ruled out is that part of the capacitative behaviour observed in the resting fibre may be anomalous, in that it may result from delayed voltage-dependent changes in permeability. This would be analogous, though opposite, to the anomalous inductive behaviour resulting from delayed K rectification in squid nerve.

Another difference between the computed and experimentally recorded curves is that the rate of fall of E_m from the peak of the spike to the plateau is smaller in the computed action potential. This might also be explained by assuming a smaller capacity during the spike. On the other hand, dE_m/dt here greatly depends on g_K and would be appreciably larger if g_{K_1} were not to fall instantaneously on depolarization of the membrane, but rather with a small delay. Such a delay might also account for the 'notch' often observed between the spike and plateau of the Purkinje fibre action potential (Draper & Weidmann, 1951, Fig. 4*b*).

Ionic fluxes in cardiac muscle

At present there is very little information available on the effect of activity on the Na fluxes in cardiac muscle, so that it is not yet possible to say how far the computed fluxes correspond to those actually occurring. In the case of potassium, experiments on different specie shave given different results. Brady (unpublished, quoted by Brady & Woodbury, 1960) found no increase in K efflux during activity in the frog ventricle. This result is consistent with the theory given here, as it is also with that given by Brady & Woodbury (1960). Wilde & O'Brien (1953) found an increase in K efflux in the turtle ventricle which they considered to be synchronous with the action potential. Weidmann (1956) has calculated that the potential changes during the action potential are themselves sufficient to account for Wilde & O'Brien's results, so that there is no need to suppose that g_K increases during the action potential. One possibility, therefore, is that potassium rectification does not occur at all in this preparation. It is difficult, however, to reconcile this suggestion with the fact that the resistance during the plateau of the turtle ventricle action potential is greater than the resting resistance (Eyster & Gilson, 1947; Cranefield, Eyster & Gilson, 1951). This observation is readily explained by supposing that g_K is reduced during the action potential (see Fig. 10). A dependence of g_K on the K electrochemical potential gradient will also explain the shortening of the turtle heart action potential produced by increasing the extracellular potassium concentration (Weidmann, 1956*a*; cf. Hoffmann & Cranefield, 1960). It does not therefore seem possible at present to reconcile Wilde & O'Brien's results with those of electrical experiments.

Comparison of theories concerning long-lasting action potentials

Various modifications of Hodgkin & Huxley's original equations have been suggested recently in order to account for long-lasting action

COMPUTED CARDIAC ACTION POTENTIALS   347

potentials, and it seems desirable briefly to summarize and compare their main features.

The modification proposed by Fitzhugh (1960) and the closely similar one of George & Johnson (1961) primarily involve a large decrease in the rate constants of the potassium current system (α_n and β_n). g_K increases on depolarization of the membrane but does so very much more slowly than in normal squid nerve. For the purpose of describing the properties of cardiac muscle this modification is not sufficient, as it does not account for the high resistance during the plateau of the action potential which has been observed in kid Purkinje fibres (Weidmann, 1951), turtle ventricle fibres (Eyster & Gilson, 1947; Cranefield *et al.* 1951) and rabbit ventricular fibres (Johnson & Tille, 1960). Thus in Fitzhugh's computations the resistance during the plateau is only about one-quarter of the resting resistance, while in cardiac muscle the plateau resistance is as great or even greater than the resting resistance. On the basis of this difference it has been suggested (e.g. Chang & Schmidt, 1960) that the Hodgkin–Huxley formulation is inadequate to account for the electrical properties of cardiac muscle and that other models such as the two-stable state hypothesis (Tasaki & Hagiwara, 1957) might be more applicable. So far as its essential features are concerned the differences between this hypothesis and that of Hodgkin & Huxley are not very great. As Fitzhugh (1960) has pointed out, Tasaki & Hagiwara's description of the transition from one stable to the other at the termination of the plateau is a good qualitative description of the way in which the modified Hodgkin–Huxley equations actually behave, and the objection based on the discrepancies between observed and computed resistance changes does not apply to the equations used in this paper (see Fig. 10).

Brady & Woodbury (1960) have recently formulated equations to describe the action potential in frog ventricle. They postulated that the principal differences between cardiac muscle and squid nerve are: (1) that g_K falls on depolarization of the cardiac fibre membrane, and (2) that the Na inactivation and activation processes in heart muscle have two components, a slow one (time constant of the order of seconds) and a fast one (time constant of the order of milliseconds, as in squid nerve). Although their equations adequately describe the frog ventricle action potential and its main properties, including Brady's failure to observe any increase of K efflux associated with activity, they had no direct experimental evidence for the modifications which they proposed. So far as Purkinje fibres are concerned, the presence of a slow component in the Na inactivation process would seem unlikely in view of Weidmann's (1955) experiments.

In the equations used in this paper the main modification involves the K current which is assumed to flow through two types of channel in the membrane, one in which g_K falls on depolarization of the membrane and another in which g_K slowly rises (Noble 1960 a, b). This hypothesis was suggested by Hutter & Noble's (1960) experiments on Purkinje fibres in Na-deficient solutions. The modification made to the Na equations is a minor one, in that it does not involve any qualitative change in the behaviour of the sodium current system, but it is necessary in order that the equations should describe the action potential shape accurately. The h equations have been unaltered, apart from the shift along the voltage axis required to make the equations fit Weidmann's (1955) experimental curve, but the m equations have been reformulated in order to make the m_∞ / E_m relation less steep than that of Hodgkin & Huxley (see Figs. 4 and 5).

The result of using the Hodgkin–Huxley m equations would be greatly to lower the position of the plateau and to decrease the magnitude of the maximum diastolic potential. This is illustrated in Fig. 15. The interrupted curves are the instantaneous K current–voltage relations shown in Fig. 9, and the thin continuous curve is the steady-state Na current–voltage relation given by the Na equations used in this paper. The thick continuous curve is a steady-state Na relation obtained by substituting Hodgkin & Huxley's m equations (equations (24) and (25)) and by setting $\bar{g}_{Na}$ at 30 mmho/cm^2, and the fraction of g_{Na} which is assumed to be independent of E_m and t at 0.25 mmho/cm^2. The peak value of the steady-state Na current now occurs at a more negative potential and the value of $\bar{g}_{Na}$ has been reduced in order that the steady-state Na current should not exceed the steady-state K current at this potential. Without integrating the equations it is clear from Fig. 15 that normal Purkinje fibre-like action and pace-maker potentials would not be obtained. The low value required for $\bar{g}_{Na}$, and the fact that the action potential would start at a potential (about -65 mV) at which h is small, mean that the rate of rise would be greatly reduced and there would be no initial spike or overshoot. The plateau would start at about -25 mV and terminate at about -45 mV. The point of maximum repolarization would be at about -75 mV.

The ability of the equations used in this paper to describe the shape of the action and pace-maker potentials fairly accurately is to a great extent due to the fact that the action potential itself was used as a source of information in obtaining the equations for m (see Methods). However, the equations also account for most of the other electrical properties of Purkinje fibres, including the impedance changes, the responses to applied current pulses, the effect of repetitive stimulation and the influence of changes in ionic permeability, which were not used in formulating the equations. Nevertheless, they are not yet based on sufficient direct experimental evidence and will probably require further detailed modification

COMPUTED CARDIAC ACTION POTENTIALS   349

when the results of voltage-clamp analysis similar to that done in squid nerve are known.

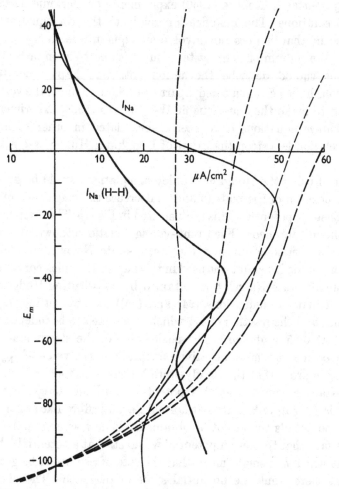

Fig. 15. Ionic current–voltage relations. Interrupted curves, instantaneous K current–voltage relations at various stages during computed action potential— same curves as in Fig. 9. Thin continuous curve, steady-state Na current–voltage relation given by Na equations used in this paper (as in Fig. 9). Thick continuous curve, Na current–voltage relation obtained by substituting Hodgkin & Huxley's *m* equations. Explanation in text.

SUMMARY

1. The equations formulated by Hodgkin & Huxley (1952 *a*, *b*, *c*, *d*) to describe the electrical activity of squid nerve have been modified to describe the action and pace-maker potentials of the Purkinje fibres of the heart.

23   Physiol. 160

350   **D. NOBLE**

2. The potassium-current equations differ from those of Hodgkin & Huxley in that K ions are assumed to flow through two types of channel in the membrane: one in which g_K falls when the membrane is depolarized and another in which g_K very slowly rises.

3. The sodium-current equations are very similar to those of Hodgkin & Huxley and are in part based on Weidmann's (1955) measurements of the properties of the h system in cardiac muscle. A method for obtaining equations for m by using some features of the action potential as an additional source of information is described.

4. The solution to these equations closely resembles the potential changes in Purkinje fibres, the only major discrepancy being that the maximum rate of depolarization is smaller in the computed action potentials than in those recorded experimentally. Possible reasons for this discrepancy are discussed.

5. The predicted changes in ionic conductances, ionic currents and fluxes are described.

6. The variation in the computed membrane impedance during the action and pace-maker potentials is very similar to that observed experimentally in Purkinje fibres.

7. Some of the effects of applied currents, repetitive stimulation and changes in ionic permeabilities have been reproduced. In general, the behaviour of the equations corresponds quite well with the observed behaviour of Purkinje fibres.

I should like to thank Professor A. F. Huxley for his comments on the manuscript of this paper, and Drs O. F. Hutter, R. Niedergerke and S. J. Hubbard for valuable discussion. I am also grateful to Dr R. A. Buckingham, Director of the University of London Computer Unit, for permission to use 'Mercury' and to Dr M. J. M. Bernal and other members of the Computer Unit Staff for advice on writing the programmes.

This work has been submitted in part fulfilment of the requirements for the degree of Ph.D. at London University.

REFERENCES

BENNETT, H. S. & PORTER, K. R. (1953). An electron microscope study of sectioned breast muscle of the domestic fowl. *Amer. J. Anat.* **93**, 61–105.

BRADY, A. J. & WOODBURY, J. W. (1960). The sodium–potassium hypothesis as the basis of electrical activity in frog ventricle. *J. Physiol.* **154**, 385–407.

CARMELIET, E. (1955a). Influence du rhythme sur la durée du potentiel d'action ventriculaire cardiaque. *Arch. int. Physiol.* **63**, 126–127.

CARMELIET, E. (1955b). Influence du rhythme sur la durée du potentiel d'action ventriculaire cardiaque. *Arch. int. Physiol.* **63**, 222–232.

CARMELIET, E. (1961). Chloride ions and the membrane potential of Purkinje fibres. *J. Physiol.* **156**, 375–388.

CARR, J. W. (1958). Error bounds for the Runge–Kutta single-step integration process. *J. Ass. comp. Mach.* **5**, 39–44.

CHANG, J. J. & SCHMIDT, R. F. (1960). Prolonged action potentials and regenerative repolarizing responses in Purkinje fibres of mammalian heart. *Pflüg. Arch. ges. Physiol.* **272**, 127–141.

COMPUTED CARDIAC ACTION POTENTIALS   351

COLE, K. S., ANTOSIEWICZ, H. A. & RABINOWITZ, P. (1955). Automatic computation of nerve excitation. *J. Soc. indust. appl. Math.* **3**, 135–172.

COLE, K. S. & CURTIS, H. J. (1941). Membrane potential of the squid giant axon during current flow. *J. gen. Physiol.* **24**, 551–563.

CORABOEUF, E. & WEIDMANN, S. (1954). Temperature effects on the electrical activity of Purkinje fibres. *Helv. physiol. acta,* **12**, 32–41.

CRANEFIELD, P. F., EYSTER, J. A. E. & GILSON, W. E. (1951). Electrical characteristics of injury potentials. *Amer. J. Physiol.* **167**, 450–456.

CURTIS, H. J. & COLE, K. S. (1940). Membrane action potentials from the squid giant axon. *J. cell. comp. Physiol.* **15**, 147–157.

CURTIS, H. J. & COLE, K. S. (1942). Membrane resting and action potentials from the squid giant axon. *J. cell. comp. Physiol.* **19**, 135–144.

DRAPER, M. H. & WEIDMANN, S. (1951). Cardiac resting and action potentials recorded with an intracellular electrode. *J. Physiol.* **115**, 74–94.

EDWARDS, G. A. & RUSKA, H. (1955). The function and metabolism of certain insect muscles in relation to their structures. *Quart. J. micr. Sci.* **96**, 151–159.

EYSTER, J. A. E. & GILSON, W. E. (1947). Electrical characteristics of injuries to heart muscle. *Amer. J. Physiol.* **150**, 572–579.

FATT, P. (1961). Transverse impedance measurements of striated muscles. *J. Physiol.* **157**, 10–12P.

FATT, P. & KATZ, B. (1951). An analysis of the end-plate potential recorded with an intracellular electrode. *J. Physiol.* **115**, 320–370.

FITZHUGH, R. (1960). Thresholds and plateaus in the Hodgkin–Huxley nerve equations. *J. gen. Physiol.* **43**, 867–896.

FRANKENHAEUSER, B. & HODGKIN, A. L. (1957). The action of calcium on the electrical properties of squid axons. *J. Physiol.* **137**, 218–244.

GEORGE, E. P. & JOHNSON, E. A. (1961). Solutions of the Hodgkin–Huxley equations for squid axon treated with tetraethylammonium and in potassium rich media. *Aust. J. exp. biol. Sci.* (in the Press).

HARRIS, E. J. & HUTTER, O. F. (1956). The action of acetylcholine on the movements of potassium ions in the sinus venosus of the heart. *J. Physiol.* **133**, 58–59P.

HODGKIN, A. L. & HOROWICZ, P. (1960). The effect of sudden changes in ionic concentration on the membrane potential of single muscle fibres. *J. Physiol.* **153**, 370–385.

HODGKIN, A. L. & HUXLEY, A. F. (1939). Action potentials recorded from inside a nerve fibre. *Nature, Lond.,* **144**, 710.

HODGKIN, A. L. & HUXLEY, A. F. (1945). Resting and action potentials in single nerve fibres. *J. Physiol.* **104**, 176–195.

HODGKIN, A. L. & HUXLEY, A. F. (1952a). Currents carried by sodium and potassium ions through the membrane of the giant axon of *Loligo. J. Physiol.* **116**, 449–472.

HODGKIN, A. L. & HUXLEY, A. F. (1952b). The components of membrane conductance in the giant axon of *Loligo. J. Physiol.* **116**, 473–496.

HODGKIN, A. L. & HUXLEY, A. F. (1952c). The dual effect of membrane potential on sodium conductance in the giant axon of *Loligo. J. Physiol.* **116**, 497–506.

HODGKIN, A. L. & HUXLEY, A. F. (1952d). A quantitative description of membrane current and its application to conduction and excitation in nerve. *J. Physiol.* **117**, 500–544.

HODGKIN, A. L., HUXLEY, A. F. & KATZ, B. (1949). Ionic currents underlying activity in the giant axon of the squid. *Arch. Sci. physiol.* **3**, 129–150.

HODGKIN, A. L. & KATZ, B. (1949). The effect of sodium ions on the electrical activity of the giant axon of the squid. *J. Physiol.* **108**, 37–77.

HOFFMANN, B. F. & CRANEFIELD, P. F. (1960). *Electrophysiology of the Heart.* New York: McGraw-Hill Book Company.

HOFFMANN, B. F. & SUCKLING, E. E. (1954). Effects of heart rate on cardiac membrane potentials and the unipolar electrogram. *Amer. J. Physiol.* **179**, 123–130.

HUTTER, O. F. & NOBLE, D. (1960). Rectifying properties of cardiac muscle. *Nature, Lond.,* **188**, 495.

HUTTER, O. F. & NOBLE, D. (1961). Anion conductance of cardiac muscle. *J. Physiol.* **157**, 335–350.

HUTTER, O. F. & TRAUTWEIN, W. (1956). Vagal and sympathetic effects on the pacemaker fibres in the sinus venosus of the heart. *J. gen. Physiol.* **39**, 715–733.

352 *D. NOBLE*

HUXLEY, A. F. (1959a). Ion movements during nerve activity. *Ann. N.Y. Acad. Sci.* **81**, 221–246.

HUXLEY, A. F. (1959b). Local activation of muscle. *Ann. N.Y. Acad. Sci.* **81**, 446–452.

HUXLEY, A. F. & TAYLOR, R. E. (1958). Local activation of striated muscle fibres. *J. Physiol.* **144**, 426–441.

JOHNSON, E. A. & TILLE, J. (1960). Changes in polarization resistance during the repolarization phase of the rabbit ventricular action potential. *Aust. J. exp. biol. Sci.* **58**, 509–513.

KEYNES, R. D. (1951). The ionic movements during nervous activity. *J. Physiol.* **114**, 119–150.

NATIONAL PHYSICAL LABORATORY (1961). *Modern Computing Methods*, 2nd ed., p. 90. London: H.M.S.O.

MOORE, J. W. (1959). Squid action potentials in isosmotic KCl. *Fed. Proc.* **18**, 107.

NOBLE, D. (1960a). Cardiac action and pacemaker potentials based on the Hodgkin–Huxley equations. *Nature, Lond.,* **188**, 495–497.

NOBLE, D. (1960b). A description of cardiac pace-maker potentials based on the Hodgkin–Huxley equations. *J. Physiol.* **154**, 64–65P.

NOBLE, D. (1961). Ionic conductance of cardiac muscle. Ph.D. Thesis. University of London.

PORTER, K. R. & PALADE, G. E. (1957). Studies on the endoplasmic reticulum. III. Its form and distribution in striated muscles. *J. biophys. biochem. Cytol.* **3**, 269–300.

ROBERTSON, J. D. (1956). Some features of the ultrastructure of reptilian skeletal muscle. *J. biophys. biochem. Cytol.* **2**, 369–380.

SCHÜTZ, E. (1936). Elektrophysiologie der Herzens bei einphasischer Ableitung. *Ergebn. Physiol.* **38**, 493–620.

SEGAL, J. R. (1958). An anodal threshold phenomenon in the squid giant axon. *Nature, Lond.,* **182**, 1370–1372.

TASAKI, I. (1959). Demonstration of two stable states of the nerve membrane in potassium-rich media. *J. Physiol.* **148**, 306–331.

TASAKI, I. & HAGIWARA, S. (1957). Demonstration of the stable potential states in the squid giant axon under tetraethylammonium chloride. *J. gen. Physiol.* **40**, 859–885.

TRAUTWEIN, W. & DUDEL, J. (1954). Aktionspotential und Mechanogramm des Warmblüter-herzmuskels als Funktion der Schlagfrequenz. *Pflüg. Arch. ges. Physiol.* **260**, 24–39.

TRAUTWEIN, W., GOTTSTEIN, U. & FEDERSCHMIDT, K. (1953). Der Einfluss der Temperatur auf den Aktionsstrom des excidierten Purkinje-fadens, gemessen mit einer intracellularen Elektrode. *Pflüg. Arch. ges. Physiol.* **258**, 243–360.

WEIDMANN, S. (1951). Effect of current flow on the membrane potential of cardiac muscle. *J. Physiol.* **115**, 227–236.

WEIDMANN, S. (1952). The electrical constants of Purkinje fibres. *J. Physiol.* **118**, 348–360.

WEIDMANN, S. (1955). The effect of the cardiac membrane potential on the rapid availability of the sodium carrying system. *J. Physiol.* **127**, 213–224.

WEIDMANN, S. (1956a). Shortening of the cardiac action potential due to a brief injection of KCl following the onset of activity. *J. Physiol.* **132**, 157–163.

WEIDMANN, S. (1956b). *Elektrophysiologie der Herzmuskelfaser*. Bern: Huber.

WILDE, W. S. & O'BRIEN, J. M. (1953). The time relation between potassium (K⁴²) outflux, action potential, and the contraction phase of heart muscle as revealed by the effluogram. *Abstr. XIX int. physiol. Congr.* pp. 889–890.

Adrenaline: Mechanism of Action on the Pacemaker Potential in Cardiac Purkinje Fibers

Abstract. *The pacemaker potential in Purkinje fibers is generated by a slow fall in potassium current which allows the inward background currents to depolarize the membrane. Adrenaline shifts the relation between activation of the potassium current and membrane potential in a depolarizing direction. Consequently, during the pacemaker potential, the potassium current falls more rapidly to lower values and the inward currents then depolarize the membrane more quickly. The shift in the potassium activation curve produced by adrenaline is large compared to that produced by calcium ions. The molecular action of adrenaline may involve either a large change in the surface charge of the membrane or a change in the dependence of the potassium permeability on the local electric field.*

Several attempts have been made to determine how adrenaline accelerates the cardiac pacemaker, but most results have been negative. Adrenaline does not greatly influence the resting conductance or the sodium conductance involved in generating the action potential (1, 2). However, the voltage-clamp technique has revealed the changes in time- and voltage-dependent current which underlie the pacemaker potential in Purkinje fibers (3–5). We have studied the action of adrenaline on these current changes.

The major time-dependent change in the pacemaker potential is a decrease in potassium current following repolarization at the end of the action potential (3–6). This fall in potassium current allows the inward background currents to depolarize the membrane to the threshold for initiating the action potential. Quantitative analysis of this potassium current (called i_{K_2} to distinguish it from i_{K_1}, which does not appear to be time-dependent) has shown that it is determined by a permeability variable, s, which is similar to the Hodgkin-Huxley permeability variables (5). Thus

$$i_{K_2} = s \cdot \overline{i_{K_2}} \qquad (1)$$

where $\overline{i_{K_2}}$ is the current when $s = 1$, and s obeys first-order kinetics:

$$ds/dt = \alpha_s(1 - s) - \beta_s s \qquad (2)$$

where α_s and β_s are functions of membrane potential (E) which have been determined empirically (5).

Adrenaline accelerates the pacemaker depolarization of spontaneously beating Purkinje fibers (1, 2, 7). To determine the mechanism of this action, we have studied the behavior of s during superfusion of a sheep Purkinje fiber with a Tyrode solution containing 5×10^{-7} g of adrenaline per milliliter (Fig. 1). The membrane potential was voltage-clamped at -80 mv, and the current changes following step depolarizations and hyperpolarizations were recorded. The current peak soon after return to -80 mv gives a measure of the degree of activation, s, since $\overline{i_{K_2}}$ is constant when E is constant. Equation 1 then becomes

$$i_{K_2} \propto s \qquad (3)$$

Figure 2 shows the variation in the steady-state value of s (that is, s_∞) with membrane potential measured in terms of i_{K_2} at -80 mv. The major effect of adrenaline is to shift the curve in the depolarizing direction by about 30 mv. The total amplitude of the curve is also reduced, but it is not yet certain whether any significance can be attached to this effect. Pronethalol (10^{-6} g/ml) restores the curve to its original position. An intermediate concentration of pronethalol (5×10^{-7} g/ml) shifted the curve by about 15 mv.

Figure 3 shows measurements of the rate of change of potassium current as a function of the membrane potential, measured in terms of the reciprocal of the time constant of current change,

$$\tau_s^{-1} = \alpha_s + \beta_s \qquad (4)$$

The symbols are the same as in Fig. 2. Adrenaline also shifts this curve in the depolarizing direction, and at -90 mv, which corresponds to the beginning of the pacemaker potential, the change in current is much faster when adrenaline is present than when it is absent or when pronethalol is applied.

Thus, adrenaline has the same effect on the properties of i_{K_2} as a hyperpolarization. A simple way in which it might do this is to alter the local electric field controlling the s kinetics, perhaps by adding its own positive charge to the surface of the membrane. However, adrenaline might also act in some more indirect manner, for example, by chemically altering the state of the membrane so that the energy levels of the s reaction are changed. At present, we cannot decide between these and various other possibilities. It is worth noting, however, that the effect of adrenaline in shifting the relation between s_∞ and membrane potential is much greater

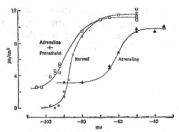

Fig. 2. Relations between steady-state degree of activation (s_∞) of slow K current and membrane potential, measured in terms of current immediately following return to -80 mv (see Fig. 1); ○, normal Tyrode; ▲, adrenaline (5×10^{-7} g/ml); and □, pronethalol (10^{-6} g/ml).

mv ‾‾50
‾80   ‾80   ‾90   ‾80   ‾80   ‾50   ‾80   ‾90   ‾80

Normal tyrode   Tyrode + 5×10⁻⁷ adrenaline

10 seconds

Fig. 1. Effect of adrenaline on voltage-clamp currents in sheep Purkinje fiber. (Left) Normal Tyrode solution. (Right) Tyrode solution containing 5×10^{-7} g of adrenaline per milliliter. Each record shows current (bottom) in response to 30 mv depolarization and 10 mv hyperpolarization from a holding potential of -80 mv. Note that in adrenaline solution no current is deactivated by hyperpolarization and that steady-state current at -80 mv is more negative than in absence of adrenaline. This change in steady-state current is attributable to shift in s_∞ (E) relation (see Fig. 2). Results shown in Fig. 2 were obtained by measuring current immediately following return to -80 mv after long-lasting depolarizations and hyperpolarizations.

916

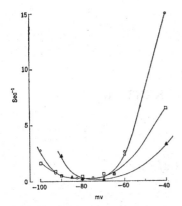

Fig. 3. Voltage dependence of rate of change of slow potassium current measured in terms of reciprocal of time constant (τ_s^{-1}) of current change. Symbols are the same as in Fig. 2.

than that of increasing the calcium ion concentration. Thus, increasing the external calcium concentration from 1.8 mM to 7.2 mM produces a shift of only 8 mv.

The physiological importance of these results lies in the fact that they account for the action of adrenaline in accelerating the pacemaker depolarization. At the beginning of the pacemaker potential the potassium current will decline more quickly, since the rate of change of i_{K_2} is faster in the presence of adrenaline (Fig. 3). Moreover, since the steady-state relation between s and membrane potential is shifted in the depolarizing direction (Fig. 2), i_{K_2} will fall toward a smaller value and so accelerate the depolarization throughout the pacemaker potential. Computer calculations have been done which show that these effects are quantitatively adequate to account for the chronotropic action of adrenaline.

O. Hauswirth
D. Noble, R. W. Tsien
*University Laboratory of
Physiology, Oxford, England*

References and Notes

1. D. G. Kassebaum, *Proc. Int. Pharmacol. Meeting 2nd* **5**, 95 (1964).
2. W. Trautwein and R. F. Schmidt, *Pfluegers Arch. Ges. Physiol.* **271**, 715 (1960).
3. K. A. Deck and W. Trautwein, *ibid.* **280**, 50 (1964).
4. M. Vassalle, *Amer. J. Physiol.* **210**, 1335 (1966).
5. D. Noble and R. W. Tsien, *J. Physiol.* **195**, 185 (1968).
6. W. Trautwein and D. G. Kassebaum, *J. Gen. Physiol.* **45**, 317 (1961).
7. M. Otsuka, *Pfluegers Arch. Ges. Physiol.* **266**, 512 (1958).
8. Supported by a Medical Research Council equipment grant. O.H. is a postdoctoral fellow of the Swiss National Fund.

12 July 1968

J. Physiol. (1968), **195**, *pp.* 185–214   185
With 14 *text-figures*
Printed in Great Britain

THE KINETICS AND RECTIFIER
PROPERTIES OF THE SLOW POTASSIUM CURRENT IN
CARDIAC PURKINJE FIBRES

By D. NOBLE and R. W. TSIEN*

From the University Laboratory of Physiology, Oxford

(*Received* 29 *September* 1967)

SUMMARY

1. The reversal potential of the slow outward current in Purkinje fibres varies with $[K]_o$ in accordance with the expected potassium equilibrium potential. It is concluded that virtually all of this current is carried by potassium ions.

2. The magnitude of the current is determined by two separable factors. The first factor is directly proportional to a variable obeying first-order voltage-dependent kinetics of the Hodgkin–Huxley type but with extremely long time constants. The time constants of this variable are extremely sensitive to temperature and the Q_{10} over the range 26–38° C is 6.

3. The second factor shows inward-going rectification with a marked negative slope in the current–voltage relation beyond about 25 mV positive to the K equilibrium potential. The current–voltage relations measured at different values of $[K]_o$ cross each other on the outward current side of the equilibrium potential.

4. The changes in slow potassium current during pace-maker activity have been calculated. It is shown that the mechanism of the pace-maker potential differs in several important respects from that described by Noble's (1962) model. The negative slope in the current–voltage relation appears to be an important factor in generating the last phase of pace-maker depolarization.

5. The role of the slow potassium current during the action potential and the consequences of the high temperature dependence of the kinetics are discussed.

INTRODUCTION

The form of the membrane currents recorded in Purkinje fibres from mammalian hearts under voltage clamp conditions depends on the range of voltages applied to the membrane. At voltages negative to the plateau

* Rhodes Scholar.

potential (approximately -20 mV) the conductance changes may be analysed in terms of three components: a fast sodium conductance, g_{Na}, similar to that observed in nerve cells (Weidmann, 1955; Deck & Trautwein, 1964; Dudel, Peper, Rüdel & Trautwein, 1967*b*); a time-independent potassium conductance, g_{K_1}, which rectifies in the inward-going direction (Hutter & Noble, 1960; Carmeliet, 1961; Hall, Hutter & Noble, 1963; Noble, 1965); and a slow time-dependent potassium conductance, g_{K_2} (Hall *et al.* 1963; Vassalle, 1966; McAllister & Noble, 1966, 1967). These components were assumed by Noble (1962) in order to reproduce the essential features of the action potential and pace-maker activity in Purkinje fibres. However, it is now evident that the behaviour of g_{K_2} differs from that assumed by Noble (1962) in several important respects (McAllister & Noble, 1966, 1967). In the work described in this paper we have therefore studied the kinetics and rectifier properties of g_{K_2} in sufficient detail to allow a reformulation of the potassium current equations.

At voltages more positive than about -20 mV the current records become more complicated and additional components appear. These components have been attributed to calcium ions (Reuter, 1966, 1967) and to chloride ions (Dudel, Peper, Rüdel & Trautwein, 1967*a*). However, since most of the important kinetic features of g_{K_2} may be studied with only moderate levels of depolarization, we shall not be concerned with these currents in the present paper.

METHODS

Short (1–2 mm) Purkinje strands taken from fresh sheep hearts obtained from the Oxford Co-operative Society Slaughterhouse were used. The voltage clamp technique was identical with that used by McAllister & Noble (1966) and is similar to that used by Deck & Trautwein (1964). A Devices Digitimer was used to obtain more accurate control over the voltage step sequences applied to the membrane. Since only very slow time course phenomena were studied, a pen recorder (4 channel, Devices) was routinely used for recording, an oscilloscope being used only to monitor the records and to check that fast time course components were normal. The pen recorder had a fairly flat frequency response up to about 50 c/s but, on many occasions, the frequency response was deliberately reduced by an RC filter to minimize noise. Checks were made to ensure that the current components being measured were not influenced by the filter. Current was continuously recorded at low and high amplification (as shown, for example, in the top records in Fig. 6) in order to obtain both the total current record and sufficiently amplified records of the important components to make accurate measurements for analysis. Temperature was continuously monitored with a copper-constantan thermocouple lying near the preparation and could be controlled over the range between room temperature and 40° C.

In order to obtain sufficient information for analysis it was usually necessary to record for several hours from the same preparation. It is difficult to eliminate completely small drifts in current and voltage recording systems over such periods of time and the absolute potential levels in these experiments are probably not accurate to within less than about 5 mV since the final voltage calibration was sometimes made a few hours after a particular experiment was performed. However, more reliance may be placed on comparisons between

CARDIAC POTASSIUM CURRENT 187

potentials during individual experiments since the records in one solution were usually obtained about half an hour after the records in the previous solution. Wherever possible, solution changes were reversed and repeated to check the results. All solutions contained 140 mM-Na$^+$, 144·6 mM-Cl$^-$, 0·5 mM-Mg^{2+}, 1·8 mM-Ca^{2+} and 1 g/l. glucose and were bubbled with oxygen. The K concentration was varied by adding various quantities of K phosphate buffer. Thus, the 4 mM-K solution contained 4 mM-K$^+$, 1·65 mM-HPO$_4^{2-}$ and 0·7 mM-H$_2$PO$_4^-$. Solutions with other K concentrations contained the same ratio of K$_2$HPO$_4$ to KH$_2$PO$_4$ but different total quantities to give total K concentrations of 2 mM, 2·7 mM or 6 mM. In all cases the buffer capacity was quite adequate to keep the pH at about 7·4 since the preparations were very small and were bathed in a large volume of solution which usually flowed over the preparation continuously.

In the region of the normal resting potential, the current changes occur with very long time constants and it was found necessary to allow up to 30 sec, or sometimes even longer, between clamp pulses in order to ensure that a steady state had been restored. In a typical experiment only about 60–100 pulses could be applied per hour. It proved very difficult therefore to obtain all the information for a full and accurate analysis of both the kinetics and rectifier properties from a single preparation. This was nearly achieved in the experiments on which the illustrations in this paper are based. In many other experiments, only partial results could be obtained. However, these results were consistent with those obtained from more complete experiments. Another difficulty arises from the fact that over some ranges of potential the slow current changes are very small and it is usually necessary to measure currents less than 10^{-8} A in order to specify the reversal potential to within a few mV. Moreover, small background current changes sometimes obscure the reversal point. However, more accurate estimates were sometimes obtained by interpolation using measurements of currents on both sides of the reversal potential (as shown, for example, in Fig. 9, bottom).

THEORY

The results and analysis to be described in this paper require fairly substantial changes in the equivalent circuit model for the K current equations and it may be helpful, therefore, to describe first the theoretical basis of the analysis.

Noble's (1962) model assumed that the potassium conductance can be represented by two parallel rectifiers (Fig. 1a). One, g_{K_1}, rectifies instantaneously in the inward-going direction, whereas the other, g_{K_2}, rectifies very slowly in the outward-going direction. g_{K_1} was described empirically as the sum of two exponentials. g_{K_2} was described by the Hodgkin–Huxley n equations (Hodgkin & Huxley, 1952) with the rate constants α_n and β_n reduced by a factor of 100. The absolute magnitude of the conductance was also greatly reduced.

The model required to account for the results described in the present paper is shown in Fig. 1b. The major differences are that g_{K_2} also rectifies in the inward-going direction for instantaneous changes in the potential (cf. Armstrong & Binstock, 1965) and that at any particular potential g_{K_2} is directly proportional to a first-order variable:

$$g_{K_2} \propto s^\gamma \quad \text{where} \quad \gamma = 1, \tag{1}$$

188   *D. NOBLE AND R. W. TSIEN*

whereas Noble used

$$g_{K_2} \propto n^{\gamma} \quad \text{where} \quad \gamma = 4. \tag{2}$$

The reasons for making these changes are that i_{K_2} is not a linear function of the driving force during sudden changes in membrane potential (McAllister & Noble, 1966) and that γ is already known to be less than 4 (the change in nomenclature from n to s for the variable obeying first-order kinetics is not important in the present context and will be justified

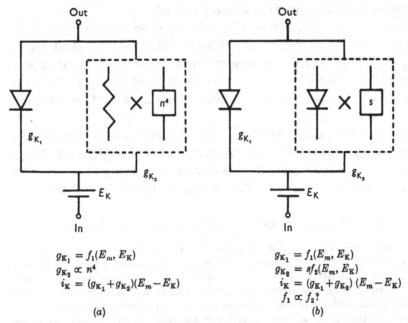

$$g_{K_1} = f_1(E_m, E_K)$$
$$g_{K_2} \propto n^4$$
$$i_K = (g_{K_1} + g_{K_2})(E_m - E_K)$$

$$g_{K_1} = f_1(E_m, E_K)$$
$$g_{K_2} = s f_2(E_m, E_K)$$
$$i_K = (g_{K_1} + g_{K_2})(E_m - E_K)$$
$$f_1 \propto f_2?$$

(a)   (b)

Fig. 1. Equivalent circuit diagrams for (a) Noble's (1962) K current equations and (b) the model obtained from the analysis of the results described in the present paper. The major changes made in the new model are that g_{K_2} also shows inward-going rectification and that the exponent on the kinetic variable is reduced to 1.

later—see Discussion). McAllister & Noble (1966, 1967) assumed $\gamma = 2$ in their analysis. However, most of their results could also be fitted fairly well with $\gamma = 1$, and the evidence against $\gamma \neq 1$ is strong enough (see Results) to conclude that McAllister & Noble's results were subject to a small systematic error in the indirect method which they used for measuring the time course of change in current. If the behaviour of Purkinje fibres is described by the model shown in Fig. 1b then they should show the following properties:

1. The instantaneous current–voltage relations for g_{K_2} should show certain resemblances to those for g_{K_1}. These relations may have a negative slope region, and relations obtained at different external K concentrations

CARDIAC POTASSIUM CURRENT   189

should cross each other when depolarizing currents are applied to the membrane, as has already been shown for g_{K_1} (Hall & Noble, 1963; McAllister & Noble, 1966). It is usually assumed that inward-going rectification occurs virtually instantaneously but the possibility of a small activation time cannot be excluded since the capacitance currents in Purkinje fibres last too long to determine the behaviour of the ionic currents at times shorter than a few msec (Fozzard, 1966). However, the changes due to inward-going rectification may be treated as instantaneous compared to the duration of the action potential.

2. *Changes* in the magnitude of g_{K_2} at any given potential should be determined entirely by changes in s. Thus, if s increases by a given factor then g_{K_2} should increase by the same factor at all potentials. Another way of expressing this property is to say that the *shape* of the $i_{K_2}(E_m)$ relation (where E_m is membrane potential) should be independent of the degree of activation of s.

3. The time constant, τ_s, for changes in s should be a unique function of E_m. Moreover, if $\gamma = 1$, the time course of current change should always be exponential and the time courses of onset and decay at a particular potential should be symmetric. This will not be true for any other value of γ.

4. The reversal potential for i_{K_2} should change as E_K changes.

The notation used in Fig. 1 differs both from Noble's notation and from that used by Armstrong & Binstock (1965). The reasons for adopting the new notation are twofold. First, it does not seem appropriate to represent a channel controlled by a Hodgkin–Huxley variable by an outward-going rectifier (as in Noble's notation), since these variables are voltage dependent rather than current dependent. Thus, the direction of current flow is determined by the appropriate ionic equilibrium potential but the voltage dependence of the permeability variable is not. Secondly, Armstrong & Binstock's use of symbols in series can be misleading since the property which needs to be represented at the macroscopic level is that the conductance factors should be multiplied, not that the resistance factors should be added. It seems more appropriate therefore to represent the Hodgkin–Huxley variable by a labelled element and to make the mathematical relation between the variables explicit in the diagram. However, the intended meaning of the 'in series' notation may still be valid since, at the single pore level, the processes of rectification and kinetics may well take place serially (Armstrong, 1966).

RESULTS

Kinetics. We have confirmed McAllister & Noble's (1967) observation that most of the slow potassium current may be activated by potentials which are too negative to activate the fast sodium conductance, and their suggestion that the steady-state value of this current at a holding potential around -80 mV is not zero. In fact, as shown in Fig. 4 below, it is sometimes possible to activate virtually all the slow current by subthreshold

190   *D. NOBLE AND R. W. TSIEN*

depolarizations. This enables the kinetics of this current to be studied
directly over an important range of membrane potentials, including the
whole range of the pace-maker potential, so that any errors which may
arise using indirect methods may be avoided.

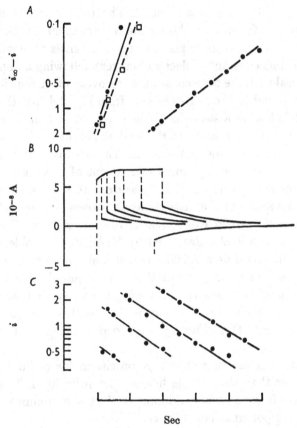

Sec

Fig. 2. Typical set of current changes in response to depolarizing and hyperpolarizing clamp pulses. Holding potential -75 mV. Preparation 40–2. $[K]_o = 4$
mM.

A. Left: continuous-line (●) shows time course of slow current change during
depolarization to -56 mV plotted on a logarithmic scale as deviation ($i_\infty - i$)
from steady-state current. Interrupted line (□) shows envelope of peak currents
following restoration of holding potential after depolarizations of various durations. The lines have virtually equal slopes so that the time constants are equal.
Right: time course of recovery of current following a hyperpolarization to -85
mV. Note that the time constant is equal to that of the changes plotted in C.

B. Superimposed tracings of currents on linear scale.

C. Decay of currents following depolarizations of various durations plotted on
logarithmic scale (for the sake of clarity only the first, second, fourth and sixth
records are plotted). Note that time constant of decay is independent of degree of
activation and is equal to the time constant of recovery of current at same potential (see right-hand plot in A).

CARDIAC POTASSIUM CURRENT   191

Figure 2*B* shows a typical set of superimposed current records in response to subthreshold rectangular depolarizations and part of a single current record in response to a rectangular hyperpolarization. In Figs. 2*A* and *C* the current changes have been plotted on logarithmic scales. It can be seen from these results that the time course of the current change is a simple exponential and is not sigmoid. The time constant of the exponential is a unique function of the membrane potential, i.e. the time constant does not depend, for example, on previous values of membrane potential, and the time constant of decay of current following a depolarizing pulse is identical with the time constant of recovery of current following a hyperpolarizing pulse. It can also be seen from Fig. 2*A* that the slow change in current following a step change in potential and the envelope of the peak currents following return to the holding potential after various intervals of time have the same time course. This means that the proportionate change in the time-dependent component of current does not depend on the potential at which it is measured. It follows from this observation that the shape of the instantaneous current–voltage relation must be independent of the degree of activation of the system (see Theory). A result similar to this was also described by McAllister & Noble (1966, Fig. 4). However, they used depolarizations which also activated the fast sodium conductance and their analysis therefore required an arbitrary choice of the position of the current scale during depolarization, since the initial part of the potassium current change is then partly obscured by the declining sodium current. This difficulty is avoided in the present experiments.

The time constants of the current change on either side of the holding potential are shorter than that at the holding potential itself. The time constant must therefore be at a maximum (τ_s^{-1} at a minimum) in the region of the holding potential (see Fig. 4).

A comparison between responses to depolarizing and hyperpolarizing pulses shows that less initial current is required to depolarize than to hyperpolarize. In the records shown in Fig. 7, for example, similar initial current changes occur in response to a 16 mV depolarization and a 8 mV hyperpolarization. This result would be expected if inward-going rectification is present. However, since g_{K_1} and g_{K_2} both contribute to these initial changes (the steady state g_{K_2} is not zero at the holding potential) it is not possible to obtain the rectifier properties of either system directly from these results. The rectifier properties must therefore be obtained by indirect means (see *Rectifier Properties* below).

These results have been repeated at a number of different values of membrane potentials and they show fairly conclusively that the kinetics of i_{K_2} are first order and that γ is 1.

192   *D. NOBLE AND R. W. TSIEN*

The question therefore arises why McAllister & Noble (1966, 1967) sometimes observed sigmoid time courses consistent with the view that $\gamma = 2$. The most likely explanation is that the currents following clamps of small duration are underestimated by the indirect method. Two possible factors might account for such an error. First, the capacity currents take a few msec to subside following a step change in potential since a substantial fraction of the membrane capacity is in series with a resistance (Fozzard, 1966). If current is measured in terms of the peak current following repolarizations, then for similar amplitudes of clamp pulse the smaller currents (i.e. those following clamps of short duration) would be reduced by the slower capacity currents to a greater proportionate extent than the larger currents following longer lasting pulses. However, it is very unlikely that this is the major factor involved since, as shown in Fig. 3, the fast transients following repolarizations are sometimes considerably longer than those following depolarizations. It is more likely, therefore,

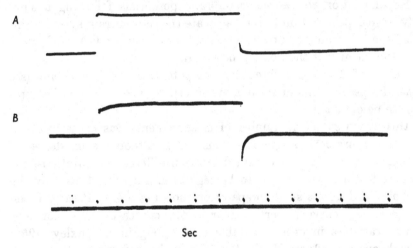

Fig. 3. Example of current record showing a much larger fast transient on repolarization than on depolarization.

A. Membrane potential. A step change 20 mV in amplitude and 6 sec in duration was applied. This depolarization is below the sodium threshold in this preparation.

B. Current record. Note that although the capacitive transients are too fast to be recorded accurately by the recording instrument, the time constant of the apparatus is short enough to record a characteristic capacity transient on depolarization. The longer-lasting fast transient on repolarization cannot therefore be due to long-lasting capacitive currents. This degree of asymmetry in the fast transients is an extreme example. Many current records (see e.g. Fig. 7) show little or no asymmetry of this kind.

that some other, relatively fast, conductance change is involved on repolarization. This phenomenon may also be responsible for the fast transients observed by Vassalle (1966), Fig. 7) on clamping back to the resting potential at various stages during the action potential plateau. Vassalle suggests that these transients might be attributable to sodium inactivation. If this interpretation is correct, the sodium conductance involved is probably not the one responsible for the spike of the action potential since, as in the record shown in Fig. 3, this transient can occur following depolarizations which are too small to activate the fast sodium conductance. Moreover, the transient is not greatly changed when pulses strong enough to activate the fast sodium conductance are applied. At present, therefore, we cannot identify

CARDIAC POTASSIUM CURRENT 193

the origin of the fast repolarizing transients. However, the slowness of the g_{K_2} changes in the region of the holding potential ensures that only very small errors can arise from this factor. Moreover, only very small errors are required to account for the deviation from exponential time courses sometimes observed with the indirect method. We have therefore continued to use this method for obtaining estimates of the time constants at potentials beyond the sodium current threshold.

In order to determine the variation in τ_s and s_∞ (the steady-state fraction of activation) with E_m, several sequences of potential change were applied to the membrane:

1. Simple rectangular steps of various durations and magnitudes from the holding potential (usually about -80 mV).

2. Repolarization to various membrane potentials following a sufficiently large depolarization to fully activate the slow current system. This enabled the time constants to be measured for a finer gradation of potentials in the vicinity of the holding potential.

3. Repeat of (1.) using different holding potentials. As shown below (see *Rectifier Properties*) this enabled more information on the rectifier properties to be obtained.

In this way a sufficient number of measurements was made to check that s_∞ and τ_s are both unique functions of E_m. Although some degree of variation in the results was obtained between different preparations, the important features were found to be regular and are illustrated by the results obtained from a single preparation shown in Fig. 4. It can be seen that the $s_\infty(E_m)$ curve is very similar to the curves obtained for conductance variables in other excitable cells (Hodgkin & Huxley, 1952; Frankenhaeuser, 1962; see Noble, 1966, for further references). The shape is sigmoid and the steepest part of the curve has a slope of 0·43/10 mV change in potential. The only other permeability variable for which this curve has been obtained in cardiac fibres is the sodium inactivation variable, h, and in this case a maximum slope of 0·5/10 mV was obtained (Weidmann, 1955).

The slow potassium current changes are extremely slow over a wide range of potentials, as indicated by the wide valley in the $\tau_s^{-1}(E_m)$ curve. Using the equation

$$\mathrm{d}s/\mathrm{d}t = \alpha_s(1-s) - \beta_s s \tag{3}$$

and the continuous lines for s_∞ ($= \alpha_s/(\alpha_s+\beta_s)$) and τ_s^{-1} ($= \alpha_s+\beta_s$), α_s and β_s were calculated and are shown as interrupted lines. It can be seen that both rate coefficients are monotonic functions of E_m. McAllister & Noble (1967) found that the β rate coefficient is sometimes non-monotonic when $\gamma = 2$ is used in the analysis. This irregularity disappears when $\gamma = 1$.

Variation of reversal potential with $[K]_0$. The slow outward current has previously been attributed to an outward movement of K ions. The best

194   *D. NOBLE AND R. W. TSIEN*

evidence for this view has come from voltage clamp experiments showing
that the current tails on repolarization reverse when the likely normal
value for E_K is exceeded (Deck & Trautwein, 1964; Vassalle, 1966;
McAllister & Noble, 1966). A typical result of this kind obtained in the
present series of experiments is shown in Fig. 5. In this case the records

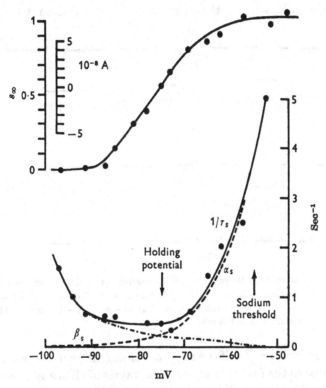

Fig. 4. Voltage dependence of kinetics of i_{K_2}.

Top: voltage dependence of fractional activation (s_∞) in the steady state
measured as the peak current change from background on return to holding
potential (-75 mV). Ordinates: peak current and s. Abscissa: membrane potential (as in bottom curves).

Bottom: points show measured values of τ_s^{-1}. Interrupted lines show α_s(----)
and β_s (—·—·—) calculated from $\tau_s^{-1} = \alpha_s + \beta_s$ and $s_\infty = \alpha_s/(\alpha_s + \beta_s)$, using continuous curves for s_∞ and τ_s^{-1}, drawn by eye through points. Arrows show position
of sodium threshold and of holding potential. Temperature 36° C. [K]$_0$ = 4 mM.

show currents in response to various hyperpolarizations from the holding
potential (-65 mV). Since an appreciable steady-state slow current is
present at this potential, the records show a slow change in current. At
-113 mV (the reversal potential in this experiment) the current change is
zero and at -129 mV the current change reverses sign.

However, it is conceivable that some other mechanism may be responsible for this result and more conclusive evidence for identifying this current with a movement of K ions would therefore be obtained by observing a shift in the reversal potential as the external K concentration is varied. We have found that the reversal potential follows the predicted shifts in E_K over a range of $[K]_o$ between 2·6 mM and 6 mM, corresponding to a 20 mV range for E_K. Below about 2·6 mM $[K]_o$, the reversal potential change appears to be smaller than expected (a change from 4 to 2 mM,

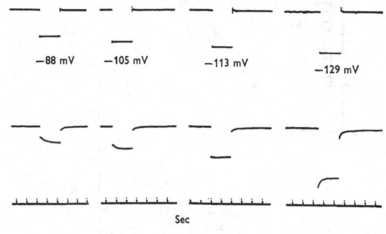

Fig. 5. Voltage clamp current records showing reversal potential determination. Preparation 46–8. $[K]_o = 2$ mM. Holding potential $= -65$ mV. The hyperpolarizing pulses deflected the membrane potential to -88 mV, -105 mV, -113 mV (reversal potential) and -129 mV.

for example, produces only half the expected change in reversal potential). This may be due to the fact that at very low values of $[K]_o$ a net loss of K from the cells will prevent $[K]_o$ immediately outside the cell membrane from falling to the value of $[K]$ in the bathing solution.

Figure 6 shows samples from three series of currents recorded during changes in K concentration. Constant hyperpolarizing clamp pulses were used as tests to show the gradual onset of the effects of changing $[K]_o$:

1. There is a marked increase in the amount of steady-state current required to achieve the same hyperpolarization when $[K]_o$ is increased. Since s_∞ is virtually zero at the hyperpolarized potential, this current increase must be attributed to changes in the current–voltage relations for g_{K_1} and any other currents. These changes are in the direction expected from the 'cross-over' effect observed previously (Hall & Noble, 1963; McAllister & Noble, 1966).

2. Changes in the amplitude of the slow current change during the hyperpolarization reflect the shifts in the reversal potential. Measuring

196   *D. NOBLE AND R. W. TSIEN*

the reversal potentials with various hyperpolarizations (as in Fig. 5) gives
-92 mV in 6 mM $[K]_o$, -99 mV in 4 mM $[K]_o$ and -111 mV in 2·7 mM
$[K]_o$. These changes are reasonably close to the 10·5 mV shifts in E_K
predicted by the Nernst equation (a few mV differences in the absolute
potential in these experiments is not very significant—see Methods):

$$E_K = 61\log([K]_o/[K]_i). \tag{4}$$

with $[K]_i = 151$ mM (Robertson & Dunihue, 1954), the calculated values
for E_K are -86 mV (6 mM), $-96·5$ mV (4 mM) and -107 mV (2·7 mM).

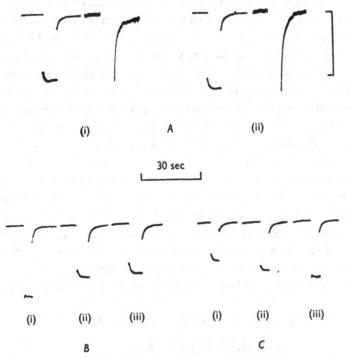

Fig. 6. Current records in response to hyperpolarizing pulses during changes in
$[K]_o$. Holding potential -68 mV. Same preparation as in Fig. 2.

    A. Records at low (left) and high (right) amplification (i) before and (ii) during
change from 4 mM $[K]_o$ to 6 mM $[K]_o$. Pulses were -28 mV in amplitude and
7 sec in duration. Note increase in total amplitude and in the current tail on return
to holding potential when $[K]_o$ is increased. This increase occurs despite a decrease
in driving force on K ions.

    B. Records at low amplification (i) before, (ii) and (iii) during, change from 6 mM
to 2·7 mM. Pulses -22 mV. In 6 mM $[K]_o$ this pulse was sufficient to deflect
potential to reversal potential. Note growth of current change during pulse as
$[K]_o$ falls. This change in $[K]_o$ corresponds to a 20 mV change in E_K.

    C. Records at low amplification (i) before, (ii) during and (iii) after change
from 2·7 mM to 4 mM. Pulses -32 mV. Time scale: 30 sec; current scale: $2·5 \times 10^{-7}$
A (low) and 5×10^{-8} A (high).

3. With increasing $[K]_o$, there is an increase in the magnitude of the slow current that is reactivated at the termination of the hyperpolarization. This change was recorded more accurately at higher gain on a second channel (see top records in Fig. 6). Provided that the same fraction of s is reactivated by the return to the holding potential in each case, as is shown below (see Fig. 10), this increase in current suggests that the instantaneous current–voltage relations, $i_{K_2}(E_m, E_K)$ obtained, in different $[K]_o$ must cross each other since the increase in current occurs despite a decrease in driving force. Further analysis of the rectifier properties of g_{K_2} confirms this suggestion (see Fig. 11).

Rectifier properties. The usual method for measuring the rectifier properties of conductance variables is to activate a certain constant fraction of the system and to plot the current–voltage relation obtained by applying step changes to various membrane potentials. This method depends on being able to establish conditions in which the conductance variable concerned controls virtually all the membrane current. It is not possible to use this method to obtain the rectifier properties of g_{K_2} in Purkinje fibres since we have found no conditions in which g_{K_2} contributes more than a fraction of the membrane conductance. This is largely due to the presence of g_{K_1} (whose absolute value is usually about twice as large as g_{K_2}), but other conductances (e.g. a resting sodium conductance) may also contribute. In this case, therefore, the shape of the instantaneous current–voltage relation for g_{K_2} must be obtained indirectly. According to the model (see Theory), the magnitude of the slow current change is given by the product of the change in s, Δs, and the rectifier function, $i_{K_2}(E, E_K)$, where E is the value of the membrane potential during the clamp pulse. In order to illustrate the analysis, two clamp records are shown in Fig. 7. Here

$$i_A = i(E, E_K, s_H + \Delta s) - i(E, E_K, s_H)$$

$$= \Delta s . f_2(E, E_K) . (E - E_K), \tag{5}$$

where i_A is the total slow current change during the pulse and s_H is the steady state value of s at the holding potential, E_H. The middle quantity in this equation expresses i_A in terms of what is actually measured, i.e. total current change during a certain change, Δs, in the degree of activation. The right hand quantity expresses i_A in terms of the model discussed in the Theory section. Now, during the slow change of current following return to the holding potential, s must return to its original value, s_H. Δs is therefore the same as during the clamp pulse. Any difference in the amplitude of the current change, i_B, must be attributed to the shape of the instantaneous current–voltage relation. Thus, in the case of the top record in Fig. 7, i_A is actually less than i_B which means that the instan-

D. NOBLE AND R. W. TSIEN

taneous current–voltage relation must have a negative slope. The shape of the current–voltage relation may be obtained as follows.

$$i_B = i(E_H, E_K, s_H + \Delta s) - i(E_H, E_K, s_H)$$

$$= \Delta s . f_2(E_H, E_K) . (E_H - E_K). \tag{6}$$

The ratio of current changes is therefore

$$\frac{i_A}{i_B} = \frac{f_2(E, E_K) . (E - E_K)}{f_2(E_H, E_K) . (E_H - E_K)} = \frac{i_{K_2}(E, E_K)}{i_{K_2}(E_H, E_K)} \tag{7}$$

Since s is now eliminated, the *shape* of the instantaneous current–voltage relation will be given by plotting the ratio i_A/i_B against E. This ratio (which we shall call the rectifier ratio) has been plotted for various values of $[K]_o$ in Fig. 8. Note that, by definition, the rectifier ratio must be equal

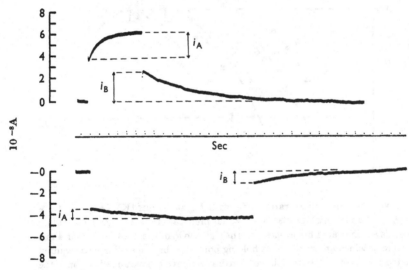

Fig. 7. Current records illustrating measurements made to obtain rectifier functions. Same preparation as Fig. 2. Holding potential -75 mV. $[K]_o = 4$ mM.

Top: current in response to depolarization to -59 mV.

Bottom: current in response to hyperpolarization to -83 mV. Time scale: sec.

Note that current change (i_A) during depolarization is actually smaller than that (i_B) following repolarization.

to 1 at E_H for all values of $[K]_o$. These plots do not enable us, therefore, to compare the absolute magnitudes of the relations at different values of $[K]_o$. However, some of the important features of the current–voltage relations are already evident. At $2\cdot7$ mM $[K]_o$ the ratio passes through a maximum near $E_m = -80$ V, so that in this range the ratio is greater than unity. Beyond this potential the relation has a negative slope, and this property may be of considerable importance in the mechanism of the

CARDIAC POTASSIUM CURRENT   199

pace-maker potential (see Discussion). The appearance of a negative slope may also be seen in the curve at 4 mM $[K]_o$. Its absence in the curve at 6 mM $[K]_o$ is probably due to the fact that an inadequate range of potentials was explored. It would, of course, be of great interest to determine the rectifier ratio for potentials more positive than the range explored in Fig. 8. Unfortunately, the current records for more positive potentials are complicated by the presence of other time-dependent currents so that further investigation of the rectifier function for g_{K_2} will require the elimination or determination of other currents.

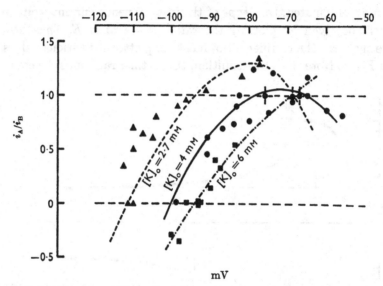

Fig. 8. Variation in rectifier ratio, i_A/i_B, with E_m at 2·7 mM $[K]_o$ (▲ – – –), 4 mM $[K]_o$ (●———) and 6 mM [K] (■ –·–·–). Same preparation as Fig. 2. The holding potentials are indicated by crosses (+) through the symbols. The results for 4 mM $[K]_o$ were obtained at two different holding potentials but no scaling was required to match the results. The curves have been drawn by eye to give an indication of the typical shape of the rectifier curves based on this and other experiments (see also Fig. 9 bottom). Note also that in these plots, the rectifier ratio has not been corrected for changes in absolute amplitude of current with changes in $[K]_o$, and that the ratio is unity at the holding potential by definition. The corrected ratios are plotted in Fig. 11.

The degree of scatter of some of the points plotted in Fig. 8 may raise the question how much reliance may be placed on our curves indicating negative slopes. In fact, the evidence for the presence of a negative slope is stronger than it might appear to be. The reason is that the relatively large scatter of points around the holding potentials at, e.g. 4 mM $[K]_o$, arises from the fact that the small pulses required to deflect the potential from the holding potential produce only small current changes. The ratios are therefore obtained by dividing one small quantity by another. The errors involved must therefore be fairly large. However, this is not true for the case of the points on which the evidence for a negative slope is based. The most positive points in the case of the 4 mM $[K]_o$ results were obtained by clamping from the most negative of the two holding potentials. These points are therefore obtained from the ratios of fairly large quantities (cf. Fig. 7, top record). The fact that these points give a ratio less than unity (in fact about 0·8) is significant since the errors in these measurements must have been considerably less than 20%. Even stronger evidence for the negative slope region is provided by the results of the experiment illustrated in Fig. 9.

Some comment may be in order here on the choice of axes for the current–voltage relations in this paper. The usual convention in voltage clamp work is to plot voltage on the abscissa. Although previous work in cardiac electrophysiology has frequently used the ordinate for voltage it may now be more appropriate to use the abscissa.

The fact that other time-dependent currents appear at more positive potentials suggests an alternative explanation for the negative slopes shown in Fig. 8. For example, a slow conductance change with a reversal potential around -50 mV (D. Noble & R. W. Tsien, unpublished) would give rise to a negative current if activated in the range of voltages studied in the present experiments. If the kinetics of this current were similar enough in time course to be included in the estimate of i_A, i_A would be reduced at more positive potentials and the effect would be to produce a negative slope in the relation obtained from the rectifier ratio, even if no negative slope exists in the true $i_{K_2}(E_m)$ relation. This possibility can be tested by determining steady-state s curves at different holding potentials. The results of such an experiment are shown in Fig. 9 (top). There are two important features of the results to which we want to draw attention:

1. The *position* of the $s_\infty(E_m)$ curve on the voltage axis is not significantly dependent on the value of the holding potential, E_H. Thus the values of E_m at which $s_\infty = 0.5$ are -83 mV ($E_H = 90$ mV), -80 mV ($E_H = -85$ mV), -86 mV ($E_H = -80$ mV) and -87 mV ($E_H = -70$ mV). This result would not be obtained either if the value of the holding potential influenced the g_{K_2} kinetics (in which case the rectifier and kinetic factors would not be strictly separable) or if another current component were being included in the estimate of i_{K_2} at more positive potentials. In the latter case the s_∞ curve would be uninfluenced only if the kinetics of

---

Legend for Fig. 9

Fig. 9. Evidence for negative slope in $i_{K_2}(E_m)$ relation based on measuring currents at different holding potentials. Preparation 47–2. [K]$_o$ = 2 mM.

Top: steady-state variation in degree of activation measured in terms of peak current deflexions on return to various holding potentials. Note that the amplitude of the curves is strongly dependent on E_H but that position on voltage axis is virtually independent of E_H.

Bottom: points show rectifier function obtained at $E_H = -70$ mV as described in text using equations (5)–(8). Vertical lines show total amplitude of steady state curve, i.e. $i(s = 1) - i(s = 0)$, as a function of E_H. Curve drawn by eye through points is a typical rectifier curve. Note that total amplitudes are also a good fit to the rectifier function. As explained in text, this result provides stronger evidence for the existence of a negative slope region than that given by the results in Fig. 8.

CARDIAC POTASSIUM CURRENT

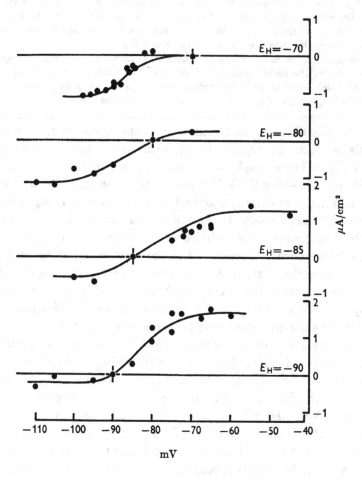

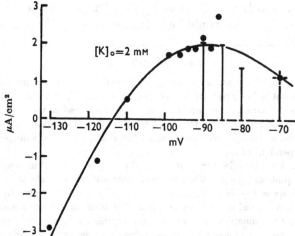

Fig. 9. For legend see opposite page.

the other current were identical with those for i_{K_2}. But in this case the reversal potential of the current controlled by s would not equal the potassium equilibrium potential.

2. The *amplitude* of the steady-state curve is strongly dependent on E_H and decreases at the most positive values of E_H. Hence $i_{K_2}(E, E_K, s = 1)$ must have a negative slope at potentials beyond about 25 mV positive to E_K (see Fig. 9, bottom). Moreover, since the shape of the current–voltage relation is independent of the value of s (see above), this result must hold for all values of s.

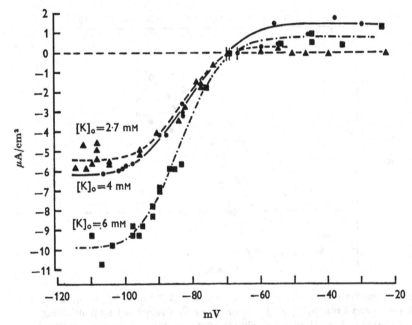

Fig. 10. Variation in peak current deflexion at the holding potential following prolonged step depolarizations. Same preparation as Fig. 2. The relations were obtained at three different K concentrations: 2·7 mM (····), 4 mM (——) and 6 mM (–·–·–). Note that position of curve on voltage axis is virtually independent of $[K]_o$ but that total amplitude increases as $[K]_o$ increases.

It is now possible to obtain the relative magnitudes of the instantaneous current–voltage relations at different values of $[K]_o$ (note that the relations shown in Fig. 8 do not give this information since they are arbitrarily scaled to give a value of 1 at E_H). The information required in order to scale the rectifier ratios to give relations in terms of K current is the value of $i(E_H, s = 1) - i(E_H, s = 0)$ at the holding potential for each value of $[K]_o$. These values may be obtained from the total amplitudes of the steady state activation curves measured (as in Fig. 4, top) in terms of the peak current on return to the holding potential. The results of an experiment in

CARDIAC POTASSIUM CURRENT   203

which these curves are determined are shown in Fig. 10. It can be seen that the position of the activation curve on the voltage axis does not depend appreciably on the value of $[K]_o$ (the small changes which occurred are no greater than the background drifts expected during experiments of this kind—see Methods). This result is further evidence for the separability of the kinetic and rectifier functions. On the other hand, the total amplitude of the curve, which gives $i(E_H, s = 1) - i(E_H, s = 0)$, changes quite substantially as $[K]_o$ is varied.

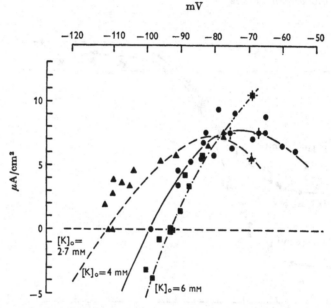

Fig. 11. Rectifier functions, $i_{K_2}(E_m, E_K, s = 1)$, at 2·7 mM $[K]_o$ (····), 4 mM $[K]_o$ (——) and 6 mM $[K]_o$ (–·–·–). The functions were calculated by multiplying ratios plotted in Fig. 8 by total amplitude of steady-state curves in Fig. 10 (see equation (8)). The relations cross each other on the positive side of the reversal potentials and show negative slopes at about 25 mV positive to the reversal potential. As in Figs. 8 and 9, the curves have been drawn by eye through points to indicate typical shape of relations.

If the rectifier ratios plotted in Fig. 8 are multiplied by the appropriate value of $i(E_H, s = 1) - i(E_H, s = 0)$ from Fig. 10, we obtain the instantaneous current–voltage relations for i_{K_2} when $s = 1$, i.e.

$$i_{K_2}(E_m, E_K, s = 1) = \frac{i_A}{i_B}[i(E_H, E_K, s = 1) - i(E_H, E_K, s = 0)]. \quad (8)$$

These relations (which we shall call the rectifier functions) have been plotted in Fig. 11. It can now be seen that the relations show a marked cross-over effect on the positive side of the reversal potentials which is similar to that already described for g_{K_1} (Hall & Noble, 1963; McAllister

D. NOBLE AND R. W. TSIEN

& Noble, 1966). This effect is expected when the conductance is a function of the driving force and not simply of the membrane potential (see Noble, 1965, Fig. 1).

In applying equations (5)–(8) to obtain the rectifier functions it is assumed that E_K remains unchanged during the passage of current across the membrane. Since McAllister & Noble (1966) have shown that E_K probably does change in a positive direction during depolarizing currents, this assumption requires justification. The justification is based on the fact that much larger depolarizations were used by McAllister & Noble (1966) than those used in the experiments described in the present paper. They showed that long-lasting depolarizations of the order of 100 mV may temporarily displace the quiescent membrane potential by about 10 mV, probably as a consequence of K accumulation in a space immediately outside the cell membrane. The changes in the present experiments, however, must be considerably smaller than this. Thus, a depolarization of the magnitude and duration shown in the top record of Fig. 7 would be expected to change the quiescent potential by less than 2 mV. Moreover, the major part of the slow current change during depolarization occurs within the first 2 sec during which time the change in quiescent potential is probably less than 1 mV. Although we cannot accurately estimate the effects which would result from such small changes, there are several reasons for thinking that the effects must be negligibly small. First, as McAllister & Noble (1966, Fig. 11) have shown, the changes in quiescent potential become very much smaller when $[K]_0$ is increased, whereas the slow current changes become much larger when $[K]_0$ is increased (see Fig. 10). Secondly, McAllister & Noble (1966) showed that the kinetics of the accumulation and depletion process are not even approximately first order, whereas the kinetics of the slow current changes are first order (see *Kinetics* above). Thirdly, the quiescent membrane potential changes observed by McAllister & Noble (1966) continue to increase with the magnitude of the preceding depolarization, the important parameter being the total charge transferred across the membrane. By contrast, the slow current change recorded on repolarization is virtually independent of the potential during the previous depolarization when potentials positive to -60 mV are applied (see Figs. 4, 10).

Temperature dependence of kinetics. The temperature dependence of the kinetics was determined by measuring the values of τ_s at various potentials as the temperature was slowly varied between 26 and 38° C. Figure 12 shows the variation of log τ_s with T at two different membrane potentials. The Q_{10} in each case is about 6. The value of $i(E_H, s = 1) - i(E_H, s = 0)$ was also found to be temperature dependent. An increase in temperature increases the total amplitude of the activation curve. However, this effect could not be expressed simply in terms of an over-all Q_{10} and more experiments will be required to establish the functional dependence of the magnitude of i_{K_2} on temperature.

DISCUSSION

Separability of kinetic and rectifier variables. The time and voltage dependence of the slow component, g_{K_2}, of the potassium conductance in Purkinje fibres is well described by a model that represents the conductance as the product of two distinct factors. The kinetic factor, s, represents the fraction of the total g_{K_2} which is activated at any potential. s depends on the past history of the membrane potential through the first-order equation (3) but it cannot change instantaneously (i.e. it is always continuous) even following step changes in membrane potential. The second factor, represented by the rectifier function, $i_{K_2}(E_m, E_K, s = s_0)$, describes the instantaneous current–voltage relation for any constant value of s, s_0. The *shape* of this relation is independent of the previous

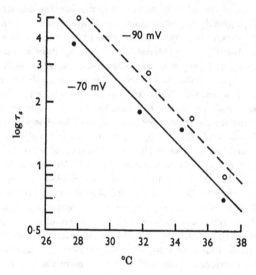

Fig. 12. Influence of temperature on rate of change of i_{K_2}. Preparation 46–5. Ordinate: $\log \tau_s$. Abscissa: temperature ° C. Open symbols show values measured at -90 mV. Filled symbols show values obtained at -70 mV. The straight lines have a slope corresponding to a sixfold change in rate for a 10° C change in temperature.

values of membrane potential and of the value of s. However, the shape does depend on the value of E_K, whereas s is independent of E_K. The shape of the relation may be constant if the current is plotted as a function of the driving force, $E_m - E_K$, but the present results are not extensive enough to justify this conclusion.

The results and analysis described in this paper show that the two factors are indeed separable (see *Rectifier properties*). Any interaction between them must be smaller than the experimental errors (see Methods). The kinetic factor is similar to the Hodgkin–Huxley permeability variables, and the $s_\infty(E_m)$, $\alpha_s(E_m)$ and $\beta_s(E_m)$ relations strongly resemble those found for other variables of this kind. The $i_{K_2}(E_m, E_K, s_0)$ function is an inward-going rectifier showing a region of negative slope conductance beyond about 25 mV positive to E_K. These results have a number of important implications which may be divided into two classes. First, although the previous analysis of the Purkinje fibre action potential and pace-maker activity involving a slow potassium conductance (Noble, 1962) is confirmed in principle, a radical revision of the equations will be required in order to reproduce the details of the potassium current described in this paper and previously (McAllister & Noble, 1966, 1967). Secondly, any explanation of the mechanism of inward-going rectification and of the Hodgkin–Huxley permeability variables must account for the possibility that these phenomena can occur together in the sense that they can control the same ionic current although the two factors themselves may behave independently. In addition to Purkinje fibres, this conjunction has also been found in TEA-treated squid nerve (Armstrong & Binstock, 1965; Armstrong, 1966) and in skeletal muscle (R. H. Adrian, W. K. Chandler & A. L. Hodgkin, personal communication). The clear separability of the two factors suggests that the K channels involved may be controlled by two physically separate gating mechanisms (cf. Armstrong, 1966).

Role of the slow K current in pace-maker activity. The spontaneous depolarization occurring during pace-maker activity in Purkinje fibres extends over the range of potentials over which s_∞ varies from nearly 0 to nearly 1. The position of the $s_\infty(E_m)$ relation is therefore considerably more negative than the slow g_{K_2} curve used by Noble (1962) to compute the time-dependent current. As a result of this, the behaviour of Purkinje fibres will differ from that of Noble's model in several important respects. These may best be indicated by calculating how s and i_{K_2} will vary during pace-maker activity. A full quantitative description of this must await a reconstruction of the action potential using the new model. However, the information on the kinetics and rectifier properties which we have obtained is sufficient to allow a preliminary calculation to be made. In order to do this we need to know how E_m varies during spontaneous activity in a cell which is known to obey the kinetics we have described. Vassalle (1966) has recorded the pace-maker potential in a cell in which he also recorded the time constant of decay of the slow current during a clamp to the maximum diastolic potential (see Fig. 1 in Vassalle's paper). His value for the time constant fits our data and it seems reasonable therefore to use his $E_m(t)$ curve. This has been traced and replotted as the upper diagram in Fig. 13. In order to calculate the variation in s, s_∞ and i_{K_2} we have used the $s_\infty(E_m)$ and $\tau_s^{-1}(E_m)$ relations shown in Fig. 4 and the rectifier function, $i_{K_2}(E_m, E_K, s = 1)$, at 2·7 mM $[K]_0$ (which corresponds to Vassalle's value) obtained from Fig. 11. s was integrated using the numerical approximation

$$\delta s_{t \text{ to } t+\delta t} = \delta t \langle \tau_s^{-1}\rangle (\langle s_\infty\rangle - \langle s\rangle), \tag{9}$$

where $\langle s_\infty \rangle$ and $\langle \tau_s^{-1} \rangle$ are average values over the step t to $t + \delta t$ and $\langle s \rangle$ is estimated from

$$\langle s \rangle = s_l + \tfrac{1}{2}\delta s_{(t-\delta t \text{ to } t)}. \tag{10}$$

A step length of 0·1 sec was used from 0 to 1·5 sec. This was reduced to 0·05 sec thereafter. This procedure gives a sufficiently accurate approximation for s as a function of time. i_{K_2} was then obtained from

$$i_{K_2} = s\, i_{K_2}(E_m, E_K, s = 1). \tag{11}$$

It was assumed that s is virtually equal to 1 at the end of repolarization (this assumption will be justified below).

At the beginning of the pace-maker potential, E_m is very negative so

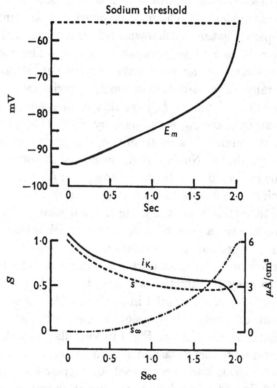

Fig. 13. Mechanism of pace-maker potential based on new model for K current.

Top: variation in membrane potential during pace-maker activity, replotted from Vassalle (1966, Fig. 1).

Bottom: s_∞ (t) relation obtained from $s_\infty(E_m)$ relation shown in Fig. 4. s and i_{K_2} were calculated from equations (9)–(11) using the rectifier function for 2·7 mM $[K]_o$ shown in Fig. 11. Note that, although s does not fall below a certain value and actually increases towards the end of the pace-maker potential, i_{K_2} falls continuously. This is a consequence of the negative slope in the rectifier function.

208　　　*D. NOBLE AND R. W. TSIEN*

that s_∞ will be nearly zero. s will therefore decline slowly (the time constants for s are longest in this range of potentials—see Fig. 4). In a full computation it would, of course, be this decline in s which is responsible for generating the pace-maker depolarization. It must be remembered that in these calculations E_m is 'forced' to follow Vassalle's recorded potential. However, the $i_{K_2}(t)$ curve obtained from the calculation is consistent with view that it may have generated this pace-maker potential. In the absence of a full reconstruction, therefore, these calculations give a useful description of the detailed mechanism of the pace-maker potential.

As the membrane depolarizes, s_∞ increases. At some point in time, therefore, s must equal s_∞. This occurs at 1·6 sec in Fig. 13. Thus, s cannot fall below a certain value during the pace-maker potential. Moreover, unless excitation occurs fairly quickly following this point, spontaneous excitation may fail to occur since any further depolarization will cause s to rise again (as in Fig. 13) so that the point at which $s = s_\infty$ could become a stable point. Whether this will happen or not may depend critically on the presence of the negative slope which we have observed in the rectifier function for i_{K_2} and possibly also on the current–voltage relation for i_{K1}. In the case of Fig. 13 the negative slope becomes apparent in the region of potential at which $s = s_\infty$. In this case, therefore, further spontaneous depolarization can occur since i_{K_2} will continue to fall even when s is rising, provided that s does not rise too quickly and that excitation does not occur too slowly. Thus, i_{K_2} declines very sharply towards the end of the pace-maker potential. In fact, most of the rapid increase in the rate of depolarization at this time is attributable to the negative slope in the instantaneous current–voltage relation rather than to activation of g_{Na} (as in Noble's model), since the Na threshold is not reached until the membrane depolarizes to about -55 mV (see Figs. 4 and 13).

The detailed mechanism of the pace-maker potential is therefore strikingly different from that in Noble's model. In this model, a point at which $n = n_\infty$ during the spontaneous depolarization following an action potential must become a stable point since the instantaneous current–voltage relation for i_{K_2} was assumed to be linear. In a spontaneously active system, therefore, n must continue to fall throughout the duration of the pacemaker potential. Other detailed differences arising from the position of the s_∞ curve have been discussed by McAllister & Noble (1967).

Role of the slow K current during the action potential. It was assumed in the previous section that s is nearly equal to 1 at the end of the action potential. This assumption is justified since the time constant, τ_s, in the region of the plateau is sufficiently short (McAllister & Noble, 1966) that s should approximate to 1 within about 200 msec. In this respect the system is similar to the behaviour of g_{K_2} in Noble's model. However, since

the K current controlled by the s system also shows inward-going rectification, and since the rectifier function is known to have a low value in the
region of the plateau (about one third of its value at -80 mV—see
McAllister & Noble, 1966, Fig. 4), the total amount of repolarizing current
supplied by K ions must be smaller than that given by Noble's model.
Even when s is fully activated, therefore, repolarization may still occur
relatively slowly. Since $s = 1$ and will remain constant between this time
and the beginning of the pace-maker potential, this later phase of repolarization may be referred to as the s-independent phase. Unless other slow
time-dependent currents also significantly contribute to the plateau
mechanism, this phase of repolarization will be determined simply by the
recharging of the membrane capacity through the non-linear resistance
given by the membrane current–voltage relation when $s = 1$.

The duration of the s-dependent phase of repolarization will, of course,
depend on the initial value of s and since s_∞ is very dependent on E_m in the
region of the resting or pace-maker potential the action potential duration
should be dependent on the initial value of E_m. A dependence of this kind
has been shown in solutions containing relatively high concentrations of
K ions by Weidmann (1956) who showed that, in these solutions, the
plateau is abolished if the action potential is initiated from the depolarized
level of membrane potential but that, if the membrane is previously
hyperpolarized back to its value in normal K solutions, a plateau could be
restored. Since an increased K concentration increases i_{K_1} as well as i_{K_2} the
abolition of the plateau cannot be attributed solely to effects on i_{K_2}. A
clearer demonstration of the role of i_{K_2} in the plateau would therefore be
obtained by varying the initial membrane potential in a fibre bathed in
normal Tyrode solution. This is shown in Fig. 14. The action potentials
were initiated (*a*) from a membrane potential at which s_∞ is nearly 1 and
(*b*) from a membrane potential at which s_∞ is nearly zero. The latter action
potential is considerably longer than the first. Van der Walt & Carmeliet
(1967) and Carmeliet & van der Walt (1968) have recently investigated
this effect more extensively and have shown that the duration of the action
potential is related to the magnitude of the initial membrane potential
according to an S-shaped relation, which would be expected since the
steady state dependence of s on membrane potential is also S-shaped (see
Fig. 4). They have also shown that the magnitude of the change in duration of the action potential depends on the duration of the preceding
change in membrane potential and lasts for some time after the end of a
hyperpolarizing pulse. Both of these effects may be attributable to the
time taken for s to change from one steady-state value to another following a sudden change in potential. These results indicate that some part of
the plateau phase of repolarization is s-dependent since our results suggest

210   *D. NOBLE AND R. W. TSIEN*

that s is the only time-dependent activation variable whose steady-state value varies appreciably over the range of potentials between -70 and -100 mV. However, it is becoming evident that the slow K current has a less important role in the mechanism of the plateau than it has in the generation of the pace-maker potential. In an action potential lasting 400 msec only about half the duration of the repolarization can be s-dependent. Moreover, the activation of the slow K current is too fast in the plateau region to be responsible for the extremely slow repolarization observed in certain conditions, e.g. low $[\text{K}]_0$ (see Noble, 1965, Fig. 4) and sometimes in Cl-free solutions (see Hutter & Noble, 1961, Fig. 5), or following strong hyperpolarization (Carmeliet & van der Walt, 1968; D. Noble & R. W. Tsien, unpublished). In these cases much slower changes must be involved.

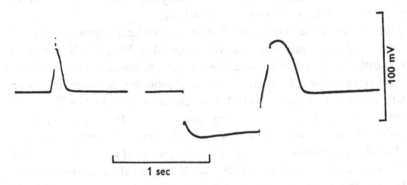

Fig. 14. Influence of initial membrane potential on duration of action potential plateau. Preparation 47–4.

Left: action potential initiated by 1×10^{-7} A depolarizing current applied for 40 msec from a potential (about -70 mV) at which s_∞ is nearly 1.

Right: action potential initiated by the break of a 8×10^{-7} A hyperpolarizing current applied for 630 msec which deflected the potential to a value (about -110 mV) at which s_∞ is nearly 0. The fast components of the action potentials, including the initial spike, are not shown. Note also that the absolute membrane potentials are uncertain for reasons explained in Methods section.

Comparison with other conductance mechanisms. The kinetics of the slow K current show some strong resemblances to the delayed K current in nerve cells. However, there are some important differences.

The s kinetics are generally two orders of magnitude slower than the n kinetics of nerve cells. It is for this reason that we have chosen the variable s rather than continuing to use the variable n as in previous work (Noble, 1962; McAllister & Noble, 1966, 1967). At present, the only other permeability variable of this kind which has been analysed and found to have kinetics as slow as those for s is the K inactivation variable, k, in nerve cells (Frankenhaeuser, 1962; Ehrenstein & Gilbert, 1966). A similar

CARDIAC POTASSIUM CURRENT   211

slow K inactivation has also been observed in skeletal muscle (Adrian, Chandler & Hodgkin, 1966). Over the voltage range and periods of time (up to tens of seconds) investigated in the present paper, the s system does not appear to be inactivated.

The second reason for using a different variable is that the presence of inward-going rectification in the same system which is controlled by the s gates makes it less likely that the s system is simply a slowed-down version of the n system found in nerve cells. This argument is not conclusive since Armstrong & Binstock (1965) interpret their results on squid-nerve to indicate that the presence of TEA introduces inward-going rectification in the n system. However, TEA does not greatly slow the n kinetics (Armstrong, 1966). Moreover, it is still possible that a much faster potassium conductance is also present in Purkinje fibres and it would, perhaps, be more consistent to use the variable n for this system than for g_{K_2}. Further experiments are required to test this possibility.

Another important difference is that the maximum current carried by g_{K_2} is very small. The maximum current around -70 mV (where $s_\infty \to 1$) is only about 10 μA/cm^2 and, at more positive potentials, the maximum current is even smaller. The s kinetics are also more temperature dependent than any other permeability variable so far described in the literature. Between 26 and 38° C the Q_{10} is 6, compared to 2–3 for the K current in nerve cells (Hodgkin, Huxley & Katz, 1952; Frankenhaeuser & Moore, 1963). Finally, g_{K_2} is determined by the first power of s, whereas a higher power (2–4, or even higher) is required for the delayed K current in nerve cells. In fact, the s system is the first known example of a system showing only simple exponential current changes; of course, the inactivation variables (h and k) in other systems also require an exponent of 1.

Influence of temperature on the electrical activity of Purkinje fibres. The extreme slowness and very high temperature dependence of the s kinetics both have important consequences for the over-all electrical activity of Purkinje fibres. Coraboeuf & Weidmann (1954) found a strikingly high temperature dependence of the rate of change of voltage during the pace-maker potential ($Q_{10} = 6.2$) and the plateau phase of repolarization ($Q_{10} = 4.5$). These results were obtained on spontaneously beating fibres and, although this means that the frequency of beating varied with temperature, it is reasonable to assume that the s system would be in equivalent states at comparable phases of the records obtained at different temperatures. This follows from the fact that the plateau and pace-maker potential account for by far the largest fraction of time during each cycle so that the total duration of the cycle changes by much the same factor as the duration of each phase. These durations are in turn to a large extent determined by the s kinetics.

On the other hand, Trautwein, Gottstein & Federschmidt (1953) using preparations stimulated at a constant frequency obtained lower values for the Q_{10}s than did Coraboeuf & Weidmann (1954). The qualitative reason for this difference is fairly clear. As the temperature is lowered, s will change more slowly and, if the frequency of stimulation is constant, the over-all variation in s during each cycle must be less than at higher temperatures. This means that on average s will deviate more from s_∞, particularly during the spontaneous phase of depolarization following the action potential, and since the rate of change of s, and hence, of membrane potential, depends on how far s deviates from s_∞, the average rate of change of s would be expected to be higher than in situations where s is allowed time to approach its steady-state value. As a consequence, the rate of change of voltage may be less temperature dependent when the preparation is stimulated at a constant frequency than when it is allowed to beat spontaneously. Whether this argument can account quantitatively for the different Q_{10}s obtained by the two methods remains to be seen.

Note added after submission. Dudel, Peper, Rüdel & Trautwein (1967c) have recently studied the potassium conductance of Purkinje fibres using the ramp voltage clamp technique. They interpret their results as evidence for the absence of significant time-dependent conductance changes and prefer to attribute changes in the potassium current–voltage relations to changes in E_K. On the basis of some experiments using the rectangular clamp technique, they estimate that less than 10 % of the outward current is time-dependent. These results may seem less surprising, however, when we consider the conditions required for observing the slow changes in i_{K_2}:

1. Even when fully activated, the time-dependent component of i_K is normally one third or less of the total potassium current, i.e. at any potential i_{K_1} is usually twice as large as the maximum value of i_{K_2}.

2. The magnitude of the slow current changes observed with rectangular clamps depends critically on the clamp potentials used. Records showing a large degree of time dependence (see Fig. 7) are obtained using depolarizations from a potential which is negative enough so that nearly all of i_{K_2} is yet to be activated (s nearly 0), and yet not so near E_K as to reduce the size of the current change on repolarization. The depolarization must also be large enough to activate a large fraction of i_{K_2}, but not so large as to allow the slow current change to be reduced as a consequence of the negative slope in the current–voltage relation.

Dudel *et al.* (1967c) also show that current–voltage relations obtained with different speeds of repolarizing ramp clamps do not intersect each other in the region of -95 mV, which they estimate as the value of E_K.

CARDIAC POTASSIUM CURRENT   213

On the basis of our results, however, we would expect that, at $[K]_0 = 2.7$ mM, the relations should intersect at a reversal potential near -110 mV (see Fig. 11) which is beyond the range of potentials studied by Dudel *et al.* (1967*c*).

This work was supported by an equipment grant from the Medical Research Council. We are grateful to Mr A. J. Spindler for valuable technical assistance and to Dr J. J. B. Jack, Dr R. E. McAllister and Dr R. S. Stein for their comments on the manuscript.

REFERENCES

ADRIAN, R. H., CHANDLER, W. K. & HODGKIN, A. L. (1966). Voltage clamp experiments in sleletal muscle fibres. *J. Physiol.* **186**, 51–52*P*.

ARMSTRONG, C. M. (1966). Time course of TEA⁺-induced anomalous rectification in squid giant axons. *J. gen. Physiol.* **50**, 491–503.

ARMSTRONG, C. M. & BINSTOCK, L. (1965). Anomalous rectification in the squid giant axon injected with tetraethylammonium chloride. *J. gen. Physiol.* **48**, 859–872.

CARMELIET, E. E. (1961). Chloride ions and the membrane potential of Purkinje fibres. *J. Physiol.* **156**, 375–388.

CARMELIET, E. E. & VAN DER WALT, J. J. (1968). Duration of the action potential as a function of the membrane potential in cardiac Purkinje fibres. *J. Physiol.* **194**, 88*P*.

CORABOEUF, E. & WEIDMANN, S. (1954). Temperature effects on the electrical activity of Purkinje fibres. *Helv. physiol. pharmac. Acta* **12**, 32–41.

DECK, K. A. & TRAUTWEIN, W. (1964). Ionic currents in cardiac excitation. *Pflügers Arch. ges. Physiol.* **280**, 63–80.

DRAPER, M. H. & WEIDMANN, S. (1951). Cardiac resting and action potentials recorded with an intracellular electrode. *J. Physiol.* **115**, 74–94.

DUDEL, J., PEPER, K., RÜDEL, R. & TRAUTWEIN, W. (1967*a*). The dynamic chloride component of membrane current in Purkinje fibres. *Pflügers Arch. ges. Physiol.* **295**, 197–212.

DUDEL, J., PEPER, K., RÜDEL, R. & TRAUTWEIN, W. (1967*b*). The effect of tetrodotoxin on the membrane current in cardiac muscle (Purkinje fibres). *Pflügers Arch. ges. Physiol.* **295**, 213–226.

DUDEL, J., PEPER, K., RÜDEL, R. & TRAUTWEIN, W. (1967*c*). The potassium component of membrane current in Purkinje fibres. *Pflügers Arch. ges. Physiol.* **296**, 308–327.

EHRENSTEIN, G. & GILBERT, D. L. (1966). Slow changes of potassium permeability in the squid giant axon. *Biophys. J.* **6**, 553–566.

FOZZARD, H. A. (1966). Membrane capacity of the cardiac Purkinje fibre. *J. Physiol.* **182**, 255–267.

FRANKENHAEUSER, B. (1962). Potassium permeability in myelinated nerve fibres of *Xenopus laevis. J. Physiol.* **160**, 54–61.

FRANKENHAEUSER, B. & MOORE, L. E. (1963). The effect of temperature on the sodium and potassium permeability changes in myelinated nerve fibres of *Xenopus laevis. J. Physiol.* **169**, 431–437.

HALL, A. E., HUTTER, O. F. & NOBLE, D. (1963). Current–voltage relations of Purkinje fibres in sodium-deficient solutions. *J. Physiol.* **166**, 225–240.

HALL, A. E. & NOBLE, D. (1963). The effect of potassium on the repolarizing current in cardiac muscle. *J. Physiol.* **167**, 53–54*P*.

HODGKIN, A. L. & HUXLEY, A. F. (1952). A quantitative description of membrane current and its application to conduction and excitation in nerve. *J. Physiol.* **117**, 500–544.

HODGKIN, A. L., HUXLEY, A. F. & KATZ, B. (1952). Measurement of current–voltage relations in the membrane of the giant axon of *Loligo. J. Physiol.* **116**, 424–448.

HUTTER, O. F. & NOBLE, D. (1960). Rectifying properties of cardiac muscle. *Nature, Lond.* **188**, 495.

HUTTER, O. F. & NOBLE, D. (1961). Anion conductance of cardiac muscle. *J. Physiol.* **157**, 335–350.

MCALLISTER, R. E. & NOBLE, D. (1966). The time and voltage dependence of the slow outward current in cardiac Purkinje fibres. *J. Physiol.* **186**, 632–662.

214   *D. NOBLE AND R. W. TSIEN*

McALLISTER, R. E. & NOBLE, D. (1967). The effect of subthreshold potentials on the membrane current in cardiac Purkinje fibres. *J. Physiol.* **190**, 381–387.

NOBLE, D. (1962). A modification of the Hodgkin–Huxley equations applicable to Purkinje fibre action and pacemaker potentials. *J. Physiol.* **160**, 317–352.

NOBLE, D. (1965). Electrical properties of cardiac muscle attributable to inward-going (anomalous) rectification. *J. cell. comp. Physiol.* **66**, suppl. 2, 127–136.

NOBLE, D. (1966). Applications of Hodgkin–Huxley equations to excitable tissues. *Physiol. Rev.* **46**, 1–50.

REUTER, H. (1966). Strom-Spannungsbeziehungen von Purkinje-Fasern bei verschiedenen extracellularen Calcium-Konzentrationen und unter Adrenalineinwirkung. *Pflügers Arch. ges. Physiol.* **287**, 357–367.

REUTER, H. (1967). The dependence of the slow inward current on external calcium concentration in Purkinje fibres. *J. Physiol.* **192**, 479–492.

ROBERTSON, W. VAN B. & DUNIHUE, F. W. (1954). Water and electrolyte distribution in cardiac muscle. *Am. J. Physiol.* **177**, 292–298.

TRAUTWEIN, W., GOTTSTEIN, U. & FEDERSCHMIDT, K. (1953). Der Einfluss der Temperatur auf den Aktionstrom des excidierten Purkinje-Fadens, gemessen mit einer intracellularen Elektrode. *Pflügers Arch. ges. Physiol.* **258**, 243–260.

VAN DER WALT, J. J. & CARMELIET, E. E. (1967). Effect of hyperpolarizing current on cardiac Purkinje fibres in K-enriched Tyrode. *Archs int. Physiol.* **75**, 139–141.

VASSALLE, M. (1966). An analysis of cardiac pacemaker potential by means of a 'voltage-clamp' technique. *Am. J. Physiol.* **210**, 1335–1341.

WEIDMANN, S. (1955). The effect of the cardiac membrane potential on the rapid availability of the sodium-carrying system. *J. Physiol.* **127**, 213–224.

WEIDMANN, S. (1956). *Elektrophysiologie der Herzmuskelfaser*. Bern: Hüber.

J. Physiol. (1969), **200**, 205–231   205
With 11 *text-figures*
Printed in Great Britain

OUTWARD MEMBRANE
CURRENTS ACTIVATED IN THE PLATEAU RANGE OF
POTENTIALS IN CARDIAC PURKINJE FIBRES

By D. NOBLE and R. W. TSIEN*

From the University Laboratory of Physiology, Oxford

(*Received* 29 *July* 1968)

SUMMARY

1. The membrane currents in Purkinje fibres under voltage clamp conditions have been investigated in the range of potentials at which the action potential plateau occurs. The results show that in this range slow outward current changes occur which are quite distinct from the potassium current activated in the pace-maker range of potentials.

2. The time course of current change in response to step voltage changes is non-exponential. At each potential the current changes may be analysed in terms of the sum of two exponential changes and this property has been used to dissect the currents into two components, i_{x_1} and i_{x_2}, both of which have been found to obey kinetics of the Hodgkin–Huxley type.

3. The first component, i_{x_1}, is activated with a time constant of about 0·5 sec at the plateau. At more positive and more negative potentials the time constants are shorter. The steady-state degree of activation varies from 0 at about -50 mV to about 1 at $+20$ mV. The instantaneous current–voltage relation is an inward-going rectifier but shows no detectable negative slope. In normal Tyrode solution ($[K]_0 = 4$ mM) the reversal potential is about -85 mV.

4. The second component, i_{x_2}, is activated extremely slowly and the time constant at the plateau is about 4 sec. The steady-state activation curve varies from 0 at about -40 mV to 1 at about $+20$ mV. The instantaneous current–voltage relation is nearly linear. The reversal potential occurs between -50 and -20 mV in different preparations.

5. It is suggested that these currents are carried largely by K ions, but that some other ions (e.g. Na) also contribute so that the reversal potentials are positive to E_K.

6. The relation of these results to previous work on delayed rectification in cardiac muscle is discussed.

* Rhodes Scholar.

206   *D. NOBLE AND R. W. TSIEN*

INTRODUCTION

Recent experiments on Purkinje fibres have shown that depolarizations to potentials within the range of the action potential plateau initiate a sequence of time and voltage dependent permeability changes which involve most of the naturally occurring ions. The first, and most rapid, changes are a substantial fall in potassium conductance (Hutter & Noble, 1960; Carmeliet, 1961; Hall, Hutter & Noble, 1963; Deck & Trautwein, 1964; Noble, 1965) and a large increase in sodium conductance (Deck & Trautwein, 1964; Dudel, Peper, Rüdel & Trautwein, 1967b) which is mostly inactivated within about 10 msec, although a small fraction of sodium inactivation requires about 100 msec to occur (Reuter, 1968). It is not yet certain whether this inactivation is complete (see Dudel *et al.* 1967b), but there are theoretical (Brady & Woodbury, 1960; Noble, 1962a) and experimental (Vassalle, 1966; Reuter, 1968; see also Rougier, Vassort & Stämpfli, 1968) reasons for supposing that incomplete inactivation contributes, together with the low potassium conductance, to maintaining depolarization during the plateau. A slow increase in permeability to calcium ions may also contribute a significant depolarizing current (Reuter, 1967).

In the present paper, and in the paper which follows it (Noble & Tsien, 1969), we shall be concerned with the question how the plateau is terminated and the membrane repolarized. Noble's (1962a) model attributed this process to activation of a potassium conductance whose subsequent decline following repolarization is responsible for the pace-maker potential. Such a potassium conductance is activated at plateau potentials (McAllister & Noble, 1966), although its kinetic and rectifier properties are substantially different from those of Noble's model (Noble & Tsien, 1968). Moreover, the time constant of activation of this conductance is very voltage-dependent. At −80 mV the time constant is of the order of a second, which is adequate to account for the pace-maker potential (Noble & Tsien, 1968). However, at −20 mV the time constant is of the order of only 50 msec (McAllister & Noble, 1966), which is too fast to account for the initiation of repolarization, except in the case of action potentials whose duration is not much longer than about 200 msec. Purkinje fibre action potentials frequently last longer than this, particularly in low K (see Noble, 1965) and Cl-free (see Hutter & Noble, 1961) solutions, which suggests that an additional mechanism may be involved in the repolarization process. We have, therefore, investigated the voltage clamp currents during long-lasting depolarizations in the region of the action potential plateau.

METHODS

The methods used have been described in detail by McAllister & Noble (1966) and Noble & Tsien (1968). The voltage clamp technique is similar to that described by Deck, Kern & Trautwein (1964). The solutions contained 140 mM-Na^+, 4 mM-K^+ 145 mM-Cl^-, 0·5 mM-Mg^{2+}, 1·8 mM-Ca^{2+}, 1·65 mM-HPO_4^{2-}, 0·7 mM-$H_2PO_4^-$ and 1 g/l. glucose. Most of the experiments were done in Cl-free solution containing 140 mM-$CH_3SO_4^-$ in place of Cl^-. About 5 mM-Cl^- remained in this solution. The solutions were saturated with oxygen and kept at a constant temperature near 35° C.

The major difficulty encountered in the experiments arises from the fact that very small current changes have to be recorded over a period of several hours in order to obtain sufficient information for a detailed analysis of the kinetics and rectifier properties. A large number of experiments were done in which a partial set of records was obtained before the preparation deteriorated or the electrodes became dislodged. These results were consistent with those of more complete experiments, although there was a significant degree of variation between fibres in the relative voltage dependency of the current components. The membrane current was continuously recorded on a pen recorder (Devices) at a constant low amplification and simultaneously at a high amplification which was varied between voltage clamp pulses to give the optimum amplification for each current record. In a typical experiment the amplification varied by a factor of 10 between the smallest and largest current records. This was essential since the current changes in the plateau region are extremely small. Noise was reduced by using a low pass filter so that only the slower current components which were being analysed were accurately recorded. An oscilloscope was used to monitor the fast components, such as the inward sodium current.

RESULTS

General characteristics of membrane currents

When long-lasting voltage clamp steps are applied to the membrane in the plateau range of voltages (0 to −40 mV) slow time and voltage dependent changes in the outward current are observed. These changes are quite distinct from the slow potassium current changes activated in the pace-maker range of potentials that have been described previously (McAllister & Noble, 1967; Noble & Tsien, 1968). This may be demonstrated in the following way. Noble & Tsien (1968) showed that the slow potassium current, i_{K_2}, whose time-dependent changes are responsible for generating the pace-maker potential, is controlled by a variable, s, which obeys first-order kinetics. The steady-state value of s, s_∞, varies from nearly 0 at −90 mV to nearly 1 at about −65 mV. Positive to −65 mV s_∞ is nearly constant. This means that when depolarizations are applied from about −65 mV (or from any potential positive to this potential) no time-dependent changes which may occur may be attributed to i_{K_2} (although some of the instantaneous variation of the current with voltage may be).

Figure 1 shows the result of an experiment in which the membrane potential was initially clamped at −60 mV. The potential was then

208   *D. NOBLE AND R. W. TSIEN*

changed in a positive direction in 10 mV steps. The first and second steps
(to −40 mV) produce only sudden changes in membrane current and it is
evident that there is a negative slope in the steady-state current–voltage
relation between −60 and −40 mV. Further depolarization steps produce
slowly increasing outward currents. Note that these currents require
several seconds to be fully activated. Their kinetics are therefore considerably slower than those of i_{K_2} in this range of potentials (see McAllister &

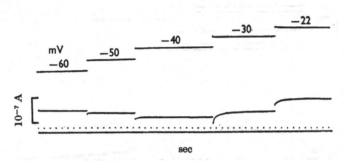

Fig. 1. Response of Purkinje fibre to 10 mV step depolarizations from −60 mV.
Top: membrane potential. Bottom: membrane current. Note that little or no slow
current change occurs on depolarization to −50 mV. A very small change occurs
on depolarization to −40 mV. Further depolarizations produce large slow current
changes in outward (upward) direction. The preparation was bathed in virtually
Cl-free Tyrode solution containing 4 mM-K and maintained at a constant temperature around 35° C.

Noble, 1966). It is also clear that in this preparation the range of potentials
which activates these slow currents is separated from the range which
activates i_{K_2} by about 20 mV so that when long-lasting potential steps are
applied there is a range within which no slow current changes occur. This
means that the slow current changes activated by the plateau range of
potentials may be investigated without interference due to changes in
i_{K_2} simply by using potentials which are positive to −60 mV. In the
experiments to be described below we have in fact chosen to use −30 mV
as the potential at which the membrane is held between pulses. We shall
refer to this potential as the holding potential. In addition to ensuring
that there is no interference from changes in i_{K_2}, holding the membrane
potential at about −30 mV also reduces interference from i_{Na} which is
largely inactivated and contributes only to the steady-state current–
voltage relations. The rapidly inactivated outward transient, identified
by Dudel, Peper, Rüdel & Trautwein (1967a) as a chloride current, is
also inactivated at −30 mV. Of course, it is not possible to completely
eliminate interference from these current components when large hyperpolarizations from the holding potential are applied, particularly at the

PLATEAU CURRENTS IN PURKINJE FIBRES 209

termination of such hyperpolarizations. The way in which this problem may be resolved will be described later.

Figure 2 shows currents recorded in response to various depolarizations and hyperpolarizations from the holding potential. It can be seen that the magnitude of the slow increase in current greatly increases as the depolarization increases so that the steady-state current–voltage relation rectifies strongly in the outward-going direction (see Fig. 10). The magni-

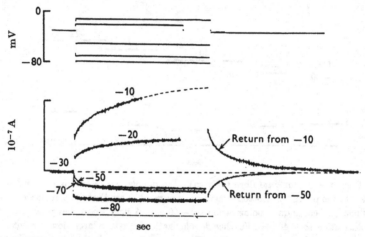

Fig. 2. Membrane currents in response to step potential changes from a holding potential (− 30 mV) in the plateau range. The currents in response to steps to − 10, − 20, − 50, − 70 and − 80 mV are shown. The response to − 10 mV was too large to be fully recorded at the amplification used and the interrupted line showing the continuation of this response was obtained from a lower amplification record (see Fig. 3). Large slow current changes occur in response to positive steps. The current changes in response to negative steps are smaller and at − 80 mV there is virtually no slow current change (Note: the slow current changes normally recorded in the pace-maker range were very small in this preparation). The records of recovery of current following returns from − 10 and − 50 mV are also shown. Note that the time courses are not symmetric and that time course of current following return from − 10 mV contains a slow component which is almost absent in the case of the recovery from − 50 mV.

tude of the current decay on repolarizing to − 30 mV also increases as the magnitude of the preceding depolarization increases (see Fig. 3). The current tails therefore probably reflect the decay of the outward current activated during the depolarization. A rigorous test of this view will be described later (see Fig. 5).

Careful inspection of the records shown in Fig. 2 reveals that at least some of the current changes are not simply exponential. Both the onset of current during depolarization and the return to the steady state following repolarization show an initial relatively fast phase of current change followed by a much slower phase. This is best shown by plotting the currents on logarithmic scales (see Fig. 4). Note also that the decay of current following depolarization is not symmetric with the recovery of current following hyperpolarization. The decay following depolarization to -10 mV shows a slow phase which is almost negligible in the recovery of current following hyperpolarization to -50 mV. The kinetics of the slow current changes in the plateau range therefore differ from those of the current changes in the pace-maker range since the latter follow simple exponential time courses and the current decays following depolarizations are symmetric in time course with the recovery of current following hyperpolarizations (Noble & Tsien, 1968, Fig. 2).

A possible hypothesis to describe current changes

Non-exponential time courses have been observed in other current systems in excitable cells. However, in those cases, the deviation from an exponential time course is in the opposite direction to that observed in the present results. Thus, in nerve cells, the sodium and potassium currents are activated slowly initially (Hodgkin & Huxley, 1952b), and this behaviour is explained in the Hodgkin–Huxley theory by making the current change proportional to a power of an exponential variable, the power being greater than 1. The physical interpretation of this behaviour is that more than one event obeying first-order kinetics must occur at each membrane site in order for it to conduct current. The power required to fit the current onsets in Fig. 2 would be less than 1, which does not appear to have any simple physical interpretation. Moreover, the current decays could not be fitted in this way. It is necessary therefore to use some other formulation.

Another possible hypothesis is that there are two or more kinetic processes occurring in parallel and controlling separate currents. In this case the time course of current change should be a sum of simple exponentials. By itself, this would not be particularly convincing evidence for the hypothesis since any decay curve can be fitted by a sum of exponentials if a sufficient number of terms are included in the sum. However, provided that sufficient experimental information is available, it is possible to test the hypothesis more rigorously in several ways. For simplicity (and because the assumption fits the experimental results over most of the range of potentials studied) we will assume that the time courses may be fitted by the sum of two exponentials. Let the slow time-dependent current in the plateau range be i_x and the two components assumed to change exponentially be i_{x_1} and i_{x_2}:

$$i_x = i_{x_1} + i_{x_2}. \tag{1}$$

Then $$\Delta i_x = \Delta i_{x_1} + \Delta i_{x_2}.$$

To reduce the number of subscripts in some equations we will refer to Δi_{x_1} and Δi_{x_2} as A and B respectively,

$$\Delta i_x = A + B.$$

During each voltage clamp pulse A and B are assumed to change exponentially with time constants τ_1 and τ_2. Hence

$$\Delta i_x = A_\infty[1 - \exp(-t/\tau_1)] + B_\infty[1 - \exp(-t/\tau_2)], \qquad (2)$$

where A_∞ and B_∞ are the steady-state values of A and B following a long step change in potential from the holding potential, E_H, to the potential during the pulse, E. On a logarithmic scale, A_∞ and B_∞ will determine the intercepts and τ_1 and τ_2 the slopes of the straight lines required to fit the experimental points (Fig. 4).

On return to the holding potential after a clamp pulse of duration b, the currents will return exponentially to their original values. However, since i_{x_1} and i_{x_2} will also be instantaneous functions of the membrane potential, there will be an initial sudden change during the potential step before A and B return exponentially to the original values. We will assume that these instantaneous changes may be represented separately from the slow kinetic changes (cf. Noble & Tsien, 1968). Let $\overline{i_{x_1}}$ and $\overline{i_{x_2}}$ be the maximum currents which may flow at each potential. Then we define

$$\overline{i_{x_1}} = i_{x_1}(E, x_1 = 1), \qquad (3)$$

$$\overline{i_{x_2}} = i_{x_2}(E, x_2 = 1), \qquad (4)$$

where x_1 and x_2 are the fractional degrees of activation of each current component. $x_1 = 1$ and $x_2 = 1$ therefore denote the fully activated states. Using this notation, sudden changes in A and B at each potential step may be attributed to changes in $\overline{i_{x_1}}$ or $\overline{i_{x_2}}$, whereas the slow changes may be attributed to changes in the degree of activation, x_1 or x_2. Note that $\overline{i_{x_1}}$ and $\overline{i_{x_2}}$ should be explicit functions of E only, whereas x_1 and x_2 are assumed to be functions of E and t.

The current change following return to the holding potential, E_H, will be given by

$$\Delta i_x = A_{t=b} \exp[(b-t)/\tau_1] + B_{t=b} \exp[(b-t)/\tau_2], \qquad (5)$$

where $A_{t=b}$, $B_{t=b}$, τ_1 and τ_2 are now all measured at E_H. $A_{t=b}$ and $B_{t=b}$ are the initial values of the currents immediately following termination of the pulse. Since it is assumed that the activation variables, x_1 and x_2, do not change instantaneously, it follows that

$$A_{E_H, t=b} \propto A_{E, t=b},$$

$$B_{E_H, t=b} \propto B_{E, t=b},$$

and the constants of proportionality will be given by the ratios of fully activated currents at each potential:

$$A_{E_H} = A_E \cdot \overline{i_{x_1,E_H}}/\overline{i_{x_1,E}}, \tag{6}$$

$$B_{E_H} = B_E \cdot \overline{i_{x_2,E_H}}/\overline{i_{x_2,E}}. \tag{7}$$

Equations (1)–(7) may be used to analyse the experimental records and the results may then be used to test the hypothesis in the following ways:

1. At any particular potential, τ_1 and τ_2 should be completely independent of the magnitude, direction and duration of the preceding polarization. In practice, the best information on this is given by the very large number of current tails recorded on return to the holding potential. These should all be fitted by exponentials with the same pair of time constants (cf. Fig. 6).

2. The magnitudes of A and B measured immediately following return to the holding potential should be simple exponential functions of the pulse duration, b, and the time constants of these exponentials should be equal to those determined from currents measured during the pulse (cf. Fig. 5).

3. τ_1 and τ_2 should be smooth simple functions of E (cf. Fig. 7) since the hypothesis would be somewhat implausible if the time constants varied with E in too complex a manner. At least it would then be worth while exploring other possible formulations.

4. $A_{b=\infty}$ and $B_{b=\infty}$, measured on return to E_H so as to keep $\overline{i_{x_1}}$ and $\overline{i_{x_2}}$ constant for all measurements, should be smooth functions of E (cf. Fig. 8). These variables will be proportional to the fraction of each component activated in the steady state and should be simple sigmoid functions of E if the variables are of the Hodgkin–Huxley type.

5. If τ_1 is sufficiently small compared to τ_2 then the decay curves for small values of b should be simple exponentials with time constant equal to the value of τ_1 at the holding potential (cf. Fig. 6).

Separation of current changes into fast and slow components

The results required to test the hypothesis described above and to obtain the important variables A_∞, B_∞, $\overline{i_{x_1}}$, $\overline{i_{x_2}}$, τ_1 and τ_2 are the membrane currents during and following polarizations of different amplitudes and various durations. In order for the tests to be fully applied and the variables calculated over a sufficiently wide range of potentials, a large number of voltage clamp pulses must be analysed. We shall therefore describe the detailed analysis of our most complete experiment. Other experiments gave similar (although usually only partial) results. Some variability between preparations was observed in the voltages at which the currents

PLATEAU CURRENTS IN PURKINJE FIBRES   213

were activated and the voltages at which the currents reverse. These differences were sometimes of the order of 20 mV between preparations but it is not yet clear what factors may be responsible for this variation.

Figure 3 shows superimposed current records in response to depolarizations and hyperpolarizations of various durations from the holding potential (-30 mV) to -80, -50, -20, -10, 0, 10, 20 and 30 mV. Hyperpolarizations to other potentials were also used but are not shown in the figure since the currents are completely inactivated by a 20 mV hyperpolarization. The important general point to note about these records is that the magnitudes of the tails of current following return to the holding potential are smooth functions of the pulse duration, b. If only one kinetic process were involved, the time course of the envelope (i.e. the maximum values of the current tails plotted as functions of b) should be identical in shape with the time course of current change during the polarization (cf. McAllister & Noble, 1966, Fig. 4; Noble & Tsien, 1968, Fig. 2). When more than one process is present, however, this will be true only if the current channels obey the same current–voltage relations. As will be shown later (Fig. 9) this is not the case for the present results and further analysis is required before comparisons may be made between the time course of current during a polarization and the time course of the envelope of the current tails.

Before discussing the analysis of the results shown in Fig. 3, it is important to explain how the current measurements were made. As will become evident later, much of the analysis depends on measurements of the peak currents immediately following return to -30 mV and it is important to ensure that these currents are measured at a time when the membrane potential is accurately controlled. This may be achieved by increasing the amplification of the feed-back circuit used in clamping the membrane voltage. Unfortunately, this also increases the tendency of the circuit to oscillate and it can be seen that some of the current records shown in Fig. 3 show rapidly damped oscillations at the beginning of each tail. When tracing the current records for analysis it is relatively easy to extrapolate the current record back through the oscillation to the time at which the clamp pulse was terminated. This extrapolation gives a better measure of the true time course of the current than would be obtained using current records at lower feed-back amplification since this would introduce unknown errors due to inadequate control of the membrane potential. In the logarithmic plots shown in Figs. 4–6 only the first point depends greatly on the accuracy of the extrapolation. The general validity of the analysis does not therefore depend very much on the extrapolation, but this factor must be taken into account when assessing the accuracy of the analysis.

Figures 4, 5 and 6 show some of the currents in response to depolarization to -10 mV plotted on logarithmic scales. The points and curves shown in Fig. 4 illustrate the way in which the analysis was done. Figure 5 shows more detailed analysis of the onset of current. The current during the depolarization was subtracted from the steady-state current to give the deviation of the current from its steady-state value, i.e.

$$[\Delta i_x (E, b = \infty) - \Delta i_x (E, t)].$$

214   *D. NOBLE AND R. W. TSIEN*

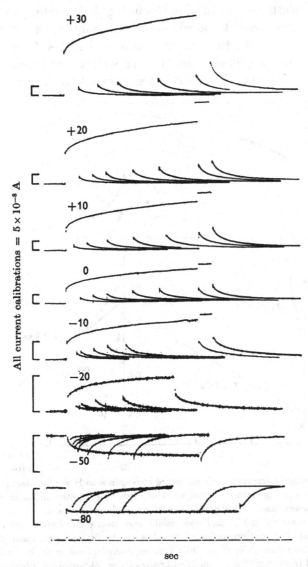

Fig. 3. Superimposed records of membrane currents in response to step depolarizations and hyperpolarizations of various magnitudes and durations from a holding potential of −30 mV. Each set of records shows a complete response to a pulse lasting 10 sec (8 sec in the case of −20 mV, 12 sec in the case of −80 mV). The tails of recovery of current following shorter pulses are shown and, in the case of depolarizations to and beyond −10 mV, the steady levels of current and tails of recovery of current following much longer pulses (20 sec in the case of −10 mV, 0 mV, +10 mV and +20 mV; 40 sec in the case of +30 mV) are also shown. These records allowed the steady-state current–voltage relations and activation curves to be obtained (see Figs. 8, 10 and 11). All current calibrations are 5 × 10⁻⁸ A. Note that current amplification was varied to give optimal amplification in each case. The steady-state current at +30 mV was too large to be shown but is plotted in Fig. 10.

PLATEAU CURRENTS IN PURKINJE FIBRES   215

The results are plotted as the filled circles in Fig. 5. The later points were
fitted by eye with a straight line whose time constant, τ_2, is 5·25 sec and
whose intercept, B_∞, with the current axis is $6·6 \times 10^{-8}$ A. This line was
then subtracted from the filled circles to give the remaining current change
which should be attributable to the fast component, A. The results are

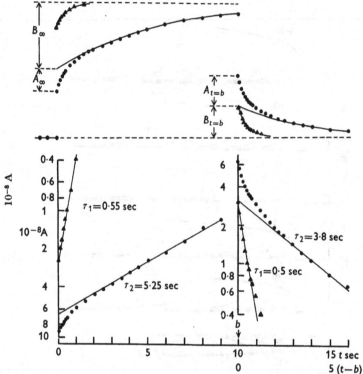

Fig. 4. Current measurements made on record in response to 10 sec depolarization
to −10 mV illustrating method of analysis. The filled circles plotted on linear scale
(top) show measurements of membrane current made on a tracing of original
record which is shown in Fig. 2. These results were then plotted on logarithmic
scales (bottom). The continuous lines were obtained by fitting logarithmic plots
with straight lines through later points and the exponential lines on the linear plot
were obtained from these lines. The filled triangles were obtained by subtracting
slow exponential component to give fast component. $A_{t=b}$, $B_{t=b}$, $A\infty$, $B\infty$, τ_1 and τ_2
refer to quantities used in equations (1)–(7).

plotted as the filled triangles in Fig. 5. Note that the points are a good fit
to a line whose time constant, τ_1, is 0·55 sec and whose intercept, A_∞, is
$2·3 \times 10^{-8}$ A. This result shows that the onset of time-dependent current
during depolarization to −10 mV is given fairly accurately by the sum of
two exponentials. Now if $A_{E_H} \propto A_E$ and $B_{E_H} \propto B_E$ (equations (6) and (7))
then the envelope of tails of current on return to the holding potential
should be fitted by exponentials with the same time constants, though not

216   *D. NOBLE AND R. W. TSIEN*

necessarily the same intercepts. The envelope of the tails has been plotted
as the open circles in Fig. 5. As can be seen the points may be fitted by the
same pair of time constants (the open triangles are obtained in the same
way as the filled triangles). The value of B_{∞, E_H} is 4.4×10^{-8} A which is
considerably smaller than $B_{\infty, E}$. This indicates that the current carried by

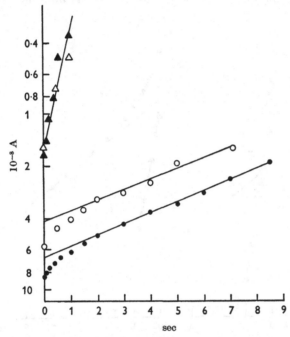

Fig. 5. Analysis of time course of current activated by depolarization to -10 mV.
The slow current change during the depolarization is measured as the deviation of
the current from the steady-state value and is plotted as filled circles ($\bullet$). The
slow component was fitted by a straight line with intercept 6.6×10^{-8} A and time
constant 5.25 sec. This line was then subtracted from the points to give the fast
component ($\blacktriangle$). This component was then fitted with a straight line with intercept 2.3×10^{-8} A and time constant 0.55 sec. The open circles were obtained by
measuring the peaks of the tails of current following return to the holding potential.
The slow component in this case has the same time constant as the slow component during the depolarization but the intercept is smaller (4.4×10^{-8} A). The
open triangles ($\triangle$) were obtained by subtraction and are fitted by the same line as
the filled triangles. The coincidence of the filled and open triangles is not significant
for the analysis of the time course of current but it does illustrate the non-linearity
of the fast current component (see Fig. 9).

the slow component is larger at -10 mV than at -30 mV for a given
degree of activation of the system. By contrast, the values of A_{∞, E_H} and
$A_{\infty, E}$ are virtually equal which means that the influence of sudden
voltage changes on the fast component of current is very small between
these two voltages (see also Fig. 9).

PLATEAU CURRENTS IN PURKINJE FIBRES   217

The tails of current on return to -30 mV after 0·2, 3·5 and 17 sec at -10 mV are plotted in Fig. 6 as filled circles. At 0·2 sec, the points are fitted by one fast exponential with a time constant equal to 0·5 sec. This is to be expected since by this time only negligible activation of the slow component will have occurred (Fig. 5). After 17 sec, by contrast, two exponentials are required. The slow component was obtained by fitting

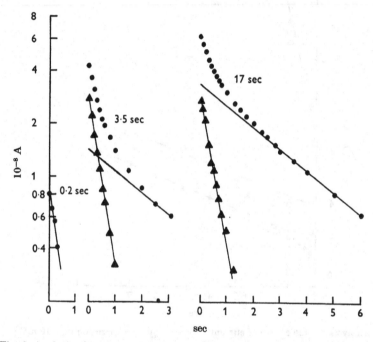

Fig. 6. Analysis of current tails following depolarizations to -10 mV lasting 0·2, 3·5 and 17 sec. The measured currents are plotted as filled circles (●). In the case of the tail following 17 sec depolarization the slow component was fitted by a straight line with intercept $3·3 \times 10^{-8}$ A and time constant 3·8 sec. The fast component (▲) was then obtained by subtraction and fitted by a line with intercept $2·7 \times 10^{-8}$ A and time constant 0·5 sec. The tail following 3·5 sec depolarization was fitted in a similar way using a slow component with the same slope as for the 17 sec tail. The 0·2 sec tail shows no slow component and has the same slope as the fast components of the 3·5 and 17 sec tails.

the later points, when the fast component may be assumed to be fully deactivated. The fast component (filled triangles) was then obtained by subtraction and is fitted by an exponential with the same time constant as that required to fit the tail following 0·2 sec. The same two exponentials fit the 3·5 sec tail. Note that the intercept of the fast component in this case is about equal to that following the 17 sec pulse. This result is expected since the fast component is fully activated by 3·5 sec (Fig. 5). By contrast, the intercept of the slow component is much smaller than that following 17 sec. The intercepts following the longest pulses give a second pair of estimates of A_∞ and B_∞ at E_H which compared well with the estimates obtained from plotting the envelope of the current peaks. For subsequent analysis the average of these two estimates of A_∞ and B_∞ have been used.

All the current records were analysed in this way and, with two exceptions, the results were entirely consistent with the hypothesis that two current components are present, each one of which is controlled by a first-order variable with voltage-dependent kinetics.

The exceptions occurred in the case of the responses to the largest polarizations. The first exception was noted at $+30$ mV. Very long-lasting depolarizations to this potential produced tails of current on return to -30 mV which required three exponentials. The third component had a time constant of 0·8 sec. There are three reasons for considering this to be a genuine third component at very strong depolarizations rather than a change in the behaviour of the other two components:

1. The other two exponentials required to fit the current tail following long-lasting depolarization to $+30$ mV had time constants equal to those of the fast and slow decays following all other polarizations.

2. Short-lasting depolarizations to $+30$ mV were followed by tails which required only the usual two components. It is evident that the third component is activated extremely slowly, as is indicated by the fact that at least 40 sec were required to achieve a steady-state current at $+30$ mV.

3. A short-lasting (2 sec) depolarization to $+50$ mV was also followed by a two-component tail. This shows that for all the potentials and durations of physiological interest, the third component is never activated. It has, therefore, been ignored in the analysis.

The second exception occurred after long-lasting hyperpolarizations beyond -70 mV. In this case, the currents on return to -30 mV differed qualitatively from all the other records. First, they showed an initial inward sodium transient. However, this component was not recorded on the pen tracings since it was too fast. It is, therefore, not shown in Fig. 3. Secondly, it can be seen that, in the case of the hyperpolarization to -80 mV, the current tails increase in amplitude up to about 5 sec but decrease slightly thereafter. This is attributable to an outward transient which reduces the size of the inward current tail at its beginning. The transient decays completely by 200 msec since after this time the current tails have the same time course as those following shorter duration hyperpolarizations. This current transient is even larger when the membrane potential is returned to potentials positive to -30 mV and has been described previously (Deck & Trautwein, 1964; Hecht, Hutter & Lywood, 1964). Dudel *et al.* (1967 a) attribute this current to chloride ions. In the present case, very little (< 5 mM) chloride was present which suggests that not all of this current is carried by chloride ions. However, this particular problem need not be resolved in order to take account of the transient in the present analysis. It has already been shown (Reuter, 1968; D. Noble & R. W. Tsien, unpublished) that at the resting potential this current recovers from inactivation only very slowly, the time constant of recovery being about 5 sec. Hence hyperpolarizations from the plateau range which do not last longer than about 2 sec should not activate very much of this current at the termination of the pulse. Since the time constants of i_x are quite short at hyperpolarized potentials, it is possible to fully analyse the behaviour of i_x without interference from the outward transient. However, it is of some importance to include this transient in the total repolarizing current when calculating the repolarization process on the basis of voltage clamp data (Noble & Tsien, 1969).

Kinetics

The results described above suggest that the slow outward current changes in the plateau range depend on two first-order processes obeying the differential equations

$$dx_1/dt = \alpha_{x_1}(1 - x_1) - \beta_{x_1} x_1, \tag{8}$$

$$dx_2/dt = \alpha_{x_2}(1 - x_2) - \beta_{x_2} x_2. \tag{9}$$

In the steady state at each potential

$$(x_1)_\infty = \alpha_{x_1}/(\alpha_{x_1} + \beta_{x_1}), \tag{10}$$

$$(x_2)_\infty = \alpha_{x_2}/(\alpha_{x_2} + \beta_{x_2}) \tag{11}$$

and, following a step change in potential, x_1 and x_2 should change exponentially with time constants given by

$$\tau_1 = 1/(\alpha_{x_1} + B_{x_1}), \tag{12}$$

$$\tau_2 = 1/(\alpha_{x_2} + \beta_{x_2}), \tag{13}$$

τ_1 and τ_2 have been determined experimentally over a range of potentials from -120 mV to $+30$ mV and their reciprocals are plotted in Fig. 7. τ_1^{-1} is a U-shaped function of membrane potential with a minimum rate at about the level of the plateau and much faster rates at hyperpolarized potentials. The rate also increases somewhat at positive potentials. Hence, x_1 should be activated fairly slowly at the plateau but should decline more quickly in the region of the resting potential. τ_2^{-1} is also at a minimum in the plateau range and becomes larger on hyperpolarization. However, we could find no evidence for an increase of τ_2^{-1} at positive potentials. We doubt whether our results are accurate enough to exclude the possibility of any increase but they probably are accurate enough to exclude an increase of the magnitude shown in the curve for τ_1^{-1}.

In order to extract the rate coefficients, the α's and β's, from these results we must also know the steady-state degree of activation of each component as a function of potential. These values are proportional to the values of A_∞ and B_∞ at E_H obtained in the analysis described in the previous section since these variables give the steady-state degree of activation of each component in terms of the current recorded immediately after return to the holding potential. The advantage of measuring the degree of activation in this way is that the results depend on current measurements which are always made at the same potential. Hence, provided that the shapes of the instantaneous current–voltage relations are constant, non-linearities in these relations will not interfere with the analysis. The values of A_∞ and B_∞ are plotted as functions of the membrane potential in Fig. 8. It

220   *D. NOBLE AND R. W. TSIEN*

can be seen that both variables are simple sigmoid functions of the membrane potential. The fast component, A_∞, is half-activated in the steady state at about -20 mV. The slow component requires more positive potentials. The curves have also been normalized and the normalized values were used as estimates of $(x_1)_\infty$ and $(x_2)_\infty$. It is now possible to use equations (10)–(13) to calculate the values of the α's and β's and the results are shown in Fig. 7 of the following paper (Noble & Tsien, 1969).

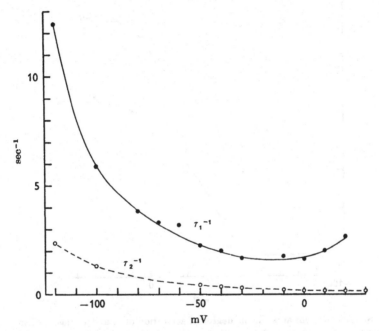

Fig. 7. Variation of rates of slow current changes, measured as reciprocal time constants (τ_1^{-1} and τ_2^{-1}), with membrane potential. The points were obtained from measurements of the slopes of the lines determined by the method explained in Figs. 4, 5 and 6. The continuous curves were drawn to fit the points by eye.

It should be emphasized that, although the curves shown in Fig. 8 are fairly typical, there is considerable variation between preparations, particularly in the position of the x_2 activation curve on the voltage axis. In some preparations the activation curve was at considerably more negative potentials and, since the reversal potential for i_{x_2} is at a fairly depolarized potential (see Fig. 9), in these preparations it was possible to activate x_2 at potentials at which i_{x_2} is negative. The result was a very slow change in current in an inward direction on depolarization of the membrane.

Current–voltage relations

In addition to determining the kinetics of the current changes, the membrane potential also has a more direct influence on the membrane current. This direct influence may be conveniently represented in terms of the current–voltage relation for a constant degree of activation of the

PLATEAU CURRENTS IN PURKINJE FIBRES   221

kinetic variable. Since it is not possible to study the current components in isolation it is necessary to use indirect methods. The information required for reconstructing the current–voltage relations is contained in the ratios A_E/A_{E_H} and B_E/B_{E_H} since these ratios give an estimate of the currents at E and E_H for the same degree of activation. If these ratios are multiplied by the total current amplitudes of the steady-state activation

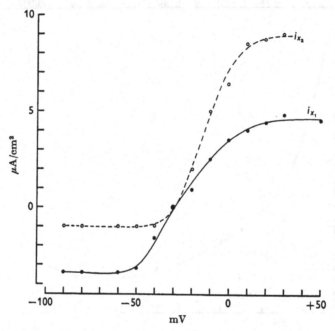

Fig. 8. Steady-state variation in the degree of activation of x_1 and x_2 measured in terms of the currents (A_∞ and B_∞) following long-lasting voltage clamp steps. All measurements are therefore made at the same potential (-30 mV) so that the properties of the fully activated current–voltage relation are eliminated. Continuous curves were fitted by eye to points.

curves (Fig. 8) we obtain the values of i_{x_1} and i_{x_2} as functions of E when x_1 and x_2 equal 1, i.e. rearranging equations (6) and (7):

$$\overline{i_{x_1,E}} = \overline{i_{x_1,E_H}}\, \frac{A_{E,\,t=b}}{A_{E_H,\,t=b}}, \tag{14}$$

$$\overline{i_{x_2,E}} = \overline{i_{x_2\,E_H}}\, \frac{B_{E,\,t=b}}{B_{E_H,\,t=b}}, \tag{15}$$

where $\overline{i_{x_1,E_H}}$ and $\overline{i_{x_2,E_H}}$ are given by the total amplitudes of the curves plotted in Fig. 8. Since these equations give the instantaneous current–voltage relations when the kinetic variable, x_1 or x_2, is fully activated, we shall refer to the resulting current–voltage relations as the 'fully activated

222   *D. NOBLE AND R. W. TSIEN*

current–voltage relations'. We have previously used the term 'rectifier
function' for this relation (Noble & Tsien, 1968, Fig. 11) but this term
seems inappropriate when it is possible for the current–voltage relation
to show virtually no rectification (as is the case for i_{x_2}—see below). Moreover, the new term indicates more clearly the condition (i.e. maximum
activation of the conducting channels at each voltage) to which the relation
applies.

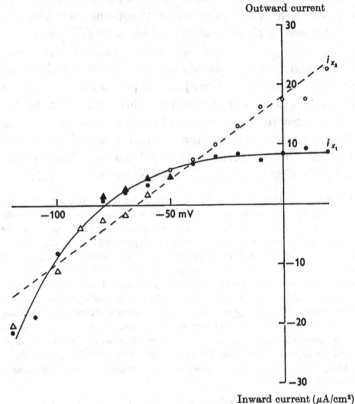

Fig. 9. Fully activated current–voltage relations for i_{x_1} (● ▲) and i_{x_2} (○ △) obtained
from the experimental results using equations (14) and (15). The circles were
obtained from results using −30 mV as the holding potential. The triangles were
obtained by using −20 mV as the holding potential. Curves fitted to points by eye.

The fully activated current–voltage relations obtained using equations
(14) and (15) are shown in Fig. 9. $\overline{i_{x_1}}$ shows inward-going rectification
although, unlike i_{K_2} (see Noble & Tsien, 1968, Fig. 9), the current–
voltage relation does not show a negative slope. The reversal potential in
this preparation is about −85 mV. This does not correspond to any known
equilibrium potential, particularly since chloride ions are virtually absent
from the solutions. Moreover, the reversal potential varies to some extent

between fibres so that the more likely explanation is that the pathways conduct mainly K ions (the K equilibrium potential is about -100 mV) with some leakage to other ions so as to give a reversal potential positive to E_K. Further experiments are required to test this hypothesis, although some preliminary experiments have shown that the reversal potential does vary with $[K]_0$ (O. Hauswirth, D. Noble & R. W. Tsien, unpublished).

$\overline{i_{x_2}}$ is fitted quite well by a straight line at all but the most negative potentials, where there is some suggestion of non-linearity. The reversal potential in this case is about -65 mV which is 20 mV positive to the reversal potential for i_{x_1}. This means that voltage pulses which deflect the membrane potential from the plateau range to a potential between the two reversal potentials (e.g. -75 mV) should produce a positive i_{x_1} component followed by a negative i_{x_2} component. This was found to be the case although it is not easy to obtain records of this kind since the currents are very small at potentials close to the reversal potentials. The reversal potential for i_{x_2} also fails to correspond to any known equilibrium potential and it seems likely that this current is also relatively non-specific.

The current–voltage relations shown in Fig. 9 were determined over a very wide range of potentials, including the range negative to -70 mV. It was unusual to obtain the relations for i_x over such a wide range since i_{K_2} normally dominates the time-dependent changes which occur negative to -70 mV. However, in some preparations (including the one used to obtain Fig. 9) i_{K_2} is relatively small and the tendency to pace-maker activity nearly absent. Since the time constants for i_x are fairly short at negative potentials, it is possible in these preparations to obtain measurements of i_x from the negative tails of current which occur before appreciable changes in i_{K_2} (which produce positive tails at potentials positive to -100 mV) occur. In preparations with a strong tendency to pace-maker activity and large changes in i_{K_2}, i_x may not be estimated accurately at very negative potentials but it is usual to observe a negative tail which is relatively fast compared to the positive tail attributable to the decline in i_{K_2}. Reuter (1968) has described some of the properties of the negative tails on repolarization and it is likely that i_x is mainly responsible for the tails which he found were not abolished by sodium removal.

The results shown in Figs. 7, 8 and 9 completely specify the behaviour of i_x. Each component may be calculated from the equations

$$i_{x_1} = x_1 \overline{i_{x_1}}, \tag{16}$$

$$i_{x_2} = x_2 \overline{i_{x_2}}, \tag{17}$$

where x_1 and x_2 obey equations (8) and (9). These equations will be used in the following paper to reconstruct the behaviour of i_x during the action potential.

Components of the steady-state current–voltage relation

In order to determine the contribution of i_x to the electrical activity of the Purkinje fibre membrane, it is also necessary to know the electrical properties of the membrane in the absence of i_x. As yet, we have no means

224   *D. NOBLE AND R. W. TSIEN*

of abolishing i_x and the only method for determining the electrical properties in the absence of i_x is to subtract i_x from the total recorded current.
This was done by plotting the total steady-state current–voltage relation
and then subtracting the calculated steady-state values of i_{x_1} and i_{x_2} given
by the equations

$$(i_{x_1})_\infty = (x_1)_\infty \, \overline{i_{x_1}}, \tag{18}$$

$$(i_{x_2})_\infty = (x_2)_\infty \, \overline{i_{x_2}}. \tag{19}$$

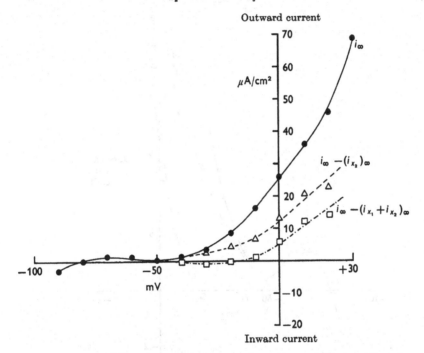

Fig. 10. Current–voltage relations illustrating the contributions of i_{x_1} and i_{x_2} to the
steady-state membrane current. The filled circles (●) show the total steady-state
current–voltage relation (i_∞). The open triangles (△) show the relation after subtraction of $(i_{x_2})_\infty$, using the experimental results and eqn. (19). The open squares
(□) were then obtained by also subtracting $(i_{x_1})_\infty$, using the experimental results
and eqn. (18).

The results are shown in Fig. 10. The filled circles and continuous curve
show the total steady-state current–voltage relation obtained from the
experimental results. As shown previously (Deck & Trautwein, 1964) this
relation is N-shaped with a negative slope conductance between − 50 and
− 70 mV. Note, however, that the chord conductance is always positive.
The open triangles and interrupted line show the current–voltage relation
after subtraction of $(i_{x_2})_\infty$. The open squares and dot-dash line show the
current–voltage relation after subtraction of $(i_{x_1})_\infty$ and $(i_{x_2})_\infty$. The latter

PLATEAU CURRENTS IN PURKINJE FIBRES   225

relation has a much more extensive negative slope conductance region and the chord conductance also becomes negative between -15 and -50 mV. These features are shown more clearly in Fig. 11 which shows the same curves plotted on a tenfold larger current scale (note that a tenfold range of current amplification was used, and found necessary, in these experiments). Some general conclusions concerning the role of i_x may be drawn immediately from Fig. 11. First, in this preparation, i_{x_2} is not required for repolarization since the subtraction of i_{x_2} does not produce a region of net

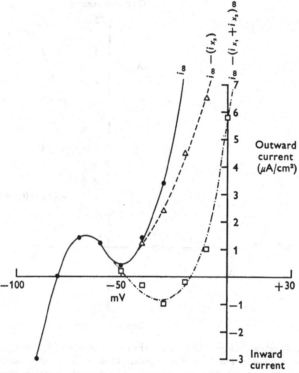

Fig. 11. Current–voltage relations on ten times large current scale. Same symbols as Fig. 10. Note that there is a region of net inward current in the current–voltage relation (dot-dash curves) when the contributions of i_{x_1} and i_{x_2} have been subtracted from the total steady-state current.

inward current. Moreover, as shown above (see *Kinetics*) the time constants of i_{x_2} are so long that it is activated extremely slowly. However, i_{x_1} is required for repolarization in this case since the current–voltage relation does not lose its region of net inward current unless i_{x_1} is at least partially activated. The question which now arises therefore is whether the properties of i_{x_1} allow the repolarization phase of the Purkinje fibre action potential to be reproduced. This question will be discussed in the following paper (Noble & Tsien, 1969).

226 *D. NOBLE AND R. W. TSIEN*

The current–voltage relation obtained after subtraction of i_x is that for all other currents (i_{Na}, i_{K_1}, i_{K_2}, i_{Ca}) in the steady state. Chloride current is not involved since the chloride concentration was extremely low. The presence of a negative current region in this relation may be accounted for if at least one of the inward currents (i_{Na}, i_{Ca}) is incompletely inactivated in the steady state. Further analysis of this current–voltage relation will be presented in another paper (R. E. McAllister, D. Noble & R. W. Tsien, in preparation).

DISCUSSION

The major conclusion of this paper is that the slow current changes in the plateau range of potentials can be accounted for quantitatively by two separate current components whose kinetics may be described by the first-order variables we have labelled x_1 and x_2. The role of these currents in electrical activity in Purkinje fibres will be discussed in later papers (Noble & Tsien, 1969; Hauswirth, Noble & Tsien, 1969). We shall therefore restrict the present discussion to the relation of our results to previous work.

Relation of results to previous work on delayed rectification

Delayed rectification was first observed in Purkinje fibres by Hutter & Noble (1960). Hall *et al.* (1963) and McAllister & Noble (1966) showed that in many preparations in Na-free solutions a delayed increase in conductance occurs when the membrane is depolarized beyond about -30 mV. Up to -30 mV the only change observed is a sudden decrease in conductance on depolarization which is similar to the inward-going (anomalous) rectification observed in skeletal muscle (Katz, 1949; Hodgkin & Horowicz, 1959; Adrian & Freygang, 1962). These properties of the membrane were both attributed to changes in potassium conductance. The fast changes due to inward-going rectification were described by assuming a non-time-dependent K current, i_{K_1}. The slow current changes responsible for delayed rectification at strong depolarizations were separately described by a time-dependent component, i_{K_2}. Together with a component of excitatory sodium current, a model incorporating these two K currents (Noble, 1962*a*) was adequate to reproduce Purkinje fibre action potential and pace-maker activity. This model explains repolarization in terms of the activation of i_{K_2} and the pace-maker potential by its deactivation. In addition to providing the large initial excitatory current, the sodium component provides a residual inward current that nearly balances the outward K current during the plateau.

Subsequent experiments showed that there are difficulties with this model. The application of the voltage clamp technique (Deck & Trautwein, 1964; Hecht *et al.* 1964; McAllister & Noble, 1966) confirmed the presence

PLATEAU CURRENTS IN PURKINJE FIBRES   227

of inward-going rectification and also revealed a region of negative slope in the current–voltage relation in sodium-free solutions. However, there has been considerable disagreement about the existence of delayed rectification and confusion concerning the voltages at which it is activated. Our results suggest a way in which these disagreements may be resolved and it will be convenient to discuss the previous results in the presence and absence of sodium ions separately.

Delayed rectification in sodium containing solutions. It is now clear that in the presence of Na ions there are *two* voltage ranges at which delayed rectification may be observed. In the pace-maker range (-90 to -60 mV) large outward current changes may be observed in response to quite small voltage steps. These currents are attributable almost entirely to K ions since the reversal potential occurs at the K equilibrium potential (Deck & Trautwein, 1964; Vassalle, 1966) and shifts with E_K when $[K]_0$ is varied (Noble & Tsien, 1968). The major differences between this current and the i_{K_2} component assumed by Noble (1962a) are that it is activated by small depolarizations (McAllister & Noble, 1967) and that the instantaneous current–voltage relation rectifies strongly in the inward-going direction (Noble & Tsien, 1968, Fig. 9). However, since this was the first time-dependent outward current to be identified in voltage clamp experiments, it was labelled i_{K_2}.

The present results show that delayed rectification also occurs at a voltage range (positive to -40 mV) which closely corresponds to the range over which Noble assumed i_{K_2} to be activated. This delayed rectification is attributable to the two components we have labelled i_{x_1} and i_{x_2}, both of which have reversal potentials which suggest that they are carried largely but not exclusively by K ions. It would clearly be confusing to change the notation yet again and we suggest that i_{K_2} be retained for the specific K current activated in the pace-maker range and that the less specific currents should be called i_{x_1} and i_{x_2}. This notation clearly distinguishes them from i_{K_2} and emphasizes the fact that they are not pure K currents.

Using the ramp voltage clamp technique, Dudel, Peper, Rüdel & Trautwein (1967c) have described results which they interpret as showing that delayed rectification is completely absent in Purkinje fibres. Their results may, however, be given a different interpretation which will be explained in the following paper (Noble & Tsien, 1969) where the role of the time-dependent outward currents in repolarization is discussed.

Delayed rectification in sodium-free solutions. Deck & Trautwein (1964) and McAllister & Noble (1966) found that in Na-free solutions the outward current changes in the pace-maker range are greatly reduced or even absent. McAllister & Noble (1966) suggested that this may be due to a large shift in the threshold for activation of i_{K_2} on removing Na ions. This explanation seemed reasonable at that time since they observed a large degree of time-dependent outward current at strong depolarizations and it was not known at that time that this current is also present and is activated in the same voltage range in Na solutions. The present results show that this hypothesis is wrong and that the delayed rectification in Na-free solutions observed by Hutter & Noble (1960), Hall *et al.* (1963) and McAllister & Noble (1966) must be attributed to i_{x_1} and possibly also to i_{x_2}. This interpretation is strengthened by the observation that sodium removal does not greatly influence the x kinetics (D. Noble & R. W. Tsien, unpublished).

This interpretation of previous work leaves two major problems unresolved. First, delayed rectification in Na-free solutions was not observed by some workers (Deck & Trautwein, 1964; Hecht & Hutter, 1965), and Hall *et al.* (1963) and McAllister & Noble (1966) did not observe it in all preparations. Hall *et al.* (1963) attributed this fact to cable complications since they were working on long fibres. However, the results obtained since 1964 have been obtained on short fibres in which cable complications are negligible (Deck *et al.* 1964). A more likely reason for failing to observe delayed rectification is that it is easily masked by the large initial outward transient identified by Dudel *et al.* (1967*a*) as a chloride current. When this current is inactivated by steady depolarization (as in Fig. 1) it is relatively easy to observe delayed rectification. The outward transient is also largely inactivated when depolarizations from the resting potential are applied at a frequency of 1 Hz or above (Reuter, 1968; D. Noble & R. W. Tsien, unpublished). This suggests that it would be easier to observe delayed rectification in preparations that are regularly pulsed at a rate more closely corresponding to the normal frequency of Purkinje fibre activity.

The second problem which remains unresolved is the effect of Na removal on i_{K_2}. Preliminary experiments (O. Hauswirth, D. Noble & R. W. Tsien, unpublished) indicate that Na removal does not greatly influence the s kinetics but simply reduces the amplitude of the slow K current changes attributable to i_{K_2}. The nature of this effect is not yet clear and more experiments will be required to resolve this problem.

Rougier *et al.* (1968) have recently succeeded in voltage-clamping frog auricular muscle and have demonstrated the presence of delayed rectification in normal and in sodium-free solutions. As yet, there is insufficient information available to allow quantitative comparison, although H. F. Brown & S. J. Noble (personal communication), using a technique similar to that of Rougier *et al.* (1968), have analysed a current component which strongly resembles the component we have called i_{x_2}.

Possible significance of instantaneous current–voltage relations

Perhaps the most surprising outcome of the present work is that there are not one but at least three separate time-dependent outward currents in Purkinje fibres. There may even be a fourth component activated at positive potentials outside the normal range (see Results). All of these currents are proportional to first-order variables of the Hodgkin–Huxley type. Apart from quantitative differences in the kinetics, which will be discussed in the following paper (Noble & Tsien, 1969, Fig. 7), the major difference between these currents is that they obey different instantaneous current–voltage relations. The expression i_{K_2} $(E, s = 1)$, which, using the notation of the present paper, is $\overline{i_{K_2}}$, is a strong inward-going rectifier with a marked negative slope (Noble & Tsien, 1968, Fig. 9). Similarly $\overline{i_{x_1}}$ is a less marked inward-going rectifier with no negative slope and $\overline{i_{x_2}}$ is nearly linear (see Fig. 9 of present paper). i_{K_2} is a pure K current; i_{x_1} and i_{x_2} are mixed currents, i_{x_1} having a reversal potential closer to E_K than i_{x_2}. It seems, therefore, that the degree of rectification becomes weaker as the specificity of the systems for K ions becomes smaller. It is tempting to speculate that each non-specific current component may be a mixture of currents passing through two kinds of channel: one highly specific for K ions and having rectification properties similar to those for i_{K_2}, and another which is a linear non-specific shunt. The only common feature of the two channels (and the only reason why they are considered together as one current component) would be that they are controlled by gating mechanisms obeying the same kinetics. This view has no concrete evidence in its favour but it is in line with the view, suggested by the separability of the kinetic and rectifier properties (Armstrong & Binstock, 1965; Armstrong, 1966; Noble & Tsien, 1968), that these properties correspond to separate molecular mechanisms since it suggests that the basic molecular mechanisms may be combined in different ways to produce a whole range of electrical characteristics.

The analysis of conductance mechanisms in terms of separable kinetic and rectifier properties is not, of course, new. It is implicit in Hodgkin & Huxley's (1952b) original notation, which uses $\bar{g}$ $(E - E_{rev})$ to specify the fully activated current–voltage relation, since it was evident that the latter could not be strictly linear at all ionic concentrations (Hodgkin & Huxley, 1952a, Fig. 7). Frankenhaeuser (1963) carried this approach a stage further by using the constant field equations to describe the instantaneous current–voltage relations in myelinated nerve. These equations describe outward-going rectification. The more recent results (Armstrong, 1966; Noble & Tsien, 1968; present paper) show that Hodgkin–Huxley kinetics may also control channels showing inward-going rectification.

230   *D. NOBLE AND R. W. TSIEN*

This work was supported by a Medical Research Council grant for equipment. We are extremely grateful to Mr A. J. Spindler for valuable technical assistance and to Mr S. J. Bergman and Dr O. Hauswirth for their helpful criticisms of the manuscript.

REFERENCES

ADRIAN, R. H. & FREYGANG, W. H. (1962). The potassium and chloride conductance of frog muscle membrane. *J. Physiol.* **163**, 61–103.

ARMSTRONG, C. M. (1966). Time course of TEA$^+$-induced anomalous rectifications in squid giant axons. *J. gen. Physiol.* **50**, 491–503.

ARMSTRONG, C. M. & BINSTOCK, L. (1965). Anomalous rectification in the squid giant axon injected with tetraethylammonium chloride. *J. gen. Physiol.* **46**, 859–872.

BRADY, A. J. & WOODBURY, J. W. (1960). The sodium–potassium hypothesis as the basis of electrical activity in frog ventricle. *J. Physiol.* **154**, 385–407.

CARMELIET, E. E. (1961). Chloride ions and the membrane potential of Purkinje fibres. *J. Physiol.* **156**, 375–388.

DECK, K. A. & TRAUTWEIN, W. (1964). Ionic currents in cardiac excitation. *Pflügers Arch. ges. Physiol.* **280**, 63–80.

DECK, K. A., KERN, R. & TRAUTWEIN, W. (1964). Voltage clamp technique in mammalian cardiac fibres. *Pflügers Arch. ges. Physiol.* **280**, 50–62.

DUDEL, J., PEPER, K., RÜDEL, R. & TRAUTWEIN, W. (1967a). The dynamic chloride component of membrane current in Purkinje fibres. *Pflügers Arch. ges. Physiol.* **295**, 197–212.

DUDEL, J., PEPER, K., RÜDEL. R. & TRAUTWEIN, W. (1967b). The effect of tetrodotoxin on the membrane current in cardiac muscle (Purkinje fibres). *Pflügers Arch. ges. Physiol.* **295**, 213–226.

DUDEL, J., PEPER, K., RÜDEL, R. & TRAUTWEIN, W. (1967c). The potassium component of membrane current in Purkinje fibres. *Pflügers Arch. ges. Physiol.* **296**, 308–327.

FRANKENHAEUSER, B. (1963). A quantitative description of potassium current in myelinated nerve fibres of *Xenopus laevis. J. Physiol.* **169**, 424–430.

HALL, A. E., HUTTER, O. F. & NOBLE, D. (1963). Current–voltage relations of Purkinje fibres in sodium-deficient solutions. *J. Physiol.* **166**, 225–240.

HAUSWIRTH, O., NOBLE, D. & TSIEN, R. W. (1969). The mechanism of oscillatory activity at low membrane potentials in cardiac Purkinje fibres. *J. Physiol.* **200**, 255–265.

HECHT, H. H. & HUTTER, O. F. (1965). The action of pH on cardiac muscle. In *Electrophysiology of the Heart*, ed. TACCARDI, B. & MARCHETTI, G., pp. 105–123. Oxford: Pergamon Press.

HECHT, H. H., HUTTER, O. F. & LYWOOD, D. W. (1964). Voltage–current relation of short Purkinje fibres in sodium-deficient solution. *J. Physiol.* **170**, 5P.

HODGKIN, A. L. & HOROWICZ, P. (1959). The influence of potassium and chloride ions on the membrane potential of frog muscle. *J. Physiol.* **148**, 127–160.

HODGKIN, A. L. & HUXLEY, A. F. (1952a). The components of membrane conductance in the giant axon of *Loligo. J. Physiol.* **118**, 473–496.

HODGKIN, A. L. & HUXLEY, A. F. (1952b). A quantitative description of membrane current and its application to conduction and excitation in nerve. *J. Physiol.* **117**, 500–544.

HUTTER, O. F. & NOBLE, D. (1960). Rectifying properties of cardiac muscle. *Nature, Lond.* **188**, 495.

HUTTER, O. F. & NOBLE, D. (1961). Anion conductance of cardiac muscle. *J. Physiol.* **157**, 335–350.

KATZ, B. (1949). Les constantes électriques de la membrane du muscle. *Archs Sci. physiol.* **3**, 285–299.

McALLISTER, R. E. & NOBLE, D. (1966). The time and voltage dependence of the slow outward current in cardiac Purkinje fibres. *J. Physiol.* **186**, 632–662.

McALLISTER. R. E. & NOBLE, D. (1967). The effect of subthreshold potentials on the membrane current in cardiac Purkinje fibres. *J. Physiol.* **190**, 381–387.

NOBLE, D. (1962a). A modification of the Hodgkin–Huxley equations applicable to Purkinje fibre action and pace-maker potentials. *J. Physiol.* **160**, 317–352.

NOBLE, D. (1962b). The voltage dependence of the cardiac membrane conductance. *Biophys. J.* **2**, 381–393.

NOBLE, D. (1965). Electrical properties of cardiac muscle attributable to inward-going rectification. *J. cell. comp. Physiol.* **66**, suppl. 2, 127–136.

NOBLE, D. (1966). Applications of Hodgkin–Huxley equations to excitable tissues. *Physiol. Rev.* **46**, 1–50.

NOBLE, D. & TSIEN, R. W. (1968). The kinetics and rectifier properties of the slow potassium current in cardiac Purkinje fibres. *J. Physiol.* **195**, 185–214.

NOBLE, D. & TSIEN, R. W. (1969). Reconstruction of the repolarization process in cardiac Purkinje fibres based on voltage clamp measurements of membrane current. *J. Physiol.* **200**, 233–254.

REUTER, H. (1967). The dependence of the slow inward current on external calcium concentration in Purkinje fibres. *J. Physiol.* **192**, 479–492.

REUTER, H. (1968). Slow inactivation of currents in cardiac Purkinje fibres. *J. Physiol.* **197**, 233–253.

ROUGIER, O., VASSORT, G. & STÄMPFLI, R. (1968). Voltage clamp experiments on frog atrial heart muscle fibres with the sucrose-gap technique. *Pflügers Arch. ges. Physiol.* **301**, 91–108.

VASSALLE, M. (1966). An analysis of cardiac pace-maker potential by means of a 'voltage-clamp' technique. *Am. J. Physiol.* **210**, 1335–1341.

WEIDMANN, S. (1955). The effect of the cardiac membrane potential on the rapid availability of the sodium-carrying system. *J. Physiol.* **127**, 213–224.

Prog. Biophys. Molec. Biol. 1975. Vol. 30, No. 2 3, pp. 99 144. Pergamon Press. Printed in Great Britain.

ANALYTICAL MODELS OF PROPAGATION IN EXCITABLE CELLS

P. J. HUNTER, P. A. McNAUGHTON and D. NOBLE

University Laboratory of Physiology, South Parks Road, Oxford

CONTENTS

INTRODUCTION

The mathematics of the propagation of impulses in excitable cells has attracted considerable interest in recent years. A number of papers have appeared in biological (Noldus, 1973; Rinzel and Keller, 1973) and mathematical journals (McKean, 1970;

P. J. HUNTER, P. A. McNAUGHTON and D. NOBLE

Hastings, 1972, 1974; Evans, 1972a, b, c; Greenberg, 1973) that follow FitzHugh (1961, 1969) and Nagumo (see Nagumo *et al.*, 1962) in using simplified analytical models for the membrane currents. These approaches contrast with that using more complex empirical descriptions such as the Hodgkin–Huxley equations (see Hodgkin and Huxley, 1952; Huxley, 1959a, b; Cooley and Dodge, 1966; Noble and Stein, 1966).

One of the most important successes of the Hodgkin–Huxley equations for ionic current in squid nerve is their ability, in combination with the cable equation, to predict the conduction velocity of the nervous impulse. Thus, Hodgkin and Huxley (1952) computed a value of 18.8 m/sec from their equations compared to 21.2 m/sec for the value obtained experimentally in squid. Hodgkin (1937) had already provided conclusive experimental evidence for Hermann's (1899) suggestion that local circuit current flow along the nerve axon is adequate to ensure the spread of excitation from active to inactive regions, but these calculations demonstrated for the first time that the ionic current flowing across the membrane is not only adequate for propagation, but also that its magnitude and kinetics are appropriate to allow the cell to generate the speed of transmission observed without making any further assumptions than those implied by Hermann's suggestion.

It is implicit in Hodgkin and Huxley's analysis that there are some quantitative relationships between the ionic currents and the conduction velocity. However, it is difficult to obtain these relationships explicitly from the Hodgkin–Huxley equations. Huxley (1959a) obtained some relations between conduction velocity and the rates and magnitudes of ionic current changes using a function (see Fig. 11 below) that depends on the rates and magnitudes. However, this function, $F(\beta)$, was obtained numerically using a series of computed action potentials corresponding to responses at different temperatures (see Huxley, 1959a, eq. 20 and Fig. 20). No simple analytical function may be used for $F(\beta)$, although some important general results may be obtained that assume knowledge of $F(\beta)$ (Huxley, 1959a, eqs. 21–23). To give a complete analysis without the use of numerical methods, it appears necessary to use models of a simpler kind and it is important to assess the degree to which such models may closely mimic the Hodgkin–Huxley equations (or the properties of real axons) whilst retaining the advantage of giving closed form solutions for the propagation process. One of the aims of this article is to explore this question using a variety of simplified models.

There is, moreover, an additional reason for investigating models simpler than the Hodgkin–Huxley equations. This is that a full voltage clamp analysis of the kinetics of ionic conductance changes is only possible in a limited number of excitable cells or over a limited range of conditions. For example, the sodium current in many kinds of cardiac muscle is difficult or impossible to analyse because of the large spatial decay of potential in even the smallest available preparations. In such cases, it becomes necessary to resort to simpler models in order to study some excitation and conduction processes in a quantitative manner.

The first model of the conduction process to give an explicit equation for conduction velocity as a function of the cable parameters and the ionic current-voltage relation was described in a classical paper by Rushton (1937). He assumed that at the threshold potential an instantaneous change in membrane electromotive force occurs. He obtained an equation that gives the conduction velocity, θ, in terms of the number of space constants travelled per time constant. This number was identified with a parameter called the "safety factor" for conduction which was defined in terms of the ratio of the active membrane e.m.f. to the threshold voltage. Offner *et al.* (1940) later modified this approach to include the large increase in conductance observed to occur during the impulse by Cole and Curtis (1939).

However, both these approaches suffer from the general deficiency of "switch-line" models: that the ionic current is assumed to change in a discontinuous manner at the threshold potential, whereas in fact the ionic current is always found to be a smooth (even if sometimes steep) function of potential, even in the threshold region. A more appropriate approach, therefore, is to use continuous functions for the ionic current.

Polynomial functions are particularly convenient as they are readily differentiated and integrated, and they may be chosen to give very good fits to current-voltage relations like those found in excitable cells without having to include more than a few powers in the polynomial equation.

The idea of using polynomials, particularly cubic equations, is not new. Van der Pol's (1928) work on the valve oscillator used a cubic function for the current flow. FitzHugh (1960) and Nagumo (Nagumo *et al.*, 1962) later used this approach to reproduce nerve excitation. More recently, Adrian *et al.* (1972) have used a cubic function to illustrate an extension of Cole's theorem for non-linear cables and Noble (1972) used another cubic function to obtain an expression for the liminal length for excitation. A more general analysis of the use of polynomial models in non-linear cable theory is given in the book by Jack *et al.* (1975, chap. 12).

Polynomial and non-polynomial functions will be used in this article to obtain analytical results for the conduction process. FitzHugh (1969) has derived an expression for the conduction velocity in a cable obeying a stationary cubic equation like that used in Van der Pol's model, using a method suggested by A. F. Huxley. A similar approach is used here although the results obtained are more general and are obtained for a much wider variety of cases than the cubic. We shall illustrate applications of the analysis to experimental results in nerve and cardiac muscle and compare the analysis with the results of Huxley (1959a) and Cooley *et al.* (1965) using the Hodgkin–Huxley equations. Models of this kind will be discussed in section I.

Although the use of polynomial functions for describing the ionic current as a function of membrane potential avoids one of the deficiencies of switch line models (inasmuch as the ionic current shows no discontinuities), there is a major deficiency that is shared by the two approaches. This is that in real cells the ionic current is not simply a function of voltage. It is also a function of time. Some of the models referred to above (e.g. Noldus, 1973) introduce time dependence by including a recovery variable that represents the processes of sodium inactivation and potassium activation (see for example, FitzHugh, 1969). However, for the purposes of reproducing the conduction process, this is an unsatisfactory modification since, at normal temperatures, the process that limits conduction velocity is primarily that of sodium activation. The recovery processes are relatively less important except at high temperatures, when the phenomenon of heat block occurs (Hodgkin and Katz, 1949; Huxley, 1959a). The unsatisfactory nature of models of the BVP kind in this connection is well illustrated by the fact that they predict a decrease in conduction velocity with temperature (Noldus, 1973), whereas real axons, and the Hodgkin–Huxley equations, show an increase in conduction velocity with temperature until the point of heat block is approached.

In sections II and III of this article we shall therefore extend the approach described in section I to include models in which the excitatory current requires time for its activation. Models of this kind (particularly those in which the sodium activation process is sigmoidal, as observed experimentally) may reproduce the real behaviour of axons very well indeed. Analytical solutions for models incorporating activation time have not been previously published, although Hodgkin (unpublished—see Hodgkin, 1975) has shown that useful analytical formulae may be obtained from such models.

Section II is concerned with models that follow the form of the Hodgkin–Huxley equations as closely as possible without sacrificing the ability to obtain analytical solutions. Section III describes a more general approach to reproducing the time dependence of the ionic current that is not restricted to equations of the Hodgkin–Huxley form. In section IV the use of non-polynomial models is described. Finally, in section V we shall list those equations that emerge from our analysis that may be of greatest use in experimental work.

All the equations discussed in the article apply to conduction in fibres of uniform geometry. The problems of conduction in non-uniform cases are discussed in a following article (Khodorov and Timin, 1975).

I. MODELS IN WHICH THE IONIC CURRENT IS A FUNCTION OF VOLTAGE ONLY

In this section we shall assume that the actual membrane current as a function of potential during a propagated action potential—the current voltage trajectory—is known. It will be shown that approximating the current-voltage trajectory with polynomial functions leads to simple expressions for the conduction velocity.

While this approach is mathematically the simplest, it is limited by the need to know—or at least to estimate—the current-voltage trajectory. This trajectory has been calculated by numerical methods (Cole, 1968) for a Hodgkin–Huxley axon at 18.5°C, although even in this case altering, say, the fully-activated conductance will alter the current-voltage trajectory in a way not simply related to the changes in conductance. We shall discuss a number of devices for estimating the trajectory of excitable tissues for which a full voltage-clamp analysis is impossible or unavailable at present.

During the rising phase of an action potential the amount of transmembrane current that flows is considerably less than the maximum possible value at each voltage attained. Early in the rising phase the activation process is too slow to keep pace with the voltage change so that incomplete activation limits the sodium current flow, while near the peak inactivation becomes important. The resulting difference between the actual current flow and the fully-activated (active state) current at each potential may be considerable. Thus, in the case of the standard Hodgkin–Huxley squid axon at 18.5°C, the actual current during a propagated action potential (Fig. 1, upper line) reaches only about one-tenth of the active state current (Fig. 1, lower line) calculated by giving the activation variable (m) its steady-state value at each potential and setting the inactivation variable (h) to its value at the resting potential.

The problem considered in this section reduces to finding functions which both provide a good fit to the current-voltage trajectory and are mathematically tractable. Odd-

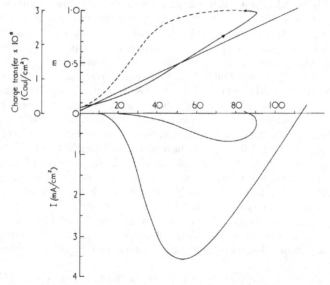

FIG. 1. Some parameters of an action potential propagating along a squid axon which conforms to the mathematical form used by Hodgkin and Huxley (1952) and Huxley (1959a, b). *Top panel*: the value of the m parameter as a function of potential during a propagated action potential. The straight line approximation to this trajectory will be used in a later section (see section II) in considering the effects of the "gating current" on the propagation velocity. The interrupted curve shows the steady-state value of the m parameter, i.e. m_∞. *Bottom panel*, *lower line*: The maximum ("active state") current available at each potential assuming that the Hodgkin–Huxley h and n variables controlling inactivation and recovery have the values they have at the resting potential, and that the m activation variable attains its steady-state value at each potential. *Upper line*: The actual current flowing across the membrane during a propagated action potential at 18.5°C. This curve was computed for us by S. B. Barton (cf. Cole, 1968, Fig. 3.14) and will be referred to as the current-voltage trajectory.

degree polynomials, with roots chosen at the resting potential, threshold and peak spike potential fortunately meet both these conditions. Cubics are the simplest polynomials, but produce too much outward current below threshold to be considered a good fit to typical trajectories. The inclusion of higher powers, however, can produce an almost perfect fit over the whole trajectory of a Hodgkin–Huxley action potential.

Evidently no power series containing a limited number of terms can provide a perfect fit to an experimental trajectory. We will derive later the conditions for considering a polynomial to be an adequate fit; briefly, these involve adjusting an overall scaling factor to make the area under the polynomial and the experimental current-voltage trajectories the same. For very accurate curve fitting the non-polynomial functions discussed in section IV are required.

1. *Ionic Current given by Cubic Functions*

We shall start with the simplest case: that in which the ionic current, I_i, is expressed as a cubic function of the membrane potential. i.e.

$$I_i = g\left\{ V\left(1 - \frac{V}{V_{th}}\right)\left(1 - \frac{V}{V_p}\right)\right\}. \tag{1.1}$$

where V is the membrane potential expressed as a deviation from the resting potential, V_{th} is the excitation threshold of the propagated action potential, and V_p is the peak or active voltage. The parameter g, with dimensions of conductance, is the slope of the cubic at the origin. The value of g should not be equated to the resting conductance of the real membrane current voltage trajectory which we seek to approximate by the cubic, as g scales the contributions of all powers of V in the expression and hence scales the amplitude of all regions of the non-linear current-voltage relation, not only the linear region close to the resting state. We shall show later that the value of g which provides the best fit to a real current-voltage trajectory is determined more by the active state conductance than by the resting conductance.

If we set $V_p = 4 + \sqrt{8}$ and $V_{th} = 8/(4 + \sqrt{8})$ we obtain the function used in a previous paper (Noble, 1972, eq. 5 and Fig. 2):

$$I_i = g(V - V^2 + (V/2)^3). \tag{1.2}$$

which gives a ratio of V_p/V_{th} equal to about 6. We shall use the general equation (1.1) to give a derivation of the conduction velocity that applies to all cubic polynomials, while using the particular equation (1.2) to illustrate the results.

The cable equation (Hodgkin and Rushton, 1946) is

$$\frac{a}{2R_i}\frac{\partial^2 V}{\partial x^2} = C_m \frac{\partial V}{\partial t} + I_i, \tag{1.3}$$

where x is distance, t is time, a is the diameter of the fibre, R_i is the axial resistivity and C_m the membrane capacitance per unit area. When we consider an impulse conducting at a uniform velocity θ at a sufficiently large distance from the initiation and termination points we may assume that the shape of the wave is constant. This is equivalent to assuming that a transposition of the wave front through distance x in space is equivalent to a transposition $-x/\theta$ in time, and therefore that

$$V(x, t) = F(x - \theta t),$$

where F describes the functional dependence of V on x and t. With this assumption

$$\frac{\partial V}{\partial x} = -\frac{1}{\theta}\frac{\partial V}{\partial t}.$$

and

$$\frac{\partial^2 V}{\partial x^2} = \frac{1}{\theta^2}\frac{\partial^2 V}{\partial t^2}.$$

so that (1.3) becomes

$$\frac{d^2V}{dt^2} = \frac{\theta^2 2\,R_i\,C_m}{a}\left(\frac{dV}{dt} + \frac{I_i}{C_m}\right) \tag{1.4}$$

(cf. Hodgkin and Huxley, 1952. eq. 30).

Now let $K = \theta^2 2R_iC_m/a$ and $\dot{V} = dV/dt$ then, using the chain rule,

$$\frac{d^2V}{dt^2} = \frac{d\dot{V}}{dt} = \frac{d\dot{V}}{dV}\frac{dV}{dt} = \frac{d\dot{V}}{dV}\,\dot{V},$$

so that (1.4) becomes

$$\frac{d\dot{V}}{dV} = K\left(1 + \frac{I_i}{\dot{V}\,C_m}\right). \tag{1.5}$$

From this equation we obtain immediately the result that when $\dot{V}$ reaches its peak value (i.e. $d\dot{V}/dV = 0$)

$$C_m\dot{V} = -I_i, \tag{1.6}$$

which is the familiar result that the capacity current $(C_m\dot{V})$ is equal and opposite to the ionic current when the action potential reaches its point of maximum rate of rise. We will refer to the voltage at which this occurs as $\bar{V}$. Thus, the trajectory $\dot{V}(V)$ passes through the point $-I_i/C_m$ at $\bar{V}$. We may also note that the trajectory must cross the voltage axis (i.e. $\dot{V} = 0$) at $V = 0$ and at $V = V_p$ since $dV/dt = 0$ at the beginning and end of the response. Thus the trajectory $\dot{V}(V)$ starts at $V = 0$, reaches a peak at $\bar{V}$ and falls again to cut the axis at V_p. These conditions suggest that the solution to equation 1.5 may be of the form of a parabola:

$$\dot{V} = \frac{gS}{C_m} V\left(1 - \frac{V}{V_p}\right), \tag{1.7}$$

where S is a scaling factor whose physical significance will be interpreted below.

Substituting (1.7) into (1.5) and using the cubic description of ionic current given by (1.1), we obtain

$$\frac{d\dot{V}}{dV} = \frac{gS}{C_m}\left[1 - \frac{2V}{V_p}\right] = K\left\{1 + \frac{[1 - (V/V_{th})]}{S}\right\}.$$

If (1.7) is to be a solution of (1.5) with a cubic I_i, the right-hand equality must hold for all values of V. Thus the coefficients of each power of V must be equal. This is satisfied if

$$\frac{gS}{C_m} = K\left(1 + \frac{1}{S}\right) \quad \text{(equating the constants)} \tag{1.8}$$

and

$$\frac{2gS}{C_m V_p} = \frac{K}{SV_{th}} \quad \text{(equating coefficients of } V). \tag{1.9}$$

Therefore (1.7) is a solution of (1.5) for the unique value of K which satisfies both (1.8) and (1.9). Solving (1.8) and (1.9) this value is found to be

$$K = \frac{g}{C_m}\left(\frac{S^2}{S+1}\right). \tag{1.10}$$

where

$$S = \frac{V_p}{2V_{th}} - 1. \tag{1.11}$$

Since $\theta^2 = Ka/2R_iC_m$ the conduction velocity of an impulse generated by a cubic current-voltage trajectory emerges from this analysis:

$$\theta^2 = \frac{ga}{2R_iC_m^2}\left(\frac{S^2}{S+1}\right). \tag{1.12}$$

This equation demonstrates the familiar dependence of the conduction velocity on the square root of the axon diameter. The velocity is also proportional to the square root of the scaling factor g—for example, increasing the amplitude of the activated current fourfold doubles the conduction velocity. Thirdly, the velocity depends on the parameter S, the *safety factor* defined by eq. (1.11).

A more intuitive basis for eq. (1.12) results if we follow Rushton's (1937) method of expressing the conduction velocity in terms of space and time constants. By analogy with the real membrane space and time constants, λ and τ_m, we can define space and time constants of the cubic approximation to the real membrane current-voltage trajectory:

$$\lambda_p = \sqrt{\left[\left(\frac{1}{g}\right)\bigg/\left(\frac{2R_i}{a}\right)\right]}$$

$$\tau_p = \frac{1}{g}C_m.$$

These polynomial space and time constants are, of course, not to be identified with true membrane constants, since, as has been emphasized above, g is not necessarily equal to the true resting membrane conductance.

Substituting λ_p and τ_p in the expression for θ we obtain

$$\theta = \frac{\lambda_p}{\tau_p}\sqrt{\left(\frac{S^2}{S+1}\right)}, \tag{1.13}$$

which approximates to

$$\theta \simeq \frac{\lambda_p}{\tau_p}\sqrt{(S-1)}. \tag{1.13a}$$

When S is large a further approximation is valid:

$$\theta \simeq \frac{\lambda_p}{\tau_p}\sqrt{S}. \tag{1.13b}$$

2. *Physical Significance of the "Safety Factor", S*

Equations (1.13)–(1.13b) show that the impulse in this model travels approximately $\sqrt{S}$ or, accurately, $\sqrt{(S^2/(S+1))}$ space constants (λ_p) per time constant (τ_p). The parameter S is an important one. It is equivalent to the "safety factor" defined by Rushton (1937) and others to indicate the extent to which the fibre's ability to become excited and conduct exceeds the minimum conditions for conduction. Thus, in this case, conduction cannot occur unless S is positive, i.e. unless

$$V_p > 2V_{th}. \tag{1.14}$$

This result is intuitively reasonable since, when $V_p = 2V_{th}$, the cubic expression for I_i is symmetric about V_{th} and the areas of inward and outward current are equal. Excitation and conduction can only occur if the area of inward current in the $I_i(V)$ relation exceeds that of outward current since this is the condition for the existence of a threshold for non-uniform excitation (cf. Noble and Hall, 1963; Noble and Stein, 1966; Adrian et al., 1972).

We may also note that the definition of S given here closely resembles that given by Rushton (1937). His equation for the safety factor becomes the same as eq. (1.10) when we note that the term RQ_B/τ in Rushton's model may be equated to V_{th} in our model (cf. Noble, 1972, Appendix). The major difference between the models lies in the fact that θ is approximately proportional to $\sqrt{S}$ in the cubic model, whereas it is proportional to S itself in Rushton's model.

Finally, we should note that, as given in eq. (1.10) the definition of the safety factor is not entirely satisfactory since it only applies to cubic models. In these models, the ratio of inward to outward current areas in the $I_i(V)$ relation is determined uniquely by the ratio V_p/V_{th}. This is also true of Rushton's model. However, it is not true of excitable cells. The area of inward current may be varied greatly by varying the sodium conductance g_{Na} with relatively little change in the voltage threshold, which is largely determined by the position of the sodium activation curve on the voltage axis. This is another way of saying that cubic equations cannot fit all shapes of $I_i(V)$ relations. To do this requires the use of polynomials of higher degree. We shall show below that this then allows the safety factor to be defined in a more satisfactory way.

3. *The Transmission Constant v*

Although we think it is useful to draw the analogy with Rushton's formula for conduction velocity by expressing the result in terms of units of λ_p travelled per τ_p, the analogy may also be misleading since these constants are not necessarily equal to λ_m and τ_m. In the rest of our treatment we shall therefore define a new variable which we shall call the *transmission constant*:

$$v = \frac{\lambda_p}{\tau_p} = \frac{1}{C_m} \sqrt{\left(\frac{ag}{2R_i}\right)}. \tag{1.15}$$

v may be regarded as a scaling parameter that enables the conduction velocity to be calculated once the shape of the current-voltage relation (which determines S) is known. Thus (1.13) may be written:

$$\theta = v \sqrt{\left(\frac{S^2}{S+1}\right)}. \tag{1.13c}$$

4. *Applications to Cubic Model*

Figure 2 shows the current-voltage diagram given by eq. (1.2) together with the trajectory $\dot{V}(V)$ which is given by eq. (1.12) with $V_p = 4 + \sqrt{8}$ and $V_{th} = 8/(4 + \sqrt{8})$.

The safety factor for this case is

$$S = \frac{(4 + \sqrt{8})^2}{16} - 1 = \sqrt{2} + 0.5 \simeq 1.914.$$

This value is positive, as required for conduction to occur, but is relatively small compared to that for most excitable cells, as we shall show later when discussing higher-degree polynomials. The conduction velocity may be obtained from eq. (1.13) to give

$$\theta = \sqrt{\left(\frac{S^2}{S+1}\right)} v = \sqrt{\left(\frac{\sqrt{2} + 2.25}{\sqrt{2} + 1.5}\right)} v = 1.121 \, v.$$

The time course of the propagating "action potential" may be obtained by integrating eq. (1.7):

$$\frac{dV}{dt} = \frac{gS}{C_m}\left(V - \frac{V^2}{V_p}\right) \tag{1.7}$$

with respect to time to obtain the solution

$$V = \frac{A}{(A/V_p) + \exp[(-gSt)/C_m]}. \tag{1.16}$$

Where A is an arbitrary constant dependent on the definition of the point $t = 0$. If we allow $V = V_p/2$ at $t = 0$ (i.e. zero time is taken arbitrarily to be at the inflection point $\bar{V}$) then

$$V = \frac{V_p}{1 + \exp(-Sgt/C_m)} \tag{1.17}$$

This equation is similar to that given by FitzHugh (1969, eqs. 6–17) for the BVP model without recovery. The equations can in fact be seen to be identical when the following

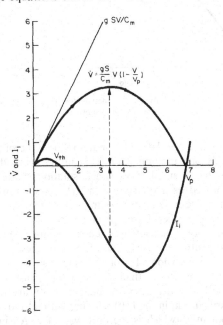

FIG. 2. The trajectory $\dot{V}(V)$ for a propagated action potential in a cable whose membrane current is given by eq. (1.2). The initial slope of the trajectory is given by $(gS/C_m)V$, where S is the safety factor for conduction. The ionic current given by eq. (1.2) is also plotted. The interrupted arrows are of equal length to show that $\dot{V} = -I_i$ when $\dot{V}$ is at its maximal value, $\dot{V}_{max}$ (eq. (1.6)).

substitutions are made to convert from FitzHugh's symbols to those used in the present paper: $V_R = 0$, $V_2 = V_p$, $S = t$, $A(V_2 - V_R) = S/RC_m$. In each case, the first symbol or expression is FitzHugh's. After substituting, FitzHugh's expression may be multiplied top and bottom by $\exp(-Sgt/C_m)$ to obtain eq. (1.17).

Finally, in Fig. 3 we show the dependence of θ on S for the cubic case, together with the relation given by $\theta \propto \sqrt{S}$. It can be seen that θ increases almost as $\sqrt{S}$ and becomes 0 when S becomes 0. Values for θ do not exist when S is negative.*

5. Ionic Current given by Higher-degree Polynomials

By including a sufficient number of powers of V, a polynomial expression can be made to fit any arbitrary current-voltage trajectory. It is therefore of great importance to the range of application to show that solutions may also be obtained for higher-degree polynomials. Thus, a quintic expression gives us two extra degrees of freedom (in the form of two additional, and arbitrary, roots) to fit the relation required. A seventh-degree polynomial gives four additional roots. Unfortunately, it can be shown that the solution for the trajectory $\dot{V}(V)$ is not generally obtainable as a simple polynomial when all

---

* This result is strictly true for a cable at rest with $V = 0$. However, propagation from a "resting" state at V_p towards V_0 is possible when S is negative. Such waves correspond to propagated repolarization waves of the kind that occur in cardiac muscle (Weidmann, 1951).

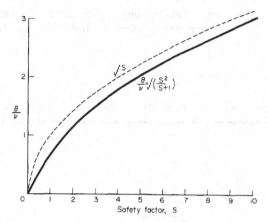

FIG. 3. Relation between conduction velocity and the safety factor. The interrupted curve shows $\sqrt{S}$. The continuous curve shows θ.

the additional roots are arbitrary (see Appendix), but $\dot{V}(V)$ may be obtained as a cubic when $I_i(V)$ is a quintic if only one additional root is arbitrary. The other root is determined by the requirement that a simple polynomial solution should exist for $\dot{V}(V)$. Similarly $\dot{V}(V)$ may be obtained as a quartic when $I_i(V)$ is seventh-degree provided that only two of the additional four roots are arbitrary.

Thus, only a limited class of higher-degree polynomials will give simple analytical solutions for $\dot{V}(V)$ and, hence, for θ and for the "action potential" shape. Nevertheless, the range of application is greatly increased since, even with these restrictions on the additional roots, we may select polynomial expressions that give a much better fit to observed ionic current-voltage trajectories than can be obtained using cubic equations.

In particular, we may use higher-degree polynomials to give larger values of S (i.e. larger relative areas of inward current in the $I_i(V)$ relation) without moving the threshold very close to the resting potential (which is the only way of increasing S using cubic models). The ratio of active to resting membrane conductances may then be made more realistic while retaining realistic values of the threshold, V_{th}.

All these approaches suffer from the same disadvantage from an experimental point of view—the actual current-voltage trajectory is rarely known, even after a full voltage-clamp analysis. It can be estimated very roughly from a voltage-clamp analysis as the current flowing at times similar to the time taken for the potential to change to the chosen clamp step, during the upstroke of the action potential. Alternatively, one point on the current-voltage relation can be deduced, as derived in eq. (1.6), by measuring the maximum rate of rise. With a reasonable guess as to the shape of the rest of the trajectory, based on the assumption that the sodium current has approximately similar characteristics in excitable cells other than the squid axon, estimates of the conduction velocity can be obtained.

This approach is similar to that adopted by Noble (1972) in obtaining equations for the liminal length for excitation. The current-voltage relation is selected fairly arbitrarily as one that applies momentarily at times similar to the time at which peak conductance occurs during strong depolarizations. The results obtained can only be approximate but they may be of use when working on excitable cells in which it is impossible to obtain a full analysis of ionic current kinetics but in which voltage clamp current-voltage relations (or even only estimates of maximum g_{Na}) are obtainable. The analysis will then be useful primarily in setting limits on the conduction velocity that may be attained and in determining whether the voltage clamp results are consistent with the excitation properties of the tissue. We shall illustrate this approximate approach by using a cubic expression that resembles the current-voltage relation given by Hodgkin *et al.* (1952, Fig. 10) for squid nerve, which is the relation used by Noble (1972) to obtain approximate estimates of the liminal length for excitation. We shall also compare the results with that for a quintic model used to reproduce Purkinje fibre conduction.

A more accurate approach will be illustrated in section IV by using non-polynomial expressions to fit the actual variation of I_i with V during a propagated action potential in the Hodgkin–Huxley model. We will show that an extremely good fit can then be obtained. The results of this approach are of interest for two reasons. First we may show that the polynomial and Hodgkin–Huxley models do in fact give the same values of θ when the same net ionic currents are reproduced. This result is of course necessary if the cable equations are applicable during action potential propagation. The second, and more important, result is that we may use such models to obtain useful working equations relating the conduction velocity to the maximum rate of rise of the propagated action potential. Finally, we may choose to reproduce the maximum current that may be activated and explicitly include time dependence of the ionic current by reproducing the activation process, m. This approach will be illustrated in sections II and III.

6. Approximate Models for Squid Nerve and Purkinje Fibres

a. Equations for a quintic model

In most excitable cells the ratio of peak inward to peak outward current is much larger than in the cubic relation plotted in Fig. 2. A more satisfactory expression for some cases is obtained by the simple device of squaring the terms corresponding to non-zero roots in eq. (1.1) to give the quintic expression:

$$I_i = gV\left(1 - \left(\frac{V}{V_{th}}\right)^2\right)\left(1 - \left(\frac{V}{V_p}\right)^2\right). \tag{1.18}$$

This is equivalent to adding two additional roots at the points $V = -V_{th}$ and $V = -V_p$. The resulting expression using the same values of V_{th} and V_p as in Fig. 2 is plotted in Fig. 4 together with the cubic expression scaled by a factor of 9 (to produce similar maximum inward currents).

It should be noted first that eq. (1.18) does satisfy the restriction on additional roots (see Appendix) required for the trajectory $\dot{V}(V)$ to be given by a cubic expression. By a derivation analogous to that given previously we may show that

$$\dot{V} = \frac{gS}{C_m}V\left(1 - \left(\frac{V}{V_p}\right)^2\right), \tag{1.19}$$

where

$$S = \left(\left(\frac{\bar{V}}{V_{th}}\right)^2 - 1\right). \tag{1.20}$$

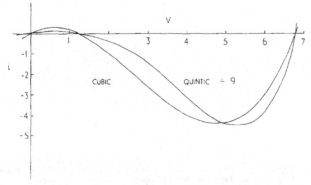

FIG. 4. Comparison between cubic and quintic polynomial representations of ionic current. The quintic relation has been divided by 9 to give nearly equal peak negative currents for the two curves.

In this case $\bar{V}$, the maximum upstroke rate occurs at the point $V_p/\sqrt{3}$ instead of $V_p/2$ (i.e. the trajectory is no longer symmetric). Hence (1.20) may also be written

$$S = \left(\frac{1}{3}\left(\frac{V_p}{V_{th}}\right)^2 - 1\right) = \left(\frac{V_p}{\sqrt{(3)}V_{th}} - 1\right)\left(\frac{V_p}{\sqrt{(3)}V_{th}} + 1\right). \qquad (1.21)$$

The first factor in this equation is slightly larger than $[(V_p/2V_{th}) - 1]$, i.e. the safety factor for the corresponding cubic case, and since the second factor must be larger than 1 the safety factor for the quintic model given by using eq. (1.18) must always be larger than that for the corresponding cubic equation (1.1). Apart from this change in the definition of the safety factor, it may be shown that the same equations for θ apply as for the cubic case. From (1.21) with $V_p = (4 + \sqrt{8})$ and $V_{th} = 8/(4 + \sqrt{8})$ we obtain a safety factor of 10.25 compared to the value of 1.914 for the cubic case with the same positive roots. The conduction velocity is given by

$$\theta = \sqrt{\left(\frac{S^2}{S+1}\right)}v = \sqrt{\left(\frac{10.25^2}{11.25}\right)}v = 3.056\,v,$$

compared to $1.121v$ for the cubic case. Thus a roughly threefold increase in θ is obtained by increasing the amplitude of the maximum inward current by a factor of about 9. This result suggests that the conduction velocity is roughly proportional to the square root of the peak magnitude of the inward current even when changes in shape are involved. The relationship is not exact since the actual ratio of θ^2 for the quintic and cubic models is 7.43. This is the amount by which the cubic expression must be scaled to give exactly the same conduction velocity as the quintic case. The value of θ is in fact more closely related to the total area of inward current under the current-voltage curve, as we shall show later.

b. *Applications to squid nerve and Purkinje fibres*

Figure 5 shows the current-voltage relations we shall use to compute approximate values of conduction velocity for squid nerve and for Purkinje fibres. The filled circles

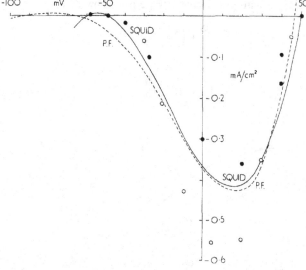

Fig. 5. Use of cubic and quintic polynomials to represent current-voltage relations in squid nerve and in cardiac Purkinje fibres. The filled circles show the current-voltage relation obtained experimentally by Hodgkin *et al.* (1952, Fig. 10). The open circles show the relation obtained by a different method (Hodgkin *et al.*, 1952, Fig. 13). The continuous curve is a cubic relation used for the squid case. The interrupted line is the quintic relation used to represent the Purkinje fibre case.

are taken from Hodgkin *et al.* (1952, Fig. 10) assuming a resting potential of -60 mV. This relation was obtained by measuring the rate of potential change at a fixed time (290 μsec) after displacements of the membrane potential by applied shocks. The ionic current was calculated from the equation

$$I_i = C_m \, dV/dt. \tag{1.22}$$

This method does not require the use of a voltage clamp and may therefore be used in circumstances that do not allow a voltage clamp analysis to be performed. The relation obtained is very similar in shape and magnitude to that obtained using the voltage clamp technique to measure ionic current at a fixed time (630 μsec) after the onset of voltage clamp steps (see Hodgkin *et al.*, 1952, Fig. 13; and the open circles in Fig. 5).

The continuous curve in Fig. 5 shows the relation given by the cubic expression

$$I_i = gV\left(1 - \frac{V}{11}\right)\left(1 - \frac{V}{110}\right) \tag{1.23}$$

with $1/g = 333$ ohms cm^2. This expression was chosen by fitting the roots (11 and 110 mV) to the threshold and peak potentials and then scaling the value of g to fit the peak inward current. V is the displacement of the potential from the resting potential.

The safety factor in this case is (see eq. (1.10))

$$S = \frac{110}{22} - 1 = 4.$$

If we use the cable constants employed by Hodgkin and Huxley (1952) for their calculation of the conduction velocity at 18.3°C, i.e. $a = 0.0238$ cm, $C_m = 1$ μF/cm^2, $R_i = 35.4 \, \Omega$ cm, we may calculate θ from (cf. eqs. (1.13) and (1.15) or (1.12)):

$$\theta = \frac{1}{C_m} \sqrt{\left(\frac{ag}{2R_i}\right)} \cdot \sqrt{\left(\frac{S^2}{S+1}\right)}.$$

which gives a value of 18 m/sec. This compares with an experimental value of the order of 20 m/sec.

It should be emphasized that this close degree of agreement is partly fortuitous. No account has been taken of the activation time and, over the temperature range 6 –30°C, the conduction velocity varies by a factor of 2 as a consequence of variations in the rate of activation (Huxley, 1959a). We shall attempt later to reproduce the effects of activation time in more complex models (see sections II and III). A calculation of this kind is significant only in giving an estimate of the order of magnitude of the conduction velocity expected.

We shall illustrate the use of this kind of calculation in applying the method to the cardiac Purkinje fibre. In this case, there is considerable doubt about the magnitude of the sodium conductance and about whether the voltage clamp technique succeeds in measuring it. The maximum rate of rise of the action potential is similar to that in squid nerve (Draper and Weidmann, 1951) and the specific membrane capacitance is similar (Weidmann, 1952; Mobley and Page, 1972; Hellam and Studt, 1974) when expressed with respect to true membrane area. This suggests that the sodium current density in those regions of membrane responsible for propagation is similar to that in squid nerve. This conclusion is reinforced by the fact that correct values for the liminal length for excitation may be obtained by making this assumption (Noble, 1972).

We may not, however, use eq. (1.23) for this case since the resting and threshold potentials are different from those in squid. The resting potential lies at -90 mV. The threshold potential lies near -65 mV which is considerably further away from the resting potential than in squid nerve. An appropriate relation for this case requires the use of the quintic expression given in eq. (1.18). This is plotted in Fig. 5 as the interrupted

The Selected Papers of Denis Noble CBE FRS

line. The values for the constants in eq. (1.18) were chosen to give $V_{th} = 23$ mV ($E_{th} = -67$ mV) and $V_p = 137$ mV ($E_p = 47$ mV). These values are in fact the same as those used to produce the quintic relation plotted in Fig. 4 with the voltage scale set to 20 mV per unit. It can be seen that the two relations give similar curves for the net inward current. The value of R was chosen as 2 kohms/cm^2 to give a peak inward current similar to that in squid membrane. Using the values $a = 40\,\mu$, $R_i = 100\,\Omega$ cm (Weidmann, 1952), $C_m = 1\,\mu$F/cm^2, we obtain $\theta = 3$ m/sec, which is very close to the values obtained experimentally for Purkinje fibres (Weidmann and Draper, 1951; Draper and Mya-Tu, 1959).

Once again, this agreement must be treated with caution. It is significant only in giving the correct order of magnitude. The agreement is, however, significant enough to reinforce the view that the voltage clamp technique as applied to Purkinje fibres either does not record all the available sodium current or that the sodium current channels are concentrated at the surface membranes of a bundle so that little or no sodium current flows across the cell membranes deep in the preparation.

This point may be illustrated by calculating the conduction velocity that would be obtained if the sodium currents recorded experimentally by Dudel and Rüdel (1969) were distributed uniformly over the cell membranes. Dudel and Rüdel obtained a value of g_{Na} which is numerically similar to that for squid nerve when expressed with respect to fibre surface area. However, since this area accounts for only 10% of the total cell area (Mobley and Page, 1972) we must divide their value by 10 to obtain a value expressed with respect to true cell membrane area. This would reduce our estimate of θ obtained above by a factor of $\sqrt{10}$ to give a value less than 1 m/sec, which is certainly too slow. Similar conclusions about the magnitude of the sodium conductance have been obtained from studies of action potentials calculated from equations of the Hodgkin–Huxley type based on voltage clamp analyses of the ionic currents (McAllister *et al.*, 1975).

c. *Higher-degree polynomials*

The trajectory occurring during a propagated action potential given by the Hodgkin–Huxley squid equations (see Fig. 1) is in fact closer to a seventh-degree polynomial than to a cubic or quintic. It is of some interest to generalize our results to higher-degree polynomials. A useful seventh-degree expression may be obtained in a manner similar to that used to obtain the quintic expression used in the previous section, i.e. the terms corresponding to the non-zero roots are raised to a power, in this case 3, to give

$$I_i = gV\left(1 - \left(\frac{V}{V_{th}}\right)^3\right)\left(1 - \left(\frac{V}{V_p}\right)^3\right). \tag{1.24}$$

As in the case of the quintic expression (1.18), eq. (1.24) automatically satisfies the restrictions on the additional roots described in the Appendix. (This is true whenever higher-degree polynomials are obtained by raising the terms containing V_{th} and V_p to powers so that all non-zero roots are equal to $\pm V_{th}$ or $\pm V_p$).

Using methods similar to those employed previously (cf. 1.11, 1.20, 1.21), the safety factor for this case can be shown to be given by

$$S = \frac{V_p^3}{4V_{th}^3} - 1 = \left(\frac{\bar{V}}{V_{th}}\right)^3 - 1. \tag{1.25}$$

The peak value of $\dot{V}$ occurs at

$$\dot{V} = \frac{V_p}{4^{1/3}} = 0.63\,V_p, \tag{1.26}$$

and the trajectory is given by the quartic expression:

$$\dot{V} = \frac{Sg}{C_m}\,V\left(1 - \left(\frac{V}{V_p}\right)^3\right).$$

Comparison of eqs. (1.10), (1.21) and (1.25) suggests the form of the general equation for S:

$$S = \frac{2}{n+1}\left(\frac{V_p}{V_{th}}\right)^{(n-1)/2} - 1.$$ (1.27)

where n is the degree of the polynomial used for $I_i(V)$. This general formula is derived in section IV (eq. (4.17) with $\gamma = (n-1)/(2)$).

7. *Relation between θ and Rate of Rise*

One of the uses to which we may put the results obtained so far is in deriving expressions relating θ to the maximum rate of rise of the action potential. These formulae may be useful experimentally in that they allow θ to be estimated from a single recorded action potential and knowledge of the fibre size. Conversely, fibre size can be estimated from knowledge of θ, without recourse to empirical formulae.

Consider first the case where membrane current is best represented by a cubic. Letting $\hat{V}$ be the maximum rate of rise (i.e. $\dot{V}_{max}$), i.e. $\dot{V}_{max}$, occurring at voltage $\bar{V}$, we have

$$I_i(\bar{V}) = -C_m\hat{V}.$$ (1.6)

Since $\hat{V}$ occurs at $V_p/2$ this becomes

$$g(V_p/2)(1 - (V_p/2V_{th}))(1 - (V_p/2V_p)) = -C_m\hat{V}.$$ (1.28)

The value of g obtained can be substituted into (1.12) to yield

$$\theta^2 = \frac{a}{2C_mR_i}\left(\frac{S}{S+1}\right)(4\hat{V}/V_p).$$ (1.29)

Higher-degree polynomials are appropriate if the maximum rate of rise occurs more than halfway up the rising phase of the action potential. For a quintic polynomial $\hat{V}$ occurs at $V_p/\sqrt{3}$ and we obtain

$$\theta^2 = \frac{a}{2C_mR_i}\frac{S}{S+1}(3\sqrt{(3)}\hat{V}/2V_p).$$ (1.30)

while for a seventh-degree polynomial $\hat{V}$ occurs at $V_p/4^{1/3}$ and the velocity is given by

$$\theta^2 = \frac{a}{2C_mR_i}\frac{S}{S+1}(4\sqrt[3]{(4)}\hat{V}/3V_p).$$ (1.31)

Similar formulae hold for higher degree polynomials. For both the quintic and the seventh-degree models the assumption $S \approx S + 1$ is often reasonable, leading to the simpler formulae:

Quintic: $$\theta^2 = 3^{3/2}a\hat{V}/4C_mR_iV_p.$$ (1.32)

Seventh-degree: $$\theta^2 = 2^{5/3}a\hat{V}/3C_mR_iV_p.$$ (1.32)

We shall discuss these relations further in section IV, where we obtain relations for non-polynomial models. Note that the conduction velocity is not critically dependent on the degree of the polynomial chosen. The degree required may be estimated by noting how far up the rising phase the maximum rate of rise occurs.

8. *Dependence of Velocity on Shape and Area of $I_i(V)$ Relation*

At several stages in our treatment we have hinted at the fact that the conduction velocity is strongly dependent on the area of the $I_i(V)$ relation and much less so on its shape. We shall now demonstrate this result in a more general way. From eq. (1.5)

$$r \cdot \frac{d\dot{V}}{dV} = K\left(\dot{V} + \frac{I_i}{C_m}\right). \tag{1.34}$$

Integrating between 0 and V_p:

$$\int_0^{V_p} \dot{V} \frac{d\dot{V}}{dV} dV = K \int_0^{V_p} \dot{V} \, dV + K \int_0^{V_p} \frac{I_i}{C_m} dV \tag{1.35}$$

Since

$$\int_0^{V_p} \dot{V} \frac{d\dot{V}}{dV} dV = \int_0^{V_p} \dot{V} d\dot{V} = \left[\frac{\dot{V}^2}{2}\right]_0^{V_p} = 0$$

we obtain

$$\int_0^{V_p} \dot{V} dV = -\frac{1}{C_m} \int_0^{V_p} I_i \, dV. \tag{1.36}$$

We shall consider an arbitrary $I_i(V)$ relation. In general this may be approximated by the odd polynomial

$$I_i(V) = gV\left(1 - \frac{V}{V_{th}}\right)\left(1 - \frac{V}{V_p}\right)\left(1 - \frac{V}{A}\right)\left(1 - \frac{V}{B}\right) \cdots, \tag{1.37}$$

$$= gP(V), \tag{1.38}$$

where $A, B, \ldots$ are real or imaginary roots outside the range $0 < V < V_p$. If $\dot{V}(V)$ is to be a polynomial satisfying eqn. (1.5) it must be of the form

$$\dot{V}(V) = \frac{gS}{C_m} V\left(1 - \frac{V}{V_p}\right)\left(1 - \frac{V}{B}\right) \cdots \tag{1.39}$$

$$= \frac{gS}{C_m} P_1(V). \tag{1.40}$$

Substituting (1.38) and (1.40) into (1.36) and using eq. (1.10)—which has general validity—we obtain

$$K = \frac{-\dfrac{1}{C_m} \displaystyle\int_0^{V_p} I_i(V) \, dV}{\displaystyle\int_0^{V_p} P_1(V) \, dV \left[1 - \dfrac{\displaystyle\int_0^{V_p} P_1(V) \, dV}{\displaystyle\int_0^{V_p} P(V) \, dV}\right]}, \tag{1.41}$$

where $P(V)$ is the ionic current polynomial and $P_1(V)$ is the trajectory ($\dot{V}(V)$) polynomial.

We now make the following approximations:

1. Since for most axons $V_{th} \ll V_p$ we have

$$\int_0^{V_p} P(V) \, dV \gg \int_0^{V_p} P_1(V) \, dV.$$

Thus the expression in square brackets in (1.41) is approximately equal to 1.

2. For many shapes

$$\int_0^{V_p} P_1(V) \, dV \simeq \text{constant}.$$

With these two approximations we have

$$\theta \propto \sqrt{K} \propto \sqrt{\left(-\int_0^{V_p} I_i \, dV \right)}.\tag{1.42}$$

which gives us the result that θ is proportional to the square root of the area of the $I_i(V)$ relation.

The errors involved in making assumptions 1 and 2 are fairly small and they tend to cancel each other. Thus the error involved in comparing cubic and quintic expressions is only about 8%. The area of a quintic must be 8% greater than that of a cubic to give the same conduction velocity. Thus, there is a small dependence of θ on the shape of the current-voltage relation.

9. *A General Equation for the Safety Factor*

The arguments in the previous section can be extended to give a more general definition of the safety factor. Simply combining (1.36), (1.38) and (1.40) we find

$$S = -\frac{\int_0^{V_p} P(V) \, dV}{\int_0^{V_p} P_1(V) \, dV}.\tag{1.43}$$

so the safety factor is equal to the ratio of the areas under the current and trajectory polynomials. Equation (1.43) is in fact a perfectly general definition since any well-behaved function can be expanded as a polynomial. In general, therefore, the safety factor is equal to the numerical value of the ratio of areas under the current *vs.* voltage and first time derivative of voltage *vs.* voltage curves, when both have been scaled to have the same gradient at the origin.

Equation (1.43) may also be used to relate S to a more intuitive definition of the safety factor. We may rewrite (1.43) as

$$S = -\frac{\int_0^{V_p} I_{i\,\text{active}} \, dV - \int_0^{V_p} I_{i\,\text{resting}} \, dV}{g \int_0^{V_p} P_1(V) \, dV}.\tag{1.44}$$

Figure 6 shows how this definition of S may be interpreted. An arbitrary current-voltage relation has been split into two components: that due to the linear term which

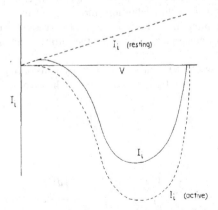

FIG. 6. Illustration of definition of safety factor in terms of "resting" (i.e. linear) and "active" (i.e. non-linear) components of total current-voltage relation.

determines the resting properties and that due to the non-linear terms, i.e. since $g = 1$,

$$I_{i\,\text{resting}} = V,$$

$$I_{i\,\text{active}} = p(V) - V.$$

The components of the area are

$$\int_0^{V_p} I_{i\,\text{resting}}\, dV = V_p^2/2, \tag{1.45}$$

$$\int_0^{V_p} I_{i\,\text{active}}\, dV = \int_0^{V_p} p(V)\, dV - V_p^2/2. \tag{1.46}$$

For the area due to $I_{i\,\text{active}}$ to exceed that for $I_{i\,\text{resting}}$,

$$\int_0^{V_p} I_{i\,\text{active}}\, dV \gg 2\,V_p^2/2 = 2\int_0^{V_p} I_{i\,\text{resting}}\, dV, \tag{1.47}$$

i.e. $\int_0^{V_p}$ of the active (non-linear) component must be larger than twice the integral of the resting (linear) component.

Using eq. (1.45) we may rewrite (1.44) as

$$S = -\frac{\displaystyle\int_0^{V_p} I_i(V)\, dV \Big/ \int_0^{V_p} I_{i\,\text{resting}}\, dV}{\dfrac{2}{V_p^2}\displaystyle\int_0^{V_p} P_1(V)\, dV}. \tag{1.48}$$

Since for many useful models we may use

$$P_1(V) = V\left(1 - \left(\frac{V}{V_p}\right)^\gamma\right) \qquad \text{(cf. eq. (4.12) below)}$$

$$\frac{2}{V_p^2}\int_0^{V_p} P_1(V)\, dV = \frac{2}{V_p^2}\left[\tfrac{1}{2}V_p^2 - \frac{1}{\gamma+1}V_p^2\right] = \frac{\gamma-1}{\gamma+1}.$$

Hence the denominator of eq. (1.48) is a constant and approaches 1 when γ is large. Equation (1.48) then approaches the form

$$S.F. = \int_0^{V_p} I_i(V)\, dV \Big/ \int_0^{V_p} I_{i\,\text{resting}}\, dV, \tag{1.49}$$

where the symbol $S.F.$ is now used for the safety factor to emphasize that it is not exactly equal to S. According to this definition, the safety factor is clearly a measure of the extent to which the integral of the ionic current exceeds the minimum required to ensure propagation. This corresponds closely to the intuitive definitions of the safety factor used by physiologists.

We may illustrate the use of these equations for S by considering an example using a cubic expression for I_i. If we set $V_{th} = 1$ and $V_p = 10$, then from eq. (1.10)

$$S = \frac{10}{2} - 1 = 4.$$

From eq. (1.43) we obtain

$$S = \frac{-\displaystyle\int_0^{10} V(1 - V)\left(1 - \frac{V}{10}\right) dV}{\displaystyle\int_0^{10} V\left(1 - \frac{V}{10}\right) dV},$$

and thus

$$S = \frac{66.7}{16.7} = 4.$$

which also gives the same value as eq. (1.10).

II. THE EFFECT OF ACTIVATION RATE ON CONDUCTION VELOCITY: EQUATIONS RESEMBLING THE HODGKIN-HUXLEY EQUATIONS

In the previous section expressions for conduction velocity were obtained by assuming that the membrane current-voltage trajectory during an action potential was known. We shall now apply a similar mathematical method to the calculation of conduction velocity under the less restrictive assumptions that only the active state current and a rate constant determining the fraction of that current activated during an action potential are known. In this case, realistic predictions of the effects on the velocity of variations in both the maximum available conductance (proportional to the density of activatable sodium channels) and in the activation rate constant will be possible.

1. Conduction Velocity with Exponential Activation

It is simplest to consider first an activation process governed by a first-order differential equation in time. Thus if m is a variable representing the fraction of the maximum possible inward current which has been activated at a given time, the future behaviour of m will be governed by

$$\frac{dm}{dt} = \alpha(m_\ell - m), \tag{2.1}$$

where α is the rate constant of activation and m_ℓ is the steady-state value of m at a given potential.

The ionic current, which we shall call I_i, is then the product of this activation variable and the maximum inward current $\bar{I}_i$:

$$I_i = m\bar{I}_i. \tag{2.2}$$

Note that I_i and m are functions of both potential and time, while $\bar{I}_i$ and m_ℓ are functions of potential only, as in Hodgkin–Huxley theory. The separation between "gating variable" m and "maximal current" $\bar{I}_i$ has been introduced to emphasize the similarity between our approach in this section and that of the Hodgkin–Huxley equations.

The total rate of change of I_i, when both potential and time are allowed to vary, is then, by the theory of partial differentials:

$$\frac{dI_i}{dt} = \frac{\partial I_i}{\partial t} + \frac{\partial I_i}{\partial V}\frac{dV}{dt}. \tag{2.3}$$

On substituting (2.2):

$$\frac{dI_i}{dt} = \bar{I}_i \frac{\partial m}{\partial t} + \left(\bar{I}_i \frac{\partial m}{\partial V} + m\frac{d\bar{I}_i}{dV}\right)\frac{dV}{dt}. \tag{2.4}$$

In Hodgkin–Huxley theory the gating variables are assumed not to change during an instantaneous variation in voltage (such as a voltage-clamp step). This is equivalent to stating that

$$\frac{\partial m}{\partial V} = 0.$$

or, equivalently, that

$$\frac{dm}{dt} = \frac{\partial m}{\partial t}.$$

With these restrictions we can substitute (2.1) in (2.4) to obtain

$$\frac{dI_i}{dt} = \bar{I}_i \alpha (m_\alpha - m) + m \frac{d\bar{I}_i}{dV} \frac{dV}{dt}. \qquad (2.5)$$

The way in which this ionic current change is composed of the sum of a change in time, with potential held constant, and a change occurring along the line passing through the inward current reversal when an instantaneous change in voltage is imposed, is illustrated in Fig. 7.

The variables m and m_α can be eliminated from (2.5) by introducing the active state current

$$I_{i_\alpha} = m_\alpha \bar{I}_i, \qquad (2.6)$$

and using eq. (2.2) to obtain

$$\frac{dI_i}{dt} = \alpha (I_{i_\infty} - I_i) + \frac{I_i}{\bar{I}_i} \frac{d\bar{I}_i}{dV} \frac{dV}{dt}. \qquad (2.7)$$

Equation (2.7) describes the way in which any current point $I_i(V)$ will move in the $I_i - V$ plane under imposed changes in potential and time. We now impose the additional restriction that the relation between I_i, V and t obeys the cable equation, namely:

$$\frac{1}{K} \frac{d^2V}{dt^2} = \frac{dV}{dt} + \frac{I_i}{C_m}, \qquad (1.4)$$

or using $dV/dt = \dot{V}$ this is written as

$$\dot{V} \left(\frac{1}{K} \frac{d\dot{V}}{dV} - 1 \right) = \frac{I_i}{C_m}. \qquad (1.5)$$

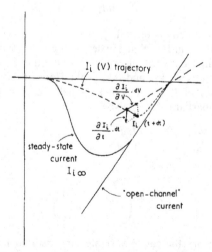

FIG. 7. The change in ionic current in a small time interval dt, expressed as the vector sum of the change $(\partial I_i/\partial t)dt = \bar{I}\alpha(m_\alpha - m)dt$ which would occur if the potential were held constant and time allowed to vary, and the change $(\partial I_i/\partial V)dV$ which would occur if potential were varied instantaneously (i.e. time held constant). Note that the change $(\partial I_i/\partial V)dV = m(d\bar{I}_i/dV)dV$ occurs along the current line passing through the inward current reversal potential (i.e. the gradient $d\bar{I}_i/dV$ of the open channel current, scaled by the fractional activation, m).

Eliminating I_i between the cable equation and eq. (2.7) and setting

$$d/dt = \dot{V}(d/dV)$$

yields

$$\frac{d}{dV}\left\{\dot{V}\left(\frac{1}{K}\frac{d\dot{V}}{dV}-1\right)\right\} = \frac{\alpha I_{i_{\star}}}{\dot{V}C_m} + \left(\frac{\dot{V}}{\bar{I}_i}\frac{d\bar{I}_i}{dV}-\alpha\right)\left(\frac{1}{K}\frac{d\dot{V}}{dV}-1\right). \tag{2.8}$$

The simplest solution to this unpalatable equation can be obtained by making the following assumptions:

(i) The "open-gate" inward current $\bar{I}_i$ is linear, as in the Hodgkin–Huxley description of the squid axon:

$$\bar{I}_i \propto -\left(1-\frac{V}{V_p}\right), \tag{2.9}$$

where V_p is the inward current reversal potential.

(ii) The active state inward current is a quartic polynomial of the form

$$I_{i_{\star}} = -gV^2\left(1-\frac{V}{V_p}\right)\left(1-\frac{V}{B}\right), \tag{2.10}$$

where B is a root to be determined and g is an overall scaling factor. Implicit in this assumption is the use of a cubic polynomial to represent m_γ (see eq. 2.6):

$$m_\gamma \propto V^2\left(1-\frac{V}{B}\right). \tag{2.11}$$

For the present treatment this approximation leads to the simplest expressions. The general form of the results obtained in this section is not altered by the more realistic approximations to m_γ considered in section IV. Note also that the quantity g has a different meaning in this part from that used in section I. While g still scales the overall size of the active state current $I_{i_{\star}}$, it is not to be identified with the gradient of I_i at the origin, which is of course zero.

(iii) The form of $\dot{V}$ is taken to be a parabola, as in section I:

$$\dot{V} = KV\left(1-\frac{V}{V_p}\right). \tag{2.12}$$

The use of K as the multiplicative constant in this equation is valid when the membrane current I_i is zero in the neighbourhood of the origin, as will be the case when polynomials with no threshold are used to represent $I_{i_{\star}}$. This result is obtained by taking the limit of eq. (1.5) as $V \to 0$, and has been described elsewhere (see e.g. Cole, 1968, p. 142). Models without thresholds have been used here to simplify the equations; incorporation of a threshold leads to complicated expressions with little change in the final results for axons like the squid axon which generate little outward current in the course of the rising phase of a normal action potential. Threshold models will be considered in section IV.

Substituting the three expressions (2.9), (2.10) and (2.12) into eq. (2.8) produces an identity in V, both sides being quadratic polynomials in V. Equating the coefficients of each power of V in turn leads to two equations in the two unknowns K and B:

$$\frac{4K^2}{V_p} + \frac{2\alpha K}{V_p} - \frac{\alpha g}{C_m} = 0 \tag{2.13}$$

(equating the coefficients of V).

Solving (2.13) for K, and discarding the negative root

$$K = \frac{\alpha}{4} \left\{ \sqrt{\left(1 + \frac{4gV_p}{C_m\alpha}\right)} - 1 \right\}.$$ (2.14)

Lastly the root B used in the definition of I_i, and m_∞ is determined by the polynomial method of solution:

$$B = \frac{g\,\alpha\,V_p^2}{4\,C_m K}.$$ (2.15)

2. *The Effects of Maximum Available Current and Rate of Activation on Velocity*

Equation (2.14) conveniently relates velocity to the two variables g (an index of the maximum conductance—proportional to the variable η used by Huxley, 1959a) and the rate of activation, α. For squid axon, a value of $g \sim 3.7$ amps/volt^2 cm^2 in eq. (2.10), together with $V_p = 114$ mV, produces a reasonable fit to the active state current, illustrated in Fig. 1, which is obtained from the Hodgkin–Huxley equations. The value of α is of the order of 10^3 sec^{-1}; taking $C_m = 1\,\mu$F/cm^2 the quantity $4gV_p/C_m\alpha$ has the value 1.7×10^3. An approximate form of (2.14) can therefore be written, dropping the 1's:

$$K \simeq \frac{\alpha}{4}\sqrt{\left(\frac{4gV_p}{C_m\alpha}\right)},$$

i.e.

$$K \simeq \sqrt{\left(\frac{g\alpha V_p}{4\,C_m}\right)}.$$ (2.14a)

And thus, by eq. (1.4)

$$\theta = \tfrac{1}{2}\,a^{1/2}\,R_i^{-1/2}\,C_m^{-3/4}\,\alpha^{1/4}\,g^{1/4}\,V_p^{1/4}.$$ (2.16)

This equation is similar to one used by Hodgkin (1975). The value of the fourth root B in (2.10) is of the order of 130 mV for the values cited above. B varies slightly with both g and α but its effect on eq. (2.16) is small for variations in α and g in physiological ranges appropriate to the squid axon. The effects in other cases would need to be investigated individually.

The actual values of K for the squid axon predicted by eq. (2.14a) should not be taken too seriously since the delayed activation observed in the squid axon and most other excitable cells is not well approximated by the exponential process assumed here. For this reason no attempt has been made to make accurate predictions of conduction velocity in the squid axon, although using the rough values quoted above the reasonable value of $\theta = 18.6$ m/sec is obtained.

a. *The time course of the rising phase*

Equation (2.11) describing the rate of rise of the action potential is identical in form to that used in section I (eq. (1.7)). The rising phase will therefore have the form derived in the cubic-trajectory model of section I, namely

$$V(t) = \frac{V_p}{1 + \exp[-gSt/C_m]}.$$ (1.17)

b. *The form of the current-voltage trajectory*

The current-voltage trajectory is obtained by substituting eq. (2.11) for $\dot{V}$ into the cable eq. (1.4). Thus

$$I_i(V) = -\frac{2KC_m}{V_p} V^2 \left(1 - \frac{V}{V_p}\right). \tag{2.17}$$

c. *Higher-degree polynomials*

The derivation outlined above can equally well be applied to higher-order polynomials. The next I_i, polynomial for which a solution is possible is

$$I_i(V) = -gV^3 \left(1 - \left(\frac{V}{V_p}\right)^2\right)\left(1 - \frac{V}{C}\right)\left(1 - \frac{V}{D}\right). \tag{2.18}$$

This form of polynomial is a much better approximation to the active state current-voltage diagram—and hence the m_i curve—of the squid axon. Solving for K, as above, we find

$$\theta = (\tfrac{1}{24})^{1/4} a^{1/2} R_i^{-1/2} C_m^{-3/4} \alpha^{1/4} g^{1/4} V_p^{1/2}.$$

Note that the dependence of θ on g and α is the same as in the quartic model, as indeed it is in all models employing an exponential activation process.

In this seventh degree model a value of $g = 20$ amps/volts³ cm² gives a reasonable fit to the squid current-voltage diagram; the value of $\theta = 14.9$ m/sec at 6.3°C is obtained using this value of g and the other values noted above. This result is closer to the true value of 12.9 m/sec than that predicted by the simple quartic approximation.

3. *Deficiencies of the Analytical Method*

From the point of view of applying the equations as developed so far to the squid axon, the chief drawback is the need to approximate the delayed activation by an exponential. The formulae are appropriate only for excitable cells showing exponential activation. There are a number of lesser approximations which should also be discussed:

(i) The activation rate constant has been assumed to have a constant value throughout the rising phase of the action potential, while the true Hodgkin–Huxley rate constant has a U-shaped dependence on potential. This should not lead to major inaccuracies, as the Hodgkin–Huxley rate constant dependence on voltage is roughly flat over a significant portion of the action potential rising phase (see Fig. 8). The dependence of velocity on the shape of the $\alpha(V)$ relation will be investigated in a later section.

(ii) The active state current-voltage relation, $I_i(V)$, does not generate any outward current and hence does not have an excitation threshold. While this has little effect in the squid axon under normal conditions, in a near-failure situation outward current becomes comparable in magnitude to the inward current. Accurate predictions in this case require the use of threshold models, which can readily be developed using the methods described above.

(iii) More serious is the need to incorporate an arbitrary root B into the polynomial describing the active state current-voltage relation. For accurate predictions of conduction velocity with, say, variable maximal conductance g, the effect of variations in B on the shape of the current-voltage diagram need to be considered. The best way to allow for these effects will be considered in the next section on delayed activation, when comparisons with numerical computations performed for the squid axon are possible.

(iv) The parameter g cannot be identified with the actual maximal inward conductance (the sodium conductance of 120 mmho/cm² in squid axon, for instance) although its use as a scaling factor to alter the overall size of the inward current is valid. The

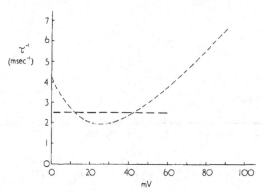

FIG. 8. The dependence of the Hodgkin–Huxley activation rate constant, τ_m^{-1}, on potential at 6.3°C. Although the deviation from a constant is considerable, the constant approximation shown in this diagram is reasonable over most of the rising phase of the action potential. A model in which the voltage dependence of the activation rate is accurately reproduced will be described in section III (see Figs. 14 and 15).

value of g must be chosen to give the best fit to the active state current-voltage relation I_{i_x} particularly (as will be discussed later) over the initial sections of the true I_{i_x} curve.

4. *Conduction Velocity with Delayed Activation*

The mathematical derivation of conduction velocity with exponential activation has been extensively described in previous sections partly because the approach developed there can be readily adapted to describe the delayed activation which is commonly observed experimentally. We have previously defined a linear open-channel current

$$\bar{I}_i \propto -(1 - V/V_p), \tag{2.9}$$

and an activation process governed by a first-order activation variable in m:

$$\frac{dm}{dt} = \alpha (m_x - m). \tag{2.1}$$

These definitions will be retained in the present approach, but the ionic current I_i will now be represented, as in Hodgkin–Huxley formalism, as the product of $\bar{I}_i$ and m raised to the power 3:

$$I_i = m^3 \bar{I}_i. \tag{2.19}$$

The active state current is therefore

$$I_{i_x} = m_x^3 I_i. \tag{2.20}$$

Thus as before

$$\frac{dI_i}{dt} = \frac{\partial I_i}{\partial t} + \frac{\partial I_i}{\partial V} \frac{dV}{dt}, \tag{2.3}$$

and on substituting (2.19) and (2.20):

$$\frac{dI_i}{dt} = \bar{I}_i \, 3m^2 \frac{dm}{dt} + \dot{V} \left(\bar{I}_i 3m^2 \frac{\partial m}{\partial V} + m^3 \frac{d\bar{I}}{dV} \right).$$

On setting $\dfrac{\partial m}{\partial V} = 0$, as before, and substituting (2.1):

$$\frac{dI_i}{dt} = 3m^2 \bar{I}_i \alpha (m_x - m) + m^3 \frac{d\bar{I}_i}{dV} \dot{V}$$

$$= (3\alpha)(I_i^{2/3} I_{i_x}^{1/3}) - (3\alpha) I_i + \frac{I_i}{\bar{I}_i} \frac{d\bar{I}_i}{dV} \dot{V}. \tag{2.21}$$

This equation is formally identical to eq. (2.7) for the case of exponential activation:

$$\frac{dI_i}{dt} = \alpha_{(exp)} I_{i_\alpha \, (exp)} - \alpha_{exp} I_i + \frac{I_i}{\bar{I}_i} \frac{d\bar{I}_i}{dV} \dot{V}, \tag{2.7}$$

where the variables appropriate to exponential activation have been subscripted (exp) to distinguish them from the delayed-activation variables.

Comparing these two equations, we see that delayed activation will produce exactly the same $I_i(V)$ trajectory—and therefore, by the results of section I, the same conduction velocity—if the exponential and delayed rate constants and active state currents are related by

$$\alpha_{delayed} = \tfrac{1}{3}\alpha_{exponential} \tag{2.22}$$

and

$$I_{i_\alpha \, delayed} = \frac{(I_{i_\alpha \, exponential})^3}{(I_i)^2}. \tag{2.23}$$

5. Comparison with the Exponentially-activated Model

In the section on exponential activation we assumed that the current-voltage trajectory was well approximated by a cubic

$$I_{i_i}(V) \simeq -\frac{2K \, C_m}{V_p} V^2 \left(1 - \frac{V}{V_p}\right), \tag{2.17}$$

and showed that under this assumption the active state current-voltage relation is a quartic:

$$I_{i_\alpha \, exp} = -gV^2 \left(1 - \frac{V}{V_p}\right)\left(1 - \frac{V}{B}\right). \tag{2.10}$$

Using the same equation (2.17) to represent the current-voltage trajectory in the delay-activated model equation (2.23) predicts that the active state current-voltage relation will be

$$I_{i_\alpha \, delayed} = -f \, V^2 \left(1 - \frac{V}{V_p}\right)\left(1 - \frac{V}{B}\right)^3, \tag{2.24}$$

where

$$f = \frac{V_p^2 g^3}{4K^2 C_m^2} \simeq \frac{V_p g^2}{\alpha C_m}, \text{ using eq. (2.14a).} \tag{2.25}$$

6. The Effect of Maximum Conductance and Activation Rate in the Delay-activated Model

One major use of eqs. (2.24) and (2.25) is to predict in a general way the effects of both maximum available conductance and activation rate constant on the conduction velocity.

In eq. (2.16) above, it was demonstrated that for exponential activation

$$\theta \propto g^{1/4},$$

where g is proportional to the maximum inward conductance.

Thus, since by eq. (2.25) $f \propto g^2$, in the delay-activated model we obtain for the dependence of conduction velocity on the inward conductance scaling factor f:

$$\theta \propto f^{1/8}. \tag{2.26}$$

Thus if a blocking experiment, with say TTX, succeeded in blocking half the sodium channels, the conduction velocity, θ, would increase to $(\tfrac{1}{2})^{1/8} \theta$, or to a value of 0.917 θ. Thus unless inactivation and recovery have a much greater effect on conduction velocity with half the inward conductance available than they do with full conductance, velocity will not be a very sensitive indicator of fractional blockage of channels. In squid axon inactivation and recovery do indeed have a marked effect when large fractions of channels are blocked but this is not necessarily true for other systems. The main point of this calculation is to show that removal of sodium channels *per se* has little effect on conduction velocity when the channel density is sufficiently large, and the major effects will be due to more rapid relative rise of inactivation and recovery processes when the rise-time has been slowed by channel blockage. In order to use conduction velocity as an index of channel blockage these inactivation and recovery processes need to be carefully quantified for the system being studied.

The dependence of conduction velocity on activation rate constant α can similarly be derived using eqs. (2.16) and (2.25):

$$\theta \propto (\alpha)^{1/4} \cdot (\alpha)^{-1/8} = (\alpha)^{3/8}. \tag{2.27}$$

Combining this with (2.26) we can compare the dependence of conduction velocity on all factors for a delay-activated system, with that exhibited in eq. (2.16) for exponential activation:

$$\theta \propto a^{1/2} R_i^{-1/2} C_m^{-5/8} \alpha^{3/8} f^{1/8} V_p^{1/8}. \tag{2.28}$$

7. The Effect of Temperature on Velocity

It is convenient to express the dependence of conduction velocity on activation rate constant, as in eq. (2.27), in a way which relates it to the more familiar notation of the Q_{10} of both θ and of α, since the most common way in which α will be altered will be by changing the temperature. Using the standard Arrhenius expression for the dependence of α on temperature

$$\alpha_T = A \exp\left(\frac{-E_A}{RT}\right),$$

where E_A is the activation energy, R and T have their usual significance and A is a constant, we obtain the approximate expression, valid for small ΔT:

$$\log_{10} \frac{\alpha_{T+\Delta T}}{\alpha_T} = \frac{\log_{10} Q_{10,\alpha}}{10} \Delta T. \tag{2.29}$$

Taking $\theta \propto \alpha^{3/8}$ we obtain

$$\log_{10} \frac{\theta_{T+\Delta T}}{\theta_T} = \frac{\log_{10} (Q_{10,\alpha})^{3/8}}{10} \Delta T.$$

Thus the relation between the Q_{10}'s of velocity and activation is:

$$Q_{10,\theta} = (Q_{10,\alpha})^{3/8}. \tag{2.30}$$

Thus the $Q_{10,\alpha}$ of 3 assumed by Hodgkin and Huxley (1952) gives a $Q_{10,\theta}$ of 1.51.

The approach described here also leads to simple formulae connecting the Q_{10} of θ with the Q_{10}'s of both channel conductance and activation rate. Using the relation (from (2.28)):

$$\theta \propto \alpha^{3/8} f^{1/8} R_i^{-1/2}$$

it can readily be shown that

$$\frac{d\theta}{\theta} = \frac{3}{8} \frac{d\alpha}{\alpha} + \frac{1}{8} \frac{df}{f} + \frac{1}{2} \frac{d(1/R_i)}{(1/R_i)}, \tag{2.31}$$

and hence that

$$Q_{10,\theta} = (Q_{10,\alpha})^{3/8} (Q_{10,f})^{1/8} (Q_{10,1/R_i})^{1/2}. \tag{2.32}$$

The success of this relation can be compared with the experimental values obtained by Chapman (1967). In squid axon he found a $Q_{10,0}$ of 1.70 between 10° and 20°, while using $Q_{10,x} = 3$, $Q_{10,f} = 1.4$ and $Q_{10,1/R_i} = 1.3$ we obtain $Q_{10,0} = 1.80$. This must, of course, be accepted with the reservation that it does not include inactivation and recovery processes. The success of this prediction, however, is a further demonstration that the major determinant of conduction velocity in the normal operating range of the squid axon is the rate of activation.

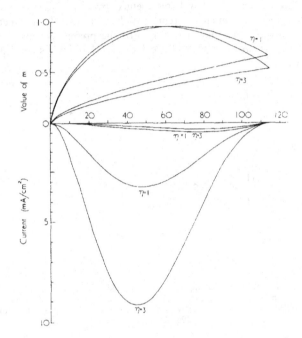

FIG. 9. Parameters of propagated action potentials at 6.3° C as used in the polynomial model. Two cases have been considered: one as close as possible to the normal squid axon parameters given by Hodgkin and Huxley (1952), labelled $\eta = 1$, and one with the maximum conductance increased by a factor of 3, labelled $\eta = 3$. *Upper panel*: Top lines show the values of m_x assumed to generate the steady-state current-voltage relations shown in the lower panel. Note that the use of a quartic expression for I_i has forced us to use m_x values which rise too steeply near the resting potential (compare the curves shown here with those shown in Fig. 1). The lower lines show the actual values of the m parameter as a function of voltage during an action potential, for the two cases described. *Lower panel*: The upper lines show actual current-voltage trajectories drawn according to eq. (2.17). The lower lines show the active state current-voltage relations drawn according to eq. (2.24).

8. *A Comparison of the Analytical Model with Some Computed Solutions*

The analytical models have been developed so far with a number of approximations. These have been introduced in the interests of bringing out general features of the equations. In this section we shall compare the approximate relations obtained above with the most accurate predictions of which the polynomial method is capable, and contrast these predictions with full computed solutions for the squid axon.

In Fig. 1 the steady-state current I_i and the current-voltage trajectory $I_i(V)$ for the squid axon were illustrated. It is interesting to see how closely eqs. (2.17) and (2.24) approximate them, and to compare the conduction velocities obtained using them.

Figure 9 shows the results of two computations—one for the normal case, and one with three times normal inward conductance. The "normal case" has been chosen so that the steady-state current I_i is as nearly like the I_i of Fig. 1 as is possible when using a sixth-degree polynomial of the type in eq. (2.24). The trajectory for this case is a cubic of the type in eq. (2.17), and using a value of $\alpha = 2.5 \times 10^3$ per sec, which is appropriate, by Fig. 8, to a temperature of 6.3°C, the trajectory has the form illustrated in Fig. 9. The velocity predicted by this trajectory is 17.8 m/sec.

The velocity in a Hodgkin–Huxley axon at 6.3°C is 12.2 m/sec (Huxley, 1957). The velocity computed by our method is evidently much too large. The inaccuracy is due to the attempt to represent the steady-state current by a sixth-degree polynomial, and hence the trajectory by a cubic. As discussed in section I, cubics do not provide a good fit to actual trajectories, and comparing Fig. 9 with Fig. 1, it is evident that the actual steady-state current is not well represented by a sixth-degree polynomial. The use of higher-degree polynomials to represent the current-voltage functions will be considered in a more complex model shortly. For the present purposes, which are primarily comparative, the easiest models of a cubic for $I_i(V)$ and a sixth-degree polynomial for $I_{i_i}(V)$ will be used.

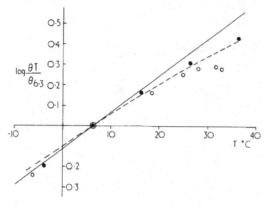

FIG. 10. A comparison of two analytical approaches, namely the approximate one of eq. (2.27) (solid line) and a full simulation (dashed line) with numerical values computed with no inactivation or recovery (filled circles), and with activation and recovery (open circles, after Huxley, 1959a).

a. *Comparison with temperature effects*

The relation

$$\theta \propto \alpha^{3/8} \tag{2.27}$$

has already been stated above. This involved two main approximations; first, some constants were neglected on deriving eq. (2.14b), and secondly the effects of the variations of the extra root B in the sixth-degree polynomial of eq. (2.26) were not considered. For Fig. 10 the first approximation has been eliminated, and the effects of the root B on the shape of the I_{i_i} relation has been allowed for as far as possible, by holding the maximum slope of the I_{i_i} relation constant as α, and hence B, varies. Some justification for this method will be found in section III.

In Fig. 10 the effects of temperature on velocity with this full method (dashed line) are compared with those predicted by eq. (2.27). Full numerical computations have also been performed using a programme devised for us by S. B. Barton, using the Hodgkin–Huxley equations (Huxley, 1959a) but with both inactivation and recovery eliminated, i.e. h and n set to their resting values and maintained there throughout the action potential. These values are plotted as the filled circles in Fig. 10, and may be compared with the full simulation performed by Huxley (1959a), replotted as open circles in Fig. 10.

Evidently the full analytical simulation provides an excellent approximation to the numerical computations with no inactivation or recovery for $T > 6.3°C$, but is poorer for $T < 6.3°C$. The approximation to the full simulation is also good for $T < 20°C$, but becomes less good at temperatures near to heat block, when inactivation and recovery are important.

b. *Comparison with Huxley's F(β) relation*

Huxley (1959a) has deduced a useful method of relating the effects of changes in maximal conductance, capacitance and rate constant of activation to one another via his $F(\beta)$ relation. He defines

$$\beta = \frac{\eta}{\gamma\phi},$$

where η scales the maximal conductance ($\eta = 1$ for the normal case)

  γ scales the capacitance ($\gamma = 1$ for the normal case)

  ϕ scales the rate constants of activation, inactivation and recovery ($\phi = 1$ at 6.3 C)

and sets

$$\frac{K}{\phi} = F(\beta),$$

the value of $F(\beta)$ being determined from numerical computations at a variety of temperatures (or of values of η or of γ).

For the purposes of illustration, Huxley's $F(\beta)$ relation is reproduced in Fig. 11, together with the values obtained in our computations with inactivation and recovery eliminated, and are compared with the approximate analytical model (see eq. 2.28) which predicts

$$\frac{K}{\phi} \propto \beta^{1/4}. \tag{2.28a}$$

Over the range of temperatures and maximal conductances close to 1 (i.e. in the normal physiological range of operation of the squid axon) the analytical method provides a good approximation, which becomes rather poorer either side of $\beta = 1$.

The full analytical simulation—scaled down to fit the activation-only computations at $\beta = 1$—fits these computations better than the approximate model for $\beta < 1$ (corresponding to $T > 6.3°C$, and to $\eta < 1$) but is surprisingly poorer than it is for $\beta > 1$.

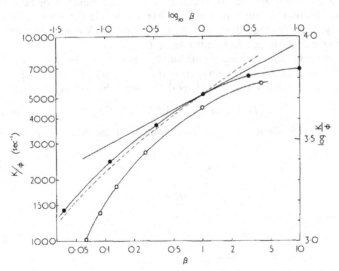

FIG. 11. A comparison of analytical and numerical methods of approximating Huxley's (1959a) $F(\beta)$ relation. O, The $F(\beta)$ relation obtained using the full Hodgkin–Huxley equations (replotted from Huxley, 1959a). ●, An $F(\beta)$ relation computed by eliminating sodium current inactivation and potassium current activation from the Hodgkin–Huxley equations. This is the relation the analytical method seeks to reproduce, as neither time-dependent sodium inactivation nor potassium activation are considered in the analytical models (computed by S. B. Barton). ––––, The approximate analytical relation $K/\phi \propto \beta^{1/4}$ (cf. eq. (2.28)). ———, A full analytical simulation of the $F(\beta)$ relation, scaled to equal the computed value at $\beta = 1$.

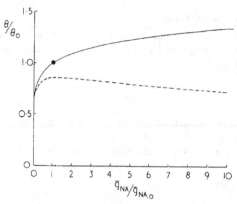

FIG. 12. The effects of gating current on impulse propagation velocity (vertical axis) as a function of sodium channel density in the squid axon. *Upper line*: The eighth-root dependence of θ on $\bar{g}_{Na}$ predicted by eq. (2.28). *Lower line*: The effect of adding gating current, as predicted by eq. (2.37).

The reason is to do with the shape changes in I_i, as β is altered to values greater than unity (see page 139).

c. *Influence of "gating current" on conduction velocity*

Recent reports (Armstrong and Bezanilla, 1973; Keynes and Rojas, 1974) have identified an asymmetrical membrane charge movement with the opening of the sodium channel. Hodgkin (1975) has used an analytical model similar to the exponentially-activated model described above to conclude that the channel density in the squid axon membrane achieves an optimum balance between the increase in velocity achieved by adding more channels and the decrease produced by adding the outward "gating current" to the membrane capacitance which must be discharged in the course of an action potential. We shall use an argument similar to Hodgkin's, but incorporating delayed activation.

We assume, as did Hodgkin, that the charge transfer as a function of potential is linear. This is not strictly true, but it is clear from Fig. 1 that it is not a bad approximation (the $m(V)$ trajectory is not too different from a straight line). We may then represent the effect of gating current by a simple capacitance C_g.

The value of this capacitance depends both on the number of channels present and on the degree of activation attained by m during the early part of an action potential. The trajectory for m in the delay activated model is given by (see eq. (2.19) *et seq.*):

$$m(V) = \sqrt[3]{[I_i/f(1 - V/B)]} = V^{2/3} \sqrt[3]{(2KC_m/f B)}, \qquad (2.33)$$

where f is the scaling factor for the density of ionic channels (i.e. f is proportional to $\bar{g}_{Na}$), i.e.

$$C_g \propto f m. \qquad (2.34)$$

Hence

$$C_g \propto f \sqrt[3]{(K/f)}. \qquad (2.35)$$

Now $K \propto \sqrt[4]{f}$, so that

$$C_g \propto f \cdot f^{-1/4} \propto f^{3/4}, \qquad (2.36)$$

or, in terms of $\bar{g}_{Na}$, the maximal sodium conductance:

$$C_g = C_{g0}(\bar{g}_{Na}/\bar{g}_{Na0})^{3/4}, $$

where C_{g0} and $\bar{g}_{Na0}$ are the values of C_g and $\bar{g}_{Na}$ in the normal axon. Thus, using eq. (2.28), the effect of a "gating capacitance" on the conduction velocity will be given by

$$\theta \propto [C_m + C_{g0}(\bar{g}_{Na}/\bar{g}_{Na0})^{3/4}]^{-5/8}(\bar{g}_{Na}/\bar{g}_{Na0})^{1/8}. \qquad (2.37)$$

Using the values of $C_m = 1 \mu F$ and $C_{q0} = 0.3 \mu F$ obtained for the squid axon (Keynes and Rojas, 1974) the effect on velocity is considerable. Figure 12 shows the eighth root dependence on increases in conductance predicted by the delay-activated model without gating capacitance (upper line) and compares it with the analytical prediction for a model incorporating gating current. The conclusion is similar to Hodgkin's (1975); the squid axon has the teleologically satisfying property of having adjusted its channel density to produce the optimum impulse velocity.

III. THE EFFECTS OF ACTIVATION RATE: MODELS USING MORE GENERAL RATE EQUATIONS

In section II we were concerned to construct models that allow closed form solutions to be obtained but which were based as closely as possible on the Hodgkin–Huxley equations. In particular, we adopted the same separation between gating and ion transfer variables and allowed the gating delay to be represented by an expression containing the third power of a first-order process. There are strong indications in the literature that the kinetics of the sodium conductance may differ from those proposed by Hodgkin and Huxley (see e.g. Goldman and Schauf, 1972). It is therefore important to determine the extent to which the results described in section II depend on the kinetic equations chosen. In this section we shall develop a model that uses a more general approach. Nevertheless, as in section II, we shall carefully compare the solutions obtained with actual results obtained in experimental work.

A convenient general form for the current-voltage trajectory is:

$$I_i = -K C_m (\gamma + 1) V \left(\frac{V}{V_p}\right)^\gamma \left(1 - \left(\frac{V}{V_p}\right)^\gamma\right), \tag{3.1}$$

where the cable equation is satisfied if

$$\dot{V} = KV \left(1 - \left(\frac{V}{V_p}\right)^\gamma\right), \tag{3.2}$$

or

$$V(t) = V_p (1 + ae^{-\gamma Kt})^{-1/\gamma}.$$

At small V this solution behaves as $V(t) = ae^{Kt}$ and K is therefore a rate constant governing the initial exponential rise of V. This suggests that an expression for the conduction velocity should be obtainable from a model able to reproduce the initial current-voltage trajectory accurately. If V is in the range where $(V/V_p)^\gamma \ll 1$, eq. (3.1) simplifies to

$$I_i = -K C_m (\gamma + 1) V \left(\frac{V}{V_p}\right)^\gamma. \tag{3.3}$$

Following the standard procedure let

$$I_i = g_{Na}(V - V_p). \tag{3.4}$$

where the conductance, g_{Na}, is both time and voltage dependent. Now, instead of following Hodgkin and Huxley kinetics in which g_{Na} is proportional to the third power of a Na-activation variable obeying a first-order equation, let g_{Na} obey a general linear nth order equation:

$$g_{Na} + \left(\frac{1}{\alpha_1} + \frac{1}{\alpha_2} + \cdots + \frac{1}{\alpha_n}\right)\frac{dg_{Na}}{dt} + \left(\frac{1}{\alpha_1\alpha_2} + \frac{1}{\alpha_1\alpha_3} + \cdots + \frac{1}{\alpha_1\alpha_n}\right)\frac{d^2 g_{Na}}{dt^2}$$
$$+ \cdots + \frac{1}{\alpha_1\alpha_2\ldots\alpha_n}\frac{d^n g_{Na}}{dt^n} = (g_{Na})_\gamma, \tag{3.5}$$

where the rate constants, $\alpha_1, \alpha_2 \ldots \alpha_n$ are voltage dependent.

From eqs. (3.4) and (3.1) (with $\gamma = 3$)

$$g_{Na} = -\frac{4K\,C_m}{(V - V_p)}V\left(\frac{V}{V_p}\right)^3\left(1 - \left(\frac{V}{V_p}\right)^3\right)$$

$$= 4\,C_m K\left(\frac{V}{V_p}\right)^4\left[1 + \frac{V}{V_p} + \left(\frac{V}{V_p}\right)^2\right].$$

The first time derivative is obtained from

$$\frac{d\,g_{Na}}{dt} = \frac{d\,g_{Na}}{dV}\frac{dV}{dt},$$

where dV/dt is given by eq. (3.2). For $(V/V_p)^3 \ll 1$,

$$\frac{dV}{dt} = KV,$$

and

$$\frac{dg_{Na}}{dt} = 4\,C_m K^2\left(\frac{V}{V_p}\right)^4\left(4 + 5\frac{V}{V_p} + 6\left(\frac{V}{V_p}\right)^2\right).$$

Similarly,

$$\frac{d^2\,g_{Na}}{dt^2} = \frac{d}{dV}\left(\frac{dg_{Na}}{dt}\right)\frac{dV}{dt} = 4\,C_m K^3\left(\frac{V}{V_p}\right)^4\left(4^2 + 5^2\frac{V}{V_p} + 6^2\left(\frac{V}{V_p}\right)^2\right),$$

and in general

$$\frac{d^n g_{Na}}{dt^n} = 4\,C_m K^{n+1}\left(\frac{V}{V_p}\right)^4\left(4^n + 5^n\frac{V}{V_p} + 6^n\left(\frac{V}{V_p}\right)^2\right).$$

Substituting g_{Na} and its derivatives into eq. (3.5) gives a $(n + 1)$th degree polynomial in K. If, at some voltage V, $(g_{Na})_r$ and all the rate constants are known, the polynomial may be solved for K. The actual order of equation required (i.e. the value n) could be chosen by comparing solutions of (3.5), under voltage clamp conditions, with the experimentally determined conductance changes following a sudden depolarization. An alternative is to examine the behaviour of K when $(g_{Na})_r$ is large. Assuming $K > 1$, the highest degree term will dominate as $(g_{Na})_r$ tends to infinity. Equation (3.5) then becomes:

$$\frac{1}{\alpha_1\alpha_2\ldots\alpha_n}4\,C_m\left(\frac{V}{V_p}\right)^4\left(4^n + 5^n\frac{V}{V_p} + 6^n\left(\frac{V}{V_p}\right)^2\right)K^{n+1} = (g_{Na})_r.$$

Hence K is proportional to the $(n + 1)$th root of $(g_{Na})_r$, or θ is proportional to the $2(n + 1)$th root of $(g_{Na})_r$.

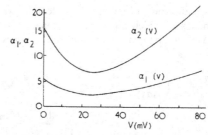

FIG. 13. The voltage dependence of the rate constants α_1 and α_2, chosen to fit eq. (3.9) to the equations used by Hodgkin and Huxley to describe their experimental conductance changes.

Since we have shown that conduction velocity varies approximately as the eighth root of (g_{Na}), when this is large (see section II), an appropriate value for n is 3, and eq. (3.5) becomes

$$g_{Na} + \left(\frac{1}{\alpha_1} + \frac{1}{\alpha_2} + \frac{1}{\alpha_3}\right)\frac{dg_{Na}}{dt} + \left(\frac{1}{\alpha_2\alpha_1} + \frac{1}{\alpha_2\alpha_3} + \frac{1}{\alpha_3\alpha_1}\right)\frac{d^2 g_{Na}}{dt^2} + \frac{1}{\alpha_1\alpha_2\alpha_3}\frac{d^3 g_{Na}}{dt^3} = (g_{Na})_r.$$

(3.6)

Substituting the various expressions for g_{Na} and its derivatives:

$$(g_{Na})_r = 4 C_m \left(\frac{V}{V_p}\right)^4 \left\{ K\left(1 + \frac{V}{V_p} + \left(\frac{V}{V_p}\right)^2\right) + K^2\left(\frac{1}{\alpha_1} + \frac{1}{\alpha_2} + \frac{1}{\alpha_3}\right)\left(4 + 5\frac{V}{V_p} + 6\left(\frac{V}{V_p}\right)^2\right) \right.$$

$$+ K^3\left(\frac{1}{\alpha_1\alpha_2} + \frac{1}{\alpha_2\alpha_3} + \frac{1}{\alpha_1\alpha_3}\right)\left(4^2 + 5^2\frac{V}{V_p} + 6^2\left(\frac{V}{V_p}\right)^2\right)$$

$$\left. + K^4 \frac{1}{\alpha_1\alpha_2\alpha_3}\left(4^3 + 5^3\frac{V}{V_p} + 6^3\left(\frac{V}{V_p}\right)^2\right)\right\}.$$

(3.7)

To test the ability of the model to predict K from knowledge of conductance changes following depolarizing voltage steps, the rate constants α_1, α_2, α_3 have been fitted to the equations used by Hodgkin and Huxley to describe their experimental conductance changes (Hodgkin and Huxley, 1952). These equations are:

$$\frac{g_{Na}}{(g_{Na})_r} = (1 - e^{-t/\tau_m})^3$$

and

where

$$(g_{Na})_r = \overline{g_{Na}} C_m m^3 h_0.$$

$$\alpha_m = 0.1(25 - V)/(e^{(25-V/10)-1}),$$

(3.8)

$$\beta_m = 4e^{-V/18},$$

$$\tau_m = 1/(\alpha_m + \beta_m)$$

and

$$m_\alpha = \alpha_m/(\alpha_m + \beta_m).$$

The Na-inactivation variable, h, is here set equal to its resting value $h_0 = 0.608$, and $\overline{g_{Na}}$ is 120 mmhos/cm^2.

The solution to eq. (3.6) following a voltage clamp step from zero to V mV is:

$$g_{Na} = (g_{Na})_r (1 + m_1 e^{-\alpha_1 t} + m_2 e^{-\alpha_2 t} + m_3 e^{-\alpha_3 t}).$$

Three initial conditions are needed and these are taken to be $(g_{Na})_{t=0} = 0$, $(d/dt)(g_{Na})_{t=0} = 0$ and $(d^2/dt^2)(g_{Na})_{t=0} = 0$. A further simplification proved possible when it was found that a good fit to the Hodgkin–Huxley equations could be obtained with $\alpha_3 = \alpha_1$; the solution at each level of depolarization is then

$$\frac{g_{Na}}{(g_{Na})_r} = 1 + \{\alpha_2(2\alpha_1 - \alpha_2 + (\alpha_1 - \alpha_2)\alpha_1 t)e^{-\alpha_1 t} - \alpha_1^2 e^{-\alpha_2 t}\}/(\alpha_1 - \alpha_2)^2.$$ (3.9)

The voltage dependencies of α_1 and α_2 are shown in Fig. 13 and the resulting solutions are shown for $V = 10$, 40 and 70 mV in Fig. 14 along with the Hodgkin–Huxley curves they are designed to reproduce.

The expressions for α_m and β_m given in (3.8) are appropriate to 6.3°C and as the conduction velocity was determined by Hodgkin and Huxley at 18.5°C they were scaled up with a Q_{10} of 3 (Hodgkin and Huxley, 1959, p. 519). A similar adjustment must be made to the values of α_1 and α_2 given in Fig. 13, that is, they are multiplied by $\phi = 3^{(T-6.3)/10}$ where $T = 18.5$.

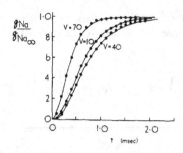

FIG. 14. Conductance changes following depolarizing steps to 10, 40 and 70 mV. The solid lines are obtained from $g_{Na}/(g_{Na})_\alpha = (1 - e^{-t + \tau^m})^3$, where τ^m is a voltage dependent time constant given by eq. (3.8). The circles (●) are obtained from eq. (3.9) with $\alpha_1(V)$ and $\alpha_2(V)$ given by Fig. 13.

Substituting the modified $\alpha_1(V)$ and $\alpha_2(V)$ into eq. (3.7) (with $\alpha_3 = \alpha_1$) and then calculating $I_\gamma(V)$ from $I_\gamma = (g_{Na})_\gamma(V - V_p)$, values of $I_\gamma(V)$ are plotted in Fig. 15 for $K = 9$, 10 and 11 msec^{-1}. Also shown is $I_\gamma(V)$ obtained from (3.8) and it is evident that a value of about $K = 10$ gives a reasonable fit to the Hodgkin-Huxley curve at small V. This results in a value of $\theta = 18.4$ m/sec and agrees well with the value of $\theta = 18.8$ m/sec obtained by numerical solution of the Hodgkin-Huxley equations. Specifying the value of $(g_{Na})_\alpha$ and solving eq. (3.7) for K at a number of voltages gives values for the conduction velocity of $\theta = 18.8$ m/sec ($V = 20$ mV), $\theta = 18.2$ m/sec (at $V = 30$ mV) and $\theta = 17.9$ m/sec (at $V = 40$ mV). The decreasing value of θ is to be expected as the error incurred by ignoring $(V/V_p)^3$ in comparison with 1 increases.

The value of 18.8 m/sec obtained for θ by fitting at $V = 20$ mV is fortuitous because the resulting I_γ curve does not exactly fit the Hodgkin-Huxley curve in this voltage range. However, the model can clearly give a value which differs from the Hodgkin-Huxley value by less than 5%.

Equation (3.7) holds for all V in the range where $(V/V_p)^3 \ll 1$. As V approaches zero, the equation may be simplified by neglecting V/V_p and $(V/V_p)^2$ in comparison with 1:

$$(g_{Na})_\alpha = 4 C_m \left(\frac{V}{V_p}\right)^4 K \left(1 + \frac{4K}{\alpha_1}\right)\left(1 + \frac{4K}{\alpha_2}\right)\left(1 + \frac{4K}{\alpha_3}\right).$$

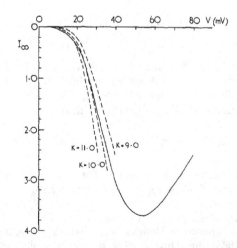

FIG. 15. Fully activated current-voltage diagram. $I_\alpha = (g_{Na})_\alpha(V - V_p)$. The solid line corresponds to the Hodgkin-Huxley description of $(g_{Na})_\alpha$ (see eq. (3.8)). The broken lines are obtained from eq. (3.7) for three values of K, with $\alpha_3 = \alpha_1$ and $\alpha_1(V)$ and $\alpha_2(V)$ given by Fig. 13.

If the rate coefficients α_1, α_2, α_3 are determined at 6.3°C, the factor ϕ_T should be included to indicate the temperature dependence:

$$(g_{Na})_r = 4 C_m \left(\frac{V}{V_p}\right)^4 K \left(1 + \frac{4K}{\phi_T \alpha_1}\right)\left(1 + \frac{4K}{\phi_T \alpha_2}\right)\left(1 + \frac{4K}{\phi_T \alpha_3}\right). \qquad (3.10)$$

The effect on conduction velocity of a change in membrane permeability is obtainable directly from this equation and is shown as the lower solid line in Fig. 16 (the permeability has been normalized to 1 at $\theta = 18.8$ m/sec).

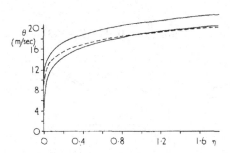

FIG. 16. The effect on conduction velocity of changes in relative membrane permeability η (expressed in terms of $(g_{Na})_r$). The lower solid line is obtained from eq. (3.10). When η is large a simplification of (3.10), given by (3.11), gives the upper solid line. Normalizing this relation to give $\theta = 18.8$ m/sec at $\eta = 1$ results in the dashed line. The similarity between the dashed line and the lower solid line (but not the upper solid line) indicates that eq. (3.11) is only applicable over the physiological range of η if it is used in a normalized sense.

When η is large a further simplification is possible and eq. (3.10) becomes:

$$(g_{Na})_r = \frac{C_m}{\phi_T^3 \alpha_1 \alpha_2 \alpha_3}\left(\frac{4KV}{V_p}\right)^4. \qquad (3.11)$$

This equation is used to give the upper line in Fig. 16. As η becomes large the two lines merge. If eq. (3.11) is also normalized such that $\eta = 1$ at $\theta = 18.8$ m/sec, the dashed line in Fig. 16 is obtained.

If a value for conduction velocity is required from given values of α_1, α_2 and $(g_{Na})_r$, eq. (3.7) should be used unless V/V_p is very small (below about 0.1) in which case eq. (3.10) will be sufficiently accurate. The similarity between the dashed line in Fig. 16 and the lower solid line suggests that the eighth root relationship between θ and $(g_{Na})_r$ is accurate when used in a normalized sense.

The effect on conduction velocity of uniform changes in the rate constants, with $(g_{Na})_r$ held constant, is also expressed in the above equations. For example, if $K_{6.3}$ refers to K at 6.3°C where $\phi_{6.3} = 1$, then (from eq. (3.11)):

$$\frac{C_m}{\phi_T^3 \alpha_1 \alpha_2 \alpha_3} \cdot \left(\frac{4K_T V}{V_p}\right)^4 = \frac{C_m}{\alpha_1 \alpha_2 \alpha_3}\left(\frac{4K_{6.3} V}{V_p}\right)^4,$$

or

$$\frac{K_T}{K_{6.3}} = \phi_T^{3/4},$$

or

$$\frac{\theta_T}{\theta_{6.3}} = \phi_T^{3/8}. \qquad (3.12)$$

The temperature dependence of conduction velocity now immediately follows by putting

$$\phi_T = 3^{(T-6.3/10)} \quad \text{(for a } Q_{10} \text{ of 3).}$$

Substituting for ϕ_T in eq. (3.12)

$$\frac{\theta}{\theta_{6.3}} = (3^{(T-6.3)/10})^{3/8}.$$

A linear relationship follows if $\log_{10} (\theta/\theta_{6.3})$ is used:

$$\log_{10} \frac{\theta}{\theta_{6.3}} = \frac{3}{80} \log_{10} 3 \, (T-6.3),$$

or

$$\log_{10} \frac{\theta}{\theta_{6.3}} = 0.0179 \, (T-6.3). \tag{3.13}$$

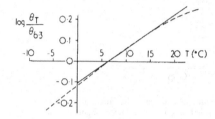

FIG. 17. The effect of temperature on conduction velocity. The dashed line shows the curve computed by Huxley (1959) (see Fig. 10) and the solid line is from eq. (3.13).

Equation (3.13) is compared with Huxley's computations (A. F. Huxley, 1959a) in Fig. 17. It can also be seen that these results compare well with those obtained in section II (eqs. (2.27)–(2.32)).

IV. NON-POLYNOMIAL MODELS

1. *Introduction*

In section I a number of expressions were derived relating conduction velocity to the safety factor, to the maximum rate of rise of voltage and other parameters. In each case the expression depended on the degree of polynomial chosen for $I_i(V)$. By including higher degree terms in the polynomial a better representation of the actual current-voltage trajectory is possible, and the resulting expressions have greater validity. However, they also become very cumbersome and the exact dependence on the shape of $I_i(V)$ is rather obscure. The same problem arises with the time-dependent case treated in section II.

In this section the expressions will be derived by assuming a simple non-polynomial form for $\dot{V}(V)$. Shape changes of the $I_i(V)$ and $I_r(V)$ relations are then equivalent to changes in a few non-integer parameters. It should be emphasized that, by choosing a form for $\dot{V}(V)$ and then deriving $I_i(V)$ from the cable equation, the approach here differs from that of sections I and II where the form of $I_i(V)$ was chosen and the corresponding $\dot{V}(V)$ derived. In the latter case equations relating the coefficients of $I_i(V)$ and $\dot{V}(V)$ gave the dependence of K on these coefficients. In this section the procedure adopted is to find K (and the other parameters) such that the expressions for $I_i(V)$ or $I_r(V)$ give the best fit to the experimental curves (or the corresponding Hodgkin–Huxley curves if the present model is to be compared with the solution of the Hodgkin–Huxley equations). When neither of these curves is known, the parameters must be fitted to any other available information, such as $\dot{V}_{max}$ (see later).

Two separate cases will be considered, one with a threshold effect and the other without. The derivation for the case without the threshold root is algebraically easier and will be dealt with first.

2. Non-threshold Case

A simple form of $\dot{V}(V)$ having an analytic solution for $V(t)$ and leading to an expression for $I_i(V)$ with approximately the right shape is:

$$\dot{V}(V) = KV\left(1 - \left(\frac{V}{V_p}\right)^\gamma\right), \tag{4.1}$$

where K is defined in section 1 and γ is a non-integer parameter. The peak voltage attained by the action potential, V_p, will be treated here as an arbitrary parameter rather than as a known constant. The solution to eq. (4.1), given that $V = 0$ at $t = -\chi$, is

$$V(t) = V_p(1 + ae^{-\gamma Kt})^{-1/\gamma}, \tag{4.2}$$

where a, the constant of integration, may be used to locate $t = 0$ at any given voltage. (For example, if $t = 0$ at $V = \frac{1}{2}V_p$, $a = 2^\gamma - 1$). From (1.5) the cable equation is

$$I_i = C_m\dot{V}\left(\frac{1}{K}\frac{d\dot{V}}{dV} - 1\right). \tag{4.3}$$

Substituting (4.1) into (4.3) gives the following non-threshold form for $I_i(V)$:

$$I_i(V) = -KC_m(\gamma + 1)V\left(\frac{V}{V_p}\right)^\gamma\left(1 - \left(\frac{V}{V_p}\right)^\gamma\right). \tag{4.4}$$

As an example of the ability of eq. (4.4) to fit a prescribed current-voltage trajectory, consider the trajectory shown in Fig. 1, found by numerical solution of the Hodgkin–Huxley equations. Apart from the membrane capacitance ($C_m = 1\,\mu F$) all the parameters in eq. (4.4) are undefined. The best fit of eq. (4.4) to this trajectory, according to a "least squares" criterion, is shown as the series of points (●) in Fig. 18 (the prescribed trajectory is reproduced as the solid line in Fig. 18). Optimal parameter values, found by the least-squares fitting program, are $\gamma = 2.37$, $V_p = 95.7\,mV$ and $K = 10.83$ per sec. This value of K corresponds to a conduction velocity of 19.1 m/sec, comparing well with the value of 18.8 m/sec numerically computed from the Hodgkin–Huxley equations.

This example proves that the chosen form of $\dot{V}(V)$ (eq. (4.1)) leads to a satisfactory form for $I_i(V)$ (eq. (4.4)). It also demonstrates the validity of considering only the rising phase of the action potential in determining the conduction velocity. In practice $I_i(V)$ may not be known experimentally and the three parameters must then be found by

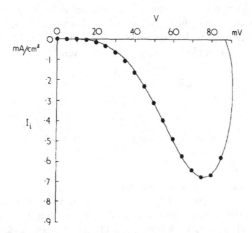

FIG. 18. The current-voltage trajectory given by eq. (4.4) (circles, ●) fitted to the trajectory found by numerical solution of the Hodgkin–Huxley equations (solid line). The optimal parameter values obtained by the fitting procedure are $\gamma = 2.37$, $V_p = 95.7\,mV$ and $K = 10.83$ m/sec.

other means, such as fitting the leading edge of the recorded action potential with eq. (4.2) or fitting the steady-state current-voltage relation found from a voltage clamp analysis (see below).

With the ionic current trajectory defined by eq. (4.4) the derivation of the steady-state current-voltage relation is straightforward. This may be done for either the Hodgkin–Huxley kinetics used in section II or for the third-order kinetics used in section III. The Hodgkin–Huxley approach will be followed here, partly because the algebra is much easier in this case. Thus, as in section II

$$I_i = m^3 \bar{g}_{Na}(V - V_p),$$ (4.5)

where the gating variable m obeys a first-order rate equation with a voltage dependent rate constant $\alpha(V)$:

$$m_x = m + \frac{1}{\alpha}\frac{dm}{dt}.$$ (4.6)

From eqs. (4.4) and (4.5) the $m(V)$ trajectory is given by

$$m(V) = \left(\frac{KC_m(\gamma + 1)}{\bar{g}_{Na}}\right)^{1/3}\left(\frac{V}{V_p}\right)^{(\gamma+1)/3}\left(1 - \frac{V}{V_p}\right)^{-1/3}\left(1 - \left(\frac{V}{V_p}\right)^{\gamma}\right)^{1/3}.$$ (4.7)

Differentiating (4.7) with respect to V and resubstituting m,

$$\frac{dm}{dV}(V) = \frac{m}{3V_p}\left[\overline{\gamma + 1}\left(\frac{V}{V_p}\right)^{-1} + \left(1 - \frac{V}{V_p}\right)^{-1} - \gamma\left(\frac{V}{V_p}\right)^{\gamma-1}\left(1 - \left(\frac{V}{V_p}\right)^{\gamma}\right)^{-1}\right].$$ (4.8)

dm/dt in eq. (4.6) may now be replaced by $(dm/dV)\dot{V}$ and the expressions for dm/dV and $\dot{V}$ substituted to give

$$m_x(V) = m\left\{1 + \frac{K}{3\alpha}\left[\frac{1 - (V/V_p)^{\gamma}}{1 - (V/V_p)} + \gamma\left(1 - 2\left(\frac{V}{V_p}\right)^{\gamma}\right)\right]\right\}.$$ (4.9)

Finally, $I_x(V)$ is recovered by substituting (4.9) into (4.5):

$$I_x(V) = I_i(V)\left\{1 + \frac{K}{3\alpha}\left[\frac{1 - (V/V_p)^{\gamma}}{1 - (V/V_p)} + \gamma\left(1 - 2\left(\frac{V}{V_p}\right)^{\gamma}\right)\right]\right\}^3,$$ (4.10)

where $I_i(V)$ is given by eq. (4.4). (Note that there are no restrictions on the form of $\alpha(V)$).

If $I_x(V)$ and $\alpha(V)$ are known from voltage clamp studies a least-squares fitting procedure will give the optimal values for γ, V_p and K. (Alternatively the values can be found by choosing three points on the experimental I_x curve.)

For example, consider the Hodgkin–Huxley I_x curve given by

$$I_{\infty} = \bar{g}_{Na} C_m h_0 \left(\frac{\alpha_m}{\alpha_m + \beta_m}\right)^3 (V - 115),$$ (4.11)

where

$$\alpha_m = 0.1(25 - V)/(e^{(25-V/10)} - 1) \times 3.83$$

and

$$\beta_m = 4e^{-V/15} \times 3.83.$$

To compare eq. (4.10) with eq. (4.11) set the Na-inactivation variable h at its resting value $h_0 = 0.608$ and let $\bar{g}_{Na} = 120$ mmhos/cm^2 (see Hodgkin and Huxley, 1952). The rate constant α in eq. (4.10) is equivalent to the sum of α_m and β_m. The solid line in Fig. 19 is the Hodgkin–Huxley curve defined by (4.11) and the points (●) are obtained from (4.10) with $\gamma = 2.5$, $V_p = 103$ mV and $K = 12.9$ m/sec (corresponding to $\theta = 20.8$ m/sec).

Analytical models of propagation in excitable cells 137

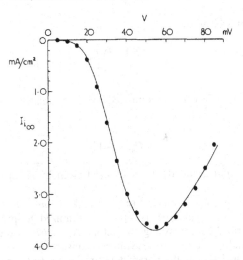

FIG. 19. The fully activated current-voltage diagram given by eqs. (4.12) and (4.4) (circles, ●) fitted to the Hodgkin–Huxley curve defined by eq. (4.13) (solid line). The optimal parameter values obtained by the fitting procedure are $\gamma = 2.5$, $V_p = 103$ mV and $K = 12.9$ m/sec.

3. *Threshold Case*

The form of $\dot{V}(V)$ chosen in the previous section (eq. (4.1)) led to an expression for the ionic current which had roots at $V = 0$ and $V = V_p$ but not at the intermediate voltage required for a threshold effect. A more general expression for $I_i(V)$ can be obtained by choosing

$$\dot{V}(V) = \frac{gS}{C_m} V \left(1 - \left(\frac{V}{V_p} \right)^\gamma \right), \tag{4.12}$$

where the parameter S is defined below in terms of the threshold root. Substituting (4.12) into the cable equation gives the following form for $I_i(V)$:

$$I_i(V) = gSV \left(1 - \left(\frac{V}{V_p} \right)^\gamma \right) \left[\frac{gS}{K C_m} \left(1 - (\gamma + 1) \left(\frac{V}{V_p} \right)^\gamma \right) - 1 \right]. \tag{4.13}$$

The threshold root occurs at

$$V = V_{th} = V_p \left(\frac{1 - (K C_m/gS)}{1 + \gamma} \right)^{1/\gamma}. \tag{4.14}$$

Solving (4.14) for K gives

$$K = \frac{gS}{C_m} \left[1 - (\gamma + 1) \left(\frac{V_{th}}{V_p} \right)^\gamma \right]. \tag{4.15}$$

and introducing this expression into (4.13) enables the ionic current to be expressed in the alternative form:

$$I_i(V) = gS \left[\frac{1}{\gamma + 1} \left(\frac{V_p}{V_{th}} \right)^\gamma - 1 \right]^{-1} V \left(1 - \left(\frac{V}{V_{th}} \right)^\gamma \right) \left(1 - \left(\frac{V}{V_p} \right)^\gamma \right). \tag{4.16}$$

Since the conductance g is by definition the coefficient of V in (4.16) (see section I), the parameter S is defined as

$$S = \frac{1}{\gamma + 1} \left(\frac{V_p}{V_{th}} \right)^\gamma - 1. \tag{4.17}$$

Thus, the ionic current is given by

$$I_i(V) = gV \left(1 - \left(\frac{V}{V_{th}} \right)^{\gamma} \right) \left(1 - \left(\frac{V}{V_p} \right)^{\gamma} \right). \tag{4.18}$$

Equations (4.15) and (4.17) may be combined to give a relation between K and S:

$$K = \frac{g}{C_m} \frac{S^2}{S+1}. \tag{4.19}$$

When S is large, $K = gS/C_m$ and the form of $\dot{V}(V)$ given by (4.12) reduces to that of (4.1).

Once the parameters V_p, V_{th}, g and γ are known, S is found from (4.17) and then K from (4.19). As before, the parameters may be optimized to produce the best fit of eq. (4.18) to a prescribed current-voltage trajectory. Alternatively the steady-state current-voltage relation and the voltage dependent rate constant may be known from voltage clamp studies, in which case the parameters are found by fitting to the following $I_x(V)$ relation:

$$I_x(V) = I_i(V) \left\{ 1 + \frac{gS}{3\alpha C_m} \left[\frac{1 - (V/V_p)^{\gamma}}{1 - (V/V_p)} - \gamma \left(\frac{V}{V_p} \right)^{\gamma} + \frac{\gamma(1 - (V/V_p)^{\gamma})}{1 - (V_{th}/V)^{\gamma}} \right] \right\}^3, \tag{4.20}$$

where $I_i(V)$ is given by eq. (4.18).

The derivation of (4.20) from (4.18) follows the derivation for the non-threshold case given in the last section (eqs. (4.5)–(4.10)).

4. Dependence of K on $\dot{V}_{max}$

We may now consider further the problem of relating K to $\dot{V}_{max}$ which we have already discussed in section I. There it was shown (eqs. (1.32) and (1.33)) that for a given polynomial chosen to fit $I_i(V)$, K is proportional to $\dot{V}_{max}$ or $\theta\alpha\sqrt{\dot{V}_{max}}$. However, comparing eqs. (1.32) and (1.33) it is clear that the constant of proportionality is moderately dependent on the degree of the polynomial. This is another way of saying that the conduction velocity is dependent on the *shape* of the action potential rising phase as well as on the maximum rate of rise.

An expression for $\dot{V}_{max}$ in terms of the parameters S, g, γ and V_p follows from eq. (4.12). When $\dot{V}$ is maximum,

$$\frac{d\dot{V}}{dV} = \frac{gS}{C_m} \left[1 - (\gamma + 1) \left(\frac{V}{V_p} \right)^{\gamma} \right] = 0$$

or

$$\frac{V}{V_p} = \left(\frac{1}{\gamma + 1} \right)^{1/\gamma}. \tag{4.21}$$

Substituting (4.21) into (4.12) gives

$$\dot{V}_{max} = \frac{gS}{C_m} V_p \frac{\gamma}{\gamma + 1} \left(\frac{1}{\gamma + 1} \right)^{1/\gamma}, \tag{4.22}$$

where gS/C_m is found in terms of K from (4.15):

$$\frac{gS}{C_m} = K \left[1 - (\gamma + 1) \left(\frac{V}{V_p} \right)^{\gamma} \right]^{-1}. \tag{4.23}$$

Introducing (4.23) into (4.22) the dependence of $\dot{V}_{max}$ on K is

$$\dot{V}_{max} = K V_p \frac{\gamma}{\gamma + 1} \left(\frac{1}{\gamma + 1} \right)^{1/\gamma} \left[1 - (\gamma + 1) \left(\frac{V_{th}}{V_p} \right)^{\gamma} \right]^{-1},$$

or the dependence of K on $\dot{V}_{max}$ is

$$K = \frac{\dot{V}_{max}}{V_p} \frac{\gamma + 1}{\gamma} (\gamma + 1)^{1/\gamma} \left[1 - (\gamma + 1) \left(\frac{V_{th}}{V_p} \right)^{\gamma} \right]. \tag{4.24}$$

As S becomes large V_{th} approaches zero and the threshold model reduces to the non-threshold model. Equation (4.24) then becomes

$$K = \frac{\dot{V}_{max}}{V_p} \frac{\gamma + 1}{\gamma} (\gamma + 1)^{1/\gamma}. \tag{4.25}$$

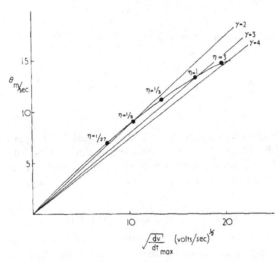

FIG. 20. An illustration of the shape changes accompanying changes in conductance. The points (●) lying on the continuous curve give the dependence of conduction velocity on the square root of the maximum rate of rise of the action potential, found by solving the Hodgkin–Huxley equations at 5 C. The straight lines give the relation between θ and $\sqrt{V}_{max}$ for the constant values of the shape parameter γ (see eq. (4.25)).

Equations (4.24) and (4.25) give us an explicit form for the dependence on shape. For given values of the shape parameters V_p, V_{th} and γ, K is proportional to $\dot{V}_{max}$. Clearly, however, V_p, V_{th} and γ also have a quite strong influence. Thus, comparing quintic ($\gamma = 2$) and seventh-degree ($\gamma = 3$) $I_i(V)$ trajectories for the non-threshold case, we obtain the values $(\gamma + 1)^{1/\gamma}(\gamma + 1)/\gamma = 3\sqrt{(3)}/2 = 2.6$, and $4\sqrt[3]{(4)}/3 = 2.12$ for the constant of proportionality.

The relevance of this result is that varying parameters such as g in a real nerve may influence γ and V_p so that the relation obtained for θ and $\sqrt{V}_{max}$ may deviate from a linear one. In the case of the Hodgkin–Huxley equations, this is indeed the case. Figure 20 shows the relation between θ and $\sqrt{V}_{max}$ obtained from these equations (S. B. Barton, unpublished). Clearly, the relation is not exactly linear and γ must be assumed to vary as η is varied.

5. *An Alternative Definition of the Safety Factor*

An intuitive definition of the safety factor is to use the ratio of the area under the actual current-voltage trajectory to the area under that trajectory required to just offset the resting current (see section I). Thus,

$$S.F. = \int_0^{V_p} I_i(V)\, dV \Big/ - \int_0^{V_p} gV\, dV.$$

140   P. J. HUNTER, P. A. MCNAUGHTON and D. NOBLE

From eq. (4.24)

$$\int_0^{V_p} I_i \, dV = \frac{\gamma}{2(\gamma + 2)} g V_p^2 \left[1 - \frac{1}{\gamma + 1} \left(\frac{V_p}{V_{th}} \right)^\gamma \right].$$ (4.26)

Introducing S from (4.17), equation (4.26) becomes

$$\int_0^{V_p} I_i \, dV = -g S V_p^2 \frac{\gamma}{2(\gamma + 2)}.$$

Since

$$\int_0^{V_p} g V \, dV = \frac{g V_p^2}{2},$$

the safety factor $S.F.$ is defined in terms of S and γ by

$$S.F. = \frac{\displaystyle\int_0^{V_p} I_i \, dV}{-\displaystyle\int_0^{V_p} I_{\text{resting}} \, dV} = \frac{\gamma}{\gamma + 2} S.$$ (4.27)

For large values of γ the safety factor, $S.F.$, is nearly equal to the parameter S. In the limit, as $\gamma \to \infty$, $S.F. = S$ and the current-voltage trajectory (eq. (4.18)) is equivalent to a "switch-line" model (see Introduction). For a more realistic shape of current-voltage trajectory (eq. (4.18) with a finite value of γ), $S.F.$ is reduced by the shape factor $\gamma/\gamma + 2$).

V. SUMMARY OF USEFUL EQUATIONS

Some of the equations we have derived in this article are directly applicable to experimental situations and it may therefore be useful to summarize those equations that should be of most use to experimental physiologists. We shall also indicate the limits on their use.

1. *Relation of θ to Basic Fibre Properties*

One of the most important relations is that between conduction velocity θ and the basic fibre properties:

$$\theta \propto a^{1/2} R_i^{-1/2} C_m^{-5/8} \tau_m^{-3/8} \bar{g}_{Na}^{1/8}.$$ (5.1)

This equation is based on eq. (2.28) with the parameters α and g simply replaced by the proportional parameters τ_m^{-1} and $\bar{g}_{Na}$. The square root dependence on fibre radius, a, and the inverse square root dependence on intracellular resistivity are well-known relations. The other relations are new and are obtained from the analysis of conduction in fibres with a delayed-activation of the sodium conductance. When the delay is represented by a third power of a first-order reaction, as in squid nerve, the conduction velocity is found to be inversely proportional to the $\frac{5}{8}$ths power of the specific capacitance, C_m and inversely proportional to the $\frac{3}{8}$ths power of the time constant of sodium activation (m). The most surprising result, however, is the very low power ($\frac{1}{8}$th) dependence on the maximum value of g_{Na}, $\bar{g}_{Na}$.

The fact that θ increases quite slowly with increase in sodium channel density or conductance has been noted by other workers. Cooley *et al.* (1965) used numerical solutions of the Hodgkin–Huxley equations and found the relation between θ and $\bar{g}_{Na}$ to be nearly flat at large values of $\bar{g}_{Na}$.

Hodgkin (1975) has recently drawn attention to the significance of this result in determining the optimum density of sodium channels. As the sodium channel density is increased, $\bar{g}_{Na}$ increases. However, the apparent membrane capacitance also increases as a result of an increased contribution of the capacitance representing the movement of the channel gating mechanisms. Since the conduction velocity is more strongly dependent on C_m than on $\bar{g}_{Na}$, and since θ decreases as C_m is increased, this leads to a decrease in θ at large values of channel density. There is therefore an optimum density that gives the fastest conduction velocity. Hodgkin (1975) shows that the optimum is similar to estimates of the channel density in squid nerve. His result was obtained using an equation for θ obtained by Huxley (unpublished) for fibres in which g_{Na} follows an exponential activation. As we have also shown (see our eq. (2.16)), this case gives a fourth root dependence of θ on $\bar{g}_{Na}$. Although this equation is less accurate than the eighth root relation for the delay-activated case, Hodgkin's general conclusion would not be very different if he had used eq. (5.1) (see section II, eqs. (2.33)–(2.37)).

These results are of considerable importance in experiments using pharmacological blocking agents. As we have noted in section II, conduction velocity will be a rather insensitive indicator of percentage blockage of ionic channels.

Finally, it is important to note that eq. (5.1) applies only when the values of g_{Na} and α_m are large enough for the recovery processes (sodium inactivation and potassium activation) to have a negligible effect. At low values of $\bar{g}_{Na}$ conduction falls very rapidly with $\bar{g}_{Na}$ and at a critical value conduction fails (see Cooley *et al.*, 1965). In this region eq. (5.1) is of no value and the use of models incorporating recovery processes is required.

2. Maximum Rate of Rise

Another important set of relations is that describing the dependence of θ on the maximum rate of rise of the action potential, $\dot{V}_{max}$, since this is a parameter that is readily obtainable experimentally. The appropriate relations have been given in eqs. (1.32), (1.33), (4.24) and (4.25). The most useful result is that obtained from (4.25):

$$\theta \propto \sqrt{\dot{V}_{max}}\sqrt{y}, \tag{5.2}$$

where y is a "shape" parameter given by $(\gamma + 1)^{(1 + 1/\gamma)}/\gamma V_p$, where γ is defined in section IV. In general, if the current-voltage trajectory is unchanged in shape, i.e. if the shape of the action potential is roughly constant, γ is constant and eq. (5.2) then predicts that θ should be proportional to the square root of $\dot{V}_{max}$. The extent to which this relation may hold experimentally was tested in Fig. 20 using the Hodgkin–Huxley equations. It can be seen that a linear relation between θ and $\sqrt{\dot{V}_{max}}$ is only approximately obeyed. A somewhat lower power than $\frac{1}{2}$ would be more appropriate in this case

We may use this approximate result and that given by (5.1) to obtain relations between $\dot{V}_{max}$ and the basic conductance parameters. Assuming γ constant, eqs. (5.1) and (5.2) give

$$\dot{V}_{max} \propto (\bar{g}_{Na})^{1/4}.(\tau_m)^{-3/4}. \tag{5.3}$$

Once again we can see that $\dot{V}_{max}$ is an insensitive measure of $\bar{g}_{Na}$. This result may appear both surprising and discouraging. It is surprising because it is generally assumed that $\dot{V}_{max}$ is *simply proportional* to I_i, the actual ionic current flowing when $\dot{V}_{max}$ occurs (see eq. (1.6)). $\dot{V}_{max}$ is therefore often thought to be a useful measure of the availability of sodium current in experimental situations where it is difficult to perform accurate voltage clamp measurements. The discouraging fact is that this is not generally correct. The reason is that the actual amount of ionic current activated during an action potential depends not only on $\bar{g}_{Na}$ (and on the initial value of h) but also on the speed with which the voltage changes. As the potential changes more rapidly so less time is available for activation and the degree of activation falls (see also Khodorov and Timin, 1975). This effect is sufficiently large to give rise to the low power dependence of $\dot{V}_{max}$ on $\bar{g}_{Na}$ given by eq. (5.3).

Unfortunately, this problem may not be solved simply by using eq. (5.3) instead of a linear relationship. The reasons for this are twofold:

1. Equation (5.3) only applies to uniformly conducted action potentials. In many experimental situations where $\dot{V}_{max}$ is measured (e.g. Weidmann, 1955) conduction is not uniform.

2. As in the case of (5.1), it is assumed that recovery processes have a negligible effect.

The actual relation between $\dot{V}_{max}$ and $\bar{g}_{Na}$ (or h_0) will therefore be even more complex than eq. (5.3) suggests. It is even conceivable that the various complicating factors may balance each other to give a result that fortuitously allows $\dot{V}_{max}$ to show similar variations as, e.g., h_0, but there is no guarantee that this will occur.

3. Influence of Temperature

The effects of temperature on conduction velocity are described by eqs. (2.32) and (3.13). As expected, the temperature dependence of θ is much lower than that of either α_m or $\bar{g}_{Na}$ since θ is proportional to low powers of these parameters. Over a range of temperatures where the recovery processes are negligible, our equations give a fairly accurate description of Huxley's (1959a) results for the squid nerve equations (see Fig. 17). The appropriate relation is (see e.g. (3.13))

$$\log_{10}(\theta/\theta_{6.3}) = 0.0179\,(T - 6.3). \tag{5.4}$$

In terms of Q_{10}'s we have (see eq. (2.32)):

$$Q_{10,\theta} = (Q_{10,\alpha_m})^{3/8} \cdot (Q_{10\bar{g}Na})^{1/8}. \tag{5.5}$$

4. The Safety Factor

The concept of the safety factor for conduction has played an important role in nerve physiology and it is natural to seek (as did Rushton, 1937) to relate S to the basic fibre properties. In this article, we have done this for a variety of models. Unfortunately, there is no single useful equation that we may give since the equations to be used depend on the nature of the model being considered. However, an approximate useful formula is given by (cf. eqs. (1.49) and (4.29))

$$S.F. = -\int_0^{V_p} I_i(V)\,dV \bigg/ \int_0^{V_p} I_{i\,resting}\,dV, \tag{5.6}$$

which corresponds closely to the intuitive idea that the safety factor is a measure of the degree to which the net inward current exceeds that minimally required to excite. According to this equation, S.F. may simply be equated with the ratio of the integral of net ionic current over the appropriate voltage range to that of the resting current (i.e. a linear extrapolation from the resting membrane resistance) over the same voltage range. As we have shown in previous sections, the more precise equations for S show some dependence on the shape of the trajectory.

APPENDIX
Restrictions on the Form of the Current-Voltage Relation
for the Time-independent Model

Equation (1.5) may be written explicitly for I_i:

$$I_i = C_m \dot{V}\left(\frac{1}{K}\frac{d\dot{V}}{dV} - 1\right), \tag{A.1}$$

where $\dot{V}(V)$ may be expressed as an Nth degree polynomial:

$$\dot{V} = a_1 V + a_2 V^2 + \cdots + a_N V^N, \tag{A.2}$$

and $I_i(V)$ as an Mth degree polynomial:

$$\frac{I_i}{C_m} = b_1 V + b_2 V^2 + \cdots + b_M V^M. \tag{A.3}$$

Substituting (A.2) into the right-hand side of (A.1) and equating the coefficients of similar powers of V in (A.1) and (A.3) gives a set of equations:

$$b_1 = a_1 \left(\frac{a_1}{K} - 1 \right)$$

$$b_2 = a_2 \left(\frac{a_1}{K} - 1 \right) + \frac{2a_1 a_2}{K}$$

$$\cdots$$

$$b_{2N-1} = N a_N^2 / K.$$

(A.4)

Any coefficients b_m with $m > 2N - 1$ are zero and therefore the degree of $I_i(V)$ is restricted to $M = 2N - 1$. For example, allowable forms of $I_i(V)$ are cubic (the trajectory $\dot{V}(V)$ is then quadratic), quintic ($\dot{V}(V)$ a cubic) and so on.

If the coefficients a_1 to a_N and the parameter K are regarded as unknowns, the set of equations (A.4) is only uniquely determinate if $(2N - 1) - (N + 1)$ of the b's are also unknowns. Thus, of the $2N - 1$ coefficients in the current-voltage relation, $N + 1$ are arbitrary parameters used to fit the experimental data and $N - 2$ are determined by the solution. This is a limitation imposed by seeking polynomial solutions and the solution is only acceptable if the final form of $I_i(V)$ is a sufficiently close approximation to the experimental data. Otherwise the non-polynomial forms discussed in section IV should be employed.

REFERENCES

ADRIAN, R. H., CHANDLER, W. K. and HODGKIN, A. L. (1972) An extension of Cole's theorem and its application to muscle. In *Perspectives in Membrane Biophysics*, (ADELMAN, Ed.), 299-309. Gordon and Breach, New York.

ARMSTRONG, C. M. and BEZANILLA, F. (1973) Currents related to movement of the gating particles of the sodium channels. *Nature, Lond.* **242,** 459–461.

CHAPMAN, R. A. (1967) Dependence on temperature of the conduction velocity of the action potential of the squid giant axon. *Nature, Lond.* **213,** 1143–1144.

COLE, K. S. (1968) *Membranes, Ions and Impulses*, University of California Press, Berkeley and Los Angeles.

COLE, K. S. and CURTIS, H. J. (1939) Electric impedance of the squid giant axon during activity. *J. Gen. Physiol.* **22,** 649–670.

COOLEY, J. W. and DODGE, F. A. (1966) Digital computer solutions for excitation and propagation of the nerve impulse. *Biophys. J.* **6,** 583–599.

COOLEY, J. W., DODGE, F. A. and COHEN, H. (1965) Digital computer solutions for excitable membrane models. *J. Cell. Comp. Physiol.* **66,** Suppl. 2, 99–109.

DRAPER, M. H. and MYA-TU, M. (1959) A comparison of the conduction velocity in cardiac tissues of various mammals. *Q. J. Exp. Physiol.* **44,** 91–109.

DRAPER, M. H. and WEIDMANN, S. (1951) Cardiac resting and action potentials recorded with an intracellular electrode. *J. Physiol.* **115,** 74–94.

DUDEL, J. and RÜDEL, R. (1970) Voltage and time dependence of excitatory sodium current in heart muscle (Purkinje fibres). *Pflügers Arch. ges. Physiol.* **315,** 136–158.

EVANS, J. W. (1972a) Nerve axon equations: I. Linear approximations. *Indiana Univ. Math. J.* **21,** 877–885.

EVANS, J. W. (1972b) Nerve axon equations: II. Stability at rest. *Indiana Univ. Math. J.* **22,** 75–90.

EVANS, J. W. (1972c) Nerve axon equations: III. Stability of the nerve impulse. *Indiana Univ. Math. J.* **22,** 577–594.

FITZHUGH, R. (1960) Thresholds and plateaus in the Hodgkin-Huxley nerve equations. *J. Gen. Physiol.* **43,** 867–896.

FITZHUGH, R. (1961) Impulses and physiological states in theoretical models of nerve membrane. *Biophys. J.* **1,** 445–466.

FITZHUGH, R. (1969) Mathematical models of excitation and propagation. In *Biological Engineering*. H. P. SCHWAN, Ed.), pp. 1–85. McGraw-Hill, New York.

FRANKENHAEUSER, B. and HUXLEY, A. F. (1964) The action potential in the myelinated nerve fibre of *Xenopus laevis* as computed on the basis of voltage clamp data. *J. Physiol.* **171,** 302–315.

GOLDMAN, L. and ALBUS, J. S. (1968) Computation of impulse conduction in myelinated fibres: theoretical basis of the velocity-diameter relation. *Biophys. J.* **8,** 596–607.

GOLDMAN, L. and SCHAUF, C. L. (1972) Inactivation of the sodium current in *Myxicola* giant axons. Evidence for coupling to the activation process. *J. Gen. Physiol.* **59,** 659–675.

HASTINGS, S. P. (1972) On a third-order differential equation from biology. *Q. J. Math., Oxford* **23,** 435–448.

HASTINGS, S. P. (1974) The existence of periodic solutions to Nagumo's equation (in preparation).

HELLAM, D. C. and STUDT, J. W. (1974) Linear analysis of membrane conductance and capacitance in cardiac Purkinje fibres. *J. Physiol.* **243,** 661–694.

144   P. J. Hunter, P. A. McNaughton and D. Noble

Hermann, L. (1899) Zur Theorie der Erregungsleitung und der elektrischen Erregung. *Pflügers Arch. ges. Physiol.* **75**, 574.
Hodgkin, A. L. (1937) Evidence for electrical transmission in nerve. *J. Physiol.* **90**, 183–210; 210–232.
Hodgkin, A. L. (1975) The optimum density of sodium channels in an unmyelinated nerve. *Phil. Trans. Roy. Soc.* **270**, 297–300.
Hodgkin, A. L. and Huxley, A. F. (1952) A quantitative description of membrane current and its application to conduction and excitation in nerve. *J. Physiol.* **117**, 500–544.
Hodgkin, A. L. and Katz, B. (1949) The effect of temperature on the electrical activity of the giant axon of the squid. *J. Physiol.* **109**, 240–249.
Hodgkin, A. L. and Rushton, W. A. H. (1946) The electrical constants of a crustacean nerve fibre. *Proc. Roy. Soc.* B **133**, 444–479.
Hodgkin, A. L., Huxley, A. F. and Katz, B. (1952) Measurement of current–voltage relations in the membrane of the giant axon of *Loligo*. *J. Physiol.* **116**, 424–448.
Huxley, A. F. (1959a) Ion movements during nerve activity. *Ann. N.Y. Acad. Sci.* **81**, 221–246.
Huxley, A. F. (1959b) Can a nerve propagate a subthreshold disturbance? *J. Physiol.* **148**, 80–81P.
Jack, J. J. B., Noble, D. and Tsien, R. W. (1975) *Electric Current Flow in Excitable Cells*, Clarendon Press, Oxford.
Keynes, R. D. and Rojas, E. (1974) Kinetics and steady-state properties of the charged system controlling sodium conductance in the squid giant axon. *J. Physiol.* **239**, 393–434.
Khodorov, B. I. and Timin, E. N. (1975) Nerve impulse propagation along non-uniform fibres. *Prog. Biophys. Molec. Biol.* **30**, 145–184.
McAllister, R. E., Noble, D. and Tsien, R. W. (1975) Reconstruction of the electrical activity of cardiac Purkinje fibres. *J. Physiol.* **251**, 1–59.
McKean, H. P. (1970) Nagumo's equation. *Adv. Math.* **4**, 209–223.
Mobley, B. A. and Page, E. (1972) The surface area of sheep cardiac Purkinje fibres. *J. Physiol.* **220**, 547–563.
Nagumo, J., Arimoto, S. and Yoshizawa, S. (1962) An active pulse transmission line simulating nerve axon. *Proc. IRE* **50**, 2061–2070.
Noble, D. (1972) The relation of Rushton's liminal length for excitation to the resting and active conductances of excitable cells. *J. Physiol.* **226**, 573–591.
Noble, D. and Hall, A. E. (1963) The conditions for initiating "all-or-nothing" repolarization in cardiac muscle. *Biophys. J.* **3**, 261–274.
Noble, D. and Stein, R. B. (1966) The threshold conditions for initiation of action potentials by excitable cells. *J. Physiol.* **187**, 129–162.
Noldus, E. (1973) A perturbation method for the analysis of impulse propagation in a mathematical neuron model. *J. Theoret. Biol.* **38**, 383–395.
Offner, F., Weinberg, A. and Young, C. (1940) Nerve conduction theory: some mathematical consequences of Bernstein's model. *Bull. Math. Biophys.* **2**, 89–103.
Rinzel, J. and Keller, J. B. (1973) Travelling wave solutions of a nerve conduction equation. *Biophys. J.* **13**, 1313–1337.
Rushton, W. A. H. (1937) Initiation of the propagated disturbance. *Proc. Roy. Soc.* B **124**, 210.
Van der Pol (1926) Relaxation oscillations. *Phil. Mag.* **2**, 978–992.
Weidmann, S. (1951) Effect of current flow on the membrane potential of cardiac muscle. *J. Physiol.* **115**, 227–236.
Weidmann, S. (1952) The electrical constants of Purkinje fibres. *J. Physiol.* **118**, 348–360.
Weidmann, S. (1955) The effect of the cardiac membrane potential on the rapid availability of the sodium-carrying system. *J. Physiol.* **127**, 213–224.

Nature Vol. 262 August 19 1976

657

Cellular basis for the T wave of the electrocardiogram

I. Cohen, W. Giles & D. Noble

University Laboratory of Physiology, South Parks Road, Oxford, UK

Differences in action potential duration in different regions of the mammalian ventricle are not systematically present when quiescent tissue is first stimulated, but develop rapidly during repetitive activity. The effects of ouabain and temperature suggest the involvement of the Na^+–K^+ exchange pump.

THE electrocardiogram (ECG) has proved to be a remarkably useful tool in the diagnosis and management of many types of malfunction or disease of the heart. But after nearly a century of use and refinement, and in spite of recent advances in cellular cardiac electrophysiology, certain features of the ECG are not well understood. One example is the T wave of the mammalian ECG. It is empirically described as arising from differences in action potential duration in different anatomical regions of the ventricle. Most attempts to account for the T wave have utilised indirect measurements of the sequence of repolarisation in ventricular tissue. For example, detailed studies of the relative refractory period in several anatomical locations in both ventricles and on the intraventricular septum have been reported[1]. And Spach and Barr[2] have used arrays of chronically implanted electrodes simultaneously to record potential distributions in the intact dog heart during excitation and repolarisation. We have used conventional microelectrode techniques to record action potentials in isolated preparations from sheep ventricle. The results show unexpected, activity-dependent changes in action potential duration which may provide further understanding of the mechanism of the T wave.

The T wave of the mammalian ECG is synchronous with the repolarisation phase of the ventricular action potential, and is therefore thought to represent the net extracellular field produced by the spreading of the repolarisation wave throughout the ventricular myocardium. But it is somewhat surprising that the T wave is found to be positive in the same ECG leads as the R wave of the QRS complex, since this complex corresponds to depolarisation of the ventricular myocardium.

Extracellular recordings of the potentials produced by the conduction of nerve impulses produce a diphasic wave, so that, if the leads are arranged such that depolarisation corresponds to a positive extracellular wave, repolarisation is represented by a negative extracellular wave. This result is necessary if depolarisation and repolarisation propagate in the same direction.

The apparent paradox in the directionality of the T wave

is usually explained by postulating that the depolarisation and repolarisation waves in at least some parts of the ventricle travel in opposite directions, so that regions that are excited latest repolarise earliest. This explanation requires that the duration of the action potential should be greater at the point of entry of the wave (in the endocardial muscle of the septum) than at later points in the conduction pathway (for example, in the epicardial surface of the apex). Early work using extracellular electrodes provided strong evidence for this conclusion (for discussion and references see ref. 3). More recently, recordings made with intracellular microelectrodes have shown that, during normal repetitive activity, the action potential at the epicardial surface of the ventricle is shorter than that at the endocardial surface[4].

An important possible function for such a difference is that the longer action potential at early points in the conduction pathway may prevent re-excitation and re-entry. It is therefore important to attempt to determine the cellular basis for this difference. The experiments we describe here show two important results. First, the duration differences responsible for the normal T wave develop as a consequence of repetitive activity; they are not present in the first action potential of a previously quiescent ventricle. Second, the effects of repetitive activity are strongly influenced by cardiac glycosides and by temperature.

Small pieces of tissue (2–4 mm square and 1 mm thick) were dissected from two regions of sheep left ventricles— the upper part (base) of the interventricular septum and the epicardial surface of the apex. The former is endocardial and the latter is epicardial. Some theories of the T wave regard the difference in action potential duration as being primarily between endocardial and epicardial surfaces. The alternative view is that a base/apex gradient exists. The two views are not, of course, mutually exclusive. Our choice of regions allows the correlate of either kind of gradient to be detected in action potential recordings. For simplicity we refer to the two regions as 'base' and 'apex', respectively.

After a suitable period for recovery (20–30 min) action potentials from each region were recorded using conventional 3-M KCl microelectrodes. All measurements of action potential duration were made from heat-sensitive pen recordings at the point of 80% repolarisation. In contrast to recordings made on the whole heart, it is possible to remain impaled in small pieces of tissue for prolonged periods. Moreover, since the preparations are quiescent until

658

Nature Vol. 262 August 19 1976

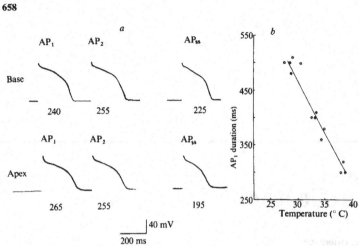

stimulated it is also possible to determine whether any differences that occur in the action potentials are dependent on the presence of repetitive activity.

First action potential

This question was answered by studying trains of action potentials following a short (30–60 s) period of rest. We found that there were no systematic differences between the first action potentials in the base and apex regions. Moreover, this similarity between the two regions occurred at all the temperatures investigated. Figure 1b shows a plot of the first action potential durations obtained from the base (filled circles) and apex (open circles) between 28 and 38 °C. There is a scatter of up to about 20 ms in the durations but it is clear that the durations are not systematically different. They can be fitted by the same relationship as a function of temperature. This result was also obtained in other experiments provided that adequate time was allowed for recovery and that the results were compared from pieces of tissue taken from the same heart.

These experiments strongly suggest that there are no substantial differences between the basic ionic conductance mechanisms that control the plateau and repolarisation phases in the two regions. If such differences exist we should have to suppose both that differences in the individual conductance mechanisms are balanced to produce identical durations and that this balance is preserved for the whole temperature range. This seems very unlikely in view of the fact that the net current flow during the plateau is approximately 1% of the magnitude of the individual inward and outward currents involved[5].

Repetitive activity and second action potential

Although there are no systematic differences between the first action potential durations, it is clear from Fig. 1a that the effect of activity is strikingly different in the two regions. At the base we invariably found that the second action potential duration (AP_2) is substantially longer than the first (AP_1) and that the steady-state action potential (AP_{ss}) is either similar to or only slightly shorter than the first. In the case illustrated, AP_{ss} is 225 ms compared with 240 ms for AP_1. In contrast, at the apex we generally found a substantial shortening so that AP_{ss} is much shorter than AP_1. This effect is particularly striking in Fig. 1a since the duration of AP_1 at the apex (265 ms) is longer than at the base (240 ms). Such a difference would result in a negative T wave, as would also be the case when the durations are identical. Nevertheless, in the apex AP_{ss} is 70 ms shorter than AP_1 and becomes 30 ms shorter than AP_{ss} at the base.

Such a difference would lead to the positive T wave observed during normal repetitive activity. We shall discuss the results for AP_2 and AP_{ss} separately.

Twin pulse studies on sheep Purkinje fibres usually show that the second action potential shortens as the interstimulus interval is reduced. This shortening is primarily caused by residual activation of i_{x1}, the outward current which speeds repolarisation[6]. But lengthening of the second action potential has also occasionally been observed in Purkinje fibres[7]. Such experiments have shown that 10^{-6} M ouabain reduces or eliminates the increase in duration of AP_2.

Fig. 1 *a*, Records obtained during trains of action potentials produced in base (endocardial) and apex (epicardial) preparations from sheep ventricle. Stimuli were applied at 150 per s after a 30–60-s quiescent period. The time to 80% repolarisation is shown in ms below each action potential. AP_1 is the first action potential in each train, AP_2 the second and AP_{ss} is taken when there are no further changes (usually 10–20 action potentials). There is a considerable difference in the effect of activity in the two regions. At the base AP_2 lengthens and AP_{ss} is only 15 ms shorter than AP_1. At the apex AP_2 shortens and AP_{ss} is 70 ms shorter than AP_1. These particular records show a 25-ms difference between AP_1 in the two regions. This difference is in the wrong direction for explaining the normal T wave and was not found to be significant (see *b*). *b*, Duration of AP_1 as a function of temperature. There is no systematic difference between AP_1 durations at the base (●) and apex (○). Similar results were found in other experiments (although the precise functional dependence of duration on temperature was not always linear).

Fig. 2 *a*, Records obtained during paired pulse stimulation at various intervals (indicated in ms). *b*, Duration ratios (AP_2/AP_1) as a function of interstimulus intervals from experiment illustrated in (*a*). The solid curves were drawn by eye. Note that the ratio is less than 1 at the apex (○) and substantially greater than 1 at the base (●).

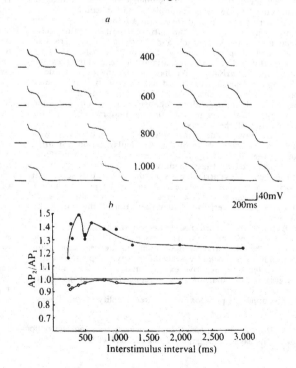

Nature Vol. 262 August 19 1976

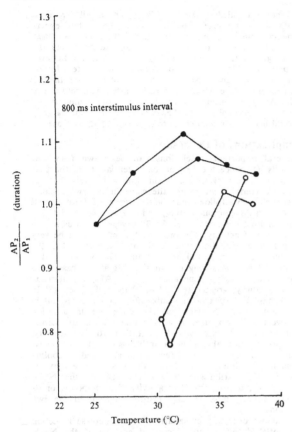

Fig. 3 Influence of temperature on the AP_2/AP_1 duration ratios at the base (●) and apex (○) at an interstimulus interval of 800 ms. Cooling has a larger influence on the apex than on the base. The effects are completely reversible in both regions.

We have studied the duration of AP_2 after application of paired stimuli at various interstimulus intervals to the base and apex regions of the ventricle. As before, a 30–60-s rest period followed each pair of stimuli. As shown in Fig. 2a, AP_2 at the base was substantially longer than AP_1, whereas the opposite result was obtained at the apex. Figure 2b shows plots of the AP_2/AP_1 ratio as a function of interstimulus interval. At all intervals, the ratio is much smaller at the apex than at the base. This result was obtained in all healthy preparations. It was important, however, to allow adequate time (about 1 h) for recovery after dissection. During recovery in apex preparations, while the resting potential was low and the first action potential was relatively short, a striking prolongation of AP_2 could be obtained. Such preparations conformed to the pattern of response shown in Fig. 2 after they reached their normal resting potential and AP_1 became normal in duration.

To test whether the prolongation of AP_2 at the base is attributable to the activity of the Na^+-K^+ exchange pump, we used two methods of blocking pump activity—cooling and application of ouabain. Figure 3 shows the results of cooling at an interstimulus interval of 800 ms. In both regions the AP_2/AP_1 duration ratio becomes smaller at low temperatures but the effect was always found to be more dramatic at the apex than at the base. At 38 °C the AP_2/AP_1 ratio is slightly larger than 1.0 in the apex but falls to about 0.8 when the tissue is cooled to 30 °C. By contrast, in the base, the ratio is still larger than 1.0 at that temperature and cooling to 25 °C was required to reduce the ratio significantly. These effects were totally reversible when the preparations were rewarmed.

Figure 4 shows the results of experiments on the effects of various concentrations of ouabain. In both regions, the AP_2/AP_1 ratio is significantly reduced by ouabain at concentrations that are thought to reduce the activity of the pump significantly. Unlike cooling, these effects were not reversible. This result is not surprising in view of the well known difficulty in reversing the inhibitory action of ouabain on the Na–K exchange pump.

Steady-state action potential

It is clear from the results described so far that very marked differences between the base and apex regions may develop

Fig. 4 The effects of ouabain on AP_2/AP_1 (○) and AP_{ss}/AP_1 (●) duration ratios. The stimulus frequency is 240 per s at the base and 150 per s at the apex. Further explanation in text.

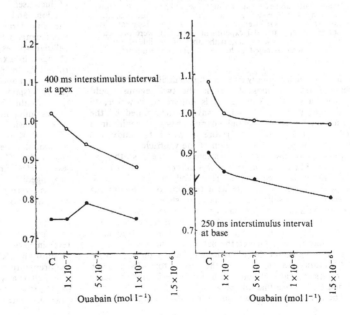

Nature Vol. 262 August 19 1976

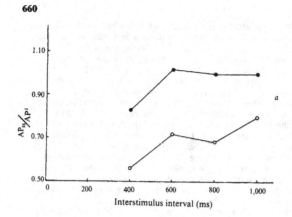

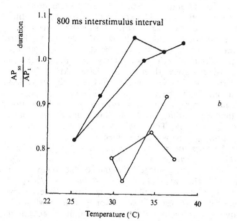

Fig. 5 *a*, AP_{ss}/AP_1 duration ratios at base (●, 38.6 °C) and apex (○, 38.0 °C) as a function of interstimulus interval. The precise influence of changes in frequency on AP_{ss} in the two regions varies somewhat between experiments but, as in this experiment, a significant difference between the two areas is found over the whole frequency range investigated. *b*, Influence of temperature on the AP_{ss}/AP_1 duration ratios in the base (●) and apex (○). These results can be compared with those for the AP_2/AP_1 ratios shown in Fig. 3. The same general phenomenon is observed, that is the apex ratio is reduced to lower levels than the base ratio by cooling. In this particular experiment, the effects were perfectly reversible at the base but not at the apex. Reversible effects on AP_{ss} were obtained in the apex in other experiments.

as a consequence of only one action potential. The difference between the durations of AP_2 in the two regions would produce a positive T wave. It is important, however, to remember that the normal T wave is determined by the steady-state action potential characteristics of maintained repetitive firing. It is therefore essential to study the durations of AP_{ss} in each region of the ventricle.

Figure 1*a* shows that AP_{ss} is longer at the base than at the apex. Figure 5*a* shows that this difference persists for a wide range of frequencies. The filled circles show the AP_{ss}/AP_1 ratio at the base as a function of interstimulus interval. The results for the apex are shown by the open circles. Although in both cases there is a tendency for the ratio to decrease at high frequencies, the difference is always substantial. This result suggests that changes in frequency *per se* should not influence the polarity of the T wave.

Figure 5*b* shows the influence of cooling on the AP_{ss}/AP_1 duration ratios. As with the results for AP_2, cooling decreases the ratio although a higher ratio is always maintained at the base than at the apex.

The influence of ouabain on the AP_{ss}/AP_1 duration ratios

is shown as filled symbols in Fig. 4. The results are more difficult to interpret than those from the cooling experiments. Thus the marked shortening of the action potentials (including AP_1) produced by inhibitory doses of ouabain may give a fairly constant AP_{ss}/AP_1 duration ratio even when both AP_{ss} and AP_1 are substantially shortened. This applies to the apex results. In the case of the base, there is a decrease in the AP_{ss}/AP_1 duration ratio. This means that the AP_{ss}/AP_1 duration ratios become more similar in the presence of ouabain. If this effect is sufficiently large it could lead to a reduction or to inversion of the T wave.

Implications of the results

Several important conclusions can be drawn from our results. (1) Since the first action potentials from the base and apex are very similar in duration it is likely that the ionic currents involved are the same in the two regions. (2) It therefore follows that the differences in duration that occur during normal beating, and that are responsible for determining the polarity of the T wave, develop as a consequence of activity. The onset of this effect must be very rapid since large differences in duration are evident in the second action potential in a train. The second action potential is always longer than the first at the base. It is sometimes longer at the apex but the AP_2/AP_1 duration ratio is always greater at the base than at the apex. (3) In both base and apex regions subsequent action potentials in a train become shorter but the difference in duration between the two areas is always maintained. At normal temperatures the base AP_{ss}/AP_1 duration ratio is near unity. At the apex it is always significantly less than 1. (4) Because the prolongation of AP_2 is reduced or abolished by cooling and by inhibitory doses of ouabain it is possible that the effects are attributable to the response of the Na–K exchange pump to repetitive activity. This response, or its consequences, must be significantly different in the two regions.

Can we offer any explanation for an increase in action potential duration after increased activity of the Na–K pump? At first sight this may seem to be opposite to the expected effect since activation of outward electrogenic pumping would by itself speed repolarisation. This suggests that the effect may arise as a secondary consequence of ionic concentration changes rather than as a direct result of increased pumping. During activity, the cells will gain Na ions and lose K ions. During a single action potential the concentration changes involved will be small unless the movements occur into very restricted spaces. Narrow extracellular spaces do exist in cardiac muscle[6–10], and rapid changes in extracellular K^+ concentration may occur in them[11,12]. In particular, $[K^+]$ in the space will increase during activity. It is possible therefore that this increase in $[K^+]$ is responsible for activating the Na–K exchange pump.

It remains, however, to explain how this might prolong the action potential. If $[K^+]$ remains elevated, the action potential will shorten as a consequence of the well known effect of $[K^+]_0$ on plateau duration[13]. To obtain prolongation we would have to suppose that the pump can transiently reduce the $[K^+]$ in the small extracellular spaces or clefts to levels below normal. The action potential would then be prolonged[14].

This interpretation of our results is clearly very tentative. We have no direct evidence that cleft $[K^+]$ is reduced below normal after the first action potential. Moreover, it may be very difficult to test the idea directly as even K^+-sensitive microelectrodes are not likely to detect the $[K^+]$ level in very narrow cleft spaces. They are more likely to monitor the $[K^+]$ level in the larger extracellular spaces in the preparation.

An alternative view is to postulate that the prolongation of AP_2 is attributable to an increased inward (for example, calcium) current, which is partly (base) or completely

Nature Vol. 262 August 19 1976 **661**

(apex) masked by the shortening effect produced by potassium accumulation. This hypothesis leaves the mechanism of the prolongation unexplained but the difference between the base and apex regions would still be attributable to differences in the capacity of the sodium pump to respond to activity.

Whatever ionic mechanisms form the basis of the effects we have observed, our results do enable us to explain various observations on the T wave. The occurrence of myocardial ischaemia or infarction leads to inversion of the T wave, as does treatment with cardiac glycosides. We can explain these effects as a result of inhibition of the Na–K exchange pump. It is also conceivable that inversion occurs as a consequence of increasing the pump's capacity to respond so that the behaviour of the apex is made to approach that of the base, rather than the opposite. Thus, warming may allow the apex to show prolongation of the action potential during activity (Fig. 3) and so in this manner become similar to the base. This phenomenon could form the basis of T-wave inversion during fever. This possibility also serves to emphasise that it may be misleading to regard T-wave inversions as having a common cause. Any manoeuvre that leads to a decrease in the difference in AP duration will reduce or invert the T wave. This may be produced either by inhibiting the property of the basal (endocardial) region that allows unexpectedly long action potentials to occur or by stimulating the apex (epicardial) areas to respond to activity in a manner resembling that of the base.

Our results also lead to the expectation that the T-wave should be reduced or inverted after a period of quiescence. This has been observed in whole hearts of frog and tortoise after a short period of quiescence produced by vagal stimulation.

We acknowledge the support of the MRC, the Canadian MRC and the Muscular Dystrophy Association of America.

Received February 23; accepted June 25, 1976.

1 Burgess, M. J., Green, L. S., Millar, K., Wyatt, R., and Abildskov, J. A., *Am. Heart J.*, **84**, 660–669 (1972).
2 Spach, M. S., and Barr, R. C., *Circulation Res.*, **37**, 243–257 (1975).
3 Mines, G. R., *J. Physiol., Lond.*, **46**, 188–235 (1913).
4 Solbery, L. E., Singer, D. H., Ten Eick, R. E., and Duffin, E. G., *Circulation Res.*, **30**, 783–797 (1974).
5 Noble, D., and Tsien, R. W., in *Electrical Phenomena in the Heart* (edit. by de Mello, W. F.), 133–161 (Academic, New York, 1972).
6 Hauswirth, O., Noble, D., and Tsien, R. W., *J. Physiol., Lond.*, **222**, 27–51 (1972).
7 Cohen, I., Daut, J., and Noble, D., *J. Physiol., Lond.* (in the press).
8 Mobley, B. A., and Page, E., *J. Physiol., Lond.*, **220**, 547–563 (1972).
9 Page, S., and Niedergerke, R., *J. cell. Sci.*, **11**, 179–203 (1972).
10 Hellam, D. C., and Studt, J. W., *J. Physiol., Lond.*, **243**, 637–660 (1974).
11 Noble, S. J., *J. Physiol., Lond.* (in the press).
12 Cohen, I., Daut, J., and Noble, D., *J. Physiol., Lond.* (in the press).
13 Weidmann, S., *J. Physiol., Lond.*, **132**, 157–163 (1956).
14 Noble, D., *J. cell. comp. Physiol.*, **66**, Suppl. 2, 127–136 (1965).

IMPLICATIONS OF THE RE-INTERPRETATION OF i_{K2} FOR THE MODELLING

OF THE ELECTRICAL ACTIVITY OF PACEMAKER TISSUES IN THE HEART

Dario DiFrancesco and Denis Noble

INTRODUCTION

Other papers in this volume and elsewhere (Brown et al., 1979; Brown and Di-Francesco, 1980; DiFrancesco and Ojeda, 1980; Yanagihara and Irisawa, 1980) have already described the properties of an inward current, i_f (or i_h), that is slowly activated during hyperpolarization beyond about -50 mV in the SA node. In its time course, its voltage range for activation/deactivation and in its response to adrenaline, this current bears many resemblances to the s-mechanism described by Noble and Tsien (1968) as controlling an outward K^+ current, i_{K2}, in Purkinje fibres (DiFrancesco and Ojeda, 1980). As this resemblance became clear, so also did an obvious puzzle; the s-mechanism is described as activating on depolarization and controls an outward current. This produces the same overall current change as an inward current activated by hyperpolarization; but had nature really developed two systems for producing this current change in the heart by such different means? It seemed rather unlikely.

For some time, therefore, it has been evident that either i_f or i_{K2} might need reinterpretation. Earlier thinking was directed towards finding ways of reinterpreting the "new" current i_f in the SA node (could it not, for example, be an i_{K2} system whose reversal potential was distorted or masked by non-uniform properties of the tissue?) In the event, however, it has been the "old" current, i_{K2}, that has needed reinterpretation. One of us (DiFrancesco, 1981) has described in a previous paper in this volume the compelling experimental reasons for thinking that i_{K2} is in fact the same as the inward current in the SA node and that it is its "reversal potential" E_{rev}, that is misleading.

Our purpose in this paper is to explore the consequences of this rein-

94   D. DiFrancesco and D. Noble

terpretation for modelling of the electrical properties of pacemaking cells in the heart. The work reported here is part of a development towards constructing a unified model of pacemaker activity that, with only quantitative differences, may be applicable to all the pacemaker regions.

Before we describe the equations we have developed, it is worth recalling the electrical properties that any model must aim to reproduce. These may be divided into specific properties of "i_{K_2}" and properties of the pacemaker potential itself.

LIST OF SYMBOLS

t	time	s
E	membrane potential	mV
E_{Na}, E_K	equilibrium potentials for Na^+ and K^+ ions	mV
K_i, K_c, K_b	internal, cleft and bulk K^+ concentrations	mM
F	Faraday's constant	96800 Coul g/ions
V	total cleft volume	μl
D	potassium diffusion constant	$\mu m^2 s^{-1}$
x	distance from fibre centre	μm
P	permeability constant for diffusion between clefts and bulk solution (3 compartment model only)	
C	membrane capacity	μF

Symbols for current (nA), fully-activated current (nA), kinetic variable, rate constants (s^{-1}), time constants (s) reversal potential (mV) respectively are indicated below for the various current components:

i_f, $\bar{i}_f$, y, α_y, β_y, τ_y, E_f	"pacemaker" current (new interpretation)
i_{fNa}, $\bar{i}_{fNa}$, y, α_y, β_y, τ_y, E_{Na}	sodium component of i_f
i_{fK}, $\bar{i}_{fK}$, y, α_y, β_y, τ_y, E_K	potassium component of i_f
i_{K2}, $\bar{i}_{K2}$, s, α_s, β_s, τ_s, E_{rev}	"pacemaker" current (old interpretation)

(note: $\alpha_s = \beta_y$, $\beta_s = \alpha_y$, $s = 1 - y$)

i_K (or i_x), $\bar{i}_K$, x, α_x, β_x, τ_x, E_K outward (delayed) current

i_{K1}^o	time-independent K^+ current
i_p	pump current
i_{Na}	fast (sodium) time-dependent current
i_{Ca}	slow (second inward) time-dependent current
i_{inb}	total inward background current
i_{mK}	total potassium current
i	total membrane current
$\beta(E, K_c)$	$= (d\bar{i}_f/dK_c)_{E, K_c}$
$\lambda(E, K_c)$	$= (d(i_{K1} + i_p)/dK_c)_{E, K_c}$

PROPERTIES OF "i_{K2}"

It must be admitted that there is some considerable surprise in the finding that i_{K2} needs reinterpretation for it was so apparently well-established as a highly-specific potassium current. The main experimental findings that gave rise to this view are:

1. On hyperpolarizing Purkinje fibres to different levels of potential, it is found that the total time-dependent current change reverses direction. In Tyrode solutions containing 4 mM K^+ this potential (E_{rev}) occurs in the range -100 to -110 mV (Vassalle, 1966; Noble and Tsien, 1968; Peper and Trautwein, 1969; Cohen et al., 1976a). This reversal is often, (see e.g. Cohen et al., 1976, Fig. 2), though not invariably, a "smooth" process that gives the impression of a single time-dependent mechanism. Even when this is not the case, it could be argued that any diphasic behaviour in the time course near E_{rev} is due to distortion by a process of K^+ depletion in the extracellular spaces (see e.g. DiFrancesco et al., 1979).

2. When the extracellular K^+ concentration is varied in the range 2-12 mM, E_{rev} shifts in the way expected for a K^+ electrode. The slope of the E_{rev} - log K_b relation is often near the expected value (60 mV/decade) for a highly selective

potassium system. This is one of the reasons for which the current was identi-fied as a pure K^+ current. It is, though, worth noting that this interpretation raised some problems. Although the slope of the Nernst relation (E_{rev} versus log K_b) is correct, the absolute values of E_{rev} are not. They are invariably too negative by about 10 mV. Cohen et al., (1976a) interpreted this to mean that the extracellular cleft K^+ concentration (K_c) was always significantly lower than the bulk concentration (K_b). This interpretation can be made self-consistent (Cohen et al., 1979, Appendix) but only at the cost of making a number of assumptions that are difficult to test experimentally. An alterna-tive explanation is that, when the K^+ depletion that occurs during hyperpolari-zations used to determine E_{rev} is taken into account, E_{rev} can be shown to be more negative than E_K (DiFrancesco et al., 1979).

3. The fully-activated current-voltage relation, $\bar{i}_{K2}$ (see Noble and Tsien, 1968) displays the phenomenon of inward-rectification characteristic of K^+ cur-rents in both skeletal and cardiac muscle. At potentials positive to about -70 mV the slope of the $\bar{i}_{K2}(E)$ relation is negative.

4. Also characteristic of K^+ rectification, the $\bar{i}_{K2}(E)$ curves at different va-lues of K_b cross each other positive to E_K. In this they resemble the proper-ties of i_{K1} (though not of the delayed K^+ rectifier, i_K (or i_x) - see discussion in Brown et al., 1980).

5. During a hyperpolarizing pulse, the slope conductance, di/dE, is reported as decreasing (Vassalle, 1966). More recently it has been observed to increase (DiFrancesco, this volume) with time. Sometimes there is very little net change in di/dE.

6. The current described as i_{K2} is absent in sodium-free solutions (McAllister and Noble, 1966).

At first sight it may seem rather unlikely that all of these properties should be compatible with an inward current hypothesis. We shall, however, show that they are indeed compatible with the new interpretation and that, in certain respects, the new hypothesis provides less complex explanations than the old one.

PROPERTIES OF THE PACEMAKER POTENTIAL

The relevant properties of the pacemaker potential are:

7. The slope conductance measured on applying small current deflections during the pacemaker depolarization decreases with time (Weidmann, 1951; 1956). This also was an important feature of the "i_{K2}" interpretation and must, clearly, be reproduced by any new model.

8. The pacemaker potential in Purkinje fibres is extremely sensitive to changes in extracellular K^+ concentration (Vassalle, 1965). While this property was not an essential feature of the "i_{K2}" model, it is of great interest to attempt to reproduce it with a new model that incorporates the effects of extracellular K^+ on the ionic currents involved in pacemaker activity.

9. Graded depolarizing and hyperpolarizing pulses have, respectively, strong negative and positive chronotropic effects on the subsequent pacemaker depolarization (Weidmann, 1951). This behaviour was extremely well reproduced by the McAllister, Noble and Tsien (1975) equations, based on the "i_{K2}" hypothesis.

DESCRIPTION OF EQUATIONS

We shall divide the description of the mathematical formulation of the model into two parts: the ionic current equations and the diffusion equations.

IONIC CURRENT EQUATIONS

In formulating the equations for ionic current we have used the McAllister et al. (1975) model where appropriate for all currents except for the new inward current, i_f. For K^+ sensitive currents such as i_{K1} and $i_K (i_x)$ we have also incorporated the modifications designed to reproduce the K^+ dependence introduced by Cohen et al. (1978) for i_{K1} and by Brown et al. (1980) for i_K. Instead of using current density units expressed per unit area of membrane, we have used absolute units of ionic current (in nA) scaled to give total currents similar to those recorded experimentally in a preparation of radius 125 μm and length 2 mm.

98   D. DiFrancesco and D. Noble

(a) THE INWARD CURRENT ACTIVATED ON HYPERPOLARIZATION

This current has so far been referred to either as i_f (Brown et al., 1979) or
as i_h (Yanagihara and Irisawa, 1980). We will continue to use the symbol i_f but
will use the activation variable, y, (see DiFrancesco, 1981) to avoid confusion
with the inactivation variable f in the calcium current equations. Then

$$\frac{dy}{dt} = \alpha_y(E) \ (1 - y) - \beta_y(E)y \tag{1}$$

The equation for $\alpha_y(E)$ was that used by McAllister et al. (1975) for $\beta_s(E)$,
while that for $\beta_y(E)$ was $\alpha_s(E)$. This gives an activation curve varying from
zero at about -60 mV to 1 at about -90 mV. Clearly, $y(E,t) = 1 - s(E,t)$.

In some preliminary work (DiFrancesco and Noble, 1980) we used a simple linear function for $\bar{I}_f$ with a reversal potential set at 0 mV. Since then, it has
become clear that i_f has both sodium and potassium components (DiFrancesco,
1980; 1981). It is not yet clear whether these components are independent,
though it is known that a component reversing at E_K exists in completely
sodium-free solutions (Hart et al., 1980). A similar result has been obtained
for the rabbit SA node (Brown et al., 1980). For simplicity, therefore, we
shall represent the components as separate. The sodium component is given by

$$\bar{I}_{fNa} = \bar{g}_{fNa} \ (E - E_{Na}) \tag{2}$$

where E_{Na} was usually set to +40 mV and $\bar{g}_{fNa}$ to 2 μs. To reproduce sodium-free
conditions $\bar{g}_{fNa}$ was set to zero.

For the K^+ component we have used two alternative formulations. In the linear case:

$$\bar{I}_{fK} = \bar{g}_{fK}(K) \ (E - E_K) \tag{3}$$

Addition of (2) and (3) gives the total current $\bar{I}_f$:

$$\bar{I}_f = \bar{g}_{fNa} (E - E_{Na}) + \bar{g}_{fK} (Kc) (E - E_K) =$$

$$= (\bar{g}_{fNa} + \bar{g}_{fK}(K_c)) \left(E - \left(\frac{\bar{g}_{fNa}}{\bar{g}_{fNa} + \bar{g}_{fK}(K_c)} E_{Na} + \frac{\bar{g}_{fK} (K_c)}{\bar{g}_{fNa} + \bar{g}_{fK}(K_c)} E_K\right)\right) \quad (4)$$

The experimental data from DiFrancesco (1981b) can be described in the general form:

$$\bar{I}_f = \bar{g}_f (K_c) (E - E_f) \quad (5a)$$

where

$$E_f = r_{Na} (K_c) E_{Na} + r_K (K_c) E_K \quad (5b)$$

The conditions under which the experimental data given by equations (5) can be described by equation (4) are deduced by comparing the two descriptions. In general, the sum of r_{Na} and r_K must be unity at any K_c and a certain relation (as deduced by equation (4)) must exist between the dependence of r_{Na} and r_K on K_c, and the dependence on K_c of the total conductance $\bar{g}_{fNa} + \bar{g}_{fK}$. Within limits the existing experimental data satisfy the above requirements.

We also sometimes used a non linear form for i_{fK}:

$$\bar{I}_{fK} = 2.6 (150 - K_c \exp (-E/25)) \quad (3a)$$

This equation is based on that for $\bar{I}_K$ (see below). This gives a non-linear relation but the total current in the pacemaker range of potentials then shows more K^+ sensitivity than for most of the linear models. Where appropriate we have checked that our more important results do not depend on the precise formulation used for i_{fK}.

Figure 1 shows the relations generated by one of our linear formulations of i_f for $K_b = 2$ and 10 mM.

(b) THE TIME-INDEPENDENT K^+ CURRENT, i_{K1}

Our first computations were done using the modification of McAllister et al. (1975) equations for i_{K1} and i_{K2} proposed by Cohen et al. (1978). The purpose of this modification was to reproduce the known dependence of potassium current on extracellular potassium. The total time-independent K^+ current is given by

$$i_{K1}(E) = i^O_{K1}(E) + \bar{i}_{K2}(E)$$

where i^O_{K1} refers to the Cohen et al. (1978) function (see DiFrancesco, 1981a). The rationale for adding $\bar{i}_{K2}$ to i^O_{K1} is to ensure that the total value of K^+ current corresponding to the state $y = 0$ (no i_K activated) in the new model corresponds to the state $s = 1$ ($\bar{i}_{K2}$ fully activated in the old model). Clearly though, the component $\bar{i}_{K2}$ is no longer controlled by a gating variable and is therefore indistinguishable from the old i_{K1}. This formulation was satisfactory for many purposes but, in the course of developing the model we have found it desirable to take account of the effects of K^+ depletion and accumulation on the measured i_{K1} relations. We have found that a better reproduction of the steady-state current voltage relations under these circumstances is given by the following equation:

$$i_{K1} = 28 \, \exp(E_K/53) \left\{ \left(\frac{\exp(0.04(E - E_K)) - 1}{\exp(0.08(E - (E_K + 50)))} \right) + \exp(0.04(E - (E_K + 50))) \right\} +$$

$$+ \, 0.016 \, (E - (E_K + 80))/(1 - \exp(-0.04(E - (E_K + 80)))) \tag{6}$$

This equation is plotted in Figure 1 for K^+ concentrations between 2 and 10 mM. It can be seen that it correctly reproduces the inward-rectification and the "cross-over" phenomenon seen experimentally.

(c) TIME-DEPENDENT (DELAYED) K^+ CURRENT

This component is also known to show inward rectification but with no negative slope conductance (Noble and Tsien, 1969) and no cross-over (DiFrancesco and McNaughton, 1979). Brown et al. (1980) used the equations derived from a rate-theory treatment (Noble, 1972) to give an equation of the form:

FIGURE 1 Top: *Current-voltage rela-*
tions for i_{K1} *at various values of* K_b
given by equation (6).
Bottom: Example of current-voltage re-
lations for i_f *at two values of* K_b.
The value of g_{fK} *in this example was*
given by $(-2007E_K)$. *We have also repre-*
sented g_{fK} *in some computations by the*
term $\sqrt{K_c}$. *Both formulations give linear*
i(E) relations with g_{fK} *increasing as*
K_c *is raised, as found experimentally.*

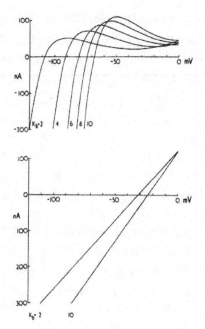

$$\bar{i}_K = 0.67\,(K_i - K_c\,\exp\,(-E/25)) \tag{7}$$

where K_i was set to the value 150 mM. Note that we have followed McDonald and Trautwein (1978), DiFrancesco et al. (1979) in using the symbol i_K for this current. Noble and Tsien (1969) called it i_x. This notation was chosen to distinguish it from i_{K2} and, since its reversal potential is much less negative than that for i_{K2}, it was thought to be a less specific channel. The new interpretation of i_{K2} however, means that the real value of E_K without accumulation or depletion (about -90 mV at K_b = 4 mM) may not be as far from the reversal potential of the delayed K^+ current as previously thought, particularly as K^+ accumulation during the large depolarizing pulses used to activate i_K may temporarily shift E_K to values less negative than -90 mV. It is now possible therefore that i_K (i_x) may be a fairly specific K^+ current.

102   D. DiFrancesco and D. Noble

We shall continue to use McAllister, Noble and Tsien's equations for the activation variable, x, controlling i_K:

$$\frac{dx}{dt} = \alpha_x (1 - x) - \beta_x x \tag{8}$$

$$\alpha_x = 0.08 \exp((E + 15)/20) \tag{9}$$

$$\beta_x = 0.08 \exp(1 - ((E + 15)/20)) \tag{10}$$

(d) INWARD BACKGROUND CURRENT, i_{inb}

As in McAllister et al. (1975) this is represented as a linear function of voltage:

$$i_{inb} = 0.2 (E - 40) \tag{11}$$

(e) THE Na^+K^+ PUMP CURRENT, i_p

This current was not included in McAllister et al.'s equations (except implicitly as an indistinguishable element of the background current). Since we now wish to develop equations for the variation of extracellular K^+ concentration with time it is necessary to include the Na^+K^+ pump explicitly in the formulation and the current, i_p, that it carries will be represented by the equation

$$i_p = 0.01 K_c VF \tag{12}$$

where V = volume of extracellular space and F is the Farady constant. At $K_b = 4$ mM, and a space of 6.4 %, this gives a pump current of 25 nA. This is less than the maximum pump current estimated by Eisner and Lederer (1980), but, as they also note, the steady state current must be less than the peak current that is activated under optimal conditions.

Strictly speaking, the dependence of i_p on K_c should be non-linear, but this complexity is not yet required in the modelling. It would, though, be relatively easy to incorporate a non-linear relation if needed in future work.

(f) i_{Na} AND i_{Ca}

When required these were represented by the same equations as those used by McAllister et al. (1975). Since, in the present paper, we are concerned only with very slow voltage changes (or with constant voltages) we have simplified the equations by setting the kinetic variables m, h, d and f to their steady state values at each voltage. We intend to remove this restriction in future work.

DIFFUSION EQUATIONS

We shall assume that the extracellular cleft space is uniformly distributed in the preparation. In this case we may use the equation for diffusion in a cylinder to represent the movement of extracellular K^+ ions:

$$\frac{dK_c}{dt} = D\left(\frac{d^2K_c}{dx^2} + \frac{1}{x}\frac{dK_c}{dx}\right) + i_{mK}/FV \qquad (13)$$

where i_{mK} is the net K^+ ion current across the cell membrane. This will, in turn, be represented by the sum of the leak due to passive ionic current carried by K^+ and the active movements due to the Na-K exchange pump:

$$i_{mK} = i_{K1} + i_K + i_{fK} - 2\,i_p \qquad (14)$$

Here i_{fK} is the K^+ component of the inward current activated on hyperpolarization. The K^+ moved by the pump is equal to twice the pump current since we are assuming a 3:2 Na:K exchange. F is the Faraday and V is the extracellular space volume.

Equation (13) was solved numerically by using an inversion procedure for a band matrix of width three (see Fox, 1961). The value of D was chosen to be equal to the free solution value, i.e. 1600 μ^2s^{-1}, adjusted by a tortuosity factor of 1.6. This assumes that on average a K^+ ion sees a cylinder of 200 μ radius instead of 125 μ, and gives an "effective" value of D of 624 μ^2s^{-1}. We have checked in our computations that the precise value of D is not in fact very critical.

Various values of V were used to correspond to Purkinje fibres with differ-

104   D. DiFrancesco and D. Noble

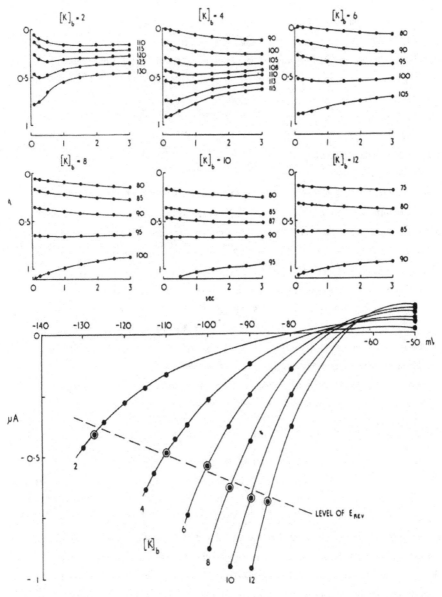

FIGURE 2 Top: *Ionic currents computed on hyperpolarizing to various potentials from -50 mV at various values of K_b. For these computations $\bar{g}_{fK}$ was made proportional to $\sqrt{K_c}$, which gives a good approximation to the K dependence of the total current carried by the i_f channels.*
Bottom: *Corresponding steady state current voltage relations. The values for E_{rev} are indicated by ringed symbols.*

ent extracellular spaces. For example, canine fibres are known to have a much larger extracellular space (about 30 %) than sheep fibres. In each case we shall indicate the value of V chosen as a percentage of the total fibre volume.

RESULTS AND DISCUSSION

(i) THE APPARENT "REVERSAL POTENTIAL" AND ITS NERNST-LIKE BEHAVIOUR

For these results we computed the total ionic currents in response to hyperpolarizations from -50 mV (where $y_\infty = 0$) to various potentials. The results, for various values of K_b, are shown in Figure 2A. It can be seen that at each value of K_b there is a potential at which the net time-dependent current changes direction. Thus, at $K_b = 4$ mM this potential is about -110 mV. At $K_b = 12$ mM, the reversal occurs at -85 mV. This change of 25 mV is close to the expected change for a Nernst potential (28.6 mV). In Figure 3A we have plotted the results for E_{rev} (●) and compared them with experimental results (□-Noble and Tsien, 1968; △-Peper and Trautwein, 1969; ⦶ -Cohen et al., 1976; X - DiFrancesco et al., 1979). We have also plotted the results for resting potentials (▲) obtained by Gadsby and Cranefield, together with the line calculated for E_K from the equation

$$E_K = \frac{RT}{F} \ln \frac{K_b}{140} \tag{15}$$

where 140 mM has been taken as the likely value for K_i. Note that E_K is a good prediction of the resting potential in the range where this varies linearly with log K_b but that, in common with the experimental results, all the values of E_{rev} are significantly more negative than E_K.

Thus, the model reproduces the apparent K^+ specificity of the "reversal potential" very well indeed. Clearly, however, this "reversal" is not a genuine reversal of a single ionic current. We can show this by reproducing DiFrancesco's result using Ba^{++} ions to block i_{K1}. As shown in Fig. 5 (below),

106   D. DiFrancesco and D. Noble

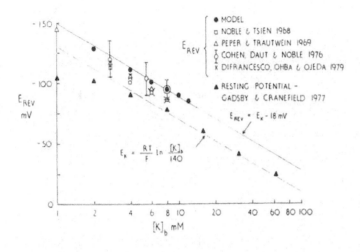

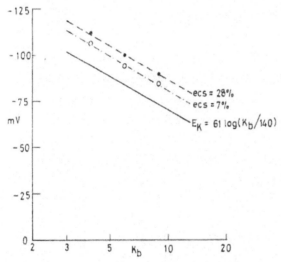

FIGURE 3 *Top:* ● *values of E_{rev} obtained from results shown in Fig. 2. Also shown are experimental results of Noble and Tsien (1968) - □, Peper and Trautwein (1969) - △, Cohen et al., (1976) - ⚨, DiFrancesco et al., (1979) - X, filled triangles show resting potentials, measured by Gadsby and Cranefield. The two straight lines are given by $E_K = (RT/F)\ln (K_b/140)$ and $E_{rev} = E_K - 18$ mV.*
Bottom: Variation of E_{rev} with K_b for extracellular space volumes of 7 and 28 %. Notice that in both cases a good fit to a 60 mV/decade slope is obtained and that increasing the space volume by this large factor (x 4) shifts the absolute values of E_{rev} by only 6 mV. For these computations, equation (3a) was used for $\overline{i}_{fk}$.

once i_{K1} is reduced (in this case to 5 % of its normal value), there is no "reversal potential" at very negative potentials.

Is the apparent K^+ specificity of E_{rev} in the model a fortuitous result or is it expected? First of all, it is not fortuitous. We have found that it does not depend critically on the formulation of the equations for i_f. Thus, DiFrancesco and Noble (1980a) used a simple linear (non K^+ dependent) equation for i_f and obtained a very similar result to that shown here. We have also repeated the computations using a much larger value for V. Increasing V by a factor of 4 to give an extracellular space of about 30 % (which would correspond to the canine Purkinje fibre) shifts the absolute values of E_{rev} by about 6 mV but still gives a nearly linear variation with log K_b with a slope of about 60 mV per decade (DiFrancesco and Noble, 1980b) - see also Figure 3.

The result then is not fortuitous. Is it expected? The answer to this question is that, to a first approximation, the K^+ dependence of E_{rev} reflects the properties of i_{K1} more than of i_f and is not very sensitive to i_f. We can derive this result as follows. The total current change during a hyperpolarization from the holding potential E_H (in this case -50 mV) to E will depend on the changes in $\bar{\imath}_f$, y, i_{K1}, i_p and K_c. If, for simplicity, we represent the extracellular space by a single compartment we obtain the equations for a 3 compartment model:

$$\Delta i(E) = \bar{\imath}_f\ (E,\ K_c).\ \ \Delta y + \left(\frac{d(i_{K1} + i_p)}{dK_c}\right)_E \Delta K_c$$

$$= \bar{\imath}_f\ (E,\ K_c).\ \ \Delta y + \lambda(E)\ \Delta K_c \qquad (16)$$

where $\Delta K_c = K_{c\infty} - K_{co}$ is the change in cleft potassium from its original value K_{co} and $y_\infty - y_0$ is the change in y. Equation (16) can be obtained by integrating equation (24) below on the assumption that $\beta/\lambda < 1$ (equation 26)).

The speed of change in K_c is then given by the continuity equation

108   D. DiFrancesco and D. Noble

$$\frac{dK_c}{dt} = \frac{1}{VF} (i_{K1} - 2 i_p + i_{fK} - FP (K_c - K_b))$$

$$= \frac{1}{VF} (i_{K1} + i_p - 3 i_p + i_{fK} - FP (K_c - K_b)) \qquad (17)$$

At potentials near E_{rev}, $i_{K1} + i_p$ is much larger than $-3 i_p$ and i_{fK}. We may therefore simplify (17) to give

$$\frac{dK_c}{dt} = \frac{1}{FV} (i_{K1} + i_p - FP (K_c - K_b)) \qquad (17a)$$

Now, in the steady state, i.e. at $t = \infty$, $dK_c/dt = 0$ and we then have

$$i_{K1} + i_p = FP (K_c - K_b) \qquad (18)$$

which simply states that the net K^+ current through the membrane must balance the net K^+ diffusion from the bulk solution. Equation (18) holds for any time t and for a first order approximation may also be written:

$$(i_{K1} + i_p) (E, K_{co}) + \lambda(E) \Delta K_c = FP \Delta K_c - FP(K_b - K_{co})$$

from which

$$\Delta K_c = \frac{(i_{K1} + i_p) (E, K_{co}) - FP (K_b - K_{co})}{FP - \lambda(E)} \qquad (19)$$

We now make use of the fact that when the membrane is initially held at a potential (say -50 mV) at which $i_{K1} + i_p$ is small, the initial value of K_c, i.e. K_{co}, will be nearly the same as K_b, so that we may neglect the term $K_b - K_{co}$. We will also assume that over a range of potentials near E_{rev}, the voltage dependence of $i_{K1} + i_p$ is well approximated by a linear function, $i_{K1} + i_p = A(E - E_K)$, where A is a constant. Then (19) becomes

$$\Delta K_c = \frac{A (E - E_K)}{FP - \lambda (E)} \tag{20}$$

Let us now define the "reversal potential" E_{rev} as the potential at which $\Delta i = 0$. (This definition will be exact when the current change near E_{rev} is monotonic, but less so when the time course is diphasic (see DiFrancesco et al., 1979). Then, from (16):

$$\bar{I}_f (E_{rev}, K_c) \Delta y = - \lambda (E_{rev}) \Delta K_c \tag{21}$$

Combining (20) and (21) and rearranging we obtain:

$$E_{rev} = E_K + \frac{\bar{I}_f (E_{rev}, K_c) \Delta y (FP - \lambda(E_{rev}))}{-A \lambda(E_{rev})} \tag{22}$$

E_{rev} will therefore differ from E_k by the value of the expression including $\bar{I}_f$, Δy, P, λ and A. Of these, P, A and Δy may be assumed constant or nearly constant for the relevant voltage range.

Notice that, since $\bar{I}_f$ and λ are always negative, E_{rev} must always be negative to E_k, as observed experimentally. We will now use numerical values for the parameters in equation (22) to see how well it reproduces the computed results. If we consider the case for $K_b = 4$ mM, at which E_{rev} is -105 mV and $E_K = -90$ mV, the values we obtain from the computed results are as follows: $\Delta K_c = -0.8$ mM, $i_{K1} + i_p (t = 0) = -366$, $i_{K1} + i_p (t = \infty) = -140$, $\bar{I}_f = -247$ nA. From these figures we may also compute $FP = 175$ nA/mM, $\lambda = -282$ nA/mM, A = 25 nA/mV. Then inserting in (22):

$$E_{rev} = E_K - 16.0 \text{ mV}$$

which, for $E_k = -90$, gives $E_{rev} = -106$ mV. Equation (22) is therefore a very

110   D. DiFrancesco and D. Noble

close approximation indeed to the numerical results.

We now ask the question, how constant will be the difference $E_{rev} - E_K$? Clearly for an exact 61 mV/decade slope for E_{rev} plotted against log K_b, the difference needs to be constant. Equally clearly, however, the difference cannot be exactly constant since $\bar{i}_f$ and λ both depend on E and K_b. Of these two, by far the most important variation is in λ, i.e. $d(i_{K1} + i_p)/dK_c$, since, as K_c is increased its influence on $i_{K1} + i_p$ decreases (see Figure 1). By differentiating equation (6) for i_{K1} as a function of E_K we may show that di_{K1}/dE_k is almost constant in the relevant voltage range and, since dE_K/dK_c is proportional to $1/K_c$, λ must vary approximately as $1/K_c$. At 8 mM it will therefore be about half its value at 4 mM. With the same figures for the other parameters this would give a difference $E_{rev} - E_K$ of about -22 mV. This value will in fact overestimate the difference since $\bar{i}_f \Delta y$ decreases, particularly when the value of E_{rev} enters the range of voltages (positive to -90) at which $\Delta y \leftarrow 1$. Nevertheless, it is interesting to note that the deviation from a constant value is in the right direction. Moreover, as expected, a larger hyperpolarizing current is then required to reach E_{rev} at high K_b than at low K_b. This is shown in Figure 2 (bottom) which shows the position of E_{rev} on the steady state current-voltage relations computed at different values of K_b. It is worth comparing these results with those obtained experimentally by Cohen et al. (1976b), for, although these authors made the assumption for theoretical purposes that the current level at E_{rev} is constant, some of their experimental results actually show the same effect as we observe in the numerical model. Thus, in figure 2 of Cohen et al.'s paper, the value of current at E_{rev} becomes significantly more negative when K_b is increased to 8 mM.

In conclusion then:

1. E_{rev} will always be negative to E_K.

2. The difference $E_{rev} - E_K$ changes by only a few mV over the relevant range of K^+ concentrations.

3. Notice also that the difference must depend on Δy. This effect has already been described experimentally by DiFrancesco et al. (1979).

(ii)"INWARD RECTIFICATION" AND "CROSS-OVER"

Noble and Tsien (1968) described $\bar{I}_{K2}$ as a set of current-voltage relations at different values of K_b which show the phenomenon of inward-going rectification and which cross each other when the net current is outward.

The new current $\bar{I}_f$ is opposite in direction to $\bar{I}_{K2}$ in the pacemaker range, and it does not resemble $\bar{I}_{K2}$ even when this difference in direction of current flow is taken into account.

Nevertheless, the new model accounts for the same net current changes as did the old model so, even though $\bar{I}_{K2}$ and $\bar{I}_f$ do not show any simple correspondence, there must be some way of deriving relations between $\bar{I}_{K2}$ and corresponding current terms in the new model. This can be done by taking account of current changes due to i_{K1} as well as those due to i_f.

In the new interpretation the current recorded on hyperpolarizing from the holding potential E_H to a potential E is

$$i(E, K_c, t) = i_f (E, K_c, t) + i_{K1} (E, K_c) + i_p(K_c) + i_{inb}(E) \qquad (23)$$

Differentiating with respect to time gives:

$$\frac{di}{dt} = \bar{I}_f \frac{dy}{dt} + \lambda \frac{dK_c}{dt} + \beta\, y \frac{dK_c}{dt} \qquad (24)$$

where $\bar{I}_f$ is the initial value of this parameter, β is $(d\,\bar{I}_f/dK_c)$ and λ, as before, is $d(i_{K1} + i_p)/dK_c$. We will assume that these parameters are independent of K_c (i.e. they depend on voltage only). Then the total current change during the hyperpolarization can be obtained by integrating (24):

$$\Delta i\,(E) = \int_0^\infty \frac{di}{dt}\, dt = \bar{I}_f\, \Delta y + \lambda\, \Delta K_c + \beta \int_0^\infty y \frac{dK_c}{dt}\, dt \qquad (25)$$

In the APPENDIX we shall give a fuller treatment of this equation. Here we will note that β is usually small compared to λ so that the final term can be neglected in a first approximation. Then

112   D. DiFrancesco and D. Noble

$$\Delta i(E) = \bar{I}_f(E)\,\Delta y + \lambda(E)\Delta K_c \qquad\qquad (26)$$

In the old model the same current change was described as

$$\Delta i(E) = \bar{I}_{K2}(E)\,\Delta s \qquad\qquad (27)$$

Since $\Delta y = -\Delta s$, it is clear that

$$-\bar{I}_{K2}(E) = \bar{I}_f(E)\,\Delta y + \lambda\,\Delta K_c/\Delta y \qquad\qquad (28)$$

and, in particular, for values of E such that $\Delta y = 1$ (full activation)

$$-\bar{I}_{K2}(E) = \bar{I}_f(E) + \lambda\,\Delta K_c \qquad\qquad (28a)$$

This "dissection" of $\bar{I}_{K2}$ into two components is illustrated in Figure 4A. Notice that the negative slope region of $\bar{I}_{K2}$ is generated by the fact that when ΔK_c is small (positive to -90 mV) $\bar{I}_{K2}$ is simply $-\bar{I}_f$. When ΔK_c is large, $\bar{I}_{K2}$ is dominated by $\Delta(i_{K1} + i_p)$ (i.e. $\lambda\Delta K_c$). We have also shown (in the points indicated by x) the effect of depletion on i_f, which corresponds to the third term in equation (26). It can be seen that the deviation of $\bar{I}_f$ from its value for $\beta = 0$ is indeed quite small, (though this effect can be seen in some of the experimental results (DiFrancesco, 1981b).

In figure 4B we have illustrated the result of increasing the extracellular K^+ concentration. This has two effects:

1. At each potential $\bar{I}_f$ is increased in magnitude (cf. Figure 1).
2. The voltage range at which Δi_{K1} becomes significant is shifted in a positive direction. This is because i_{K1} becomes large and negative at a less negative range of potentials (see Fig. 1).

Clearly the total current will display a "cross-over" at about -85 mV. Thus the "negative slope" and the "cross-over" are necessary consequences of considering as one current alone the superimposition of the inward component i_f and the effects of K^+ depletion on i_{K1} during strong hyperpolarizations, even if neither of these characteristics feature as properties of $\bar{I}_f$.

Reinterpretation of IK   113

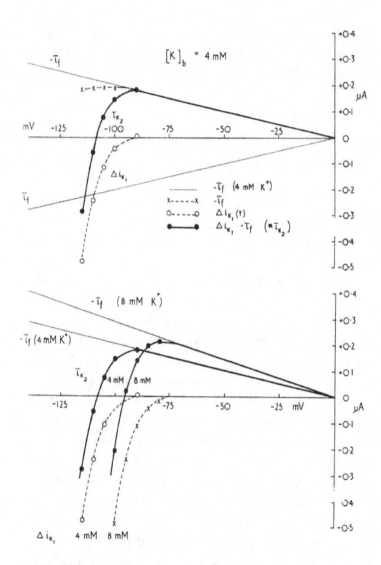

FIGURE 4 Top: *Diagram showing how $\bar{i}_{K_2}$ is generated by new model as the alge-braic sum of $-i_f$ and $\Delta(i_{K_1} + i_p)$. We have also shown the small effect of de-pletion on i_f (xxx). This corresponds to the last term in equation (25) and can for most purposes be neglected.*
Bottom: *Effect of increasing K_b. For strong depolarizations, where $\bar{i}_f$ domi-nates the total current, i_{K_2} is increased. At more negative potentials where $\Delta(i_{K_1} + i_p)$ dominates, the opposite effect occurs. This generates the "cross-over" phenomenon observed experimentally (Noble and Tsien, 1968). In both diagrams the very small values of Δi_{K_1}, when this term is positive have been neglected.*

114   D. DiFrancesco and D. Noble

(iii) SLOPE CONDUCTANCE MEASUREMENTS

When small voltage steps (δE) are applied the membrane current will change. The current displacement (δi (E, t)) is given by

$$\delta i \ (E, \ t) = \delta (i_{K1} + i_p) + \delta i_f + \delta i_{inb}$$

$$= \left(\frac{d(i_{K1} + i_p)}{dE} \right)_{E,t} \delta E + y(E, t) \left(\frac{di_f}{dE} \right) \delta E + \left(\frac{di_{inb}}{dE} \right) \delta E$$

The conductance ($\delta i / \delta E$) measured in this way is given by

$$\frac{\delta i}{\delta E} (E, \ t) = \lambda(E, t) + y(E, t) \ \beta(E) + g_{inb} \ (E) \tag{29}$$

It is worth noting that in equation (29) the restriction on the dependence of λ on K_c has been removed and λ is considered to be changing even during the relatively small K_c changes occurring during hyperpolarization. As we shall show below, the decrease in λ with time is of crucial importance since, while λ is decreasing, y increases with time. Clearly the net result will depend on the relative magnitudes of these two time-dependent changes in conductance.

In Figure 5 we show the result of applying repetitive 5 mV hyperpolarizations in the numerical model during clamp pulses to various voltages. Notice that at -95, -100 and -105 mV the slope conductance decreases as a function of time. At -95 mV the change in g is about 10 %. At -100 mV it is 23 % and at -105 mV it becomes 33 %. Vassalle (1966) reported approximately a 20 % decrease during a voltage clamp applied at the level of the maximum diastolic potential. Clearly, this result is not inconsistent with the new model.

When i_{K1} is made very small (reduced to 5 %) to reproduce the effect of Ba^{++}, the conductance increases with time, as shown experimentally by DiFrancesco (1981b).

It should be noted that, although the model can reproduce Vassalle's (1966) result, this result is not the only one expected. If i_{K1} is reduced (though not necessarily as much as in the presence of barium), or if a larger extracellular

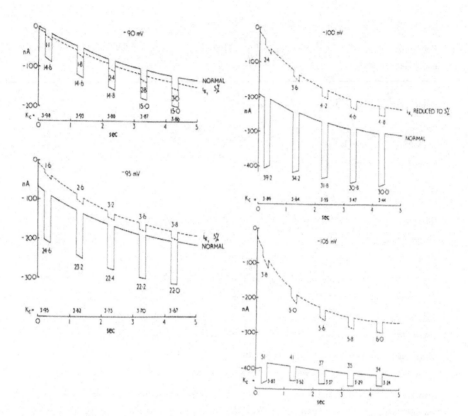

FIGURE 5 *Reconstruction of Vassalle's (1966) slope conductance measurements. For these calculations we set K_b = 4 mM and the extracellular space was 7 %. The membrane was hyperpolarized from -50 mV to the voltages shown. 5 mV test pulses were then applied (1 mV in the case of the pulse in normal solution to -105 mV). The figures on each test pulse are the value of slope conductance in μS. At -90 mV there is almost no net change in g. At all other voltages the slope conductance decreases as a function of time. When i_{K1} is reduced to 5 % of its normal value (to reproduce the effect of barium) large increases in slope conductance occur at each voltage and the net current no longer displays a reversal at -105 mV. The figures at the bottom of each graph show the mean value of the cleft K concentration, K_c, at the time of each test pulse. Notice that the slope in normal solution can be dominated by the properties of i_{K1} (and the effects of changes in K_c on its slope conductance) even though the total current change is dominated by i_f. The reason for this is that, in general for this range of voltages, di_{K1}/dE is much larger than di_f/dE - see figure 1. (Note also that the slope conductance values given here are the mean values obtained by calculating the current jumps at the "on" and "off" of each pulse after correcting for the change in net current with time. This calculation was performed automatically by the computer programme from the formula $\delta i = (i_{t + \delta t} - i_t) - (i_t - i_{t - \delta t})$ where t is the time of onset of the voltage change and δt is the integration step length).*

116   D. DiFrancesco and D. Noble

space is used, or if a function for i_f is chosen for which $d\bar{i}_f/dE$ is larger, the balance can be shifted in favour of the second term in equation (29). We then obtain a small increase in conductance with time even in normal conditions. This is the result reported by DiFrancesco (this volume). It is important to note that while Vassalle's (1966) results are consistent with both old and new models, DiFrancesco's results in both normal and barium solutions cannot be reproduced by the old i_{K2} explanation except in the range of voltages where the negative slope characteristic is found.

(iv) SODIUM-FREE SOLUTIONS

The "sodium-dependence" of "i_{K2}" has always been a puzzle. Since, in the new theory, a major part of i_f is carried by sodium ions it is to be expected that a large change will occur in sodium-free solutions. The existence of a K^+ component in i_f may nevertheless seem to pose a problem. It is therefore important to check on the effect of removing the sodium component alone on the total ionic current. This has been done in the calculations shown in Figure 6. Here we have chosen to use the large extracellular space of 30% to correspond to the situation found in canine Purkinje fibres.

As can be seen, the effect of removing the Na^+ component of i_f is the apparently complete disappearance of "i_{K2}", even though a large K^+ component of i_f is still present. The reason for this result is that the K^+ component of i_f is large and negative only in the range of voltages (negative to about 10 mV from E_K) over which i_{K1} is also large and negative. Hence, as i_f grows on applying stronger hyperpolarizations, so does i_{K1} and this generates sufficient Δi_{K1} to mask i_f. Only when i_{K1} is reduced, by applying Ba^{++}, is it possible to record the K^+ component of i_f in completely sodium-free solutions (Hart et al., 1980). In this connection it is interesting to note that, even in normal Na^+ containing Tyrode solution, results like those in Na-free solution occasionally occur, i.e. there appears to be no "i_{K2}". Exposure of such preparations to Ba^{++} then reveals that i_f is nevertheless present (Hart, Noble and Shimoni, unpublished). It is likely therefore that in such preparations, either the Na component of i_f is already relatively weak or the magnitude of i_{K1} is sufficient to generate a depletion component, i_{K1}, that masks i_f even in normal solutions.

One consequence of the explanation for results in Na^+-free solutions given here is that the time-dependent current change obtained cannot be attributed un-

Reinterpretation of IK   117

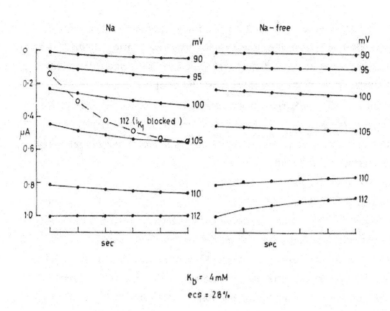

FIGURE 6 *Example of reconstruction of the effect of sodium-free solution on net ionic currents.*
Left: The net ionic current shows a reversal at -112 mV. At this voltage, the change in i_f is shown by the interrupted line (i_{K_1} blocked). The K component of i_f was represented by equation (3a). The extracellular space was set to 28 %.
Right: After removing i_{fNa}, the current records are almost flat at potentials between -90 and -105 mV. At -112 mV the current change is now reversed and appears to represent a simple depletion process. In fact, a substantial current change due to i_{fK} is still present.
We have carried out a number of such computations for various formulations of i_f and for various values of K_b (for K_b = 6 mM see DiFrancesco and Noble, 1980b) and extracellular space. The precise quantitative results depend on the equations and conditions used, but in all cases the apparent disappearance of "i_{K2}" in Na-free solutions is well reproduced. The reason is that, at potentials close to E_K, i_{fK} is too small to be evident (it reverses at E_K) and at more negative potentials it is always effectively masked by the effect of depletion on i_{K1}. The result is therefore a necessary one given the known voltage and K^+ dependence of i_{K1} and of i_{fK}.

iquely to K^+ depletion.

(v) THE SLOPE CONDUCTANCE DURING THE PACEMAKER POTENTIAL

Weidmann's (1951) experiment showing that the voltage change produced by applying small current pulses increases during the pacemaker depolarization in Purkinje fibres has been of seminal importance in the study of pacemaker mechanisms.

118   D. DiFrancesco and D. Noble

The simplest interpretation of the result is that a time-dependent decay of, e.g., K^+ conductance is responsible for pacemaker activity. It has, however, also been clear since the discovery of inward-going rectification that the decrease in conductance recorded in Weidmann's experiment might also (or entirely) be a consequence of the depolarization rather than its cause (see Hutter and Noble, 1960). The relation between slope conductance and total ionic conductance is in fact quite complex (Noble and Tsien, 1972) and it has already been shown that the slope conductance change during the plateau of the cardiac action potential may not reflect the total conductance change.

It is clearly very important to check whether the new model can reproduce Weidmann's result in the pacemaker range of potentials. It might be thought that we have already done this in principle in our reconstruction of Vassalle's result (see Figure 5). However, this is by no means the case. Thus, at -90 mV we found that $\delta i/\delta E$ is almost constant. At -85 mV it in fact slightly increased (not shown in figure 5). It is, however, important to note that in these calculations the voltage level at which the conductance is measured is constant (apart from the small voltage step used to make the measurement). No voltage-dependent conductance changes were therefore involved.

To reproduce Weidmann's result we have let the voltage change spontaneously by using the equation:

$$\frac{dE}{dt} = \frac{-\Sigma i_i}{C} \tag{30}$$

where Σi_i was set equal to the sum of all ionic currents calculated at every point in the preparation. The capacitance, C, was set to 0.05 µF, which is a typical value obtained for Purkinje fibre preparations of the size we have assumed. As explained in an earlier part of the paper, the sodium and calcium currents were set equal to their steady state values at each potential. This means that the calculation is restricted for all practical purposes to the slow pacemaker depolarization.

Figure 7 shows one of our results. The initial conditions were chosen to correspond to the end of an action potential: E = -90 mV, y = 0. Since the model was quiescent at this value of K_b, we shifted the y curve in the depolar-

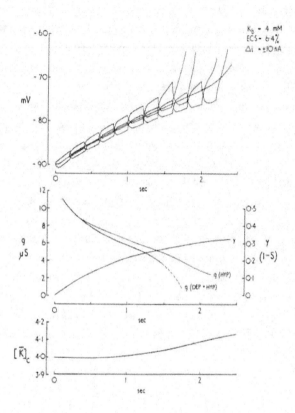

FIGURE 7 *Reconstruction of Weidmann's (1951) measurement of slope conductance during the pacemaker potential in Purkinje fibres.*
Top: pacemaker depolarization with and without ± 10 nA repetitive current pulses. Note that voltage deflection increases with time.
Middle: variation in calculated slope conductance with time and of activation, y, of i_f with time. The slope conductance was calculated either for depolarizing and hyperpolarizing pulses combined (g DEP + HYP) or for hyperpolarizing pulses only (g(HYP)). Note that g varies in the opposite direction to y. The slope conductance is therefore dominated by the voltage-dependent decrease in i_{K1} slope conductance, not by the time-dependent increase in g_f.
Bottom: Variation in mean value of K_c with time. There is a small (0.12 mM) increase in K_c. This would slightly increase g for i_{K1}. It is not therefore responsible for the decrease in g with time. This means that the explanations for Weidmann's g measurements and those of Vassalle (see figure 5) are quite different.

120   D. DiFrancesco and D. Noble

izing direction by 5 mV. This produced the pacemaker potential shown in the diagram. We then repeated the calculation with small (10 nA) repetitive current
pulses added to Σi_i. The slope conductance was calculated either as the mean
value of $\delta i/\delta E$ for both hyperpolarizing and depolarizing pulses or as the value
for hyperpolarizing pulses alone. In both cases a very substantial decrease in
$\delta i/\delta E$ was obtained, as observed by Weidmann.

   Note, however, that this result cannot be due to changes in K_c. First of
all, the change in the mean value of K_c was very small (only 0.1 mM). Secondly,
the change in K_c is in the wrong direction: it slightly increases during the
later part of the pacemaker depolarization which by itself would increase the
slope conductance.

   The decrease in conductance computed here is in fact produced by the effect
of voltage on i_{K1}. It is therefore a consequence of the pacemaker depolarization, not its primary cause.

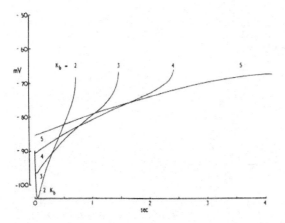

FIGURE 8 *Pacemaker depolarizations computed at various values of K_b between 2
and 5 mM. At 5 mM the "fibre" becomes quiescent. As in Vassalle's (1965) experimental results, the maximum diastolic potential becomes more negative and
the slope of the pacemaker potential greatly increases as K_b is reduced. For
these computations the voltage dependence of y was shifted by 10 mV in a positive direction. Without this shift, the "fibre" is also quiescent at $K_b = 4$ mM.*

(vi) INFLUENCE OF EXTRACELLULAR K⁺ ON PACEMAKER ACTIVITY

In 1965 Vassalle described some important experiments designed to clarify the
extreme sensitivity of pacemaker activity in Purkinje fibres to the level of the

extracellular K^+ concentration. Having constructed a model which, for the first time, takes account of the effects of K^+ on ionic currents and of the extracellular space it seemed possible that we should be able to reproduce Vassalle's results. This is indeed the case, as Figure 8 shows. We computed the pacemaker depolarization in our model for values of K_b = 2, 3, 4, 5, 6 and 8 mM. It can be seen that high K_b suppresses pacemaker activity and that at low K_b the maximum diastolic potential becomes more negative while the slope of the pacemaker depolarization greatly increases. These computations reproduce all the main features of Vassalle's experimental results.

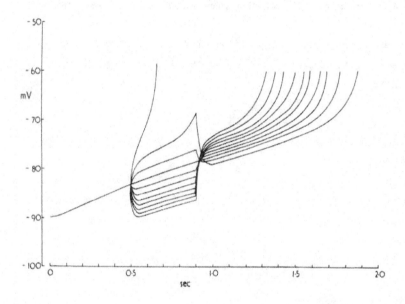

FIGURE 9 *Reconstruction of Weidmann's (1951) experiment showing effect of short hyperpolarizing and depolarizing current pulses applied at about the middle of the pacemaker depolarization. The current used were: depolarizing: 10, 20, and 30 nA.*
hyperpolarizing: 10, 20, 30, 40, 50, 60, 70 nA
Weidmann's result is well reproduced by the new model, just as it was reproduced by McAllister, Noble and Tsien's (1975) equations.

(vii) INFLUENCE OF GRADED CURRENT PULSES

In Figure 9 we show the results of calculations designed to reproduce Weidmann's experiment using positive and negative current pulses of varying amplitude applied at about the middle of the pacemaker depolarization. As in Weidmann's experiment hyperpolarizations are followed by accelerated depolarization and more rapid firing, whereas depolarizations (unless they induce firing themselves) are followed by a less rapid approach to threshold. This result was also well reproduced by the McAllister, Noble and Tsien equations. Clearly, the new interpretation of i_{K2} does not influence this result. The reason is that the result depends only on the fact that the activation curve for the current controlling the rate of depolarization varies steeply in the pacemaker range. A less dramatic result, or even the opposite result, would be obtained if the time-dependent current involved had its activation curve outside the pacemaker range.

CONCLUSIONS

The computations we have described in this paper show that all the major features of the ionic currents in Purkinje fibres that led so apparently conclusively to the "i_{K2}" interpretation receive full and natural explanations with the new "i_f" interpretation. In some respects the new interpretation offers less complex explanations, as, for example, for the behaviour of the ionic current in sodium-free solution (for which the "i_{K2}" hypothesis offered no explanation) and the relatively negative level of E_{rev} (for which the "i_{K2}" hypothesis required more complex explanations - see Cohen et al., 1979; Appendix, and DiFrancesco et al., 1979).

This conclusion is both reassuring and disturbing. It is reassuring inasmuch as it is no longer necessary to suppose that the different pacemaker regions possess fundamentally different ionic current mechanisms. Yet it is also disturbing. For all the standard criteria for a pure K^+ current in the heart had been so fully satisfied that, without the uneasy analogy with the SA node and, even more, the striking and unexpected results of blocking i_{K1} with barium, no-one would have seen any reason to replace i_{K2} with a simple, non-specific linear conductance that carries current in the opposite direction.

Any lessons here for the philosophy of science? Perhaps; and if so, they are fairly obvious: the troughs and peaks are not so very far apart -

Reinterpretation of IK   123

*Ara vos prec, per aquela valor
Que vos guida al som de l'escalina:
Sovenhatz vos a temps de ma dolor!

(Dante, Divina Commedia, Purgatorio,
end of song XXVI)

APPENDIX

Equation (25) may be further treated by noting that the integral

$$\int_0^\infty y \, \frac{dK_c}{dt} \, dt$$

can be obtained as follows. In general

$$y(E, t) = (y_0 - y_\infty) \exp(-t/\tau_f) + y_\infty$$

or, if $y_0 = 0$

$$y(E, t) = y_\infty(1 - \exp(-t/\tau_f)) \tag{A1}$$

Hence

*Now, preie you, by that power whiche not in vayn
up this high montaigne-staire hath lad you sure,
bethynke you in due sesoun of my payne!

(Translation into archaic English of Dante's Occitan)

124   D. DiFrancesco and D. Noble

$$\int_0^\infty y \, \frac{dK_c}{dt} \, dt = y_\infty \int_0^\infty \frac{dK_c}{dt} \, dt - y_\infty \int_0^\infty \exp(-y/\tau_f) \, \frac{dK_c}{dt} \, dt \qquad (A2)$$

Now,

$$\int_0^\infty e^{-pt} \frac{dK_c}{dt} \, dt$$

is, by definition, the Laplace transform of dK_c/dt and

$$\int_0^\infty e^{-pt} \frac{dK_c}{dt} \, dt = \mathcal{L}\left\{\frac{dK_c}{dt}\right\} = -K_{co} + p\bar{K}_c(p)$$

where $\bar{K}_c(p)$ is the Laplace transform of $K_c(t)$.
Therefore

$$\int_0^\infty e^{-t/\tau_f} \frac{dK_c}{dt} \, dt = -K_{co} + \frac{1}{\tau_f} \bar{K}_c\left(\frac{1}{\tau_f}\right) \qquad (A3)$$

Suppose the decay of K is exponential, then

$$K_c(t) = (K_{co} - K_{c\infty}) \exp(-t/\tau_K) + K_{c\infty}$$

and

$$\bar{K}_c(p) = \frac{K_{c\infty}}{p} + \frac{K_0 - K_{c\infty}}{p + 1/\tau_K}$$

which leads to

$$\int_0^\infty e^{-t/\tau_f} \frac{dK_c}{dt}\, dt = -K_{co} + \frac{1}{\tau_f}\left(\tau_f K_{c\infty} + \frac{K_{co} - K_{c\infty}}{(1/\tau_f + 1/\tau_K)}\right)$$

$$= -K_{co} + K_{c\infty} + (K_{co} - K_{c\infty})\, \frac{1/\tau_f}{(1/\tau_f + 1/\tau_K)}$$

$$= \Delta K_c - \Delta K_c \left(\frac{1/\tau_f}{(1/\tau_f + 1/\tau_K)}\right) \tag{A4}$$

So, from (A2)

$$\int_0^\infty y\, \frac{dK_c}{dt}\, dt = y_\infty \Delta K_c \left(\frac{1/\tau_f}{1/\tau_f + 1/\tau_K}\right) \tag{A5}$$

In the limiting case when $y_\infty = 1$ (i_f fully activated by strong hyperpolarization) and τ_f if short compared to τ_K, this expression simplifies to ΔK_c. For all other cases it will be less than ΔK_c.

ACKNOWLEDGEMENTS

We are grateful to the Medical Research Council and the British Heart Foundation for financial support. We are also indebted to the Wellcome Trust for providing funds to enable one of us (D.D.) to visit Oxford to complete this work.

126   D. DiFrancesco and D. Noble

REFERENCES

Brown, H.F. and DiFrancesco, D.: Voltage clamp investigations of membrane currents underlying pacemaker activity in rabbit sino-atrial node. J. Physiol. (London), 308: 331-351, 1980.

Brown, H.F., DiFrancesco, D., Noble, D. and Noble S.J.: The contribution of potassium accumulation to outward currents in frog atrium. J. Physiol. (London), 306: 127-149, 1980.

Brown, H.F., DiFrancesco, D. and Noble, S.J.: Cardiac pacemaker oscillation and its modulation by autonomic transmitters. J. Exp. Biol., 81: 175-204, 1979.

Brown, H.F., Kimura, J. and Noble, S.J.: Evidence that the current i_f in sino-atrial node has a potassium component. J. Physiol. (London), 308: 33P, 1980.

Cohen, I., Daut, J. and Noble, D.: The effects of potassium and temperature on the pacemaker current, i_{K2}, in Purkinje fibres. J. Physiol. (London), 260: 55-74, 1976a.

Cohen, I., Daut, J. and Noble, D.: An analysis of the actions of low concentrations of ouabain on membrane currents in Purkinje fibres. J. Physiol. (London), 260: 75-103, 1976b.

Cohen, I., Eisner, D. and Noble, D.: The action of adrenaline on pacemaker activity in cardiac Purkinje fibres. J. Physiol. (London), 280: 155-168, 1978.

Cohen, I., Noble, D., Ohba, M. and Ojeda, C.: Actions of salicylate ions on the electrical properties of sheep cardiac Purkinje fibres. J. Physiol. (London), 297: 163-185, 1979.

DiFrancesco, D.: The pacemaker current, i_{K2}, in Purkinje fibres is carried by sodium and potassium. J. Physiol. (London), 308: 32P, 1980.

DiFrancesco, D.: A new interpretation of the pacemaker current in Purkinje fibre. J. Physiol. (London), 314: 359-376, 1981a.

DiFrancesco, D.: A study of the ionic nature of the pacemaker current in Purkinje fibres. J. Physiol. (London), 314: 377-393, 1981b.

DiFrancesco, D.: The current "i_{K2}" in Purkinje fibres reinterpreted and identified with the pacemaker current i_f in the SA node. (this volume), 1981c.

DiFrancesco, D. and McNaughton, P.A.: The effects of calcium on outward membrane currents in the cardiac Purkinje fibre. J. Physiol. (London), 289, 347-373, 1979.

DiFrancesco, D. and Noble, D.: If "i_{K2}" is an inward current, how does it display potassium specificity? J. Physiol. (London), 305: 14-15P, 1980a.

DiFrancesco, D. and Noble, D.: Reconstruction of Purkinje fibre currents in sodium-free solution. J. Physiol. (London), 308: 35P, 1980b.

DiFrancesco, D., Noma, A. and Trautwein, W.: Kinetics and magnitude of the time-dependent K-current in the rabbit SA node: effect of external potassium. Pfluegers Arch., 381: 271-279, 1979.

DiFrancesco, D., Ohba, M. and Ojeda, C.: Measurement and significance of the reversal potential for the pacemaker current (i_{K2}) in sheep Purkinje fibres. J. Physiol. (London), 297: 135-162, 1979.

DiFrancesco, D. and Ojeda, C.: Properties of the pacemaker current i_f in the sinoatrial node of the rabbit compared with those of the current i_{k2} in Purkinje fibres. J. Physiol. (London), 308: 353-367, 1980.

Eisner, D.A. and Lederer, W.J.: Characterisation of the electrogenic sodium pump in cardiac Purkinje fibres. J. Physiol. (London), 303, 441-474, 1980.

Fox, L.: In: Modern Computing Methods, Chapter 12. H.M.S.O. London, 1961.

Gadsby, D.C. and Cranefield, P.F.: Two levels of resting potential in cardiac Purkinje fibres. J. Gen. Physiol., 70: 725-746, 1977.

Hart, G., Noble, D. and Shimoni, Y.: Adrenaline shifts the voltage dependence of the Na^+ and K^+ components of i_f in sheep Purkinje fibres. J. Physiol. (London), 308: 34P, 1980.

Hutter, O.F. and Noble, D.: Rectifying properties of heart muscle. Nature, 188: 495, 1960.

McAllister, R.E. and Noble, D.: The time and voltage dependence of the slow outward current in cardiac Purkinje fibres. J. Physiol. (London), 186: 632-662, 1966.

McAllister, R.E., Noble, D. and Tsien, R.W.: Reconstruction of the electrical activity of cardiac Purkinje fibres. J. Physiol. (London), 251: 1-59, 1975.

McDonald, T.F. and Trautwein, W.: The potassium current underlying delayed rectification in cat ventricular muscle. J. Physiol. (London), 274: 217-246, 1978.

128   D. DiFrancesco and D. Noble

Noble, D.: Conductance mechanisms in excitable cells. Biomembranes 3, Kreuzer,
    F. and Slegers, J.F.G., eds., Plenum Press, New York, pp. 427-447, 1972.

Noble, D. and Tsien, R.W.: The kinetics and rectifier properties of the slow
    potassium current in cardiac Purkinje fibres. J. Physiol. (London), 195:
    185-214, 1968.

Noble, D. and Tsien, R.W.: Outward membrane currents activated in the plateau
    range of potentials in cardiac Purkinje fibres. J. Physiol. (London),
    200, 205-231, 1969.

Noble, D. and Tsien, R.W.: The repolarization process of heart cells.   In:
    Electrical Phenomena in the Heart, De Mello, W.C., ed., Academic Press, New
    York, pp. 133-161, 1972.

Peper, K. and Trautwein, W.: A note on the pacemaker current in Purkinje fi-
    bres. Pfluegers Arch., 309: 356-361, 1969.

Vassalle, M.: Cardiac pacemaker potentials at different extracellular and in-
    tracellular K concentrations. Am. J. Physiol., 208: 770-775, 1965.

Vassalle, M.: Analysis of cardiac pacemaker potential using a "voltage clamp"
    technique. Am. J. Physiol., 210: 1335-1341, 1966.

Weidmann, S.: Effect of current flow on the membrane potential of cardiac mus-
    cle. J. Physiol. (London), 115: 227-236, 1951.

Weidmann, S.: Electrophysiologie der Herzmuskelfaser. Huber, Bern, 1956.

Yanagihara, K. and Irisawa, H.: Inward current activated during hyperpolariza-
    tion in the rabbit sinoatrial node cell. Pfluegers Arch., 385: 11-19,
    1980.

Phil. Trans. R. Soc. Lond. B **307**, 353–398 (1985)   [353]
Printed in Great Britain

A MODEL OF CARDIAC ELECTRICAL ACTIVITY INCORPORATING IONIC PUMPS AND CONCENTRATION CHANGES

By D. DiFrancesco[1] and D. Noble, F.R.S.[2]

[1]*Dipartimento di Fisiologia e Biochimica Generali, Sez Elettrofisiologia, Via Celoria, 26, 20133 Milano, Italy*
[2]*University Laboratory of Physiology, Parks Road, Oxford, OX1 3PT, U.K.*

(*Received 8 February* 1984)

CONTENTS

354   D. DiFRANCESCO AND D. NOBLE

Equations have been developed to describe cardiac action potentials and pacemaker activity. The model takes account of extensive developments in experimental work since the formulation of the M.N.T. (R. E. McAllister, D. Noble and R. W. Tsien, *J. Physiol., Lond.* **251**, 1–59 (1975)) and B.R. (G. W. Beeler and H. Reuter, *J. Physiol., Lond.* **268**, 177–210 (1977)) equations.

The current mechanism i_{K2} has been replaced by the hyperpolarizing-activated current, i_f. Depletion and accumulation of potassium ions in the extracellular space are represented either by partial differential equations for diffusion in cylindrical or spherical preparations or, when such accuracy is not essential, by a three-compartment model in which the extracellular concentration in the intercellular space is uniform. The description of the delayed K current, i_K, remains based on the work of D. Noble and R. W. Tsien (*J. Physiol., Lond.* **200**, 205–231 (1969a)). The instantaneous inward-rectifier, i_{K1}, is based on S. Hagiwara and K. Takahashi's equation (*J. Membrane Biol.* **18**, 61–80 (1974)) and on the patch clamp studies of B. Sakmann and G. Trube (*J. Physiol., Lond.* **347**, 641–658 (1984)) and of Y. Momose, G. Szabo and W. R. Giles (*Biophys. J.* **41**, 311a (1983)). The equations successfully account for all the properties formerly attributed to i_{K2}, as well as giving more complete descriptions of i_{K1} and i_K.

The sodium current equations are based on experimental data of T. J. Colatsky (*J. Physiol., Lond.* **305**, 215–234 (1980)) and A. M. Brown, K. S. Lee and T. Powell (*J. Physiol., Lond.* **318**, 479–500 (1981)). The equations correctly reproduce the range and magnitude of the sodium 'window' current.

The second inward current is based in part on the data of H. Reuter and H. Scholz (*J. Physiol., Lond.* **264**, 17–47 (1977)) and K. S. Lee and R. W. Tsien (*Nature, Lond.* **297**, 498–501 (1982)) so far as the ion selectivity is concerned. However, the activation and inactivation gating kinetics have been greatly speeded up to reproduce the very much faster currents recorded in recent work. A major consequence of this change is that Ca current inactivation mostly occurs very early in the action potential plateau.

The sodium–potassium exchange pump equations are based on data reported by D. C. Gadsby (*Proc. natn. Acad. Sci. U.S.A.* **77**, 4035–4039 (1980)) and by D. A. Eisner and W. J. Lederer (*J. Physiol., Lond.* **303**, 441–474 (1980)). The sodium–calcium exchange current is based on L. J. Mullins' equations (*J. gen. Physiol.* **70**, 681–695 (1977)). Intracellular calcium sequestration is represented by simple equations for uptake into a reticulum store which then reprimes a release store. The repriming equations use the data of W. R. Gibbons & H. A. Fozzard (*J. gen. Physiol.* **65**, 367–384 (1975b)). Following Fabiato & Fabiato's work (*J. Physiol., Lond.* **249**, 469–495 (1975)), Ca release is assumed to be triggered by intracellular free calcium. The equations reproduce the essential features of intracellular free calcium transients as measured with aequorin.

The explanatory range of the model entirely includes and greatly extends that of the M.N.T. equations. Despite the major changes made, the overall time-course of

MODEL OF CARDIAC ELECTRICAL ACTIVITY   355

the conductance changes to potassium ions strongly resembles that of the M.N.T. model. There are however important differences in the time courses of Na and Ca conductance changes. The Na conductance now includes a component due to the hyperpolarizing-activated current, i_f, which slowly increases during the pacemaker depolarization. The Ca conductance changes are very much faster than in the M.N.T. model so that in action potentials longer than about 50 ms the primary contribution of the fast gated calcium channel to the plateau is due to a steady-state 'window' current or non-inactivated component. Slower calcium or Ca-activated currents, such as the Na–Ca exchange current, or Ca-gated currents, or a much slower Ca channel must then play the dynamic role previously attributed to the kinetics of a single type of calcium channel. This feature of the model in turn means that the repolarization process should be related to the inotropic state, as indicated by experimental work.

The model successfully reproduces intracellular sodium concentration changes produced by variations in $[\text{Na}]_0$, or Na–K pump block. The sodium dependence of the overshoot potential is well reproduced despite the fact that steady state intracellular Na is proportional to extracellular Na, as in the experimental results of D. Ellis *J. Physiol., Lond.* **274**, 211–240 (1977)).

The model reproduces the responses to current pulses applied during the plateau and pacemaker phases. In particular, a substantial net decrease in conductance is predicted during the pacemaker depolarization despite the fact that the controlling process is an increase in conductance for the hyperpolarizing-activated current.

The immediate effects of changing extracellular [K] are reproduced, including: (i) the shortening of action potential duration and suppression of pacemaker activity at high [K]; (ii) the increased automaticity at moderately low [K]; and (iii) the depolarization to the plateau range with premature depolarizations and low voltage oscillations at very low [K].

The ionic currents attributed to changes in Na–K pump activity are well reproduced. It is shown that the apparent K_m for K activation of the pump depends strongly on the size of the restricted extracellular space. With a 30% space (as in canine Purkinje fibres) the apparent K_m is close to the assumed real value of 1 mM. When the extracellular space is reduced to below 5%, the apparent K_m increases by up to an order of magnitude. A substantial part of the pump is then not available for inhibition by low $[\text{K}]_b$. These results can explain the apparent discrepancies in the literature concerning the K_m for pump activation.

DEFINITION OF SYMBOLS

Voltages are in millivolts, concentrations in millimoles per litre, currents in nanoamperes.

t	time (seconds)
E_m	membrane potential
E_{Na}	sodium equilibrium potential
E_{Ca}	calcium equilibrium potential
E_K	potassium equilibrium potential
i_{tot}	total membrane ionic current flow
C	membrane capacitance (microfarads)
a	radius of preparation (micrometres)
l	length of preparation (micrometres)
x	radial distance (micrometres)
D	K$^+$ ion diffusion constant

356   D. DiFRANCESCO AND D. NOBLE

V	total volume of preparation (microlitres)
V_i	total intracellular volume (microlitres)
V_e	total extracellular volume (microlitres)
V_{up}	volume of sarcoplasmic reticulum (s.r.) uptake store
V_{rel}	volume of store of releasable calcium (note: no assumptions are made on whether these stores are physically distinct)
V_{ecs}	fraction occupied by extracellular space
F	Faraday constant
$[Na]_o$, $[Na]_i$	extra- and intracellular Na concentrations (millimoles per litre)
$[K]_b$, $[K]_c$, $[K]_i$	bulk, cleft and intracellular K concentrations
$[Ca]_o$, $[Ca]_i$	extra- and intracellular Ca concentrations
$[Ca]_{up}$, $[Ca]_{rel}$	Ca concentrations in s.r. uptake and release stores
$[\overline{Ca}]_{up}$	maximum concentration in s.r. uptake store
$i_{b,Na}$	sodium background current
$g_{b,Na}$	sodium background conductance
i_p	sodium–potassium exchange pump current
$\overline{i}_p$	maximum value of i_p
$K_{m,K}$	K_m for K activation of Na–K pump
$K_{m,Na}$	K_m for Na activation of Na–K pump
i_{NaCa}	Na–Ca exchange current
k_{NaCa}	scaling factor for i_{NaCa}
E_{NaCa}	reversal potential for i_{NaCa}
n_{NaCa}	stoichiometry of Na–Ca exchange (Na:Ca)
γ_{NaCa}	position of energy barrier controlling voltage-dependence of i_{NaCa}
d_{NaCa}	denominator constant for i_{NaCa}
$i_{b,Ca}$	calcium background current
$g_{b,Ca}$	calcium background conductance
i_{Na}	TTX sensitive fast sodium current
g_{Na}	conductance of i_{Na} channels
m, α_m, β_m	activation gate and rate coefficients
h, α_h, β_h	inactivation gate and rate coefficients
E_{mh}	Reversal potential for sodium channel
i_{si}	total TTX-insensitive inward current (the 'second inward current')
$i_{Ca,f}$	fast calcium current (first component of i_{si})
$\overline{i}_{Ca,f}$	fully-activated value of $i_{Ca,f}$
$i_{Ca,s}$	slow calcium current (third component of i_{si})
$i_{si,Ca}$, $i_{si,Na}$, $i_{si,K}$	Ca, Na and K components of $i_{Ca,f}$
d, α_d, β_d	activation gating and rate coefficients for $i_{Ca,f}$
f, α_f, β_f	inactivation gating and rate coefficients for $i_{Ca,f}$
$f_2, \alpha_{f2}, \beta_{f2}$	Ca_i dependent inactivation of $i_{Ca,f}$
$i_{m,Na}$, $i_{m,K}$, $i_{m,Ca}$	net membrane fluxes expressed as currents
E_{rev}	'reversal potential' for i_{K2}
$K_{m,Ca}$	K_m for Ca binding to release site
r	number of Ca ions required to bind to activate release
i_{up}	Ca uptake into s.r. expressed as a current

i_{tr}	Ca transferred into releasable form
i_{rel}	Ca release
p	variable controlling transfer of Ca to release sites
τ_{up}	time constant for s.r. uptake of calcium
τ_{rep}	time constant for repriming release store
τ_{rel}	time constant for Ca release (note: these time constants are not necessarily the overall time constants: see equations (42) to (51) for more details)
i_f	hyperpolarizing-activated Na–K current (nearest equivalent to i_{K2} in M.N.T. model)
$\bar{i}_f$	fully-activated value of i_f
$K_{m,f}$	K_m for extracellular K activation of i_f
$g_{f,K}$	K conductance of i_f channels
$g_{f,Na}$	Na conductance of i_f channels
y, α_y, β_y	gating variable and rate coefficients for i_f
i_K	delayed K current (equivalent of i_x in M.N.T. model)
$\bar{i}_K$	fully activated value of i_K
$i_{K,max}$	maximum outward current carried by i_K (at $[K]_i = 140$ mM)
x, α_x, β_x	gating variable and rate coefficients for i_K
i_{K1}	background K current (inward rectifier)
$K_{m,K1}$	K_m for K activation of i_{K1}
i_{to}	transient outward current
$K_{m,to}$	K_m for $[Ca]_i$ activation of i_{to}

INTRODUCTION

In 1975, McAllister *et al.* published a model of Purkinje fibre electrical activity. This model (which in the present paper we shall refer to as the M.N.T. model) represented the ionic currents using gating equations of the Hodgkin–Huxley form, and was based on Noble & Tsien's (1968, 1969 *a*, *b*) experimental analysis of slow ionic current mechanisms together with Beeler & Reuter's (1970 *a*, *b*) work on the second inward current. Beeler & Reuter (1977) subsequently developed a similar model for ventricular activity.

The very substantial delay between the experimental and theoretical papers reflects, in part, the difficulties involved. Detailed experimental information on some of the important currents (i_{Na} in particular) was scanty, and it was a matter for judgement to decide when a worthwhile model had been developed. That was bound to be a difficult judgement given the nature of the arguments on the use of voltage clamp techniques in the heart (Johnson & Lieberman 1971; Attwell & Cohen 1977; Beeler & McGuigan 1978).

However useful the M.N.T. model may have been, it has now outlived that usefulness, and for a variety of reasons. First, one of the major elements of the model, that is, the i_{K2} system, has recently been radically re-interpreted (DiFrancesco 1981 *a*, *b*; DiFrancesco & Noble 1980 *a*, 1981, 1982). Secondly, much better experimental information on the sodium current in the heart (Lee *et al.* 1979; Ebihara *et al.* 1980; Brown *et al.* 1981) and in Purkinje fibres in particular (Colatsky 1980) is now available. Thirdly, it has become increasingly important to take account

358   D. DiFRANCESCO AND D. NOBLE

of intracellular and extracellular ion concentration changes and, therefore, of the influence of ionic pumps, exchange mechanisms and of restricted diffusion. Good experimental information is also now available on the sodium pump in Purkinje fibres (Isenberg & Trautwein 1974; Ellis 1977; Deitmer & Ellis 1978; Gadsby 1980; Eisner & Lederer 1980) and on the influence of extracellular potassium ions on potassium and potassium-dependent currents (DiFrancesco & McNaughton 1979; DiFrancesco et al. 1979b; Brown et al. 1980).

Some information is also available on the sodium–calcium exchange process (Horackova & Vassort 1979; Chapman & Tunstall 1980; Coraboeuf et al. 1981; Fischmeister & Vassort 1981; Sheu & Fozzard 1982; Mentrard & Vassort 1982), and the possible equations for an electrogenic Na–Ca exchange have recently been reviewed by Mullins (1977, 1981). We have incorporated this information together with modelling of the Ca sequestration and release mechanisms based on the data given by Chapman (1979) and on the calcium-induced calcium release hypothesis of Fabiato & Fabiato (1975). Important changes have also occurred in the description and analysis of the second inward current (see review by Noble 1984).

Initially, our work was directed towards the question whether all the properties of 'i_{K2}' and of the pacemaker potential that had led, apparently so conclusively, to the i_{K2} hypothesis were compatible with the new interpretation of this mechanism as an inward, largely sodium, current i_f that is activated by hyperpolarization. The answer to that question is that these properties, including the 'Nernstian' behaviour of the reversal potential (E_{K2}) (DiFrancesco & Noble 1980a), inward-going rectification, the 'cross-over' phenomenon, and the slope conductance changes are indeed fully compatible with the new interpretation, and that some other properties, such as the disappearance of 'i_{K2}' in sodium-free solutions (McAllister & Noble 1966; DiFrancesco & Noble 1980b) and the otherwise anomalous conductance measurements reported by DiFrancesco (1981a), now receive natural explanations that were not within the scope of the i_{K2} hypothesis or the M.N.T. model. A full account of this work has recently appeared in the Amsterdam symposium on cardiac rate and rhythm (DiFrancesco & Noble 1982). In the present paper we shall refer only fairly briefly to the relevant results presented in that paper using an earlier and much simpler version of the equations.

The work for the Amsterdam paper was limited to answering a particular and pressing question, but it clearly formed the basis for the more ambitious undertaking to develop a model that incorporates the full explanatory range of the M.N.T. equations and the greatly extended range that is now possible with the newer results referred to above. It is this development that we report in this paper and in a subsequent paper (DiFrancesco et al. 1985). Accompanying papers (Noble & Noble 1984; Brown et al. 1984a, b) describe the extension of the model to the mammalian s.a. node and its application to experimental results in that tissue.

DESCRIPTION OF EQUATIONS

We have chosen to use absolute units of current (in nanoamperes) scaled to give currents similar to those recorded experimentally in a Purkinje strand of length 2 mm and radius 50 μm. The reason for choosing this convention rather than using current density is that in many of the calculations current density varies as a function of position in the preparation (to take account of concentration profiles in the extracellular space). With regard to K-dependent currents therefore a single current density might be a misleading parameter. The magnitudes were sometimes scaled up or down to give currents for larger or smaller preparations. The surface area of our standard fibre is 0.0063 cm². Assuming that the total cell membrane area is ten times larger than the cylinder surface (Mobley & Page 1972), the total cell surface would be 0.063 cm². Thus, to convert our figures to nanoamperes per square centimetre the currents should be multiplied by a factor of about 15. We have assumed a membrane capacitance of 12 µF cm⁻² of cylinder surface (Weidmann 1952) or 1.2 µF cm⁻² of cell surface, which gives a value of 0.0756 µF for our standard preparation.

The differential equation for the variation of membrane potential, E_m, is

$$dE_m/dt = -i_{tot}/C \tag{1}$$

where C is the membrane capacitance and i_{tot} is the total current:

$$i_{tot} = i_f + i_K + i_{K1} + i_{to} + i_{b,Na} + i_{b,Ca} + i_p + i_{NaCa} + i_{Na} + i_{Ca,f} + i_{Ca,s} + i_{pulse}. \tag{2}$$

Each of these current components will now be explained in turn.

(a) Hyperpolarizing-activated current, i_f

The experimental evidence (DiFrancesco 1981 a) shows that the fully activated current–voltage relation for this channel is nearly linear. Some of the deviation from linearity, particularly at extreme negative potentials, might be attributed to residual K ion depletion in the extracellular space, although the presence of outward-going rectification at high K⁺ concentrations (DiFrancesco 1982) argues in favour of it being in part a genuine channel property. Nevertheless, a linear i_f function is a good approximation in the pacemaker range of potentials where i_f has its most important functional role. The behaviour of the reversal potential is consistent with the view that the total current is composed of relatively independent Na⁺ and K⁺ components and that, at normal K⁺ and Na⁺ concentrations, the contributions of these two ions to the total conductance are approximately equal. The net reversal potential in normal physiological solutions is then around -20 mV. At high values of external bulk potassium $[K]_b$, the current is greatly increased (DiFrancesco 1981 b). This property suggests that the channel is activated by external potassium. We have assumed a simple first-order binding process for this activation. The experimental value for $K_{m,f}$ (that is, the value of $[K]_b$ for half activation) is 45 mM (DiFrancesco 1982). In Na-free solutions, only the K⁺ component is present. This then shows a reversal potential close to the expected value for E_K (Hart *et al.* 1980).

The equation we shall use for the fully activated current, i_f, is therefore:

$$i_f = ([K]_c/([K]_c + K_{m,f})) \{g_{f,K}(E - E_K) + g_{f,Na}(E - E_{Na})\}. \tag{3}$$

Suitable experimental values for the constants in this equation are $g_{f,Na} = 3$ µS, $g_{f,K} = 3$ µS, $K_{m,f} = 45$ mM (DiFrancesco 1981 b, 1982).

The gating mechanism controlling i_f is the s process described by Noble & Tsien (1968), except that activation occurs on hyperpolarization, not depolarization. The fully activated state in our model therefore corresponds to the fully deactivated state in Noble & Tsien's analysis. We have chosen the variable y to represent the degree of activation of i_f. So, $y = 1 - s$. The equations for α_y and β_y are those in the M.N.T. model for β_s and α_s respectively:

$$dy/dt = \alpha_y(1-y) - \beta_y y \tag{4}$$

where:

$$\alpha_y = 0.025 \exp(-0.067(E+52)), \tag{5}$$

$$\beta_y = 0.5 (E+52)/(1 - \exp(0.2(E+52))), \tag{6}$$

$$(\beta_y)_{E=-52} = 2.5. \tag{6a}$$

The net current is then given by:

$$i_f = y i_f. \tag{7}$$

It should be noted that, while these equations assume first-order voltage-dependent kinetics for the gating parameter, y, the most recent experimental data (DiFrancesco & Ferroni 1983; Hart 1983; DiFrancesco 1984) shows that the onset of i_f is in fact sigmoid: there is a delay in the time course which can be removed by conditioning hyperpolarizations. This property, which is of course important for detailed modelling of channel properties, does not have much importance in reconstructing the pacemaker potential since in the relevant voltage range the current is very slow and an initial small delay not too important. For simplicity we have retained the M.N.T. first-order kinetics, though these could readily be substituted in the program by more complex equations without significant change in the results computed here.

(b) Time-dependent (delayed) K⁺ current, i_K

A considerable amount of new experimental information has appeared on this current since the M.N.T. equations were formulated. First, it has been shown in a variety of preparations (Purkinje fibres: DiFrancesco & McNaughton 1979; frog atrium: Brown *et al.* 1980; ventricle: McDonald & Trautwein 1978; Rabbit s.a. node: DiFrancesco *et al.* 1979) that, while the instantaneous current–voltage relation shows inward-going rectification without a negative slope conductance region (as first shown by Noble & Tsien 1969a), it does *not* show the cross-over phenomenon, that is, at all potentials, the current is a monotonic function of $[K]_c$. In this respect, the current differs quite markedly from i_{K1}. The absence of the cross-over effect allows us to use a very simple formulation both for the rectification property and for the K⁺ dependence of the current. This is based on using rate theory, assuming that the major energy barrier for ion movement in the electric field is situated at the inner surface of the membrane (Noble 1972; Jack *et al.* 1975). This gives the equation:

$$i_K = i_{K,\,max} \{[K]_i - [K]_c \exp(-E/25)\}/140. \tag{8}$$

The usual value used for the 'maximum' current (actually the maximum outward current at positive potentials when $[K]_i = 140$ mM) is 180 nA. $[K]_i$ was usually set to 140 mM (Lee & Fozzard 1975; Miura *et al.* 1977). These parameters give outward currents similar to the delayed outward current recorded by Noble & Tsien (1969a).

Notice that, following McDonald & Trautwein (1978), we have chosen the symbol i_K for this current rather than the symbol i_x used by Noble & Tsien (1969a). The justification for this change is that, in the M.N.T. model, E_{K2} is regarded as the true value of E_K. Since this was considerably *negative* to the reversal potential for the delayed current activated in the plateau range of potentials, it was concluded that the latter was a less specific channel. The new interpretation of E_{K2} as a mixed 'reversal' potential means that the true value of E_K is almost certainly 10–20 mV positive to E_{K2} (see DiFrancesco & Noble (1982) for an equation relating E_{K2} to the true value of E_K), so that the reversal potential for the plateau-activated current is much closer to E_K than in the M.N.T. equations. We have therefore regarded it as a specific K⁺ current for which it is more natural to use the symbol i_K. While this current should clearly *not* be confused with i_{K2} in the M.N.T. model, it *does* correspond to the g_{K2} system first described by Hall *et al.* (1973) which was used in the 1962 model (Nobel 1962). In several respects, our formulation of the equations for K⁺ currents closely resembles the 1962 model and its development (Noble 1965) to account for extracellular K⁺ effects.

The second aspect of this system that has been investigated further experimentally is the fact, also first observed by Noble & Tsien (1969a), that at least two and sometimes three exponential terms are required fully to describe the time course of the current following voltage step changes. This feature has been confirmed in all the multicellular preparations investigated so far with a wide variety of different voltage clamp techniques, though the detailed kinetics sometimes differ from those of Noble & Tsien (see, for example, Brown *et al.* 1972) even in Purkinje fibres (R. H. Brown and D. Noble, unpublished). The question that arises is whether this reflects a genuine property of the gating process or whether it is produced partly or even wholly by perturbations due to ion concentration changes. The most complete analysis of this problem (Brown *et al.* 1980; DiFrancesco & Noble 1980a) shows that the slowest exponential term, when present, is indeed due to a K^+ accumulation process but that, although this necessarily perturbs the time course of i_K (Attwell *et al.* 1979b), this perturbation does not account for the biexponential time course of the remaining components, whose time constants are not seriously perturbed. We are therefore left with the problem faced by Noble & Tsien (1969a) that a single Hodgkin–Huxley type gating reaction does not account for the current time course. We will adopt the same solution as Noble & Tsien (1969a) that is, to note, like them, that only one of the components is of significant importance during repolarization (Noble & Tsien 1969b). For simplicity, we shall drop subscripts and use the gating symbol x for the controlling reaction:

$$\mathrm{d}x/\mathrm{d}t = \alpha_x(1-x) - \beta_x x, \tag{9}$$

$$\alpha_x = 0.5 \exp\ (0.0826\ (E+50))/(1+\exp\ (0.057\ (E+50))), \tag{10}$$

$$\beta_x = 1.3 \exp\ (-0.06\ (E+20))/(1+\exp\ (-0.04\ (E+20))) \tag{11}$$

where the equations for α_x and β_x are those used in the M.N.T. model. The total current is given by

$$i_K = x i_K. \tag{12}$$

(c) *Time-independent (background)* K^+ *current,* i_{K1}

The M.N.T. model represented this current by a purely empirical function describing inward-going rectification with a negative slope over a range of potentials positive to the resting potential. Since this current is obtained by measuring the current that remains when other identifiable components have been subtracted (in this respect it is exactly analogous to the leak current in the original Hodgkin–Huxley (1952) analysis), it has always been evident that it must include currents other than the true i_{K1}, such as the pump and exchange currents. In the new model we represent these currents by separate equations (see below). This is one reason why our i_{K1} cannot correspond exactly to that in the M.N.T. model. Furthermore, since the state $y = 0$ in our model corresponds to the state $s = 1$ in the M.N.T. model, a term corresponding to i_{K2} in that model now becomes indistinguishable from i_{K1} (for a further explanation of the mapping between these aspects of the two models see DiFrancesco & Noble (1982)).

These changes simply add or subtract to the magnitude and slightly change the form of i_{K1}. A more radical question is whether the basic form of the i_{K1} function is correct or whether it is possible that features such as inward-going rectification are not properties of a single mechanism, but reflect rather our ignorance of some unidentified component. This question acquires added force since we have ourselves shown that the 'inward-going rectification displayed by i_{K2} is not a genuine property of a single mechanism (DiFrancesco & Noble 1982, figure 4).

The most direct way of answering this question is to measure K^+ fluxes as a function of potential. This was first done by Haas & Kern (1966) who showed that the radioactive flux was consistent with the presence of inward-going rectification. Recently, Vereecke *et al.* (1980) have used a much improved technique to show not only that the K^+ efflux is consistent with the presence of inward-going rectification but also that the negative slope region is a genuine characteristic. Clear-cut evidence of the inward-rectifying property comes also from experiments where the i_{K1} (E) relation is measured by substracting the time-independent curves in the presence and absence of Ba^{2+} ions (DiFrancesco 1981 *b*). Finally, recent work with patch-clamp techniques (Momose *et al.* 1983; Sakmann & Trube 1984) shows the presence of i_{K1} and has provided valuable data on its kinetics and $[K]_o$-activation at potentials negative to E_K.

In place of the purely empirical formulation of the M.N.T. model we have chosen to use Hagiwara & Takahashi's (1974) equation. This is also empirical but it is a simple formulation which closely resembles the curves generated by the more complex equations for the blocking particle model of Hille & Schwarz (1978; see their comparison in figure 9 of that paper). We have also incorporated the fact that the channel is K^+ activated (cf. the development of the M.N.T. equations by Cohen *et al.* (1978) and the patch clamp data of Sakmann & Trube (1984)). Our equation is:

$$i_{K1} = g_{K1}([K]_c/([K]_c + K_{m,1})) \{(E - E_K)/(1 + \exp((E - E_K + 10) 2F/RT))\}. \quad (13)$$

$K_{m,1}$ was set to 210 mM (Sakmann & Trube 1984, figure 5) and the maximum conductance (which is the maximum conductance reached during strong hyperpolarizations) was set to 920 µS. We shall show later that this reproduces the main experimental features of the current–voltage relations attributable to i_{K1}.

Carmeliet (1982) has recently raised the question whether i_{K1} is strictly instantaneous. The patch-clamp work indeed shows that there is time-dependent inactivation (Sakmann & Trube 1984) but since this time-dependence becomes important only at very negative potentials we have not used equations for this process. If needed, they could easily be incorporated into the program.

(d) The transient outward current, i_{to}

It has been known since the first studies of i_{K1} that, beyond about -20 mV, the inward rectifier is either masked by a rapidly activated outward rectifier or that the i_{K1} channel itself shows outward rectification positive to -20 mV. The experimental evidence (including the action of blocking agents like Ba^{2+} and Cs^{2+} on inward but not outward-going rectification) favours the first interpretation (see Isenberg 1976; Carmeliet 1980) which is why our equation for i_{K1}, unlike that in the M.N.T. model, describes inward-going rectification only.

Flux measurements by Vereecke *et al.* (1980) favour the view that the outward-rectification, instantaneous and transient, is also largely carried by K^+ ions. The current is very sensitive to external K^+ ions (Hart *et al.* 1982), and is largely, but not entirely, blocked by 4-aminopyridine (Boyett 1981 *b*; Coraboeuf & Carmeliet 1982).

Originally, a Hodgkin–Huxley type model was used for this current which was attributed to Cl^- ions (see Dudel *et al.* 1967 *a*; Fozzard & Hiraoka 1973; McAllister *et al.* 1975). There are, however, serious difficulties with this interpretation. The time constants are in fact relatively independent of voltage and 'envelope' tests (cf. Noble & Tsien 1968) do not work (Hart *et al.* 1982). Moreover, Siegelbaum & Tsien (1980) have shown that the activation is $[Ca]_i$-dependent.

MODEL OF CARDIAC ELECTRICAL ACTIVITY   363

We have therefore represented i_{to} as an outward rectifier that is $[Ca]_i$-activated and which depends on $[K]_o$. As McAllister, Noble & Tsien (1975) have shown, the precise inactivation process for i_{to} is not important during a single action potential, though repriming of the process is important during repetitive firing (for example, Hauswirth *et al.* 1972; Boyett 1981 *a*; Boyett & Jewell 1980). Moreover, the inactivation process is not well understood.

Our equation for the K^+ activation, Ca^{2+} activation and rectification properties of i_{to} is

$$i_{to} = 0.28((0.2+[K]_c)/(K_{m,1}+[K]_c))\,([Ca]_i/(K_{m,to}+[Ca]_i))$$
$$\times\{(E+10)/(1-\exp(-0.2\,(E+10)))\}\{([K]_i\exp(0.02\,E)-[K]_o\exp(-0.02\,E)\}. \quad (14)$$

The first term in this equation represents activation by external $[K^+]$ which saturates at about 30 mM (Hart *et al.* 1982). The second term represents $[Ca]_i$ activation. We usually set $K_{m,to}$ to 1 μM, which allows the normal $[Ca]_i$ transient to activate the current with the correct magnitude and speed to reproduce Siegelbaum & Tsien's (1980) experimental results. The third term represents the voltage dependence (this term could be replaced by a gating process if desired). This term was set to 5 at $E = -10$. The final term is obtained from rate theory assuming that the energy barrier is placed at the centre of the membrane, which generates a moderate degree of outward rectification.

The inactivation process was described by a first order equation fitted to Fozzard & Hiraoka's (1973) data:

$$\alpha_r = 0.033\exp(-E/17) \quad (15)$$

$$\beta_r = 33/(1+\exp(-(E+10)/8)) \quad (16)$$

$$dr/dt = \alpha_r(1-r)-\beta_r r \quad (17)$$

$$i_{to} = ri_{to}. \quad (18)$$

(It should be noted that these equations represent the main features of i_{to} but they do not fully represent the multicomponent nature of i_{to}. This will be dealt with by DiFrancesco *et al.* (1985).)

(e) Background sodium current, $i_{b,Na}$

As in the M.N.T. model, the resting sodium flux is represented by a linear relation:

$$i_{b,Na} = g_{b,Na}(E-E_{Na}). \quad (19)$$

Setting $g_{b,Na}$ equal to 0.18 μS gives a resting sodium influx that is both sufficient to account for the deviation of E from E_K and for the rate of increase in intracellular sodium when the sodium pump is blocked (Ellis 1977).

When varying $[Na]_o$ in the computations, we have assumed that the fraction of $g_{b,Na}$ that is carried by sodium is proportional to $[Na]_o$. In effect, this assumption allows for the fact that common Na^+ substitutes like choline are known to permeate the membrane. The assumption of a linear dependence of the Na current on $[Na]_o$ is the simplest we could make but it turns out to be adequate for the present purposes. Equation (19) then becomes:

$$i_b = ([Na]_o/[Na]_{o,c})\,g_{b,Na}(E-E_{Na})+i_{b,Ch} \quad (20)$$

where $[Na]_{o,c}$ is the control level of $[Na]_o$ (usually 140 mM) and $i_{b,Ch}$ is the background current due to choline or another Na substitute. This equation assumes that Na and other ions move independently through the background channel. A further test of the value chosen for $g_{b,Na}$

is whether it allows accurate prediction of the rate of change of internal sodium following external sodium concentration changes. This is the case (see figure 9).

During the development of this model, Colquhoun *et al.* (1981) published patch-clamp studies of a linear non-specific cation channel activated by Ca^{2+} ions. It was initially tempting to conclude that this channel might account for the resting background conductance. This possibility was incorporated into the computer program by allowing an option to use a background conductance equally permeable to Na^+ and K^+ ions and which is Ca^{2+} activated. There are, however, serious difficulties in using this option (see Conclusions) and we have not used it in the present paper.

(f) Na–K exchange pump current, i_p

The Na–K exchange pump in Purkinje fibres has been extensively studied recently (Ellis 1977; Deitmer & Ellis 1978; Gadsby 1980; Gadsby & Cranefield 1979; Eisner & Lederer 1980; Eisner *et al.* 1981). The results agree in showing that the pump is directly electrogenic with a probable stoichiometry of 3:2 (Na:K). At rest, therefore, there must be an outward pump current, i_p, equal to one third of the net sodium influx generated by Na conducting channels and by the Na–Ca exchange process (see (g) below).

For simplicity, we have assumed that the pump is activated by external K^+ and by internal Na^+ by first-order binding processes:

$$i_p = i_p([K]_c/(K_{m,K}+[K]_c))([Na]_i/(K_{m,Na}+[Na]_i)). \qquad (21)$$

where i_p is the maximum pump current, $K_{m,K}$ is the value of $[K]_c$ for half-activation and $K_{m,Na}$ is the value of $[Na]_i$ for half-activation. The experimental evidence (Eisner *et al.* 1981) shows that over the whole range of values so far explored (up to about 20 mM) the pump rate is linearly dependent on $[Na]_i$. This means that $K_{m,Na}$ must be considerably larger than 20 mM. We have chosen to use 40 mM.

At first sight, there is considerable disagreement on the value of $K_{m,K}$. Gadsby (1980) obtained 1 mM in canine Purkinje fibres, whereas Eisner & Lederer (1980) obtained 4–5 mM in the sheep. Deitmer & Ellis (1978) obtained an even higher value (around 10 mM). We shall show in this paper that this variation is in fact compatible with a single value of $K_{m,K}$ provided that effects due to the restricted extracellular space are taken into account. On this view, the best value for $K_{m,K}$ is the lowest one obtained in the species with the largest extracellular space. We shall therefore use 1 mM for this parameter. This does in fact correspond well with the values in other tissues. With these values for the activation parameters, a maximum current of 125 nA gives a resting pump current of about 20 nA when $[K]_c = 4$ mM and $[Na]_i = 9$ mM. This current is similar to that estimated by extrapolating the $i_p([Na]_i)$ function of Eisner *et al.* (1981).

(g) Na–Ca exchange current, i_{NaCa}

The evidence that this exchange mechanism is directly electrogenic has recently been reviewed by Mullins (1981) who has also proposed that the current generated, which we will call i_{NaCa}, may replace some of the currents already identified in cardiac electrophysiology. We will discuss elsewhere the extent to which our results support this suggestion (see DiFrancesco *et al.* 1985 and also Brown *et al.* 1984*a*, *b*). Fischmeister & Vassort (1981) have recently incorporated the Na–Ca exchange current into the M.N.T. model (for a comparison, see DiFrancesco *et al.* 1985).

MODEL OF CARDIAC ELECTRICAL ACTIVITY   365

The equations for i_{NaCa} are based on the assumption that the only energy available to the process is that of the Na and Ca ion gradients and the membrane potential. Two alternative expressions have been used in our work. The simplest assumes that the current is a hyperbolic sine function of the energy gradient expressed in millivolts:

$$i_{NaCa} = k_{NaCa} \{\exp((E - E_{NaCa})F/RT) - \exp((-(E - E_{NaCa})F/RT)\}/2 \quad (22)$$

where

$$E_{NaCa} = (n_{NaCa}E_{Na} - 2E_{Ca})/(n_{NaCa} - 2), \quad (23)$$

$$E_{Na} = (RT/F) \ln([Na]_o/[Na]_i), \quad (24)$$

$$E_{Ca} = (RT/2F) \ln([Ca]_o/[Ca]_i) \quad (25)$$

and n_{NaCa} is the stoichiometry of the exchange. We have used either $3:1$, as suggested by some of the experimental literature in the heart, or $4:1$ as suggested by work on squid nerve (Mullins (1981) figure 4.3). Most of the results in this paper use $3:1$ and the question whether $4:1$ would equally well fit the results will be treated by DiFrancesco *et al.* (1985).

Equation (22) is given by Mullins (1977, 1981) as a simplification for his more general model. It may apply moderately well for sudden small voltage changes at fixed ion concentrations. There is however no reason to suppose that it will be at all accurate when large ion concentration changes are involved. In fact, the variations in $[Ca]_i$ may be one or two orders of magnitude during normal electrical activity and it is then important to use a more realistic function that reproduces the expected $[Ca]_i$ dependence of the exchange process. The full equations for the Mullins model are however very complex and many of the rate coefficients are unknown. We have therefore used an intermediate version based on the fact that sodium concentration changes are fairly small, at least during a few action potentials. Some of the terms in Mullins full equations are then constant and we obtain (26):

$$i_{NaCa} = k_{NaCa} (\exp(\gamma(n_{NaCa} - 2)EF/(2RT))[Na]_i{}^n[Ca]_o$$

$$- \exp(-(1-\gamma)(n_{NaCa} - 2)EF/(2RT))[Na]_o{}^n[Ca]_i)/$$

$$(1 + d_{NaCa}([Ca]_i[Na]_o{}^n + [Ca]_o[Na]_i{}^n.) \quad (26)$$

This equation would require further refinement (replacing 1 in the denominator by a function of the sodium concentrations) to take proper account of $[Na]_i$ and $[Na]_o$ changes. The variable γ was set to 0.5 in the standard model. This parameter represents the shape or position of the energy barrier in the electrical field and is exactly analogous to similar parameters used in rate theory to describe current–voltage relations (see, for example, Noble 1972). Some of the computations were run with values of γ set to the extreme values of 0 or 1. It was found that this produces some quantitative changes in the precise time course of i_{NaCa} during an action potential, but does not seriously change the qualitative aspects of the results.

Some of the variables in these equations are either fixed (for example, $[Na]_o$ is usually 140 mM, $[Ca]_o$ is usually 2 mM), or can be computed from the model (for example, $[Na]_i$ and $[Ca]_i$), or can be determined once other model parameters are fixed). Thus, k_{NaCa}, which scales the exchange current for a given energy gradient, can be determined as the value required to ensure that, in the steady state, all the calcium entering the cells is eventually pumped out. A suitable value for k_{NaCa} when $[Na]_i$ is in the range 5–10 mM and $[Ca]_i$ is in the range 0.05–0.1 μM is 20, when (22) is used. For (26) appropriate values are $k_{NaCa} = 0.02$ and $d_{NaCa} = 0.001$.

366   D. DiFRANCESCO AND D. NOBLE

Finally, to keep the resting calcium in this range (as suggested by experimental results with aequorin and Ca electrodes – see Marban *et al.* 1980; Sheu & Fozzard 1982), we require a resting Ca^{2+} leak (cf. Fischmeister & Vassort 1981):

$$i_{b, Ca} = g_{b, Ca}(E - E_{Ca}). \tag{27}$$

A value of $g_{b, Ca}$ that satisfies the above conditions is 0.02 μS.

(h) *The fast sodium current, i_{Na}*

Major experimental advances have been made recently in measuring this current, the most important being the use small synthetic ventricular strands (Ebihara *et al.* 1980), rabbit Purkinje fibres (Colatsky & Tsien 1979; Colatsky 1980) and of single ventricular cells (Brown *et al.* 1981). These studies have provided more reliable information on the kinetics which are significantly different from those used in the M.N.T. model. Another approach has been to measure the steady-state properties by determining the TTX-sensitive steady state ('window') current (Attwell *et al.* 1979*a*).

The data that is most relevant for our purposes is that obtained on Purkinje fibres by Colatsky (1980). The major disadvantage of this data is that it was obtained in cooled fibres, which means that the speeds of the gating reactions must be adjusted to 37 °C. It is also possible that the inactivation curve shifts in a negative direction on the voltage axis at low temperature, which would reduce the overlap of the activation and inactivation curves. The single cell data at 37 °C does indeed show more overlap. Colatsky (1980) even concluded that there was no overlap in his experiments. We shall show that this is too strong a conclusion. Even with his data, we can reconstruct fairly easily the observed 'window' current (see below).

The equations we have used are:

$$i_{Na} = m^3 h \{g_{Na}(E - E_{mh})\}, \tag{28}$$

$$E_{mh} = (RT/F) \ln (([Na]_o + 0.12 [K]_e)/([Na]_i + 0.12 [K]_i)) \tag{29}$$

that is, the sodium channel is assumed to show a 12 % permeability to K^+ ions (Chandler & Meves 1965)

$$dm/dt = \alpha_m (1 - m) - \beta_m m, \tag{30}$$

$$dh/dt = \alpha_h (1 - h) - \beta_h h, \tag{31}$$

$$\alpha_m = 200(E + 41)/(1 - \exp (0.1 (E + 41)), \tag{32}$$

$$\alpha_{(m)E=-41} = 2000, \tag{33}$$

$$\beta_m = 8000 \exp (-0.056(E + 66)), \tag{34}$$

$$\alpha_h = 20 \exp (-0.125 (E + 75)), \tag{35}$$

$$\beta_h = 2000/\{320 \exp (-0.1(E + 75))\}. \tag{36}$$

The value we have used for g_{Na} is 750 μS. This generates a maximum depolarization rate similar to that recorded experimentally. The maximum inward current on depolarizing to 0 mV is then about 3000 nA which, using the scaling factor of 15 for conversion to current density (see above) gives about 500 μA cm^{-2}, that is, the value recorded experimentally (Colatsky 1980).

The *m* equations used here are in fact that of Hodgkin & Huxley (1952) shifted on the voltage axis to give a steady state value of 0.5 for m^3 at -30 mV, which fits Colatsky's data – see below. The rate constants were then scaled to give a time constant, τ_m, of about 100 μs at $E = 0$ mV (Brown *et al.* 1981). It can be seen (see figure 4) that this gives an activation curve that is somewhat less steep than that obtained in Colatsky's experiments. Our reason for choosing a less steep function is that this fits better the experimental data of Brown *et al.* (1981) which was obtained in more favourable conditions. The greater steepness of Colatsky's curve could be due to a small degree of voltage non-uniformity in a multicellular preparation which would be minimized in a single cell clamp.

The h equations were fitted to Colatsky's data to give $h_\infty = 0.5$ at $E = -70$ mV. This in fact corresponds well to one of Colatsky's published curves but it is worth noting that his half-inactivation potential is usually nearer -75 mV, which may well be due to cooling the fibres. The absolute values for the rate constants were adjusted to give τ_h values of about 50 ms at -80 mV, decreasing to 0.7 ms at 0 mV. Brown *et al.* (1981) also found a steep voltage dependence for τ_h between the resting potential, where τ_h is very large, and 0 mV, where it becomes very small. This means that, during a normal action potential, the inactivation is considerably faster than in the M.N.T. model.

These equations do not reproduce slower components of Na inactivation and recovery. Gintant *et al.* (1984) and E. Carmeliet (personal communication) have very recently shown that such a process does exist and that the 'window' current is considerably larger at the beginning of the plateau than at its end, that is, a small but significant component of Na current inactivates with a time course of several hundred milliseconds. In this connection it is worth noting that a persistent problem in our computations has been the presence (though not in the particular computations illustrated in this paper) of a small bump on the repolarization process which is due to the 'window' current. Introducing slow inactivation would be one way of eliminating this problem.

(i) *The second inward current, i_{si}, and its components*

Considerable advances have been made in studying this current since the formulation of the M.N.T. and B.R. equations.

First, Reuter & Scholz (1977) showed that the reversal potential for i_{si} requires that some K^+ ions should cross the channel in addition to Ca^{2+} ions. This view, has been confirmed in the work of Lee & Tsien (1982) using a perfusion electrode clamp of single guinea-pig ventricular cells. Reuter & Scholz also concluded that Na^+ ions cross the channel. This conclusion is now doubtful (see Mitchell *et al.* 1983; Noble 1984). Our computer program allows for this possibility but we have not used this facility in most of our computations.

Second, the work with isolated cells shows that the kinetics of the largest component of i_{si} are *very much* faster than in the M.N.T. and B.R. models. Activation peaks occur within 2–3 ms and the inactivation time constant lies in the range 10–20 ms (see review by Noble 1984). These figures are at least an order of magnitude faster than previously supposed.

Third, the peak amplitude of the calcium current is considerably greater than the multi-cellular work suggested (see discussion in Mitchell *et al.* 1983).

Finally, there is evidence for two or three different components of i_{si}. In addition to the fast component, which we will call $i_{Ca,f}$, a Cd^{2+} and Mn^{2+} resistant channel has been found in single guinea-pig ventricular cells (Lee *et al.* 1984a) and in single frog atrial cells (Hume & Giles 1983). This component is very slowly and, at some voltages, only partly inactivated.

In some ways, therefore, it may play a role similar to the non-inactivated component of i_{si} in the M.N.T. equations for which experimental evidence was recently presented by Kass & Wiegers (1982).

However, there is also another component that may play this role. This is strongly correlated with contraction and may, therefore, be $[Ca]_i$-activated. It has been found and called $i_{si, 2}$ in the mammalian s.a. node (Brown *et al.* 1983, 1984*a*) and in single guinea-pig ventricular cells (Lee *et al.* 1983, 1984*b*). One interpretation of this is that it is carried by the Na–Ca exchange process for which we have given equations in a previous section. One of the purposes of our model is to explore the extent to which these equations reproduce the properties of $i_{si, 2}$ in Purkinje fibres, the s.a. node and in single ventricular cells.

For the fast component, $i_{Ca, f}$, we have followed Reuter & Scholz (1977) in using a constant field type formulation for the individual ion movements (though see Attwell & Jack 1978) for an important critique of this approach).

$$i_{Ca, f} = dff2 \, (i_{si, Ca} + i_{si, K}), \tag{37}$$

$$i_{si, Ca} = 4 P_{si} (E-50) \, (F^2/RT)/(1 - \exp(-(E-50) \, 2F/RT))$$
$$\times \{[Ca]_i \exp(100F/RT) - [Ca]_o \exp(-2(E-50) \, F/RT)\}, \tag{38}$$

$$i_{si, K} = 0.01 P_{si} (E-50) \, (F^2/RT)/(1 - \exp(-(E-50) \, F/RT))$$
$$\times \{[K]_i \exp(50F/RT) - [K]_c \exp(-(E-50) \, F/RT)\}. \tag{39}$$

If required, an equation similar to (39) was used for describing a sodium component.

Note that, in these equations, we do not use an explicit equation for the reversal potential. When required (for example, for calculations of conductance), this was computed either by an iterative procedure or by solving the quadratic equation given by Attwell & Jack (1978).

We now require a description of the gating kinetics (d and f). The original Beeler & Reuter (1977) equations used in the M.N.T. model describe an activation gate, d, with a time constant of about 22 ms at about 0 mV and an inactivation gate with a very long time constant (about 300 ms). This was a very important feature of the M.N.T. and Beeler–Reuter models since the process of i_{si} inactivation is then strongly implicated in controlling the duration of the action potential plateau. More recent work shows that both activation and inactivation occur very much more quickly than in the M.N.T. model. In Purkinje fibres, the most direct evidence on this question comes from the experiments of Siegelbaum & Tsien (1980) who injected EGTA to abolish the internal $[Ca]_i$ transient and so record i_{si} in the absence of currents (such as i_{NaCa} and a component of i_{to}) dependent on $[Ca]_i$. The time constants for i_{si} in single ventricular cells have also been found to be very short, typical values being about 2–5 ms for activation and 10–20 ms for inactivation (Powell *et al.* 1981; Isenberg & Klöckner 1982; Lee & Tsien 1982; Mitchell *et al.* 1982; Mitchell *et al.* 1983). The equations we have used are:

$$dd/dt = \alpha_d (1-d) - \beta_d \, d, \tag{40}$$

$$\alpha_d = 30(E+24)/(1 - \exp(-(E+24)/4)), \tag{41}$$

$$(\alpha_d)_{E=-24} = 120, \tag{41a}$$

$$\beta_d = 12(E+24)/(\exp((E+24)/10) - 1), \tag{42}$$

$$(\beta_d)_{E=-24} = 120. \tag{42a}$$

These equations describe an activation process that has a 'threshold' near -35 mV, half activation at -24 mV and a peak time constant of about 5 ms.

$$\mathrm{d}f/\mathrm{d}t = \alpha_f(1-f) - \beta_f f, \tag{43}$$

$$\alpha_f = 6.25\,(E+34)/(\exp\,((E+34)/4)-1), \tag{44}$$

$$(\alpha_f)_{E=-34} = 25, \tag{44a}$$

$$\beta_f = 50/(1+\exp\,(-(E+34)/4)). \tag{45}$$

These equations describe an inactivation process that is half-maximal at -34 mV (cf. Reuter *et al.* 1982) and has a peak time constant of about 20 ms. With these kinetics, i_{si} reaches a peak in less than 5 ms and is largely inactivated by 50 ms.

For the description of the Ca-dependent inactivation (see Brown *et al.* 1984*a*) we have used a formulation similar to that used recently by Standen & Stanfield (1982):

$$\mathrm{d}f_2/\mathrm{d}t = \alpha_{f2}(1-f_2) - \beta_{f2}[\mathrm{Ca}]_i f_2 \tag{46}$$

which represents Ca-inactivation as occurring via a first-order binding reaction to the channel. In this equation, the speed of recovery from inactivation is determined by α_{f2}, its reciprocal being the time constant of recovery, which we usually set to 0.1 s. At the steady-state the degree of inactivation (that is, $1-f_2$) is given by:

$$1-f_2 = [\mathrm{Ca}]_i/([\mathrm{Ca}]_i + K_{\mathrm{m},f2}) \tag{47}$$

where
$$K_{\mathrm{m},f2} = \alpha_{f2}/\beta_{f2}.$$

The value usually used for $K_{\mathrm{m},f2}$ was 1 μm which gives negligible inactivation at resting levels of $[\mathrm{Ca}]_i$ but appreciable inactivation during the $[\mathrm{Ca}]_i$ transient, as required if the experimental results are to be reproduced. An important result that is reproduced by this formulation is that i_{Ca} inactivation and recovery have quite different time constants even when measured at the same potential (see Brown *et al.* 1984*a*, figure 3).

We will show that, together with the equations for the exchange current, $i_{\mathrm{Na,Ca}}$, the equations reproduce the fast and slow components of i_{si} in Purkinje fibres (see figure 5 below) and in the s.a. node (see Brown *et al.* 1984*a*).

The question, though, remains whether there exists also a component corresponding to $i_{\mathrm{Ca,s}}$ in Purkinje fibres. This question will be explored in another paper (DiFrancesco *et al.* 1984).

(j) Intracellular sodium concentration

If we assume negligible binding of Na$^+$ ions the change in $[\mathrm{Na}]_i$ will be given by:

$$\mathrm{d}[\mathrm{Na}]_i/\mathrm{d}t = -(i_{\mathrm{Na}} + i_{\mathrm{b,Na}} + i_{\mathrm{f,Na}} + i_{\mathrm{si,Na}} + 3\,i_{\mathrm{p}} + (n_{\mathrm{NaCa}}/(n_{\mathrm{NaCa}}-2))\,i_{\mathrm{NaCa}})/V_i F \tag{48}$$

where V_i is the intracellular fluid volume.

Note that, strictly speaking, i_{Na} is not pure Na movement since we have assumed a 12 % permeability to K$^+$ for the Na channel. The error this introduces is however very small. It would make a difference of less than 4 % to the overall Na flux during an action potential. The reason for this is that the Na–Ca exchange process is at least as much involved in sodium entry as is the sodium current (DiFrancesco *et al.* 1985).

370   D. DiFRANCESCO AND D. NOBLE

(k) *Intracellular calcium concentration*

Here we encounter the major difficulty in developing the model. It is clearly incorrect to assume that intracellular calcium is not bound. In fact, most of it is sequestered and the processes of sequestration are both complex and not very well understood. The approach we have adopted is to use the simplest possible equations to represent the essential features of the uptake and release processes from an electrophysiological point of view. Our aim has been to produce computed $[Ca]_i$ transients that show a time course similar to that recorded in recent experiments (Allen & Kurihara 1980). Our assumptions are (see figure 1):

(i) The main sequestration store (which we shall refer to as the uptake store) is the sarcoplasmic reticulum. It is assumed that this occupies about 5% of the intracellular fluid volume and can sequester Ca^{2+} up to a concentration of 5 mM (Chapman 1979).

(ii) A fraction of the stored calcium is either transferred to a separate release store or is converted into a releasable form by a repriming process (cf. Hodgkin & Horowitz 1960). This process may be voltage dependent with a time constant of the order of a second or more at -80 mV (Gibbons & Fozzard 1975 a,b).

(iii) Release of Ca^{2+} from the release store is induced by calcium (Fabiato & Fabiato 1975). With these assumptions, the equations are:

$$i_{up} = \alpha_{up}[Ca]_i([\overline{Ca}]_{up} - [Ca]_{up}) - \beta_{up}[Ca]_{up}, \tag{49}$$

$$i_{tr} = \alpha_{tr}\,p([Ca]_{up} - [Ca]_{rel}), \tag{50}$$

$$i_{rel} = \alpha_{rel}[Ca]_{rel}([Ca]_i{}^r/([Ca]_i{}^r + K_{m,Ca})) \tag{51}$$

where $[\overline{Ca}]_{up}$ is the maximum value of $[Ca]_{up}$, r is the number of Ca^{2+} ions assumed to bind to the release site (usually set to 2) and

$$dp/dt = \alpha_p(1-p) - \beta_p\,p. \tag{52}$$

This equation represents the time- and voltage-dependence of the exchange between storage and release sites. For the rate coefficients we used the same equations as for f slowed by a factor of 10. This gives the required steady state voltage dependence for the repriming process, which is similar to that for Ca current inactivation and reavailability. Then:

$$d[Ca]_{up}/dt = (i_{up} - i_{tr})/2V_{up}F, \tag{53}$$

$$d[Ca]_{rel}/dt = (i_{tr} - i_{rel})/2V_{rel}F, \tag{54}$$

$$d[Ca]_i/dt = -(i_{si,Ca} + i_{b,Ca} - \{2i_{NaCa}/(n_{NaCa} - 2)\} + i_{up} - i_{rel})/2V_iF, \tag{55}$$

where V_{up} and V_{rel} are the volumes of the uptake and release stores respectively. The general features of this model are represented in figure 1.

The usual values used for the constants in these equations were as follows:

$[\overline{Ca}]_{up} = 5$ mM. This corresponds to the known Ca^{2+} sequestering ability of the sarcoplasmic reticulum (Chapman 1979).

$V_{up} = 0.05\,V_i$, the reticulum is assumed to occupy 5% of the intracellular volume (Chapman 1979). Chapman's figure is for ventricular muscle. We have used the same parameter for the Purkinje model but it would clearly be desirable to replace this with an experimental value for Purkinje fibres.

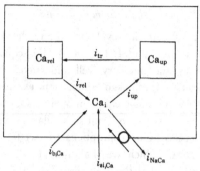

FIGURE 1. Diagram summarizing the processes assumed to control Ca^{2+} movements within the cell and across the cell membrane. An energy-consuming pump is assumed to transport calcium into the sarcoplasmic reticulum uptake store which then reprimes a release store. This may either be a physically distinct store or a releasable state of calcium within the same store. Release is assumed to be activated by cytoplasmic calcium ions. Ca^{2+} ions enter the cell through a background leak channel and through a gated channel. Ca leaves the cell through the Na–Ca exchange. Rarely it may enter through the exchange (for example, when $[Ca]_i$ is very low and the voltage very positive). These are the minimum assumptions required to model the $[Ca]_i$ transient. The model would need further development if it were thought necessary to add voltage-dependent Ca release, energy consuming surface membrane calcium pump, other calcium sequestration processes (such as binding to the contractile proteins), or further compartmentation of intracellular calcium.

$V_{rel} = 0.02\ V_i$. This figure is arbitrary and was chosen to give roughly the correct quantity of releasable calcium.

$K_{m,Ca} = 0.001$ mM when $r = 1$, 0.001^2 when $r = 2$. This figure is also somewhat arbitrary. Clearly, it cannot be as low as 0.0001 since resting calcium levels do not trigger release. 0.001 is sufficient to allow the quantity of calcium entering during an action potential to release stored calcium. The precise value of $K_{m,Ca}$ was not found to be important in the computations described in this paper.

The value of r was set to 1 or 2. The standard value was 2 since this gave oscillatory release (see DiFrancesco *et al.* 1983) more readily.

The rate coefficients were computed by the program from values set to the time constants of the processes involved. The release time constant, τ_{rel}, was set to 50 ms to enable $[Ca]_i$ to rise to a peak within 50 to 100 ms (Wier & Isenberg 1982; Allen & Kurihara 1980). The repriming time constant, τ_{rep}, was set to 2 s at -80 mV (Gibbons & Fozzard 1975a, b). The uptake time constant, τ_{up}, was set to 25 ms to allow uptake to occur sufficiently rapidly to reproduce the falling phase of the measured $[Ca]_i$ transients. This value also allows the s.r. to accumulate Ca^{2+} ions up to a concentration (about 2 mM) near half the maximum value (assumed to be 5 mM). These conditions are appropriate for a situation where the larger part of the $[Ca^{2+}]_i$ transient is due to internal cycling.

The values for the time constants were then used by the computer program to compute the rate coefficients using the relations:

$$\alpha_{up} = 2FV_i/(\tau_{up}\,[\overline{Ca}]_{up}), \tag{56}$$

$$\alpha_{tr} = 2FV_{rel}/\tau_{rep}, \tag{57}$$

$$\alpha_{rel} = 2FV_{rel}/\tau_{rel}. \tag{58}$$

We should emphasize that this part of the modelling is not thought to be very secure. There are too many arbitrary factors and, in any case, the major issue of whether Ca^{2+} release is Ca^{2+}-induced or voltage-induced (or, perhaps, both) is still controversial. Our purpose here is therefore largely limited to reproducing the known $[Ca]_i$ transient time course. We have succeeded in doing this reasonably well, although we have not found it possible to reproduce the biphasic feature found by Wier & Isenberg (1982). We suspect that this would require further or different assumptions about intracellular calcium location and diffusion.

Despite the very tentative nature of the modelling of the $[Ca]_i$ transient, this feature of the model greatly extends its explanatory range since it is essential to model the $[Ca]_i$ transient in equations for Ca-dependent currents, like $i_{Na, Ca}$, $i_{Ca, f}$ and i_{to}. Even a primitive model, here, is much better than no model at all. An important consequence is that activity computed in this model is dependent on the inotropic state. This will be explored more fully in a subsequent paper (DiFrancesco *et al.* 1985).

(*l*) *Extracellular potassium concentration*

We assume that K^+ ions diffuse freely in the extracellular space so that we may use the free solution diffusion constant. In some calculations we have also assumed that there is a restriction factor that determines the degree to which free diffusion may be impeded.

The equation for diffusion in a cylinder where, at any point, ions may also cross the cell membrane, is

$$\partial[K]_c/\partial t = D\{\partial^2[K]_c/\partial x^2 + (1/x)\,\partial[K]_c/\partial x\} + i_{m, K}/V_e F \qquad (59)$$

where
$$i_{m, K} = i_{K, 1} + i_K + i_{f, K} + i_{si, K} + i_{b, K} - 2i_p \qquad (60)$$

and V_e is the extracellular space volume. For a cylinder this would be:

$$V_e = V_{ecs}\, a^2 l \quad \text{and, similarly,} \quad V_i = (1 - V_{ecs})\, V_e \qquad (61)$$

where V_{ecs} is the fractional extracellular space (usually set to 5%), a is the radius and l the length of the preparation.

These are the equations we have used for calculations of $[K]_c$ in a Purkinje preparation when it has seemed important to represent the non-uniform distribution of extracellular K^+.

For a spherical preparation we have used the equation:

$$\partial[K]_c/\partial t = D\{\partial^2[K]_c/\partial x^2 + (2/x)\,\partial[K]_c/\partial x\} + i_{m, K}/V_e F \qquad (62)$$

in place of (59).

Finally, for many purposes, we have found that the results are little affected by assuming a homogeneous K^+ concentration in a three-compartment model:

$$d[K]_c/dt = -P([K]_c - [K]_b) + i_{m, K}/V_i F \qquad (63)$$

(cf. Attwell *et al.* 1979*b*), where $[K]_b$ is the bulk extracellular K^+ concentration and P is the rate constant for exchange between the bulk and cleft space. For most calculations we used values for P between 0.2 and 1.0 s^{-1}. This range of values was determined using a comparison between calculations using the cylindrical and three-compartment equations.

(m) Intracellular potassium concentration

This was computed by using:

$$d[K]_i/dt = i_{m, K}/V_i F. \tag{64}$$

Finally, it should be noted that we have used concentrations as synonymous with activities. This assumes that the intracellular and extracellular activity coefficients are very similar.

Methods

The set of equations (1–64) is extremely stiff since the range of time constants is exceedingly large. The sodium activation equation time constant is only of the order of 0.1 ms, whereas the equation for $[Na]_i$ has a time constant of the order of 5 min, a ratio of over a million. It is not therefore practical to use exactly the same numerical approach for all computations. With this and other requirements in mind a very general computer program, HEART, has been written which varies the computation methods to suit a wide variety of possible experimental situations. The ordinary differential equations were integrated using the methods described by Plant (1979). The partial differential equations, when used, were integrated separately using a method for inverting a band matrix of width three (see Modern computing methods 1961). Figure 2 shows a flow diagram of the main part of the program. The original programming language used for development was a version of Algol60 suitable for running on small machines. The program has subsequently been translated into Pascal. These languages were chosen for their superior logical structure compared with the Fortran IV available on PDP11 computers. The advantages that this gives in very large programs with extensive use of nested control loops were found to be very important in building-in the extreme flexibility which is one of the major features of the program. This readily permits new versions of the model (for example, for preparations other than the Purkinje fibre) to be incorporated as parameter procedures that set the constants and determine the pathway through the nested control loops. New control loops can also be added with ease since they do not refer to fixed labels. We have successfully run the Pascal version of the program using the RT11SJ monitor on PDP11/34 and PDP11/23 computers and the VMS monitor on a VAX computer. The program is extensively documented and no knowledge of Algol or Pascal is required unless substantial developments are envisaged, in which case an appropriate compiler will be required. A Pascal compiler for RT11 and other DEC systems is available from Oregon Software. The Pascal used is very close to the international standard, so that the program should easily transfer to other computers. Enquiries about the availability and use of the software should be addressed to Dr Noble.

Results and discussion

(a) Current–voltage relations

The steady state current–voltage relations given by the model can be analysed in the same way as those in the M.N.T. model (see McAllister *et al.* 1975, figures 2 and 3). The results obtained are not in general very different and will not be repeated here. Instead we shall describe new features that were not within the scope of the M.N.T. model.

First, we may now correctly describe the influence of extracellular potassium ions on the

374   D. DiFRANCESCO AND D. NOBLE

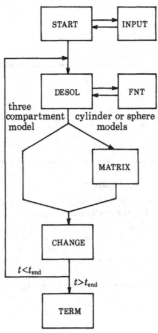

FIGURE 2. Flow diagram of the main features of the program. A procedure START either calls specific INPUT procedures
containing parameters relevant to each preparation or version of the model, or reads a separate INPUT file. START
then computes a variety of parameters that are used repeatedly in the computations and organizes output files.
Each integration step then involves a call of the integration control procedure DESOL (see Plant 1979) which
calls a number of other procedures, including FNT. The latter contains all the model equations and can readily
be modified to produce new versions of the model. On exit from DESOL, procedure MATRIX is called to solve
the diffusion equations when these are used. The three-compartment model bypasses this procedure. Procedure
CHANGE controls the time changes of concentrations, currents, voltage clamp protocols etc. When t_{end} is
reached, procedure TERM terminates the computation and tidies up the output files. This diagram shows only
the main overall features. The program also contains about 20 other procedures not shown here which control
an almost infinitely large number of modes of operation that can be tailored to the requirements of particular
problems. The program is extensively annotated to enable these facilities to be operated without requiring any
significant understanding of the program language.

current–voltage relations since all the known K^+-dependent processes are represented. Figure
3 shows the results of computing the quasi-instantaneous current–voltage relations at values
of $[K]_b$ between 1 mM and 40 mM. It can be seen that the major features of the experimental
results (see, for example, Dudel *et al.* 1967*b*; Sakmann & Trube 1984) are reproduced,
including: (i) the presence of inward-going rectification with a negative slope region; (ii) the
crossover of current–voltage relations at different values of $[K]_b$; (iii) the fact that at very low
$[K]_b$ the net current–voltage relation becomes almost flat over a wide range of potentials; (iv)
the presence of a net inward current region at low values of $[K]_b$.

The last feature was an important part of Noble & Tsien's (1969*a*) results and of their
reconstruction of the plateau (Noble & Tsien 1969*b*).

The relations shown in figure 3 do not include the steady-state sodium current since the
experiments of Dudel *et al.* (1967*a*,*b*) were performed in sodium-free (choline-substituted)
solutions. It is however of interest to compare the results obtained including the steady-state
sodium current since this has recently been measured experimentally by substrating current–

MODEL OF CARDIAC ELECTRICAL ACTIVITY   375

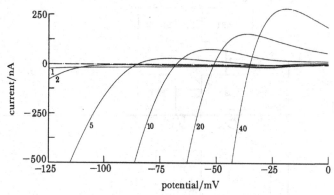

FIGURE 3. Steady-state current–voltage relations computed at various values of extracellular potassium concentration, $[K]_b$, from 1 to 40 mM. At each value of $[K]_b$ the model was clamped at -50 mV. Current–voltage relations were then computed assuming that the gating mechanisms m, h, d and f are held at their steady-state values at each potential, and that there are no significant variations in the Na–Ca exchange current. The last assumption is justified by our finding that the exchange system only carries large currents transiently and that these transients are quite fast when $[Ca]_i$ is in the diastolic range. This kind of result has been partly reconstructed by previous models (Noble 1965; Cohen *et al.* 1978). This is the first, though, to incorporate all the known $[K]_o$-dependent processes (i_{K1}, i_p and, to a lesser extent, i_K and i_{to}) with detailed experimental parameters.

voltage relations in the presence and absence of TTX (Attwell *et al.* 1979*a*; Colatsky & Gadsby 1980). Figure 4 (bottom) shows the 'window' current obtained from the model. This curve reproduces the experimental results fairly well. Attwell *et al.* (1979*a*) obtained a mean peak current value of -20 nA. The model gives -23 nA. The 'range' of the 'window' is about -60 to -20 mV which is closer to the experimental results than was the M.N.T. model. It is important to note that the 'window' current is well-reproduced even though our i_{Na} equations are based on Colatsky's (1980) results. The top part of figure 4 shows how our equations for h and m^3 fit Colatsky's data. There is only very little overlap between m^3 and h but this is sufficient to generate a 'window' current that only needs to be less than 1 % of the peak i_{Na}.

(b) Reconstruction of voltage clamp currents

Figure 5 shows the extent to which the equations can reproduce the voltage clamp results obtained in Purkinje fibres with regard to the fast calcium current, slower inward current and the transient outward current. Traces (*a*) to (*f*) show currents computed on voltage clamping from -80 mV to the potentials shown. In each case the current was computed for the standard case with g_{Na} set to zero (that is, TTX block of g_{Na} is assumed), and then with i_{to} and i_{NaCa} set to zero to eliminate the current dependent on the $[Ca]_i$ transient. This was done to mimic the situation in Siegelbaum & Tsien's (1980) experiments where the $[Ca]_i$ transient was eliminated by EGTA injection. Record (*f'*) shows the computed $[Ca]_i$ transient corresponding to (*f*). Record (*a'*) shows the result of changing from -50 mV to -40 mV. The reason for this additional record will become clear later. Finally, records (*g*) and (*h*) show experimental records chosen for comparison with computed records (*a*) and (*e*), (*f*).

First, it is worth noting that the amplitudes and speeds of i_{si} and i_{to} at potentials near 0 mV correspond well with those in Siegelbaum & Tsien's (1980) results. Moreover, when i_{to} is blocked the peak inward current level is increased and the current record becomes much

376   D. DiFRANCESCO AND D. NOBLE

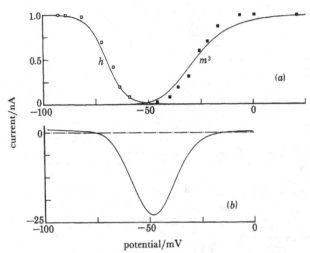

FIGURE 4. (a) Points show experimental data (Colatsky 1980) on the activation (■) and inactivation (□) of the sodium current in Purkinje fibres. The continuous lines show the steady state values of m^3 and h given by the model. Note that we have fitted the h data fairly accurately, but have chosen a somewhat less steep function for m^3. In this choice we were influenced by the data on single cells (Brown *et al.* 1981) showing an activation curve similar in steepness to our equations. This choice also allows a better reconstruction of the 'window' current. (b) 'Window' current computed by subtracting current–voltage relations obtained before and after setting g_{Na} to zero. This curve is similar to that obtained experimentally. The small outward current shift negative to −75 mV is due to a small change in the Na gradient when g_{Na} is blocked. Note that the 'overlap' region shown in (a) appears to be a very small region near −50 mV. The curve in (b) shows that, when dealing with very small currents, the overlap is more extensive at values of m^3 and h that are too small to appear significantly different from zero in the top curves.

simpler. The records before and after removing $[Ca]_i$-dependent currents cross each other as they do in the experimental results. In the model this is due to the presence of a long-lasting small inward current caused by Na–Ca exchange. The main difference between the computed and experimental results here is that the difference persists to much longer time in the experimental results. This might be due to the presence of a genuine slow Ca current, $i_{Ca, s}$, (Lee *et al.* 1984*a*) if that current is $[Ca]_i$-dependent. We will return to these differences later in discussing figure 6.

Turning now to the voltages below the range of activation of i_{to}, it is clear that near −40 mV a very slow inward transient occurs that lasts about 500 ms. Its amplitude and duration are similar to those of the current recorded at −40 mV by Eisner *et al.* (1979) – see also Lederer & Eisner (1982) – which is shown as record (g). Also shown is the effect of caffeine at a level thought to discharge the s.r. This removes the current, as does removal of transient changes in i_{NaCa} in the model. Of course, caffeine should first itself induce an inward current while the stores are being discharged. Clusin *et al.* (1983) have recently described just this effect in embryonic heart cells. They also attribute the current to the Na–Ca exchange process.

The main differences between traces 5a and 5g is that the computed response has a sharper onset compared to its decay. It is worth noting that this may also occur experimentally (see, for example, Lederer & Eisner 1982, figure 2). In our equations, this feature depends on the current magnitude: the onset is faster the larger the current (see also Brown *et al.* 1984*a*). Another feature worth noting is that Siegelbaum & Tsien's (1980) results do *not* show this very slow current in the region of −40 mV. The reason may be that they used a holding potential

MODEL OF CARDIAC ELECTRICAL ACTIVITY

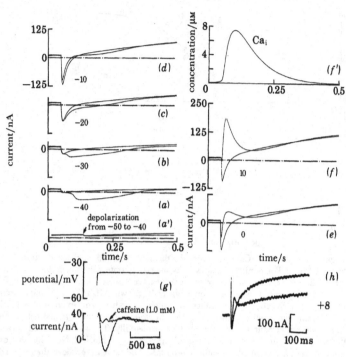

FIGURE 5. Voltage clamp currents computed from the model. In records (*a*) to (*f*), the voltage was stepped from −80 mV to the potentials indicated, first using the full equations (with g_{Na} set to zero) and then with i_{to} and i_{NaCa} also set to zero to mimic the expected result of eliminating the currents dependent on [Ca]$_t$. Record (*f'*) shows the intracellular Ca transient computed during record (*f*). Record (*a'*) shows the effect of changing the holding potential to −50 mV. The very slow inward current seen on clamping to −40 or −30 from −80 mV is then no longer seen. Records (*h*) show superimposed experimental records from Siegelbaum & Tsien (1980). They clamped from −45 mV to, in this case, +8 mV. Record (*g*) shows experimental records from Eisner *et al.* (1979). Note that the time scales for the experimental records are not the same as those for the computed records. See text for further description.

around −45 mV. It is important to note (see record (*a'*)) that in the model also, holding at, in this case, −50 mV eliminates the slow component. This is because little Ca release occurs on depolarizing from −50 to −40 mV.

In single guinea-pig ventricular cells a similar situation is found to occur. Depolarizations from −80 mV to −50 mV produce a slow component of current (see Lee *et al.* 1983) whereas depolarizations from −50 mV to −40 mV or −30 mV fail to trigger this component.

This is a suitable point at which to comment on the diversity of the experimental information concerning the slower components of i_{si} (see also review by Noble 1984). The range is so wide that, in some experiments, currents like that shown in figure 5*a* and 5*g* are apparently not observed at all. It is important to note that this is quite consistent with the system of equations we have described. To ensure the apparent absence of the slow component, it is sufficient to reduce a little the sensitivity of the Ca-release mechanism to intracellular calcium (by increasing $K_{m,Ca}$). Transients due to i_{NaCa} are then always greatly masked by the activation of the much larger $i_{Ca,t}$, combined with the fact that the onset of i_{NaCa} is then also much faster. An example of this behaviour is shown using the sinoatrial node version of the model in Brown *et al.* (1984*a*, figure 10*a*), where the computed time course of i_{si} decay is clearly monotonic.

(c) *Standard action and pacemaker potentials*

Figure 6 shows the standard action potential and intracellular Ca²⁺ transient computed at $[K]_b = 4$ mM. To induce pacemaker activity the y variable was shifted 10 mV in a positive direction (cf. Hauswirth *et al.* 1968). The intracellular Ca²⁺ transient rises to a peak within about 50 ms and decays well before the faster phase of repolarization begins. This corresponds

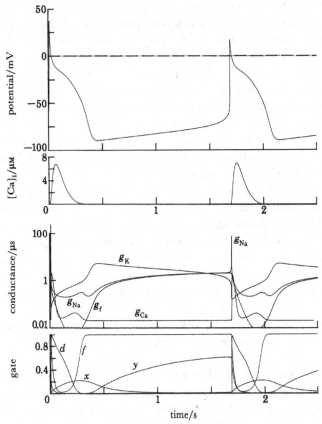

FIGURE 6. Standard action potential, pacemaker potential, intracellular calcium transient, conductances (on a logarithmic scale) and gating variables computed for $[K]_b = 4$ mM. Description in text.

well to the experimental results with aequorin, except that we do not find the biphasic response seen by Wier & Isenberg (1982) in Purkinje fibres. In principle, two peaks are possible in the model since the Ca²⁺ transient is made up of two components: a smaller one due to the calcium current and a larger one due to internal release. In practice, these fuse together as they appear to do experimentally in ventricular muscle. We do not know whether any of the electrophysiological phenomena dependent on intracellular calcium depend on the biphasic response, but we suspect that the time course reproduced in figure 6 is a good first approximation.

The lower part of figure 6 shows the computed conductance changes plotted on a logarithmic scale. In this diagram 'g_K' includes the conductances (computed as chord conductances) due

MODEL OF CARDIAC ELECTRICAL ACTIVITY   379

to i_{K1}, i_K and $i_{t,K}$; 'g_{Na}' includes the conductances due to i_{Na}, and $i_{t,Na}$; 'g_{Ca}' includes those due to $i_{Ca,f}$ and $i_{b,Ca}$, while 'g_t' is the sum of $g_{t,Na}$ and $g_{t,K}$. Our reason for using these combinations is that, apart from g_t, they correspond most closely to the equivalent parameters in the M.N.T. model.

It can be seen that the variations in the equivalent conductances show some resemblance to those in the M.N.T. model. In particular, the K$^+$ conductance time course is almost identical to that of the M.N.T. model. The other conductances, however, show significant differences. The main differences are: (i) the decay of g_{Ca} is very much faster, and (ii) the onset of g_t during the pacemaker depolarization is a new feature that was not present in the M.N.T. model which represented the equivalent process as the decay of a specific K$^+$ conductance. The reason for the close resemblance for the remaining terms included in 'g_K' is that by far the largest factors in the time course of 'g_K' in the M.N.T. model and in the new model are the voltage-dependent variations in $i_{K,1}$ and the time and voltage dependent variations in i_K (the formulations of which are very similar in the two models). Another way of demonstrating these features of the model is to measure 'slope' conductance as it would be measured experimentally by applying small repetitive voltage pulses (as shown in DiFrancesco & Noble 1982). The results reproduce Weidmann's (1951) data despite the radical re-interpretation of i_{K2} (for further discussion of this and related results see DiFrancesco & Noble (1982)).

The importance of the hyperpolarizing-activated current, i_f, in the pacemaker depolarization may be demonstrated by computing the effects of shifting the voltage dependence of the gating variable to reproduce the effects of adrenaline (Hauswirth *et al.* 1968; Tsien 1974; Hart *et al.* 1980). With $[K]_b = 2.7$ mM, a 15 mV shift is sufficient to double the firing frequency. A 30 mV shift leads to substantial depolarizaton. If g_{si} is increased the result is very rapid pacemaker activity of the kind seen experimentally after strong doses of adrenaline. The results are so similar to those illustrated by McAllister *et al.* (1975) that we have not shown them as a figure in this paper.

Figure 7 shows the time courses of the main current components. For clarity, the fast sodium current has been omitted (its time course can be estimated from the conductance changes plotted in figure 7).

While i_f is the main time-dependent gated current that contributes to pacemaker activity, other currents also contribute substantially. The net increase in i_f during the pacemaker depolarization in figure 8 is -14 nA. By comparison, i_K shows a fall of 5 nA during the pacemaker potential; $i_{b,Na}$ carries a roughly constant -26 nA, $i_{b,Ca}$ carries about -10 nA, i_{NaCa} carries about -4 nA and the sodium–potassium exchange pump carries about 17 nA. The difference is made up by i_{K1} which carries about 31 nA.

(d) *Influence of external* [K] *on action potentials and pacemaker activity*

Figure 8 shows the influence of varying the bulk extracellular K$^+$ concentration. At 12 and 20 mM the action potential is of fairly brief duration and a stable resting potential is established immediately following repolarization. Decreasing $[K]_b$ to 8 or 6 mM lengthens the action potential, hyperpolarizes the membrane and, in consequence, activates i_f to produce a pacemaker depolarization, though at these concentrations the depolarization is insufficient to reach the action potential threshold. At 4 mM slow repetitive firing occurs. Further reduction to 2.7 mM lengthens the action potential even further and the pacemaker potential becomes much steeper. These are the well-known effects of external [K] on action potentials and

380 D. DiFRANCESCO AND D. NOBLE

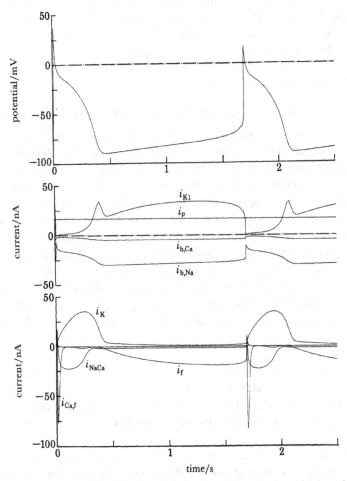

FIGURE 7. Continuation of figure 6. This shows ionic currents (except for i_{Na} which is too large for the current scale used here).

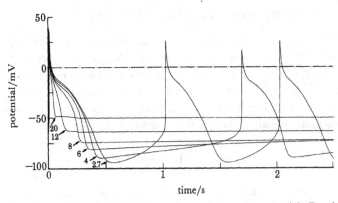

FIGURE 8. Influence of extracellular [K] on action and pacemaker potentials. Description in text.

pacemaker activity in Purkinje fibres (Weidmann 1956; Vassalle 1965). At values below 2.7 mM, the behaviour depends critically on the value assumed for the background sodium conductance $g_{b, Na}$. With this conductance set at 0.02 μS, the model fails to repolarize at very low [K] and, after a damped oscillation, the membrane potential settles at -40 mV. This effect is shown in Noble (1984, figure 4) and corresponds to the well-known depolarizing effect of very low [K] in Purkinje fibres (Weidmann 1951; Gadsby & Cranefield 1977).

(e) *Influence of external* [Na] *on action potentials, pacemaker activity and intracellular sodium*

In 1951, Draper & Weidmann described the influence of $[Na]_o$ on the overshoot and pacemaker activity in Purkinje fibres. The overshoot potential was found to follow closely the behaviour of a sodium electrode, while the duration of the plateau and the rate of pacemaker depolarization were both greatly reduced in low $[Na]_o$. At the time these results appeared they were taken at face value as strong support for the application of the Na hypothesis (Hodgkin & Katz 1949) to the heart, and as support for a role of Na ions in the pacemaker depolarization.

More recent experiments, however, have made Draper & Weidmann's work seem less simple than when it first appeared. When $[Na]_o$ is changed $[Na]_i$ changes fairly rapidly, the time constant of change being about 3 min (Ellis 1977; Sheu & Fozzard 1982). Moreover, the change in $[Na]_i$ is almost linearly proportional to the change in $[Na]_o$ with the consequence that E_{Na} changes by very much less than 61 mV per tenfold change in $[Na]_o$. In fact a tenfold decrease in $[Na]_o$ would be expected to produce less than 30 mV change in E_{Na}. This raises the question how Draper & Weidmann could possibly have obtained such an apparently simple result for the overshoot potential.

The answer may be provided by the second complication, which is that the value of E_{Na} predicted from intracellular Na measurements is about 30–40 mV positive to the observed overshoot potential. Thus, with $[Na]_i$ in the range 4–10 mM (which is fairly typical) and $[Na]_o$ at 140 mM, E_{Na} is expected to be about 70–100 mV, whereas the overshoot is only about 30–40 mV.

The explanation for the last result is fairly obvious: the 'Na' channel may not exclude other ions. Indeed, in our equations, we have allowed for this by using the result obtained in squid nerve (Chandler & Meves 1965) showing a 12% permeability to K^+ ions in the 'Na' channel. The reversal potential is then given by equation (29), which since 0.12 $[K]_c$ is very small compared to $[Na]_o$ simplifies to:

$$E_{mh} = (RT/F) \ln \left([Na]_o/([Na]_i + 0.12 [K]_i)\right) \tag{65}$$

and, since $[K]_i \gg [Na]_i$, E_{mh} would be expected to be relatively insensitive to $[Na]_i$.

First, we checked whether the equations can reproduce the $[Na]_o$-dependence of $[Na]_i$. The results are shown in figure 9. When $[Na]_o$ is reduced, $[Na]_i$ falls in an almost exponential manner with a time constant (3.3 min) that is very close to Ellis's (1977) experimental value (see also Sheu & Fozzard 1982; Chapman *et al.* 1983). Moreover, over a wide range of concentrations, $[Na]_i$ is almost linearly proportional to $[Na]_o$, as shown in figure 10. We then used these values of concentrations to investigate the $[Na]_o$-dependence of the computed overshoot potential. The results are shown in figure 11 and clearly closely follow Draper & Weidmann's results. Moreover, as found by them, the 'fibre' becomes inexcitable below about 15 mM $[Na]_o$. We also found the pacemaker depolarization to be less evident and the action potential duration reduced (not shown here). The latter effect is in part attributable to the contribution of the

382   D. DiFRANCESCO AND D. NOBLE

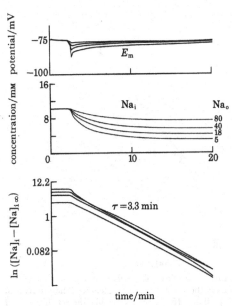

FIGURE 9. Influence of $[Na]_o$ on membrane potential and on intracellular $[Na]_i$. $[Na]_o$ was reduced from 140 mM at time 2 min to 80, 40, 18 or 5 mM at time 2.5 min. There is a transient hyperpolarization similar in amplitude and duration to that seen experimentally (Ellis 1977). In the model this is attributed to a reduction in i_{NaCa} while the Na gradient is reduced. Note that this effect is largely transient. The bottom diagram shows the $[Na]_i$ changes plotted on a semilogarithmic scale. $[Na]_i$ falls exponentially with a mean time constant of 3.3 min.

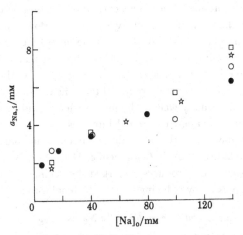

FIGURE 10. Steady-state variation of $a_{Na,i}$ with $[Na]_o$. The open symbols show results replotted from Ellis (1977). The closed symbols show the model's predictions using an activity coefficient of 0.75. In both experimental and computed results there is a roughly linear variation of $[Na]_i$ with $[Na]_o$. (See also Chapman *et al.* 1983.)

Na 'window' current to the plateau and in part to entry of Na by the Na–Ca exchange mechanism.

The suppression of pacemaker activity in low $[Na]_o$ requires further comment. It might be thought that this represents the contribution of the sodium background current to the pacemaker depolarization. This is not so since we have assumed (see above) that the Na

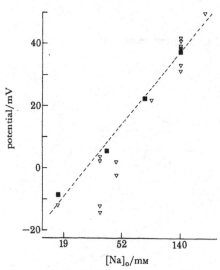

FIGURE 11. Variation of overshoot potential with $[Na]_o$. The open triangles are results replotted from Draper & Weidmann (1951). The closed squares show the model's predictions. The interrupted line shows a 61 mV variation per decade change in $[Na]_o$. Below 15 mM the model, like the real fibres, is inexcitable.

replacement can also pass through the background channel (Na replacement by choline, for example, does not greatly alter the resting potential in Purkinje fibre – see Hall *et al.* 1963). The reduction in the rate of the pacemaker depolarization is in fact attributable to the fact that, in the pacemaker range of potentials, i_f is largely carried by Na ions. Draper & Weidmann (1951) actually gave as one of their explanations the view that 'the slow depolarization during diastole...depends on the entry of sodium'. For the Purkinje fibre, on the new interpretation of i_{K2}, this is entirely correct even during the early phase of the pacemaker depolarization. As in the M.N.T. model, the later part of the pacemaker depolarization is also dependent on a small degree of activation of the fast sodium current. All the conclusions drawn by McAllister *et al.* (1975) on this point apply equally well to the new model, including their explanation for the influence of surface charge changes due to calcium ions.

One way of demonstrating the role of the fast sodium current is to compute the effects of reducing g_{Na}. This is shown in figure 12. As in Coraboeuf & Deroubaix (1978) and Colatsky's (1982) recent experimental work, this produces a marked shortening of the action potential and pacemaker activity is suppressed by reducing the rate of depolarization in the later phase.

(f) Ionic current changes due to the Na–K pump

The current carried by the Na–K pump has been extensively investigated recently. A standard method (used both by Gadsby and by Eisner & Lederer) has been to place a preparation in a K-free solution for several minutes to reduce pump activity and so to increase $[Na]_i$. The preparation is then returned to a K^+-containing solution. An outward current transient is then recorded as the increased internal sodium stimulates the pump. An example of this kind of experiment and its reconstruction is shown in figure 13. The top records are reproduced from Gadsby (1980) and show currents in a dog Purkinje fibre following various periods of K-free superfusion for up to 3 min. The middle record shows the computed result from the model with the variations in cleft [K] and $[Na]_i$ shown below. The computed variation

384   D. DiFRANCESCO AND D. NOBLE

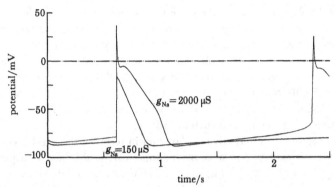

FIGURE 12. Influence of decreased g_{Na} on action potential and pacemaker activity. g_{Na} was reduced from 2000 μS
    to 150 μS. This abolishes the spike of the action potential and eliminates pacemaker activity. The effect on
    the action potential duration illustrates the role of the sodium 'window' current in the plateau.

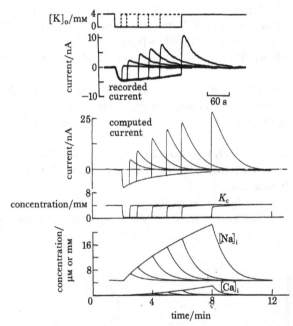

FIGURE 13. (*a*) Experimental records of changes in ionic currents in response to changes in $[K]_b$ in a canine Purkinje
    fibre (from Gadsby 1980). (*b*) Computed variations in ionic current and in $[K]_c$. (*c*) Computed variations in
    $[Na]_i$ (in millimoles per litre) and in $[Ca]_i$ (in micromoles per litre).

in $[Na]_i$ corresponds well to Ellis (1977) and Deitmer & Ellis's (1978) measurements showing
that in K^+-free medium $[Na]_i$ doubles in a period of about 5 min. The computed increase in
$[Ca]_i$ (also shown) corresponds well to the fact that tonic tension is known to increase over
this period of time, and Sheu & Fozzard (1982) have recently recorded $[Ca]_i$ with a calcium
electrode showing changes comparable to those computed here.

   We think, therefore, that we can have some confidence in the model's predictions concerning
the intracellular concentration changes during K-free inhibition of the pump.

   We turn now to the reconstruction of Gadsby's ionic current measurements. It can be seen that the computed results show a very similar pattern. Not only does the model correctly reproduce the outward current transients on return to K^+-containing solution; it also reproduces the slow upward current creep that occurs while the preparation stays in the K^+-free medium. The model provides a possible explanation for this phenomenon, which is that, although we have assumed a large extracellular cleft space (30 % in this case) the cleft K^+ concentration does not fall to the bulk K^+ concentration since it takes time for diffusion to occur. This allows a residual degree of K^+ activation of the pump, which is then further activated as $[Na]_i$ increases, so producing the upward current drift.

Mullins (1981) has proposed an alternative explanation in terms of the Na–Ca exchange current. We can exclude this explanation since on the time scale of this kind of experiment the Na–Ca exchange system will be close to its steady-state activity at nearly all times. If the background Ca influx remains constant and small, there is no reason why the Na–Ca exchange current should vary greatly.

This is perhaps a suitable point at which to emphasize a general result we have found with the model: this is that the Na–Ca exchange current is nearly always very small (about 4–5 nA, that is, much smaller than the Na–K pump current) in the steady-state. Large currents are carried by the exchange process only as transients. When $[Ca]_i$ is very low (less than 0.1 μM) these transients are very rapid (a few milliseconds); when $[Ca]_i$ is large (for example, 5 μM) the transients can last several hundred milliseconds (as during a computed action potential – see figure 7).

The magnitude of the upward current drift is somewhat larger in the model than in Gadsby's result. This amplitude is strongly dependent on the size of the extracellular space and on the time constant for cleft-bulk space diffusion. A shorter time constant for the diffusion process would give a smaller current creep.

We conclude that the model does accurately reproduce current changes due to Na–K pump activity. We will now use the model to investigate two other kinds of experiment in which the influence of pump changes has been measured.

Figure 14 shows the influence of the Na–K pump activity on the duration of the action potential. This computation is designed to reproduce Gadsby's (1982) measurements of the

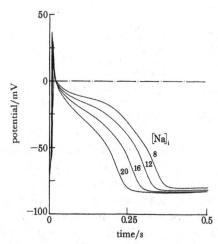

FIGURE 14. Action potentials computed at various values of $[Na]_i$ between 8 and 20 mM.

386   D. DiFRANCESCO AND D. NOBLE

shortening of the action potential on return to K^+ containing solutions after a period of several minutes in K^+-free solution. We computed the standard action potential in 6 mM K^+ at various values of $[Na]_i$ between the normal level of 8 mM and up to 20 mM. This is the range of $[Na]_i$ increase expected during several minutes exposure to K^+-free solution. The shortening of the action potential is similar to that recorded experimentally. Notice also the small hyperpolarization in the resting state, which is also seen experimentally.

The computations on the influence of the Na–K pump described so far were done with large extracellular space volumes appropriate to the known structure of canine Purkinje fibres. We now turn to the possible effects of more restricted spaces such as are found in sheep Purkinje fibres.

Figure 15 shows the results computed on return to a range of external activator cation

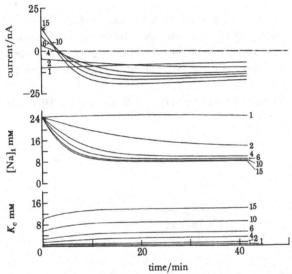

FIGURE 15. (a) Computed ionic currents following reactivation of the Na–K pump by various concentrations of external activator cation (1–15 mM) after allowing $[Na]_i$ to rise to 25 mM blocking Na–K pump. (b) Corresponding variation in $[Na]_i$. (c) Corresponding variations in $[K]_c$.

concentrations between 1 and 15 mM after a period of K-free superfusion leading to an increase in $[Na]_i$ to 25 mM. For these computations we used a cleft space volume of 5% and a diffusion time constant of 5 s. These parameters are interrelated. A smaller cleft volume together with a shorter diffusion time constant would give similar results. We have also run some computations using the full diffusion equations for a two-dimensional cylindrical space. The results are similar to those shown in figure 16, but each case takes much longer (several hours instead of a few minutes) to compute. Finally, since these computations were designed to reproduce the experimental conditions investigated by Eisner & Lederer (1980), in which Rb was used in place of K as the activator cation to reduce the effects of K depletion by reducing the inward rectifier current, we reduced g_{K1} to 5% of its usual value.

The top traces in figure 15 show the net ionic current changes. It can be seen that at 1 mM Rb (Rb and K are roughly equipotent activators of the Na–K pump) there is virtually no current transient. Eisner & Lederer (1980) also found only a very small current at 1 mM. A

MODEL OF CARDIAC ELECTRICAL ACTIVITY   387

comparable computation using a 30% cleft volume gave a nearly 50% activation of the pump (as expected since the 'true' K_m assumed in these computations is 1 mM). 2 mM produces a small slowly declining current transient. As much as 4 mM is required to activate 50% of the maximum current and speed of current change, which is not reached until the activator cation concentration is increased above 10 mM. This corresponds quite closely to Eisner & Lederer's curve for activation of the pump current by external cation, giving an apparent K_m in the region of 4–5 mM, despite the fact that the true value is 1 mM.

We found the apparent K_m value to be strongly dependent on the extracellular space size and the assumed diffusion time constant. It is easy to obtain apparent K_m values as high as 10 mM by halving the space size or increasing the diffusion time constant. With cleft spaces less than about 1–2% it becomes almost impossible to deactivate the pump in low $[K]_b$. Thus the high apparent value of K_m is attributable to the 'inertia' of the cleft system in relation to bulk $[K]$ changes.

There may appear to be a difficulty, though, with this explanation. This is that Eisner *et al.* (1981) were careful in their experiments to check that the pump current change is linearly dependent on $[Na]_i$. This is the result they found over the range of $[Na]_i$ values between 8 and 16 mM. As they point out, a large effect of external cation depletion on the pump activity might upset this linearity.

Nevertheless, we don't find this effect to be very significant. Figure 16 shows our results

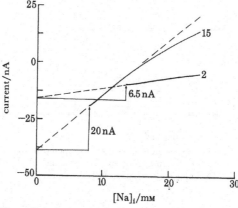

FIGURE 16. The results of figure 15 are replotted as current–$[Na]_i$ relations as they change during pump reactivation at various concentrations of activator cation. When the activator cation is 2 mM there is a nearly linear relation over the whole range. At 15 mM the result is nearly linear over the range of $[Na]_i$ between 8 and 16 mM. The extrapolated lines show how the current at $[Na]_i = 0$ can be estimated, as done by Eisner *et al.* (1981) to estimate the resting pump current.

replotted as current–$[Na]_i$ curves. With activation by 2 mM external cation, the result is close to linear over the whole range. With 15 mM external cation concentration the curve is close to linear over the range 8–16 mM but deviates from linearity above this range. Thus, over the relevant range of the experiments, a nearly linear relation is obtained. Furthermore, the computed deviation from linearity above 16 mM is almost entirely attributable to the arbitrary value of 40 mM assumed for the K_m for internal Na activation of the pump. If this is increased to, say, 100 mM the results are linear up to much larger Na^+ concentrations.

Further results related to i_p and consequential changes in $[Na]_i$, $[K]_c$ and in current–voltage relations have already been published by Hart *et al.* (1983).

(g) *Current changes formerly attributed to i_{K2}*

We have already given a fairly complete treatment of this question using an earlier, and much simpler, version of the model (DiFrancesco & Noble 1982). Here we will restrict ourselves to showing that essentially the same results are obtained with the more complete version described in the present paper.

Figure 17 shows ionic currents and mean cleft K^+ concentrations computed using the

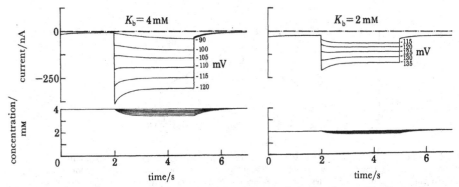

FIGURE 17. Examples of currents computed in response to hyperpolarizations from -50 mV to various potentials in the activation range for i_f. The extracellular cleft space was set to 10% and the full cylindrical diffusion equations were used to estimate the K concentration profiles as a function of radial distance. Below each set of current records we show the mean values of $[K]_c$. At $[K]_b = 2$ mM there is a reversal of total time-dependent current between -125 and -130 mV. At $[K]_b = 4$ mM the reversal occurs at about -110 mV. Similar results have been obtained for extracellular space volumes between 0.5 and 30%. For further analysis of the influence of extracellular space volume, see DiFrancesco & Noble (1982).

diffusion equations for a cylindrical space. The space volume was set to 10% and step hyperpolarizations were imposed from -50 mV to the potentials indicated. A variety of bulk extracellular K^+ concentrations was used. The results for 2 mM and 4 mM are illustrated. Note that at each value of $[K]_b$ there exists a potential at which the net time-dependent current change changes direction. As shown in our previous work (DiFrancesco & Noble 1980c, 1982) this reversal, although it often gives the appearance of a simple single component (which is what led to its identification in experimental work as a true ionic channel reversal potential) is in fact attributable to a balance between an inward current change due to the activation of i_f during hyperpolarizations and an outward current change due to a decrease in inward-flowing i_{K1} during depletion of K^+ ions from the cleft space. The time constants for these two processes are sufficiently close under most circumstances to produce the impression that a single component (perhaps slightly perturbed by depletion) is responsible. It is noteworthy that the amounts of K^+ depletion required to produce this effect are very small. Typically a reduction of only 0.5 mM in the mean $[K]_c$ is sufficient to generate a change in i_{K1} sufficient to mask the opposite change in i_f. This decrease in mean $[K]_c$ only represents a change of 10% at $[K]_b = 5$ mM. The reduction in the total i_f conductance – which is

MODEL OF CARDIAC ELECTRICAL ACTIVITY 389

K-activated (see DiFrancesco 1982) – during large hyperpolarizations may further contribute to the observed reversal effect.

There is some argument about the precise size of the extracellular space (see, for example, Cohen *et al.* 1983). We have therefore repeated these computations over the range 0.5–30%. The same kind of result is obtained in all cases. The influence of space size on E_{rev} is treated fully in DiFrancesco & Noble (1982).

Figure 18 shows the variation in reversal potential as a function of external [K] on a

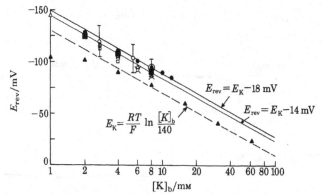

FIGURE 18. Variation of E_{rev} for 'i_{K2}' with $[K]_b$ given by the model and by various experimental results. We also show the results on measurements of resting potentials and the predictions of the Nernst equation for potassium (interrupted line) and of (66) (solid lines) for two values of ΔE. ● Model 1; ■ model 2; □ Noble & Tsien 1968; △ Peper & Trautwein 1969; ○ Cohen *et al.* 1976; × DiFrancesco *et al.* 1979*b*; ▲ resting potential (Gadsby & Cranefield 1977).

logarithmic scale. The filled square symbols show the results for the present model, while the filled round symbols show the results for the earlier version (DiFrancesco & Noble 1982). The open symbols show the results of various experiments, while the filled triangles show the variation in resting potential obtained by Gadsby & Cranefield (1977). The interrupted line shows the value for E_K computed by assuming that $[K]_i = 140$ mM. This is clearly a good fit to the resting potential results for values of $[K]_b$ above about 8 mM. Equally clearly, all the reversal potential estimates, experimental and theoretical lie significantly negative to the estimated values of E_K. To a first approximation, the results fit an equation of the form:

$$E_{rev} = E_K - \Delta E \tag{66}$$

where ΔE is nearly a constant. For the early version of the model the best value of ΔE is 18 mV. For the present version it lies at about 14 mV. The theoretical derivation of and justification for this surprisingly simple equation has been given already in DiFrancesco & Noble (1982). All we need to add to what was shown in that paper is that we have now checked this result with numerical computations in about eight different versions of the same basic model with various formulations of i_{K1} and i_f. Any lingering suspicion that the result is fortuitous can now be laid firmly to rest. Given the properties of i_{K1} and its strong K-dependence at negative potentials and the similar time constants for the y gating reaction and the K^+ depletion process an approximate equation of the form of (59) is far from fortuitous: it is rather a necessary consequence of the given properties of the ionic currents and geometries involved.

390   D. DiFRANCESCO AND D. NOBLE

Nevertheless, there are some significant variations. First, as shown by the comparison between the two versions of our model, the best value for ΔE can vary. Among the variables concerned in determining this parameter is the size of the extracellular cleft space (DiFrancesco & Noble 1982). Secondly, it is worth noting that the precise shape of the ionic current record near the reversal potential varies with the detailed characteristics assumed. Sometimes the current record remains virtually monotonic (cf. Noble & Tsien 1968, figure 5, and the results plotted in figure 18). Sometimes, it is clearly biphasic (cf. Cohen *et al.* (1976), figure 2B, and the computed results shown in DiFrancesco & Noble (1980c)). It is even sometimes impossible to obtain a reversal potential (see, for example, Cohen *et al.* (1976), figure 2C). This is of course the natural situation in the mammalian s.a. node where i_{K1} is too weak to produce sufficient depletion dependent current change to mask i_f. It is therefore significant that the case showing absence of reversal published by Cohen *et al.* (1976) is from a Purkinje fibre in which the instantaneous current jumps attributable primarily to i_{K1} were very small indeed. It is easy to produce this behaviour in the model by reducing i_{K1} (DiFrancesco & Noble 1982).

Recently, Clay & Shrier (1981a, b) have recorded an ionic current change in spherical aggregates of embryonic ventricular cells which strongly resembles i_f or i_{K2}. In their analysis they use the Noble & Tsien (1968) i_{K2} hypothesis. We therefore thought it important to check the extent to which their results are also compatible with an i_f hypothesis. To reproduce their experimental situation we made the following modifications to the model: (i) the equations for K$^+$ diffusion in a spherical space were used instead of the cylindrical equations; (ii) the sphere was assumed to have a radius of 100 µm with an extracellular space volume of 4 % (the values given by Clay & Shrier (1981a, b)); (iii) the ionic currents were all scaled down by a factor of 10 to give absolute values similar to those recorded in Clay & Shrier's experiments. We have in fact repeated the computations for a variety of other parameter sets (see, for example, DiFrancesco & Noble (1981) for an example that uses Clay & Shrier's kinetics). The results all resemble those shown in figure 19 which shows currents computed in response to

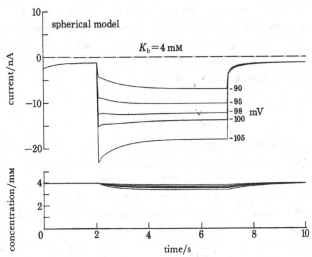

FIGURE 19. Computed variations in ionic current in a spherical model in response to various hyperpolarizations from −50 mV to the potentials shown. The extracellular cleft space was set to 4 %. The full diffusion equations for a three-dimensional spherical space were used. The mean values of [K]$_c$ are plotted below. [K]$_i$ was set to 110 mM. This gives a reversal potential at −98 mV.

hyperpolarizations from -50 mV with the bulk [K] set at 4 mM. To obtain a 'reversal' potential at about -98 mV (near the value found by Clay & Shrier) we used a value of 110 mM for $[K]_i$. The results clearly closely resemble those of Clay & Shrier. A feature of their results which they feel strongly supports the i_{K2} hypothesis is that the current records at the reversal potential are very flat. The result computed in figure 19 shows only 2 % variation in current level at the reversal potential. This figure depends naturally on the precise parameters assumed. With other possible parameters consistent with the experimental data, this figure for the current variation at E_{rev} could be higher or lower. Our own view is that this is not the crucial argument for distinguishing between the hypothesis. The more important one is to ask, first, what is the minimum plausible magnitude of the depletion process during hyperpolarization to E_{rev} and, second, would the change in i_{K1} expected from such a change in $[K]_c$ be within, say, less than 1 or 2 % of the total current. The answer to the first question is already provided in figure 19. It should be noted that in our computations using the full diffusion equations (in this case for a three-dimensional spherical space) we have used the free diffusion coefficient either with no restriction factor, or with a restriction factor of 0.5 to represent possible slowing of diffusion in the extracellular space either by the cells or by the external matrix. The computations were very similar for both situations since only the K concentration very near the surface was found to depend strongly on the diffusion coefficient. For a 4 % space this gives a mean depletion of about 0.5 mM at the reversal potential (in the region of -98 mV). Now in this range of potentials the observed variation of ionic current with external K^+ is very large: from Clay & Shrier (1981 *a*, figure 2) we estimate of the order of 5–7 nA mM^{-1}, or about 2.5–3.5 nA for 0.5 mM change in $[K]_c$. Clearly such a current change is much larger than 1–2 % of the total current; it is more like 20–30 %. On this argument, a truly flat current record at about -95 mV requires that some other process (such as a slow activation of an inward current) should also occur, rather than being evidence for a single component. Put another way, to reduce the predicted cleft K^+ depletion to values (say less than 0.025 mM) sufficiently small to produce a less than 1–2 % variation in ionic current we would have to increase the cleft space volume by at least a factor of 10 to about 40–50 %. This is far from the value given by Clay & Shrier and, we suspect, much larger than an extracellular space size could possibly be in a tight-fitting cell aggregate. Our conclusion here, therefore, is that it is quantitatively implausible to hold that depletion is negligible in a 100 μm radius sphere conducting strongly K^+-dependent ionic currents of the magnitudes recorded by Clay & Shrier.

Conclusions

We have discussed most of our results together with their presentation since it is not possible in a paper of this kind to defer all the discussion to a separate section. In this concluding section we shall therefore restrict ourselves to discussion of a more general nature.

In one sense, our model is conceived in a manner similar to previous ones. In other ways it is a radical departure from them. The sense in which it resembles previous cardiac models is that it uses the experimental data on individual ionic current mechanisms to construct a mathematical description that acts as a convenient quantitative catalogue of the relevant results. While being primarily descriptive, this function of a model is nevertheless important and its importance grows as the number of separate mechanisms increases. Cardiac electrophysiology has long ago passed the stage at which numerical predictions on the basis of known experimental data are sufficiently obvious not to require a proper overall formulation. Even from this point of view our model is a major advance on the previous ones, and on the M.N.T. model in particular, since we have taken the opportunity to incorporate a very large range of new experimental information. Moreover, even at this simply descriptive level, it has already proved very useful in, for example, exploring the consequences of the very much faster kinetics determined for the calcium channel for the role of this channel in the action potential plateau, in reassessing the variations in ionic conductances during the action potential and pacemaker potential, and in reconstructing the influence of extracellular potassium ions on electrical activity. Viewed simply as an up-dating of the numerical catalogue, the model clearly replaces the M.N.T. model for the kinds of purpose for which that model was constructed.

Nevertheless, up-dating the M.N.T. model was not our initial or even primary aim. This was, rather, to begin to construct a model that, for the first time, fully integrates the electrophysiological description of gated channels in the heart with a description of the ionic pump and sequestering processes. The present state of development of the field clearly requires a model of this kind since it is no longer plausible to ignore either the direct contributions of ionic pumps and exchange mechanisms or the indirect effects arising from ion concentration changes. Doubtless, these underlie the well-known fact that cardiac muscle electrical activity changes in quite complex ways with time, and over a time scale that must involve changes in intracellular and extracellular ion concentrations. In addition to the examples provided by the computations described in the present paper, good examples of uses of the model that exploit this integration are the complete 'mapping' of the old i_{K_2} hypothesis onto the new i_f hypothesis which we described in a previous paper (DiFrancesco & Noble 1982) and the use by Hart *et al.* (1983) to account for the transient nature of some of the electrical correlates of perturbation of the Na–K pump by low concentrations of cardiotonic steroids. Further examples are also provided by the extensive use of the mammalian s.a. node version of the model (Noble & Noble 1984) to provide plausible explanations for a variety of otherwise puzzling results obtained recently in experiments on this tissue (Brown *et al.* 1984*a, b*). We shall give further examples in DiFrancesco *et al.* (1985) which relate to longer-term changes and to possible interrelations between inotropic state and electrical properties. This also is an area that no useful model of electrical activity could now properly ignore. Our own initial involvement in the need to take account of ionic concentration changes was of course due to the requirement to investigate the theoretical consequences of potassium depletion processes in the extracellular spaces; it is an obvious and logical step to extend this approach to intracellular spaces.

While it was relatively easy to carry out this extension in principle, we have found it difficult to make some of the choices we found were necessary. It is extremely unlikely that our representation of the Ca-sequestering processes or of the Na–Ca exchange mechanism or of other Ca-activated currents (such as i_{to}) will remain among the best available for very long. Yet, our own experience (like that of McAllister *et al.* (1975)) is that the development of an overall model for the heart is a tedious process requiring at least two or three years and indefinite amounts of computer time. It was largely for this reason that we decided to program the model in a high structured language (Algol) that readily allows future developments. Many of the possible future developments are already built-in to the program and, as noted in the Methods section, we have translated the program into the closely related language, Pascal. Our hope is that those who wish to build onto the structure we have created will be able to do so relatively easily.

Finally, some comments are appropriate on a few choices in the development of the model that we could have made but didn't. First of all, we were inclined at an early stage to conclude that it would be most economical to assume that all or a major part of the background Na current is carried by non-specific (as between Na and K ions) Ca-activated channels of the kind described recently by Colqhoun *et al.* (1981) in the heart and which have also been observed in a wide variety of other tissues. This facility exists in the program and we spent several months investigating its consequences. While it is perfectly possible to construct a Purkinje fibre model in which this assumption is made, the assumption created fairly severe difficulties in extending the model to other tissues such as the s.a. node (Noble & Noble 1984) and ventricle (DiFrancesco *et al.* 1985). The reason is fairly simple. In a tissue in which the plateau potential is near the reversal potential of the non-specific channel the channel carries little current even when strongly Ca-activated and so does not greatly influence the action potential shape. By contrast, in preparations with fairly positive plateau potentials, the channel would carry fairly substantial outward currents. The result is in all cases to deform the repolarization process so that it resembles the Purkinje fibre repolarization process. Niedergerke & Page (1982) have recently shown that incorporating this channel mechanism into the M.N.T. model or the Beeler–Reuter model produces just this effect and that this may explain the shape of frog action potentials in high calcium at higher frequencies. This kind of repolarization waveform may also accompany what is usually called the 'rested state' contraction. In both cases, the contraction, and therefore the $[Ca]_i$ transient, are very large. Our results would fully confirm Niedergerke & Page's conclusions, but clearly this process cannot be significantly involved in action potentials from nodal or ventricular tissue when they do not show this particular repolarization waveform. Our conclusion here is that the full role and significance of this channel remains to be clarified. It may well be activated during unusually large $[Ca]_i$ transients, but it cannot be significantly activated during normal ventricular action potentials of the type in which the net repolarizing current is at its minimum during the $[Ca]_i$ transient.

Another area in which we initially explored some unsatisfactory formulations is the description of the Na–Ca exchange process. While there is now little doubt that this process is electrogenic, there are many ways in which its dependence on ionic concentrations and membrane potential might be formulated. We have satisfied ourselves that the simple hyperbolic sine function (see Mullins (1981), p. 42) is unsatisfactory except for a very restricted range of purposes when the only significant variable is membrane potential. In practice this is hardly ever the case since calcium concentration changes are nearly always involved. At the least, therefore, a better description of the Ca-dependence of the current is required. Yet, a complete version (whether that of Mullins (1977) or any other plausible model) of the equations for Na–Ca exchange would be so complex and use so many arbitrary coefficients that it would be cumbersome to formulate and would be of doubtful validity. We eventually opted for a compromise: a version that does incorporate a plausible description of the Ca-dependence of the exchange process but which does not fully represent the Na-dependence. This was achieved by representing a number of the Na-dependent terms in Mullins' model by a constant. We draw attention to this so as to warn other users of the model that, if substantial changes in Na concentrations are involved and the Na–Ca current is very significant, then they may have to develop the equations further than we have done in this direction. The possible roles of the Na–Ca exchange current have been quite extensively discussed recently. Our model may allow some of the questions raised to be put to some quantitative tests.

394 D. DiFRANCESCO AND D. NOBLE

We should like to acknowledge the support of the Medical Research Council and of the British Heart Foundation. We are particularly grateful to the Wellcome Trust for providing support for Dr D. DiFrancesco, and to the Medical Research Council for the award of a Research Fellowship for Academic Staff to Dr D. Noble.

We should like to thank Dr F. Edwards for his invaluable help in setting up the PDP computer systems we used. We are also grateful to Dr G. Hart, Dr H. F. Brown, Dr K. S. Lee, Dr S. J. Noble, Dr T. Powell and Dr A. Taupignon for their comments on the manuscript of this paper.

REFERENCES

Allen, D. G. & Kurihara, S. 1980 Calcium transients in mammalian ventricular muscle. *Eur. Heart J.* **1**, 5–15.

Attwell, D. & Cohen, I. 1977 The voltage clamp of multicellular preparations. *Prog. Biophys. molec. Biol.* **31**, 201–245.

Attwell, D., Cohen, I., Eisner, D., Ohba, M. & Ojeda, C. 1979a The steady-state TTX sensitive ('window') sodium current in cardiac Purkinje fibres. *Pflügers Arch. Eur. J. Physiol.* **379**, 137–142.

Attwell, D., Eisner, D. A. & Cohen, I. 1979b Voltage clamp and tracer flux data: effects of a restricted extracellular space. *Q. Rev. Biophys.* **12**, 213–261.

Attwell, D. & Jack, J. J. B. 1978 The interpretation of current–voltage relations: a Nernst–Planck analysis. *Prog. Biophys. molec. Biol.* **34**, 81–107.

Beeler, G. W. & McGuigan, J. A. S. 1978 Voltage clamping of multicellular myocardial preparations: capabilities and limitations of existing methods. *Prog. Biophys. molec. Biol.* **34**, 219–254.

Beeler, G. W. & Reuter, H. 1970a Membrane calcium current in ventricular myocardial fibres. *J. Physiol., Lond.* **207**, 165–190.

Beeler, G. W. & Reuter, H. 1970b The relation between membrane potential, membrane currents and activation of contraction in ventricular myocardial fibres. *J. Physiol., Lond.* **297**, 211–229.

Beeler, G. W. & Reuter, H. 1977 Reconstruction of the action potential of ventricular myocardial fibres. *J. Physiol., Lond.* **268**, 177–210.

Boyett, M. R. 1981a A study of the effect of the rate of stimulation on the transient outward current in sheep cardiac Purkinje fibres. *J. Physiol., Lond.* **139**, 1–22.

Boyett, M. R. 1981b Two transient outward currents in cardiac Purkinje fibres. *J. Physiol., Lond.* **320**, 32P.

Boyett, M. R. & Jewell, B. R. 1980 Analysis of the effects of changes in rate and rhythm upon electrical activity in the heart. *Prog. Biophys. molec. Biol.* **36**, 1–52.

Brown, A. M., Lee, K. S. & Powell, T. 1981 Sodium in single rate heart muscle cells. *J. Physiol., Lond.* **318**, 479–500.

Brown, H. F., Clark, A. & Noble, S. J. 1972 Analysis of pacemaker and repolarization currents in frog atrial muscle. *J. Physiol., Lond.* **258**, 547–577.

Brown, H. F., DiFrancesco, D., Noble, D. & Noble, S. J. 1980 The contribution of potassium accumulation to outward current in frog atrium. *J. Physiol., Lond.* **306**, 127–149.

Brown, H. F., Kimura, J., Noble, D., Noble, S. J. & Taupignon, A. 1983 Two components of 'second inward current' in the rabbit SA node. *J. Physiol., Lond.* **334**, 56P.

Brown, H. F., Kimura, J., Noble, D., Noble, S. J. & Taupignon, A. 1984a The slow inward current, i_{si}, in the rabbit sino-atrial node investigated by voltage clamp and computer simulation. *Proc. R. Soc. Lond.* B **222**, 305–328.

Brown, H. F., Kimura, J., Noble, D., Noble, S. J. & Taupignon, A. 1984b The ionic currents underlying pacemaker activity in rabbit sino-atrial node: experimental results and computer simulation. *Proc. R. Soc. Lond.* B **222**, 329–347.

Carmeliet, E. 1980 Decrease of K efflux and influx by external Cs ions in cardiac Purkinje and muscle cells. *Pflügers Arch. Eur. J. Physiol.* **383**, 143–150.

Carmeliet, E. 1982 Induction and removal of inward-going rectification in sheep cardiac Purkinje fibres *J. Physiol., Lond.* **327**, 285–308.

Chandler, W. K. & Meves, H. 1965 Voltage clamp experiments on internally perfused giant axons. *J. Physiol., Lond.* **180**, 821–836.

Chapman, R. A. 1979 Excitation-contraction coupling in cardiac muscle. *Prog. Biophys. molec. Biol.* **35**, 1–52.

Chapman, R. A., Coray, A. & McGuigan, J. A. S. 1983 Sodium/calcium exchange in mammalian ventricular muscle: a study with sodium-sensitive microelectrodes. *J. Physiol., Lond.* **343**, 253–276.

Chapman, R. A. & Tunstall, J. 1980 The interaction of sodium and calcium ions at the cell membrane and the control of contractile strength in frog atrial muscle. *J. Physiol., Lond.* **305**, 109–123.

Clay, J. R. & Shrier, A. 1981a Analysis of subthreshold pacemaker currents in chick embryonic heart cells. *J. Physiol., Lond.* **312**, 471–490.

Clay, J. R. & Shrier, A. 1981*b* Developmental changes in subthreshold pace-maker currents in chick embryonic heart cells. *J. Physiol., Lond.* **312**, 491–504.

Clusin, W. T., Fischmeister, R. & deHaan, R. L. 1983 Caffeine-induced current in embryonic heart cells: time course and voltage dependence. *Am. J. Physiol.* **254**, H528–H532.

Cohen, I., Daut, J. & Noble, D. 1976 The effects of potassium and temperature on the pacemaker current i_{K_2} in Purkinje fibres. *J. Physiol., Lond.* **260**, 55–74.

Cohen, I., Eisner, D. A. & Noble, D. 1978 The action of adrenaline on pacemaker activity in cardiac Purkinje fibres. *J. Physiol., Lond.* **280**, 155–168.

Cohen, I. S., Falk, R. T. & Mulrine, N. K. 1983 Actions of barium and rubidium on membrane currents in canine Purkinje fibres. *J. Physiol., Lond.* **338**, 589–612.

Colatsky, T. J. 1980 Voltage clamp measurements of sodium channel properties in rabbit cardiac Purkinje fibres. *J. Physiol., Lond.* **305**, 215–234.

Colatsky, T. J. 1982 Mechanisms of action of lidocaine and quinidine on action potential duration in rabbit cardiac Purkinje fibres. *Circ. Res.* **50**, 17–27.

Colatsky, T. J. & Gadsby, D. C. 1980 Is tetrodotoxin block of background sodium channels in canine Purkinje fibres voltage-dependent? *J. Physiol., Lond.* **306**, 20P.

Colatsky, T. J. & Tsien, R. W. 1979 Electrical properties associated with wide intercellular clefts in rabbit Purkinje fibres. *J. Physiol., Lond.* **290**, 227–252.

Colquhoun, D., Neher, E., Reuter, H. & Stevens, C. F. 1981 Inward channels activated by intracellular Ca in cultured heart cells. *Nature, Lond.* **294**, 752–754.

Coraboeuf, E. & Carmeliet, E. 1982 Existence of two transient outward channels in sheep cardiac Purkinje fibres. *Pflügers Arch. Eur. J. Physiol.* **392**, 352–359.

Coraboeuf, E. & Deroubaix, E. 1978 Shortening effect of tetrodotoxin on action potentials of the conducting system in the dog heart. *J. Physiol., Lond.* **280**, 24P.

Coraboeuf, E., Gautier, P. & Guiraudou, P. 1981 Potential and tension changes induced by sodium removal in dog Purkinje fibres: role of an electrogenic sodium–calcium exchange. *J. Physiol., Lond.* **311**, 605–622.

Deitmer, J. W. & Ellis, D. 1978 The intracellular sodium activity of cardiac Purkinje fibres during inhibition and reactivation of the sodium–potassium pump. *J. Physiol., Lond.* **284**, 241–259.

DiFrancesco, D. 1981*a* A new interpretation of the pace-maker current i_{K_2} in Purkinje fibres. *J. Physiol., Lond.* **314**, 359–376.

DiFrancesco, D. 1981*b* A study of the ionic nature of the pace-maker current in calf Purkinje fibres. *J. Physiol., Lond.* **314**, 377–393.

DiFrancesco, D. 1982 Block and activation of the pace-maker channel in calf Purkinje fibres: effects of potassium, caesium and rubidium. *J. Physiol., Lond.* **329**, 485–507.

DiFrancesco, D. 1984 Characterization of the pacemaker (i_f) current kinetics in calf Purkinje fibres. *J. Physiol., Lond.* **348**, 341–367.

DiFrancesco, D. & Ferroni, A. 1983 Delayed activation of the cardiac pacemaker current and its dependence on conditioning pre-hyperpolarizations. *Pflügers Arch. Eur. J. Physiol.* **396**, 265–267.

DiFrancesco, D., Hart, G. & Noble, D. 1982 Ionic current transients attributable to the Na–Ca exchange process in the heart: computer model. *J. Physiol., Lond.* **328**, 15P.

DiFrancesco, D., Hart, G. & Noble, D. 1983 Demonstration of oscillatory variations in $[Ca]_i$ and membrane currents in a computer model of Ca-induced Ca release in mammalian Purkinje fibre and ventricular muscle. *J. Physiol., Lond.* **334**, 8P.

DiFrancesco, D., Hart, G. & Noble, D. 1985 (In preparation.)

DiFrancesco, D. & McNaughton, P. A. 1979 The effects of calcium on outward membrane currents in the cardiac Purkinje fibre. *J. Physiol., Lond.* **289**, 347–373.

DiFrancesco, D. & Noble, D. 1980*a* The time course of potassium current following potassium accumulation in frog atrium: analytical solutions using a linear approximation. *J. Physiol., Lond.* **306**, 152–173.

DiFrancesco, D. & Noble, D. 1980*b* Reconstruction of Purkinje fibre currents in sodium-free solution. *J. Physiol., Lond.* **308**, 35P.

DiFrancesco, D. & Noble, D. 1980*c* If 'i_{K_2}' is an inward current, how does it display potassium specificity? *J. Physiol., Lond.* **305**, 14.

DiFrancesco, D. & Noble, D. 1981 A model of cardiac electrical activity incorporating restricted extracellular spaces and the sodium potassium pump. *J. Physiol., Lond.* **320**, 25P.

DiFrancesco, D. & Noble, D. 1982 Implications of the re-interpretation of i_{K_2} for the modelling of the electrical activity of pacemaker tissues in the heart. In *Cardiac rate and rhythm* (ed. L. N. Bouman and H. J. Jongsma), pp. 93–128. The Hague: Martinus Nijhoff.

DiFrancesco, D., Noma, A. & Trautwein, W. 1979*a* Kinetics and magnitude of the time-dependent K current in the rabbit SA node: effect of external potassium. *Pflügers Arch. Eur. J. Physiol.* **381**, 271–279.

DiFrancesco, D., Ohba, M. & Ojeda, C. 1979*b* Measurement and significance of the reversal potential for the pacemaker current i_{K_2} in sheep Purkinje fibres. *J. Physiol., Lond.* **297**, 135–162.

Draper, M. H. & Weidmann, S. 1951 Cardiac resting and action potentials recorded with an intracellular electrode. *J. Physiol., Lond.* **115**, 74–94.

396   D. DiFRANCESCO AND D. NOBLE

Dudel, J., Peper, K., Rüdel, R. & Trautwein, W. 1967*a* The dynamic chloride component of membrane current in Purkinje fibres. *Pflügers Arch. Eur. J. Physiol.* **295**, 197–212.

Dudel, J., Peper, K., Rüdel, R. & Trautwein, W. 1967*b* The potassium component of membrane current in Purkinje fibres. *Pflügers Arch. Eur. J. Physiol.* **296**, 308–327.

Ebihara, L., Shigeto, N., Lieberman, M. & Johnson, E. A. 1980 The initial inward current in spherical clusters of chick embryonic heart cells. *J. gen. Physiol.* **75**, 437–456.

Eisner, D. A. & Lederer, W. J. 1980 Characterization of the sodium pump in cardiac Purkinje fibres. *J. Physiol., Lond.* **303**, 441–474.

Eisner, D. A., Lederer, W. J. & Noble, D. 1979 Caffeine and tetracaine abolish the slow inward current in sheep cardiac Purkinje fibres. *J. Physiol., Lond.* **293**, 76P.

Eisner, D. A., Lederer, W. J. & Vaughan-Jones, R. 1981 The dependence of sodium pumping and tension on intracellular sodium activity in voltage-clamped sheep Purkinje fibres. *J. Physiol., Lond.* **317**, 167–187.

Ellis, D. 1977 The effects of external cations and ouabain on the intracellular sodium activity of sheep heart Purkinje fibres. *J. Physiol., Lond.* **274**, 211–240.

Fabiato, A. & Fabiato, F. 1975 Contractions induced by a calcium-triggered release of calcium from the sarcoplasmic reticulum of single skinned cardiac cells. *J. Physiol., Lond.* **249**, 469–495.

Fischmeister, R. & Vassort, G. 1981 The electrogenic Na/Ca exchange and the cardiac electrical activity. 1. Simulation on Purkinje fibre action potential. *J. Physiol., Paris* **77**, 705–709.

Fozzard, H. & Hiraoka, M. 1973 The positive dynamic current and its inactivation properties in cardiac Purkinje fibres. *J. Physiol., Lond.* **234**, 569–586.

Gadsby, D. C. 1980 Activation of electrogenic Na^+/K^+ exchange by extracellular K^+ in canine cardiac Purkinje fibres. *Proc. natn. Acad. Sci. U.S.A.* **77**, 4035–4039.

Gadsby, D. C. 1982 Hyperpolarization of frog skeletal muscle fibres and canine cardiac Purkinje fibres during enhanced K^+–Na^+ exchange: extracellular K^+ depletion or increased pump current? *Curr. Top. Memb. Transport.* **16**, 17–34.

Gadsby, D. C. & Cranefield, P. F. 1977 Two levels of resting potential in cardiac Purkinje fibers. *J. gen. Physiol.* **70**, 725–746.

Gadsby, D. C. & Cranefield, P. F. 1979 Electrogenic sodium extrusion in cardiac Purkinje fibers. *J. gen. Physiol.* **73**, 819–837.

Gibbons, W. R. & Fozzard, H. A. 1975*a* Relationships between voltage and tension in sheep cardiac Purkinje fibers. *J. gen. Physiol.* **65**, 345–365.

Gibbons, W. R. & Fozzard, H. A. 1975*b* Slow inward current and contraction of sheep cardiac Purkinje fibers. *J. gen. Physiol.* **65**, 367–384.

Gintant, G. A., Datyuner, N. B. & Cohen, I. 1984 Slow inactivation of a tetrodotoxin-sensitive current in canine cardiac Purkinje fibres. *Biophys. J.* (In the press.)

Haas, H. G. & Kern, R. 1966 Potassium fluxes in voltage clamped Purkinje fibres. *Pflügers Arch. Eur. J. Physiol.* **291**, 69–84.

Hagiwara, S. & Takahashi, K. 1974 The anomalous rectification and cation selectivity of the membrane of a starfish egg cell. *J. membrane Biol.* **18**, 61–80.

Hall, A. E., Hutter, O. F. & Noble, D. 1963 Current–voltage relations of Purkinje fibres in sodium-deficient solutions. *J. Physiol., Lond.* **166**, 225–240.

Hart, G. 1983 The kinetics and temperature dependence of the pacemaker current i_f in sheep Purkinje fibres. *J. Physiol., Lond.* **337**, 401–416.

Hart, G., Noble, D. & Shimoni, Y. 1980 Adrenaline shifts the voltage dependence of the sodium and potassium components of i_f in sheep Purkinje fibres. *J. Physiol., Lond.* **308**, 34P.

Hart, G., Noble, D. & Shimoni, Y. 1982 Analysis of the early outward currents in sheep Purkinje fibres. *J. Physiol., Lond.* **326**, 68P.

Hart, G., Noble, D. & Shimoni, Y. 1983 The effects of low concentrations of cardiotonic steroids on membrane currents and tension in sheep Purkinje fibres. *J. Physiol., Lond.* **334**, 103–131.

Hauswirth, O., Noble, D. & Tsien, R. W. 1968 Adrenaline: mechanism of action on the pacemaker potential in cardiac Purkinje fibres. *Science, Wash.* **162**, 916–917.

Hauswirth, O., Noble, D. & Tsien, R. W. 1969 The mechanism of oscillatory activity at low membrane potentials in cardiac Purkinje fibres. *J. Physiol., Lond.* **200**, 255–265.

Hauswirth, O., Noble, D. & Tsien, R. W. 1972 The dependence of plateau currents in cardiac Purkinje fibres on the interval between action potentials. *J. Physiol., Lond.* **222**, 27–49.

Hille, B. & Schwartz, W. 1978 Potassium channels as multi-ion single-file pores. *J. gen. Physiol.* **72**, 409–442.

Hodgkin, A. L. & Horowitz, P. 1960 Potassium contractures in single muscle fibres. *J. Physiol., Lond.* **153**, 386–403.

Hodgkin, A. L. & Huxley, A. F. 1952 A quantitative description of membrane current and its application to conduction and excitation in nerve. *J. Physiol., Lond.* **117**, 500–544.

Hodgkin, A. L. & Katz, B. 1949 The effect of sodium ions on the electrical activity of the giant axon of the squid. *J. Physiol., Lond.* **108**, 37–77.

Horackova, M. & Vassort, G. 1979 Sodium–calcium exchange in regulation of cardiac contractility. Evidence for an electrogenic voltage-dependent mechanism. *J. gen. Physiol.* **73**, 403–424.

MODEL OF CARDIAC ELECTRICAL ACTIVITY   397

Hume, J. R. & Giles, W. R. 1983 Ionic currents in single isolated bullfrog atrial cells. *J. gen. Physiol.* **81**, 153–194.

Isenberg, G. 1976 Cardiac Purkinje fibres: caesium as a tool to block inward rectifying potassium currents. *Pflügers Arch. Eur. J. Physiol.* **365**, 99–106.

Isenberg, G. & Klöckner, V. 1982 Calcium currents of isolated bovine ventricular myocytes are fast and of large amplitude. *Pflügers Arch. Eur. J. Physiol.* **195**, 30–41.

Isenberg, G. & Trautwein, W. 1974 The effect of dihydroouabain and lithium ions on the outward current in cardiac Purkinje fibres. Evidence for electrogenicity of active transport. *Pflügers Arch. Eur. J. Physiol.* **350**, 41–54.

Jack, J. J. B., Noble, D. & Tsien, R. W. 1975 *Electric current flow in excitable cells.* Oxford: Clarendon Press. (Paperback edition, 1983.)

Johnson, E. A. & Lieberman, M. 1971 Heart: excitation and contraction. *A. Rev. Physiol.* **33**, 499–532.

Kass, R. S. & Wiegers, S. E. 1982 The ionic basis of concentration related effects of noradrenaline on the action potential of calf cardiac Purkinje fibres. *J. Physiol., Lond.* **322**, 541–558.

Kenyon, J. L. & Gibbons, W. R. 1979 4-aminopyridine and the early outward current of sheep cardiac Purkinje fibres. *J. gen. Physiol.* **73**, 139–157.

Lederer, W. J. & Eisner, D. A. 1982 The effects of sodium pump activity on the slow inward current in sheep cardiac Purkinje fibres. *Proc. R. Soc. Lond.* B **214**, 249–262.

Lee & Fozzard, H. A. 1975 Activities of potassium and sodium ions in rabbit heart muscle. *J. gen. Physiol.* **65**, 695–708.

Lee, E., Lee, K. S., Noble, D. & Spindler, A. J. 1983 A very slow inward current in single ventricular cells. *J. Physiol., Lond.* **345**, 6P.

Lee, E., Lee, K. S., Noble, D. & Spindler, A. J. 1984a A new, very slow inward Ca current in single ventricular cells of adult guinea-pig. *J. Physiol., Lond.* **346**, 75P.

Lee, E., Lee, K. S., Noble, D. & Spindler, A. J. 1984b Further properties of the very slow inward currents in isolated single guinea-pig cells. *J. Physiol., Lond.* (In the press.)

Lee, K. S. & Tsien, R. W. 1982 Reversal of current through calcium channels in dialyzed heart cells. *Nature, Lond.* **297**, 498–501.

Lee, K. S., Weeks, T. A., Kao, R. L., Akaike, N. & Brown, A. M. 1979 Sodium current in single heart muscle cells. *Nature, Lond.* **278**, 269–271.

McAllister, R. E. & Noble, D. 1966 The time and voltage dependence of the slow outward current in cardiac Purkinje fibres. *J. Physiol., Lond.* **186**, 632–662.

McAllister, R. E., Noble, D. & Tsien, R. W. 1975 Reconstruction of the electrical activity of cardiac Purkinje fibres. *J. Physiol., Lond.* **251**, 1–59.

McDonald, T. F. & Trautwein, W. 1978 The potassium current underlying delayed rectification in cat ventricular muscle. *J. Physiol., Lond.* **274**, 217–246.

Marban, E., Rink, T. J., Tsien, R. W. & Tsien, R. Y. 1980 Free calcium in heart muscle at rest and during contraction measured with Ca^{++}-sensitive microelectrodes. *Nature, Lond.* **286**, 845–850.

Mentrard, D. & Vassort, G. 1982 The Na–Ca exchange generates a current in frog heart cells. *J. Physiol., Lond.* **334**, 55P.

Mitchell, M. R., Powell, T., Sturridge, M. F., Terrar, D. A. & Twist, V. W. 1982 Action potentials and second inward current recorded from individual human ventricular muscle cells. *J. Physiol., Lond.* **332**, 51P.

Mitchell, M. R., Powell, T., Terrar, D. A. & Twist, V. W. 1983 Characteristics of the second inward current in cells isolated from rat ventricular muscle. *Proc. R. Soc. Lond.* B **219**, 447–469.

Miura, D. S., Hoffman, B. F. & Rosen, M. R. 1977 The effect of extracellular potassium on the intracellular potassium ion activity and transmembrane potentials of beating canine cardiac Purkinje fibres. *J. gen. Physiol.* **69**, 463–474.

Mobley, B. A. & Page, E. 1972 The surface area of sheep cardiac Purkinje fibres. *J. Physiol., Lond.* **220**, 547–563.

Modern computing methods 1961 London: Her Majesty's Stationery Office.

Momose, Y., Szabo, G. & Giles, W. R. 1983 An inwardly rectifying K^+ current in bullfrog atrial cells. *Biophys. J.* **41**, 311a.

Mullins, L. J. 1977 A mechanism for Na/Ca transport. *J. gen. Physiol.* **70**, 681–695.

Mullins, L. J. 1981 *Ion transport in the heart.* New York: Raven Press.

Niedergerke, R. 1963 Movements of Ca in beating ventricles of the frog. *J. Physiol., Lond.* **167**, 551–580.

Niedergerke, R. & Page, S. 1982 Changes of frog heart action potential due to intracellular calcium ions. *J. Physiol., Lond.* **328**, 17–18P.

Noble, D. 1962 A modification of the Hodgkin–Huxley equations applicable to Purkinje fibre action and pacemaker potentials. *J. Physiol., Lond.* **160**, 317–352.

Noble, D. 1965 Electrical properties of cardiac muscle attributable to inward-going (anomalous) rectification. *J. cell. comp. Physiol.* **66**, 127–136.

Noble, D. 1972 Conductance mechanisms in excitable cells. In *Biomembranes* 3 (ed. F. Kreuzer and J. F. G. Slegers), pp. 427–447. New York: Plenum Press.

Noble, D. 1979 *The initiation of the heartbeat.* 2nd edn. Oxford University Press.

Noble, D. 1984 The surprising heart: A review of recent progress in cardiac electrophysiology. *J. Physiol., Lond.* (In the press.)

Noble, D. & Noble, S. 1984 A model of sino-atrial node electrical activity based on a modification of the DiFrancesco–Noble (1984) equations. *Proc. R. Soc. Lond.* B **222**, 295–304.

Noble, D. & Tsien, R. W. 1968 The kinetics and rectifier properties of the slow potassium current in cardiac Purkinje fibres. *J. Physiol., Lond.* **195**, 185–214.

Noble, D. & Tsien, R. W. 1969a Outward membrane currents activated in the plateau range of potentials in cardiac Purkinje fibres. *J. Physiol., Lond.* **200**, 205–231.

Noble, D. & Tsien, R. W. 1969b Reconstruction of the repolarization process in cardiac Purkinje fibres based on voltage clamp measurements of the membrane current. *J. Physiol., Lond.* **200**, 233–254.

Plant, R. E. 1979 The efficient numerical solution of biological simulation problems. *Computer Progr. in Biomed.* **10**, 1–15.

Peper, K. & Trautwein, W. 1969 A note on the pacemaker current in Purkinje fibres. *Pflügers Arch. Eur. J. Physiol.* **309**, 356–361.

Powell, T., Terrar, D. A. & Twist, V. W. 1981 The effect of noradrenaline on slow inward current in rat ventricular myocytes. *J. Physiol., Lond.* **319**, 82P.

Reuter, H. 1967 The dependence of slow inward current in Purkinje fibres on the extracellular calcium concentration. *J. Physiol., Lond.* **192**, 479–492.

Reuter, H. & Scholz, H. 1977 A study on the ion selectivity and the kinetic properties of the calcium dependent slow inward current in mammalian cardiac muscle. *J. Physiol., Lond.* **264**, 17–47.

Reuter, H., Stevens, C. F., Tsien, R. W. & Yellen, G. 1982 Properties of single calcium channels in cardiac cell culture. *Nature, Lond.* **297**, 501–504.

Rougier, O., Vassort, G., Garnier, D., Gargouil, Y.-M. & Coraboeuf, E. 1969 Existence and rôle of a slow inward current during the frog atrial action potential. *Pflügers Arch. Eur. J. Physiol.* **308**, 91–110.

Sakmann, B. & Trube, G. 1984 Conductance properties of single inwardly rectifying potassium channels in ventricular cells from guinea-pig heart. *J. Physiol., Lond.* (In the press.)

Sheu, S. S. & Fozzard, H. A. 1982 Transmembrane Na^+ and Ca^{2+} electrochemical gradients in cardiac muscle and their relationship to force development. *J. gen. Physiol.* **80**, 325–351.

Siegelbaum, S. A. & Tsien, R. W. 1980 Calcium-activated transient outward current in calf cardiac Purkinje fibres. *J. Physiol., Lond.* **299**, 485–506.

Siegelbaum, S., Tsien, R. W. & Kass, R. S. 1977 Role of intracellular calcium in the transient outward current in calf Purkinje fibres. *Nature, Lond.* **269**, 611–613.

Sjodin, R. A. 1980 Contribution of Na/Ca transport to the resting membrane potential. *J. gen. Physiol.* **76**, 99–108.

Standen, N. B. & Stanfield, P. R. 1982 A binding site model for calcium channel inactivation that depends on calcium entry. *Proc. R. Soc. Lond.* B **217**, 101–110.

Tsien, R. W. 1974 Effects of epinephrine on the pacemaker potassium current of cardiac Purkinje fibres. *J. gen. Physiol.* **64**, 293–319.

Vasalle, M. 1966 Cardiac pacemaker potentials at different extra- and intracellular K concentrations. *Am. J. Physiol.* **208**, 770–775.

Vassalle, M. 1966 Analysis of cardiac pacemaker potentials using a 'voltage-clamp' technique. *Am. J. Physiol.* **210**, 1335–1341.

Vereecke, J., Isenberg, G. & Carmeliet, E. 1980 K efflux through inward rectifying K channels in voltage clamped Purkinje fibres. *Pflügers Arch. Eur. J. Physiol.* **384**, 207–217.

Weidmann, S. 1951 Effect of current flow on the membrane potential of cardiac muscle. *J. Physiol., Lond.* **115**, 227–236.

Weidmann, S. 1952 The electrical constants of Purkinje fibres. *J. Physiol., Lond.* **118**, 348–360.

Weidmann, S. 1956 *Elektrophysiologie der Herzmuskelfaser.* Bern: Huber.

Wier, W. G. 1980 Calcium transients during excitation-contraction coupling in mammalian heart: aequorin signals of canine Purkinje fibres. *Science, Wash.* **207**, 1085–1087.

Wier, W. G. & Isenberg, G. 1982 Intracellular [Ca] transients in voltage-clamped cardiac Purkinje fibres. *Pflügers Arch. Eur. J. Physiol.* **392**, 284–290.

Proc. R. Soc. Lond. B **240**, 83–96 (1990)
Printed in Great Britain

A model of the single atrial cell: relation between calcium current and calcium release

By Y. E. Earm and D. Noble, F.R.S.

University Laboratory of Physiology, Parks Road, Oxford OX1 3PT, U.K.

(*Received 6 July* 1989 – *Revised 24 November* 1989)

The hypothesis that calcium release from the sarcoplasmic reticulum in
cardiac muscle is induced by rises in free cytosolic calcium (Fabiato 1983,
Am. J. Physiol **245**) allows the possibility that the release could be at
least partly regenerative. There would then be a non-linear relation
between calcium current and calcium release. We have investigated this
possibility in a single-cell version of the rabbit-atrial model developed by
Hilgemann & Noble (1987, *Proc. R. Soc. Lond.* B **230**). The model
predicts different voltage ranges of activation for calcium-dependent
processes (like the sodium–calcium exchange current, contraction or
Fura-2 signals) and the calcium current, in agreement with the experimental results obtained by Earm *et al.* (1990, *Proc. R. Soc. Lond.* B **240**)
on exchange current tails, Cannell *et al.* (1987, *Science, Wash.* **238**) by
using Fura-2 signals, and Fedida *et al.* (1987, *J. Physiol., Lond.* **385**) and
Talo *et al.* (1988, *Biology of isolated adult cardiac myocytes*) by using
contraction. However, when the Fura-2 concentration is sufficiently high
(greater than 200 μm) the activation ranges become very similar as the
buffering properties of Fura-2 are sufficient to remove the regenerative
effect. It is therefore important to allow for the buffering properties of
calcium indicators when investigating the correlation between calcium
current and calcium release.

Introduction

Earm *et al.* (1989, 1990) have shown that the late phase of the action potential in
isolated single rabbit atrial cells is maintained by a slowly decaying inward current
that displays all the properties required for its identification as the sodium–calcium
exchange current. These results are similar to those obtained for the rat ventricular
cell (Mitchell *et al.* 1983, 1984, 1987; Schouten & Ter Keurs 1985), which displays
a very similar action potential. They are also consistent with the experimental
results and theory of calcium movement in the rabbit atrium given by Hilgemann
& Noble (1987).

The identification of the exchange current during the action potential and its
separation from the calcium current in voltage-clamp experiments enabled Earm
et al. to make the unexpected observation that the voltage dependence of the
exchange current tails does not match that of the calcium current. The tail current
increases rapidly with potential towards a maximum at about −20 mV and then
stays relatively constant with further depolarization, whereas the calcium current
displays the typical current–voltage relation for a voltage-activated channel, with

[83]

84   Y. E. Earm and D. Noble

activation over the range $-35\,\text{mV}$ to $+10\,\text{mV}$, followed by a nearly linear approach to the reversal potential at $+60\,\text{mV}$. Because the exchange current is proportional to $[\text{Ca}^{2+}]_i$ and should therefore be a good indicator of the intracellular calcium transient (Egan *et al.* 1989), this implies that calcium release is not proportional to calcium entry.

This is a controversial issue and there are experimental results that support both sides of the controversy. On the one hand (Barcenas-Ruiz & Wier 1987; Callewaert *et al.* 1988; Näbauer *et al.* 1989), there are results supporting a simple correlation between calcium entry and calcium release. On the other hand, some workers have found calcium release to occur below the apparent threshold for the main calcium-current channel in cardiac muscle. Thus Cannell *et al.* (1987) found that the voltage dependence of the Fura-2 signal showed that the calcium signal increases with voltage before the calcium current, whereas Fedida *et al.* (1987) found that some contraction and a slow inward current that they attributed to sodium–calcium exchange, could be recorded below the threshold for the calcium current and Talo *et al.* (1988) showed that a calcium current that is a relatively small percentage of the maximum can initiate a nearly maximal contraction.

At first sight, the calcium-induced calcium-release hypothesis might lead one to expect a simple correlation between calcium entry and release. But this is not necessarily so. In some circumstances, for example during the generation of the oscillations of internal calcium that underly the transient inward current, calcium release occurs in a regenerative manner without being triggered by calcium entry. Moreover, in principle there should be regenerative behaviour because, once the release process is started, it may contribute to the calcium pool that is controlling release. Whether this occurs will depend on the location of this calcium pool (is it the general cytosol calcium, or is it a special region accessed by at least part of the calcium current?), and on the gain of the loop between intracellular free calcium and calcium release.

The Hilgemann–Noble (Hilgemann & Noble 1987) model of the atrium uses the experimental data on calcium-induced release, and is perhaps the most extensive model of this process so far constructed. It is certainly the only model that enables this process to be coupled to the action potential and surface membrane currents. We therefore thought it would be worthwhile to convert this multicellular model into a single-cell model and then to test whether it can reproduce the voltage-clamp results of Earm, Ho and So.

METHODS

Single-cell model

This was developed by scaling down the multicellular model developed by Hilgemann & Noble (1987). The cell length was set to 80 μm, and the cell diameter to 8 μm. As noted in the preceding paper (Earm *et al.* 1990) the exact cell volume is uncertain as the assumption of an exact cylinder will overestimate the volume. Some of the calculations shown here were done with smaller cell volumes, and we note in the text how this influences the results. The cell capacitance was assumed to be 40 pF. These values are based on the measurements of Earm *et al.* (1990).

The conductances were then all scaled down by a factor of 100. This gives almost exactly the correct amplitude of single-cell currents, implying that the Hilgemann–Noble model must have represented about 100 cells. The constants determining the membrane currents were: $g_{Na} = 0.5$ nS, $P_{Ca} = 0.05$, $g_{to} = 0.01$ nS, $g_{b,Na} = 0.00012$ nS, $g_{b,Ca} = 0.00005$ nS, $\bar{\iota}_p = 0.14$ nA, $K_{NaCa} = 0.0001$, $\bar{g}_{K1} = 0.017$ nS, $g_{b,K} = 0.0017$ nS.

For the calcium-release process, we had to make some choices within the Hilgemann–Noble formulation (see their equations (19–(27)). The first was whether to keep the combined voltage and calcium-induced release, or whether to make release uniquely calcium dependent, by setting E_{on} to zero. For most of the calculations we decided to keep the original formulation as we found that it correctly predicts the results of Earm *et al.* (1990) in showing that the calcium dependent current remains high even following steps to very positive potentials at which i_{Ca} becomes small or even reverses. It is worth noting, however, that this is not the result obtained by Callewaert *et al.* (1988) by using Fura-2 signals in rat ventricular cells. To model their situation correctly at positive potentials it would be necessary to set E_{on} in equation (21) to zero. Later calculations (see figure 4) explore this case.

The second choice concerns the value of the binding constant for calcium binding to the release site. This was set arbitrarily to 0.5 µm in the Hilgemann–Noble equations (see their equation (20)). We found that values in the range 0.4–0.45 µm gave the best results for the difference in voltage dependence of calcium current and release. In the results shown here we used 0.4 µm, and we modified HEART to allow the input variable *KMCA* to be used for this constant (see the example input files in Appendix 2). This modification is included in versions 2.2 and 3.0 of HEART. If version 2.1 is used, then the program would need to be edited to give the correct value for the binding constant in equation (20). (See Appendix 2.)

The transient outward current in the Hilgemann–Noble equations (see their equations (34) and (35)) shows activation and deactivation only. This is perfectly sufficient for action-potential calculations, when deactivation is the primary means by which i_{to} is switched off. For voltage-clamp reconstructions an inactivation process is required. For this purpose we added the expressions used by DiFrancesco & Noble (1985, see their equations (15)–(18)). As the DiFrancesco–Noble Purkinje-fibre model was the starting point for the sinus-node model from which the atrial model was developed, it is worth noting here that, since the appearance of the DiFrancesco–Noble (1985) paper, several discrepancies have been noted between the expressions as given in their text and those actually used in the computer program OXSOFT HEART. In Appendix 1, we summarize these differences. This enables anyone who wishes to program the equations again to do so.

Computations

These were done by using the Unix version of OXSOFT HEART (version 2.2), compiled via the Berkeley Pascal compiler, and running on a SUN microsystems 3/50 computer. All the calculations shown here were done by using the full equations for calcium buffering introduced by Hilgemann & Noble (1987). Given the importance of avoiding possible errors, we did not use the faster 'steady-state'

86   Y. E. Earm and D. Noble

calculation incorporated as an option in the HEART program. To enable users of OXSOFT HEART to reproduce our results directly, we include two examples of data input files for the calculations (Appendix 2).

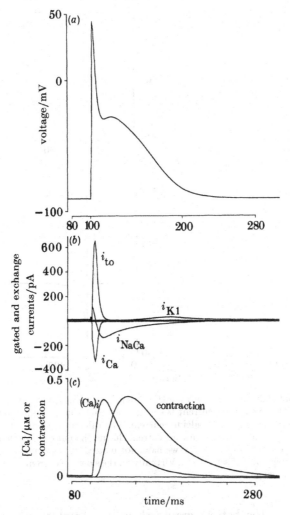

FIGURE 1. Calculation of response of single atrial cell model to a current pulse of 1.3 nA applied for 2 ms. (a) Membrane potential; (b) currents carried by the transient outward, inward rectifier and calcium channels, and by the sodium–calcium exchange; (c) variation in free intracellular calcium and contraction.

RESULTS

Figure 1 shows the response of the single-cell model to a stimulus of 1.3 nA applied for 2 ms at time 100 ms. The shape of the action potential is similar to that recorded experimentally (see Earm *et al.* (1990), figure 1). The middle graphs show the computed currents attributable to i_{to}, i_{Ca}, i_{K1}, and the sodium–calcium

Atrial cell model   87

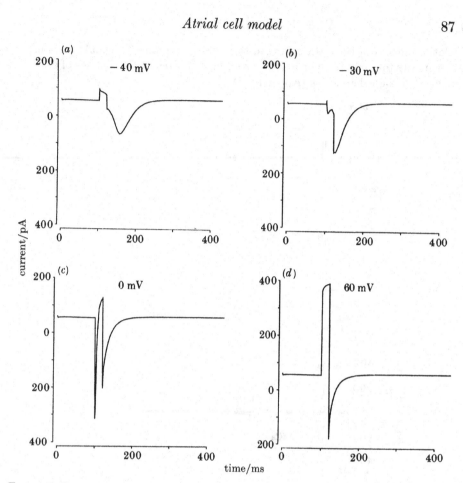

FIGURE 2. Currents calculated during and following 20 ms depolarizations from -70 mV to the
    potentials indicated. The inward tail current is substantially activated by pulses (to -30
    and -40 mV) that activate very little calcium current. The calcium current activated at
    -30 mV is insufficient to generate a net inward current during the pulse. Note that as the
    deactivation of i_{Ca} is very rapid at -70 mV we have not plotted the calcium current and
    sodium–calcium exchange current separately. Virtually all the slow tail current is generated
    by sodium–calcium exchange.

exchange current, i_{NaCa}. The bottom graphs show the computed intracellular
calcium transient and contraction.

Figure 2 shows the computed currents during and following voltage-clamp steps
from -70 mV to the potentials indicated for a period of 20 ms. To simplify the
records, we set the sodium and transient outward currents to zero. It is clear that
the step to -40 mV, which produces a very small calcium current (insufficient to
produce net inward current during the clamp step) nevertheless initiates a substantial slow tail current, which takes nearly 100 ms to reach its peak inward value.
At -30 mV the calcium current is large enough to produce a net inward-current
change (though not an absolute net inward current) during the pulse, but the slow
tail current, representing sodium–calcium exchange is still much larger than the

88   Y. E. Earm and D. Noble

calcium current. At 0 mV, the situation is reversed. Finally, at 60 mV, there is no calcium current (this is close to the calcium-channel reversal potential), but there is a large inward tail current. These results are qualitatively very similar to those of Earm *et al.* (1990).

Figure 3 (top left) shows the normalized calcium and tail currents plotted against the pulse voltage. The tail current 'activates' over a voltage range approximately 20 mV more negative than the calcium current.

Because the model reproduces, at least semiquantitatively, the results of Earm *et al.* (1990), Cannell *et al.* (1987), Fedida *et al.* (1987) and Talo *et al.* (1988) it is

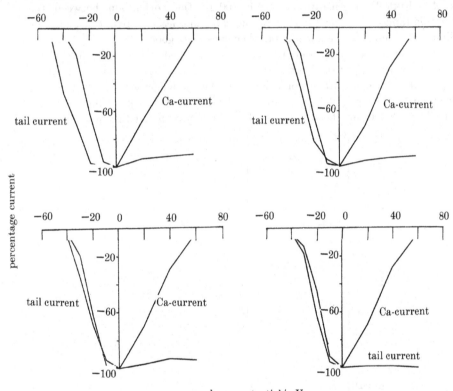

membrane potential/mV

FIGURE 3. Peak calcium current and peak inward tail current plotted as a function of the potential during the 20 ms voltage-clamp pulse. The calcium current is calculated as the peak achieved during the pulse, whereas the tail current is calculated as the peak achieved following the pulse. In each plot the currents have been normalised to emphasize the correlation or lack of correlation. (*a*) In the standard model, the tail current activates over a voltage range 20 mV more negative than the calcium current. A similar result is obtained when the intracellular calcium or contraction is plotted (we have not shown these as in the model, the exchange current is proportional to intracellular calcium and is therefore a good indicator of the calcium level). (*b–d*) The same computations were done with increasing intracellular concentrations of Fura-2 by using experimental buffering kinetics published by Jackson *et al.* (1987) and incorporated in HEART by Noble & Powell (1990). Above 200 μM Fura-2 the additional buffering is sufficient to linearize the relation between calcium current and release over the negative range of potentials. ((*a*) Fura-2 = 0 mM; (*b*) Fura-2 = 0.1 mM; (*c*) Fura-2 = 0.2 mM; (*d*) Fura-2 = 0.4 mM.)

Atrial cell model   **89**

worth asking what would need to be modified to produce the results of Callewaert *et al.* (1988) showing good correlation between the activation range of the calcium release and that of the calcium current. The major difference between the results of Earm *et al.* and those of Callewaert *et al.* is that the latter were obtained with up to 400 μM Fura-2 inside the cell. Fura-2 is itself a buffer with known calcium-binding properties (Jackson *et al.* 1987) and these kinetics have recently been introduced into the HEART program by Noble & Powell (1990). We therefore repeated the computations of figure 2 by using various concentrations of Fura-2 in the model cell. The results are shown in the other panels of figure 3. It is clear that for Fura-2 concentrations above 200 μM the correlation between the activation range of the calcium current and calcium release (here indicated by the sodium–calcium-exchange current tail) becomes very good.

It is important to determine whether the large calcium release at positive potentials, where i_{Ca} is small or even reversed, is attributable to the assumed component of direct voltage dependence in the model or whether it could be attributed to calcium entry via sodium–calcium exchange operating in reverse mode. The calculations shown in figure 4 were done to answer this question. Two

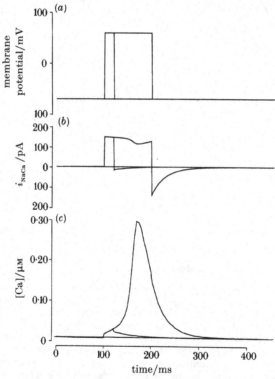

FIGURE 4. Calculations of sodium–calcium exchange current (*b*) and intracellular calcium (*c*), during 20 ms and 100 ms depolarizations to +60 mV from −70 mV when the model is modified to remove direct voltage dependence of calcium release. The 20 ms pulse then fails to trigger release (all the small rise in [Ca]ᵢ is attributable to calcium entry via reversed sodium–calcium exchange). During the 100 ms pulse this calcium entry becomes sufficient to trigger calcium release from the sarcoplasmic reticulum after about 50 ms.

90   Y. E. Earm and D. Noble

voltage pulses to $+60$ mV were applied, one of 20 ms duration, the other 100 ms. In both calculations the directly voltage-dependent component of release was removed by setting E_{on} (ALPHA [26] in the HEART program) to zero. The results show that for the 20 ms pulse this almost completely eliminates calcium release, and the sodium–calcium-exchange tail on repolarizing to -70 mV is very small. Note, however, that there is some calcium entry via the sodium–calcium exchange during the pulse (i.e. the exchange current is outward). When the pulse is prolonged to 100 ms, the entry becomes sufficient to fully trigger release from the reticulum.

Discussion
Nonlinearity between calcium current and calcium release

The results clearly show that the calcium-induced release hypothesis, as expressed mathematically by Hilgemann & Noble, can explain the experimental results of Earm *et al.* (1990) in a semi-quantitative way. They are also consistent with the results of Cannell *et al.* (1987) showing that the voltage dependence of calcium release is steeper than that of the calcium current so that the release even appears to occur below the calcium current threshold. Our calculations show that this effect may be apparent rather than real as *some* calcium current, however small, does flow in all the cases where the model induced calcium release. The reason why the release may appear to occur below the calcium-current threshold is that, for very small values of i_{Ca}, the current can easily be masked by other voltage-dependent current changes. If we had plotted net current rather than the calcium current itself in figure 3, we would have obtained the same results, i.e. release at voltages apparently below the calcium-current threshold.

, This result is encouraging because it shows that the calcium-induced release hypothesis is quite capable of explaining release that is not proportional to the calcium current. Indeed, as noted in the Introduction, this kind of nonlinearity is to be expected in a system that should display some positive feedback between calcium current. Indeed, as noted in the Introduction, this kind of non-linearity is

Nevertheless, there are some quantitative differences between the computed and experimental results. The most important of these concern the timescale. In the model results, the timecourse of release, and therefore of the sodium–calcium-exchange current is strongly voltage dependent (see figure 2). Moreover, at negative voltages, both reach a peak very slowly. Whereas, in the experimental results, the peak inward tail current is reached fairly quickly and the time to peak is not strongly voltage dependent (see Earm *et al.* (1990), fig. 3). Finally, we needed to use 20 ms pulses in the model, whereas in the experimental results, 2–5 ms pulses were sufficient to induce release, even at negative voltages.

All these differences could be attributed to the same potential failing in the model. It is assumed, for lack of further evidence to the contrary, that the calcium entering through the surface-membrane channels, and that released from intracellular stores is immediately distributed throughout the cytosol, ready to be buffered by calmodulin and troponin. In fact, of course, there must be at least some diffusion restrictions, and it is even possible that the space into which the current triggering release flows, and perhaps the release itself, is restricted by more than ordinary diffusion. There is no doubt that, if we were to introduce such complexities, the speed of the regenerative process linking rise in $[Ca]_i$, to calcium release, to further rise in calcium, etc. would be faster. We considered the possibility of determining how small such a space would need to be to obtain the results described by Earm *et al.* (1990) but decided that it is premature to introduce such complexity without further experimental information. There are simply too many ways in which the timecourse of release could be affected by diffusion or compartmentation. We did, however, find that the speed of release could be greatly increased by reducing the cell size.

Moreover, exactly the same deficiencies could be related to the question whether direct voltage dependence of release is essential to explain large release at very positive potentials when i_{Ca} is small or reversed. The results shown in figure 4 show that calcium entry via reversed sodium–calcium exchange *can* play the same role when the depolarization lasts long enough. Clearly, therefore, if calcium entering through the cell membrane were to accumulate in a subsarcolemmal space, the duration of the pulse required to ensure release triggered by calcium entry via sodium–calcium exchange would be reduced. We checked this point by rerunning calculations with the same ionic currents but with a smaller cell volume. These showed that a restricted space equivalent to about 10 % of the cell volume would be adequate to enable the results of Earm *et al.* (1990) at positive voltages to be obtained without assuming a directly voltage dependent component of release. Hilgemann & Noble (1987, p. 174) anticipated the possibility of a large, local calcium transient functioning in place of direct voltage dependence when formulating their equations for the release process. There is clearly an urgent need for more experimental data on intracellular compartmentation of calcium to clarify this matter, particularly as the detailed properties may depend on the cell type used. Thus, although rabbit atrial cells show a rapid and large release (as monitored by subsequent activation of sodium–calcium-exchange current tails) at very positive potentials, in guinea-pig ventricular cells the response in the absence of i_{Ca} is slow. Isenberg *et al.* (1988) show that in these cells calcium entering via reversed sodium–calcium exchange (as shown by its dependence on the patch pipette sodium concentration) can produce only a slow increase in cytosolic calcium.

Influence of calcium probes on calcium release

The question that now arises is how Callewaert *et al.* (1988) and Näbauer *et al.* (1989) were able to obtain a simple quantitative relation between calcium current and calcium release. The results shown in figure 3 show that it is possible that, at high enough concentrations, the buffering properties of Fura-2 may be strong enough to reduce the positive feedback between calcium and calcium release sufficiently to produce a simpler relation between calcium current and calcium release. It is worth noting that the concentration range of intracellular Fura-2 used by Näbauer *et al.* was 200–400 μM. It should be noted that this effect of Fura-2 and similar calcium probes does not depend on where the calcium initiating release comes from. Not only is the positive feedback between cystosol calcium and release reduced, but more calcium current is required to initiate a given quantity of release.

92   Y. E. Earm and D. Noble

Increasing the buffering capacity of the cytosol is not the only way to reduce the nonlinearity. We also found that reducing the calcium released from the reticulum achieved the same effect. The degree of loading of the sarcoplasmic reticulum and the extent of inactivation of the release process will also determine whether the relation is linear or nonlinear.

In this paper, we have concentrated on the different voltage dependence of calcium current and release in the model and in the experimental results. It is worth noting, however, that the features of the Hilgemann–Noble equations that enable us to reproduce the experimental results of Earm *et al.* are the same features that enabled Egan *et al.* (1989) to use the model to reproduce their experimental finding that, in the guinea-pig ventricular cell, rapid modulation of the calcium current within a single beat has relatively little effect on the sodium–calcium exchange current tail (see Egan *et al.* (1989) figs 7 and 12).

We acknowledge the British Heart Foundation for supporting this work with a visiting research fellowship for Y. E. Earm, and ICI for providing a fellowship for travel and research expenses.

Appendix 1

Comparison between OXSOFT HEART *and the DiFrancesco–Noble* (1985) *paper*

This Appendix lists the differences that have been reported between the original model equations as described in DiFrancesco & Noble (1985) and the expressions used in OXSOFT HEART.

Note on origin of OXSOFT HEART

Some of the differences between the DiFrancesco–Noble paper (referred to here as D–F) and HEART arise from the way in which HEART developed from the early versions used for the original computations for the D–F paper. These versions were first written in ALGOL-60. For distribution the software package was completely translated into PASCAL. Ninety-five percent of this translation was done with an OXSOFT translator program. The rest was done by hand. This is where most of the important differences crept in. The opportunity was taken to clean up the program formatting, remove the more inelegant parts, and to make the standard model procedures conform to the then current models.

The general rule to follow is that, where there is a difference between D–F and HEART, HEART should be taken as the definitive source.

Differences in equations

Equations (5) *and* (6)

The rates for the y reaction in HEART are twice those given in D–F. So the constant 0.025 should be 0.05 and the constant 0.5 should be 1. The D–F paper used the original ALGOL listing but failed to note that SPEED(Y) was set to 2 in the procedure PURKINJE. This inelegance was removed during the translation to PASCAL.

Atrial cell model 93

Equation (8)

The constant 25 should be replaced by RT/F. The ALGOL program used 25, which is correct at only one temperature.

Equation (14)

The constant $K_{m,1}$ should be replaced by the constant $K_{m,\text{to}}$, which has the value 10 mM. This allows the K^+ activation of i_{to} to saturate near 30 mM as stated in the text. (Note: the constant $K_{m,1}$ is a constant in the equation for i_{Ki} and should not be present in the equation for i_{to}.)

The constant $K_{m,\text{to}}$ should be replaced by the constant K_{act4}, which was normally set to 0.5 μM. This constant determines the calcium concentration for half activation of i_{to} when calcium activation is incorporated into the model. (In HEART this occurs when AMODE is set to 4 or 9.)

Equation (26)

The term $2RT$ should be replaced by RT in each case. The constant $2RT$ was that used in the ALGOL program that was written before introducing the partition parameter, γ. In the standard model, this parameter is set to 0.5, which is equivalent to dividing ER/RT by 2. The error was missed in proof-reading the D–F paper.

Equation (32)

The constant 0.1 should be -0.1. The minus sign was missed during proof-correcting.

Equation (36)

The denominator should read $1+320\ldots$ not $320\ldots$

Equation (37)

The variable $f2$ should read f_2.

Equations (38) *and* (39)

F^2 should read F. Note that this means that the permeabilities, PCA etc, are the constant field permeabilities *multiplied by F*.

Equation (49)

The variable β_{up} is set to zero in HEART. This is because, for any reasonable values for the uptake and leak to and from the uptake store, the leak was found to be negligible compared to transfer of calcium to the release store.

Equation (52)

As stated in the text, the equations for α_p and β_p were the same as those used for f, but slowed by a factor of 10. This enables repriming of Ca release to occur more slowly than repriming of the calcium current, as observed experimentally.

94   Y. E. Earm and D. Noble

Equation (60)

The equation for variation in external [K] does not contain a contribution from i_{to} either in the D–F paper or in HEART. For greater accuracy, this should be included, and will be incorporated into a new version of HEART.

Equation (61)

The constant π is missing in the D–F paper. Also, the standard value of V_{ecs} is 10% not 5%. The difference does not make very much difference to the general results of the computations of external potassium.

Equation (63)

V_i should be V_e. The standard value of P was 0.7 s^{-1}.

Equation (63)

A minus sign is missing in the D–F paper.

Initial values and constants

Some of these were omitted in the D–F paper. They are: $P_{si} = 15$; $[Ca]_{up} = 2$ mM; $[Ca]_{rel} = 1$ mM; $[Ca]_i = 50$ nM (0.00005 mM); $y = 0.2$; $x = 0.01$; $d = 0.005$; $f = 1$; $m = 0.01$; $h = 8.0$.

The stimulus used to generate the Purkinje fibre response was usually -250 nA applied for 5 ms. The initial voltage was usually set to -75 mV.

APPENDIX 2

This Appendix gives two of the data input files used with OXSOFT HEART to generate the results shown in this paper. These input files were used with version 2.2 of the HEART program. If version 2.1 is used then, instead of setting the value of KMCA in the input file, edit the module CONCEN.PAS so that the expression Z1: = (Y[4]/(Y[4]+0.0005)) under CASE SRMODE 2 is replaced by Z1: = (Y[4]/(Y[4]+0.0004)).

Single atrial cell model
Compared with multicellular model all current parameters have been scaled down by 100

$**** PREP: HATRIUM*******

```
OUT = 9
IPULSESIZE = -1.3
ON = 0.2
OFF = 0.202
REP = 0.3
NAI = 6.48
SPACE = 4
TOMODE = 5   GTO = 0.01   Q = 0      R = 1          IDISP = 8
GNA = 0.5    PCA = 0.05   IKM = 0
```

Atrial cell model   95

```
GBNA = 0.00012          GFK = 0    GFNA = 0    GBCA = 0.00005
PUMP = 0.14  KNACA = 0.0001        GK1 = 0.017

                                              GBK = 0.0017

KMCA = 0.0004
E = −91.6
PREPLENGTH = 0.08
DX = 0.4
CAPACITANCE = 0.00004
TIMESCALE = 1000
ISCALE = 1000
TEND = 0.49
```

$$

Single atrial cell model
Double-pulse voltage clamps

$****PREP: HATRIUM******

```
OUT = 4
MODE = 2
E = −70
E2 = 0      T2 = 0.10
E3 = −70    T3 = 0.12
E4 = 20     T4 = 0.42
E5 = −70    T5 = 0.44
NAI = 6.48
SPACE = 4
BUFFAST = 1
TOMODE = 5  GTO = 0       Q = 0      R = 1        IDISP = 8
GNA = 0        PCA = 0.05   IKM = 0
GBNA = 0.00012              GFK = 0    GFNA = 0    GBCA = 0.00005
PUMP = 0.14 KNACA = 0.0001            GK1 = 0.017  GBK = 0.0017
KMCA = 0.0004
PREPLENGTH = 0.08
DX = 0.4
CAPACITANCE = 0.00004
TIMESCALE = 1000
ISCALE = 1000
TEND = 0.59
```

$$$

96   Y. E. Earm and D. Noble

REFERENCES

Barcenas-Ruiz, L. & Wier, W. G. 1987 Voltage dependence of intracellular $[Ca^{2+}]_i$ transients in guinea-pig ventricular myocytes. *Circ. Res.* **61**, 148–154.

Callewaert, G., Cleeman, L. & Morad, M. 1988 Epinephrine enhances Ca^{2+} current regulated Ca^{2+} release and Ca^{2+} reuptake in rat ventricular myocytes. *Proc. Natn. Acad. Sci. U.S.A.* **85**, 2009–2013.

Cannell, M. B., Berlin, J. R. & Lederer, W. J. 1987 Effects of membrane potential changes on the calcium transient in single rat cardiac muscle cells. *Science, Wash.* **238**, 1419–1423.

DiFrancesco, D. & Noble, D. 1985 A model of cardiac electrical activity incorporating ionic pumps and concentration changes. *Phil. Trans. R. Soc. Lond.* B **307**, 353–398.

Earm, Y. E., Ho, W. K. & So, I. S. 1989 An inward current activated during late low-level plateau phase of the action potential in rabbit atrial cells. *J. Physiol., Lond.* **410**, 64P.

Earm, Y. E., Ho, W. K. & So, I. S. 1990 Inward current generated by Na–Ca exchange during the action potential in single atrial cells of the rabbit. *Proc. R. Soc. Lond.* B **240**, 61–81 (Preceding paper.)

Egan, T. M., Noble, D., Noble, S. J., Powell, T., Spindler, A. J. & Twist, V. W. 1989 Sodium–calcium exchange during the action potential in guinea-pig ventricular cells. *J. Physiol., Lond.* **411**, 639–661.

Fabiato, A. 1983 Calcium-induced release of calcium from the cardiac sarcoplasmic reticulum. *Am. J. Physiol.* **245**, C1–C14.

Fedida, D., Noble, D., Shimoni, Y. & Spindler, A. J. 1987 Inward currents and contraction in guinea-pig ventricular myocytes. *J. Physiol., Lond.* **385**, 565–589.

Hilgemann, D. W. & Noble, D. 1987 Excitation–contraction coupling and extracellular calcium transients in rabbit atrium: reconstruction of basic cellular mechanisms. *Proc. R. Soc. Lond.* B **230**, 163–205.

Isenberg, G., Spurgeon, H., Talo, A., Stern, M., Capogrossi, M. & Lakatta, E. 1988 The voltage dependence of the myoplasmic calcium transient in guinea-pig ventricular myocytes is modulated by sodium loading. In *Biology of isolated adult cardiac myocytes* (ed. W. A. Clark, R. S. Decker & T. K. Borg), pp. 354–357. New York: Elsevier.

Jackson, A. P., Timmerman, M. P., Bagshaw, C. R. & Ashley, C. C. 1987 The kinetics of calcium binding to fura-2 and indo-1. *FEBS Lett.* **216**, 35–39.

Mitchell, M. R., Powell, T., Terrar, D. A. & Twist, V. W. 1983 Characteristics of the second inward current in cells isolated from rat ventricular muscle. *Proc. R. Soc. Lond.* B **219**, 447–469.

Mitchell, M. R., Powell, T., Terrar, D. A. & Twist, V. W. 1984 The effects of ryanodine, EGTA and low-sodium on action potentials in rat and guinea-pig ventricular myocytes: evidence for two inward currents during the plateau. *Br. J. Pharmacol.* **81**, 543–550.

Mitchell, M. R., Powell, T., Terrar, D. A. & Twist, V. W. 1987 Calcium-activated inward current and contraction in rat and guinea-pig ventricular myocytes. *J. Physiol., Lond.* **391**, 545–560.

Näbauer, M., Callewaert, G., Cleeman, L. & Morad, M. 1989 Regulation of calcium release is gated by calcium current, not gating charge, in cardiac myocytes. *Science, Wash.* **244**, 800–803.

Noble, D. & Noble, S. J. 1984 A model of S.A. node electrical activity using a modification of the DiFrancesco–Noble (1984) equations. *Proc. R. Soc. Lond.* B **222**, 295–304.

Noble, D. & Powell, T. 1990 The attenuation and slowing of calcium signals in cardiac muscle by fluorescent indicators. *J. Physiol.* (In the press.)

Schouten, V. J. A. & ter Keurs, H. E. D. J. 1985 The slow repolarization phase of the action potential in rat heart. *J. Physiol.* **360**, 13–25.

Talo, A., Spurgeon, H. A. & Lakatta, E. G. 1988 The relationship of calcium current, sarcoplasmic reticulum function and contraction in single rat ventricular myocytes excited in the resting state. In *Biology of isolated adult cardiac myocytes* (ed. W. A. Clark, R. S. Decker & T. K. Borg), pp. 362–365. New York: Elsevier.

Wier, W. G., Cannell, M. B., Berlin, J. R., Marban, E. & Lederer, W. J. 1987 Cellular and subcellular heterogeneity of $[Ca]_i$ in single heart cells revealed by fura-2. *Science, Wash.* **235**, 325–328.

Proc. R. Soc. Lond. B **230**, 163–205 (1987)
Printed in Great Britain

Excitation–contraction coupling and extracellular calcium transients in rabbit atrium: reconstruction of basic cellular mechanisms

By D. W. Hilgemann and D. Noble, F.R.S.

*Department of Physiology, American Heart Association, Greater Los Angeles
Affiliate Cardiovascular Research Laboratory, University of California,
Los Angeles, California 90024, U.S.A.
and University Laboratory of Physiology, Parks Road, Oxford OX1 9PT, U.K.*

(*Received 2 May* 1986)

CONTENTS

Interactions of electrogenic sodium–calcium exchange, calcium channel and sarcoplasmic reticulum in the mammalian heart have been explored by simulation of extracellular calcium transients measured with tetramethylmurexide in rabbit atrium. The approach has been to use the simplest possible formulations of these mechanisms, which together with a minimum number of additional mechanisms allow reconstruction of action potentials, intracellular calcium transients and extracellular calcium transients. A 3:1 sodium–calcium exchange stoichiometry is assumed. Calcium-channel inactivation is assumed to take place by a voltage-dependent mechanism, which is accelerated by a rise in intracellular calcium; intracellular calcium release becomes a major physiological regulator of calcium influx via calcium channels. A calcium release mechanism is assumed, which is both calcium- and voltage-sensitive, and which undergoes prolonged inactivation. 200 μM cytosolic calcium buffer is assumed. For most simulations only instantaneous potassium conductances are simulated so as to study the other mechanisms independently of time- and calcium-dependent outward current. Thus, the model

[163]

164   D. W. Hilgemann and D. Noble

reconstructs extracellular calcium transients and typical action-potential configuration changes during steady-state and non-steady-state stimulation from the mechanisms directly involved in trans-sarcolemmal calcium movements. The model predicts relatively small trans-sarcolemmal calcium movements during regular stimulation (ca. 2 μmol kg^{-1} fresh mass per excitation); calcium current is fully activated within 2 ms of excitation, inactivation is substantially complete within 30 ms, and sodium–calcium exchange significantly resists repolarization from approximately -30 mV. Net calcium movements many times larger are possible during non-steady-state stimulation. Long action potentials at premature excitations or after inhibition of calcium release can be supported almost exclusively by calcium current (net calcium influx 5–30 μmol kg^{-1} fresh mass); action potentials during potentiated post-stimulatory contractions can be supported almost exclusively by sodium–calcium exchange (net calcium efflux 4–20 μmol kg^{-1} fresh mass). Large calcium movements between the extracellular space and the sarcoplasmic reticulum can take place through the cytosol with virtually no contractile activation. The simulations provide integrated explanations of electrical activity, contractile function and trans-sarcolemmal calcium movements, which were outside the explanatory range of previous models.

Introduction

Progress in understanding cardiac excitation–contraction coupling is hindered by two major obstacles in spite of very impressive progress towards an understanding of individual mechanisms which regulate electrical and mechanical activity:

1. The integrated function of sarcolemmal mechanisms with internal calcium stores and buffers is clearly of such complexity that overall function cannot readily be derived even from well-characterized isolated mechanisms.

2. Available data on individual mechanisms may allow a range of theoretical explanations and/or objections, which cannot yet be proven or eliminated by reliable experimentation.

One potentially useful means of addressing these problems is to reconstruct the underlying mechanisms in simulations, and analyse their interactions. Although a variety of simulations of cardiac excitation–contraction coupling have in fact been made (Manring & Hollander 1971; Opiz 1973; Wussling & Szymanski 1973; Kauffmann *et al.* 1974; Hilgemann 1980; Adler *et al.* 1985; Schouten 1985) even a superficial reading of that literature reveals striking gaps between the mechanisms simulated, experimental evidence for the mechanisms stimulated and the biophysical plausibility of individual formulations. Significantly, none of these simulations allows quantitative treatment of sarcolemmal mechanisms in excitation–contraction coupling in relation to electrical activity.

The recent simulations of DiFrancesco & Noble (1985) appear to provide the first versatile basis for such a unification. This article presents some first steps towards this goal from the perspective of new data on the magnitude, timing and overall characteristics of trans-sarcolemmal calcium movements in rabbit atrium, as revealed by extracellular calcium transient measurements in rabbit atrium (Hilgemann 1986*a*, *b*). The pursuit of these simulations seemed especially useful, because a range of data obtained with very different methods has supported the same general interpretations:

1. During the course of a normal contraction cycle in mammalian heart, nearly the same amount of calcium would be extruded by electrogenic 3:1 sodium–calcium exchange as entered previously by fast calcium channels.

2. A calcium dependence of calcium-channel inactivation would be an important means of regulating calcium influx by negative feedback to internal calcium, thus preventing not only calcium overload but also sodium overload.

3. These two mechanisms together would play a major role in shaping the cardiac action potential and its changes of configuration.

Several new formulations used in the present work will be described below, together with basic reasons for the choices made. It may be mentioned that detailed simulations of the extracellular calcium transients were possible with quite modest modifications of the published model. Primary attention was focused on the description of the function of the sarcoplasmic reticulum, the implications of intracellular calcium buffering, and the formulation of calcium-channel inactivation. Only those sarcolemmal mechanisms essential for excitation–contraction coupling are included in the present simulations. These are: (1) sodium channel, (2) electrogenic sodium pump, (3) background sodium and calcium conductances, (4) one calcium channel, (5) electrogenic sodium–calcium exchange mechanism, and (6) instantaneous potassium conductances. Time- and calcium-dependent potassium conductances were not generally simulated in order to study the isolated influences of calcium channels and sodium–calcium exchange on action-potential configuration. The role of time-dependent (delayed) change in K^+ conductance is, in any case, thought to be very small in action potentials of the kind simulated here. A sarcolemmal calcium pump was treated as an optional mechanism, and sodium–calcium exchange is the sole mechanism of calcium efflux in all simulations presented except one. For the sake of brevity, extensive citations concerning the individual mechanisms simulated will not be given here; the wealth of available literature is, however, readily accessible from the major articles cited.

After performing the original simulations, multiple formulations were made and tested for each of the model changes described. For calcium-induced calcium release more than 20 different formulations were examined. Wide ranges of parameter settings were explored, particularly in those cases where available data left a large degree of uncertainty about the settings chosen. It will be argued that, for a rough simulation of the processes of interest, enormously simplifying assumptions can be adequate. Naturally, the important question arises as to when and if these simplifications critically jeopardize the goals of the simulations and the conclusions that may be reached. The approach taken to this problem was to pursue each central aspect of the simulations at a higher level of sophistication than necessary for the simulations presented and by simulating alternative mechanisms. This has been carried out for the descriptions of cytosolic calcium buffering, calcium diffusion (treated as instantaneous mixing in the simulations presented), sodium–calcium exchange and the calcium pump of the sarcoplasmic reticulum. Thus, the simulations presented are one of several sets of simulations, which have all clearly supported the general interpretations given. Those points

6

166   D. W. Hilgemann and D. Noble

will be stressed at which consideration of alternatives and/or more sophisticated formulations leave doubts about significant details, and in fact raise important unresolved questions. One of the criteria for choice of the simulations presented was that they illustrated specific problems which were not apparent in other simulations. New experimental results are presented in those cases where significant problems in the original simulations could be addressed by further experimentation.

METHODS AND EXPERIMENTAL FORMULATIONS FOR SIMULATION

The integration methods given previously (DiFrancesco & Noble 1985) were also used in the present simulations. The original simulations of extracellular calcium transients (Hilgemann & Noble 1986) were performed on Minc and PDP 11/23 computers. With one exception, the simulations selected for presentation here were performed on Compaq Deskpro and IBM AT personal computers equipped with the INTEL 80287 mathematics coprocessor, in the TURBO Pascal language. The relative loss of calculation speed on IBM-compatible personal computers was far overshadowed by the advantages of interactive graphics and the versatility of the TURBO Pascal programming environment (Borland International, Scotts Valley, California).

The simulations were developed as a modification of simulations made of the rabbit sinus node preparation (Noble & Noble 1984). The 'preparation' is assumed to be 1.6 nl in volume and is assumed to have a 40% extracellular space, from measurements made in much larger rabbit left atrium preparations (Hilgemann 1986a). The cytosolic space was also assumed to be 40% of total preparation volume. This value underestimates the volume of mitochondria and nucleoli, but has the advantage that extracellular and intracellular calcium concentration changes can be directly compared during trans-sarcolemmal calcium movements and calcium release (i.e. calcium leaves and enters equal volumes during the calcium current). Reduction of intracellular volume to 25% of total volume had effects on the simulations essentially identical to decreasing the cytosolic buffer capacity by 30%.

Average cell diameter of cat atrial cells is 5–6 µm according to McNutt & Fawcett (1969); Hume & Uehara (1985) give a larger value of 11 µm for atrial myocytes isolated from guinea pig atrium. Assuming an average diameter of 6 µm and an average cell length of 125 µm, the average cell volume would be 15 pl. Accordingly, the simulated currents reflect the sum currents of 60 cells. A capacitance of 6 nF was used for the simulations (or 0.1 nF per cell); this is twice the capacitance expected if the cells were simple cylinders with $1\ \mu F\ cm^{-2}$ capacitance. This reflects the possibility that atrial cells may have substantial membrane invaginations and/or t-tubules (see, for example, Sommer & Johnson 1979); values obtained by Giles & Van Ginneken (1985) in crista terminalis cells of rabbit heart were in close agreement with these assumptions.

(a) Previously published mechanisms

In all simulations presented, the formulations given by DiFrancesco & Noble (1985) were used unchanged for the fast sodium channel, electrogenic sodium

Ca²⁺ movements and cardiac E–C coupling **167**

pump, background sodium and calcium conductances, sodium–calcium exchange and instantaneous potassium conductances. Unless specified otherwise the parameter settings involving time are related to seconds and those relating to concentrations are in millimoles per litre.

(i) *Sodium channel*

The sodium conductance was selected to generate action potential upstrokes of a velocity comparable to experimental results; in all simulations presented, peak membrane potential is achieved within 1.5 milliseconds of displacing the membrane potential to -55 mV. The conductance selected results in a net increment of intracellular sodium of 4.2 μM per action potential, and reflects a sodium influx of about 2.5 μmol kg^{-1} fresh mass. This relatively conservative estimate of sodium influx by fast sodium channels becomes important for maintaining sodium homeostasis with the sodium pump rate used in these simulations.

(ii) *Sodium pump*

Particular attention was given here to the problem of maintaining sodium homeostasis with a maximum sodium pump rate of 6 mmol kg^{-1} fresh mass min^{-1}, which corresponds to the nearly maximal net sodium extrusion rate measured in atrial muscle (Glitsch *et al.* 1976). This value was not exceeded under essentially optimal pump conditions in recent measurements of sodium pump current in single ventricular myocytes (Gadsby *et al.* 1985). It also correlates with extrapolations from Purkinje fibres via sodium electrode measurements (Eisner *et al.* 1981). Assuming that normal intracellular sodium is in the midrange of the activation curve for internal sodium, this number limits the possible net sodium influx by all mechanisms to around 2.5 mmol kg^{-1} fresh mass min^{-1}. With the simulation parameter settings used here, internal sodium increases by 4.5 mM between the resting-state and steady-state simulation at 2 Hz. This is substantially more than found in Purkinje fibres (see, for example, Ellis 1985); comparable data from atrial and ventricular muscle is not available. This suggests that internal sodium may be regulated by more mechanisms than are currently represented in the model. DiFrancesco & Noble (1985) came to this same conclusion in considering the fact that the model overestimates the rise of internal sodium found on inhibition of the sodium pump, particularly when sodium exceeds 20 mM.

(iii) *Sodium–calcium exchange*

In the simulations presented, no changes of parameters related to sodium–calcium exchange were made from those given previously, except to study the effect of the overall rate of exchange on simulations. It is, however, worth noting that, since the equations were originally formulated, strong new experimental evidence on the sodium–calcium exchange has appeared in the work of Kimura *et al.* (1986), Mechmann & Pott (1986) and Fedida *et al.* (1987). The results of Kimura *et al.* show current–voltage relations very similar to those used in the model; the results of Fedida *et al.* show dynamic behaviour of a slow inward current very similar to the predictions of the present modelling. Recently, simulations have been performed with several alternative formulations of exchange, based on different possible models (sequential, empty-carrier-return type, and a model with competitive binding of sodium and calcium to a carrier with binding sites on each side of the membrane). For the overall interpretations given by the present simulations, changes resulting from the alternative formulations *per se* were not critical (D. W. Hilgemann, unpublished). However, the general shape of *I–V* relations, which can be changed in each formulation, can indeed have a large influence on details of simulations.

(iv) *Potassium conductances*

For reasons already given, the goal of the present simulations was not to recreate rabbit atrial action potentials, and the great majority of simulations were performed with only instantaneous potassium conductances. A linear conductance and a conductance with inward rectification were used. To permit rough simulations of action potentials (as described in Results), the inward rectifier was shifted 30 mV to more positive potentials from the values used in the DiFrancesco–Noble (1985) simulations. The linear conductance was set large enough to repolarize roughly at the speeds found in experiments in the positive range to mid-voltage range of the action potential, and the rectifying conductance was set large enough to terminate the action potential at a realistic rate. Two simulations are given in the Results in which a transient outward current has been added, as this current plays a major role in rabbit atrium. Since most of the simulations presented here were performed, it has become known that two potassium conductances are probably of primary importance in rabbit atrium, a transient outward current with the characteristics of an 'A-current' and a linear background conductance (W. Giles, personal communication of data from rabbit atrial cells, similar to descriptions in crista terminalis, indicating the existence of only small 'i_{K1}' and delayed ('i_K') K$^+$ conductances).

(v) *Background sodium and calcium conductances*

Background sodium conductance was adjusted to bring resting internal sodium to 3.5 mm with the pump rate selected, and a background calcium conductance was set to increase resting sodium further to 5 mm via sodium–calcium exchange. Internal sodium concentration in the simulations given is 7 mm, which corresponds to the steady-state value reached during 0.5 Hz stimulation.

(b) *New formulations*

(i) *Calcium-channel inactivation*

Only one calcium channel was simulated in the present work. The activation formulation was not changed from that given by DiFrancesco & Noble (1985), but it was accelerated by a factor of 2.5 to reproduce the rapid onset of extracellular calcium depletion. Several formulations of inactivation were explored. Briefly, the formulation used in all results presented was made with the following problems in mind, assuming that inactivation is in some way related to intracellular calcium (see, for example, Mentrard *et al.* 1984; Mitchell *et al.* 1984): (1) Pure calcium-mediated inactivation is probably an impossibility in the heart, since the calcium current would reactivate upon relaxation of internal calcium; the formulation should therefore result in inactivation which is maintained for the duration of

Ca^{2+} *movements and cardiac E–C coupling* 169

depolarization, but allow a rapid recovery from inactivation upon repolarization. (2) The formulation should be consistent with several studies which implicate a role of calcium together with voltage in the inactivation process (Kass & Saguinetti 1984; Lee *et al.* 1983). (3) The formulation should potentially explain extremely prolonged extracellular calcium depletion responses found under some circumstances (Hilgemann 1986*b*) as well as extremely prolonged calcium currents found in some atrial myocytes with strong internal EGTA loading (Bechem & Pott 1985).

The formulation used here assumes that calcium binding to a regulatory site results in an inactivated state; inactivation is 'preserved' by an inherently voltage-dependent inactivation reaction, which takes place only slowly in the absence of calcium. This formulation minimizes, but does not always eliminate, reactivation of calcium current during relaxation of intracellular calcium. For simplicity, the calcium binding reaction is regarded as instantaneous. Accordingly, four channel states are possible with respect to inactivation:

$$F_{c0} = F_c(1 - F_v), \tag{1}$$

$$F_{0v} = (1 - F_c)F_v, \tag{2}$$

$$F_{00} = (1 - F_c)(1 - F_v), \tag{3}$$

and
$$F_{cv} = F_c F_v, \tag{4}$$

where F_c is the fraction of channels with regulatory site occupied by calcium, and F_v is the fraction of channels inactivated by the voltage-dependent process.

Now
$$F_c = [Ca]_i/(0.001 + [Ca]_i) \tag{5}$$

and
$$dF_v/dt = (F_{00} + 120F_{c0})\beta - F_v\alpha. \tag{6}$$

Here
$$\alpha = 6.25(E_m + 34)/(e^{\frac{1}{4}(E_m + 34)} - 1), \tag{7}$$

$$\beta = 2.5/(1 + e^{-\frac{1}{4}(E_m + 34)}). \tag{8}$$

For all of the simulations presented, the fraction of conducting channels was defined as the fraction which had undergone activation (see DiFrancesco & Noble 1985) multiplied by the F_{00} fraction; that is to say, the F_{c0}, F_{0v} and F_{cv} fractions of the total calcium-channel population were assumed to be zero conductance states. Treatment of F_{c0} and F_{0v} as partially conducting states gives the formulation a high degree of flexibility (results not shown). It seems noteworthy enough to mention that with moderately high intracellular calcium (e.g. with moderately high internal sodium), many of the formulations tested for inactivation, including this formulation, resulted in small oscillations of internal calcium independent of the sarcoplasmic reticulum (i.e. with no sarcoplasmic reticulum; voltage range of about 0 to −40 mV).

(ii) *Sarcoplasmic-reticulum calcium pump*

A primary concern with regard to the sarcoplasmic-reticulum calcium pump was the extent to which complete reaction schemes might be critical for the simulations, and for particular details the simulations indeed appear to force a relatively

170   D. W. Hilgemann and D. Noble

complete formulation. Five different formulations of sarcoplasmic-reticulum calcium pump were used, the simplest being the expression of calcium uptake and loss from the sarcoplasmic reticulum as linear rates (DiFrancesco & Noble 1985). Increasing complexity was added, starting with a formulation as an uptake process proportional to calcium binding to a single site, then to two sites with cooperativity, then with translocation processes and stimulation by a calcium-dependent reaction, and finally, for some simulations, with six steps of the calcium ATPase cycle as suggested by Inesi *et al.* (1980). For the simulations presented a very simple formulation was used, which nevertheless expresses a presumably important kinetic aspect of calcium ATPase for this work, namely that calcium uptake rates will decrease with increased calcium loading by the sarcoplasmic reticulum as enzyme conformations accumulate which are not available for the forward pump reaction:

$$[X_c]_{cy} \underset{\beta}{\overset{\alpha}{\rightleftharpoons}} [X_c]_{sr}$$

$$k_{cyca} \Big\updownarrow \qquad\qquad \Big\updownarrow k_{srca}$$

$$c + [X]_{cy} \xleftarrow[k_{xcs}]{} [X]_{sr} + c$$

The calcium binding reactions and the translocation of the empty enzyme were treated as instantaneous reactions, assuming that the summed reactions translocating calcium would be rate-limiting. Accordingly, the fractions (X) of enzyme in the four conformations are determined solely by the free calcium concentrations of the cytosol ($[Ca]_i$) and the lumen of the sarcoplasmic reticulum ($[Ca]_{sr}$). The following equations were derived algebraically to calculate the concentrations of the four X fractions:

$$K_1 = k_{cyca}\,k_{xcs}/k_{srca}, \tag{9}$$

$$K_2 = [Ca]_i + ([Ca]_{sr}\,K_1) + k_{cyca}\,k_{xcs} + k_{cyca}, \tag{10}$$

$$[X_c]_{cy} = [Ca]_i/K_2, \tag{11}$$

$$[X_c]_{sr} = [Ca]_{sr}\,K_1/k_2, \tag{12}$$

$$[X]_{cy} = k_{cyca}/K_2, \tag{13}$$

$$[X]_{sr} = k_{cyca}\,k_{xcs}/k_2. \tag{14}$$

The value of k_{cyca} in the present simulations was 0.0003 mM, k_{srca} was 0.5 mM, and k_{xcs} was 0.4. The forward reaction rate constant, α, was set high enough in simulations to give the desired time course of relaxation of intracellular calcium, and the back reaction rate, β, was set to keep these reactions also in a ratio of 0.4 ($= k_{xcs}$). These settings would have the implication that investment of energy in the reaction is reflected solely in the change of calcium-binding affinity upon translocation.

Calcium movement by the pump, simulated in terms of a change of cytosolic calcium concentration, was taken as:

$$Ca_{srup} = -[X_c]_{cy}\,\alpha + [X_c]_{sr}\,\beta. \tag{15}$$

It may be mentioned that the kinetics of calcium transport simulated here need not be very different from simulations involving two Ca binding sites and several sequential steps, if the rate-limiting step is a reaction closely related to binding of the first calcium ion. As described later, calcium uptake is assumed to take place into a store of small volume, which equilibrates rapidly with a release store of 10 times greater functional volume (i.e. with calcium buffering).

(iii) *Sarcolemmal calcium pump*

A sarcolemmal calcium pump was included in some simulations, both with and without electrogenicity. For the single simulation presented with this assumption, extrusion was simply treated as a rate proportional to calcium binding to a single site without electrogenicity, voltage-dependence or secondary calcium regulation:

$$\mathrm{Ca}_{\mathrm{slpump}} = k_{\mathrm{pump}}[\mathrm{Ca}]_i/([\mathrm{Ca}]_i + 0.0002). \tag{16}$$

(iv) *Cytosolic calcium buffers*

Cytosolic calcium buffers were treated as independent calcium-binding sites with the following general formulation for the calcium-occupied site (X_c), concentrations being related to the cytosol:

$$\mathrm{d}X_c/\mathrm{d}t = (X_{\mathrm{total}} - X_c)[\mathrm{Ca}]_i\,\alpha - X_c\,\beta. \tag{17}$$

Although the kinetics of calcium binding to isolated troponin, calmodulin and the calcium pump of the sarcoplasmic reticulum have been characterized experimentally, physiological kinetics may well be different (see, for example, Fabiato 1983). Given the uncertainties involved (e.g. possibilities such as changes of buffer characteristics during the contraction; see Allen & Kentish (1985) for recent review), simulations were performed with a very wide range of buffer characteristics to examine their influence on the individual mechanisms simulated. One characteristic of buffering, in particular, appeared to raise serious problems for the overall simulations presented. Substantial buffering with slow off-rates of less than about 50 s⁻¹ would not permit loading of the sarcoplasmic reticulum by a premature excitation; intracellular calcium would still be high upon repolarization and a large portion of the calcium would be extruded by the exchanger. In initial simulations, three buffers were simulated with kinetic characteristics resulting in affinities equivalent to those summarized by Fabiato (1983) for troponin C, the sarcoplasmic calcium pump and calmodulin. Then, for simplicity (and considering the uncertainty about physiological characteristics), only two buffers were included.

For off-rates of approximately 150 s⁻¹ and greater (in the range of isolated troponin C), it was found that the kinetic simulation of buffers was of very little consequence to the overall simulations, and deviations of free-calcium waveforms from total-calcium waveforms were relatively small. Therefore, for routine purposes, the buffer equations were solved for the steady state, which had the great advantage of decreasing simulation times manyfold, particularly during the diastolic period. This was in fact essential for testing a number of other mechanisms, over a large parameter range during series of excitations, with available

172   D. W. Hilgemann and D. Noble

computer facilities. In those simulations where buffers were set to steady state, free calcium was either calculated by a linear iterative procedure or a modified Newton procedure to an accuracy of greater than 0.02 %. The differential equation for total cytosolic calcium was

$$d[Ca]_{cytot}/dt = Ca_{ICa} + Ca_{Inaca} - Ca_{srup} + Ca_{rel} - Ca_{slpump}, \qquad (18)$$

where Ca_{ICa} is calcium influx by the calcium channel, Ca_{Inaca} is trans-sarcolemmal calcium movement by the exchanger, Ca_{srup} is calcium movement by the calcium pump, and C_{rel} is calcium movement through the calcium release pore, described below. Conversion of the sarcolemmal current-generating mechanisms to calcium concentration changes was given previously (DiFrancesco & Noble 1985). When simulating the buffers kinetically, the differential equation becomes that for free calcium after addition of the differential equations for the buffer sites.

For three reasons, a high degree of cytosolic calcium buffering was selected for the simulations presented:

1. In the original simulations of DiFrancesco & Noble, cytosolic calcium buffering was not simulated, except as uptake by the sarcoplasmic reticulum. It will be shown here that the essential conclusions reached in those simulations about the role of sodium–calcium exchange are not greatly affected by addition of a high degree of buffering.

2. Yue *et al.* (1986) have recently presented evidence that free calcium during steady state tension development may be very substantially less than estimated in experiments with 'skinned' fibres. Furthermore, calibrations of aequorin transients during normal twitches were lower than previous estimates (*ca.* 2 μM free calcium at maximal twitch). The buffer characteristics chosen for simulation predict these values.

3. Although it is not essential to the simulations, higher cytosolic buffering made certain aspects of the simulations more flexible to parameter changes. Alternative assumptions, which allow equally good simulations of action-potential configuration changes, will be mentioned in the Discussion.

For the simulations presented, 180 μM cytosolic calcium buffer is assumed with a K_m of 2 μM (on- and off-rates of 10^8 M^{-1} s^{-1} and 200 s^{-1}, respectively, which are the approximate characteristics of isolated troponin-C (Robertson *et al.* 1982)). This buffer will be referred to as X_1. Of the 180 μM total buffer, 140 μM is assumed to represent troponin-C (assuming a troponin concentration of 70 μM or 24 μmol troponin kg^{-1} fresh mass). This is equivalent to the concentration used by Canell & Allen (1984) for skeletal muscle. The remaining 40 μM of X_1 is assumed to represent other calcium binding sites with moderately high affinity (primary possibilities are sarcoplasmic reticulum pump and membrane binding sites). A second binding site, X_2, is assumed to represent calmodulin (or other higher-affinity binding sites). X_2 was simulated with an affinity four times higher, resulting from an off-rate four times slower, and therefore X_2 buffers relatively more calcium at low cytosolic calcium concentrations. A 20 μM cytosolic concentration of X_2 was used in all simulations presented (or 8 μmol kg^{-1} fresh mass). The largest total cytosolic calcium concentration reached in the simulations presented is 120 μM, which results in a free calcium concentration of 2.6 μM with these buffer charac-

Ca^{2+} *movements and cardiac* E–C *coupling* 173

teristics. This magnitude of release, which corresponds to 30 µmol calcium per kilogram fresh mass, would according to Fabiato (1983) result in about 50 % of maximum tension developed by cardiac muscle, but nearly maximal twitch in an intact muscle. The free-calcium concentrations of the present experiments are approximately twofold lower than those suggested by Fabiato and threefold lower than would be projected from the majority of aequorin measurements (see, for example, Allen & Orchard 1983).

(v) *Calcium release mechanism*

Calcium-induced release of calcium has been for many years the most widely accepted mechanism of calcium release in cardiac muscle, and recent work of Fabiato has provided a great deal of quantitative data on this process in skinned cardiac myocytes (Fabiato 1985 *a, b, c*). However, proposals about the mechanistic and molecular details of release are still lacking. Therefore, it was of considerable interest to attempt to simulate the basic results of calcium-induced calcium-release experiments. Fabiato has suggested a qualitative treatment in terms of activation and inactivation processes controlled by calcium, which might be analogous in principle to Hodgkin–Huxley channel regulation by voltage and suggests the possibility of modelling the release mechanism in similar fashion. This was attempted with the hope of first modelling the basic experimental results of Fabiato, and then placing the mechanism in the framework of the overall model. By working along these lines, parallel activation and inactivation reactions were modelled with control by calcium-binding reactions of variable rates and stoichiometries; gating was also assumed to be of the Hodgkin–Huxley type with variable numbers of activation and inactivation gates. Extended attempts along these lines were not successful with minimum requirements for recreating Fabiato's results and for stability in an intact cell: (1) The formulation should reproduce the basic experimental result that a brief application of calcium in the range of 1 µM effectively initiates release, while continued application of 1 µM results in suppression of release at the same release cycle. (2) The release mechanism must be able to operate over at least a fivefold range of calcium-current magnitudes (i.e. it should not fail at 150 µM extracellular calcium). (3) The mechanism should not become unstable with an approximately fivefold rise of internal calcium (e.g. with small to moderate rises of internal sodium). The formulation of an inactivation process, controlled by calcium and operating parallel to the activation process, was a major cause of instabilities (especially during relaxation) no matter how powerful or steeply dependent on calcium. In this context, it was found that a calcium permeability of 2 % of the permeability needed at the peak of release was sufficient to short-circuit the SR calcium pump substantially during relaxation. This result underscores the problem of inactivation, if calcium uptake and release sites are not physically separate, and calcium release does not deplete the calcium store.

In general, much greater success was achieved with inactivation taking place in some way through the activated state. This allows purely time-dependent inactivation, which can be very powerful already in the absence of calcium. Individual formulations allowed partial but not complete reconstruction of the

174   D. W. Hilgemann and D. Noble

calcium and time-dependences of release described by Fabiato. The formulations
used in simulations presented here are broadly consistent with work of Fabiato
and also reflect the possible involvement of a second messenger in calcium release
as suggested for skeletal muscle (Vergara *et al.* 1985). One of the definite successes
of this formulation is that it remains stable without assuming physical separation
of uptake and release stores. A voltage dependence of the release process was finally
assumed in order to overcome a serious problem of release failure with reduction
of calcium current in a buffered cytosolic environment. This could be substituted
for a large local calcium transient, if release were initiated at special sarcolemmal
sites where sarcoplasmic reticulum membranes would come in very close apposition
to calcium channels at the surface. Thus, the formulations chosen meet minimum
requirements for simulation, but cannot be defended on the basis of available
experimental data.

A voltage-dependent function (E_{on}) initiates an activation process, formally
thought of as the production of a second messenger (F_2) from an inactive precursor
(F_1), but which could also be thought of as the activity of an enzyme, for example.
The activation process is also stimulated by calcium via occupation of two
regulatory calcium binding sites (R_c), making the calcium dependence relatively
steep. The activator (F_2) is broken down rapidly to an inactive product (F_3), which
is regenerated slowly to the precursor (F_1). Activation of the release channel takes
place in dependence on the binding of two activator molecules; this adds a further
steepness to the activation processes. The possibility of adding a calcium leak from
the release channel was included to simulate the effects of agents thought to act
by activating the release channel. The following equations are now obtained for
the different variables:

Voltage dependence of release (E_{on}):

$$E_{on} = \exp\left(0.08(E_m - 40)\right). \tag{19}$$

Regulatory calcium binding site (R_c):

$$R_c = ([Ca]_i/([Ca]_i + 0.0005))^2. \tag{20}$$

Activation rate (K_{act}):

$$K_{act} = 600(E_{on} + R_c). \tag{21}$$

Inactivation rate (K_{inact}):

$$K_{inact} = 100 + 1000R_c. \tag{22}$$

Precursor fraction (F_1):

$$dF_1/dt = 0.6F_3 - F_1 K_{act}. \tag{23}$$

Activator fraction (F_2):

$$dF_2/dt = F_1 K_{act} - F_2 K_{inact}. \tag{24}$$

Product fraction (F_3):

$$dF_3/dt = F_2 K_{inact} - 0.6F_3. \tag{25}$$

Open release channel fraction (F_r):

$$F_r = (F_2/(F_2 + 0.25))^2. \tag{26}$$

Calcium release (Ca_{rel}):

$$Ca_{rel} = (F_r KM_{Ca} + K_{leak}) [Ca]_{rel} KM_{Ca}, \tag{27}$$

where KM_{Ca} is the maximal release rate.

Maximum calcium release in the present simulations results in a reduction of calcium in the release store by about 55 %.

(vi) *Calcium 'translocation' from uptake to release store*

A translocation step has been included in the present simulations as a very rapid process, which could be accounted for by diffusion through the longitudinal sarcoplasmic reticulum to terminal cisternae (with a time constant of 20 ms, not the 1 s constant which would be needed to account for 'restitution' of contractility). This process influences only one of the simulations presented, namely the oscillatory release pattern in the final figure (figure 15) at high internal sodium. The reason is that the rapid translocation process adds an initial steepness to the recovery of the ability to release calcium, which favours the generation of oscillatory release patterns. For all other simulations this translocation step can be treated as instantaneous with almost no effect on simulations. Ca_{trans} is the concentration change of the uptake store due to 'translocation':

$$Ca_{trans} = ([Ca]_u - [Ca]_{rel}) \, k_{trans}, \tag{28}$$

where $[Ca]_u$ is the concentration of calcium taken up. The differential equations for calcium concentrations in the uptake and release stores are then

$$d[Ca]_u/dt = Ca_{srup}/f_{srup} - Ca_{trans}, \tag{29}$$

$$d[Ca]_{rel}/dt = Ca_{trans} \, f_{srup}/f_{srrel} - Ca_{rel}/f_{srrel}, \tag{30}$$

where f_{srup} and f_{srrel} are the fractional volumes of the uptake and release stores in relation to cytosolic volume. The release store is assumed to represent a functional volume, larger than the actual terminal cisternae, due to the presence of low-affinity calcium buffering (casequestrin). The significance of the 10 times larger release volume than uptake volume is that, to reproduce many results, it is necessary to assume that large fractions of the total calcium store are released at single excitations.

A note on internal calcium diffusion. Internal calcium diffusion is treated as instantaneous mixing in the present simulations. The possible implications of subsarcolemmal inhomogeneities of calcium for the simulations are being studied at present with representation of radial diffusion. The most important results for the present work will be mentioned in the Discussion.

(vii) *Extracellular calcium transients*

The extracellular calcium transients described in this article were determined exactly as described previously (Hilgemann 1986*b*). The derivative signals presented with the transients were determined as follows: a second-order polynomial was fitted to 11–21 data points of the averaged signals, which were acquired at 12 kHz. The first derivative of the polynomial was calculated at the midpoint of the fitted data and was assigned to that point. The procedure was repeated throughout the data of interest.

For the simulations presented, calcium equilibration between the assumed extracellular space of the muscle (calcium concentration = $[Ca]_o$) and the assumed muscle bath (calcium concentration = $[Ca]_b$) was represented by a simple rate equation with a time constant of 5 min. This is comparable to estimates of calcium exchange in such preparations in experiments, and means that calcium exchange to the muscle bath plays no role in the simulations presented:

$$\mathrm{d}[Ca]_o/\mathrm{d}t = ([Ca]_b - [Ca]_o)\,k_{\mathrm{diff}} - (Ca_{ICa} + Ca_{Inaca} - Ca_{slpump})/f_{exsp}, \quad (31)$$

where k_{diff} is the rate constant of exchange between extracellular space and the muscle bath and f_{exsp} is the fractional volume of the extracellular space in relation to the cytosolic space (unity in the present simulations). As a check on the accuracy of simulations of calcium movements, k_{diff} was set to zero and simulations were run for 48 h, corresponding to about 500 excitation cycles. Total calcium in the preparation was calculated before and after the simulation period from the concentrations in the different compartments and their respective volumes. Conservation was maintained to an accuracy of better than 1 %.

RESULTS

(a) *Systolic calcium movements during steady-state stimulation*

Figure 1 shows the basic extracellular calcium transient obtained in rabbit atrium with its corresponding derivative and the calcium-independent motion artefact at 680 nm (1 Hz steady stimulation, 150 μM total calcium, 100 μM free extracellular calcium). As was typical for more than 200 observations during regular stimulation at frequencies of 0.5–2 Hz, extracellular calcium depletion begins extremely rapidly on excitation and in mammalian preparations re-equilibrates to nearly the pre-stimulatory level by the end of relaxation. The precise time course of the onset of depletion has still not been accurately resolved, except to document confidently that maximum depletion rates are achieved within 2 ms of the initial onset of depletion (often within 3 ms of excitation by field stimulation). Filtering of the data by the polynomial fitting procedure, used to calculate derivatives, clearly determined this time course with reduction of the number of points used in the calculation down to the lowest reasonable signal:noise ratio. The initial depletion rates and depletion magnitudes were both reduced to less than 20 % of control by dihydropyridines (nifedipine at 1 μM and nitrendepine at 0.3 μM: no examples are given, for brevity). Furthermore, initial depletion rates were never found to increase convincingly (but often decreased) with interventions thought to enhance internal sodium moderately (more than 20 observations with heart glycosides, low external potassium and veratridine). These observations are the primary reasons for assuming that depletion reflects calcium-channel activity, primarily a single type of channel. Regarding this assumption, it should be mentioned that in the presence of high concentrations of nifedipine and threefold increased extracellular calcium, very brief depletions were found (*ca.* 15 ms from onset at excitation to termination), which were insensitive to both adrenergic stimulation and the interventions acting on internal sodium. These characteristics are consistent with the existence of the additional calcium channel described by Bean (1985) in atrial muscle, but its relative contribution to calcium influx under normal circumstances would be very small.

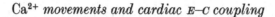

Ca^{2+} *movements and cardiac* E–C *coupling*   **177**

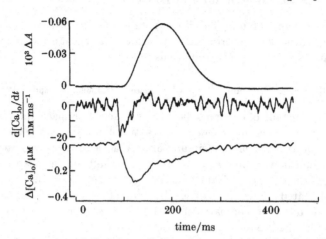

FIGURE 1. A representative extracellular calcium transient, its first derivative, and the corresponding motion artefact at a calcium-independent wavelength (680 nm), determined in rabbit atrium (1 Hz regular stimulation; 150 µM total extracellular calcium; 100 µM free calcium; field stimulation at twice threshold via electrodes mounted outside the light path). Five hundred consecutive signals were averaged for these composite records. The time from initial downward deflection of the derivative signal to attainment of the maximal rate of depletion is less than 2 ms, and is determined by the filter characteristics.

As described previously (Hilgemann 1986*b*), the depletion rates would correspond to a calcium current of 5 nA at about 3 mM extracellular calcium if the preparation were made up of 50 pl cells and the relationship of external calcium to calcium current were linear; assuming average atrial cell volumes of 15 pl, peak current magnitudes at 2 mM calcium (simulated here) would be 1.2 nA. The rate of extracellular calcium depletion decreases rapidly, and approaches zero early during the rise of contraction. The maximum rate of replenishment is about 15 % of the initial rate of calcium depletion. Thus, if 3:1 sodium–calcium exchange is the only mechanism of calcium efflux, the peak exchange current achieved during the action potential would be about 7 % of the fast calcium current.

Figure 2 shows the basic simulation of extracellular calcium transients during continuous stimulation, whereby the simulated signals project results obtained at low extracellular calcium to 2 mM calcium. The selection of parameter settings for simulations will be described over the course of the results presented, which should give a reasonably complete picture of the implications of each setting. Emphasis will be placed on those aspects most directly related to the extracellular calcium transients. In this figure, and most later figures, the simulated action potentials are given with the simulated extracellular calcium transient and the total cytosolic calcium concentration. The simulated free intracellular calcium transients are not presented, because they have very similar time course to total calcium. In this one case, a simulated contraction signal is also given as a reminder that the simulated cytosolic calcium signals would be related to contraction by subsequent reactions, resulting in a very substantial displacement in time of the two signals. To simulate a contraction, the calcium bound to two of the low-affinity binding sites (X_1) and one high-affinity site (X_2) was used as the driving function for two further saturable

178   D. W. Hilgemann and D. Noble

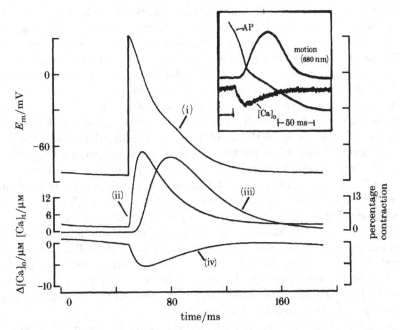

FIGURE 2. Basic behaviour of the model during regular stimulation; experimental results are
given here and in all subsequent figures as an inset. The experimental results were obtained
at 37 °C and a stimulation frequency of 0.5 Hz (Hilgemann 1985). The simulation projects
results obtained at low external calcium to 2 mM extracellular calcium. (See text for details
of the simulations.) Note that even with the small magnitude of trans-sarcolemmal calcium
movements simulated here (maximal depletion of 5 μM extracellular calcium) calcium
extrusion by sodium–calcium exchange significantly slows repolarization between approximately −30 and −80 mV (upper simulated curve). Curve (ii) corresponds to total cytosolic
calcium (calibration is given at left); curve (iii) shows contraction, expressed as a percentage
of maximal contraction (see text for details). The extracellular calcium transient (iv) reflects
almost exclusively calcium influx by calcium channels and calcium efflux by sodium–
calcium exchange; overlap of the two processes is minimized by rapid calcium-channel
inactivation. (See text for complete details.)

reactions (R_1 and R_2) with time constants of 10 ms and 25 ms. The two reactions
would grossly represent (but are not defended as) the rates of light-chain
conformation changes and sum crossbridge reactions, respectively. The 'contraction' signal corresponds to R_2:

$$dR_1/dt = (X_1^2 X_2 (1-R_1) k_1) - (R_1 k_2), \tag{32}$$

$$dR_2/dt = (R_1 (1-R_2) k_3) - (R_2 k_4). \tag{33}$$

The on-rates were selected to keep the reactions at approximately 30% of
saturation during regular stimulation; this percentage corresponds to about
12 μmol bound internal calcium per kilogram fresh mass.

The calcium-channel conductance was selected to give a peak single-cell current
of 0.8 nA, or about 45 nA in the preparation assumed to be made up of 60 cells
of 15 pl volume. This value turned out to be very close to values obtained in crista

Ca²⁺ *movements and cardiac* E–C *coupling*   **179**

terminalis of rabbit heart (Giles & Van Ginneken 1985). From the standpoint of the extracellular calcium depletions, this extrapolation assumed that the relation between calcium current and extracellular calcium would be somewhat less than linear. The calcium-channel inactivation parameters were set to terminate the calcium current largely by inactivation even within the course of the short action potential simulated here. Deactivation with repolarization is, however, significant with such short action potentials, and lengthening of the action potential by reducing the background potassium conductance results in continued calcium current at a small magnitude. In the simulation, the maximal depletion achieved is 6 μM or 2.4 μmol calcium influx per kilogram fresh mass; the maximal depletion achieved in the experimental result at 20 times lower free calcium was 0.4 μM or 0.08 μmol kg⁻¹ fresh mass. Thus, of the 28 μM increment of cytosolic calcium at activation, about 75 % comes from the internal release. The peak of cytosolic calcium is achieved at 13 ms from excitation. The rapidity of these signals reflects the temperature chosen for this experiment (37 °C), and the simulation rates are correspondingly high. Details of the rabbit atrial action potential will be considered in the final section of the results, but it should be noted here that the sharply biphasic form of the action potential given in the inset very probably reflects deactivation of an early outward current (I_{to}) with terminal repolarization being determined by a smaller background conductance.

The replenishment of extracellular calcium via sodium–calcium significantly slows repolarization in this simulation from roughly the midpoint of the action potential, even though the exchange current is very small in relation to the calcium current and free intracellular calcium is relatively low in these simulations. Regarding selection of the exchange rate, the crucial factors were: (1) that the rate of exchange was slow enough so that it contributed less than 15 % of the total calcium influx (consistent with experimental findings to date); and (2) that it nevertheless became activated as an extrusion mechanism capable of moderately strong competition with the sarcoplasmic reticulum during short action potentials.

The rate of calcium uptake by the sarcoplasmic reticulum is the primary determinant of the rate of fall of internal calcium in the simulation. It thereby is a major determinant of the time course of the sodium–calcium exchange current and time course of equilibration of extracellular calcium. A larger rate of uptake abbreviates the extracellular calcium transient (and the terminal repolarization of the action potential), and in the course of about six contractions results in a steady-state calcium release of correspondingly greater magnitude. Increasing the rate of sodium–calcium exchange (over a factor of 2) potently decreases the steady-state calcium release. In the steady state, small changes of exchange rate have little effect on the extracellular calcium transient in comparison with the magnitude of calcium release. The fundamental principle is that changes of the pump or exchange rates result in net calcium movements over a few excitations until calcium efflux again matches influx in steady state. A prerequisite for this simulation, consistent with all extracellular calcium transients obtained in the mammalian heart, is that calcium exchange between the sarcoplasmic reticulum and the cytosol is slow during diastole. A more subtle, but important, characteristic effect of changing the exchange rate is that a decrease of exchange rate can

180   D. W. Hilgemann and D. Noble

ultimately shift calcium extrusion into a more positive potential range, as the
magnitude of calcium release increases.

(b) Systolic calcium movements during non-steady-state stimulation

Figures 3 and 4 show experimental results, typical for more than 50 observations,
obtained during continuous paired and triple pacing (regular stimulation at 0.5 Hz
with the addition of one or two premature excitations at a 250 ms interval after

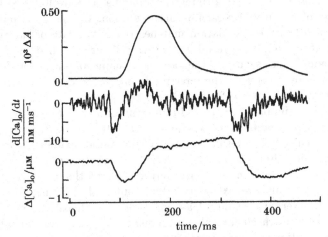

FIGURE 3. A typical extracellular calcium transient, its first derivative, and the corresponding
motion signal at 680 nm obtained during continuous paired-pulse stimulation (basal rate
of 0.5 Hz; 250 ms paired-pulse separation). The maximal rate of extracellular calcium
accumulation at the initial contraction is approximately 60 % of the initial rate of calcium
depletion. These results were obtained in the presence of 1.5 mM 4-aminopyridine.

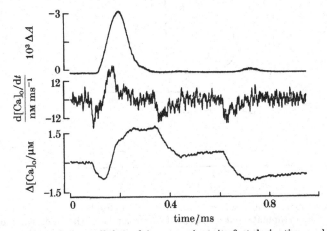

FIGURE 4. A typical extracellular calcium transient, its first derivative, and the corresponding
motion signal at 680 nm obtained during continuous triple-pulse stimulation (0.5 Hz
basal rate; 250 ms triple-pulse separation). The maximal rate of extracellular calcium
accumulation exceeds the initial rate of extracellular calcium depletion. Note the small
magnitude of motion signals at the premature excitations, which result in net extracellular
calcium depletion. These results were obtained in the presence of 1.5 mM 4-aminopyridine.

Ca²⁺ *movements and cardiac* E–C *coupling*   181

each regular excitation). Characteristically, this pattern results in a strong potentiation of the initial contraction in mammalian heart; the premature contractions are of very small magnitude. The experimental results are in the presence of 1.5 mM 4-aminopyridine to minimize the influence of the transient outward current on action potential duration. Note that the initial depletion signals are similar at all excitations: at least at low calcium the stimulation pattern results in no detectable changes of the maximal rate of depletion. However, the depletions are abbreviated at the potentiated contractions, and are followed by a pronounced phase of extracellular calcium accumulation. In the triple pulse sequence the rate of calcium accumulation matches the initial rate of calcium depletion. The premature excitations result in net extracellular calcium depletion. In the case of the paired-pulse sequence, the baseline calcium concentration does re-equilibrate considerably towards the prestimulatory level between paired stimulations (see Hilgemann (1986 a) for complete details).

Figures 5, 6 and 7 show simulations of paired-pulse stimulation. For these figures the background potassium conductance was reduced by 20 % in relation to the previous simulation. This change results in steady-state action-potential forms similar to those found in rabbit atrium at intermediate frequencies (i.e. often a nearly triangular form with smooth repolarization). In figure 5 the quasi-steady-state cycle was followed by an early excitation after an interval of 250 ms. At this second excitation the internal calcium release mechanism is almost completely inactivated, so that the intracellular calcium transient is small. Accordingly, calcium-channel inactivation is slow in comparison with the initial excitation and the action potential shows a much more pronounced plateau. Sodium–calcium

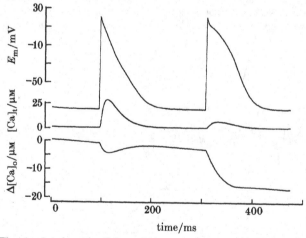

FIGURE 5. The simulated extracellular calcium transient, total cytosolic calcium and action potential during regular stimulation at 0.5 Hz (first excitation) and with interjection of a premature stimulus (second excitation). The premature excitation results in prolongation of the action potential and an enhancement of calcium influx by about a factor of 4. Note that the premature excitation results in essentially complete net uptake of the calcium influx. Baseline extracellular calcium is shifting downward, because the sarcoplasmic reticulum is loading up between excitations; genuine steady-state stimulation at 0.5 Hz results in net calcium efflux, albeit less pronounced than in subsequent figures.

182   D. W. Hilgemann and D. Noble

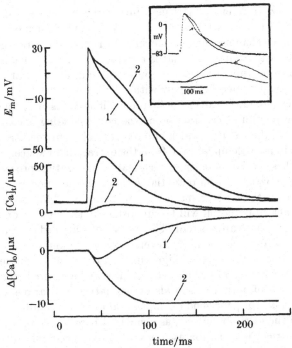

FIGURE 6. Continuation of figure 5 after one intervening paired-pulse sequence. The simulated
results for paired stimulation have been superimposed, whereby the signals marked 1
correspond to the initial excitation and signals marked 2 correspond to the premature
excitation. The inset gives results of Schouten & Ter Keur (1985) from rat ventricle; signals
marked by arrows correspond to potentiated contraction and unmarked signals were
obtained during decay of potentiation. (See text for complete details.)

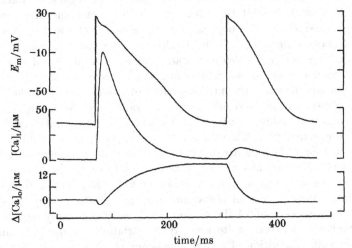

FIGURE 7. Continuation of figures 5 and 6 after five additional paired-pulse sequences; this is
essentially the steady state for the paired-pulse sequence. The initial excitation results in
a net Ca efflux of ca. 5 µmol kg^{-1} fresh mass; the premature excitation results in a net Ca
influx of ca. 5 µmol kg^{-1} fresh mass. (See text for complete details.)

exchange does not favour calcium extrusion during this time, so that calcium entering the cell can be taken up almost entirely by the sarcoplasmic reticulum. Note that intracellular calcium does not continue to rise through the simulated depletion during the early excitation. That is because the rate of calcium uptake by the sarcoplasmic reticulum, needed to remove calcium released at a normal excitation, is large enough to nearly keep up with calcium entrance by the calcium channel. Thus only calcium influx over the first few milliseconds of depolarization is relevant to the activation of contraction at the corresponding excitation. On repolarization, most of the calcium which entered the cell is in the sarcoplasmic reticulum, resulting in net calcium uptake from the extracellular space. Although not pronounced in these action potentials, terminal repolarization proceeds more rapidly during the premature action potential without calcium release.

It should be stressed that, at premature excitations in rabbit myocardium, prolongation of the action potential will be supported by the slow recovery from inactivation of the early outward current. That is to say, outward current will be smaller at the premature excitation. However, a very pronounced broadening effect is still found in the presence of high concentrations of 4-aminopyridine (Wohlfart 1982; Hilgemann 1986*a*). Furthermore, it may be pointed out that the pronounced broadening of action potentials at premature excitations in rabbit atrium (see Tanaka *et al.* 1967 for good examples) cannot be accounted for either by the behaviour of the calcium current, as simulated here, or by the behaviour of I_{to}, alone. The critical reasons for assuming a pronounced role of calcium-mediated inactivation will be summarized in the Discussion.

The paired-pulse protocol was then repeated with 1.5 s rest periods, time enough for the calcium release mechanism to recover from inactivation virtually completely. Figure 6 shows the third paired-pulse sequence, with signals superimposed. The initial excitation (1) now results in a much larger intracellular calcium transient, roughly a factor of two larger than during regular excitation. Simulation of the premature excitation (2) is very similar to that shown in figure 5. Action potentials show crossing over during this sequence at about -35 mV. The underlying reasons are that: (1) calcium-channel inactivation is fast at the excitation with large internal calcium release; (2) even with this degree of potentiation, sodium–calcium exchange does not strongly resist repolarization in the positive potential range; and (3) in the negative potential range the exchanger results in a pronounced slowing of repolarization or 'trailing off'. Accordingly, the potentiated contraction results in net calcium efflux and extracellular calcium cumulation. These action potential configuration changes come close to those found in rabbit atrium in the presence of a high concentration of 4-aminopyridine (Hilgemann 1986*b*). They are also similar to those found in rat ventricle during similar protocols. The inset shows results of Schouten & Ter Keur (1985) from rat ventricle, obtained during decay of poststimulatory potentiation; these results were obtained at a frequency of 0.1 Hz, so that the I_{to} has presumably recovered essentially completely between excitations. With variations of the potassium conductances (including addition of a time-dependent conductance) quite good simulations of the corresponding action potential configuration changes in rabbit ventricle (Wohlfart (1982) with 4-aminopyridine) and in guinea-pig atrium (D. W. Hilgemann, unpublished observations) have been obtained. It should be noted

184   D. W. Hilgemann and D. Noble

that the crossover point of action potentials in this sequence is in no way related
to the reversal potential of sodium–calcium exchange. Note that calcium efflux
(= exchange) is fully activated in a positive potential range at the potentiated
contraction, in spite of the high degree of cytosolic calcium buffering assumed in
these simulations.

Figure 7 shows the final steady-state sequence, after an additional six paired-
pulse excitations. The initial excitation now results in a 110 μM peak of total
cytosolic calcium concentration, which corresponds to 2.6 μM peak free calcium.
Note that the action potential at the initial excitation now shows negligible
shortening in the positive potential range in comparison to the subsequent action
potential. The initial excitation results in an approximately 14 μM increment of
extracellular calcium, which corresponds to extrusion of about 12 % of the calcium
released. The premature excitation results in net calcium influx of approximately
equal magnitude. The very large magnitude of calcium release here is dependent
on the fact that in the simulated model there is very little re-equilibration of
extracellular calcium between excitations. Experimentally, the re-equilibration
between paired pulses can in fact be considerable (for example, in figure 3 about
50 % of the depletion associated with the premature excitation is replenished
during the 2 s diastolic period). A number of possibilities can correct this aspect
of the modelling: a slow cytosolic calcium buffer is one. Note that the initial
depletions at potentiated contractions in the simulations are more brief than
those obtained experimentally at low calcium. The assumption here is that
calcium-channel inactivation will be more rapid and complete at the much greater
contractions occurring with 2 mM external calcium. As particularly apparent in
figure 5, extracellular calcium is decreasing slowly over the course of the simulation
period. With the rates of calcium uptake by the sarcoplasmic reticulum, needed
to remove cytosolic calcium in the time course of rapid calcium transients, the
uptake carries on into the diastolic period and over the course of 20 s quiescence
comes to a high steady state.

Figure 8 shows the behaviour of the model after a long rest period, whereby
three excitations were simulated, equivalent to the triple pulse sequence. The
background potassium conductance is set as in figure 2 (i.e. slightly larger than
in the previous figures); this results in larger net calcium efflux relative to total
calcium released in the previous simulations. The reason for the exact choice of
outward current magnitude was to illustrate that action potentials supported
almost entirely by inward exchange current may appear only insignificantly
different from action potentials supported almost entirely by calcium current. The
simulated records of calcium current, sodium–calcium exchange current, intracellular sodium changes, intracellular calcium and extracellular calcium are
presented to illustrate behaviour of all major variables during the sequence. The
extracellular calcium transients are essentially similar to those described
previously (compare the simulated wave form to that obtained experimentally in
figure 4). It is mentioned that continued rapid pacing results in further depletion
of extracellular calcium over about eight excitations; this result is consistent with
experimental observations (Hilgemann 1986*a*). The action potential changes come
quite close to those found in guinea-pig atrium (unpublished observations) for this
sequence.

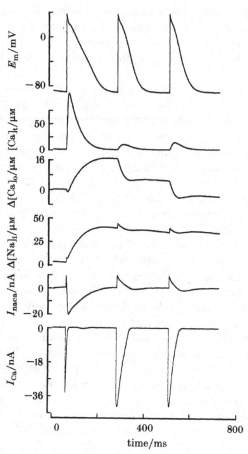

FIGURE 8. Simulated results for triple-pulse stimulation from the final rested state of the model; the results come close to steady state for repetitive triple-pulse stimulation at a basal rate of 0.4 Hz. The figure presents, from bottom to top, the simulated behaviour of calcium current, sodium–calcium exchange current, intracellular sodium, extracellular calcium, total intracellular calcium and membrane potential. This simulation was made with a slightly larger background potassium conductance than previous simulations. Note that the action potential at the initial excitation with large intracellular calcium release is supported almost exclusively by sodium–calcium exchange, whereas the premature action potentials are supported almost exclusively by calcium current. Compare the simulated extracellular calcium transient with those obtained in intact muscle, given in figure 4.

Note that the brief calcium current associated with the rested-state contraction shows only a very small degree of reactivation during the action potential; with several other formulations of calcium-mediated calcium-channel inactivation, the reactivation of calcium current during such an action potential was very substantial. Thus the action potential at the potentiated contraction is supported almost entirely by sodium–calcium exchange inward current, while the premature action potentials are supported almost entirely by calcium current. Note at the early excitations the small magnitude of the outward sodium–calcium exchange current, in relation to the calcium current; the outward exchange current continues, however, over a relatively long period at the early excitations with little calcium release, and therefore results in extrusion of essentially the amount of sodium needed for the upstroke. If the exchange rate is increased to allow extrusion of a greater portion of the calcium released at the potentiated contraction (which would be consistent with many observations on the decay of potentiation over 1–3 excitations), the magnitude of sodium extrusion due to calcium influx at the premature excitations can bring internal sodium back to the resting level over four premature excitations. With simulation of quadruple pacing (three premature excitations; not presented here for brevity) the ion movements at the potentiated contraction can be increased to 24 μM extracellular calcium increment and 74 μM sodium increment at the potentiated contraction. These values are somewhat less than the maximal exchange per beat, which would be expected by linear extrapolation of extracellular calcium transients at highly potentiated contractions with 150 μM free calcium (6 μM increment of extracellular calcium maximum).

A major question in the present simulations is whether a substantial component of calcium efflux might be accounted for by a sarcolemmal calcium pump rather than sodium–calcium exchange. One argument in favour of sodium–calcium exchange raised previously (Hilgemann 1986a) was that the exchanger could strongly favour calcium extrusion at potentiated contractions, but at a small premature excitation could allow calcium entering the cell to be taken up by the sarcoplasmic reticulum. A sarcolemmal calcium pump would compete with the sarcoplasmic reticulum for calcium independent of membrane potential. This argument was examined in simulations, and the basic result is shown in figure 9. A triple excitation sequence was simulated (exchange rates set so that the exchanger does not contribute more than 15% of calcium influx at the premature excitations; curves marked 1) and then the exchange rate was reduced by a factor of 10 with substitution by an electroneutral sarcolemmal calcium pump as described in Methods. Starting with the same calcium release conditions as in the number 1 signals, the extrusion rate was increased until the same amount and rate of extrusion was achieved at the potentiated contraction (signals 2); the difference in action potentials corresponds therefore to the role of sodium–calcium exchange inward current. At the premature excitations, calcium influx is decreased by about 40%. There is a 26% reduction in the rate of *net* calcium influx, because calcium extrusion is simultaneous with the calcium-channel current.

In this modelling the dependence of contractility on the trans-sarcolemmal sodium gradient is lost. It could be partially regained if sodium–calcium exchange resulted in a substantial calcium influx at excitation, but available evidence does not support this assumption. Further, a powerful sarcolemmal calcium pump is not consistent with the fact that the loss of potentiated contractility during non-steady-state stimulation (and the accompanying accumulations of extracellular calcium) are indeed suppressed by reduction of extracellular sodium (Hilgemann 1986a). In sheep Purkinje fibres the sodium-dependence of such contractile staircases has been examined in a more extreme fashion, because this preparation survives well after complete sodium removal. Figure 10 shows the effect of complete removal of extracellular sodium on negative contraction staircases found on stimulation from the rested state (15 min rest). The results shown are before

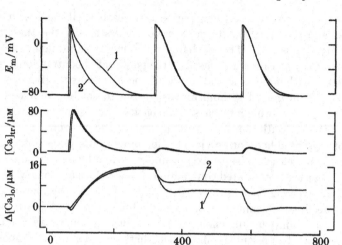

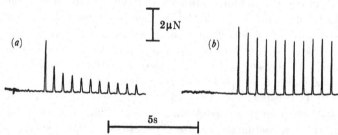

FIGURE 9. The effect of substituting a sarcolemmal calcium pump for sodium–calcium exchange in the triple-pulse sequence. The curves marked 1 were with sodium–calcium exchange, as in figure 8; the curves marked 2 are with a powerful sarcolemmal calcium pump. Note that the primary manifestation of the sodium–calcium exchange in action potentials is at the initial potentiated contraction. In extracellular calcium transients, the primary difference is that a calcium pump would significantly extrude calcium simultaneous with calcium influx, and thereby reduce net calcium uptake at premature excitations. A small part of the reduction of net calcium influx at premature excitations is due to the inhibition of the sodium–calcium exchange *per se.* (See text for complete details.)

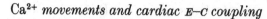

FIGURE 10. Experimental results on negative contractile staircases in sheep Purkinje fibres. In (*a*), voltage-clamp pulses were applied from -80 to 0 mV for 250 ms at a frequency of 0.5 Hz from the rested state. Note the very large negative contractile staircase. In (*b*) the same voltage protocol was applied after complete substitution of extracellular sodium for N-methylglucamine. Note the almost complete suppression of the negative staircase.

and after complete sodium removal for 3 h with N-methylglucamine replacement. A two-microelectrode voltage clamp is used for voltage control throughout, and depolarizing pulses in zero sodium solutions result in normal contraction form. The results shown were obtained in collaboration with R. D. Vaughan-Jones and C. Bountra with methods as described in Vaughan-Jones *et al.* (1985).

Figure 10(*a*) shows the control response in which voltage-clamp pulses were applied from -80 to 0 mV for 200 ms at 0.5 Hz from the rested state. Typically for this preparation (as well as atrial muscle), contractility is lost in two to four excitations to quite a low level, and one to two minutes are needed for complete recovery during rest. In other experiments (results not shown) it was found that the negative staircase is essentially abolished when the voltage pulses are made to $+40$ mV. Figure 10(b) shows the same protocol as in (a) after sodium removal; the rested state contraction is increased in magnitude by about 25%, and contractility is nearly fully maintained from one contraction to the next. Thus, it appears by that in Purkinje fibres contractile viability can be maintained without sodium–calcium exchange (or other sodium-dependent mechanisms) and that resting calcium can be maintained low enough to prevent contractile activation. By inference from work with atrium, however, large calcium movements resulting in loss of contractility at post-rest stimulation are not possible.

The simulations described thus far were all performed with instantaneous potassium conductances simply selected to give action potentials of realistic duration for atrial muscle. After establishing that the model predicts well the extracellular calcium transients and basic action-potential configuration changes, the modelling was further pursued with more realistic outward-current formulations. Figure 11 describes basic results for rabbit atrium, with a simple formulation of transient outward current. No changes of any other parameters were made in relation to the previous figures. The insert in the upper part of figure 11 shows action potentials and contractions obtained in rabbit atrium at 1.3 mm calcium; the first seven excitations are given on stimulation at 0.2 Hz from the rested state. Note the large negative staircase of developed tension and the late action potential over just three excitations. There is little or no change of the early repolarization because the stimulation frequency was selected to be low enough to allow (presumably) complete recovery from inactivation of the transient outward current. At the same time, the 5 s interval in this preparation results in only minor recovery of contractility after the rested state contraction. A typical extracellular calcium transient, obtained at a highly potentiated contraction with comparably fast repolarization, is shown below the other experimental results. (These results are from Hilgemann (1986a, b).)

In the lower panel the simulations are presented. Steady-state activation of transient outward current was taken from the data of Giles & Van Ginneken (1985) in crista terminalis of rabbit heart:

$$f_{ss} = 1/(1 + \exp\{-\tfrac{1}{5}(E_m + 4)\}). \tag{34}$$

Inactivation was assumed to be negligible during a single post-rest potential, and the activation/deactivation reactions were assumed to take place with a time constant of 3 ms; at normal temperatures these rates often cannot be resolved, and for the purpose of this simulation a more detailed formulation of the current is of no consequence. The only important feature is that the current is not instantaneous, so that the transient outward current results in repolarization to a potential negative to its activation threshold. The fractional activation of the I_{to} conductance (f_{act}) was simulated as

$$df_{act}/dt = (f_{ss} - f_{act}) \times 333, \tag{35}$$

and a linear instantaneous current–voltage relation was assumed. In addition to

Ca^{2+} *movements and cardiac* E–C *coupling*   189

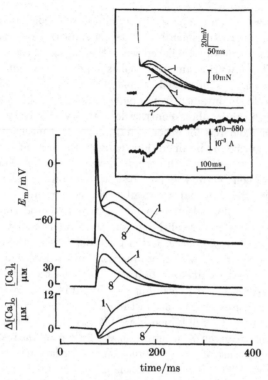

FIGURE 11. Post-rest stimulation at low frequency in rabbit atrium. In the inset, the bottom record shows the extracellular calcium transient, marked 1, obtained at a potentiated post-rest contraction in rabbit atrium (low extracellular calcium). The top two records of the insert show the developed tension and action potentials obtained during post-rest stimulation in rabbit atrium at normal calcium (0.2 Hz stimulation; otherwise same experimental conditions as extracellular calcium transient). See text for details of the simulations. There is a discrepancy here between the muscle behaviour and the simulated behaviour in that the muscle comes essentially to steady state within three excitations, whereas the simulated responses need about eight excitations to attain steady state.

the I_{to}, the simulation includes a linear background potassium conductance responsible for terminal repolarization, but no other potassium conductance.

The lower panel of the figure shows the first, fourth and eighth action potentials during 0.6 Hz stimulation from the rested state in the simulated model, together with the corresponding internal and external calcium transients. Note that the timing of the extracellular calcium accumulation is almost exactly predicted by the simulation, maximum efflux rate being achieved before the development of the late action-potential 'hump', and being complete well before termination of the action potential. The fact that rabbit atrial muscle has only a small or negligible I_{k1} conductance is presumably an important reason why late action-potential configuration changes with changes of contractility are more pronounced than in other preparations. Finally, it should be noted that if a calcium-activated non-specific cation channel were present in this preparation, it would result in still larger late action-potential changes than simulated here; up to now patch clamp

190   D. W. Hilgemann and D. Noble

studies have failed to identify the activation of non-specific channels during normal activity, as distinct from the calcium-overloaded state.

Without altering other simulation parameters, logical changes of the outward current formulations resulted in quite good simulations of action potentials of other preparations with a rapid contraction, in particular rat ventricle. Figure 12 presents simulation of rather striking rat action-potential changes described by Schouten & Ter Keur (1985). For the simulation, an I_{k1} conductance was simply substituted for the background potassium conductance. The magnitude of the I_{to} conductance was reduced in relation to the previous figure, and a moderate degree of inward rectification was added to the instantaneous current–voltage relation (see results of Giles & Van Ginneken (1985)); the rectifying property of the I_{to} conductance determines importantly the shape of repolarization in the positive potential range. Again, details of gating characteristics of the I_{to} were of no consequence to the simulation, except that the channel must deactivate in the course of 3–6 ms on repolarization beyond the activation threshold. The result of the formulations is that with moderate I_{to} magnitudes a low-conductance 'window' is formed between about −35 and −55 mV during repolarization. If an inward current is flowing at this time, the I_{to} deactivates fully before terminal repolarization can be brought about by the I_{k1} conductance, and a late low-level plateau is formed. The sodium–calcium exchange current has precisely the kinetic characteristics to promote generation of the plateau. During the plateau itself, net inward current can be very small, and the duration of the plateau need not reflect at all the time course of the inward exchange current.

The action potential marked 1 was obtained during low frequency stimulation,

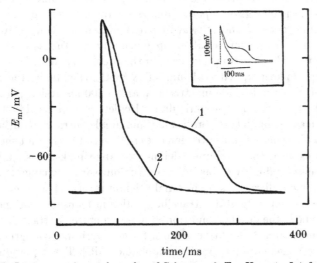

FIGURE 12. Inset: experimental results of Schouten & Ter Keur (1985) from rat ventricle; simulated results are given below. Curves marked 1 correspond to low-frequency excitations (i.e. large contractions); curves marked 2 are after substitution of extracellular sodium for lithium, which results in a pronounced positive inotropic effect. For the simulation, only the effective sodium concentration for the sodium–calcium exchanger was reduced. (See text for complete details.)

therefore at a relatively high rat ventricular contraction. The action potential marked 2 was obtained soon after partial substitution of external sodium for lithium. This was simulated by simply reducing the external sodium concentration in the sodium–calcium exchange equation, under the assumption that lithium would substitute for sodium in the upstroke and background sodium conductance. Essentially the same action-potential configuration change (1 to 2) is obtained when the calcium release by the sarcoplasmic reticulum is turned off, and correspondingly, ryanodine is known to reduce the late action potential of rat ventricular myocytes. (See Mitchell *et al.* (1984) for description of action potential changes related to internal calcium in myocytes.)

(c) *Trans-sarcolemmal calcium movements without contractile activation*

A major implication of the present modelling is that only a rapid increment of intracellular calcium can activate a contraction, because otherwise calcium uptake and extrusion effectively hinders cytosolic calcium cumulation. Correspondingly, calcium movements to and from the sarcoplasmic reticulum can be very substantial under certain circumstances with little or no active force development. This conclusion is highly relevant to the interpretation of a number of basic contractile phenomena and inotropic interventions. Therefore a few examples are presented.

Figure 13 shows an example relevant to the action of ryanodine. For the modelling of the ryanodine state, the sarcoplasmic reticulum is either made leaky by adding a continuous small activation of the release channel or the release mechanism is turned off; the 'leak' predicts more accurately findings in extracellular calcium transients (e.g. rapid re-equilibration of cumulative depletions); a leak was used in the modelling presented in the left panel of the figure. For the simulation presented, the background potassium conductance was reduced to just a point where calcium current and outward current match at about −10 mV. Calcium entering the sarcoplasm is taken up efficiently by the sarcoplasmic reticulum, and the rate of calcium uptake used in the previous simulations turns out to be fast enough to keep up with calcium influx by the calcium current. The calcium current undergoes only little inactivation, and a 600 ms action potential can strongly load the sarcoplasmic reticulum. Depletion here is 140 μM or 60 μmol kg⁻¹ fresh mass, enough to generate a half-maximal contraction, although the peak total cytosolic calcium concentration does not exceed 10 μM (not shown). Note in this circumstance that sodium–calcium exchange makes a significant contribution to calcium influx, whereas normally the outward exchange current is stopped within a few milliseconds by internal calcium release. The right panel of the figure shows the experimental extracellular calcium transient obtained in rabbit left atrium under the special condition of a rested-state contraction after ryanodine pretreatment and the addition of 2 mM 4-aminopyridine to lengthen the action potential (see Hilgemann (1986*b*) for complete details). The depletion was estimated at 15 μM of 200 μM total extracellular calcium concentration. It is mentioned that this figure is from a different set of simulations with slightly different parameter settings for calcium-channel inactivation and a different formulation of calcium release. With the calcium-channel parameters used in the other simulations, calcium current would in fact decrease substantially during such

192   D. W. Hilgemann and D. Noble

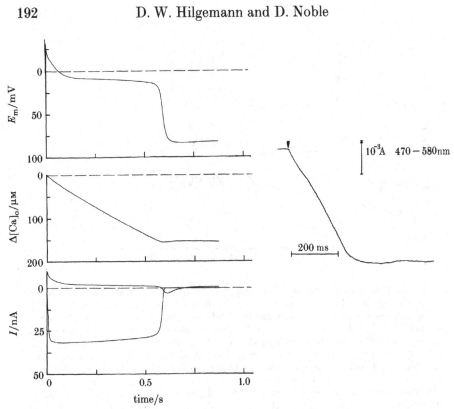

FIGURE 13. Left, simulated responses of calcium current, sodium–calcium exchange current, extracellular calcium current and membrane potential at a post-rest excitation with no calcium release and with reduction of outward current to just the value needed to generate a prolonged plateau. Calcium release is negligible because of the addition of a calcium leak via the release mechanism. Right, the extracellular calcium depletion obtained in rabbit atrium in the 'ryanodine state', and additionally with 4-aminopyridine to slow repolarization. Note that the waveform of depletion is predicted accurately by the simulation, and that extracellular calcium depletion can continue over 200 ms with only little decrease of rate.

a long action potential. However, the figure illustrates the basic possibility of a sustained calcium current, which the inactivation formulation predicts. This possibility appears realistic in relation to calcium currents described by Bechem & Pott (1985) in guinea-pig atrial cardioballs, to extracellular calcium transients such as that shown (Hilgemann (1986b); also found in ventricular preparations), and could explain the extremely high net calcium influx described by Lewartowski *et al.* (1982) in guinea-pig ventricle at single rested-state contractions.

Diastolic calcium movements, as well as systolic movements, via sodium–calcium exchange can either load or unload the sarcoplasmic reticulum in the course of a few seconds with little or no contractile activation under certain circumstances. These calcium movements are ultimately controlled by the sarcoplasmic reticulum; figures 14 and 15 illustrate the basic possibilities. Figure 14 is essentially a continuation of figure 13, in which the sarcoplasmic-reticulum leak was increased further to result in loss of stored calcium almost completely within

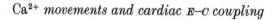

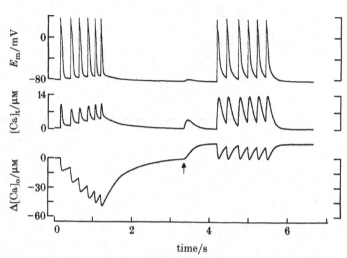

FIGURE 14. Simulations relevant to the actions of ryanodine and caffeine on extracellular calcium transients. Initially, six excitations were made with a calcium-channel release leak large enough to allow re-equilibration of extracellular calcium within 2 s of quiescence. Note that 20 μmol Ca kg⁻¹ fresh mass can return to the extracellular space within 1.3 s with only a slight trailing-off of membrane potential when the efflux takes place 'slowly' as here during quiescence. At the arrow the calcium pump was turned off; note that the sarcoplasmic reticulum is not entirely empty. If the rate of translocation between uptake and release stores were on the order of 1 s, as in most simulations, and not a few milliseconds, as in these simulations, the result of turning off the calcium pump (one of several actions of caffeine) would be a very large contracture. In the last part of the records, six rapid excitations were again simulated. Note that these excitations result in no net depletion of extracellular calcium. (See text for further details.)

2 s. Six short action potentials were simulated, which result in intracellular calcium transients of about 7 μM magnitude total calcium. The series of action potentials results in a 50 μM depletion of extracellular calcium, enough to activate a moderately large contraction, if it were maintained in the store. Note the small rise of calcium in the cell between excitations, which would correspond to the small rise of rest tension found during rapid stimulation in the ryanodine state (see, for example, Hilgemann 1986a). After the last stimulation, extracellular calcium replenishes within 2 s, and the effect on membrane potential is only about a 3 mV shift in the positive direction. At the arrow in the figure, the calcium pump was turned off, and the calcium remaining in the sarcoplasmic reticulum generated a small transient rise of internal calcium as it was lost from the sarcoplasmic reticulum and extruded by sodium–calcium exchange. Note that even with a large calcium leak, the sarcoplasmic reticulum contains a substantial amount of calcium which can be released by inhibiting calcium uptake; this result is relevant to the finding of substantial caffeine-induced contractures even after pretreatment with ryanodine in cardiac muscle (see, for example, Marban & Weir (1985)). If the calcium 'translocation' process took place with a 1 s time constant, the caffeine-induced contracture could be of much larger magnitude.

The following six excitations result in depletions which re-equilibrate during the

194   D. W. Hilgemann and D. Noble

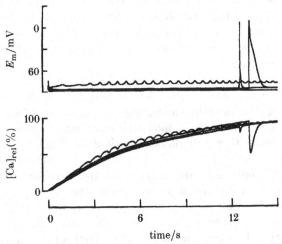

FIGURE 15. Time course of calcium-store loading during quiescence. Four superimposed simulations of percentage calcium-store load and membrane potential. One record corresponds to the cumulation of calcium by the sarcoplasmic reticulum upon stopping regular stimulation; the three subsequent records are with increasing concentrations of intracellular sodium (10, 14 and 18 mM) introduced as a step change. The results with spontaneous oscillations are at 18 mM sodium. In spite of very different calcium-store calcium concentrations and different magnitudes of calcium uptake (which were reflected in the extracellular space as depletion) the time courses are all determined by the back rate of calcium movement from the sarcoplasmic reticulum to the cytosol. (See text for further details.)

action potentials themselves. Intracellular calcium transients are somewhat larger, and they relax more slowly. These results are very similar to those found with high concentrations of caffeine in extracellular calcium transients (Hilgemann *et al.* (1983); fast extracellular calcium transients with caffeine have not been published). The almost complete inhibition of cumulative extracellular calcium depletions by caffeine (assuming one of its actions to be inhibition of calcium uptake) is an additional argument that most calcium exchange found in experiments reflects exchange to the sarcoplasmic reticulum. It is stressed that this is not the case in frog heart, where the calcium store and/or buffer sites responsible for calcium uptake during cumulative depletions are not sensitive to caffeine or ryanodine (D. W. Hilgemann, unpublished results).

Figure 15 illustrates briefly calcium loading by the sarcoplasmic reticulum after regular stimulations as occurs normally in the simulations during quiescence, and as would occur if the trans-sarcolemmal sodium gradient could be decreased instantaneously. A series of four superimposable results is presented in the figure, whereby the change of calcium concentration in the release store has been scaled and plotted as percent of maximum change: (1) the time course of calcium loading by the sarcoplasmic reticulum during quiescence was examined after regular stimulation; (2) the time course of calcium-store loading with a step increase of sodium from 7 to 12 mM was examined; (3) 7–14 mM; and (4) 7–18 mM. Only the 18 mM sodium curve shows significant deviations from the other curves, because of spontaneous calcium release. Total calcium movement over the 15 s shown corresponded to a 52 μM extracellular calcium depletion in (1), 94 μM depletion in (2), 160 μM depletion in (3), and 162 μM depletion in the experiment with oscillatory calcium release. The time course given would correspond to the recovery of contractility in atrial muscle after rested state contractions (termed 'NIEA decay' by Blinks & Koch-Weser (1961)).

These results are relevant to the general finding of rapid contractile responses on lowering external sodium, even during a quiescent period. From the standpoint of the present modelling, this provides the basis for the inotropic effect of heart glycosides. By increasing internal sodium, internal calcium-store loading increases without having to assume either a pronounced activation of calcium influx at excitation or a pronounced inhibition of calcium efflux by the increased internal sodium concentration. The only necessary assumption is that diastolic free calcium will be determined largely by the exchange mechanism. The exchanger need not operate in the reverse mode during calcium-store loading at rest, if the background calcium leak is substantial. However, with the routine simulation parameter settings, the exchanger is indeed operating in the calcium influx mode during these sequences. These results support the hypothesis that sodium–calcium exchange will ultimately regulate contractility by regulating calcium influx (an idea proposed for many years by G. A. Langer; see, for example, Langer (1974)) albeit with the twist that diastolic calcium-store loading rather than systolic calcium influx is the underlying basis for the inotropic effect.

DISCUSSION

Action-potential configuration changes in relation to extracellular calcium transients

It is now just 35 years since Weidmann (1951) published his milestone experiments designed to characterize conductance changes during the cardiac action potential. It is perhaps not surprising that basic conclusions from squid axon were not readily applicable to cardiac muscle, but surely no one at the time would have expected that three and a half decades later fundamental issues about the nature of the cardiac action potential would remain highly problematic. Furthermore, who would have guessed that electrophysiological techniques of unimaginable sophistication 35 years ago (primarily whole-cell and single-channel patch-clamp techniques) would not quickly and definitively resolve the integral function of basic mechanisms underlying the physiological cardiac action potential? The simulations presented in this article have examined hypothetical interactions of two of those mechanisms, calcium channel and sodium–calcium exchange, with intracellular calcium-regulating mechanism. A major motivation for these simulations has been the fact that both of these mechanisms are inextricably tied to intracellular calcium regulation, which in turn must be disrupted to study their isolated current-generating characteristics. In both cases, this means intracellular dialysis, substitution of ions and addition of EGTA to the internal solution. Although the new data on these mechanisms from patch-clamp work reflects a greatly enhanced experimental control of cellular and methodological variables, it is simply not obvious how the new data can be extrapolated

196   D. W. Hilgemann and D. Noble

to normal function. Therefore, a strong point of the present simulations is that extracellular calcium transients measured in intact muscle could be used as a set of guide posts, which have favoured certain possibilities and appear to exclude others.

After completing one of the early descriptions of myocyte calcium currents (Isenberg & Klöckner 1982), Isenberg (1982) concluded that the speed and magnitude of the currents recorded were appropriate for direct activation of contraction, but that the currents recorded did not contradict general concepts about the role of an internal calcium store, as positive contractile staircases could still be evoked without an increase of the current (i.e. the current primed an intracellular store). Those currents, which reflected an increase of intracellular calcium concentration of around 100 μM, did in fact very seriously contradict available concepts, since no established mechanism could be proposed to extrude such quantities of calcium. Given the generally short term nature of such experiments, it remains a question whether the calcium is ever extruded; in fact, the peak current magnitudes in those studies (greater than 10 nA) have seldom, if ever, been duplicated by other authors.

One tempting explanation of the current magnitudes was that removal of the glycocalyx (Isenberg & Klöckner 1980) increased the apparent conductivity of the calcium channels without altering their activation properties. However, the major possibility, which relaxed requirements for calcium extrusion in intact muscle, is the one simulated here, calcium-mediated inactivation of calcium current. From the perspective of the present simulations the inactivation process is a pivotal regulatory point for both calcium and sodium homeostasis. It allows calcium influx during regular stimulation to be no larger than original estimates of calcium influx by calcium-flux and calcium-current analysis (see Chapman 1983). Nevertheless, it allows calcium influx to be of sufficient magnitude to activate a moderate contraction when calcium release is negligible (albeit the next contraction). Potentially, it explains such problematic findings as a net calcium influx of more than 100 μmol kg^{-1} fresh mass during single excitations in guinea pig ventricle (Lewartowski *et al.* 1982). Finally, the calcium-dependence of calcium-channel inactivation may resolve an important paradox, which arises from the assumption of sodium–calcium exchange as the primary calcium efflux mechanism. During regular contraction, any mechanism which increases calcium influx via calcium channels (e.g. increasing extracellular calcium) would necessarily increase sodium influx by the exchanger and thereby generate positive feedback in the excitation–contraction coupling system (i.e. would finally inhibit calcium efflux). The calcium-dependence of calcium-channel inactivation would prevent such a positive feedback, and the relation between extracellular calcium and calcium influx per beat can be quite flat (results not presented).

Given the calcium-dependence of calcium current inactivation, as simulated here, a smaller calcium release at a premature excitation and a larger release at poststimulatory excitations will lengthen and shorten the duration of the plateau via negative feedback of calcium release to the calcium current. However, this effect can be offset by the activation of sodium–calcium exchange in the positive potential range. Even with a high degree of cytosolic calcium buffering, as in the present simulations, the exchanger can support an action potential whose configuration need not be greatly different from an action potential supported entirely by calcium current (figure 8). The teleological sense of these explanations is that the calcium-dependence of calcium-channel inactivation minimizes (or eliminates) simultaneous calcium influx and efflux. With the calcium buffering used in the present experiments, the inward exchange current supports action-potential trailing-off and low-level action-potential plateaux at intermediate levels of contractility (figures 6, 11 and 12).

It must be stressed again here that the details of the simulated action-potential configuration changes depend on the assumption of intracellular free calcium transients approximately one-third of those generally inferred from aequorin measurements. This assumption further limits the possibility of simultaneous calcium influx and efflux, because calcium efflux by exchange is forced into the negative potential range during steady stimulation. To simulate the basic action-potential changes with less internal calcium buffering, which results in greater inward exchange currents in the positive potential range, it appears essential to introduce a calcium-activated outward current as the cause of action-potential shortening at potentiated contractions, and at least in bovine cardiac Purkinje cells a calcium-dependent potassium channel has finally been identified (Callewaert *et al.* 1986). Without this assumption, early action-potential plateaux are increased by inward exchange current during potentiated contractions, and this result is unrealistic. Therefore, several possibilities to limit exchange current at positive potentials with free calcium in the range 5–10 µM were examined (D. W. Hilgemann, unpublished simulations):

1. The voltage-dependence of exchange in the calcium efflux mode might tend to limit inward exchange current in positive potential range; analysis of the data of Kimura *et al.* (1986) shows some tendency in this direction, but not of adequate magnitude.

2. Calcium inhomogeneities may develop from the cell membrane to the interior of the cell during efflux, resulting in lower free-calcium concentrations for exchange during calcium efflux than mean free cytosolic calcium. Although subsarcolemmal calcium fluctuations probably do not markedly affect the driving force for calcium currents (Fischmeister & Horackova 1983; Simon & Llinas 1986), relatively small cytosolic inhomogeneities of free calcium could have a substantial effect on exchange activity. Estimations of the possible magnitude of inhomogeneity during rapid extrusion after a large calcium release have been carried out (D. W. Hilgemann, unpublished simulations); with 120 µM fixed binding sites in a cell of 6 µm diameter, a factor of about 2 was obtained from the cell centre to the cell membrane.

3. The Donnan effect of fixed negative charge in the cytosol may complicate the interpretation of mean free cytosolic calcium measurements.

4. Calcium efflux by exchange might be a much less than linear function of internal calcium. For example, the calcium-unbinding reaction appears to rate-limit the exchange reaction in the mechanism simulated by Johnson & Kootsey (1985; see also Sanders *et al.* (1984)). In sequential models, the translocation of bound sodium could, for example, be rate-limiting at positive potentials. A tendency of exchange rates to saturate with increasing internal calcium would

198   D. W. Hilgemann and D. Noble

also possibly explain why poststimulatory potentiation is often lost within one beat at low external calcium (e.g. in extracellular calcium transients) but at normal extracellular calcium usually takes several excitation–contraction cycles (D. W. Hilgemann, unpublished observations in guinea-pig atrium, rabbit atrium and rabbit ventricle).

The question as to the relative involvement of a sarcolemmal calcium pump in calcium efflux is not yet resolved with certainty. A significant involvement would have the advantage of lessening the cellular sodium load produced by exchange. A high degree of activity would have the consequences described in figure 9: the pump would extrude calcium simultaneous with calcium influx, and it would negate the sodium-dependence of cardiac contractility (assuming that sodium–calcium exchange is not massively involved in calcium influx). For the time being, available evidence very clearly supports a major involvement of sodium–calcium exchange in calcium efflux, and is consistent with the idea that the sarcolemmal calcium pump is a slow back-up calcium efflux mechanism:

1. The simulations presented here are consistent with the magnitude of exchange currents found in myocytes (i.e. maximal inward exchange currents with 1 μM internal free calcium are approximately 40 % as large as calcium currents (Kimura *et al.* 1986)).

2. Fedida *et al.* (1987) have recently recorded currents in single ventricular myocytes, closely consistent with the mechanisms simulated here. This was achieved by using relatively high-resistance electrodes with no calcium buffer in the pipette to reduce dialysis. Slow tail currents were found using double pulse and repetitive pulse protocols. The current was closely related to myocyte movement, and the dynamic behaviour of the slow tail current fitted closely with calcium accumulation in extracellular calcium measurements. Also, prolongation of calcium current was found at premature excitations.

3. It is virtually impossible to evoke net calcium efflux upon excitation after reduction of external sodium by 50 % (Hilgemann 1986*a*).

4. The contractile patterns closely associated with calcium efflux (i.e. negative staircase over 1–4 excitations) are essentially abolished after complete sodium removal (figure 10) in sheep Purkinje fibres.

Interval-strength mechanisms in relation to extracellular calcium transients

Details of simulations related to non-steady-state stimulation are given first as a brief summary of the underlying mechanisms:

1. The loss of contractility at premature excitations is accounted for by inactivation of the calcium-release mechanism.

2. Almost complete net uptake of calcium entering the cardiac cell at premature excitations is possible in mammalian heart because sodium–calcium exchange does not generally favour calcium efflux without internal calcium release.

3. Calcium influx at a premature excitation can be much larger than during steady-state stimulation because the calcium current is prolonged, and accordingly the potentiating effect of single premature excitations on later excitations can be large.

4. Calcium efflux at potentiated contractions can be larger than expected

Ca^{2+} *movements and cardiac E–C coupling* 199

alone from the enhanced internal calcium concentration, because a more rapid repolarization favours calcium efflux by sodium–calcium exchange.

5. Calcium efflux necessarily takes place primarily during systole, because calcium taken up by the sarcoplasmic reticulum during relaxation of internal calcium is not available to sarcolemmal calcium-extrusion mechanisms during diastole (i.e. passive calcium leak across the SR is minimal). Apart from these interpretations, several problems arose in modelling details of interval-strength mechanisms, which were in part overcome by other formulations of sarcoplasmic-reticulum function, but clearly demand further attention.

In the present simulations the sarcoplasmic reticulum loads to a high level during rest; this result is consistent with the high rested-state contraction of atrial muscle at normal calcium levels (see, for example, Kruta & Stejskalova 1960). In ventricular muscle the contractile potentiation produced by a few rapid excitations is lost over the course of several minutes during quiescence (Wood *et al.* 1969) and, correspondingly, cumulative extracellular calcium depletions related to the production of potentiation can also take minutes to re-equilibrate during rest (Hilgemann & Langer 1984). This is also the case in atrial muscle at low external calcium (Hilgemann 1986*a*). Rather surprisingly, it is not readily possible to simulate a slowly dissipating store during quiescence, which the experimental results presumably reflect (1–3 min time constants are often found). The basic problem which arises is that, with the rates of calcium uptake needed to remove most cytosolic calcium within about 200 ms, calcium uptake remains fairly active even at the low cytosolic calcium concentrations during quiescence. As the free-calcium concentrations in the sarcoplasmic reticulum occurring in the present simulations are in the range 0.3–1.3 mM, the thermodynamic capabilities of a two-calcium per ATP pump model are not approached (see Kodama 1985) and the problem therefore appears to reflect control of the rate-limiting steps in calcium translocation. The problem becomes still more severe if resting free calcium is not close to the thermodynamic equilibrium of the exchanger, as in the present simulations, but approaches 0.1 μM as indicated by calcium electrode measurements (note that this same problem arose in simulations of Cannell & Allen (1984) in skeletal muscle). In the present work the problem has been relatively minimized by a simple pump formulation, which results in pump inhibition as calcium accumulates in the sarcoplasmic reticulum, but it remains a serious problem.

A rather obvious solution tested was to make calcium uptake dependent on calcium binding to two pump sites with equal or variable affinities, consistent with evidence for a relatively steep uptake function (two calcium pumped per cycle). These simple formulations had the desired effect of greatly reducing the steady-state calcium-store load, but resulted in extreme prolongation of intracellular calcium transients with small internal release (or low external calcium) and long tails of the normal intracellular calcium transient. Furthermore, these formulations did not allow the sarcoplasmic reticulum to take up calcium efficiently at premature excitations or in the 'ryanodine state' when cytosolic calcium is relatively low. For these reasons, a favoured possibility at present is that the calcium pump undergoes a quite large transitory stimulation with each contraction cycle, which is maintained through relaxation and then decays rapidly during diastole. Two mechanistic possibilities were considered: stimulation of the calcium pump via a calmodulin-dependent (Klee & Banaman 1982) phosphorylation reaction (LePeuch *et al.* 1980) during each contraction cycle; and the relatively slow formation and decay of an enzyme state, intermediate between the two calcium-binding reactions in the forward pump reaction cycle, as proposed by Inesi and colleagues (1980). Both solutions allow simulation of the basic ventricular frequency–force relation with slow loss of contractility during quiescence (D. W. Hilgemann, unpublished results). It would obviously be important to know whether calmodulin-dependent phosphorylation is a beat-to-beat process.

Another closely related point, where present formulations of calcium exchange by the sarcoplasmic reticulum are not adequate, concerns the development of the high rested-state contraction in atrial muscle under normal conditions. The time course of recovery of the high rested-state contraction after one to several post-rest excitations ('decay of the negative inotropic effect of activation') is complex and shows a fast and a slow phase (Blinks & Koch-Weser 1961; Hilgemann 1980). The fast phase (1–2 s) would correlate with the recovery from inactivation of the release mechanism; the slow phase would correlate with the reloading of calcium by the sarcoplasmic reticulum, which was lost primarily to the extracellular space at the previous excitations. Assuming that sodium–calcium exchange acts as quite a rapid cytosolic calcium clamp during quiescence, it was concluded previously that the time course of the slow phase would be determined ultimately by the back rate constant of calcium movement from the sarcoplasmic reticulum to the cytosol (Hilgemann 1980). This consideration was tested in the context of the present modelling, and was verified in figure 15; the approximate half-life of 10 s for calcium uptake by the sarcoplasmic reticulum is identical to the rate at which calcium is lost from the sarcoplasmic reticulum if it has been loaded above the rested-state equilibrium (results not shown). What is not predicted by the present modelling is that the slow phase of recovery is accelerated markedly by all interventions, thought to increase resting cytosolic calcium (heart glycosides, low potassium, low external sodium and high external calcium) from time constants of 1 min and more down to just 2–3 s (Koch-Weser & Blinks 1962; Vierling & Reiter 1975; Hilgemann 1980). When the time of recovery approaches 2 s, the recovery curve shows damped oscillatory behaviour (Hilgemann 1980), which is the prelude to continuous oscillatory behaviour during quiescence. A possible explanation, not included in the present simulations, is that the back rate of calcium movement from the sarcoplasmic reticulum to the cytosol reflects reversal of the calcium pump, which would depend on the binding of two calcium ions to pump sites facing the lumen of the sarcoplasmic reticulum (Mensing & Hilgemann 1981; Hilgemann 1980). This readily predicts the acceleration of 'NIEA decay' with increase of diastolic calcium, but a transition to oscillatory recovery has not been reproduced. Alternative explanations of 'NIEA decay' did not arise in the course of the present work.

A problem of more general interest concerns the kinetics with which potentiated contractility is lost at post-rest stimulations, as corresponding experiments have often been used to draw conclusions about the fractional extrusion of calcium released in relation to re-uptake by the sarcoplasmic reticulum (see, for example,

Morad and Goldman 1973; Wohlfart 1979; Schouten 1985). Briefly, the experimental tactic is to make a series of excitations from the potentiated state (either after rapid loading excitations or from the rested state of atrial muscle) at intervals long enough, so that the rapid 'restitution of contractility' is complete, and then to analyse the loss of contractility over several excitations (see figures 10 and 11). The fractional decay of potentiation from one beat to the next would in the simplest case correspond to the fractional extrusion of calcium and would allow calculation of the relative contributions of calcium influx and release to activation of contraction. Such experiments in mammalian ventricle, for example, give estimates of fractional extrusion from nearly 90 % in rabbit ventricle (Wohlfart 1979), 60 % in many mammalian ventricular preparations (see, for example, results of Wood *et al.* (1969)), to 10 % in rat ventricle (see, for example, Schouten (1985); but note that loss of potentiation can also take place in just two excitations in rat ventricle). As already pointed out, trans-sarcolemmal calcium movements in non-steady-state will usually be much larger than during steady-state stimulation. Therefore, projection of such simple calculations to the steady state is almost certainly invalid.

It is, however, a different question as to whether or not the fractional decay of potentiation from one beat to the next indeed reflects the fractional extrusion of cytosolic calcium during non-steady-state contractions (i.e. whether loss of potentiation in 2–3 beats indeed reflects sarcolemmal extrusion of about 50 % of calcium released per beat). Extracellular calcium transients have favoured this interpretation qualitatively but not quantitatively; the amount of calcium appearing in the extracellular space at potentiated contractions is less than expected for the negative inotropic effect on the basis of Fabiato's estimates of calcium requirements for activation. For this reason, one of the goals in formulating calcium-release mechanisms was to look for a hypothetical mechanism, which would release progressively smaller fractions of calcium in the store with progressive depletion of the store. This approach was not successful, and correspondingly potentiation in the model decays over approximately 9 excitations rather than the 1–3 excitations routinely found in experiments with rabbit myocardium.

The most simple adjustment of the model to lessen this discrepancy is to allow release of essentially all calcium in the store at a fully 'restituted' contraction, as proposed for mammalian myocardium many years ago (Antoni *et al.* 1969). This brings the decay of potentiation down to 4–5 beats, but not less. The discrepancy could also be lessened if major internal redistributions of calcium took place at the potentiated contraction (e.g. to mitochondria) but it then becomes difficult to account for a dependency of the loss of potentiation on sodium. Finally, the discrepancy could be lessened if the calcium requirements for activation by a rapid release of calcium were in fact not larger than those originally suggested by Solaro *et al.* (1974) for example, because cooperativity between adjacent troponin molecules in activating contraction is greater in intact muscle than in skinned preparations. In any case, the decay kinetics of poststimulatory potentiation remains a problem of considerable importance. A closely related problem arises for these simulations in recent studies of rabbit myocyte contractility after rapid

202 D. W. Hilgemann and D. Noble

changes of external solution by Langer & Rich (1987). These authors have found large changes of contraction magnitude, on changing external sodium, which are complete within one contraction cycle. Both the magnitude and kinetics of these effects would be consistent with a more profound role of sodium–calcium exchange in that preparation than simulated in the present work.

In summary, we are well aware of a number of experimental findings which can not be readily explained by the simulations in their present form, and we do not exclude the possibility of important missing mechanisms in the simulation scheme. Several important open questions, such as the role of sodium–calcium exchange in a long, physiological ventricular action potential, have not been addressed specifically by these simulations. Nevertheless, we feel that the present simulations give a detailed picture of how available data on calcium current, sodium–calcium exchange, and sarcoplasmic reticulum can be projected to physiological function during steady- and non-steady-state stimulation in the mammalian atrium. On the basis of present knowledge, substantial revisions of the magnitudes, mechanisms and timings of trans-sarcolemmal calcium movements do not appear possible.

We thank Dr Charanjit Bountra and Dr Richard Vaughan-Jones for their generous collaboration in one of the experiments. We thank the British Heart Foundation, the Wellcome Trust and the Medical Research Council for financial support. D.H. thanks Dr Jan Koch-Weser for his support at an early stage (1980–81) in this work, Leota Green for secretarial assistance, and Dr W. R. Gibbons for helpful comments on this manuscript.

D.H. is a Senior Research Fellow of the American Heart Association, Greater Los Angeles Affiliate. This work was supported by a Grant-in-Aid from the American Heart Association (no. 85099600) and a Distinguished Visitor Award from the British Heart Foundation to D.W.H.

Questions about any further details of the formulations used for simulation (selection of parameter settings, alternative simulations, etc.) should be addressed to D.H. A complete listing of equations and a copy of the program will be made available on request. The program is currently being prepared for publication, together with recent developments of individual aspects.

REFERENCES

Adler, D., Wong, A. Y. K., Mahler, Y. & Klassen, G. A. 1985 Model of calcium movements in the mammalian myocardium: interval–strength relationship. *J. theor. Biol.* **113**, 379–394.

Allen, D. G. & Kentish, J. C. 1985 The cellular basis of the length-tension relation in cardiac muscle. *J. molec. cell. Cardiol.* **17**, 821–840.

Allen, D. G. & Orchard, C. H. 1983 Intracellular calcium concentration during hypoxia and metabolic inhibition in mammalian ventricular muscle. *J. Physiol., Lond.* **339**, 107–122.

Antoni, H., Jakob, R. & Kaufmann, R. 1969 Mechanische Reaktionen des Frosch- und Säugetiermyokards bei Veränderung der Aktionspotential-Dauer durch konstante Gleichstromimpulse. *Pflügers Arch. Eur. J. Physiol.* **306**, 33–57.

Bean, B. P. 1985 Two kinds of calcium channels in canine atrial cells. Differences in kinetics, selectivity, and pharmacology. *J. gen. Physiol.* **86**, 1–30.

Bechem, M. & Pott, L. 1985 Removal of Ca current inactivation in dialysed guinea-pig atrial cardioballs by Ca chelators. *Pflügers Arch. Eur. J. Physiol.* **404**, 10–20.

Blinks, J. R. & Koch-Weser, J. 1961 Analysis of effects of changes of rhythm upon myocardial contractility. *J. Pharmac. exp. Ther.* **134**, 373–389.

Callewaert, G., Vereecke, J. & Carmeliet, E. 1986 Existence of a calcium-dependent potassium channel in the membrane of cow cardiac Purkinje cells. *Pflügers Arch. Eur. J. Physiol.* **406**, 424–427.

Cannell, M. B. & Allen, D. G. 1984 Model of calcium movements during activation in the sarcomere of frog skeletal muscle. *Biophys. J.* **45**, 913–925.

Chapman, R. A. 1983 Control of cardiac contractility at the cellular level. *Am. J. Physiol.* **245**, H535–H552.

DiFrancesco, D. & Noble, D. 1985 A model of cardiac electrical activity incorporating ionic pumps and concentration changes. *Phil. Trans. R. Soc. Lond.* B **307**, 353–398.

Eisner, D. A., Lederer, W. J. & Vaughan-Jones, R. D. 1981 The dependence of sodium pumping and tension on intracellular sodium activity in voltage-clamped sheep Purkinje fibres. *J. Physiol., Lond.* **317**, 163–187.

Ellis, D. 1985 Effects of stimulation and diphenylhydantoin on the intracellular sodium activity in Purkinje fibres of sheep heart. *J. Physiol., Lond.* **284**, 241–259.

Fabiato, A. 1981 Myoplasmic free calcium concentration reached during the twitch of an intact isolated cardiac cell and during calcium-induced release of calcium from the sarcoplasmic reticulum of a skinned cardiac cell from the adult rat or rabbit ventricle. *J. gen. Physiol.* **78**, 457–497.

Fabiato, A. 1983 Calcium-induced release of calcium from the cardiac sarcoplasmic reticulum. *Am. J. Physiol.* **245**, C1–C14.

Fabiato, A. 1985*a* Rapid ionic modifications during the aequorin-detected calcium transient in a skinned canine cardiac Purkinje cell. *J. gen. Physiol.* **85**, 189–246.

Fabiato, A. 1985*b* Time and calcium dependence of activation and inactivation of calcium-induced release of calcium from the sarcoplasmic reticulum of a skinned canine cardiac Purkinje cell. *J. gen. Physiol.* **85**, 247–289.

Fabiato, A. 1985*c* Simulated calcium current can both cause calcium loading in and trigger calcium release from the sarcoplasmic reticulum of a skinned canine cardiac Purkinje cell. *J. gen. Physiol.* **85**, 291–320.

Fedida, D., Noble, D., Shimoni, Y. & Spindler, A. J. 1987 Slow inward currents related to contraction in guinea-pig ventricular myocytes. *J. Physiol., Lond.* **385** (In the press.)

Fischmeister, R. & Horackova, M. 1983 Variation of intracellular Ca²⁺ following Ca²⁺ current in heart. *Biophys. J.* **41**, 341–348.

Gadsby, D. C., Kimura, J. & Noma, A. 1985 Voltage dependence of Na/K pump current in isolated heart cells. *Nature, Lond.* **315**, 63–65.

Giles, W. R. & Van Ginneken, A. C. G. 1985 A transient outward current in isolated cells from the crista terminalis of rabbit heart. *J. Physiol., Lond.* **368**, 243–264.

Glitsch, H. G., Pusch, H. & Venetz, K. 1976 Effects of Na and K ions on the active Na transport in guinea-pig auricles. *Pflügers Arch. Eur. J. Physiol.* **365**, 29–36.

Glitsch, H. G., Grabowski, W. & Thielen, J. 1978 Activation of the electrogenic sodium pump in guinea-pig atria by external potassium ions. *J. Physiol., Lond.* **276**, 515–524.

Hilgemann, D. W. 1980 New perspectives on the interval–strength relationship of guinea pig atrium. Dissertation, Faculty of Biology, University of Tübingen.

Hilgemann, D. W. 1985 Extracellular calcium transients at single excitations in rabbit atrium. (Abstract.) *Biophys. J.* **47**(2, pt 2), 376a.

Hilgemann, D. W. 1986*a* Extracellular calcium transients and action potential configurations changes related to post-stimulatory potentiation in rabbit atrium. *J. gen. Physiol.* **87**, 675–706.

Hilgemann, D. W. 1986*b* Extracellular calcium transients at single excitations in rabbit atrium. Measured with tetramethylmurexide. *J. gen. Physiol.* **87**, 707–735.

Hilgemann, D. W., Delay, M. J. & Langer, G. A. 1983 Activation-dependent cumulative depletions of extracellular free calcium in guinea pig atrium measured with antipyrylazo III and ttramethylmurexide. *Circulation Res.* **53**, 779–793.

204 D. W. Hilgemann and D. Noble

Hilgemann, D. W. & Langer, G. A. 1984 Transsarcolemmal calcium movements in arterially perfused rabbit right ventricle measured with extracellular calcium-sensitive dyes. *Circulation Res.* **54**, 461–467.

Hilgemann, D. & Noble, D. 1986 Simulation of extracellular Ca transients in rabbit cardiac muscle. *J. Physiol., Lond.* **371**, 195P.

Hume, J. R. & Uehara, A. 1985 Ionic basis of the different action potential configurations of single guinea-pig atrial and ventricular myocytes. *J. Physiol., Lond.* **368**, 525–544.

Inesi, G., Kurzmack, M., Coan, C. & Lewis, D. E. 1980 Cooperative calcium binding and ATPase activation in sarcoplasmic reticulum vesicles. *J. biol. Chem.* **255**, 3025–3031.

Isenberg, G. & Klöckner, U. 1980 Glycocalyx is not required for slow inward calcium current isolated rat heart myocytes. *Nature, Lond.* **284**, 358–360.

Isenberg, G. & Klöckner, U. 1982 Calcium currents of isolated bovine ventricular myocytes are fast and of large magnitude. *Pflügers Arch. Eur. J. Physiol.* **395**, 30–41.

Isenberg, G. 1982 Ca entry and contraction as studied in isolated bovine ventricular myocytes. *Z. Naturforsch* **37c**, 502–512.

Johnson, E. A. & Kootsey, J. M. 1985 A minimum mechanism of Na^+–Ca^{++} exchange: net and unidirectional Ca^{++} fluxes as functions of ion composition and membrane potential. *J. Membrane Biol.* **86**, 167–187.

Kass, R. & Sanguinetti, M. C. 1984 Inactivation of calcium channel current in the calf cardiac Purkinje fiber. Evidence for voltage-and calcium-mediated mechanisms. *J. gen. Physiol.* **84**, 705–726.

Kauffmann, R., Bayer, R., Fürniss, T., Krause, H. & Tritthart, H. 1974 Calcium movement controlling cardiac contractility. II. Analogue computation of cardioac excitation-contraction coupling on the basis of calcium kinetics in a multicellular compartment model. *J. molec. cell. Cardiol.* **6**, 543–559.

Kimura, J., Noma, A. & Irisawa, H. 1987 Na–Ca exchange current in mammalian heart cells. *Nature, Lond.* **319**, 596–597.

Klee, C. B. & Banaman, T. C. 1982 Calmodulin. *Adv. Protein Chem.* **35**, 213–321.

Koch-Weser, J. & Blinks, J. R. 1962 Analysis of the relation of the positive inotropic effect of cardiac glycosides to the frequency of contraction of heart muscle. *J. Pharmacol. exp. Ther.* **136**, 305–317.

Kodama, T. 1985 Thermodynamic analysis of muscle ATPase mechanisms. *Physiol. Rev.* **65**, 467–551.

Kruta, V. & Stejskalová, J. 1960 Allure de la contractilité et fréquence optimale du myocarde auriculaire chez quelques mammifères. *Arch. int. Physiol.* **68**, 152–164.

Langer, G. A. 1974 Ionic movements and the control of contraction. In *The mammalian myocardium* (ed. G. A. Langer & A. J. Brady), pp. 193–218. New York: John Wiley and Sons.

Langer, G. A. & Rich, T. L. 1987 Functional demonstration of two calcium pools contributing to contraction in cardiac myocytes. (In preparation.)

LePeuch, C. J., LePeuch, D. A. M. & Demaille, J. G. 1980 Phospholamban, activator of the cardiac sarcoplasmic reticulum calcium pump. Physicochemical properties and diagonal purification. *Biochemistry, Wash.* **19**, 3368–3373.

Lee, K. S., Marban, E. & Tsien, R. W. 1985 Inactivation of calcium channels in mammalian heart cells: Joint dependence on membrane potential and intracellular calcium. *J. Physiol., Lond.* **364**, 395–411.

Lewartowski, B., Pytkowski, B., Prokopczuk, A., Wasilewska-Dziubinska, E. & Otwinowski, W. 1982 Amount and turnover of calcium entering the cells of ventricular myocardium of guinea pig heart in a single excitation. In *Advances in myocardiology*, vol. 3 (ed. E. Charov, V. Smirnov & N. S. Dhalla), pp. 345–357. New York: Plenum Medical Book Company.

Manring, A. & Hollander, P. B. 1971 The interval-strength relationship in mammalian atrium: A calcium exchange model. *Biophys. J.* **11**, 483–501.

Marban, E. & Weir, W. G. 1985 Ryanodine as a tool to determine the contributions of calcium and calcium release to the calcium transient and contraction of cardiac Purkinje fibers. *Circulation Res.* **56**, 133–138.

McNutt, A. & Fawcett, D. W. 1969 The ultrastructure of the cat myocardium. II. Atrial muscle. *J. Cell Biol.* **42**, 46–67.

Ca^{2+} *movements and cardiac* E–C *coupling* 205

Mechmann, S. & Pott, L. 1986 Identification of Na–Ca exchange current in single cardiac myocytes. *Nature, Lond.* **319**, 597–599.

Mensing, H. J. & Hilgemann, D. W. 1981 Inotropic effects of activation and pharmacological mechanisms in cardiac muscle. *Trends pharmacol. Sci.* **2**, 303–307.

Mentrard, D., Vassort, G. & Fischmeister, R. 1984 Calcium-mediated inactivation of the calcium conductance in Cesium-loaded frog heart cells. *J. gen. Physiol.* **83**, 105–131.

Mitchell, M. R., Powell, T., Terrar, D. A. & Twist, V. W. 1984 Ryanodine prolongs Ca-currents while suppressing contraction in rat ventricular muscle cells. *Br. J. Pharmac.* **83**, 13–15.

Morad, M. & Goldman, Y. 1973 Excitation–contraction coupling in heart muscle: Membrane control of development of tension. *Proc. Biophys. molec. Biol.* **27**, 257–313.

Noble, D. & Noble, S. 1984 A model of sino-atrial node electrical activity based on a modification of the DiFrancesco–Noble (1984) equations. *Proc. R. Soc. Lond.* B **222**, 295–304.

Opiz, H. 1973 Die Dynamik der Frequenzinotropie am Säugerherzen. *Nova Acta Leopoldina* **38**, 91–107.

Robertson, S. P., Johnson, J. D., Holroyde, M. J., Kranias, E., Potter, J. D. & Solaro, R. J. 1982 The effect of TnI phosphorylation on static and kinetic calcium binding by cardiac TnC. *J. biol. Chem.* **257**, 260–263.

Schouten, V. J. A. 1985 Excitation–contraction coupling in heart muscle. Dissertation, University of Utrecht.

Schouten, V. J. A. & Ter Keur, H. 1985 The slow repolarization phase of the action potential in rat myocardium. *J. Physiol., Lond.* **360**, 13–25.

Sheu, S. S. & Fozzard, H. A. 1982 Transmembrane Na$^+$ and Ca^{2+} electrochemical gradients in cardiac muscle and their relationship to force development. *J. gen. Physiol.* **80**, 325–351.

Simon, S. M. & Llinas, R. R. 1986 Compartmentalization of the submembrane calcium activity during calcium influx and its significance in transmitter release. *Biophys. J.* **48**, 485–498.

Sommer, J. R. & Johnson, E. A. 1979 Ultrastructure of cardiac muscle. In *Handbook of physiology*, Section 2, *The cardiovascular system*, vol. 1, *The heart* (ed. R. M. Berne, N. Sperelakis & S. Geiger), pp. 113–186. Bethesda, Maryland: American Physiological Society.

Tanaka, I., Tosaka, T., Saito, K., Shin-mura, H. & Saito, T. 1967 Changes in the configuration of the rabbit atrial action potential after various periods of rest. *Jap. J. Physiol.* **17**, 487–504.

Vaughan-Jones, R. D., Eisner, D. A. & Lederer, W. J. 1985 The effects of intracellular Na on contraction and intracellular pH in mammalian cardiac muscle. *Adv. Myocardiol.* **5**, 313–330.

Vergara, J., Tsien, R. Y. & Delay, M. 1985 Inositol 1,4,5-triphosphate: A possible chemical link in excitation–contraction coupling in muscle. *Proc. natn. Acad. Sci. U.S.A.* **82**, 6352–6356.

Vierling, W. & Reiter, M. 1975 Frequency–force relationship in guinea pig ventricular myocardium as influence by magnesium. *Nauyn-Schmiedeberg's Arch. Pharmacol.* **289**, 111–125.

Weidmann, S. 1951 Effect of current flow on the membrane potential of cardiac muscle. *J. Physiol., Lond.* **115**, 227–236.

Wohlfart, B. 1979 Relationships between peak force, action potential duration and stimulus interval in rabbit myocardium. *Acta physiol. scand.* **106**, 395–409.

Wohlfart, B. 1982 Analysis of mechanical altenans in rabbit papillary muscle. *Acta physiol. scand.* **115**, 405–414.

Wood, E. H., Heppner, R. L. & Weidman, S. 1969 Inotropic effects of electric currents. *Circulation Res.* **24**, 409–445.

Wussling, M. & Szymanski, G. 1973 Ein zwei-Ca-Speicher-Model zur qualitativen Beschreibung von Potentiationserscheinungen am Kaninchenpapillarmuskel. *Nova Acta Leopoldina* **38**, 141–173.

Yue, D. T., Marban, E. & Wier, W. G. 1986 Relationship between force and intracellular [Ca^{2+}] in tetanized heart muscle. *J. gen. Physiol.* **87**, 223–242.

J. Physiol. (1984), **353**, *pp. 1–50*
With 16 text-figures
Printed in Great Britain

1

REVIEW LECTURE*

THE SURPRISING HEART: A REVIEW OF RECENT PROGRESS IN CARDIAC ELECTROPHYSIOLOGY

By DENIS NOBLE

From the University Laboratory of Physiology, Parks Road, Oxford, OX1 3PT

(*Received 25 October 1983*)

INTRODUCTION

My aim in this lecture will be to review recent progress in our understanding of cardiac pace-maker mechanisms. In doing so I shall also try to explain the major changes that have occurred recently in the interpretation of two of the important ionic current mechanisms involved: i_{K2} (the Purkinje fibre 'pace-maker' current) and i_{si} (the 'second inward' or 'slow inward' current which is known to be important in rhythmic activity in the sino-atrial (s.a.) node).

The earliest ionic current models for pace-maker activity in the heart were based on those for repetitive activity in nerve. My own first involvement in modelling the cardiac action potential and pace-maker activity (Noble, 1960, 1962a) was greatly influenced by the fact that Huxley (1959) had just shown that the Hodgkin–Huxley (1952) equations could reproduce the repetitive firing observed in nerve axons in Ca-deficient solutions. The mechanism involved in the 1962 model was one of those proposed by Weidmann (1951) on the basis of his resistance measurements during the cardiac pace-maker potential: that a delayed increase in K conductance, g_K, that may occur during the action potential slowly decays after repolarization so allowing any background permeability, e.g. to Na ions, to drive the membrane potential away from the K equilibrium potential, E_K. I will call this hypothesis the g_K decay hypothesis.

It is worth noting that in 1960 the evidence for such a delayed K conductance change in the heart was not too strong. Hall, Hutter & Noble (1963) described a delayed K conductance change (which they called g_{K2}) superimposed on a background K current showing inward rectification (which they called g_{K1}). But the date of that paper (coming *after* the 1962 model paper) is significant: we were partly convinced that it was worth looking for the delayed K current by the fact that the modelling work seemed to require it. I spent some considerable time convincing myself that a Na conductance plus a K ion inward rectifier was theoretically not sufficient. Experimentally though, the delayed K change was not always easy to find. We now know some of the reasons why: in many cases it is obscured by a large transient outward current, i_{to} (see below).

* Given at the Meeting of the Physiological Society held in Edinburgh on 23 September 1983 as the Society's Annual Review Lecture.

PHY 353

2   *D. NOBLE*

Moreover, in some Purkinje fibres (as in some ventricular preparations) the delayed K change is in fact small or even negligible. It was later found to be much more prominent in atrial and nodal tissues. This was a case therefore of theory leading experiment. We were also not to know at that time that the delayed current change *during* depolarization in Purkinje fibres is not generated by the same mechanism that generates the greater part of the current change during the pace-maker depolarization *following* repolarization, nor that Na-deficient solutions would eliminate the latter.

The changes that have occurred since that early work may conveniently be divided into two main phases. First, there was a period, following the introduction of the voltage clamp to the heart by Deck, Kern & Trautwein (1964), of increasing complexity within an existing framework of interpretation. During this period, more conductance mechanisms were found but the basis for analysis of rhythmic activity remained the g_K decay hypothesis in one form or another. During this period (roughly from 1964 to 1977) the second inward (Ca) current was recorded, the delayed current changes were separated into several components and a transient outward current was analysed.

The second period, roughly from 1978 to the present time, is rather more revolutionary. Currents (such as those generated by ion pumps or ion exchange mechanisms) that were neglected in earlier work have now assumed a considerably greater importance and some existing currents, i_{K2} and i_{si} in particular, have been substantially reinterpreted. Moreover, the introduction of single-cell- and patch-clamp methods has revolutionized the techniques employed and led to the discovery of even more components of ionic current, such as the non-specific current channel described by Colquhoun, Neher, Reuter & Stevens (1981) and the components of the second inward current that I shall describe later in this review.

The result is a bewildering array of ionic current mechanisms that naturally provokes the question whether one really does need that many (for an interesting theoretical essay in doing without most of them, see Johnson, Chapman & Kootsey; 1980). There is also a sense of confusion concerning the theory of the field for, if what was by all the standard criteria firmly established as a specific K ion current can turn out to be carried largely by Na ions then it becomes important both to tighten up the theory (by taking account of previously neglected processes) and, in the light of this, to carefully reassess our conclusions concerning other ionic identities and reversal potentials. Finally, there is an urgent need to reassess the pace-maker mechanisms in the heart not only because the g_K decay hypothesis no longer applies to normal Purkinje fibre rhythm but also because, for the other main pace-maker mechanism, the s.a. node, different groups have come up recently with at least four different answers to the question – what is the main 'pace-maker' current?

I am therefore very grateful to the Physiological Society for inviting me to give this review lecture at such an opportune time and I will try to rise to the challenge: to explain the recent changes to a wider audience and to try to answer some of the questions and deal with the important issues I have just raised.

If I start with a little history that is in part to provide a perspective from which to view the impact of the new work, but also because this is a somewhat personal review lecture: the period I cover is naturally co-extensive with my own involvement and I presume that to give such an account is one of the purposes of the Physiological Society review lectures.

THE SURPRISING HEART   3

The 1962 cardiac model differed from its neuronal parent in only two essential features.

(1) The leak current was separated into a background Na current, $i_{b,Na}$, and a K current mechanism, i_{K1}, showing inward-going rectification (see Hutter & Noble, 1960; Carmeliet, 1961; Hall, Hutter & Noble, 1963).

(2) The speed and magnitude of the delayed K current change (called g_{K2}) were greatly reduced compared to the relevant parameters for nerve fibres (see Hall *et al.* 1963).

Subsequent experimental work using the voltage-clamp technique led to three main developments that required additional modifications to the Hodgkin–Huxley type of model.

(i) Evidence for a role of Ca ions in generating inward current had already come from action potential experiments (Niedergerke, 1963; Orkand & Niedergerke, 1966) and flux measurements (Winegrad & Shanes 1962). In 1967, Reuter succeeded for the first time in recording a net inward current that flowed even in the absence of Na ions and which was dependent on extracellular Ca ions. The ionic current recorded in Reuter's experiment (see Fig. 1 *A*) achieved a peak inward level after about 100 ms and required several hundred milliseconds for its inactivation. This observation was important for two reasons. First, because the entry of Ca during the flow of this current could be responsible for triggering contraction and secondly, because a current with this time course would play an important role in maintaining the plateau of the action potential. The 'slow inward' or 'second inward' current has since been recorded by many workers in a wide variety of cardiac cells. However, the kinetics show a bewildering and important diversity, with inactivation time constants, for example, varying from 20 ms (see, e.g. Rougier, Vassort, Garnier, Gargouil & Coraboeuf, 1969) to more than 1 s. The importance of this diversity has, I think, been underestimated. It is of functional importance since the kinetics of the current change determine its role in the action potential mechanism. It is bewildering because, even in the same tissue (the Purkinje fibre), the variations are large. Siegelbaum & Tsien (1980) and Marban & Tsien (1982) recorded currents with much faster activation and inactivation kinetics in EGTA-injected or Cs-loaded preparations (see Fig. 1 *B* – I will comment further on the significance of this result later in this review), whereas at the other extreme Lederer & Eisner (1982) and Eisner, Lederer & Noble (1979) recorded a current that differs from those shown in Fig. 1 *A* and *B* in showing almost symmetric activation and inactivation time courses. This current (shown in Fig. 1 *C*) is best obtained over a fairly narrow range of voltages (in the region of −40 mV) and in its time course it strongly resembles another slow inward current: the transient inward current, i_{TI}, first described by Lederer & Tsien (1976) in preparations in which the Na–K exchange pump has been blocked by cardiac glycosides. An example of this kind of inward current is shown in Fig. 1 *D*. It should be noted that the voltage protocol for obtaining i_{TI} is rather different. It is usually recorded on hyperpolarizing back from a depolarizing pulse, but it can also be recorded during depolarization and I shall discuss the significance of this fact later. My reason for presenting this diverse array of 'second inward' current records together in a single Figure will also become evident later when I have reviewed more recent work;

4   *D. NOBLE*

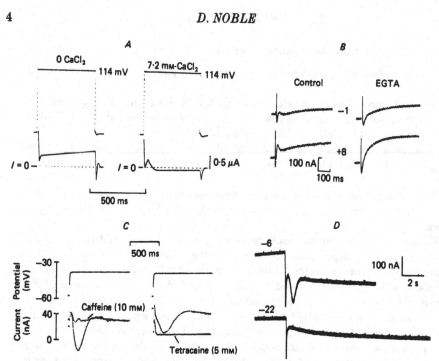

Fig. 1. Slow inward currents recorded in multicellular Purkinje fibre preparations. *A*, this shows the first slow inward current recorded by Reuter (1967). The preparation was depolarized from −80 to +34 mV in Ca-free, Na-free solution (left) and solution containing 7·2 mM-Ca ions (right). The peak inward current occurs at about 100 ms after the application of the pulse and is not fully inactivated even after 500 ms. *B*, this shows the influence of intracellular EGTA injection to remove currents (like the transient outward current) that are activated by the [Ca]$_i$ transient (Siegelbaum & Tsien, 1980). The membrane was depolarized from −46 to −1 and +8 mV. The control records on the left show the same kind of initial complex time course as in *A*. After EGTA injection, the current record becomes simpler and shows an inward current that activates and inactivates fairly rapidly. The record approaches that obtained in single cells (see Fig. 6). *C*, these records show slow inward currents with almost symmetric activation and inactivation time courses when a Purkinje fibre is clamped from −58 to −40 mV (Eisner *et al.* 1979; see also Lederer & Eisner, 1982). It is likely that this current is activated by intracellular Ca since it is abolished by caffeine (which is thought to discharge the intracellular Ca stores) and tetracaine (which abolishes the Ca release and tension response). *D*, the transient inward current, i_{TI}, recorded by Lederer & Tsien (1976) in pump-blocked Purkinje fibres. The membrane was clamped back to −72 mV following 5 s depolarizations to −6 and −22 mV in a fibre that had been exposed to 1 μM-strophanthidin. Note the similar time course of the inward currents shown in *C* and *D*, even though the voltage protocols are quite different. The thesis explored in this review is that the currents recorded in *C* and *D* may be produced by the same mechanism and that this is quite different from the mechanisms that produce the majority of the currents shown in *A* and *B*. On this view, depolarization activates two or three separate Ca or Ca-dependent inward current systems (see eqns. (2) and (3)).

(ii) In 1968 and 1969 Tsien and I (Noble & Tsien, 1968, 1969*a,b*) analysed the kinetics of the slow permeability changes attributed to K ions. The main outcome of that work was the discovery that, in contrast to the 1962 model, there are at least two distinct time-dependent mechanisms. The closest equivalent of g_{K_2}, the delayed

THE SURPRISING HEART 5

K current in the 1962 work, is a current (or currents) activated in the plateau range of potentials (-40 to $+20$ mV). The other mechanism has a gating range at much more negative potentials in the pace-maker range (between -100 and -60 mV) and can be distinguished by its kinetic properties even when both components are present (Hauswirth, Noble & Tsien, 1972). More recently, it has been found that the two

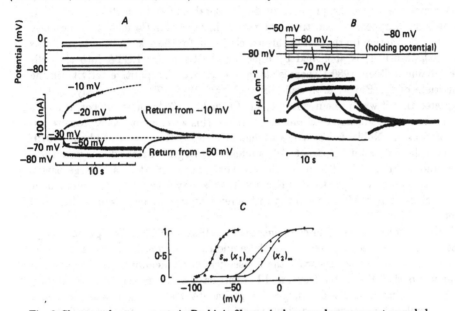

Fig. 2. Slow membrane currents in Purkinje fibres. *A*, slow membrane current recorded in the plateau range of membrane potentials by Noble & Tsien (1969). These currents have been, and still are, attributed to slow activation and deactivation of a K current. The fibre chosen for analysis here showed little or no slow current change in the pace-maker range. More usually, the plateau current needs to be separated from the pace-maker range current (see Hauswirth *et al.* 1972). *B*, slow membrane current recorded in the pace-maker range of potentials (Noble & Tsien, 1968). This was originally also attributed to K ions but is now thought to be carried by non-specific channels activated by hyperpolarization into the pace-maker range (see Fig. 3). *C*, voltage ranges of activation of these two currents as analysed by Noble & Tsien (1968, 1969). The plateau current divided into two components (labelled x_1 and x_2). The current activated in the pace-maker range (labelled *s*) was analysed as *deactivating* as the membrane is hyperpolarized between -60 and -90 mV. The analysis proposed by DiFrancesco (1981) is that the channels *activate* over this voltage range but, since they are presumed to carry inward rather than outward current, the net current *change* is still as shown in the records in *B*.

systems respond quite differently to divalent cations: the plateau current (which I will call i_K, but which was originally called i_x (Noble & Tsien, 1969a) or g_{K2} (Hall *et al.* 1963; McAllister & Noble, 1966) in previous work) is blocked by Ba ions, as is the K background current, i_{K1} (DiFrancesco, 1981a), whereas the 'pace-maker' current (originally called i_{K2} by Noble & Tsien (1968) is blocked by Cs ions (Isenberg, 1976; Carmeliet, 1980; DiFrancesco, 1982). Fig. 2 summarizes the kinetic properties of these currents.

(iii) The third additional current mechanism is the transient outward current, i_{to}.

6 D. *NOBLE*

This current greatly resembles the current i_A in molluscan neurones. In the case of the heart, it was first found in Purkinje fibres and is responsible, in part, for the characteristic notch in the Purkinje fibre action potential. Like the current, i_A, a large fraction of i_{to} is blocked by 4-aminopyridine, 4-AP (Kenyon & Gibbons, 1979; Boyett, 1981). There is, however, a small component (which in the calf is the only component) that remains even in the presence of 4-AP (Coraboeuf & Carmeliet, 1982). This component disappears if the transient rise in intracellular Ca ions that normally initiates contraction (in the rest of this article I will, for the sake of brevity, call this the $[Ca]_i$ transient) is abolished by injecting the Ca buffer EGTA into the cells (Siegelbaum & Tsien, 1980.) There is also a strong effect of repetitive activity on the amplitude of i_{to} (Boyett & Jewell, 1980; Boyett 1981). The early analysis of i_{to} suggested that it was a Cl current (see, e.g. Carmeliet, 1961; Dudel, Peper, Rudel & Trautwein, 1967; Fozzard & Hiraoka, 1973) and this was how it was represented in the McAllister, Noble & Tsien (1975) model (MNT). Kenyon & Gibbons (1979) however showed that its apparent Cl dependence is to a great extent due to variations in extracellular free Ca. The best evidence now suggests that it is in large part a $[Ca]_i$-activated and $[K]_o$-activated K current. This may account for the component that is blocked by 4-AP. The identity of the remaining component will be discussed later.

Thus, in addition to the four components of the 1962 model (i_{Na}, i_K, i_{K1}, and $i_{b,Na}$), these results revealed the presence of three additional currents (i_{si}, i_{K2} and i_{to}). It was for the purpose of exploring the consequences of including these additional components that McAllister, Tsien and I constructed a new model (McAllister *et al.* 1975). Being based on a wider range of experimental results, the new model had a greater explanatory range. In particular it could in principle account for the fact that Purkinje fibre rhythmic activity can occur in two different potential ranges: the normal Purkinje pace-maker range (-90 to -60 mV) and a depolarized range (-60 to -40 mV). The oscillations in the latter range are fairly similar to the natural oscillations of the s.a. node (Hauswirth, Noble & Tsien, 1969) and an early 'model' of the s.a. node rhythm was based on the view that i_{si} and i_K were the main currents involved. As we shall see later, this is close to one of the present views of s.a. node pace-maker activity: indeed I shall show that this is the only form of pace-maker activity that still conforms at all closely to the original g_K decay hypothesis.

In the MNT (McAllister *et al.* 1975) model, both kinds of pace-maker depolarization are attributable to (different) g_K decay processes. Before I start to review the more recent developments it is worth summarizing the weight of evidence that had accumulated in favour of the g_K decay hypothesis by 1975.

(a) In 1951 Weidmann performed a series of experiments that have been of seminal importance in cardiac electrophysiology. He injected small pulses of current at various times during the cardiac cycle in an attempt to measure the variations in total membrane conductance. This work confirmed the presence of all-or-nothing repolarization during the plateau (which was one of the reasons I was certain in 1960 (see also Noble, 1962*b*; Noble & Hall, 1963) that there had to be a slow time-dependent current: i_{Na}, i_{K1} and $i_{b,Na}$ were not sufficient by themselves to reproduce Weidmann's result). It also showed a slow decline in membrane slope conductance during the pace-maker depolarization. Important though these experiments were (all models

THE SURPRISING HEART 7

must take them into account and reproduce them) their interpretation has proved quite difficult. The general reasons were reviewed by Noble & Tsien (1972) who gave the equation for the slope conductance as a function of individual ionic currents and their voltage dependence. This equation shows that there is no reason, *a priori*, for assuming that the underlying membrane conductance change responsible for a particular potential change should vary in the same direction as the total slope conductance. Weidmann (1956) clearly realized the importance of this argument in relation to the plateau slope conductance measurements. We shall see later that this general problem has assumed great importance in interpreting the conductance measurements during pace-maker potentials and voltage clamps in the pace-maker range of potentials.

(b) In relation to the pace-maker slope conductance measurements, when Hutter and I (1960) first described our work on inward-going rectification we drew attention to the possibility that a decrease in slope conductance during the pace-maker depolarization could be merely a secondary consequence of the depolarization (via its effect on i_{K1}) rather than a reflexion of the primary cause of the depolarization, which Weidmann's pulse experiments showed must involve a voltage- and time-dependent change. This kind of problem was met by Vassalle's (1966) work in which he measured the change in slope conductance under voltage-clamp conditions. The change he observed was small (about 20 %) compared to Weidmann's measurements, which confirms that a large part of the slope conductance change during the pace-maker potential is voltage-dependent rather than time-dependent, but he still found a decrease in conductance with time under voltage-clamp conditions.

(c) The obvious and then standard interpretation of this conductance measurement was reinforced by the fact that all the measurements of the reversal potential E_{rev} for i_{K2} showed that it varies with $[K]_0$ in the manner expected for a very *specific* K current, i.e. there is a 60 mV change in E_{rev} for a 10-fold change in $[K]_0$ (Noble & Tsien, 1968; Peper & Trautwein, 1969; Cohen, Daut & Noble, 1976). It requires only a fairly small leak in a Goldman-type equation for this slope to fall to, say, 50 or 40 mV.

(d) Also consistent with the view that a pure K current was involved is the fact that the instantaneous $i(V)$ relations for i_{K2} (dissected using the ratio method that Tsien and I described in 1968) display both inward-going rectification and the 'cross-over' phenomenon (in which the net current at different values of $[K]_0$ changes in different directions depending on the voltage range and K concentrations) that are so typical of some K conductance mechanisms (e.g. i_{K1} in the heart, the inward rectifiers in skeletal muscle and in starfish egg cells).

In fact, I think it is fair to say that the weight of evidence for the g_K decay hypothesis in Purkinje fibres was so strong that i_{K2} was regarded as perhaps the most fully analysed current in the heart and served as a model for other systems. One of the surprises to which the title of this review refers is that this interpretation of the pace-maker current in the Purkinje fibre should not only turn out to be deeply incorrect but that a very different interpretation should be able to account for these results both accurately and with no need to suppose that any of the original experimental results is incorrect. Indeed, as I shall show, some important and awkward details of the experimental results then receive very natural explanations.

8   *D. NOBLE*

The second element of surprise implied by my title concerns the analysis of the slow inward current i_{si}. The standard analysis of this current in multicellular preparations is that of Reuter & Scholz (1977). Though this work appeared too late to be incorporated into the MNT model, it nevertheless is in accord with that model in giving i_{si} inactivation kinetics a crucial role in determining the time course of the plateau phase of the action potential. Another important feature of Reuter & Scholz's (1977) analysis is the measurement of the relative contributions of Ca, Na and K ions in carrying the current. They estimated the relative permeability coefficients to be 1:0·01:0·01 (Ca:Na:K) which, given the high concentration of Na and K compared to Ca ions means that as much as 30–40% of the current carried might be carried by ions other than Ca. There are both theoretical and experimental reasons for reassessing these conclusions.

The third element of surprise is one that contributes strongly to the other two: this is that changes in ion concentrations, intracellular and extracellular, have turned out to be much more important, and more rapid, than previous work allowed. A model, like the MNT model, that assumes fixed concentrations is now therefore severely limited, as will become apparent as the story unfolds.

THE SECOND STAGE: THE RECENT REVOLUTION

I call this stage a revolution because it does indeed turn things inside out and upside down, but also because the changes have occurred with great speed. At the end of 1979 it was still reasonable to use models of the MNT kind. By mid-1980 it was already clear that on at least three counts, the model's usefulness was in question. Moreover, the new results with single cells and patch-clamp methods have appeared with great rapidity over the last 3 years. This will be the longest section of my review so, for convenience, I shall divide it into five parts.

1. *The 'pace-maker' current, i_{K2}, turns upside down*

Increasing difficulties with the i_{K2} model

Although the analysis of this current as a pure K current seemed very secure, it is worth recalling that there were some awkward details evident in the results from the very beginning. Eric McAllister and I (McAllister & Noble, 1966) had found that the current disappears in the pace-maker range in Na-free solutions (this is also evident in Deck & Trautwein's (1964) analysis), which is fairly odd behaviour for a highly specific K mechanism. The best explanation we could propose was that the current was perhaps dependent on $[Ca]_i$ which is known to increase greatly in Na-free solutions. Then Dick Tsien and I (Noble & Tsien, 1968) found that the absolute values of the reversal potentials (E_{rev}) were always a few millivolts more negative than expected from the Nernst equation for K ions using a likely value for $[K]_i$. We noted the result and suggested that the problem might lie in trusting intracellular electrode measurements of absolute potential to within 5–10 mV. Peper & Trautwein (1969) also found this result (their Nernst plot requires $[K]_i$ to be 200 mM which is 50 mM greater than has been measured with K-sensitive electrodes (Miura, Hoffman & Rosen, 1977)).

Cohen, Daut and I took up the story again in 1976 (Cohen *et al.* 1976) and fully

THE SURPRISING HEART 9

confirmed that E_{rev} is always 5–15 mV negative to the calculated E_K. Instead of postulating a systematic error or a high $[K]_i$ we opted for the view that the extracellular cleft K concentration, $[K]_c$, is lower than the bulk extracellular concentration, $[K]_b$, by about 1 mм.

This hypothesis however creates theoretical difficulties, some of which are treated in a review by Attwell, Eisner & Cohen (1979). It is possible to construct a consistent model with a steady-state extracellular concentration difference (see Appendix to Cohen, Noble, Ohba & Ojeda, 1979) but only at the cost of making assumptions about non-uniform distribution of channels and ion pump sites that would be very difficult to test experimentally.

But the necessity for such hypotheses was soon removed and, paradoxically, the work that achieved that result was initially aimed at exploring the contributions of cleft concentration changes in more detail *within* the framework of the i_{K2} hypothesis.

It would require a full length review of its own to explore fully the recent work on extracellular ion accumulation and depletion in the heart. For the present purpose it will suffice to take a short cut with two propositions that I think are now well established.

(1) Given the size of the extracellular space and the magnitudes of the currents flowing in Purkinje fibres it is impossible to avoid the conclusion that, at the i_{K2} reversal potential (about -100 mV), about 0·5–1 mм of extracellular K depletion must occur during prolonged voltage-clamp pulses.

(2) The instantaneous rectifier, i_{K1}, is extremely sensitive to $[K]_0$ at negative potentials. Thus, at the least, there must be considerable contamination of i_{K2} by changes in i_{K1} due to K depletion.

This was the stage of the story investigated by DiFrancesco, Ohba & Ojeda (1979) who showed that the experimental results on E_{rev} for i_{K2} conform fully to the predictions of these propositions. But that was the opening of the way to the more fundamental move, which was made by DiFrancesco when he returned to work in Italy: if K-dependent changes in i_{K1} can distort i_{K2} to that extent, how do we know anything at all about i_{K2} near its supposed reversal potential? Perhaps it does not reverse at all in that range! Maybe it is not a specific K ion current after all.

DiFrancesco's reinterpretation

And so, i_{K2} was turned upside down. From being a specific K ion channel generating outward current in the pace-maker range it became a non-specific channel generating *inward* current in this range largely carried by Na ions. The hypothesis in its quantitative form (see DiFranceso & Noble, 1982, 1984) demonstrates that what was measured as i_{K2} in fact consists of two components:

$$i_{K2}(t) = \Delta i_{K1}([K]_c) - i_f(t), \tag{1}$$

i.e. a current component due to the inward rectifier, i_{K1}, that varies with time (t) because the cleft K ion concentration varies with time, and a genuine time-dependent gated current, i_f. It so happens that the time constants of K depletion and the gating of i_f are of the same order of magnitude so that the net current records are often, though not always, monotonic and so give the appearance of a single component. Moreover, the voltage range at which Δi_{K1} becomes significant and can mask i_f to

10   *D. NOBLE*

produce a 'reversal potential' varies almost exactly in parallel with E_K since the voltage range of i_{K1} rectification varies in parallel with E_K.

Fig. 3 shows one of DiFrancesco's crucial experiments. First, he obtained i_{K2}-like behaviour (see records on the left). Then he used Ba ions to block i_{K1} (and so massively reduce the problems of K ion depletion). He then obtained the records

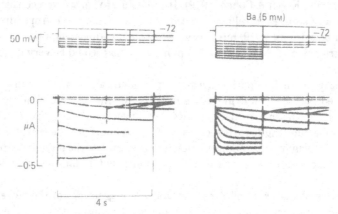

Fig. 3. DiFrancesco's (1981) experiment showing that the 'reversal potential' for the current activated in the pace-maker range of potentials is abolished when i_{K1} is blocked by Ba ions. The records on the left show what happens when the voltage protocol shown in Fig. 2*B* is extended to more negative potentials. The time-dependent current change becomes smaller and eventually reverses. In this case, the reversal potential is situated at -127 mV. This reversal potential is negative to the expected value of E_K but it changes accurately in parallel with E_K when $[K]_o$ is varied (for a review of these results and their explanation see DiFrancesco & Noble, 1982). The records on the right show the currents obtained when 5 mM-Ba is added to the solution. Even though the voltage protocol (see above) is extended to much more negative potentials (down to -165 mV in this case), no reversal potential is observed. DiFrancesco's conclusion was that the reversal depended on K-dependent changes in i_{K1} rather than in the slow time-dependent current (see eqn. (1)).

shown on the right: no reversal potential is then found near -100 mV. He telephoned me from Milano in January 1980 to tell me this result and the same night I was able to use a computer program he and I had developed together to show that his new interpretation of i_{K2} as a non-specific inward current i_f could give a full and accurate theoretical account of the i_{K2} results (we presented this theory at the March 1980 meeting of the Physiological Society (DiFrancesco & Noble, 1980) and later developed it into a full paper (DiFrancesco & Noble, 1982)). Meanwhile, DiFrancesco went on to determine the ionic species (almost equally Na and K) carrying the current i_f (DiFrancesco 1981*a*, *b*) and to analyse the kinetics, which turn out to be significantly different from those obtained when changes in i_{K1} due to K ion depletion are present (Hart, 1983; DiFrancesco, 1984).

DiFrancesco's reinterpretation has been amply confirmed by other experimental work, notably Callewaert, Carmeliet, Van der Heyden & Vereecke's (1982) demonstration that in isolated *single* Purkinje cells, with no restricted intercellular spaces, i_{K2} behaves like i_f even without using a blocker such as Ba ions to eliminate i_{K1}.

THE SURPRISING HEART 11

The background to the reinterpretation

This reinterpretation of i_{K_2} was, without doubt, a big surprise and even more so is the fact that it did not need to cast doubt on *any* of the previous experimental results: on the contrary, it has provided a better explanation for those results since the new theory accommodates the previously awkward results quite naturally. Nevertheless, some of the controversy surrounding the reinterpretation has implied that, at least originally in 1980, it was based on inadequate reasons or on the basis of a single result crucially dependent on a correct interpretation of the action of Ba ions.

I have never felt that, which was why I so readily accepted the full implications of Dario's telephone call from Milano. The reasons are very relevant to the theme of this review so I shall explain them in some detail.

First of all it is important to recall that the current i_f was not discovered first in Purkinje fibres (except under its disguise as i_{K_2}). It was first discovered, as a hyperpolarizing activated current, in the s.a. node and since a lot of the work involved was done in the Oxford laboratory it was natural that I and my colleagues should be greatly influenced by the close resemblance to i_{K_2} (this resemblance was in fact noted in the 2nd edition of *The Initiation of the Heartbeat* (Noble, 1979, p. 102). Seyama (1976) described inward-going rectification in the mammalian s.a. node and drew attention to the fact that a lot of it was time dependent and was reduced in Cl-free solutions. Brown, Giles & Noble (1977) then described a hyperpolarizing activated current in the frog sinus venosus, though they did not actually prove that this was the correct interpretation. The convincing proof that it was correct came with measurements of conductance changes in the mammalian node (DiFrancesco & Ojeda, 1980) showing that the membrane conductance *increases* with time on hyperpolarization. Noma, Yanagihara & Irisawa (1977) also found the current (which they called i_h) in rabbit s.a. node and Yanagihara & Irisawa (1980) showed that its activation range is very similar to that for the supposed deactivation of i_{K_2} in Purkinje fibres, i.e. from -60 to -90 mV. Now the relevant interesting feature of the s.a. node is that, by comparison with the Purkinje fibre, i_{K_1} is very weak, i.e. the s.a. node naturally provides precisely the conditions for which Ba was used in the Purkinje fibre. So i_f could be recorded as such (and not under the guise of i_{K_2}) even without using blockers. Moreover, like i_{K_2} or i_f in Purkinje fibres, i_f in the s.a. node is blocked by low concentrations (1 mM) of Cs ions and is almost eliminated in Na-free solutions (Kimura, 1982). Brown & DiFrancesco (1980) and DiFrancesco & Ojeda (1980), in their papers on the s.a. node, summarized these similarities very effectively even before the reinterpretation of i_{K_2}. It is a sign of how much has changed that it is now almost inconceivable that some should have reacted to those papers by finding the analogy far-fetched and implausible.

So, my laboratory had been living already for 3 years with the tensions generated by the uncanny resemblance of the nodal i_f with i_{K_2}, and with the increasing theoretical difficulties posed by the i_{K_2} model itself. In fact, before Dario and I used the computer program we had developed to test the i_f theory as an explanation for i_{K_2}, we had been using it for internal consistency checks on the i_{K_2} hypothesis. While he was carrying out the Ba experiment in Milano, I had already satisfied myself that

the i_{K_2} hypothesis is internally inconsistent: one cannot have *both* the Cohen *et al.*
(1976) explanation for the negative level of the reversal potential *and* the possibility
even of measuring a true reversal potential. (This point resembles and is greatly
influenced by an argument presented by David Attwell – see Attwell *et al.* 1979; it
differs from his argument largely in that Dario and I carried out full computations

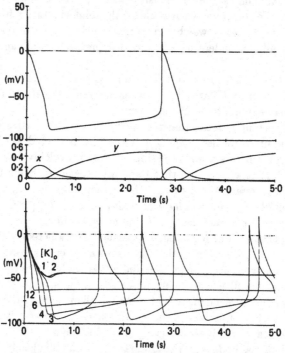

Fig. 4. Reconstruction of Purkinje fibre pace-maker activity using the DiFrancesco–Noble
(1984) equations which incorporate DiFrancesco's reinterpretation of the current in the
pace-maker range. Top: computed action potentials at 4 mM-external [K] together with
the computed variations in the gates controlling i_K (x) and i_f (y). The largest variation
during the pace-maker potential is clearly attributable to i_f. This diagram in fact
over-emphasizes the role of i_K since at the beginning of the pace-maker depolarization the
potential is close to E_K so that i_K is very small even when activated. By contrast, i_f is
large in this range of potentials. Near E_K little K flows through the i_f channels so that
the major current change is due to Na ion flow through these channels. Bottom: influence
of [K]$_o$ on computed pace-maker activity in Purkinje fibres. The figures on each record
show the value of [K]$_o$. The model accurately reproduces the K dependence of Purkinje
fibre pace-maker activity (Vassalle, 1966) including the acceleration and eventual
depolarization in low [K]$_o$ and the suppression of pace-maker activity above 6 mM. This
high K sensitivity is attributable to the actions of [K]$_o$ on i_{K1} (cf. Noble, 1965). Thus,
while i_f is the main current that *initiates* the pace-maker depolarization, other currents
can also play a major role in determining the frequency.

of the influence of a restricted extracellular space with parameters appropriate to a
Purkinje fibre, whereas Attwell's argument was presented in general terms of the
relevant differential equations.) I had in fact reached almost the same conclusion in
some rough calculations discussed with Dick Tsien in 1977 showing that there were

THE SURPRISING HEART 13

not enough K ions in the extracellular space to carry all the current that flows negative to the supposed reversal potential without postulating massive replacement by diffusion.

DiFrancesco and I never published the theoretical work on the internal inconsistency of the i_{K_2} hypothesis because we became too fully occupied with his new hypothesis. But in some ways that is a pity: I think some people would have been less surprised by the revolution and its apparent suddenness if they had fully appreciated the weight of this theoretical evidence and of the sinus node results. Seen in this light, Dario's 1980 experiment came not as an awkward fact destroying a pet theory but rather as a great source of relief: an internally consistent theory could be reconstructed and the s.a. node – Purkinje fibre comparison fell fully into place.

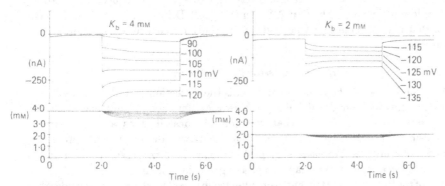

Fig. 5. Reconstruction of the 'reversal potential' for i_{K_2} using the DiFrancesco–Noble (1984) model. Results for two values of $[K]_b$ are shown (2 and 4 mm). The model membrane was clamped to the potentials shown from a holding potential of -50 mV. Note reversal of time-dependent current change at -110 when $[K]_b = 4$ mm and -125 when $[K]_b = 2$ mm. The lower traces show the computed mean $[K]_b$ in the extracellular space. Note that only very moderate quantities of K ion depletion are required to produce the result. The change in [K] at the reversal potential is only about 10% of the bulk extracellular concentration.

Role of i_f in Purkinje fibre pace-maker activity

This is not the place to review fully all the experimental work on i_f. For that the reader is referred to DiFrancesco's recent papers (DiFrancesco 1981 a, b, 1982, 1984). I shall deal rather with the implications for the reinterpretation of pace-maker mechanisms. In the case of the Purkinje fibre these are profound. Fig. 4 shows a reconstruction of Purkinje fibre pace-maker activity using the DiFrancesco–Noble (1984) equations. This Figure is based on Figs. 6 and 8 of DiFrancesco & Noble (1984) showing both the large variation of i_f (carried largely by Na ions in this range of potentials) and the large influence of external K concentration on Purkinje fibre pace-maker activity (see Vassalle, 1965; Noble, 1965).

Normal Purkinje fibre pace-maker activity is therefore attributable to the slow onset of a largely Na current during the pace-maker depolarization. Fig. 5 shows computations of ionic current during voltage-clamp protocols of the kind used to analyse i_{K_2}. It can be seen that the 'reversal' potential and its K dependence are nicely

14   *D. NOBLE*

reproduced by the new model. For further details on the theoretical comparisons between the i_t and i_{K2} hypotheses the reader is referred to DiFrancesco & Noble (1982). It is however worth noting here that two of the important experimental results that are naturally explained by the new model are Weidmann's (1951) and Vassalle's (1966) conductance measurements. Thus we arrive at the conclusion that the underlying conductance change here *is* in the opposite direction to the over-all measured slope conductance change.

i_t in other parts of the heart

In addition to Purkinje fibres and the s.a. node, i_t has also been found in the mammalian atrium (Earm, Shimoni & Spindler, 1983) and in frog atrium (Bonvallet, personal communication). A current behaving like i_{K2} has also been analysed in embryonic ventricular cells by Clay & Shrier (1981 *a,b*). DiFrancesco and I (1984) have shown that it is possible that this is also i_t partially masked by changes in i_{K1} due to depletion.

2. How many 'second inward' currents?

The second pillar of the recent revolution is the dissection of and finer resolution of the kinetics and amplitude of i_{si}. There are some rough analogies with the reinterpretation of i_{K2}. First, what was initially described as a single current mechanism turns out to be a composite of at least two, and even three, processes. Secondly, one of these processes very probably may involve substantial variations in ion concentrations rather than in voltage-gated ion channels. And thirdly, the kinetics of the largest current mechanism involved in recent work are very considerably different from those of the largest component usually detected in previous work. But there are also some important differences: the main current component involved is still thought to be carried by Ca moving inwards across the cell membrane. The extreme view that most or all of i_{si} is due to processes like the Na–Ca exchanger or Ca-activated channels is clearly incorrect: though I shall show that such processes may contribute significantly, the major part of i_{si} continues in one form or another to consist of Ca inflow through membrane channels. I have already shown (see Fig. 1) the great diversity of forms that i_{si} has taken in experimental work even before the introduction of single cell work. The inactivation time course has been variously estimated as having a time constant as brief as 20 ms (Rougier *et al.* 1969) and as long as 500 ms (Beeler & Reuter, 1970). Sometimes inactivation appears to be much slower than activation: sometimes (as in Fig. 1 *C*) the current appears to take as long to activate as to inactivate. Could all these diverse wave forms really be manifestations of the same mechanism?

My approach here will be, first, to review the recent evidence on i_{si} obtained in work on the multicellular s.a. node preparation and on single cells from various sources. I will then reinterpret previous work in the light of these more recent experiments. The approach will be rather more piecemeal than for the i_t/i_{K2} story since, as yet, there is no generally accepted theoretical structure within which *all* the new results can be accommodated. The DiFrancesco–Noble (1984) equations, however, do provide a partial reinterpretation and I shall use these where I think they help to clarify the situation. I should though emphasize the more speculative nature of some of what I am going to say in this section of the review.

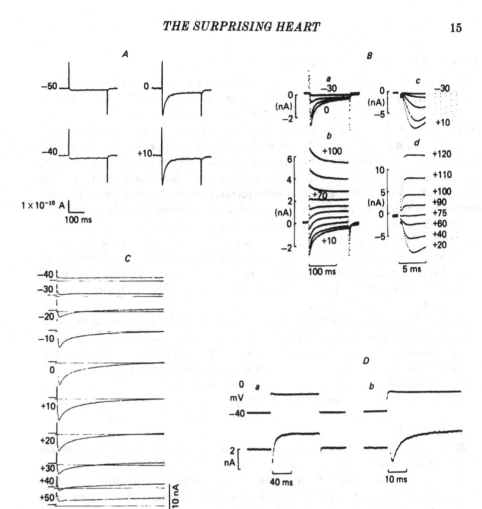

Fig. 6. 'Fast' second inward currents recorded in isolated single cardiac cells. *A*, atrial cells from the bull-frog (Hume & Giles, 1983). The membrane was clamped from -80 mV to the potentials shown. The first inward current is a *maintained* inward current at -50 and -40 mV. This current is resistant to the fast Ca channel blocker Cd and will be discussed, together with a similar current found in guinea-pig ventricular cells, in a later section of this review. Stronger depolarizations (examples here at 0 and $+10$) show a fast current that reaches a peak in less than 5 ms and a time constant of inactivation around 20 ms. The maintained current is also still present. *B*, perfused guinea-pig ventricular cells (Lee & Tsien, 1982). The records in *a* show currents recorded up to 0 mV and those in *b* show the currents at positive potentials. Note that a reversal potential occurs at $+70$ mV beyond which outward (probably K) current flows through the channel. The expanded records in *c* and *d* show that the current reaches a peak at about 2–3 ms. The time constant of inactivation is about 20 ms. *C*, bovine ventricular cells (Isenberg & Klockner, 1982). The potential was stepped from -50 mV to the potentials shown. Peak activation time is around 2 ms and inactivation time constant near 20 ms. *D*, rat ventricular cells (Mitchell *et al.* 1983). The potential was stepped to 0 from -40 mV. The inward current peaks at 2–3 ms and inactivates with a time constant near 10 ms. Thus, in a wide range of cell types, the kinetics of the fast Ca current (called $i_{Ca,f}$ in this review) are very similar. This suggests that the enormous diversity of time courses found in previous multicellular work is *not* attributable to large natural variations in the speed of the same mechanism.

16   *D. NOBLE*

The main component of i_{si} is very fast

It is convenient to start with the single-cell work. In any case, this is likely to be more reliable since voltage-clamp control is more rapid and probably more uniform in single cells than it is in multicellular preparations, even if very small as in the case of the s.a. node. The striking and uncontrovertible feature of the single-cell results is that the main part of what is *now* described as i_{si} is far from being a *slow* inward current. It is *much* faster and considerably larger than previous work suggested (see, e.g. discussion in Mitchell, Powell, Terrar & Twist, 1983, for details) and these facts have considerable importance for the reinterpretation of the role of the second inward current in normal electrical activity. Fig. 6 shows a selection of results taken from various laboratories using the single-cell technique to study i_{si}. It is clear that peak activation of the current occurs within 2–5 ms rather than 20–100 ms and that the inactivation time constant is in the range 10–20 ms rather than hundreds of milliseconds or even seconds. Thus these figures give kinetics at least an order of magnitude faster than most of the multicellular work suggested. This immediately raises several important questions.

(1) Were the much slower currents recorded in multicellular preparations distorted versions of the current recorded in single cells, or is it rather that the single-cell work has fully revealed a fast component that was greatly masked in the multicellular work, and that there are also slower components which formed a proportionately greater part of the recorded current in multicellular preparations?

(2) If there are several components, are they all gated channels or could some of the current be attributed to carrier processes and their variation with ion concentrations?

(3) What are the implications for the role of i_{si}, or its components, in the generation of the plateau, in repolarization and in pace-maker activity?

Before we proceed it is important to establish some consistent terminology. I shall now use the symbol i_{si} to refer to the *total* slow inward current. This convention is consistent with the fact that most previous work uses i_{si}, though usually without recognizing that it may not be a single component (though it is worth noting that the MNT model used two components, one of which was a maintained, non-inactivated, current). Moreover, even in the single-cell work we still need a term for the over-all current since it is not always possible to dissect the components. For the *fast* component of current identified in the single-cell work I shall use the symbol $i_{ca,f}$. This terminology has not been used before but as my review progresses it will become obvious why I need a separate term. We shall also require new terminology for any slower components. I shall introduce this later.

First, then, we deal with the question whether slower currents recorded in multicellular work could be artifactually slow recordings of the fast current, $i_{Ca,f}$. A possible mechanism for this would be slowing of the current record due to the presence of a series resistance between the cell membranes and the current collecting electrode (Isenberg & Klöckner, 1982; Noble & Powell, 1983). This approach looks plausible initially since Isenberg & Klöckner (1982) tried the effect of inserting a series resistance of up to 10 MΩ on their single-cell current recordings. This does indeed increase the time to peak but not to much more than 10 ms. Powell and I (Noble & Powell, 1983) used the DiFrancesco–Noble (1984) equations, which use kinetics for i_{si} based on the fast single-cell kinetics. We found that increasing the series resistance to 10 MΩ could delay the peak to about 6 ms but the amplitude would then become very small (only about 20 % of the true value). A peak delayed by much more than this would become negligible. Now it is indeed the case that the peak amplitude of i_{si} is much larger in single cells than it is in multicellular preparations (when due allowance has been made for the different membrane areas) (see Isenberg & Klöckner, 1982; Mitchell *et al.* 1983). It is likely therefore that i_{si} (and in particular $i_{Ca,t}$) has been seriously underestimated in the older work.

Another factor that would lead to an underestimate of the amplitude of i_{si} is masking by the transient outward current. This would be particularly serious in Purkinje fibres where Siegelbaum & Tsien (1980) were able to show that much faster and larger inward currents could be recorded when i_{to} had been removed by injecting the preparation with EGTA to buffer the Ca ion concentration and so eliminate the $[Ca]_i$ transient. This important experiment (Fig. 1*B*) shows that a large and fast part of i_{si} is indeed masked by a Ca-activated outward transient. It may also be the case that the Ca buffering may have eliminated slower components of i_{si}.

It is probable, I think, that both series resistance 'blunting' and outward current 'masking' are at work. But it is also, I think, clear that this cannot be the full explanation of relatively slow and small inward currents in multicellular preparations. There are two reasons for this view. First, the EGTA injection experiments of Siegelbaum & Tsien show that, even if inward currents quite as fast as in single cells cannot be obtained in a multicellular preparation, nevertheless, once i_{to} is removed, the inward current is fairly fast, and reaches a peak in about 5 ms. Moreover, the inactivation time constant is also fairly short. In other words, with the $[Ca]_i$ transient removed by EGTA, the inward current record *approaches* that obtained in the single-cell work. There is no reason to suppose that EGTA injection should alter the extracellular series resistance. Series resistance problems cannot therefore be responsible for much slower currents with times to peak as long as 50 ms. The more reasonable explanation is that such currents also exist, are small compared to $i_{Ca,t}$ and are obtained most readily when most of $i_{Ca,t}$ is masked by the transient outward current. The second reason for this view is that the computations that Trevor Powell and I did show that substantial slow inward currents peaking at about 50 ms could not be generated from $i_{Ca,t}$ by series resistance artifacts since the series resistance required would be implausibly large and the current amplitude much too small.

A very slow transient component of i_{si}

But if there really do exist additional components of inward current then why not look for them directly ? This has been done both in single cells and in the multicellular s.a. node preparation. The work on the s.a. node was part of a project to assess quantitatively the contribution of different ionic currents to the generation of the normal rhythm of the heart. i_{si} is one of the currents that has been proposed as the main 'pace-maker' current so it seemed important to analyse its behaviour during the pace-maker depolarization. It was during such an analysis that the discovery was made that if one applies a voltage clamp at the potential reached towards the end of the pace-maker depolarization (about −45 mV) the inward current recorded

18   *D. NOBLE*

sometimes shows the time course illustrated in Fig. 7. This very clearly resembles the Purkinje current recorded at a similar voltage in Fig. 1*C*, i.e. activation and inactivation take almost equal periods of time and the peak inward current is not reached until 50–100 ms after the onset of the clamp. Here there is no question of the delay being produced by the time taken for the voltage step to occur via current flow through a series resistance since there is no voltage step. The preparation had

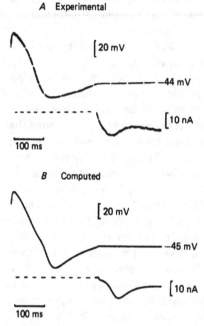

Fig. 7. The current, $i_{si,2}$, in rabbit s.a. node. The top records show one way of obtaining this current with little contamination from $i_{Ca,f}$. The membrane potential is allowed to follow the natural time course of the pace-maker depolarization until a potential near -40 to -50 mV is reached. The voltage is then clamped at this point. There is no voltage step and no transients attributable to a step. In some experiments, as here, a very slow inward current develops. The time course resembles that of Fig. 1*C* since activation and turn-off take almost equal periods of time. The bottom records show a reconstruction of this type of current record using the s.a. node version (Noble & Noble, 1984) of the DiFrancesco–Noble (1984) equations. The current here is computed using a Mullins-type model for the Na–Ca exchanger assuming a stoicheiometry of 3:1. (From Brown *et al.* 1983, 1984*a*.)

already naturally achieved the voltage required. A further argument for the view that such current records are not artifactual is that it is sometimes possible in voltage-clamp experiments in the s.a. node to distinguish two separate peaks of inward current. An example of this is shown in Fig. 8*A*.

It is of importance to note that the separation is only rarely as complete as this and that there is only a fairly narrow voltage range over which two peaks can clearly be seen. This is typically in the range -50 to -30 mV. At more negative potentials, no inward currents are recorded (i_{Na} has been blocked with tetrodotoxin (TTX)). At more positive potentials, the slower component speeds up and fuses with the faster

THE SURPRISING HEART 19

one to give the appearance of a single component. In some experiments it does this at all potentials.

This behaviour is not specific to the mammalian s.a. node. In the presence of 2 mM-4-AP to block i_{to}, two peaks of inward current can also be recorded in this voltage range in Purkinje fibres (G. Hart & D. Noble, unpublished). But perhaps the

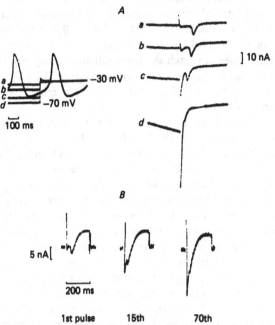

Fig. 8. This shows two other protocols that sometimes reveal $i_{si,s}$ in the rabbit s.a. node. *A*, left: voltage steps superimposed on a record of natural pace-maker activity. Right: current records. Note that as the holding potential is made more negative, progressively more $i_{Ca,f}$ is activated. Following this a very slow current similar in time course to that shown in Fig. 7 occurs. The time delay between $i_{Ca,f}$ and $i_{si,s}$ decreases as more $i_{Ca,f}$ is activated. Our interpretation of this result is that the speed of onset of $i_{si,s}$ depends on the speed of Ca entry and subsequent intracellular release This record is unusual in showing such a complete separation between the two components. More usually, $i_{si,s}$ appears as a slow tail following on $i_{Ca,f}$, as in the record in *d*. This is also how we think it usually occurs during natural pace-maker activity (see computer reconstruction in Fig. 12). (From Brown *et al.* 1984*a*.) *B*, two components of i_{si} recorded during a train of depolarizations (from -65 to -30 mV) following a long period of rest. The total current shows the 'staircase' phenomenon (cf. Noble & Shimoni, 1981). As in *A*, when $i_{Ca,f}$ becomes larger, the delay between this and $i_{si,s}$ decreases and the two components fuse together. (Reprinted with permission from Brown *et al.* 1982, 1984*a*.)

most convincing evidence comes from work on isolated single guinea-pig ventricular cells (Lee, Lee, Noble & Spindler, 1983), where it is possible not only to record the two peaks but also to show that when the first peak, i.e. $i_{Ca,f}$ is rapidly blocked by 0·2 mM-Cd ions (Lee & Tsien, 1982), the second peak remains for a few minutes. The second peak is strongly correlated with the contractile activity and, when, as is sometimes the case, this is oscillatory, the very slow inward current is also

20 *D. NOBLE*

oscillatory. In its time course and in the fact that it is sometimes oscillatory, this current strongly resembles the transient inward current, i_{TI}, first described in Purkinje fibres by Lederer & Tsien (1976). i_{TI} has also been shown to correlate strongly with the contractile properties of the preparation (see review by Tsien, Kass & Weingart, 1979). Lederer & Tsien used preparations in which the Na–K exchange pump was blocked by cardiotonic steroids to create the conditions (raised intracellular Na and, therefore, raised intracellular Ca) required to generate the after-contractions with which the current is correlated. Eisner & Lederer (1979) subsequently showed that Na-pump blockade produced by using K-free solutions was equally effective. But, of course, during a normal twitch the intracellular Ca may rise to values even greater than that occurring during after-contractions produced by Na-pump blockade. There is therefore no surprise in finding that a similar current may be activated during normal electrical activity.

What produces it? So far there are two theories, both of which were discussed by Kass, Lederer, Tsien & Weingart (1978). The first is that i_{TI} is the inward current generated by an electrogenic Na–Ca exchange process extruding Ca ions (which explains why high levels of intracellular Ca are required) in exchange for an electrically greater quantity of Na ions (which would explain why i_{TI} disappears in Na-free solutions). The second theory is that the current flows through non-specific (as between Na and K ions) channels activated by intracellular Ca ions. Such a channel has been found in patch-clamp studies (Colquhoun *et al.* 1981). I shall refer to the Na–Ca exchange current as i_{NaCa} and the non-specific channel current as i_{NaK}. When wishing to remain neutral as between these hypotheses (and any other hypotheses) I shall use the symbol $i_{si,2}$ for the current recorded experimentally. While using this neutral terminology I do nevertheless think that the resemblance of the current recorded in the s.a. node and in single guinea-pig ventricular cells to the original i_{TI} is so strong that a provisional identification of the currents as the same mechanism is justified.

How can we distinguish between the proposed mechanisms? There are several possible lines of argument here.

(1) If $i_{si,2}$ (or i_{TI}) is generated by the non-specific current, i_{NaK}, it should display a reversal potential near 0 mV, where the Na and K gradients are nearly equal and opposite and where the current i_{NaK} has been found to reverse in patch-clamp studies. The early work on i_{TI} in Na-pump-blocked preparations does describe such a reversal potential (Kass, Tsien & Weingart, 1978). By contrast, i_{NaCa} should show more complex behaviour depending on the way in which voltage and the [Ca]$_i$ transient combine to control current flow via the exchanger. This is complicated, but it is possible to show, using the DiFrancesco–Noble (1984) equations, that, at the least, i_{NaCa} should become smaller as the voltage is made more positive (see Fig. 16 and D. DiFrancesco, G. Hart & D. Noble, in preparation). A genuine reversal though of the time-dependent current should not occur since the direction of *change* of current should depend on the direction of change of [Ca]$_i$ at all voltages (this point needs to be clearly distinguished from the fact that i_{NaCa} itself must have an absolute reversal potential, which might not be too different from that for i_{NaK} at resting Ca levels (see Mullins, 1981). The problem is that we are not only dealing with elevated Ca levels, we are dealing with 'reversal' of the current *change* with time. Thus, an outward exchange current might still generate an 'inward' *change* of current level

THE SURPRISING HEART 21

if $[Ca]_i$ reduces, as it should, the level of such an outward i_{NaCa}, whether or not it succeeds in making it become absolutely inward. The absolute instantaneous reversal potential for i_{NaCa} is therefore irrelevant to the present discussion.

At first sight, the existence of a reversal potential for i_{TI}, therefore favours i_{NaK} as the mechanism. But here I will be frank and reveal my prejudice: I suspect that the 'reversal potential' for i_{TI}, is not always a genuine channel reversal potential. There are several reasons for my suspicion: (a) Karagueuzian & Katzung (1982) and Per Arlock & Katzung (1982) failed to find a reversal potential for i_{TI} in work on mammalian ventricular muscle; (b) even in Purkinje fibres, the reversal is sometimes fairly odd, as shown for example in the current–voltage diagrams for i_{TI} published recently by Hennings & Vereecke (1983): the current falls towards zero between -40 and -30 mV and then *stays at zero* for about 30 mV before showing a very small reversal. This looks much more like a voltage *range* over which i_{TI} is negligible or non-oscillatory (cf. Fig. 16) rather than a specific channel current reversal potential; (c) in work on i_{TI} in the s.a. node we have sometimes found that as the range of the presumed reversal potential is approached the oscillation shows a phase shift (Brown, Noble, Noble & Taupignon, 1984) which would imply that the phase of the $[Ca]_i$ oscillation had changed. A 'reversal' in such an experiment might then be no more than a combination of (i) a voltage range over which the oscillatory *change* in current is very small, and (ii) a phase shift that enables an outward phase of one oscillation at positive potentials to appear coincident in time with the inward phase of an oscillation at negative potentials.

At the least, therefore, I think the argument based on the existence of a reversal potential for i_{TI} is equivocal.

I should though make it clear that I am not proposing that i_{TI} *never* shows a specific reversal potential. The original experiments of Kass *et al.* (1978) clearly show such a reversal and this result, even if it is not that always obtained, needs to be explained. I will return to this problem below.

2. If i_{TI} is indeed activated (as $i_{si,2}$) during normal electrical depolarizations (as the s.a. node work and the results on isolated guinea-pig ventricular cells suggests) then it is relevant to ask which mechanism would be most compatible with the known behaviour of the action potential. This question is easy to answer. i_{NaCa} is perfectly compatible with the generation of normal ventricular action potentials in which the rate of repolarization is *smallest* at a time (e.g. about 50–100 ms) when the $[Ca]_i$ transient is maximal since the $[Ca]_i$ transient will either activate an inward i_{NaCa} or reduce an outward i_{NaCa}, in both cases assisting the depolarization and therefore the plateau of the action potential. By contrast i_{NaK} activation would distort the repolarization phase in exactly the same way as the end-plate current distorts the skeletal muscle repolarization phase, by always driving the potential towards the current reversal potential (Fatt & Katz, 1951).

There is evidence that at high $[Ca]_o$ and high frequencies this does indeed happen in frog ventricle (Niedergerke & Page, 1982) but it can hardly happen in ventricular action potentials showing the *lowest* rate of repolarization at the time of the *peak* $[Ca]_i$ transient. This argument has been given a quantitative form recently in the DiFrancesco–Noble (1984) equations in which i_{NaCa} supports a normal action potential while i_{NaK} distorts it.

This is a convenient point at which to return to the i_{TI} reversal potential measured by Kass *et al.* (1978). Their current–voltage diagram differs from that obtained by Per Arlock & Katzung (1982), by Hennings & Vereecke (1983) and in our own work since they found a *specific* reversal potential rather than an extensive *range* of zero current potentials or phase shifts. Moreover, in their experiment they excluded the phase shift argument I have presented above since they recorded the contraction, which, like the current, did not show a significant phase shift. Clearly then, two or three significantly different types of result can be obtained. I think the simplest explanation for this is that, in addition to activating the Na–Ca exchanger, the Ca transient also sometimes activates the non-specific current channels. i_{TI} might then be a composite current, which would be entirely dominated by the non-specific current over the voltage range at which the exchanger current would be expected to fall towards zero. This happens to be in the range of the reversal potential for the non-specific current channel. In conclusion then, the voltage dependence of i_{TI} shows three forms: reversal at a specific membrane potential, a voltage range of zero current change, and a voltage range of combined low amplitude and phase shift of the oscillation. I doubt whether this diversity can succumb to a single component explanation and I therefore propose that the best explanation is that i_{NaCa} and i_{NaK} are activated by internal Ca to different extents in different preparations and experimental conditions. When i_{NaK} is very small or absent, there will be no specific reversal potential.

(3) The third line of argument is more purely theoretical. It is easy to show that if i_{NaCa} is the main mechanism by which Ca leaves the cell then for exchange ratios of 4:1 or 3:1 (the currently favoured Na:Ca exchange ratios) the time integral of i_{NaCa} must be either of the same order of magnitude as that for i_{Ca} or about half this magnitude. This argument is most readily appreciated for a 4:1 exchange since in that case the inward current carried by the excess Na ions must then equal the charge on the Ca ions transported. Now I think it is significant that the slow current $i_{si,2}$ recorded in the s.a. node and in guinea-pig ventricular cells is of the right order of magnitude. Fig. 7 *B* shows that the DiFrancesco–Noble (1984) equations, as modified for the s.a. node (Noble & Noble, 1984) can readily account quantitatively for $i_{si,2}$. This model does assume that the Na–Ca exchanger is the main route by which Ca ions leave the cell and that the exchange ratio is 3:1. The Purkinje version of the equations also succeed in reproducing the very slow inward current recorded (see, e.g. Fig. 1 *B*) in Purkinje fibres (DiFrancesco & Noble, 1984).

My conclusion, therefore, is that, at present the arguments favour the view that, in *normal* electrical activity, $i_{si,2}$ (or i_{TI}) is largely i_{NaCa} rather than largely i_{NaK}, but I would readily admit that this view could greatly benefit from more direct ways of demonstrating the existence of i_{NaCa} in a way analogous to the elegant experiments of Gadsby (1980), Eisner & Lederer (1980) and Daut & Rüdel (1982) showing the existence of the Na–K exchange pump current, i_p, in the heart. Some recent evidence on i_{NaCa} has been produced by Horackova & Vassort (1979), Mentrard & Vassort (1983) and by Jacob, Murphy & Lieberman (1983) using experimental conditions in which blockers or substituted solutions are used to eliminate all other possible currents.

There is a final point about $i_{si,2}$ which needs to be made to complete the picture as I now see it. I have so far described it as a very slow transient because that is how it appears when it is dissectable from $i_{Ca,f}$ in the voltage range between -60 and

-30 mV. It is important, however, to realize that it is very easy to miss seeing the current for two reasons. First, to obtain a large slow current in this voltage range it is usually, though not always, necessary to use a sufficiently negative holding potential. We used -80 mV in our single-cell work, which is much more negative than is usually used in studies of i_{si}. This requirement may depend on repriming of the internal Ca-release mechanism, which is known to be voltage dependent (Gibbons & Fozzard, 1975). The second reason is that, in both the s.a. node work and the single ventricular cells, as the membrane voltage is made more positive, and $i_{\text{Ca,f}}$ is more strongly activated, $i_{\text{si,2}}$ becomes considerably faster and then fuses more completely in its time course with $i_{\text{Ca,f}}$ to give the appearance of a single component. This behaviour is consistent with a mechanism attributable to i_{NaCa} and is seen also in the behaviour of $i_{\text{si,2}}$ in the DiFrancesco–Noble (1984) equations. The mechanism is that as $i_{\text{Ca,f}}$ is increased, the Ca transient that is supposed to activate i_{NaCa} rises more rapidly. Depolarization also reduces the magnitude of i_{NaCa} by approaching its reversal potential. Thus, the protocols for separating $i_{\text{si,2}}$ and $i_{\text{Ca,f}}$ have to be chosen with care and they may well vary in different cells and in different experimental conditions. Amongst the latter, one that we have found to be very important is the condition of the isolated cells. If this is not good enough for them to display action potentials with long plateaus, the slower components of current are difficult to detect. This is not suprising in view of the likely functional importance of the slow components (see below).

A Cd-resistant maintained component of i_{si}

This discussion of additional slow inward currents may already have exhausted the reader, but it certainly does not exhaust the mechanisms that have been found. I referred earlier to the fact that $i_{\text{si,2}}$ (the i_{TI}-like current) has been found in single guinea-pig ventricular cells (Lee *et al.* 1983) and that it can be observed for a period of time even when $i_{\text{Ca,f}}$ has been blocked by Cd ions. Eventually, after a minute or two in Cd solution, the contractile behaviour changes and $i_{\text{si,2}}$ also disappears. In these conditions though, there *still* exists an inward current which is totally abolished when Ca ions are removed (Lee, Lee, Noble & Spindler, 1984*a*). This current, whose peak amplitude is only about 20 % of $i_{\text{Ca,f}}$, shows very slow inactivation and at some potentials (around -50 to -30 mV) it hardly inactivates at all. We have called this component $i_{\text{Ca,s}}$ since we think it is a very slowly inactivated Ca current. Although much smaller than $i_{\text{Ca,f}}$ it is just large enough to support a prolonged action potential with a very slow rate of rise and no overshoot (see Fig. 9).

An inward current very similar to $i_{\text{Ca,s}}$ in the guinea-pig ventricle cell has also been recorded recently in isolated single frog atrial cells by Hume & Giles (1983). Like $i_{\text{Ca,s}}$ it starts to activate in the region of -60 mV, is resistant to Cd ion block, shows very little inactivation at some potentials and carries a peak current equal to about 20 % of the total i_{si}. There is however an important difference, which is that, whereas in guinea-pig cells $i_{\text{Ca,s}}$ promptly disappears in Ca-free solution, the persistent inward current in atrial cells does not (see Hume & Giles, 1983, Fig. 12). It requires many minutes of Ca-free superfusion to be eliminated. Hume & Giles note the resemblance of their persistent inward current to persistent Ca currents found in *Helix* neurones (Eckert & Lux, 1975; Akaike, Lee & Brown, 1978) and vertebrate motoneurones (Schwindt & Crill, 1980) and do not exclude the possibility that their current is also

24 *D. NOBLE*

a Ca current since brief superfusion with Ca-free solution may not remove all the Ca that may be involved.

The existence of $i_{Ca,s}$ should not come as too great a surprise. Ca currents of different inactivation speeds also exist in other cells (see recent review by Tsien, 1983) and the work on multicellular preparations (e.g. Beeler & Reuter, 1970) shows that some component of Ca current must exist that requires several hundred milliseconds for inactivation. Indeed it seems likely that a very large part of the 'slow inward current' recorded in multicellular preparations must consist of currents like $i_{si,2}$ and $i_{Ca,s}$ rather than $i_{Ca,t}$ as I have defined it in this review.

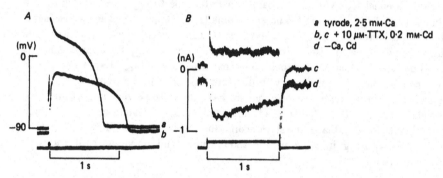

Fig. 9. The very slow Cd-resistant current in single guinea-pig cells. *A*, this shows action potentials recorded before and after application of 0·2 mM-Cd. Note that, even when $i_{Ca,t}$ is blocked by Cd, a long plateau at a lower membrane potential can still be obtained. *B*, this shows ionic current in Cd solution before and after removal of Ca ions. The action potential plateau is also then completely abolished (not shown here). The persistent inward current (which we have called $i_{Ca,s}$) differs markedly from $i_{Ca,t}$ in its voltage range (threshold for $i_{Ca,s}$ is 20–30 mV more negative than for $i_{Ca,t}$), time course (very slow), resistance to Cd, absence of effect of isoprenaline, and absence of Ca-induced inactivation (at some voltages the current is hardly inactivated at all). Together with $i_{si,2}$ (which is also found in single guinea-pig cells (Lee *et al.* 1983)), $i_{Ca,s}$ is thought to generate the very long plateau. (From Lee, Lee, Noble & Spindler, 1984*a*.)

To conclude this rather difficult, and doubtless more controversial part of my review, I propose that i_{si} is composed of three components, which, using completely neutral terminology, might be referred to as $i_{si,1}$, $i_{si,2}$ and $i_{si,3}$. Provisionally, at least, the evidence that the first and third components are Ca channels leads me to replace this neutral terminology with $i_{Ca,t}$ and $i_{Ca,s}$. We then arrive at the dissection:

$$i_{si} = i_{Ca,t} + i_{si,2} + i_{Ca,s}, \qquad (2)$$

and it will also be clear that my prejudice on the composition of the second component would lead to the hypothesis of eqn. (3):

$$i_{si} = i_{Ca,t} + \Delta i_{NaCa} + i_{Ca,s}. \qquad (3)$$

The idea that there may be such a mixture of currents would also readily explain the diversity of opinion on the extent to which Na and Ca ions are responsible. $i_{Ca,s}$

THE SURPRISING HEART 25

seems to be purely Ca (it completely disappears in Ca-free solution) while currents like i_{NaCa} (or i_{NaK} if this should turn out to be the correct explanation for $i_{si,2}$) should disappear in Na-free solutions.

Correlations with multicellular work

At this point I am tempted to speculate on the correlation between this dissection of i_{si} based on the single-cell and s.a. node work and the currents recorded in previous work. In some cases, I think we can be fairly certain of the correlations. Thus, it is likely that the first i_{si} recorded by Reuter (1967) in Purkinje fibres (see Fig. 1 A) contained virtually no $i_{Ca,f}$ since it reached a peak at a time when most of the fast current would be already inactivated. From which one would conclude that it was largely components like $i_{Ca,s}$ and $i_{si,2}$. Similarly, i_{TI}-like currents of the kind shown in Fig. 1 C are likely to be largely $i_{si,2}$, with some persistent contribution from $i_{Ca,s}$. The much faster currents (Fig. 1 B) recorded, e.g. by Siegelbaum & Tsien (1980) and by Marban & Tsien (1982) must have contained much more $i_{Ca,f}$ and, with intracellular Ca buffered, we may guess that $i_{si,2}$ was absent. Inward current with fast ($\tau = 20$ ms) inactivation time constants was also recorded in multicellular work on frog atrial trabeculae (Rougier *et al.* 1969). And so one could go on through the past literature. It would, however, be disingenuous of me to suggest that drawing out these correlations is all that easy. There are also some awkward problems. For example, the voltage range of the activation curve for $i_{Ca,s}$ in single cells is appreciably more negative (about -60 mV) than the accepted 'threshold' (about -40 mV) for activating i_{si} in multicellular preparations (Reuter, 1967; Beeler & Reuter, 1970; Reuter & Scholz, 1977). All I think we can say for certain is that there would have been an even greater problem if the single cell work had *not* shown the existence of slower components of inward current.

There is another reason for being cautious about the interpretation of the work on multicellular preparations in the light of that on single cells, particularly when comparing the presumed charge carriers for various currents. This is that it is far from clear what the relative permeabilities for various ions should be. Reuter & Scholz (1977) attempted to measure these by using reversal potentials interpreted using a form of Goldman equation. The equation they used is in fact incorrect, as is the 'corrected' version they published later (see Attwell & Jack (1978) for a correct derivation of equations for channels conducting divalent and monovalent ions simultaneously, and for an important discussion on the ambiguity of using constant-field type models for defining permeabilities in such cases). The conclusion that the i_{si} channels carry Na and K ions as well as Ca ions needs reassessment, both for this theoretical reason and because it is now important to assess the ion carriers *separately* for each of the components of i_{si}. In the case of $i_{Ca,f}$, Lee & Tsien (1982) have shown that outward current (presumably carried by K ions) can be recorded positive to the reversal potential. By itself, given the problems that may arise from current 'masking', this would not be sufficient. They were therefore careful to show that this outward current flow is blocked by Ca current blockers and that there is no change in current level at the reversal potential during such a block. Hume & Giles (1983) made the same observation on current recorded beyond the reversal potential in the case of the fast channel in frog atrial cells. They also investigated the contribution

26   *D. NOBLE*

of Na ions and concluded that Na$^+$ does not contribute to the fast or persistent inward currents. Mitchell *et al.* (1983) have also concluded that Na ions do not contribute to the fast Ca current in single rat ventricular cells. Yet, the work on i_{si} in multicellular preparations repeatedly suggests that Na ions are involved. If neither $i_{Ca,t}$ nor $i_{Ca,s}$ conducts Na ions, what does? Mitchell, Powell, Terrar & Twist (1984) describe a 'late' plateau current in rat ventricular cells that is Na sensitive. Whether this is identical with what I have called $i_{si,2}$ in this review remains to be seen. Certainly, if $i_{si,2}$ were to be carried either by the Na–Ca exchanger or by the Ca-activated non-specific channels, it would be Na sensitive. Another problem in this connexion is that Ca ion channels may only conduct Na ions in Ca-free solutions.

Before I leave the question of the components of i_{si}, there are some semantic problems that arise. If the 'dissection' I have proposed here is correct (and it is by no means the only 'dissection' that has been proposed – see Isenberg (1983) for an interesting alternative based on distinguishing between sarcolemmmal i_{si} and current that may be carried by internal membrane systems), then it is clear that until very recently the single-cell work has concentrated on what I have called $i_{Ca,t}$. Because it was by no means evident initially whether i_{si} was a single or multicomponent current, the symbol i_{si} tends to be used *both* for the fast current recorded in single cells *and* for the slower currents seen until recently only in multicellular preparations. The impression given is that the same current is being referred to. This is doubly unfortunate from a semantic point of view. First, because it is likely, as I have discussed in this review, that the current that contributed *least* to the early records of i_{si} in multicellular preparations is $i_{Ca,t}$, which has nevertheless tended to acquire the label i_{si} in the single-cell work. This is likely to give rise to just as much confusion as was generated when the symbol i_{K2} was used (by Tsien and me) to describe the current that now turns out to *least* resemble the g_{K2} system originally described by Hall *et al.* (1963). Whether or not my suggested terminology is adopted, I think it is important to drop the non-specific label i_{si} when referring to the fast Ca current in single cells. The second reason why the confusion is unfortunate is that it is difficult to decide correctly what component(s) (if any) should be described as 'new'. Taking the admittedly very short history of the single-cell work as a basis, $i_{si,2}$ and $i_{Ca,s}$ are the 'new' currents, since they are not the already well-established fast current. But, of course, all of them must have been flowing across the cell membranes in the multicellular preparations (at some semantic level, no currents are 'new'!). The problem, as I have indicated above, is that it is not always easy to decide which component contributed to the total i_{si} in particular cases. What is certainly new is the much greater resolution provided by the single-cell-clamp work which makes it more plausible to dissect the components with some confidence. Currents that might have been dismissed as 'abominable notches' due to clamp escape at the surface membrane in multicellular preparations cannot be so dismissed in single cells when they carry a current only a small fraction of currents (like i_{Na} and the fast Ca current) which can clearly be clamped successfully.

THE SURPRISING HEART   27

Functions of the various components of i_{si}

Leaving these semantic problems aside, let us turn to function. How might the various components now be thought to contribute to the action potential and to rhythmic activity? I shall be discussing pace-maker activity later in this review. So far as the action potential is concerned, there are some very significant changes to be made to the picture given by the MNT and BR models. The fast component of i_{si} is sufficiently rapid (peak current is reached in 2–3 ms) that it would fuse in with the later part of the Na current peak. More important, its inactivation is so fast that it can hardly contribute much current (except in the form of a 'window' current) to the later parts of the action potential plateau. If we take the inactivation time constant to be 10–20 ms then inactivation would be expected to be largely over by 40–80 ms. The role of such a current in the plateau then comes to resemble more closely that of the Na current. Another important consequence of the increased speed is that the Ca ions required to trigger the contraction process are made available more quickly. If the Ca release from the sarcoplasmic reticulum is Ca induced then the speed of $i_{Ca,f}$ may be important in accounting for the speed at which $[Ca]_i$ has been observed to rise in experiments with aequorin (Allen & Kurihara, 1980; Wier, 1980). In attempting to reproduce these transients in our equations (which assume Ca-induced Ca ion release), DiFrancesco and I were greatly helped by the increased speed and intensity we were able to give to the Ca current. But the correlate of this speed for activation and inactivation of $i_{Ca,f}$ is that the maintenance of the plateau must depend on other components. As shown above (Fig. 9), $i_{Ca,s}$ is capable of maintaining a plateau, albeit lower than normal, in Cd-blocked cells. Together with $i_{si,2}$, it may well carry enough current to maintain the normal ventricular plateau in its later phases. Mitchell *et al.* (1983, 1984) and ter Keurs & Schouten (1984) give a similar role to the Na-sensitive current (? $i_{si,2}$) in rat ventricular cells. Hume & Giles (1983) also propose a late plateau role for the persistent inward current (? $i_{Ca,s}$) which they find in single frog atrial cells.

Finally, I should like to draw attention to a number of other interesting features of slow inward currents in the heart that have been described recently. Since they have been dealt with fairly fully in other recent reviews or papers I shall mention them only briefly with regard to their importance to work on reconstructing pace-maker activity in the heart.

Ca-entry-dependent inactivation

This phenomenon has been described in the rabbit s.a. node (Brown, Kimura & Noble, 1981; Brown, Kimura, Noble, Noble & Taupignon, 1984*a*), in frog atrium (Fischmeister & Vassort, 1981; Mentrard, Vassort & Fischmeister, 1983), in single frog atrial cells (Hume & Giles, 1982) and in Purkinje fibres (Marban & Tsien, 1981). These findings are, of course, important in relating work on the heart to the already well-established existence of this phenomenon in other cells. Is it of importance in normal cardiac electrical activity? I think the answer to that question must be yes. If the inactivation of $i_{Ca,f}$ were purely voltage dependent then it can be shown that the 'window' current that is necessarily generated by a Hodgkin–Huxley type of model is always too large and this results in a 'hump' of varying size during the repolarization phase. In the s.a. mode modelling (Noble & Noble, 1984) we managed to keep this 'hump' to a minimum: it is only really evident on the plot of dV/dt (see Noble & Noble, 1984, Fig. 1). Nevertheless, the fact is that such a hump is not usually seen at all in normal s.a. nodal activity. We have subsequently modified the model to include Ca-dependent inactivation of $i_{Ca,f}$ and it is then relatively easy to eliminate the hump almost completely, particularly if the recovery from inactivation is assumed to take several hundred milliseconds, as shown experimentally in the s.a. node (Brown *et al.* 1984*a*) and in frog atrium (Mentrard *et al.* 1983).

The 'staircase' phenomenon

Another phenomenon that may be of importance in rhythmic activity is that during repetitive depolarizations i_{si} shows slow changes in amplitude that resemble the well-known tension 'staircases' in the heart (S. J. Noble & Shimoni, 1981*a*, *b*; Shimoni, 1981). Like the tension staircase, this sometimes leads to i_{si} *increasing* during repetitive depolarizations, whereas incomplete recovery would lead only to a decrease in amplitude. This phenomenon was first described in frog atrial trabeculae but it has also been observed in the rabbit s.a. node (Brown *et al.* 1984*a*).

This is a convenient point at which to comment on the correlation between i_{si} and tension changes in the heart. Broadly speaking, there are two possible schools of thought on this correlation. The first holds that it is by changing i_{si} and hence the quantity of Ca entering the cell, that the change in tension is brought about. This is the way in which adrenaline is thought to increase the force of contraction since it is well known that adrenaline increases i_{si} (Reuter, 1967; Reuter & Scholz, 1977). The alternative view is that changes in i_{si} *reflect* changes in $[Ca]_i$ that cause the changes in tension. This view is best exemplified by the accepted view of i_{TI} which is seen as being *induced by* the changes in $[Ca]_i$, not as causing those changes. But if, as on the interpretation I have given earlier in this review, i_{TI} is also activated (as $i_{si,2}$) during normal depolarizations to form a part of i_{si} then some changes, particularly those involving slow components of i_{si}, might be induced by $[Ca]_i$ changes rather than directly causing them. My suspicion is that some of the changes in i_{si} produced by Na–K pump inhibition (Lederer & Eisner, 1982) and by drugs like caffeine and tetracaine (see Fig. 1*C*) may be in this category. It is significant that these changes may run closely in parallel with the tension changes, rather than preceding them, as do the changes induced by catecholamines (see Lederer & Eisner, 1982).

But are the 'staircase' effects also $[Ca]_i$ induced rather than contributing to $[Ca]_i$ changes? This question is still open, but I presently favour the view that they may at least partly reflect rather than cause $[Ca]_i$ changes. There are two main lines of evidence for this view.

1. Using the s.a. node version of the DiFrancesco–Noble (1984) equations, Brown *et al.* (1984*a*) were able to reproduce the current 'staircase' by allowing changes in $[Ca]_i$ transients to induce changes in i_{NaCa}. These computations show, at least, that such a process could be *sufficient* to explain the phenomena.

2. Preliminary experiments on $i_{Ca,f}$ in single cells (E. Lee, K. S. Lee, D. Noble & A. J. Spindler, unpublished) have failed to show the 'staircase' effect. This would be compatible with the view that only the slower components, like $i_{si,2}$, are involved.

THE SURPRISING HEART 29

This fact must though be interpreted with care at this stage. It is important first to check that the tension 'staircase' really does exist in single cells since an alternative explanation for a failure to observe the current 'staircase' in single cells would be that intracellular Ca cycling may not be normal in isolated cells. A good test for this would be the existence of tension 'staircases'.

It is, of course, possible to develop a hybrid hypothesis: that some changes in i_{si} are Ca-induced *and* that they then contribute to $[Ca]_i$ changes (this would require that either $i_{Ca,t}$ or $i_{Ca,s}$ should be changed). This is the position favoured by Marban & Tsien (1982) in their work on cardiotonic steroids who point out that this would lead to the existence of a positive feed-back in the control of i_{si}. A small increase in the $[Ca]_i$ transient would increase i_{si} and (if this involves $i_{Ca,t}$ or $i_{Ca,s}$) in turn further increase the $[Ca]_i$ transient.

One possible difficulty with this view is that it remains to reconcile it with the view that $[Ca]_i$ *inactivates* $i_{Ca,t}$ (see previous section). Could $[Ca]_i$ both *inactivate* $i_{Ca,t}$ and *increase* the transient amplitude? Perhaps, but we really do not have any idea of the mechanisms that might be involved.

One of my reasons for introducing the influence of repetitive activity on i_{si}, or its components, is that any such effects may be important in analysing the contribution of i_{si} to the generation of repetitive activity. After all the heart never rests.

3. *Other current components*

The potassium currents, i_{K1} and i_K

By comparison with the developments concerning i_t/i_{K2} and i_{si}, the interpretation of the currents described as i_{K1} (the inward rectifier) and i_K (the delayed rectifier) has not changed so much. I shall therefore give only a brief treatment of the most important developments before proceeding to discuss the quantitative assessment of the contributions of these currents and of i_t and i_{si} to natural pace-maker activity in the heart.

The analysis of the inward rectifier has until recently tended to be relegated to a description of the current that remains when other (primarily time-dependent) currents have been analysed. This was the approach used in the development of the MNT model. But, this approach must incorporate the Na–K exchange pump current, i_p, and the steady-state value of the Na–Ca exchange current, i_{NaCa}, in the description of what functions in much the same way as the leak current, i_l, in the Hodgkin–Huxley model.

We can, however, now be more direct in approach. There are at least three lines of approach that have been used to characterize i_{K1} more specifically: (a) to use a blocker, such as Ba ions, to perform a subtraction of time-independent $i(E)$ relations in the presence and absence of the blocker (see DiFrancesco, 1981b); (b) to measure ^{42}K fluxes as a function of membrane potential (see Vereecke, Isenberg & Carmeliet, 1980), and (c) to use patch-clamp methods to study i_{K1} channels in isolation (Momose, Szabo & Giles, 1983; Sakmann, Noma & Trautwein, 1983). All these approaches confirm that i_{K1} is a strong inward rectifier and that it is very sensitive to $[K]_o$. This sensitivity to external concentration of K ions has two important consequences that are relevant to this review: first it allows large depletion current changes to occur during hyperpolarization. I have already discussed the significance of this fact in relation to the reinterpretation of i_{K2}. Secondly, it means that any pace-maker system, such as the Purkinje fibre, that conducts i_{K1} fairly strongly, will be very sensitive to $[K]_0$, such that increasing $[K]_0$ will tend to abolish pace-maker activity. I have already referred to this property of Purkinje fibres in Fig. 4. These developments have largely served to confirm the properties of i_{K1} as assumed in previous work. There is however one development that introduces a significant difference. This is the demonstration that at very negative potentials i_{K1} rectification may show some time dependence. This was proposed by Carmeliet (1982) and has been confirmed recently in patch-clamp work on single channels (Kameyama, Kiyosue, Soejima & Noma, 1983). Moreover, Ba ion block of i_{K1} has been shown to be time dependent, which means that in the presence of Ba ions a form of pace-maker activity may develop in ventricular cells that is controlled by the time dependence of i_{K1} rather than of i_K or i_t (Carmeliet, Van der Heyden & Vereecke, 1983). This observation is important in relation to a later discussion in this review on s.a. node rhythm since the presence of nodal rhythm even in Ba-containing media has been used as an argument against the importance of i_K in s.a. nodal rhythm.

The delayed K current, i_K, has been studied in a number of cardiac preparations since its analysis in Purkinje fibres (McAllister & Noble, 1966; Noble & Tsien, 1969a). These include frog atrium (Brown, DiFrancesco, Noble & Noble, 1980), mammalian ventricle (McDonald & Trautwein, 1978), rabbit s.a. node (Katzung & Morgenstern, 1977; DiFrancesco, Noma & Trautwein, 1979), single frog atrial cells (Hume & Giles, 1983) and single sinus venosus cells from the frog (Shibata & Giles, 1983). The results all agree in showing that the voltage range of activation is from about -50 mV to 0 mV, that the channels show instantaneous rectification, but that they do not show the extreme sensitivity to $[K]_0$ displayed by i_{K1}. The cross-over phenomenon does not therefore occur so that a pace-maker mechanism strongly dependent on i_K should *not* necessarily be quiescent at high $[K]_0$. The s.a. node model based on the g_K decay hypothesis using i_K (Noble & Noble, 1984) does indeed show only a very moderate sensitivity to $[K^+]_0$ compared to that displayed by Purkinje fibres.

A comment on terminology is appropriate here. The first description of i_K used the symbol g_{K2} (Hall *et al.* 1963). This became confusing when the i_{K2} system emerged and Noble & Tsien (1969a) used the symbol i_{x1} in the belief that the delayed K current was a less pure K channel than i_{K2}. This view, of course, no longer carries any force since it is very probable that E_K is not after all so far away from the reversal potential for the delayed rectifier as was thought when the 'reversal potential' for i_{K2} was thought to be at E_K. I think the simplest way to resolve the terminological difficulties is to follow McDonald & Trautwein's (1978) use of the symbol i_K. This has been done in formulating the DiFranceso–Noble (1984) equations and I shall also adopt this terminology here.

The sodium current, i_{Na}

There have been very considerable advances made in the analysis of the Na current in the heart. This has in part resulted from better voltage clamping of multicellular preparations (e.g. Colatsky, 1980; Ebihara, Shigeto, Lieberman & Johnson, 1980), in part from work on isolated single cells using whole cell clamps (Lee, Weeks, Kao, Akaike & Brown, 1979; Brown, Lee & Powell, 1981; Lee, Hume, Giles & Brown, 1981; Bodewei, Hering, Lemke, Rosenshtraukh, Undrovinas & Wollenberger, 1982) and using patches (Cachelin, De Peyer, Kokubun & Reuter, 1983). The kinetics determined show considerable quantitative differences from those used in the MNT model and much of the new data was used in formulating new equations for this current in the DiFrancesco–Noble (1984) equations. This formulation gives a good reconstruction of the steady–state ('window') current (Attwell, Cohen, Eisner, Ohba & Ojeda, 1979), which has been shown to contribute to the maintenance of the plateau in Purkinje fibres and which may be involved in low voltage pace-maker activity in these cells.

The sodium pump current, i_p

It has been clear since Isenberg & Trautwein's (1974) work that a significant outward current may be carried in the heart by the Na–K exchange pump. In a very elegant series of experiments Gadsby (1980), Eisner & Lederer (1980) and Daut & Rüdel (1982) have succeeded in characterizing this current sufficiently well for a quantitative formulation to be possible for the first time. This has been incorporated into the DiFrancesco–Noble equations. The Na–K pump current has also been detected recently in isolated frog atrial cells (Shibata, Momose & Giles, 1983).

4. The natural pace-maker: the s.a. node

I have already discussed the fact that normal pace-maker activity in Purkinje fibres does *not* conform to the g_K decay hypothesis. What about the natural pace-maker, the s.a. node?

Alternatives to the g_K decay hypothesis

Until about 1976 the s.a. node itself was thought to be too difficult for voltage-clamp experiments. The best available evidence therefore came from analogies with partially depolarized Purkinje fibres (Hauswirth *et al.* 1969) or atrial trabeculae (Brown, Clark & Noble, 1972, 1976). In both cases, the suggested mechanism was a decay of K current transported by the delayed rectifier followed by activation of the second inward current, i.e. a version of the g_K decay hypothesis. This hypothesis was fully supported by the Brown *et al.* (1977) study of the frog sinus venosus. Although they found i_f in this tissue, they also pointed out that its activation range was usually too negative for it to be involved in generating the pace-maker depolarization. Yet the threshold for activation of i_{si} was situated towards the last 30–50% of the pace-maker depolarization. These observations virtually eliminated i_f and i_{si} as the primary cause of the pace-maker depolarization, leaving i_K decay as the only remaining viable hypothesis. Since then, the main object of study has been the mammalian s.a. node, and in particular the small dissected rabbit s.a. node preparation pioneered by Noma & Irisawa (1976a). Four theories have emerged:

1. Noma & Irisawa and their colleagues (see e.g. Yanagihara & Irisawa, 1980) have proposed that the main pace-maker current is i_{si} and that i_K decay or i_f onset are not important or are very small compared to the changes in i_{si}.

2. Maylie, Morad & Weiss (1981) also proposed that i_K decay was not important but that i_f onset was involved.

3. Pollack (1976) has proposed that the rhythm depends on intercellular mechanical interactions that are implicated in propagation in nodal tissues.

4. Brown *et al.* (1982) have proposed that the original g_K decay hypothesis is correct.

The field is now somewhat confused (or, at least, confusing to the uninitiated) because some of these authors have explicitly or implicitly abandoned their original views and I suspect that there may soon be a degree of unanimity on some form of the g_K decay hypothesis. Regardless of authorship, or ownership, though, the arguments for the alternatives are interesting in their own right and it is partly because I think that it is instructive to weigh them up that, in this review, I have chosen to discuss the four theories in turn.

First, then, the i_{si} hypothesis. There is, of course, no denying the fact that i_{si} (but which components?) is of crucial importance to s.a. node activity. S.a. node pace-maker activity is not abolished by TTX and does not therefore require i_{Na}. But, to take an analogous situation, the fact that the upstroke in Purkinje fibres is generated by i_{Na} does not lead to its being identified as the 'pace-maker' current in this tissue; so the fact that the upstroke is generated by i_{si} in the s.a. node does not identify it as the 'pace-maker' current. The crucial question is not what generates the upstroke but what brings the membrane potential towards the threshold for generating the upstroke. Now this *could* be the same mechanism as that which generates the upstroke provided that its 'threshold' is very broadly spread over the voltage range of the pace-maker depolarization. Voltage-dependent conductances of the Hodgkin–Huxley type do not of course have a strictly defined threshold. What is meant here is the spread on the voltage axis of the foot of the activation curve. Thus, in the case of the Na conductance this is sufficient to allow i_{Na} to contribute a significant fraction of the depolarizing current to the last 30 % or so of the pace-maker depolarization in Purkinje fibres (this is a feature both of the MNT (1975) and the DiFrancesco–Noble (1984) equations). But the extent of the foot is not large enough to enable i_{Na} to *initiate* the pace-maker depolarization. Rather, it increasingly helps the depolarization, once it is already well developed, to succeed in developing spontaneously into a fully-fledged action potential.

I think the experimental evidence suggests a similar role for i_{si} in the case of the s.a. node pace-maker depolarization. Measurements of the threshold for i_{si}, both in frog sinus venosus (Brown *et al.* 1977, Fig. 13) and in the rabbit s.a. node (Brown *et al.* 1984a) show that it is not significant at the beginning of the depolarization and that it is largely during the last third of the depolarization that i_{si} becomes significant. The question which component(s) of i_{si} are then significant is hard to answer with certainty. The inward current that dominates the record when a voltage clamp is imposed at the end of the pace-maker depolarization activates very slowly indeed and is i_{TI}-like in its time course (this is the component I have earlier described as $i_{si,2}$, see Fig. 7). But, of course, if $i_{Ca,f}$ is also activated it would already be so, and be close to its steady-state value, in such an experiment since the rate of change of voltage during the pace-maker depolarization is so slow compared to what we now know are the very fast kinetics of activation of $i_{Ca,f}$.

How then did Noma and Irisawa and their colleagues come to the view that i_{si} is more important than this analysis suggests? I think the answer to that lies in an analysis of the dynamics of i_K during a pace-maker depolarization. Their measurements of the kinetic properties of i_K are similar to those of the Oxford group (and to those published by DiFrancesco *et al.* (1979) – I shall show later that this *requires* i_K decay to be the main initiator of the pace-maker depolarization). But, in their computations of pace-maker activity using these kinetics they found that i_K remains fairly constant during the pace-maker depolarization (see Yanagihara, Noma & Irisawa, 1980). How is this consistent with the g_K decay hypothesis? To answer this question it is important to note that it is g_K not i_K that matters. To clarify this point consider the following argument. The rate of change of potential during a pace-maker depolarization is small and nearly constant so that we can, to a first approximation, regard the net ionic current, which will be $-C\,dV/dt$ (where C is the membrane capacity), as nearly constant. Suppose, for the sake of argument that the background inward current, $i_{b,in}$ does not vary greatly over the pace-maker range of potentials so that this may also be regarded as nearly constant. Then, approximately,

$$i_K = -i_{b,in} - C\,dV/dt \simeq \text{constant.} \tag{4}$$

Thus i_K will be nearly constant even if g_K is varying in precisely the way required to be the prime cause of the change in potential. This argument was illustrated graphically in Brown *et al.* (1977, Fig. 12). Thus, if g_K is decaying, the depolarization must increase the driving force sufficiently for i_K itself to remain nearly constant. What is certain (since the instanteous $i(E)$ relation for i_K does not show a negative slope region) is that i_K will change by much less than g_K does. I will return to a more quantitative version of this argument when I have considered the other theories.

   Maylie *et al.* (1981) used a potentially powerful and ingenious approach based on measuring $[K]_0$ changes with an ion-sensitive electrode. Their argument is that the rate of change of $[K]$, $d[K]/dt$, should reflect, in part at least, the K flux across the membrane. If a large part of the current relaxation that occurs when one uses a voltage clamp to stop the pace-maker depolarization is due to i_K decay then the rate of decay of $[K]$ should initially be slow and should then increase as i_K decays. But their experimental results showed that, even when the total current relaxation is large $d[K]/dt$ is nearly constant. Their conclusion was that i_K decay is not important and that the majority of the current relaxation must be attributable to the onset of i_t (which they call i_p).

   If the conditions of the experiment were ideal, i.e. if the extracellular K electrode was sensing an undisturbed $[K]$ deep in the restricted spaces of the preparation it would be difficult to find fault with this argument. And, indeed, I was myself initially very puzzled by this result since, whatever its quantitative role in pace-maker activity, g_K does switch on significantly during each action potential and must decay again afterwards. When developing the DiFrancesco–Noble (1984) equations into a version for the s.a. node (Noble & Noble, 1984) we therefore made use of the fact that the model represents non-uniform distribution of K ions in the extracellular space with appropriate partial differential equations. Our computations (see Brown *et al.* 1984*b*, Fig. 4) confirm that the argument of Maylie *et al.* (1981) would indeed be valid for an undisturbed space but they also show that it would not be valid for a near surface or for a disturbed space. In such cases it is then quite easy to compute a relatively constant $d[K]/dt$ even when i_K varies substantially.

   Very recently, Noma, Morad & Irisawa (1983) have published a paper using Ca block of i_t to show that it plays only a minor role in the s.a. node pace-maker potential, though they still give the most important role to i_{si} (see my review of this view above).

   The third theory mentioned above is that of Pollack who argues that intercellular mechanical forces are important in pace-maker activity. Part of the basis for this view is that intercellular electrical connexions (nexuses) are fairly rare in s.a. node tissue and that conduction may itself require other forms of intercellular interaction (see Pollack, 1976). There are two main arguments against this line of approach. Perhaps

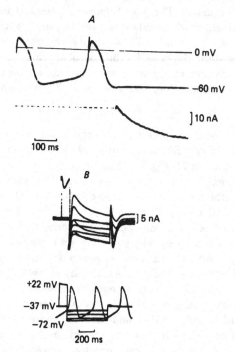

Fig. 10. Estimation of relative contributions of i_K and i_f to the pace-maker depolarization in rabbit s.a. node. *A*, this shows the current relaxation recorded when the voltage clamp is imposed at the maximum diastolic potential (in this case -60 mV). About 20 nA of current relaxation in the inward direction occurs during the period of time (about 200 ms) taken up by the pace-maker depolarization. *B*, this shows how the likely relative contributions of i_K decay and i_f onset to the current relaxation can be estimated. First, a depolarizing pulse (in this case to $+22$ mV for 100 ms) is found that activates a current relaxation very similar to that activated by an action potential. Not surprisingly, this requires a square pulse similar in amplitude and duration to the action potential. The current relaxations at various potentials produced by this pulse are then compared with those produced by simple hyperpolarization from a potential (in this case -37 mV) chosen to be at the foot of the i_K activation curve (so that no i_K decay contributes to the relaxations). This voltage is also positive to the foot of the i_f activation curve. Subtraction of the two relaxations at each potential then gives an estimate of the contribution of i_K decay. It can be seen that, at all potentials within the pace-maker range, i_K decay is much larger than i_f onset. Only when E_K is approached (towards -80 mV) does i_f decay dominate. (Brown *et al.* 1984*b*.)

the most powerful is that isolated single cells show normal pace-maker activity. Shibata & Giles (1983) have analysed the ionic currents underlying pace-maker activity in isolated frog sinus venosus cells and their work not only shows that rhythmic activity occurs in single node cells but also that it conforms to the g_K decay hypothesis, with i_{si} playing a role towards the end of the pace-maker depolarization.

THE SURPRISING HEART 35

Isolated mammalian s.a. node cells also show perfectly normal rhythmic activity (Irisawa, Nakayama, Kurachi & Noma, 1983). The other line of argument is a theoretical one. Astonishingly, few nexus channels are required to allow synchronization of pacing cells (de Haan, 1982). In fact, a *single* nexus channel per cell could be sufficient (see Noble, 1982a). This density would be even lower than that resolvable by electron microscopical techniques!

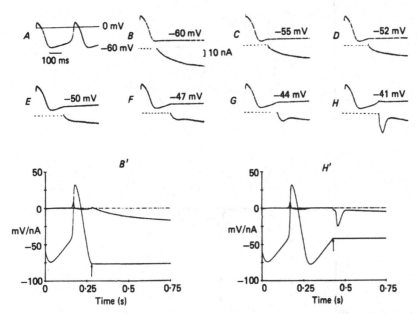

Fig. 11. Progressive change in current relaxations as the s.a. node pace-maker potential develops. The first change is that the current changes due to i_K decay and i_t onset (see Fig. 10) become smaller. This happens at about the speed expected for the i_K decay process, whose time constant varies only slightly over this narrow range of potentials. Then, beginning at about -50 mV the relaxation speeds up again and by -47 mV it is quite clear that an additional inward transient is superimposed on the remaining decay of i_K. This transient becomes larger and faster during the last few millivolts before the action potential upstroke occurs. Records B' and H' show computer reconstructions of traces similar to those shown experimentally in B and H. The slow transient inward current here is the component we have called $i_{si,2}$. See text for discussion of presence and role of $i_{Ca,t}$. (From Brown *et al.* 1984b.)

Recent evidence for the g_K decay hypothesis

It will already be evident from this discussion that I favour the g_K hypothesis in the case of the s.a. node, despite having abandoned it in the case of the Purkinje fibre. The time has come to review carefully the evidence in favour of this view. Some of this evidence has already been discussed in rejecting the alternative views. But there remains a body of evidence which is more directly aimed at substantiating the hypothesis. This has been described recently in Brown *et al.* (1982) and Brown *et al.* (1984b).

A key feature of the work described in these papers is to perturb the preparation and its normal pace-maker activity as little as possible. Natural pace-maker activity is allowed to develop and the voltage clamp is applied at various times during the depolarization. The results are then supplemented with the results of more

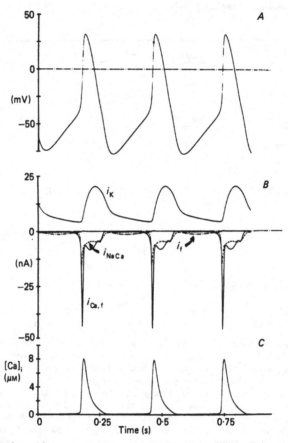

Fig. 12. Computed records summarizing our present picture of normal rhythmic activity in the s.a. node. *A*, computed variation in membrane potential. *B*, computed variations in $i_{Ca,t}$, $i_{si,2}$ (reconstructed in the model as the Na–Ca exchanger current, i_{NaCa}), i_K and i_f. These results resemble those computed by Noma & Irisawa's group, except for the presence of $i_{si,2}$, which is not represented in their model. *C*, computed variations in $[Ca]_i$. (Based on Brown *et al.* 1984*b*.)

conventional voltage-clamp pulse protocols. This approach is not new. It was pioneered by Vassalle (1966) in his work on the Purkinje fibre pace-maker depolarization. The advantage of this protocol is that it enables quantitative assessments to be made of the magnitudes of the various current components flowing at various times during the pace-maker depolarization. The disadvantage is that these currents are very small, as indeed they must be given the very small rate of change of potential during the pace-maker potential.

Fig. 10 shows the result of such an experiment on the rabbit s.a. node. It is clear that when the clamp is imposed at the beginning of the pace-maker depolarization there is a current relaxation that could be either decay of i_K or onset of i_f. Fig. 10B shows how the relative contributions can be assessed. First, the record at the beginning of the pace-maker depolarization was reproduced by recording the current

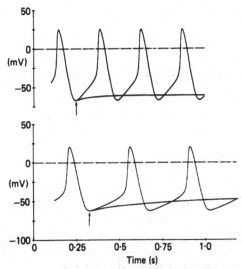

Fig. 13. Quantitative demonstration of the essential role of g_K decay in generation of s.a. node pace-maker potential. The top records show the behaviour of our model, while the lower records show the behaviour of the 'Japanese' model when the decay of the gating variable controlling i_K is stopped at the time of the maximum diastolic potential. The two models then agree in showing only a very small pace-maker depolarization which is attributable to the small influence of i_f.

following a short depolarizing pulse that activates a similar current relaxation as did the naturally occurring action potential. A second voltage-clamp trace is then obtained by omitting the depolarizing pulse. The key feature of this experiment is that the initial holding potential is chosen to be at the bottom of the i_K activation curve. Thus, no i_K will be present in the second trace and any remaining current relaxation may be attributed to i_f. Clearly, i_K decay forms by far the largest component of the total current relaxation. This approach may be extended to determine how quickly i_K decay occurs during the pace-maker depolarization. It is clear from Fig. 11 that when the pace-maker potential has reached the last third or so of its time course the current recorded ceases to be dominated by a monotonic relaxation and is increasingly dominated by a transient inward current whose time course identifies it as $i_{si,2}$. It would be tempting from this to conclude that the component of i_{si} involved is a very slow component, such as would be produced by i_{NaCa}. And, indeed, the model results, which incorporate a description of i_{NaCa}, do reproduce the very slow current (see Fig. 11). However, if the model is a good reconstruction then this tempting conclusion would be incorrect. The component of i_{si} that is involved in the last third of the pace-maker depolarization in the model is the fast component, which I have called $i_{Ca,f}$. The reason it is not seen either in

38   *D. NOBLE*

the experimental or model results is that, at each voltage, it is already activated. On this time scale, the speed of the fast component of i_{si} ensures that it is almost at its steady-state level at each voltage. There are two further reasons for thinking that the i_{TI}-like component of i_{si} is not crucial for pace-making. First, if it is induced by the $[Ca]_i$ transient that produces contraction then contraction would begin during

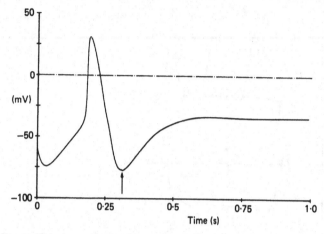

Fig. 14. This shows the complementary calculation to that shown in Fig. 13. Here, instead of stopping the i_K decay, we blocked the Ca current $i_{Ca,f}$. It is clear that the first two-thirds or so of the pace-maker depolarization is unaffected by this change. I have not shown the equivalent behaviour of the Yanagihara *et al.* (1980) model since, as the authors note (see their p. 854) the model fails to reproduce the effect of Ca current block. The difference lies in the fact that the 'Oxford' model distinguishes between the D-600-sensitive Ca current and the background inward currents, whereas the 'Japanese' model does not. This difference does though serve to emphasize the importance of background inward current in pace-maker activity in the heart. These background channels remain the least characterized of the current components.

the pace-maker depolarization. This is not the case. The second reason is that the results shown in Fig. 10 are not found in all experiments. In some cases, the i_{TI}-like component is not evident until one clamps at a time and voltage which would more properly be counted as part of the upstroke of the action potential. I think this reflects natural variation in the membrane depolarization or the level of $[Ca]_i$ at which the regenerative release of Ca (which is what in turn generates i_{NaCa} in the model) occurs.

These findings are summarized in Fig. 12 which shows our present interpretation of the variations in $i_{Ca,f}$, i_K, i_f and $i_{si,2}$ (here attributed in the model to i_{NaCa}) during natural pace-maker activity in the s.a. node.

There are two further uses to which the model may be put that are relevant to testing the g_K decay hypothesis.

Fig. 13 shows the result of 'freezing' the i_K gating process at the time of the maximum diastolic potential. It is clear that this abolishes pace-maker activity. The small depolarization that remains illustrates the small calculated contribution of i_f to the pace-maker potential. In this Figure I have shown the result of this computation both on the s.a. node model based on the DiFrancesco–Noble (1984) equations and on the model developed by Yanagihara *et al.* (1980). It is clear that the results are essentially identical, as indeed they should be since the kinetic description of i_K is fairly similar in the two models.

The second use of the model is shown in Fig. 14. This shows the result of blocking the fast Ca current, as would occur in the presence of D-600. It can be seen that the great majority of the pace-maker depolarization still occurs and the membrane settles down at a fairly depolarized level (around -35 mV). This is also what is found experimentally.

Contribution of i_f to natural pace-maker activity

Can we produce more direct evidence for the small contribution made by i_f? This has been done experimentally by using Cs ions which at low concentrations (1 mM) is a fairly selective blocker of i_f. This usually slows the pace-maker rhythm by about 10–20 % (Brown *et al.* 1982; Brown *et al.* 1984*b*) which corresponds well to the expected contribution based on the model computations. Noma, Morad & Irisawa (1983) have also recently described the use of Cs in a paper in which they propose that i_f is not involved at all. Their experimental results though show a small slowing (see their figs. 1 and 5) of the same order of magnitude as in our experiments.

The reason I have emphasized this small role of i_f is that it is still an open question whether i_f might not play a larger role in some circumstances. For example, it is increased by adrenaline and it may be that some part of the acceleration induced by adrenaline may be attributable to i_f, particularly since the s.a. node is not at all uniform in its electrophysiology (see review by Brown, 1982). Shifts of the dominant pace-maker region are known to occur in adrenaline, possibly to cells in which i_f makes more contribution. In this connexion it is interesting to note that adrenaline accelerates the depolarization *throughout* the duration of the pace-maker potential (Hutter & Trautwein, 1956). If it acted only by increasing i_{si} one would expect the acceleration to occur only during the second half of the pace-maker potential.

i_{K1} in the s.a. node

I have dealt with the role of i_K, i_f and i_{si} in s.a. node pace-maker activity. What about i_{K1}? The significant feature here is that it is nearly absent. This has very important consequences since it is the sensitivity of i_{K1} to $[K]_o$ that allows very moderate changes in $[K]_o$ to modulate greatly the frequency of the pace-maker rhythm in Purkinje fibres. By contrast the s.a. node rhythm is much less sensitive to $[K]_o$. This lack of sensitivity is well reproduced by the model equations (see Noble & Noble, 1984, Fig. 3).

5. Abnormal rhythms and the possible roles of i_{NaCa} or i_{NaK}

It will already be clear from this review that, in clear contrast to the situation only a few years ago, there is considerable diversity in the pace-maker mechanisms of the heart. So far, I have reviewed three mechanisms: g_K decay (in the s.a. node), i_f –mainly Na current – onset (in normal Purkinje fibre rhythm) and time-dependent changes in i_{K1} (ventricular rhythm induced by Ba ions). These mechanisms do nevertheless have one important feature in common: they all depend on surface membrane voltage-induced current relaxations. Hence, the underlying oscillations of membrane

40   *D. NOBLE*

current are completely suppressed in voltage-clamp conditions. The fourth mechanism does *not* share this feature. This is the rhythm generated when the Na–K pump is blocked. Lederer & Tsien (1976) showed that this rhythm depends on an inward current, i_{TI}, that is not identified with variations in voltage-dependent currents normally implicated in cardiac rhythm. Moreover, oscillations in this current can continue even in voltage-clamp conditions and have therefore been attributed to current changes induced by intracellular oscillations in [Ca] (see review by Tsien, Kass & Weingart, 1979). My colleagues and I have recently investigated a possible theoretical model for such oscillation based on three main assumptions.

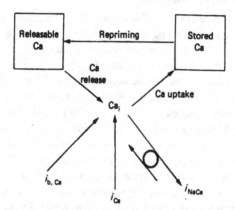

Fig. 15. Diagram summarizing the processes assumed in the DiFrancesco–Noble model to control Ca ion movements. An energy-consuming pump is assumed to transport Ca into the sarcoplasmic reticulum which then reprimes a release store. Release is activated by cytoplasmic free Ca (Ca_i). Ca leaves the cell via the Na–Ca exchanger and may enter the cell through the voltage-gated Ca channel i_{Ca} or through a background (leak) channel $i_{b, Ca}$. Rarely, it may enter the cell through the exchanger (when $[Ca]_i$ is very low and the voltage very positive). These were the minimum assumptions necessary to model the $[Ca]_i$ transient and to model oscillatory variations at raised $[Ca]_i$. There are though some important Ca movements that are neglected in this simple model, such as binding to contractile proteins, to calmodulin, buffering by mitochondria, and transport via a surface ATP-driven pump. The purpose of the model is restricted to demonstrating some minimal assumptions required to model the behaviour of $[Ca]_i$ for electrophysiological purposes.

1. That Ca ion release from the sarcoplasmic reticulum is induced by Ca ions (the Ca-induced Ca release hypothesis of Fabiato & Fabiato, 1975).

2. That, following release, the release store is slowly reprimed by a voltage-dependent process (this idea was first proposed for skeletal muscle by Hodgkin & Horowicz, 1960 and Gibbons & Fozzard (1975) have described very similar results in cardiac muscle).

3. That variations in $[Ca]_i$ induce variations in the current carried by the Na–Ca exchange mechanism. For this we have used a model based on that of Mullins (1977, 1981) – see DiFranceso & Noble (1984).

These main assumptions are illustrated diagrammatically in Fig. 15. When incorporated into a model of cardiac electrical activity that represents the other membrane currents, it is possible to show that Na–K pump inhibition is then a sufficient condition to induce oscillations even under voltage-clamp conditions. An

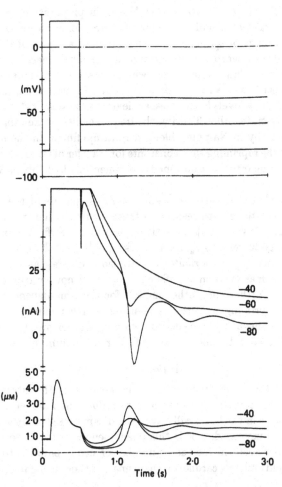

Fig. 16. Use of the DiFrancesco–Noble equations to reproduce oscillatory current changes and variations in [Ca]$_i$ when the Na–K exchange pump is partially blocked. In this particular set of computations the pump was reduced to 25 % of its normal capacity, which allows [Na]$_i$ to rise from its normal value of about 8 mM to about 20 mM. This is a sufficient condition for the generation of oscillatory current changes. Top: voltage-clamp steps. Middle: computed net membrane currents. Superimposed on the onset of i_f and decay of i_K (which are responsible for the general downward current relaxation at the end of each step) there is an inward current whose onset and decay are almost symmetric (cf. Fig. 1C), followed by a damped oscillation. The peak amplitude is voltage dependent and the steepness of this dependence on voltage depends on the function chosen for the voltage dependence of the Na–Ca exchanger. In this case a steep sinh function (cf. Mullins, 1981) was used and this reproduces the kind of experiment (cf. Hennings & Vereecke, 1983) in which i_{TI} falls towards zero at about -40 mV. A small phase shift also occurs (cf. Brown *et al.* 1984). Bottom: computed variations on [Ca]$_i$. Note that, before the oscillatory variations, the Ca concentration falls below the steady-state level. (D. DiFrancesco, G. Hart & D. Noble, in preparation; see DiFrancesco, Hart & Noble, 1983.)

42 *D. NOBLE*

example (from D. DiFrancesco, G. Hart & D. Noble, in preparation) is shown in Fig. 16. There are several features of this result that are worth noting. First, as in the experimental recordings of i_{TI} (or of $i_{si,2}$ – see previous sections of this review) the onset and decay of the inward current are almost symmetric. Secondly, they are generated by oscillations in $[Ca]_i$ that occur following repolarization. These oscillations (like those recently described experimentally using aequorin – see Allen, Eisner, Lab & Orchard, 1983) lead to Ca levels *below* those achieved in the steady state as well as levels above the steady-state value. Thirdly, the frequency lies in the range 1–2 Hz, as also seen experimentally in Na-pump-blocked Purkinje fibres. In the model this frequency depends on the repriming time constants for transfer of Ca to the releasable state. We used figures similar to those obtained experimentally by Gibbons & Fozzard (1975).

I should emphasize that there may be nothing very unique about the particular model used in these equations with respect to these results. Other models for the current involved (e.g. allowing $[Ca]_i$ to activate the non-specific channels, i_{NaK}, rather than the exchange current, i_{NaCa}) might work equally well, though there would be some differences since i_{NaCa} *contributes* to variations in $[Ca]_i$ as well as being activated by them, whereas the non-specific current would not contribute directly to variations in $[Ca]_i$. Furthermore, other models for the time dependence of the release process and its repriming are perfectly plausible. Further clarification of these abnormal rhythm mechanisms will depend on obtaining further precise evidence on the composition of i_{TI} and on the mechanisms of internal calcium release.

Conclusions

It will be clear from this review that anyone who hoped that the introduction of single-cell and patch-clamp methods would help to reduce the number of separate ionic current mechanisms in the heart will have been disappointed. On the contrary, the number has further increased. Nor is this phenomenon restricted to the heart. In a recent review I discussed the increasing number of strong parallels to be found between many of the old and new cardiac mechanisms and those being found in nerve cells (Noble, 1983).

It is interesting to compare the number of separate mechanisms postulated in the 1962, 1975 and 1984 Purkinje fibre models. I have done this in Table 1.

The horizontal and vertical lines and organization of this Table are all significant. Horizontal lines are used to classify the current mechanisms according to the types listed at the left, while vertical lines are used to indicate where reinterpretations have taken place. This allows mechanisms that have been reinterpreted to be classified both under their original category and in their new category. At the right I have listed the main activators and examples of known blockers.

It is sincerely to be hoped that this is not a case of indefinite exponential growth or, by the year 2000, we shall all have great difficulty in explaining cardiac excitation to ourselves, let alone to students with formidable memories!

Can we really be sure that so many separate components exist? It is worth noting that, of the mechanisms that could appear as current jumps in patch-clamp studies (clearly, pump and exchange mechanisms are well beyond resolution at single-current jump level), the majority have already been identified in such studies. These include

THE SURPRISING HEART 43

TABLE 1. Summary of ionic current mechanisms in the heart

Ionic current mechanisms in heart

Row	1962	1975	1984	Activator	Blocker
Na current	i_{Na}	i_{Na}	i_{Na}	Depol.	TTX
K currents	i_{K1}	i_{K1}	i_{K1}	Hyperpol.	Ba
	i_{K2}	i_x	i_K	Depol.	Ba
Cl current		i_{qr}	i_{to}	$[Ca]_i$ depol.	4-AP
		i_{K2}	i_f	$[K]_o$ hyperpol.	Cs
Non-specific currents			i_{NaK} ($?i_{TI}$)	$[Ca]_i$	?
Ca currents			$i_{Ca,f}$	Depol. (Ca inact)	Cd D-600
		i_{si}	$i_{Ca,s}$	Depol.	?
			$i_{si,2}$ ($?i_{TI}$)	$[Ca]_i$	?
Exchange and pump currents			i_{NaCa}	$[Ca]_i$ $[Na]_o$	?
			i_p	$[K]_o$ $[Na]_i$	Ouabain
Passive background currents			$i_{b,Na}$		
	i_b	$i_{b,in}$	$i_{b,Ca}$		
	i_{Cl}		i_{Cl}		

The Table lists the current components assumed in the 1962, 1975 and 1984 Purkinje fibre models and classifies them according to their ionic composition. Where this has been reinterpreted, a vertical line is used. Where the terminology has changed, e.g. from i_{K2} to i_x to i_K, this is indicated by arranging the relevant terms on the same horizontal row. Even where the terminology has not changed though (e.g. as with i_{K1}) there may have been significant changes in the formulation of the equations for the component in the different models. The main uncertainty in this Table now lies in the area between the Ca currents, the exchange pump current and the Ca-activated non-specific current. Our interpretation of the second component of i_{si} ($i_{si,2}$) is that it may be the same mechanism as the transient inward current, i_{TI} and that both are to a large extent carried by the Na–Ca exchanger as i_{NaCa}, but another possibility is that these components are carried by the non-specific current, i_{NaK}, activated by intracellular Ca. i_{TI} is therefore included in the Table together with a query in both the appropriate places.

44   *D. NOBLE*

to my knowledge: i_{Na}, i_{K1}, i_K, i_{to} (probably – at least Ca-activated K currents have been seen), i_{NaK}, $i_{Ca,f}$. Of the components that remain: i_p has been seen in single cells; $i_{Ca,s}$ has also been seen (indeed, identified as such) in single cells, but may be very difficult to find in patch studies of single channels since either the density of channels is much lower (only about 20 %) than for $i_{Ca,f}$ or the single channel current is much smaller. Thus we are left only with the background currents and $i_{si,2}/i_{NaCa}$. Could we, at least, eliminate these?

In theoretical work with DiFrancesco and Hart, I have tried very hard to do so. For, when Colqhoun *et al.* (1981) identified their Ca-activated non-specific channels in patch-clamp studies it became very tempting to conclude that here, at last, were the background current channels. And, of course, if they were significantly activated even at resting levels of $[Ca]_i$, then *a fortiori* they would be activated during $[Ca]_i$ transients so there would be no need to postulate a separate mechanism for $i_{si,2}$ and indeed no particular need to include the exchanger current, i_{NaCa}.

The opposite position has been very forcefully expounded by Mullins in his recent book *Ion Transport in Heart* (Mullins, 1981) where i_{NaCa} (called i_c by Mullins) has been proposed as responsible for a wide variety of already identified current components in the heart. I have already expressed my reasons for doubting some of Mullins' conclusions (see Noble, 1982*b*). But, on his essentially important conclusion, that the exchanger current may contribute to the second inward current, I think I must agree. My reasons are primarily theoretical, but they are as well researched as were the reasons in the early 1960s for being convinced, on theoretical grounds, that i_K (or something like it) must exist (I have discussed this point at the beginning of this article). They also illustrate what I see to be one of the main purposes of theoretical models in this field: to test the *quantitative* plausibility of various hypotheses.

And it is on grounds of quantitative implausibility that I find it very difficult to give a major role to the non-specific current, i_{NaK}, in normal electrical activity. I have reviewed the arguments in earlier sections of this review and they lead me to conclude that, if we are to reduce Table 1 at all it will be by identifying $i_{si,2}$ as largely i_{NaCa}. We could then make the Table much neater by closing the uncertain gap between Ca currents and exchange currents. This, or some alternative rationalization of the problems I have discussed here, must be one of the major aims of single cell work in the immediate future.

This review lecture was written during the tenure of a Research Fellowship for academic staff awarded by the Medical Research Council.

I should like to acknowledge valuable discussions with Jean Banister, Hilary Brown, Dario DiFrancesco, Wayne Giles, George Hart, Junko Kimura, Kai Lee, Susan Noble, Trevor Powell, Tony Spindler and Anne Taupignon.

Note added in proof. There are, of course, surprising and exciting developments that could not be covered in this review. One of the most important of these (the recording of single channels) has been dealt with in a valuable review by Reuter (*A. Rev. Physiol.* 1984, in the Press).

THE SURPRISING HEART 45

REFERENCES

AKAIKE, N., LEE, K. S. & BROWN, A. M. (1978). The calcium current of *Helix* neuron. *J. gen. Physiol.* **71**, 509–531.

ALLEN, D. G., EISNER, D. A., LAB, M. J. & ORCHARD, C. H. (1983). Oscillations of intracellular [Ca^{2+}] in ferret ventricular muscle. *J. Physiol.* **336**, 64–65P.

ALLEN, D. G. & KURIHARA, S. (1980). Calcium transients in mammalian ventricular muscle. *Eur. Heart J.* **1**, suppl. A, 5–15.

ARLOCK, P. & KATZUNG, B. G. (1982). Effects of sodium substitutes on ouabain induced transient inward current. *Proc. West. Pharmacol. Soc.* **25**, 57–60.

ATTWELL, D., COHEN, I., EISNER, D. A., OHBA, M. & OJEDA, C. (1979). The steady-state TTX sensitive ('window') sodium current in cardiac Purkinje fibres. *Pflügers Arch.* **379**, 137–142.

ATTWELL, D., EISNER, D. A. & COHEN, I. (1979). Voltage clamp and tracer flux data: effects of a restricted extracellular space. *Q. Rev. Biophys.* **12**, 213–261.

ATTWELL, D. & JACK, J. J. B. (1978). The interpretation of current–voltage relations: a Nernst–Planck analysis. *Prog. Biophys. molec. Biol.* **34**, 81–107.

BEELER, G. W. & REUTER, H. (1970). Membrane calcium current in ventricular myocardial fibres. *J. Physiol.* **207**, 165–190.

BODEWEI, R., HERING, S., LEMKE, B., ROSENSHTRAUKH, L. V., UNDROVINAS, A. I. & WOLLEN-BERGER, A. (1982). Characterization of the fast sodium current in isolated rat myocardial cells: simulation of the clamped membrane potential. *J. Physiol.* **325**, 301–315.

BOYETT, M. R. (1981). A study of the effect of the rate of stimulation on the transient outward current in sheep cardiac Purkinje fibres. *J. Physiol.* **319**, 1–22.

BOYETT, M. R. & JEWELL, B. R. (1980). Analysis of the effects of changes in rate and rhythm upon electrical activity in the heart. *Prog. Biophys. molec. Biol.* **36**, 1–52.

BROWN, A. M., LEE, K. S. & POWELL, T. (1981). Sodium current in single rat heart muscle cells. *J. Physiol.* **318**, 479–500.

BROWN, H. F. (1982). Electrophysiology of the sinoatrial node. *Physiol. Rev.* **62**, 505–530.

BROWN, H. F., CLARK, A. & NOBLE, S. J. (1972). Pacemaker current in frog atrium. *Nature, New Biol.* **235**, 30–31.

BROWN, H. F., CLARK, A. & NOBLE, S. J. (1976). Identification of the pace-maker current in frog atrium. *J. Physiol.* **258**, 521–545.

BROWN, H. F. & DiFRANCESCO, D. (1980). Voltage-clamp investigations of membrane currents underlying pace-maker activity in rabbit sino-atrial node. *J. Physiol.* **308**, 331–351.

BROWN, H. F., DiFRANCESCO, D., NOBLE, D. & NOBLE, S. J. (1980). The contribution of potassium accumulation to outward currents in frog atrium. *J. Physiol.* **306**, 127–149.

BROWN, H. F., GILES, W. R. & NOBLE, S. J. (1977). Membrane currents underlying activity in frog sinus venosus. *J. Physiol.* **271**, 783–816.

BROWN, H. F., KIMURA, J. & NOBLE, S. J. (1981). Calcium entry dependent inactivation of the slow inward current in the rabbit sino-atrial node. *J. Physiol.* **320**, 11P.

BROWN, H. F., KIMURA, J. & NOBLE, S. J. (1982). The relative contributions of various time-dependent membrane currents to pacemaker activity in the sino-atrial node. In *Cardiac Rate and Rhythm*, ed. BOUMAN, L. N. & JONGSMA H. J., pp. 53–68. Martinus Nijhoff: The Hague.

BROWN, H. F., KIMURA, J., NOBLE, D., NOBLE, S. J. & TAUPIGNON, A. (1983). Two components of 'second inward current' in the rabbit SA-node. *J. Physiol.* **334**, 56–57P.

BROWN, H. F., KIMURA, J., NOBLE, D., NOBLE, S. J. & TAUPIGNON, A. (1984a). Mechanisms underlying the slow inward current, i_{si} in the rabbit sino-atrial node investigated by voltage-clamp and computer simulation. *Proc. R. Soc.* B (in the Press).

BROWN, H. F., KIMURA, J., NOBLE, D., NOBLE, S. J. & TAUPIGNON, A. (1984b). The ionic currents underlying pace-maker activity in rabbit sino-atrial node: experimental results and computer simulations. *Proc. R. Soc.* B (in the Press).

BROWN, H. F., NOBLE, D., NOBLE, S. & TAUPIGNON, A. (1984). Transient inward current and its relation to the very slow inward current in the rabbit SA node. *J. Physiol.* **349**, 47P.

CACHELIN, A. B., DE PEYER, J. E., KOKUBUN, S. & REUTER, H. (1983). Sodium channels in cultured cardiac cells. *J. Physiol.* **340**, 389–401.

CALLEWAERT, G., CARMELIET, E., VAN DER HEYDEN, G. & VEREECKE, J. (1982). The pace-maker current in a single cell preparation of bovine cardiac Purkinje fibres. *J. Physiol.* **326**, 66–67P.

46   *D. NOBLE*

CARMELIET, E. (1961). Chloride ions and the membrane potential of Purkinje fibres. *J. Physiol.* **156**, 375–388.

CARMELIET, E. (1980). Decrease of K efflux and influx by external Cs ions in cardiac Purkinje and muscle cells. *Pflügers Arch.* **383**, 143–150.

CARMELIET, E. (1982). Induction and removal of inward-going rectification in sheep cardiac Purkinje fibres. *J. Physiol.* **327**, 285–308.

CARMELIET, E., VAN DER HEYDEN, G. & VEREECKE, J. (1983). Spontaneous activity in cardiac ventricular cells. *Proc. Int. Union physiol. Sci.* **15**, 119.

CLAY, J. R. & SHRIER, A. (1981*a*). Analysis of subthreshold pace-maker currents in chick embryonic heart cells. *J. Physiol,* **312**, 471–490.

CLAY, J. R. & SHRIER, A. (1981*b*). Developmental changes in subthreshold pace-maker currents in chick embryonic heart cells. *J. Physiol.* **312**, 491–504.

COHEN, I., DAUT, J. & NOBLE, D. (1976). The effects of potassium and temperature on the pace-maker current, i_{K_2}, in Purkinje fibres. *J. Physiol.* **260**, 55–74.

COHEN, I., NOBLE, D., OHBA, M. & OJEDA, C. (1979). Action of salicylate ions on the electrical properties of sheep cardiac Purkinje fibres. *J. Physiol.* **297**, 163–185.

COLATSKY, T. J. (1980). Voltage clamp measurement of sodium channel properties in rabbit cardiac Purkinje fibres. *J. Physiol.* **305**, 215–234.

COLQHOUN, D., NEHER, E., REUTER, H. & STEVENS, C. F. (1981). Inward current channels activated by intracellular Ca in cultured heart cells. *Nature, Lond.* **294**, 752–754.

CORABOEUF, E. & CARMELIET, E. (1982). Existence of two transient outward currents in sheep cardiac Purkinje fibres. *Pflügers Arch.* **392**, 352–359.

DAUT, J. & RÜDEL, R. (1982). The electrogenic sodium pump in guinea-pig ventricular muscle: inhibition of pump current by cardiac glycosides. *J. Physiol.* **330**, 243–264.

DECK, K. A., KERN, R. & TRAUTWEIN, W. (1964). Voltage clamp technique in mammalian cardiac fibres. *Pflügers Arch.* **280**, 50–62.

DECK, K. A. & TRAUTWEIN. (1964). Ionic currents in cardiac excitation. *Pflügers Arch.* **280**, 65–80.

DE HAAN, R. L. (1982). In vitro models of entrainment of cardiac cells. In *Cardiac Rate and Rhythm*, ed. BOUMAN, L. N. & JONGSMA, H. J., pp. 323–359. Martinus Nijhoff: The Hague.

DIFRANCESCO, D. (1981*a*). A new interpretation of the pace-maker current i_{K_2} in Purkinje fibres. *J. Physiol.* **314**, 359–376.

DIFRANCESCO, D. (1981*b*). A study of the ionic nature of the pace-maker current in calf Purkinje fibres. *J. Physiol.* **314**, 377–393.

DIFRANCESCO, D. (1982). Block and activation of the pace-maker channel in calf Purkinje fibres: effects of potassium, caesium and rubidium. *J. Physiol.* **329**, 485–507.

DIFRANCESCO, D. (1984). Characterization of the pace-maker current kinetics in calf Purkinje fibres. *J. Physiol.* **348**, 341–367.

DIFRANCESCO, D., HART, G. & NOBLE, D. (1983). Demonstration of oscillatory variations in [Ca]$_i$ and membrane currents in a computer model of Ca-induced Ca release in mammalian Purkinje fibre and ventricular muscle. *J. Physiol.* **334**, 8–9P.

DIFRANCESCO, D. & NOBLE, D. (1980). If 'i_{K_2}' is an inward current, how does it display potassium specificity? *J. Physiol.* **305**, 14–15P.

DIFRANCESCO, D. & NOBLE, D. (1982). Implications of the re-interpretation of i_{K_2} for the modelling of the electrical activity of pace-maker tissues in the heart. In *Cardiac Rate and Rhythm*, ed. BOUMAN, L. N. & JONGSMA, H. B., pp. 93–128. Martinus Nijhoff: The Hague.

DIFRANCESCO, D. & NOBLE, D. (1984). A model of cardiac electrical activity incorporating ionic pumps and concentration changes. *Phil Trans. R. Soc. B* (in the Press).

DIFRANCESCO, D., NOMA, A. & TRAUTWEIN, W. (1979). Kinetics and magnitude of the time-dependent K current in the rabbit s.a. node: effect of external potassium. *Pflügers Arch.* **381**, 271–279.

DIFRANCESCO, D., OHBA, M. & OJEDA, C. (1979). Measurement and significance of the reversal potential for the pace-maker current (i_{K_2}) in sheep Purkinje fibres. *J. Physiol.* **297**, 135–162.

DIFRANCESCO, D. & OJEDA, C. (1980). Properties of the current i_f in the sino-atrial node of the rabbit compared with those of the current i_{K_2} in Purkinje fibres. *J. Physiol.* **308**, 353–367.

DUDEL, J., PEPER, K., RUDEL, R. & TRAUTWEIN, W. (1967). The dynamic chloride component of membrane current in Purkinje fibres. *Pflügers Arch.* **295**, 197–212.

THE SURPRISING HEART 47

EARM, Y. E., SHIMONI, Y. & SPINDLER, A. J. (1983). A pace-maker-like current in the sheep atrium and its modulation by catecholamines. *J. Physiol.* **342**, 569–590.

EBIHARA, L., SHIGATO, N., LIEBERMAN, M. & JOHNSON, E. A. (1980). The initial inward current in spherical clusters of chick embryonic heart cells. *J. gen. Physiol.* **75**, 437–456.

ECKERT, R. & LUX, H. D. (1975). A non-inactivating inward current recorded during small depolarizing voltage steps in snail pace-maker neurones. *Brain Res.* **83**, 486–489.

EISNER, D. N. & LEDERER, W. J. (1979). Inotropic and arrythmogenic effects of potassium-depleted solutions on mammalian cardiac muscle. *J. Physiol.* **294**, 255–277.

EISNER, D. A. & LEDERER, W. J. (1980). Characterization of the sodium pump in cardiac Purkinje fibres. *J. Physiol.* **303**, 441–474.

EISNER, D. A., LEDERER, W. J. & NOBLE, D. (1979). Caffeine and tetracaine abolish the slow inward current in sheep cardiac Purkinje fibres. *J. Physiol.* **293**, 76P.

FABIATO, A. & FABIATO, F. (1975). Contractions induced by a calcium-triggered release from the sarcoplasmic reticulum of single skinned cardiac cells. *J. Physiol.* **249**, 469–495.

FATT, P. & KATZ, B. (1951). An analysis of the end-plate potential recorded with an intracellular electrode. *J. Physiol.* **115**, 320–370.

FISCHMEISTER, R. & VASSORT, G. (1981). The electrogenic Na/Ca exchange and the cardiac electrical activity. 1. Simulation on Purkinje fibre action potential. *J. Physiol., Paris.* **77**, 705–709.

FOZZARD, H. A. & HIRAOKA, M. (1973). The positive dynamic current and its inactivation properties in cardiac Purkinje fibres. *J. Physiol.* **234**, 569–586.

GADSBY, D. C. (1980). Activation of electrogenic Na^+/K^+ exchange by extracellular K^+ in canine cardiac Purkinje fibres. *Proc. natn. Acad. Sci. U.S.A.* **77**, 4035–4039.

GIBBONS, W. R. & FOZZARD, H. A. (1975). Relationships between voltage and tension in sheep cardiac Purkinje fibres. *J. gen. Physiol.* **65**, 345–365.

HALL, A. E., HUTTER, O. F. & NOBLE, D. (1963). Current–voltage relations of Purkinje fibres in sodium-deficient solutions. *J. Physiol.* **166**, 225–240.

HART, G. (1983). The kinetics and temperature dependence of the pace-maker current, i_f, in sheep Purkinje fibres. *J. Physiol.* **337**, 401–416.

HAUSWIRTH, O., NOBLE, D. & TSIEN, R. W. (1969). The mechanism of oscillatory activity at low membrane potentials in cardiac Purkinje fibres. *J. Physiol.* **200**, 255–265.

HAUSWIRTH, O., NOBLE, D. & TSIEN, R. W. (1972). Separation of the pace-maker and plateau components of delayed rectification in cardiac Purkinje fibres. *J. Physiol.* **225**, 211–235.

HENNINGS, B. & VEREECKE, J. (1983). Effects of Cs and Ba on the transient inward current of sheep cardiac Purkinje fibres. *J. Physiol.* **345**, 149P.

HODGKIN, A. L. & HOROWICZ, P. (1960). Potassium contractures in single muscle fibres. *J. Physiol.* **153**, 386–403.

HODGKIN, A. L. & HUXLEY, A. F. (1952). A quantitative description of membrane current and its application to conduction and excitation in nerve. *J. Physiol.* **117**, 500–544.

HORACKOVA, M. & VASSORT, G. (1979). Sodium–calcium exchange in regulation of cardiac contractility. Evidence for an electrogenic, voltage-dependent mechanism. *J. gen. Physiol.* **73**, 403–424.

HUME, J. R. & GILES, W. R. (1982). Turn-off of a TTX-resistant inward current 'i_{Ca}^{++}' in single bullfrog atrial cells. *Biophys. J.* **37**, 240A.

HUME, J. R. & GILES, W. R. (1983). Ionic currents in single isolated bullfrog atrial cells. *J. gen. Physiol.* **81**, 153–194.

HUTTER, O. F. & NOBLE, D. (1960). Rectifying properties of cardiac muscle. *Nature, Lond.* **188**, 495.

HUTTER, O. F. & TRAUTWEIN, W. (1956). Vagal and sympathetic effects on the pace-maker fibres in the sinus venosus of the heart. *J. gen. Physiol.* **39**, 715–733.

HUXLEY, A. F. (1959). Ion movements during nerve activity. *Ann. N.Y. Acad. Sci.* **81**, 221–246.

IRISAWA, H., NAKAYAMA, T., KURACHI, Y. & NOMA, A. (1983). Ionic current density of the single pace-maker cell isolated from rabbit SA and AV nodes. *Proc. Int. Union physiol. Sci.* **15**, 50.

ISENBERG, G. (1976). Cardiac Purkinje fibres: caesium as a tool to block inward rectifying potassium currents. *Pflügers Arch.* **365**, 99–106.

ISENBERG, G. (1983). Isolated mammalian ventricular myocytes: Ca release may contribute to I_{Ca}. *Proc. Int. Union physiol. Sci.* **15**, 74.

48 *D. NOBLE*

ISENBERG, G. & KLOCKNER, U. (1982). Calcium currents of isolated bovine ventricular myocytes are fast and of large amplitude. *Pflügers Arch.* **395**, 30–41.

ISENBERG, G. & TRAUTWEIN, W. (1974). The effect of dihydro-ouabain and lithium ions on the outward current in cardiac Purkinje fibres. Evidence for electrogenicity of active transport. *Pflügers Arch.* **350**, 41–54.

JACOB, R., MURPHY, E. & LIEBERMAN, M. (1983). Electrogenic aspects of sodium–calcium exchange in cultured chick embryo heart cells. *J. Physiol.* **345**, 28P

JOHNSON, E. A., CHAPMAN, J. B. & KOOTSEY, J. M. (1980). Some electrophysiological consequences of electrogenic sodium and potassium transport in cardiac muscle: a theoretical study. *J. theor. Biol.* **87**, 737–756.

KAMEYAMA, M., KIYOSUE, T., SOEJIMA, M. & NOMA, A. (1983). I_{K1} channel in the rabbit ventricular cell is time- and voltage-dependent. *Proc. Int. Union physiol. Sci.* **15**, 50.

KARAGUEUZIAN, H. S. & KATZUNG, B. G. (1982). Voltage-clamp studies of transient inward current and mechanical oscillations induced by ouabain in ferret papillary muscle. *J. Physiol.* **327**, 255–271.

KASS, R. S., LEDERER, W. J., TSIEN, R. W. & WEINGART, R. (1978). Role of calcium ions in transient inward current and after contractions induced by strophanthidin in cardiac Purkinje fibres. *J. Physiol.* **281**, 187–208.

KASS, R. S., TSIEN, R. W. & WEINGART, R. (1978). Ionic basis of transient inward current induced by strophanthidin in cardiac Purkinje fibres. *J. Physiol.* **281**, 209–226.

KATZUNG, B. G. & MORGENSTERN, J. A. (1977). Effects of extracellular potassium on ventricular automaticity and evidence for a pace-maker current in mammalian ventricular myocardium. *Circulation Res.* **40**, 105–111.

KENYON, J. L. & GIBBONS, W. R. (1979). 4-aminopyridine and the early outward current of sheep cardiac Purkinje fibres. *J. gen. Physiol.* **73**, 139–157.

KIMURA, J. (1982). Electrophysiology of the mammalian SA node. D. Phil. thesis, Oxford University.

LEDERER, W. J. & EISNER, D. A. (1982). The effects of sodium pump activity on the slow inward current in sheep cardiac Purkinje fibres. *Proc. R. Soc.* B **214**, 249–262.

LEDERER, W. J. & TSIEN, R. W. (1976). Transient inward current underlying arrhythmogenic effects of cardiotonic steroids in Purkinje fibres. *J. Physiol.* **263**, 73–100.

LEE, E. W., LEE, K. S., NOBLE, D. & SPINDLER, A. J. (1983). A very slow inward current in single ventricular cells. *J. Physiol.* **345**, 6P.

LEE, E., LEE, K. S., NOBLE, D. & SPINDLER, A. J. (1984*a*). A new, very slow inward Ca current in single ventricular cells of adult guinea-pig. *J. Physiol.* **346**, 75P.

LEE, E., LEE, K. S., NOBLE, D. & SPINDLER, A. J. (1984*b*). Further properties of the very slow inward currents in isolated single guinea-pig ventricular cells. *J. Physiol.* **349**, 48P.

LEE, K. S., HUME, J. R., GILES, W. R. & BROWN, A. M. (1981). Sodium current depression by lidocaine and quinidine in isolated ventricular cells. *Nature, Lond.* **278**, 269–271.

LEE, K. S. & TSIEN, R. W. (1982). Reversal of current through calcium channels in dialysed single heart cells. *Nature, Lond.* **297**, 498–501.

LEE, K. S., WEEKS, T. A., KAO, R. L., AKAIKE, N. & BROWN, A. M. (1979). Sodium current in single heart muscle cells. *Nature, Lond.* **278**, 269–271.

MCALLISTER, R. E. & NOBLE, D. (1966). The time and voltage dependence of the slow outward current in cardiac Purkinje fibres. *J. Physiol.* **186**, 632–662.

MCALLISTER, R. E., NOBLE, D. & TSIEN, R. W. (1975). Reconstruction of the electrical activity of cardiac Purkinje fibres. *J. Physiol.* **251**, 1–59.

MCDONALD, T. F. & TRAUTWEIN, W. (1978). The potassium current underlying delayed rectification in cat ventricular muscle. *J. Physiol.* **274**, 217–246.

MARBAN, E. & TSIEN, R. W. (1981). Is the slow inward calcium current of heart muscle inactivated by calcium? *Biophys. J.* **33**, 143a.

MARBAN, E. & TSIEN, R. W. (1982). Enhancement of calcium current during digitalis inotropy in mammalian heart: positive feed-back regulation by intracellular calcium? *J. Physiol.* **329**, 589–614.

MAYLIE, J., MORAD, M. & WEISS, J. (1981). A study of pace-maker potential in the rabbit sino-atrial node: measurement of potassium activity under voltage-clamp conditions. *J. Physiol.* **311**, 161–178.

THE SURPRISING HEART   49

MENTRARD, D. & VASSORT, G. (1983) The Na–Ca exchange generates a current in frog heart cells. *J. Physiol.* **334**, 55P.

MENTRARD, D., VASSORT, G. & FISCHMEISTER, R. (1983). Calcium-mediated inactivation of the calcium conductance in caesium-loaded frog heart cells. *J. gen. Physiol.* (in the Press).

MITCHELL, M. R., POWELL, T., TERRAR, D. A. & TWIST, V. W. (1983). Characteristics of the second inward current in cells isolated from rat ventricular muscle. *Proc. R. Soc.* B **219**, 447–469.

MITCHELL, M. R., POWELL, T., TERRAR, D. A. & TWIST, V. W. (1984). Possible association between an inward current and the late plateau of action potentials in ventricular cells isolated from rat heart. *J. Physiol* (in the Press).

MIURA, D. S., HOFFMAN, B. F. & ROSEN, M. R. (1977). The effect of extracellular potassium ion activity on the intracellular potassium ion activity and transmembrane potentials of beating canine cardiac Purkinje fibres. *J. gen. Physiol.* **69**, 463–474.

MOMOSE, Y., SZABO, G. & GILES, W. R. (1983). An inwardly rectifying K^+ current in bullfrog atrial cells. *Biophys. J.* **41**, 311a.

MULLINS, L. J. (1977). A mechanism for Na/Ca transport. *J. gen. Physiol.* **70**, 681–695.

MULLINS, L. J. (1981). *Ion Transport in the Heart.* New York: Raven Press.

NIEDERGERKE, R. (1963). Movements of Ca in beating ventricles of the frog. *J. Physiol.* **167**, 551–580.

NIEDERGERKE, R. & PAGE, S. (1982). Changes of frog heart action potential due to intracellular calcium ions. *J. Physiol.* **328**, 17–18P.

NOBLE, D. (1960). Cardiac action and pace-maker potentials based on the Hodgkin-Huxley equations. *Nature, Lond.* **188**, 495–497.

NOBLE, D. (1962a). A modification of the Hodgkin–Huxley equations applicable to Purkinje fibre action and pacemaker potentials. *J. Physiol.* **160**, 317–352.

NOBLE, D. (1962b). The voltage dependence of the cardiac membrane conductance. *Biophys. J.* **2**, 381–393.

NOBLE, D. (1965). Electrical properties of cardiac muscle attributable to inward-going (anomalous) rectification. *J. cell. comp. Physiol.* **66**, suppl. 2, 127–136.

NOBLE, D. (1979). *The Initiation of the Heartbeat*, 2nd edn. Oxford: The Clarendon Press.

NOBLE, D. (1982a). In discussion following DE HAAN, R. L. (1982). In *Cardiac Rate and Rhythm*, ed. BOUMAN, L. N. & JONGSMA, H. J., pp. 359–361. Martinus Nijhoff: The Hague.

NOBLE, D. (1982b). Book review. *Q. Rev. Biol.* **57**, 501–503.

NOBLE, D. (1983). Ionic mechanisms of rhythmic firing. In *Symp. Soc. exp. Biol.* **37**, 1–28.

NOBLE, D. & HALL, A. E. (1963). The conditions for initiating 'all-or-nothing' repolarization in cardiac muscle. *Biophys. J.* **3**, 261–274.

NOBLE, D. & NOBLE, S. J. (1984). A model of s.a. node electrical activity using a modification of the DiFrancesco-Noble (1984) equations. *Proc. R. Soc.* B (in the Press).

NOBLE, D. & POWELL, T. (1983). The effects of series resistance on second inward current recorded under voltage clamp in ventricular muscle. *J. Physiol.* **345**, 7P.

NOBLE, D. & TSIEN, R. W. (1968). The kinetics and rectifier properties of the slow potassium current in cardiac Purkinje fibres. *J. Physiol.* **195**, 185–214.

NOBLE, D. & TSIEN, R. W. (1969a). Outward membrane currents activated in the plateau range of potentials in cardiac Purkinje fibres. *J. Physiol.* **200**, 205–231.

NOBLE, D. & TSIEN, R. W. (1969b). Reconstruction of the repolarization process in cardiac Purkinje fibres based on voltage clamp measurements of the membrane current. *J. Physiol.* **200**, 233–254.

NOBLE, D. & TSIEN, R. W. (1972). The repolarization process of heart cells. In *Electrical Phenomena in the Heart*, ed. DE MELLO, W. C., pp. 133–161. London: Academic Press.

NOBLE, S. J. & SHIMONI, Y. (1981a). The calcium and frequency dependence of the slow inward current 'staircase' in frog atrium. *J. Physiol.* **310**, 57–75.

NOBLE, S. J. & SHIMONI, Y. (1981b). Voltage-dependent potentiation of the slow inward current in frog atrium. *J. Physiol.* **310**, 77–95.

NOMA, A. & IRISAWA, H. (1976a). Membrane currents in the rabbit sinoatrial node as studied by the double microelectrode method. *Pflügers Arch.* **364**, 45–52.

NOMA, A. & IRISAWA, H. (1976b). A time- and voltage-dependent potassium current in the rabbit sinoatrial node cell. *Pflügers Arch.* **366**, 251–258.

NOMA, A., MORAD, M. & IRISAWA, H. (1983). Does the 'Pacemaker Current' generate the diastolic depolarization in the rabbit s.a. node cells? *Pflügers Arch.* **397**, 190–194.

50   *D. NOBLE*

Noma, A., Yanagihara, K. & Irisawa, H. (1977). Inward current activated during hyper-polarization in the rabbit sinoatrial node cell. *Pflügers Arch.* **385**, 11–19.

Orkand, R. K. & Niedergerke, R. (1966). The dual effect of calcium on the action potential of the frog's heart. *J. Physiol.* **184**, 291–311.

Peper, K. & Trautwein, W. (1969). A note on the pace-maker current in Purkinje fibres. *Pflügers Arch.* **309**, 356–361.

Pollack, G. H. (1976). Intercellular coupling in the atrioventricular node and other tissues of the rabbit heart. *J. Physiol.* **255**, 275–298.

Reuter, H. (1967). The dependence of slow inward current in Purkinje fibres on the extracellular calcium concentration. *J. Physiol.* **192**, 479–492.

Reuter, H. & Scholz, H. (1977). A study on the ion selectivity and the kinetic properties of the calcium dependent slow inward current in mammalian cardiac muscle. *J. Physiol.* **264**, 17–47.

Rougier, O., Vassort, G., Garnier, D., Gargouil, Y.-M. & Coraboeuf, E. (1969). Existence and role of a slow inward current during the frog atrial action potential. *Pflügers Arch.* **308**, 91–110.

Sakmann, B., Noma, A. & Trautwein, W. (1983). Acetylcholine activation of single muscarinic K^+ channels in isolated pace-maker cells of the mammalian heart. *Nature, Lond.* **303**, 250–253.

Schwindt, P. G. & Crill, W. E. (1980). Properties of a persistent inward current in normal and TEA-injected motoneurones. *J. Neurophysiol.* **43**, 1700–1724.

Seyama, I. (1976). Characteristics of the rectifying properties of the sino-atrial node cell of the rabbit. *J. Physiol.* **255**, 379–397.

Shibata, E. F. & Giles, W. R. (1983). Ionic currents in isolated cardiac pace-maker cells from bullfrog sinus venosus. *Proc. Int. Union physiol. Sci.* **15**, 76.

Shibata, E. F., Momose, Y. & Giles, W. R. (1983). Measurement of an electrogenic Na^+/K^+ pump current in individual bullfrog atrial myocytes. *Proc. Int. union physiol. Sci.* **15**, 51.

Shimoni, Y. (1981). Parameters affecting the slow inward channel repriming process in frog atrium. *J. Physiol.* **320**, 269–291.

Siegelbaum, S. A. & Tsien, R. W. (1980). Calcium-activated transient outward current in calf cardiac Purkinje fibres. *J. Physiol.* **299**, 485–506.

ter Keurs, H. E. D. J. & Schouten, V. J. A. (1984). The slow repolarization phase of the action potential in rat myocardium. *J. Physiol.* (in the Press).

Tsien, R. W. (1983). Calcium channels in excitable cell membranes. *A. Rev. Physiol.* **45**, 341–358.

Tsien, R. W., Kass, R. S. & Weingart, R. (1979). Cellular and subcellular mechanism of cardiac pace-maker oscillations. *J. exp. Biol.* **81**, 205–215.

Vassalle, M. (1965). Cardiac pace-maker potentials at different extra- and intracellular K concentrations. *Am. J. Physiol.* **208**, 770–775.

Vassalle, M. (1966). Analysis of cardiac pace-maker potential using a 'voltage clamp' technique. *Am. J. Physiol.* **210**, 1335–1341.

Vereecke, J., Isenberg, G. & Carmeliet, E. (1980). K efflux through inward rectifying K channels in voltage clamped Purkinje fibres. *Pflügers Arch.* **384**, 207–217.

Weidmann, S. (1951). Effect of current flow on the membrane potential of cardiac muscle. *J. Physiol.* **115**, 227–236.

Weidmann, S. (1956). *Elektrophysiologie der Herzmuskelfaser*. Bern: Huber.

Wier, W. G. (1980). Calcium transients during excitation-contraction coupling in mammalian heart: aequorin signals of canine Purkinje fibres. *Science, N.Y.* **207**, 1085–1087.

Winegrad, S. & Shanes, A. M. (1962). Calcium flux and contractility in guinea-pig atria. *J. gen. Physiol.* **45**, 371–394.

Yanagihara, K. & Irisawa, H. (1980). Potassium current during the pace-maker depolarization in rabbit sino-atrial node cell. *Pflügers Arch.* **388**, 255–260.

Yanagihara, K., Noma, A. & Irisawa, H. (1980). Reconstruction of sino-atrial node pace-maker potential based on the voltage clamp experiments. *Jap. J. Physiol.* **30**, 841–857.

Reciprocal role of the inward currents $i_{b,Na}$ and i_f in controlling and stabilizing pacemaker frequency of rabbit sino-atrial node cells

D. NOBLE[1], J. C. DENYER[1], H. F. BROWN[1] AND D. DiFRANCESCO[2]

[1] *University Laboratory of Physiology, Parks Road, Oxford OX1 3PT, U.K.*
[2] *Dipartimento di Fisiologia e Biochimica Generali, Via Celoria 26, 20133 Milano, Italy*

SUMMARY

Experiments and computations were done to clarify the role of the various inward currents in generating and modulating pacemaker frequency. Ionic currents in rabbit single isolated sino-atrial (SA) node cells were measured using the nystatin-permeabilized patch-clamp technique. The results were used to refine the Noble–DiFrancesco–Denyer model of spontaneous pacemaker activity of the SA node. This model was then used to show that the pacemaker frequency is relatively insensitive to the magnitude of the sodium-dependent inward background current $i_{b,Na}$. This is because reducing $i_{b,Na}$ hyperpolarizes the cell and so activates more hyperpolarizing-activated current, i_f, whereas the converse occurs when $i_{b,Na}$ is increased. The result is that i_f and $i_{b,Na}$ replace one another and so stabilize nodal pacemaker frequency.

1. INTRODUCTION

Patch-clamp recording from mammalian isolated sino-atrial node (SA) cells has led to a greater understanding of pacemaker mechanisms, but the question of the relative contributions of various conductance mechanisms to determining the overall rate of pacemaker activity is still largely unresolved.

One reason for this is that the net current flowing in a single cell may be as little as 3 pA (calculated for a cell of 30 pF with a diastolic depolarization rate of 0.1 V s^{-1} and using the relation $I_{net} = -C dV/dt$; see DiFrancesco (1991)). It can, indeed, be directly demonstrated that injecting a very small amount of current into an SA node cell will cause a marked change in membrane potential: thus 5 pA of injected current causes about 5 mV hyperpolarization (Denyer 1989). All the known conductance mechanisms involved can contribute currents vastly in excess of the small net current, which therefore represents a very fine balance between much larger individual currents. In each case the current contributed depends on the lowest part (foot) of the relevant activation curve, and this is the region most subject to experimental error. This is precisely the kind of problem where a combined experimental and theoretical approach is required. A model based as closely as possible on the experimental data can be used to determine what would happen as a result of perturbations in the parameters that are most uncertain. That is the approach used in the present paper.

By using the nystatin-permeabilized patch method of whole-cell recording (Horn & Marty 1988) that avoids cell dialysis, we have collected new data from several isolated SA node cells and updated the Noble–DiFrancesco–Denyer model (Noble *et al.* 1989) to bring the model parameters as close as possible to those of normal cells.

By using this model we have investigated the way in which pacemaker frequency varies on changing either one of the two inward currents involved in the first part of the pacemaker depolarization: the sodium-dependent inward background current ($i_{b,Na}$) or the hyperpolarization-activated inward current (i_f). The results suggest that i_f has an extremely important role in stabilizing the pacemaker frequency against changes induced by altering other conductances, whatever the magnitude of its own contribution to the depolarizing current. In this study we have restricted ourselves to consideration of changes in these two membrane currents. Other inward currents, such as the calcium currents (L and T), contribute particularly to the last third of the pacemaker depolarization and to the action potential upstroke, but we have not in this paper attempted any systematic study of their role in modulating pacemaker frequency.

2. METHODS

The cell isolation procedure, the recording method (whole-cell patch-clamp technique) and the solutions for cell isolation used have been fully described in Denyer & Brown (1990*a*). In the present experiments, all the recordings were made using the nystatin-permeabilized patch technique to reduce the rate of cell rundown (Horn & Marty 1988).

The Tyrode solution contained (in millimoles per litre): 140 NaCl, 5.4 KCl, 1.8 CaCl$_2$, 0.5 MgSO$_4$, 5 HEPES 10 glucose; pH adjusted to 7.4 (at 37 °C) by titrating with 4 N

NaOH solution. The pipette solution (nystatin permeabilized patch) contained (in millimoles per litre): 140 KCl, 3 $MgSO_4$, 11 HEPES, 0.14 mg ml^{-1} nystatin, DMSO 0.28%; pH adjusted to 7.2 by titrating with 1 N KOH solution. All experiments were done at 37 °C.

The numerical computations were done using OXSOFT HEART version 3.2 or 3.4 running on PC-AT-type computers. The starting point was the single-cell model described by Noble *et al.* (1989), which uses equations for i_f originally based on experiments described in DiFrancesco & Noble (1989). The conductances and activation curves were adjusted until the voltage-clamp results predicted by the model fitted those obtained in the present experiments.

Other results on i_f have also recently appeared (see, for example, van Ginneken & Giles 1991; Frace *et al.* 1992), but we have found that the computations described in the present paper do not depend strongly on the particular equations chosen for i_f from the several formulations now available. Similarly, the precise formulation used for $i_{b, Na}$ is not very important. Hagiwara *et al.* (1992) describe a moderate degree of nonlinearity (of the constant-field type) in $i_{b, Na}$, whereas the equation we have used is linear. The degree of nonlinearity described by Hagiwara *et al.* in the pacemaker range of voltages is, in any case, almost negligible.

Examples of data input files for use with the HEART program are given in the appendices.

3. RESULTS

Figure 1 *a–c* shows typical current records obtained from an SA node cell during voltage-clamp pulses from a holding potential of −40 mV. Depolarizing pulses of 1 s duration (figure 1 *a, b*) activated inward L-type calcium current ($i_{Ca, L}$) and outward delayed rectifier potassium current (i_K). Hyperpolarizing pulses of 800 ms duration (figure 1 *c*) elicited first an instantaneous inward background current jump followed by slow activation of the hyperpolarization-activated inward current (i_f). Cell capacitance (13 cells), peak $i_{Ca, L}$ (11 cells), peak i_K tail current (13 cells), isochronal i_f (at 800 ms: 10 cells) and the absolute instantaneous value of i_b (10 cells) were measured and the mean current densities and their s.e.m.s calculated. The current density–voltage relations of these data are plotted in figure 1 *g* as solid symbols. All the cells used in this study showed vigorous spontaneous activity under current clamp conditions ($I = 0$).

The experimental results were fitted by modification of the Noble–DiFrancesco–Denyer model. The following input parameters were reset: PCA (permeability of L-type calcium channel) was reduced to 0.075, IKM (maximum outward i_K) reduced to 0.25 nA, SHIFTY (parameter for shifting the voltage dependence of i_f) set to −6 mV, PUMP (maximum value of i_p) reduced to 0.25 nA, KNACA (scaling factor for i_{NaCa}) set to 0.000088, GBNA ($i_{b, Na}$ conductance) reduced to 0.00018 µS and CAPACITANCE set to 23.2 pF (see also Appendix I for a listing of input parameters). Hagiwara *et al.* (1992) have recently measured the value of $g_{b, Na}$ in rabbit sinus node cells. They found about 0.0004 µS for a 40 pF cell. The value assumed here is therefore close to that found experimentally when allowance is made for cell membrane area. Two values for KMCA (constant for activation of calcium release) were required to fit the experimental data

accurately. A value of 0.002 mM was used for simulations of action potentials and for the depolarizing voltage clamp pulses: values greater than 0.002 mM caused distortion of the repolarization phase of the action potential. A higher KMCA value of 0.2 mM was used for the hyperpolarizing voltage-clamp simulations: this prevented the transient calcium release and activation of i_{NaCa} which otherwise occurred.

The need to vary KMCA is related to a problem in these models already noted and discussed by Hilgemann & Noble (1987) and Earm & Noble (1990). This is that the models represent cytosol calcium as instantaneously uniform. This is extremely unlikely to be the case (for recent evidence of inhomogeneity of calcium release in atrial cells see Lipp *et al.* (1990)). The SR release site may therefore detect calcium concentrations that are transiently very different from the general cytosol calcium concentration. In the model developed here the only way to recognize this problem is to use different values for KMCA for different conditions, with the lowest value being required when fast release by relatively low concentrations is required. It requires much more extensive modelling to represent intracellular calcium diffusion and inhomogeneity. This has been done by incorporating a small subsarcolemmal space into the models (D. Noble, unpublished work) and, as expected, this does completely overcome the problem discussed here. However, the resulting computations are very much slower. Moreover, there is no evidence to suggest that the results of the present work would be significantly affected: SA node pacemaker activity does not obviously differ when internal calcium is strongly buffered and we have not found the normal pacemaker behaviour of the sinus node models to depend at all strongly on the equations used to represent internal calcium release.

Simulations of the experimental patch-clamp data are shown in figure 1 *d–f*. The computed current densities are plotted as hollow symbols in figure 1 *g*. Note that the simulation is based on the mean current densities obtained experimentally, and that the current records shown in figure 1 *a–c* were obtained from an individual cell. The differences in peak values of $i_{Ca, L}$ and i_K are therefore not significant. The alternative approach would be to fit accurately the experimental data from one particular cell, but this would have the disadvantage that general conclusions would be less likely to be valid: any one cell can deviate quite substantially from the average, either in overall current amplitudes or in exact action potential shape. Thus the cell illustrated in figure 1 *a–c* shows less i_K than the mean, but a larger and more rapid $i_{Ca, L}$. We will comment later on the significance of these features. Note also that, although the fit of i_f to the mean current is relatively poor negative to −80 mV, the physiologically important range of i_f (positive to −75 mV) is very closely fitted indeed. Our priority, given the purpose of this paper, was to fit very accurately the mean gated ionic currents between −70 mV and 0 mV.

There is some discrepancy between the experimental and computed values of the net instantaneous background current, i_b. The value of GBNA used in the

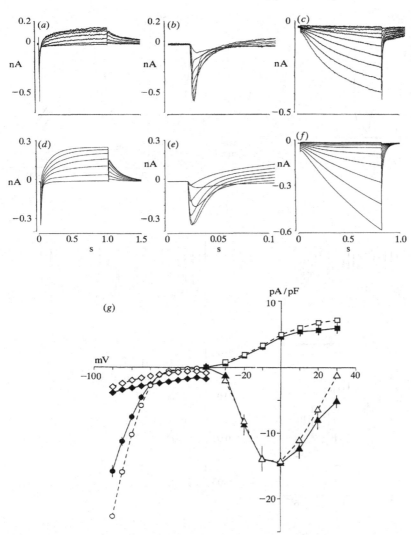

Figure 1. (*a–c*) Example of experimentally recorded currents recorded under voltage-clamp conditions in one particular cell. (*d–f*) Computed currents obtained under voltage-clamp conditions using the parameters fitting the mean currents as in (*g*). (*a*) A cell was held at −40 mV and given depolarizing pulses to +30 mV in successive 10 mV steps of 1 s duration at a frequency of 0.2 Hz. Current traces are shown superimposed. Voltage traces not shown. (*b*) First 100 ms of currents in (*a*) shown on an expanded timescale. (*c*) Current records obtained during hyperpolarizing pulses. The cell was held at −40 mV and given successive −5 mV steps of 800 ms duration down to −90 mV at a frequency of 0.2 Hz. (*d–f*) Computer simulations using the same protocols as in (*a*), (*b*) and (*c*), respectively. (*g*) Current density–voltage relations for the experimentally recorded (solid symbols) and computed (open symbols) currents shown in (*a–c*). The peak i_K tail current (squares) and the peak inward calcium current (triangles) were measured with respect to the holding current; i_f (circles) was measured as the difference between the instantaneous current level attained upon hyperpolarization and the current at 800 ms; i_b (diamonds; net background current which includes $i_{b,Na}$) was measured with respect to the zero current level.

simulations was chosen to set the maximum diastolic potential (MDP) at about −63 mV, close to the mean MDP recorded in the experiments (−60.5 ± 2.0 mV; seven cells). However, the value of GBNA required to fit the experimental current–voltage relation for i_b (0.00055) depolarized the MDP to about −20 mV, terminating activity. One possible reason for this difference between the experimental and the computed results is that experimentally i_b is very difficult to measure accurately. Unless capacity transients can be cancelled exactly and leakage current compensated, it is likely that significant errors will be made. The calculated leakage current can be as large as $i_{b,Na}$ (DiFrancesco & Noble 1989). It is also worth noting

202   D. Noble and others   *Modulation of pacemaker rhythm*

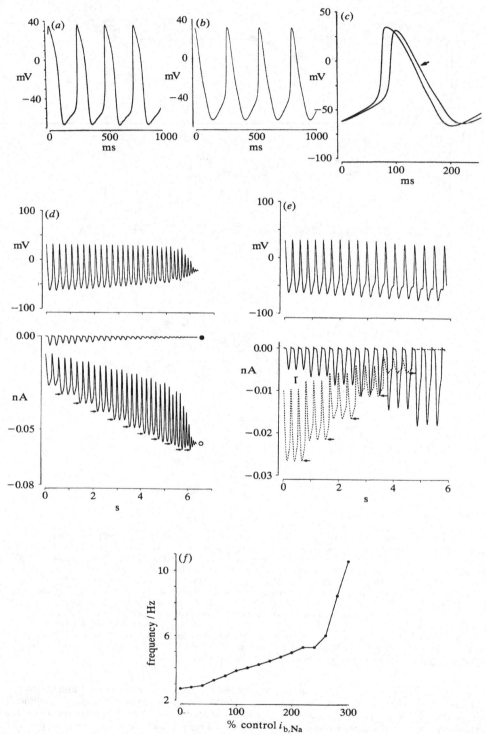

Figure 2. (*a*) Spontaneous activity recorded in current clamp mode in the SA node cell voltage clamped in figure 1. (*b*) Computed spontaneous activity using the mean current parameters obtained experimentally. (KMCA value of 0.002 mM.) (*c*) Computed activity for a single cycle comparing the behaviour of the standard model (as in (*b*);

that, in other SA node cells, the measured net background current may even be outward in the diastolic range positive to -65 mV (DiFrancesco 1991; see also the Discussion).

Figure 2a shows the spontaneous electrical activity recorded under current clamp conditions ($I = 0$) in the same SA node cell used in figure 1a–c. Figure 2b shows a simulation of spontaneous activity using the new version of the model (KMCA value of 0.002 mM). The frequency of beating is very similar in the real cell and in the model. Again the minor differences between the experimental and computed results can be explained by cell-to-cell variation. It is interesting that the sharper rate of rise of the action potential in the patch-clamped cell correlates with the above-average amount of $i_{Ca, L}$ also recorded in this cell (figure 1a, b). These features of individual cells (including in this case a more marked 'plateau' preceding repolarization) can be reproduced in the model by fine tuning of the relevant parameters, such as the amplitude and timecourse of the calcium and sodium–calcium exchange currents, as shown in figure 2c. We have also investigated the effect of including $i_{Ca, T}$ in the calculations (for details see Appendix 1). This succeeds in reproducing the small contribution of this current to the rate of depolarization during the pacemaker potential described by Hagiwara et al. (1988, figure 10, top). However, including this component, or modifying $i_{Ca, L}$ and the sodium–calcium exchange current even by as much as 100%, does not significantly alter the conclusions of the present paper concerning the relative roles of i_b and i_f. The model used for figure 2b was therefore used as the control for all the remaining calculations because it represents the behaviour of equations fitted carefully to the means of the experimental results.

Figure 2d, e shows the results of varying the amplitude of the background sodium conductance which constitutes the majority of the background inward conductance in the model. In figure 2d the effect of increasing $i_{b, Na}$ from 0.000 18 pS to 0.000 54 pS (100–300% control) is shown (20% increase at each arrow), and figure 2e illustrates the effect of decreasing $i_{b, Na}$ from 0.000 18 pS to 0 pS (100–0% control: 20% decrease at each arrow). The result is quite remarkable: despite the fact that in the standard version of this model (i.e. using the value of background conductance chosen to fit the data) the background current contributes four times as much depolarizing current to the pacemaker depolarization as does i_f, there is only a rather moderate variation in frequency as $g_{b, Na}$ is varied over this enormous range until, at very depolarized levels, rapid unstable oscillations are produced. This is further illustrated in figure 2f, in

which the frequency of firing as a function of $g_{b, Na}$ (percentage of control value) is plotted. Thus, doubling control $g_{b, Na}$ only increases the frequency from about 4 Hz to about 5 Hz, whereas totally blocking the background sodium-dependent current only reduces the frequency to just under 3 Hz. There is therefore less than a twofold variation in frequency over the whole of this range. In the Noble–DiFrancesco–Denyer model (1989) a similar resistance to frequency change is observed.

Figure 2e provides the explanation for this unexpectedly small change in frequency. As $g_{b, Na}$ is reduced, the maximum diastolic potential becomes more negative and, as a result, i_f is activated more strongly. This process continues until the additional i_f activated balances the reduced background inward current. The dynamics of this process are extremely well tuned to that, in the extreme case, when the background conductance is zero, the variations in i_f almost exactly match the former variations in $i_{b, Na}$. This means that, apart from a few millivolts difference in maximum diastolic potential, there is very little obvious difference in the activity of the model when running almost entirely on i_f as the depolarizing current and when running very largely on $i_{b, Na}$.

A similar computation to that in figure 2e but with i_f blocked produces a totally different result. Whereas before block of i_f a 50% reduction in $i_{b, Na}$ only reduced the frequency from 4.43 Hz to 4.32 Hz, when i_f is blocked the same reduction in $i_{b, Na}$ reduced the frequency from 4.03 Hz to 2.78 Hz (plots not shown).

Similar computations were done to investigate the frequency dependence of the SA node cell activity on i_f. Figure 3 shows the results of shifting the voltage dependency of i_f to (a) more negative potentials or (b) more positive potentials. As summarized in the graph in (c), there is very little effect on overall frequency of beating when i_f is shifted negative to the control position. Shifting the voltage dependence of i_f to more positive potentials, however, significantly increases spontaneous rate: a $+50$ mV shift almost doubles the rate of firing (b). Positive shifts of more than $+50$ mV produce unstable rapid oscillations.

4. DISCUSSION

The model described in this paper is based on experimental data obtained from SA node cells isolated in conditions which leave them as close as possible to their normal shape, and with a recording method that eliminates dialysis. Hopefully, therefore, the model is very close indeed to a reconstruction of normal pacemaker activity in these cells. With presently

arrowed) with that obtained after increasing the calcium channel current and the sodium–calcium exchange current by 50%. (d) Computed effect of increasing $i_{b, Na}$ on the spontaneous activity in an SA node cell. Top, membrane potential; bottom, i_f (filled circles) and $i_{b, Na}$ (open circles). At each arrow, GBNA is increased by a 20% increment from 100% of the control value to 300% of the control value. (e) Computed effect of decreasing $i_{b, Na}$ on the spontaneous activity in an SA node cell. Top, membrane potential; bottom, i_f (solid line) and $i_{b, Na}$ (broken line). At each arrow, GBNA is decreased by a 20% decrement from 100% of the control value to 0% of the control value. (f) Frequency dependence of spontaneous activity on $i_{b, Na}$, plotted from the data shown in (d) and (e).

204   D. Noble and others   *Modulation of pacemaker rhythm*

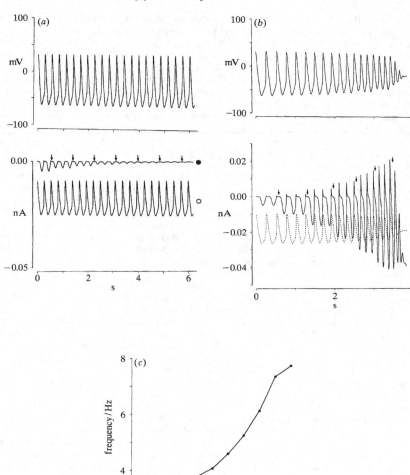

Figure 3. (*a*) Computed effect of negative shifts of the voltage dependence of i_f on the spontaneous activity in an SA node cell. Top, membrane potential; bottom, i_f (filled circles) and $i_{b, Na}$ (open circles). The voltage dependence of i_f is shifted negatively at the points indicated by arrows from the control position (half-activation point at -61 mV) to -35 mV more negative in -5 mV steps. (*b*) Computed effect of positive shifts of the voltage dependence of i_f on the spontaneous activity in an SA node cell. Top, membrane potential; bottom, i_f (solid line) and $i_{b, Na}$ (broken line). The voltage dependence of i_f is shifted positively at the points indicated by arrows from the control position (half-activation point at -61 mV) to $+60$ mV more positive in 10 mV steps. (*c*) Frequency dependence of spontaneous activity on the voltage shift in i_f activation, plotted from the data shown in (*a*) and (*b*).

available methods it is difficult to know from which region of the SA node the cells come. It seems likely from the appearance of the voltage traces that they were cells from the periphery rather than from the centre of the node.

The first problem we have tackled with this model is whether it is possible to determine the relative quantitative roles of the sodium-dependent background inward current $i_{b, Na}$ and the hyperpolarization-activated inward current, i_f, in generating the pacemaker depolarization. Different approaches to this

problem have given different answers varying from ascribing a mainly modulatory role to i_f (the block of i_f with 2 mM caesium only causes a 19–30 % reduction in frequency (Denyer & Brown 1990 b)) to regarding i_f as essential to the pacemaker depolarization (Di-Francesco 1991; see also the Appendix to DiFrancesco (1991) by D. Noble, J. C. Denyer & D. DiFrancesco). The present results show that, whatever the answer to that quantitative question, the dynamic role of i_f is of major physiological significance, as the properties of this current are finely tuned to enable it to protect SA node cells from frequency changes induced even by massive changes in other depolarizing currents. This is almost certainly functionally important because, although the background currents are not yet well understood in cardiac cells, it is possible that they represent a whole range of transport mechanisms that underly other purposes (e.g. the possible transport of amino acids (Earm *et al.* 1989, 1990).

Our results also show that the pacemaker frequency is very insensitive to the amount of $i_{b,Na}$ in the cell (figure 2). This means that cells with genuinely different amounts of $i_{b,Na}$ (possible examples are cells isolated under differing biochemical conditions or those which come from different regions of the SA node (see, for example, Kodama & Boyett 1985; Honjo & Boyett 1992)) could still pacemake in an apparently similar way with regard to frequency and voltage waveforms. In the present paper, experimental results suggest the presence of a net inward background current in the diastolic depolarization range. However, in cells used in a different study which showed almost identical spontaneous activity (DiFrancesco 1991), the measured net background current positive to -65 mV was outward, which implies that in these cells the major depolarizing current contribution was from i_f. It is significant, therefore, that these latter cells also showed a sufficiently negative maximum diastolic potential to activate i_f strongly. In the present computations these two conditions (net outward background current and very negative maximum diastolic potential) are satisfied when $i_{b,Na}$ is reduced to around one third of the control value in the model.

Is there any way, therefore, of quantifying the relative amplitudes of contribution of $i_{b,Na}$ and i_f to the pacemaker depolarization? Clearly a specific blocker of i_f would provide the answer because, unlike the frequency response to changes in background current, the frequency response to change in i_f does give a reasonable guide to the relative amplitude of i_f involved because there is no 'buffering' effect on frequency by another current system. This was the basis on which Denyer & Brown (1990*a, b*) used block of i_f with 2 mM caesium to estimate a contribution from i_f of around 30% in the isolated beating cells in their experiments. Our present results show that this fraction would depend very steeply on the value of maximum diastolic potential attained, and this does vary by at least 10 mV, according to both the experimental conditions and the area of the sinus node from which they are isolated. Moreover, this estimate also depends on how specific is the action of caesium. All we can say with certainty at present is that even if 2 mM caesium does have other actions in addition to i_f block it does not change the background current level in the appropriate direction to counter block of i_f (Denyer & Brown 1990*b*). Nevertheless, the development of a completely specific blocker of i_f is of great importance.

Although, in this paper, we have illustrated the 'buffering' role of i_f with changes in $i_{b,Na}$, it is worth noting that changes in other currents, such as the sodium pump current (which would change as a consequence of $i_{b,Na}$ changes) and the acetylcholine-activated K current, would also be subject to this

effect, so that larger changes in these currents would be required to achieve a particular modulation of frequency than would be the case in the absence of i_f.

A possible role for i_f buffering *in vivo* might be to counteract the slowing influence of electrotonic spread of hyperpolarizing current from atrial cells to the sinus node. This interaction is very strong and is known to shift the site of origin of pacemaker depolarization from the periphery to the centre of the node (Kirchof *et al.* 1987; Winslow *et al.* 1992). However, it has not yet been determined how much i_f contributes to this interaction.

These results and computations emphasize how the pacemaker system of the SA node cell is designed to compensate for changes in net background current so that pacemaking will continue. They also highlight the role of i_f both in 'buffering' the frequency changes when inward currents alter and in modulating pacemaker frequency.

This study was supported by the B.H.F, the M.R.C. and the Wellcome Trust.

APPENDIX 1.

This Appendix gives the basic input file used for the computations in this study. The command PREP:SINUS selects the original sinus node model in the HEART program. The other parameters determine the standard single-cell model developed in the present paper. In each case we have included a comment on the significance, and, where relevant, the units of each parameter. In HEART 3.4 and later versions, these comments can actually be included in the data file for future reference.

$****PREP: SINUS****

CA12 = 3.9609 (SR calcium concentration/mM)
CA13 = 0.3006 (release store calcium/mM)
CAI = 0.002 220 ([Ca]$_i$ mM)
CAO = 1.8 ([Ca]$_o$ mM)
CAPACITANCE = 0.000 023 2 (capacitance 23.2 pF)
D = 0.9973 (calcium-channel activation gate)
DT = 0.001 (integration step length 1 ms)
DX = 1 (cell radius = $10 \times DX = 10$ μm)
E = 30.9948 (membrane potential mV)
F = 0.3180 (calcium-channel inactivation gate)
GBCA = 0.0001 (background calcium conductance 100 pS)
GBK = 0 (background potassium conductance zero)
GBNA = 0.000 18 (background sodium conductance 180 pS)
GFK = 0.06 (K component of i_f 60 nS)
GFNA = 0.06 (Na component of i_f 60 nS)
GK1 = 0.0075 (maximum value of g_{K1} 7.5 nS)
GNA = 0.0125 (g_{Na} 12.5 nS)
H = 0.0 (sodium-channel inactivation gate)
IKM = 0.25 (maximum current carried by i_K 250 pA)
ISCALE = 1000 (convert plotted currents from nA to pA)
KB = 5.4 (external [K] mM)
KC = 5.4 (external [K] mM)
KI = 142.9889 (internal [K] mM)
KMCA = 0.002 (binding constant for calcium-induced calcium release)
KMODE = 4 (use Noble–DiFrancesco–Denyer equations for i_K)
KNACA = 0.000 008 8 (scaling factor for sodium–calcium exchange)
M = 0.9976 (sodium activation gate)
MODE = 1 (action potential mode)
NAI = 5.5736 ([Na]$_i$ mM)

206   D. Noble and others   *Modulation of pacemaker rhythm*

NAO = 140.0 ([Na]$_o$ mM)
PCA = 0.075 (calcium channel permeability × F)
LENGTH = 0.028 (cell length 28 μm)
PRIM = 0.0994 (SR repriming parameter)
PUMP = 0.25 (maximum pump current 0.25 nA)
SHIFTD = 5.0 (calcium activation shifted 5 mV)
SHIFTY = −6 (i_f activation shifted −6 mV)
SPACE = 4 (single cell conditions for external space)
TABT = 0.001 (tabulation interval 1 ms)
TEND = 10.0 (compute for 10 s)
TIMESCALE = 1000 (plot in ms not s)
X = 0.1602 (i_K activation gate)
Y = 0.0002 (i_f activation gate)
YMODE = 20 (use Noble–DiFrancesco–Denyer i_f equations
   with square (delayed) kinetics
\$\$\$\$ (row of dollar signs ends data file).

For the calculations in figure 2c we increased both PCA (to 0.11) and KNACA (to 0.000015) to represent the larger relative calcium movements found experimentally in the cell illustrated in figure 1. For the calculation of possible contribution from $i_{Ca,T}$ referred to in the Results section, we used the above data file together with the following additional parameters used to set the properties of $i_{Ca,T}$:

PCA2 = 0.075 (set permeability of T channels similar to that for L)
SPEED32 = 2 (activation rate double that for L channel)
SPEED33 = 2 (inactivation rate double that for L channel)
SHIFT32 = −20 (activation curve shifted −20 mV compared with L)
SHIFT33 = −50 (inactivation curve shifted −50 mV compared with L)
Y[32] = 0 (initial value of activation)
Y[33] = 0.1 (initial value of inactivation).

This data is valid for HEART version 3.4 and later versions. It reproduces quite accurately the experimental data of Hagiwara *et al.* (1988).

APPENDIX 2.

For completeness, the basic input file used for the computations shown in the Appendix to DiFrancesco (1991) is given below. In this case, we have not annotated the data as the units and comments are the same as in Appendix 1.

CA12 = 2.391
CA13 = 0.221
CAI = 0.000056
CAO = 2.0
CAPACITANCE = 0.00006
D = 0.0
DISP = 4
DT = 0.003
DX = 1.000
E = −69.186
F = 0.9997
GBCA = 0.0
GBK = 0.00013
GBNA = 0.00008
GFK = 0.06
GFNA = 0.06
GK1 = 0.0075
GNA = 0.0125
H = 0.197
IKM = 0.4
ISCALE = 1000
KB = 5.4
KC = 3.0
KI = 140.0

KMCA = 0.004
KMODE = 4
KNACA = 0.00002
M = 0.9976
MODE = 1
NAI = 7.5
NAO = 140.00
OUT = 5
PCA = 0.12
PREPLENGTH = 0.028
PRIM = 0.237
Q = 1.0
R = 0.103
SHIFTD = 12.0
SHIFTF = 11.0
SHIFTX = 10.0
SHIFTY = 8
SPACE = 4
TABT = 0.005
TEND = 0.8
TIMESCALE = 1000.0
X = 0.1231
Y = 0.0822
YMODE = 20

REFERENCES

Denyer, J. C. & Brown, H. F. 1990*a* Rabbit sino-atrial node cells: isolation and electrophysiological properties. *J. Physiol., Lond.* **428**, 405–424.

Denyer, J. C. & Brown, H. F. 1990*b* Pacemaking in rabbit isolated sino-atrial node cells during Cs⁺ block of the hyperpolarization-activated current, i_f. *J. Physiol., Lond.* **429**, 401–409.

DiFrancesco, D. 1991 The contribution of the 'pacemaker' current (i_f) to generation of spontaneous activity in rabbit sinoatrial node myocytes (with an Appendix by D. Noble, J. C. Denyer and D. DiFrancesco). *J. Physiol., Lond.* **434**, 23–40.

DiFrancesco, D. & Noble, D. 1989 The current i_f and its contribution to cardiac pacemaking. In *Cellular and neuronal oscillators* (ed. J. W. Jacklet), pp. 31–57. New York: Dekker.

Earm, Y. E., Noble, D., Noble, S. J. & Spindler, A. J. 1989 Taurine activates a sodium-dependent inward current in isolated guinea-pig ventricular cells. *J. Physiol., Lond.* **417**, 56*P*.

Earm, Y. E., Spindler, A. J., Noble, S. J. & Noble, D. 1990 Conductance changes induced by amino acids in isolated guinea-pig ventricular cells. *J. Physiol., Lond.* **425**, 55*P*.

Earm, Y. & Noble, D. 1990 A model of the single atrial cell: relation between calcium current and calcium release. *Proc. R. Soc. Lond.* B **240**, 83–96.

Frace, A. M., Maruoka, F. & Noma, A. 1992 Control of the hyperpolarization-activated current by external anions in rabbit sino-atrial node cell. *J. Physiol., Lond.* **453**, 307–318.

Hagiwara, N., Irisawa, H. & Kameyama, M. 1988 Contribution of two types of calcium currents to the pacemaker potentials of rabbit sino-atrial node cells. *J. Physiol., Lond.* **395**, 233–253.

Hagiwara, N., Irisawa, H., Kasanuki, H. & Hosoda, S. 1992 Background current in sino-atrial node cells of the rabbit heart. *J. Physiol., Lond.* **448**, 53–72.

Hilgemann, D. & Noble, D. 1987 Excitation–contraction coupling and extracellular calcium transients in rabbit atrium: reconstruction of basic cellular mechanisms. *Proc. R. Soc. Lond.* B **230**, 163–205.

Honjo, H. & Boyett, M. R. 1992 Correlation between action potential parameters and the size of single sino-atrial node cells isolated from the rabbit heart. *J. Physiol., Lond.* **452**, 128*P*.

Horn, R. & Marty, A. 1988 Muscarinic activation of ionic currents measured by a new whole-cell recording method. *J. gen. Physiol.* **92**, 145–159.

Irisawa, H., Brown, H. F. & Giles, W. R. 1993 Cardiac pacemaking in the sinoatrial node. *Physiol. Rev.* (In the press.)

Kirchof, C. J. H. J., Bonke, F. I. M., Allessie, M. A. & Lammers, W. J. E. P. 1987 The influence of the atrial myocardium on impulse formation in the rabbit sinus mode. *Pflügers Arch. Eur. J. Physiol.* **410**, 198–203.

Kodama, I. & Boyett, M. R. 1985 Regional differences in the electrical activity of the rabbit sinus node. *Pflügers Arch. Eur. J. Physiol.* **404**, 214–226.

Lipp, P., Pott, L., Callewaert, G. & Carmeliet, E. 1990 Simultaneous recording of Indo-1 fluorescence and Na^+/Ca^{2+} exchange current reveals two components of Ca^{2+}-release from sarcoplasmic reticulum of cardiac atrial myocytes. *FEBS Lett.* **275**, 181–184.

Noble, D., DiFrancesco, D. & Denyer, J. C. 1989 Ionic mechanisms in normal and abnormal cardiac pacemaker activity. In *Cellular and neuronal oscillators* (ed. J. W. Jacklet), pp. 59–85. New York: Dekker.

Van Ginneken, A. & Giles, W. R. 1991 Voltage clamp measurements of the hyperpolarization-activated inward current i_f in single cells from rabbit sino-atrial node. *J. Physiol., Lond.* **434**, 57–83.

Winslow, R., Kimball, A., Varghese, A. & Noble, D. 1992 Simulating cardiac sinus and atrial network dynamics on the Connection Machine. *Physica D.* (In the press.)

Received 26 August 1992; accepted 21 September 1992

Physica D 64 (1993) 281–298
North-Holland

Simulating cardiac sinus and atrial network dynamics on the Connection Machine

Raimond L. Winslow[a,b1], Anthony L. Kimball[c],
Anthony Varghese[b] and Denis Noble[d]

[a]*Department of Physiology, The University of Minnesota, Minneapolis, MN 55455, USA*
[b]*Army High Performance Computing Research Center, The University of Minnesota, Minneapolis, MN 55455, USA*
[c]*Thinking Machines Corporation, Cambridge, MA 02142, USA*
[d]*University Laboratory of Physiology, Oxford University, Oxford, UK*

Received 21 August 1991
Revised manuscript received 4 August 1992
Accepted 8 August 1992
Communicated by A.V. Holden

Computational methods for simulating biophysically detailed, large-scale models of mammalian cardiac sinus and atrial networks on the massively parallel Connection Machine CM-2, and techniques for visualization of simulation data, are presented. Individual cells are modeled using the formulations of Noble et al. Models incorporate properties of voltage-dependent membrane currents, ion pumps and exchangers, and internal calcium sequestering and release mechanisms. Network models are used to investigate factors determining the site of generation and direction of propagation of the pacemaker potential. Models of the isolated sinus node are used to show that very few gap junction channels are required to support frequency entrainment. When cell membrane properties in the isolated sinus node models are modified to reproduce regional differences in oscillation properties, as described by the data of Kodama and Boyett, an excitatory wave is generated in the node periphery which propagates towards the node center. This agrees with activation patterns measured in the isolated sinus node by Kirchoff. When the model sinus node is surrounded by a region of atrial cells, the site of pacemaker potential generation is shifted away from the periphery towards the node center. This is in agreement with activation patterns measured by Kirchoff in the intact sinus node of the rabbit heart, and demonstrates the importance of sinus node boundary conditions on shaping the site of generation and direction of propagation of the pacemaker potential.

1. Introduction

The rate of contraction of cardiac muscle is controlled by a region of specialized cells located near the apex of the heart called the sinus node. Cells within the node function as the clock of the heart, generating a rhythmic electrical wave of depolarization which propagates outward from

[1]Current Address (and address for correspondence): Dr. Raimond L. Winslow, Department of Biomedical Engineering, The Johns Hopkins University School of Medicine, Traylor Research Building, 720 Rutland Ave., Baltimore, MD 21205, 410-550-5090 (Office), 410-955-0549 (FAX).

the node center into the surrounding atria. Specialized cells known as Purkinje fibers also conduct this wave of excitation into the ventricles. Atrial and ventricular cells are normally quiescent. However, when excited by this wave, they undergo rapid depolarization (an action potential). This rapid depolarization in turn opens calcium (Ca) channels in the cell membrane, allowing Ca ions to flow into the cells. When intracellular levels of Ca reach a sufficiently high level, additional Ca is released from intracellular organelles. It is this massive internal release of Ca during each action potential that leads to the coordinated contraction of cardiac muscle.

Development of cell dissociation methods, whole-cell voltage-clamp recording, and optical techniques for measuring intracellular ion concentration have provided a wealth of detailed information on the biophysical mechanisms that give rise to oscillation of membrane potential in single sinus cells, and generation of action potentials in atrial cells [11,35,16,26,8,6,12,32,7]. This in turn has led to the formulation of several biophysically detailed models of isolated sinus and atrial cells [1,41,4,12,30,10,27,29]. The models of Noble and his colleagues are unique in that they include quantitative descriptions of not only voltage-dependent membrane currents, but also the sodium–potassium pump current, sodium–calcium exchanger current, changes in intracellular ion concentrations over time, intracellular Ca sequestering and release mechanisms, and cytosolic Ca buffering. In this paper, we report the incorporation of these single cell models into large-scale, two-dimensional models of the cardiac sinus node and atria implemented on the massively parallel Connection Machine CM-2. Building on the work of Jalife and Michaels [17,24,25,23] we use these models to investigate the mechanisms by which networks of coupled sinus cells entrain to a common oscillation frequency, and the ways in which alteration of sinus node boundary conditions, by embedding the node within a large network of model atrial cells, effects the site of initiation, frequency, and direction of propagation of the pacemaker potential [39,38,40].

2. Methods

2.1. Single cell and network models

Models of single cells employed in this study are those described previously by Noble and his colleagues [27,30,31], and implemented in the OXSOFT Heart simulation package [27]. The

models are based on data obtained from whole-cell voltage-clamp studies of sinus cells isolated from the rabbit sinus node [7,6], and have been modified to account for new membrane current data obtained using the nystatin patch-clamp technique [15,29]. In this method, nystatin in the patch-clamp electrode is used to induce conductance in the patch membrane to access the cell interior electrically, thus preventing dialysis of the cell by the electrode solution. This method reduces calcium channel "rundown" and shift of the hyperpolarizing-activated current I_f to a more negative activation range, as observed when conventional whole-cell recording methods are used [29]. OXSOFT Heart input files defining properties of central and peripheral sinus cells are included in appendix A. The single atrial cell model is based on data obtained in cells isolated from rabbit atrium, and has been described elsewhere [12,10].

Features of both these models are illustrated schematically in fig. 1, and include: (a) time- and voltage-dependent membrane currents; (b) membrane ion pumps and exchangers; (c) intracellular Ca sequestering and release mechanisms; (d) cytosolic Ca buffering systems; and (e) variation of intracellular sodium (Na), potassium (K), and Ca concentrations. Time and voltage dependent properties of membrane currents are modeled using equations of the form originally proposed by Hodgkin and Huxley [14], with

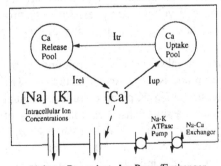

Fig. 1. The generic cardiac cell model.

parameters adjusted to fit single cardiac cell voltage-clamp data. An expression for the kth voltage-dependent outward membrane current $I_k(V_t)$ is

$$I_k(V_t) = (V_t - E_k)G_k^{\max} m_k(V_t)^{p_k} h_k(V_t)^{q_k}, \quad (1)$$

$$\dot{m}_k(V_t) = \alpha_k^m(V_t)[1 - m_k(V_t)] - \beta_k^m(V_t)\, m_k(V_t), \quad (2)$$

$$\dot{h}_k(V_t) = \alpha_k^h(V_t)[1 - h_k(V_t)] - \beta_k^h(V_t)\, h_k(V_t), \quad (3)$$

where V_t is trans-membrane potential (intracellular minus extracellular potential), E_k is the reversal potential of the ion(s) to which the kth membrane conductance is permeable (given by the Nernst or Goldman–Hodgkin–Katz equations), and $G_k^{\max}$ is the peak conductance of the kth voltage-dependent membrane conductance. The peak conductance is multiplied by a factor m_k, raised to the p_kth power. This factor assumes values between 0 and 1, and models activation of the kth membrane current. $G_k^{\max}$ is also scaled by a factor h_k, raised to the q_kth power. This factor ranges from 1 to 0, and models inactivation of the kth membrane current. Activation and inactivation variables m_k and h_k are solutions of the nonlinear ordinary differential equations (2), (3). The functions $\alpha_k^{m,h}(V_t)$ and $\beta_k^{m,h}(V_t)$ in these equations are forward and backward voltage-dependent rate constants for the activation and inactivation variables m_k and h_k. Whole-cell voltage-clamp experiments provide the data required to estimate $G_k^{\max}$, $a_k^{m,h}(V_t)$, and $B_k^{m,h}(V_t)$.

The models also include descriptions of the Na–K pump and Na–Ca exchange currents. The energy requiring Na–K pump extrudes 3 sodium for every 2 potassium ions. The stoichiometry of the Na–Ca exchanger is 3:1. In each case, outward current through the kth membrane ion pump/exchanger is modeled as an instantaneous function of the intra- and extra-cellular concentrations ($[C_i^{m,n}]$ and $[C_o^{m,n}]$, respectively) of the

relevant ionic species m and n (either Na and K or Na and Ca), as well as of membrane potential V_t:

$$I_k^{\text{pump}} = f_k([C_o^{m,n}], [C_i^{m,n}], V_t). \quad (4)$$

Intracellular concentrations of both Na, Ca, and K are allowed to vary over time. The time rate-of-change of the concentration of the kth ionic species is

$$[\dot{C}_k] = -\sum_k I_k^{\text{total}}/z_k F V_{\text{eff}}, \quad (5)$$

where $-I_k^{\text{total}}$ is the total inward current associated with the kth ionic species, V_{eff} the effective cell volume, F Faraday's constant, and z_k the valence of the kth ionic species. Extracellular K concentration is held constant, unlike the original Purkinje and sinus cell models derived from multicellular preparations [9,31].

Sequestration and release of Ca from intracellular organelles is modeled using a set of three first order differential equations. These equations specify the net Ca current I_{up} pumped into organelles (uptake pool), the net Ca current I_{rel} released from intracellular stores (release pool) when the cytosolic level of Ca exceeds a threshold value, and the net Ca current I_{tr} corresponding to transfer of Ca from the uptake to the release pool. A more detailed description of this model is given in DiFrancesco and Noble [9,27]. Cytosolic buffering of Ca by troponin and calmodulin is modeled as described by Hilgemann and Noble [12].

The structure of a two-dimensional cardiac network model is shown in fig. 2. The model is a simple rectangular mesh of interconnected cells with dimension $N \times N$. Each interior cell is modeled as having four nearest neighbors to which it is connected by linear conductors, labeled G_c, representing gap junctions. Measurements in coupled cardiac cell pairs support the model assumption of linearity for the gap junction coupling conductance [33,34]. Additional as-

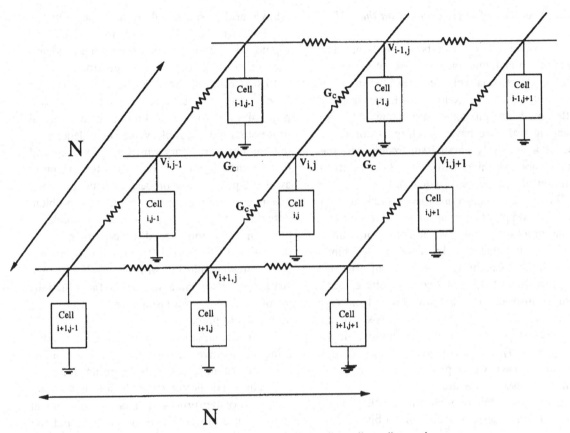

Fig. 2. Structure of the network models. G_c is cell-to-cell coupling conductance.

sumptions inherent in the network model are: (a) membrane potential and ionic concentrations are uniform within each cell; (b) coupling is uniform throughout the array, with value G_c; and (c) flux of ions through gap junctions is small, leading to negligible concentration changes from cell to cell.

A set of state equations for the rate of change of membrane potential of the ijth cell is derived by application of Kirchoff's current law at node ij:

$$\dot{V}_t^{ij} = \left(-\sum_k I_k^{ij} - \sum_k {}^{ij}I_k^{\text{pump}} \right.$$

$$\left. + (S_{ij} - N_{ij}V_t^{ij})G_c \right) \Big/ C_m , \qquad (6)$$

where S_{ij} is the sum of the membrane potentials

of the cells neighboring cell ij, N_{ij} is the number of such neighbors, and C_m is the membrane capacitance of each cell. The first term in the above equation is inward current through voltage-dependent membrane conductances, the second is total inward current carried by membrane ion pumps/exchangers, and the third is total inward current through gap junctions. Additional state equations are given by the time rate-of-change of model activation/inactivation variables (eqs. (2), (3)) and intracellular ion concentrations (eq. (5)) in each cell. This brings the total number of state variables N_{eq} to 14 and 17 for each sinus and atrial cell, respectively. Network properties are simulated by numerical integration of the $N_{eq}N_{tot}$ nonlinear differential equations defining the mesh, where $N_{tot} = N^2$ is the total number of cells in the mesh.

2.2. Integration of state equations on the CM-2

The value of N_{tot} can be quite large. For example, simulating the response of a 1024×1024 mesh of atrial cells requires integration of more than 17 million coupled nonlinear ordinary differential equations, a computationally demanding task. We have therefore explored the use of a massively parallel supercomputer, the Connection Machine CM-2, in the numerical solution of this system of equations.

The CM-2 is a single-instruction multiple-data (SIMD) computer with up to 65 536 (64 k) bit-serial processors and 2048 floating point units (FPU's), manufactured by Thinking Machines Corporation, Cambridge, MA [13,36]. Briefly, all processors of the CM-2 concurrently execute a single instruction stream issued by a front end host computer. Each processor consists of an arithmetic logic unit and memory (up to 1 Mbyte per processor). Sixteen bit-serial processors are present on each CM-2 processor chip. Pairs of CM-2 processor chips are components in a section, the basic unit of replication in the CM-2. Each section includes a 32- or 64-bit Weitek FPU, interprocessor communication hardware, memory (64 Kbyte to 1 Mbyte 23-bit words per section), and a number of bit matrix transposers. These transposers convert 32-bit serial data to 32-bit parallel format for input to the FPU (and vice versa). CM-2 processor chips are interconnected to form a $\log_2 m$ dimensional hypercube network, where m is the number of processor chips in the CM-2. There are also specialized communication paths for nearest neighbor interconnection of processors on the same processor chip.

Problems defined on n-dimensional regular grids, in which each grid element has 2^n nearest neighbors, are particularly well suited for solution using the Connection Machine. In this case, grid elements may be embedded in the hypercube so that neighboring elements occupy neighboring processors. This topology is called the *NEWS* grid. Local communication in the NEWS grid is supported by high-speed special purpose hardware. For $n = 2$, the NEWS grid topology is a 2-dimensional square mesh, identical to that of the cardiac cell network model.

Virtual processor sets can be defined with up to r virtual processors being emulated by each physical processor; r is known as the virtual processor ratio. Possible values of r depend on the amount of memory which must be allocated to each virtual processor out of the total memory available per physical processor. Thus, the obvious mapping of the cardiac cell network problem onto the CM-2 is a one-to-one correspondence between cells and virtual processors on a 2-dimensional NEWS grid. This is the implementation strategy employed, with the result that each virtual processor integrates all of the equations for one cell in a serial fashion, in synchrony with other virtual processors.

The model equations are integrated in parallel using an adaptive step size fourth-order Runge–Kutta algorithm [21]. In this algorithm, the global OR line (a dedicated circuit which allows rapid reduction of per-processor data into one global value) is used to determine whether or not the estimated local error at the current step size in each virtual processor is below a specified tolerance. If so, the integration step is accepted; otherwise a new, smaller step is selected in each processor and the local error re-computed. A more detailed presentation of techniques used to model sinus, atrial, and combined sino-atrial networks is given in appendix B.

2.3. Visualization of simulation data

Analyses of simulation data poses unique problems since very large data sets are generated, ranging in size from 1 to 100 Gbytes depending on network size. We have developed highly interactive real-time graphics tools to help us deal with these problems. Briefly, following a CM-2 simulation, data stored on the Connection Machine DataVault (a parallel disk array) is transferred to a 50 Gbyte IPI-2 disk farm attached to a Silicon Graphics Inc. workstation. Disks are configured four in parallel (stacked five deep), with each data word being distributed between the four disk stacks to speed access, a procedure known as disk striping. The resulting high data transfer rates (18 Mbytes/s) make it possible to animate large data sets stored on disk using full 24 bit color at high frame rates. Movies showing the time-evolution of selected state-variables or membrane currents in all cells of the mesh are generated using interactive software developed in our laboratory. Time evolution of any state variable or current, as well as instantaneous oscillation frequency of any cell within the mesh can be displayed. Instantaneous oscillation frequency is computed at each time t by setting a threshold voltage equal to the membrane potential $V(t)$, and computing the time at which the last threshold crossing of the same sign occurred. A variety of animations can be generated, including "top-down" views of the mesh in which the time-evolution of displayed-variables is color-coded, and "perspective views", in which the time-evolution of the selected variables appear as a color-coded surface rising above a plane. The visualization tool is highly interactive in that the surface may be enlarged or contracted, and rotated in real time to be viewed from any perspective. It is possible to select any cell, or set of cells, within the mesh for more detailed analysis. Upon selecting a cell, a new graphics window is opened in which the time evolution of the selected variable is displayed. Responses of single cells, or sub-grids of cells, can be extracted from the data set and written to disk for subsequent analysis.

3. Results

3.1. CM-2 performance analysis

Two-dimensional cardiac network models ranging in size from 64×64 to 1024×1024 cells have been simulated, computing responses as

long as 60 seconds. Performance statistics, measured in Mflops (millions of floating point operations per second), are shown in fig. 3. CM-2 performance is compared to that obtained using single processors on Cray X-MP/4-16 and Cray-2/4-512 supercomputers.

In order to best utilize the CM-2, the front-end computer must issue instructions in rapid sequence, and the communication link between the front-end computer and the CM-2 must operate as quickly as possible. It takes longer for the

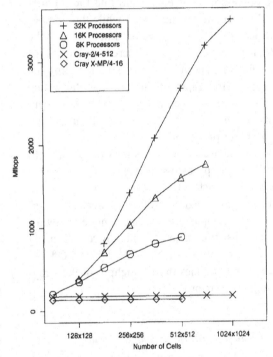

Fig. 3. Execution rate (Mflops, ordinate) versus number of cells simulated (abscissa) for the sinus network simulation code on the Connection Machine CM-2 and Cray supercomputers. See Key for the number of CM-2 processors utilized in each benchmark. The virtual processor ratio can be determined by dividing the number of cells simulated by the number of physical processors used. Performance was evaluated using the 32 768 processor CM-2 at The Army High Performance Computing Research Center, The University of Minnesota, configured with 4 Gbytes of memory, 64-bit Weitek floating point units, and Sun 490 front-ends. CM-2 codes were written in CM Fortran slicewise compiler Version 1.2. Single processor Cray results are from CFT77 implementations running under UNICOS on the Cray X-MP/4-16 and Cray-2/4-512 supercomputers of The Minnesota Supercomputer Center.

CM-2 to execute a single instruction at a high virtual processor ratio since each physical processor must execute the same instruction r times, once for each virtual processor. This results in more efficient utilization of the CM-2 by increasing the ratio of computation to front-end communication time. This is shown in the data of fig. 3. The abscissa is the number of cells simulated, and the ordinate is throughput in units of Mflops. Solid lines connecting data points for the CM-2 were computed using either 8K, 16K, or 32K processors. For each solid line the number of cells simulated is increasing while the number of processors is fixed, therefore implying that the virtual processor ratio is increasing. Each of these lines has positive slope, demonstrating the increased throughput of the CM-2 as the virtual processor ratio is increased. This stands in contrast to the approximately constant throughput of the Cray processors.

Actual measured performance shows, for sufficiently large problems, a speed advantage of more than a factor of five for 8K processors of a CM-2 versus a single Cray-2 processor. The maximum throughput on a sinus node network model consisting of over 1 million cells utilizing all 32K processors of the CM-2 is 3.5 Gflops, more than five times the throughput of the Cray-2 using all four processors.

3.2. Cardiac network simulation results

3.2.1. Sinus node models

Initial efforts have focussed on development of a model of the mammalian sinus node [39,28,40], primarily because there are sufficient experimental data from these cells to permit accurate modeling of their oscillation properties [30,27,31,29]. We have identified a minimum set of membrane parameters which, when varied, account for regional differences of intrinsic cell characteristics (as seen in central, transitional, and peripheral sinus cells) such as oscillation frequency, minimum and maximum oscillation amplitude, up-

stroke velocity, and sensitivity to external potassium concentration [20]. The parameters which must be varied to model central to peripheral cells are described in appendix A. Briefly, in order to fit the data of Kodama and Boyett [20], it is necessary to increase (in peripheral cells) the peak conductance of: (a) the hyperpolarizing activated I_f current [5]; (b) the fast inward Na current (for increased upstroke velocity); (c) the second inward Ca current (to increase overshoot potential); and (d) the instantaneous outward K current (to obtain the required sensitivity to external K concentration). Model central, transitional, and peripheral cells are generated by scaling each of these parameter values by a factor between zero and one, with a scale factor of one corresponding to peripheral cells, and a scale factor of zero to central cells. The scaling function employed is

$$S = 1.0 - e^{-d^2/\sigma^2}, \tag{7}$$

where d is distance from the center of the node (in number of cells) and σ is standard deviation (also in units of cells). The value of σ determines the rate at which a transition is made from central to peripheral cell oscillation properties. In 128×128 cell sinus network models, setting σ equal to 32 yields model cell oscillation properties which change with relative distance from the node center in the same way as do the data of Kodama and Boyett [20]. Examples of the oscillation waveform of central, transitional, and peripheral cells are shown in fig. 4. Cell oscillation parameters such as overshoot potential, maximum diastolic potential, and period, match the Kodama–Boyett data to within 10%.

Two types of numerical experiments have been performed. In the first, parameters of sinus cells within the mesh are varied randomly to produce model networks in which each cell has a slightly different oscillation frequency and waveform. This is done by choosing the scaling factor S for each cell in the mesh from a uniform distribution on the interval [0, 1]. The resulting

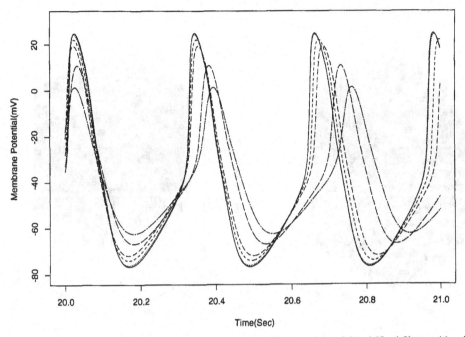

Fig. 4. Membrane potential (ordinate, mV) versus time (abscissa, seconds) for model peripheral ($S = 1.0$), transitional ($S = 0.89$, 0.71, 0.43, 0.13), and central cells ($S = 0$). S is defined using eq. (7).

assortment of cells have oscillation periods in the range of 318 to 370 ms. Coupling conductance is held constant throughout the mesh, a uniform initial condition is imposed on each cell, and the state equations describing the dynamics of the network are integrated in time on the Connection Machine.

Figure 5 shows a perspective view of membrane potential in 128×128 sinus cell network models with random network parameters. The x- and y-axes in each plot specify the coordinates of cells within the mesh. Displacement along the z-axis represents depolarization of membrane potential. Four images are shown. Moving from left to right, and top to bottom, images show membrane potential throughout the same mesh when coupling conductance is set to 0, 160 pS, 1 nS, and 10 nS, respectively. In each image, membrane potential is displayed at the same instant of time, after integrating network activity for 2.515 seconds. This particular time was selected since it is a time at which many cells in the

mesh are undergoing the rapid upstroke phase of the cardiac cycle. The top left image shows that with zero coupling between cells, activity is completely asynchronous, with some cells maximally depolarized and others maximally hyperpolarized at the same instant of time. The image shown at the top right was computed using a coupling conductance of 160 pS. Dispersion of cell membrane potential is reduced; network voltage rises and falls in a more synchronized fashion. At the third coupling value (1 nS; lower left) membrane potential oscillation exhibits even more synchrony. The lower right image in fig. 5 shows membrane potential in the grid with cell coupling set to 10 nS. In this case, membrane potential throughout the mesh is approximately uniform, even during the rapid upstroke phase of the pacemaker cycle.

While demonstrating that increased cell-to-cell coupling increases the synchrony of membrane potential oscillation in network cells, these particular images do not adequately demonstrate

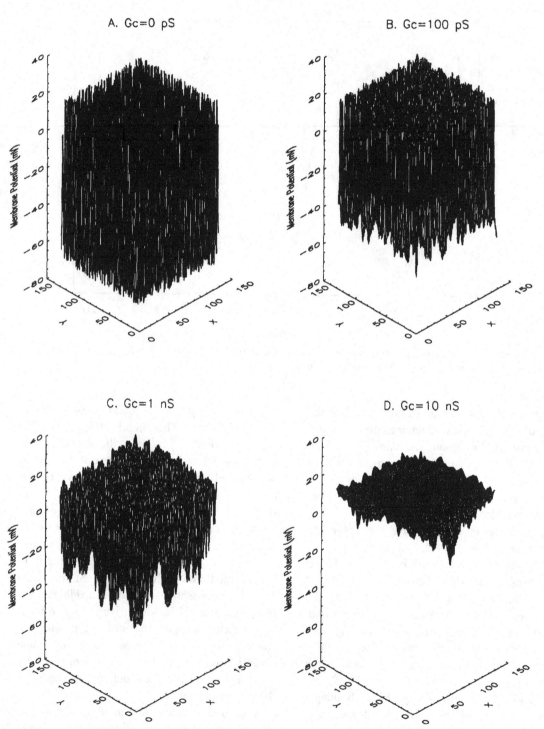

Fig. 5. Membrane potential (z-axis) in a 128×128 mesh of sinus cells at time $t = 2.515$ seconds. Cell membrane parameters are randomly distributed so that each cell has a different oscillation frequency and waveform (see section 3.2.1). Cell-to-cell coupling is uniform throughout the mesh at values of: (A) 0 pS; (B) 160 pS; (C) 1000 pS; (D) 10 000 pS.

290   *R.L. Winslow et al. / Cardiac network dynamics*

the extent of frequency entrainment in the networks. Figure 4 shows the reason for this. Peripheral cells have a much faster upstroke velocity than do central cells. Thus, even if the peripheral and central cells reach threshold at the same time, this increased upstroke velocity means that the peripheral cell may fire an action potential prior to the central cell. Membrane parameters are distributed randomly in the network model, resulting in a random distribution

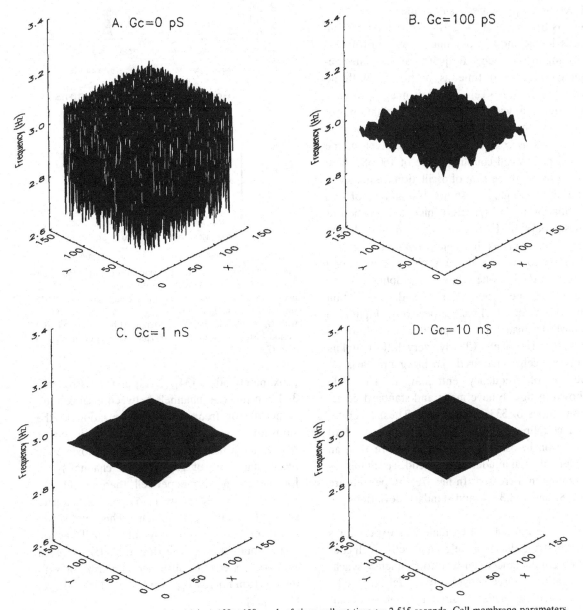

Fig. 6. Instantaneous frequency (z-axis) in a 128×128 mesh of sinus cells at time $t = 2.515$ seconds. Cell membrane parameters are randomly distributed so that each cell has a different oscillation frequency and waveform (see section 3.2.1). The parameter distribution is the same as for the network described in fig. 5. Instantaneous frequency is computed as described in the text (section 2.3). Cell coupling is uniform throughout the mesh at values of: (A) 0 pS; (B) 160 pS; (C) 1000 pS; (D) 10 000 pS.

of upstroke velocities. These random differences in upstroke velocity manifest themselves as dispersion of membrane potential during the rapid upstroke phase.

To overcome this problem, we have developed a measure we refer to as instantaneous frequency (section 2.3). Figure 6 shows plots of instantaneous frequency at $t = 2.515$ s computed over all cells in the mesh. Four images are shown; displaying instantaneous frequency at the same coupling values and time as in fig. 5. With no coupling between cells (top left image), there is of course a wide dispersion of oscillation frequencies throughout the mesh. This dispersion decreases as coupling is increased. Even at the small cell-to-cell coupling value of 160 pS, there is very little dispersion of oscillation frequency in the mesh. Figure 7 shows histograms of cell oscillation period (abscissa, ms) as a function of mesh coupling. Histograms were computed from the instantaneous frequency data at time $t = 2.515$ seconds. Figure 7a shows the estimated density function when the mesh coupling is set to zero. The mean period and standard deviation are 335.1 and 13.1 ms, respectively. Figure 7b shows the histogram computed when cell-to-cell coupling is 160 pS. Clearly, very little coupling between cells is required to achieve a remarkable degree of frequency entrainment. The data shown in fig. 7b have mean and standard deviation values of 333.2 and 1.2 ms. Thus, a cell-to-cell coupling value of 160 pS reduces standard deviation in oscillation period by more than an order of magnitude. This trend continues as coupling in increased. In figs. 7c, 7d, periods are 332.8, and 332.3 ms, and standard deviations are 0.4 and 0.11 ms.

The results of these numerical experiments demonstrate that relatively small cell-to-cell coupling conductances can lead to frequency synchronization in large networks of model sinus cells. The coupling value required for frequency entrainment is certainly less than 1 nS and seems to be as small as 160 pS. The conductance of a single gap junction channel is known to be ap-

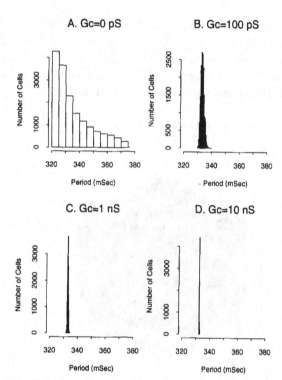

Fig. 7. Estimated oscillation period histograms. Plots show number of cells per period bin (ordinate) versus oscillation period (abscissa, msec). Histograms computed at $t = 2.515$ seconds from the data shown in fig. 6. (A) Period histogram with $G_c = 0$ pS. (B) Period histogram with $G_c = 160$ pS. (C) Period histogram with $G_c = 1000$ pS. (D) Period histogram with $G_c = 10\,000$ pS.

proximately 50 pS [34]. Thus, as few as roughly 3–4 gap junction channels between each cell are sufficient for frequency synchronization in the sinus node. Since there are 50–100 gap junctions per nexus (a nexus is an ordered, approximately hexagonal array of gap junction channels) and four nearest neighbors per cell, there need be no more than a single nexus region per cell to assure strong frequency entrainment. This number of nexus regions would occupy only a small fraction of the total surface area (less than 1%) of each sinus cell, a conclusion which is consistent with the electron microscopic studies of Masson-Pevet et al. [22].

We have performed a second class of numerical experiments in which properties of the sinus node have been modeled with a greater degree of realism. In these models, the scaling factor S controlling the degree to which the oscillation pattern of sinus cells resemble those of peripheral or central cells is varied as a function of distance from the center of the node according to eq. (7). In the simulations shown here, grid size was 128×128, $\sigma = 32$ cells, and cell-to-cell coupling was set to 10 nS. As noted before, this choice yields model sinus cell maximum diastolic potentials, overshoots, and oscillation periods in central, transitional, and peripheral sinus cells matching the data of Kodama and Boyett [20].

When properties of the resulting network are simulated, it is seen that an excitatory wave is generated first in the peripheral regions. This wave propagates inwards toward the center of the sinus node model. This result is demonstrated in fig. 8. The top and bottom images in this figure show the distribution of membrane potential within a 128×128 network with Gaussian distributed membrane parameters at two different times (top: $t = 3.575$ s; bottom: $t = 3.725$ s). As shown in the top image, depolarization first occurs in the network periphery. Propagation of the excitatory wave towards the center region can be seen by comparing the location of peak depolarization within the network at time $t = 3.575$ s (top image) with that 150 ms later (bottom image). This numerical result demonstrates the same behavior as seen experimen-

3575 milliseconds frame 715

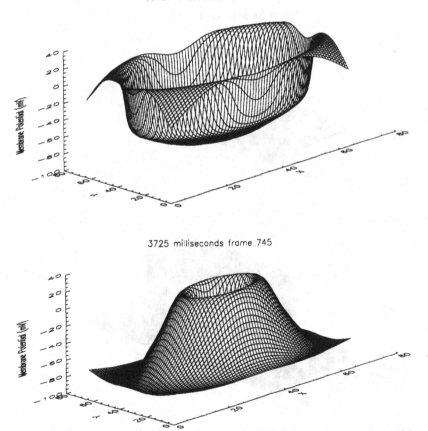

3725 milliseconds frame 745

Fig. 8. Membrane potential (z-axis) in a 128×128 sinus network incorporating regional variation of cell membrane parameters, as described in section 3.2.1. Top image shows network potential at time $t = 3.575$ s. Bottom image shows potential distribution at time $t = 3.725$ s.

tally in the rabbit sinus node [19]; when the node is separated from the atrium (as it is in this model), the pacemaker potential is initiated in peripheral regions, and propagates towards the node center. This stands in marked contrast to behavior in the intact sinus node. In such preparations, the pacemaker potential is generated in more central regions of the node, and propagates outwards towards the node periphery [19,2]. Thus, the Kirchoff data indicate that altering the boundary conditions on the sinus node by coupling the node to atrial cells shifts the site of pacemaker potential generation from the node periphery towards the center, and reverses the direction of pacemaker potential propagation.

3.2.2. Sinus node – atrial network models

We have investigated the effects of sinus node boundary conditions on pacemaker potential generation in a third set of numerical experiments [40]. In these studies, a circular central region of sinus cells was surrounded with a mesh of normally quiescent atrial cells. For the models shown here, the radius of the sinus node was set to 63 cells, with the total mesh size 256×256. This gives an atrial : sinus cell ratio of 4.3 to 1. Cell membrane parameters were varied in the sinus node as described above. Figure 9 illustrates results of this experiment. The top image in fig. 9 shows the initial site of generation of the pacemaker potential at time $t = 190$ ms. The cen-

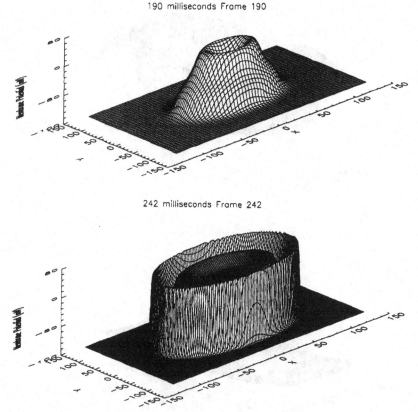

Fig. 9. Pacemaking and atrial conduction in a sinus node–atrial network model. Model structure described in section 3.2.2. Sinus node was modeled as a circular central region of cells of radius 63 cells. The node was surrounded by a mesh of atrial cells with total dimension 256×256 cells. $G_c = 100$ nS. Top image shows action potential initiation at time $t = 190$ ms after simulation onset. Bottom image shows propagation of this action potential into, and excitation of, the atrium at time $t = 242$ ms.

ter of the sinus node is at $X = 0$ and $Y = 0$; the node extends to approximately $X = \pm 63$ and $Y = \pm 63$. These data therefore show that the site of generation of the pacemaker potential is shifted from the peripheral region of the sinus node towards the center of the node. This is clearly a loading effect of the atrium, and is consistent with the data of Kirchoff [19]. At time $t = 242$ ms (bottom image), the pacemaker potential has propagated into the atrium, evoking an excitatory travelling wave shown by the outer ring of activity.

4. Discussion

4.1. The computational approach

We have described an integrated approach to modeling of large two-dimensional networks of cardiac cells. Our approach makes use of the fact that the topology of the massively parallel Connection Machine CM-2 is in exact correspondence with that of the network models. Because of this correspondence, we are able to simulate properties of these networks significantly faster on the CM-2 than on shared memory vector supercomputers, without sacrificing biophysical detail present in the single cell models. This approach has been extended to three-dimensional volumes of tissue.

The enormous computational throughput of the Connection Machine would be of little value in the absence of software tools for analyzing simulation data. The visualization system which we have described enables us to animate extremely large data sets in an interactive fashion, providing immediate visual feedback on the results of the supercomputer simulations.

4.2. Sinus node models

Our studies of the dynamics of large-scale sinus node models build on the work of Michaels and his colleagues [25,23]. These investigators

have studied properties of sinus network models in which the basic network element is a region of tissue with total surface area one cm^2. Their goal was to test the hypothesis that frequency entrainment in the sinus node is a result of phase-dependent interactions between node elements. In one series of experiments, network elements were assigned random oscillation frequencies by varying the amount of current injected into each element. Network dynamics were then studied as a function of coupling conductance [25]. At low coupling values, network element oscillation was unsynchronized. Leading pacemaker regions organized about more rapidly beating elements were observed as coupling was increased. At still higher coupling conductances, oscillation frequency became entrained throughout the network. In another series of experiments, oscillation frequency was made highest in the sinus node center and lowest in the node periphery, again, by variable current injection into network elements as a function of location within the node. A region of elements with reduced excitability was also positioned to one side of the node center. Under these conditions, the authors reported sinus node activation patterns similar to those measured in intact heart [2].

These previous models have provided valuable insight into mechanisms of frequency entrainment within the sinus node. We have built on this work by incorporating additional features into the network models. We have used as the basic building blocks of our model, network elements with physical dimension, membrane surface area, and membrane conductances matching those of individual isolated sinus cells. We have also increased the size of the network models. This is of potential significance, since model boundary conditions will have a greater effect on network dynamics when model dimensions are small relative to the network space constant. Finally, we have incorporated differences in cell oscillation properties as a function of location within the sinus node.

Our purpose in developing such a model was

to determine whether the data of Kodama and Boyett, when incorporated into a model of the sinus node, are consistent with measured activation times in the node [2]. Simulation of the model demonstrates that an excitatory wave of activity propagates from peripheral to central regions, in contradiction to the data of Bleeker et al. [2]. These properties are consistent, however, with Kirchoff's data [19] which show that in the intact sinus node, depolarizing excitatory waves are generated in the central region of the node and propagate to the periphery, consistent with the data of Bleeker et al. [2]. However, following isolation of the sinus node from the atrium, the excitatory wave is generated in peripheral sinus regions, and propagates towards the node center, precisely as predicted by our network model.

The experimental results of Kirchoff [19], as well as the results of our own numerical experiments, therefore demonstrate two major points. First, large-scale sinus network models incorporating data on regional variation of sinus cell membrane properties predict the same site of initiation, frequency, and direction of propagation of the pacemaker potential as observed experimentally when the sinus node is isolated from the atrium. This demonstrates consistency of the Kodama and Boyett single cell data, our sinus node network modeling results, and the experimental data of Kirchoff obtained in isolated sinus node preparations. Second, Kirchoff's data shows that "loading" of the sinus node by the surrounding atrial cells is critical in shifting the site of generation and reversing the direction of propagation of the pacemaker potential. Network models of the sinus node which do not include models of the surrounding atria therefore should not replicate the activation data of Bleeker et al. [2], as these data only apply under conditions of atrial loading. Rather, they should reproduce the Kirchoff data obtained in the isolated sinus node.

Our studies of frequency entrainment demonstrate that extremely small cell-to-cell coupling values (as few as three gap junction channels between each cell) lead to a remarkable degree of frequency entrainment [37] in large cell networks. This finding suggests that a modification to the network models may be necessary. Given that a few gap junction channels between each cell can give rise to frequency entrainment, it is likely that the probabilistic opening and closing of these channels effect the entrainment process. Properties (mean channel open/close times) of single gap junction channels in neonatal rat heart myocytes have been measured [34]. A future objective will to incorporate these data into the network models in order to investigate their effects at low gap junction channel densities.

4.3. Sinus node – atrial network models

Given the above results, our present objective is to develop a model of the sinus node which is consistent with both the known regional variation of sinus cell properties, activation times, and propagation patterns in the sinus node of the intact rabbit heart. We believe two factors will be of critical importance in replicating the Kirchoff data: (a) loading of the sinus node cells by surrounding atrial cells, as shown directly in Kirchoff's experiments; and (b) spatial variation of gap junction coupling in the atrium and sinus node. Our simulations of sinus node–atrial networks demonstrate that in network models with uniform coupling in both atrium and the sinus node, the site of pacemaker potential generation is in fact shifted to more central regions as the coupling magnitude is increased. We hasten to add that uniform coupling in atrium and sinus node is not a physiologically realistic assumption, since increased coupling in the atrium is required to model higher conduction velocities observed in this region. The use of uniform coupling merely represents a first attempt at identifying all factors that bring the model network activation data into closer agreement with the data of Bleeker et al. [2]. Further, we have not been able to shift the site of pacemaker potential generation directly into the node center by use of uniform coupling.

There are now competing views as to the nature of gap junction coupling within the sinus node itself. The "classical" theory holds that there is a gradation of coupling, with coupling strength increasing from the sinus node center to the periphery [2,3]. More recent results obtained using an antibody stain to gap junction proteins has called this view into question [18]. These new results suggest the presence of uniform, sparse coupling within the sinus node, and stronger coupling within the atria. There appear to be interdigitating strands of heavy and light staining at the border between sinus node and atria. There is no evidence of significant coupling variation in the node itself. Of course, these results must be interpreted with great caution. As noted above, our numerical studies of frequency entrainment in large sinus node network models indicate that as few as three gap junction channels between each cell gives rise to near complete frequency entrainment in the node. Given the small number of channels required to achieve frequency entrainment, it is unclear whether gap junction staining techniques can resolve small variations in channel density that could be physiologically relevant. Numerical experiments investigating properties of the sinus node–atrial network model will therefore provide a powerful tool for the quantitative evaluation of hypotheses on the nature of gap junction coupling, since the models must reproduce well known propagation patterns and conduction velocities in rabbit sinus node and atria [2,3,19].

Acknowledgements

Drs. Jane Denyer and Hilary Brown of the Oxford University Laboratory of Physiology, Oxford, England, assisted in development of the cardiac network models. Dongming Cai provided help in generating plots of network data. The Army High Performance Computing Research Center (AHPCRC) of the University of Minnesota Institute of Technology provided support for all phases of this research, including Connection Machine time. The Director of the AHP CRC Graphics and Visualization Laboratory, Paul Woodward, provided valuable advice on design of graphics software as well as implementation of the Cray network codes. Ken-Chin Purcell and Wes Barris of the AHPCRC assisted in the production of graphics and animations during the course of this research. Thinking Machines Corporation, Cambridge, MA provided technical assistance, use of the Connection Machine Network Server, and video production facilities. 3M Corporation, St. Paul, MN, provided financial support. The Minnesota Supercomputer Center, Minneapolis, MN, provided both technical assistance and access to the Cray X-MP and Cray-2 supercomputers.

Appendix A

The following are OXSOFT Heart input files for simulating central and peripheral sinus cells. Oscillation properties have been adjusted to fit the data of Kodama and Boyett [20]. In addition to these parameter changes, the voltage-dependent time constant for the delayed rectifier current I_k is slowed by a factor of 2 and 4 (relative to the current OXSOFT value) in peripheral and central cells.

A.1. Central cell model

$****PRES:ESINUS*****

CONTINUE = 0; PREPLENGTH = 0.028; DX = 1; TIMESCALE = 1000; ISCALE = 1000; disp = 0; yrange3 = 0.05; silent = 1; grafitot = 0; grafik = 1; grafib = 1; grafik1 = 1; grafica = 0; grafinc = 0; CANmode = 0; TEND = 5.0; kb = 4.0; Kc = 4.0; GBNA = 0.00091; GK1 = 0.015; PCA = 0.078; IKM = 1.6; SHIFTY = −4.0.

A.2. Peripheral cell

$*****PREP:ESINUS*****

CONTINUE = 0;   PREPLENGTH = 0.028; DX = 1; TIMESCALE = 1000; ISCALE = 1000; disp = 0; yrange 3 = 0.05; silent = 1; grafitot = 0; grafik = 1;   grafib = 1;   grafik1 = 1;   grafica = 0; grafinc = 0; CANmode = 0; TEND = 5.0; kb = 4.0;   Kc = 4.0;   GFK = 0.12;   GFNA = 0.12; GK1 = 0.0225;   GNA = 0.375;   PCA = 0.30; IKM = 1.76; SHIFTY = −20.0.

Appendix B

We have simulated three different classes of cardiac networks: (a) homogeneous networks in which all cells are of similar type (either sinus or atrial) and have identical membrane parameters; (b) nonhomogeneous sinus networks in which all cells are of similar type, but have different membrane parameters; and (c) sinus node–atrial network models which include both cell types as well as variation in individual cell parameters. The mapping of cells to virtual processors is simple for cases (a) and (b); each processor updates the state equations for a small subgrid of cells in lockstep with every other processor. This is possible since all cells in the network are of similar type. Case (b) poses no particular problem since differences in membrane properties of individual cells are incorporated by storage of membrane parameters in the local memory of each virtual processor. Nearest neighbor communication is required to update membrane potential in every cell (the current into each cell from its immediate neighbors is proportional to the difference in voltage between these neighbors, see eq. (6)). Thus, when updating membrane potential, each cell in the mesh must send its membrane potential to north, south, east, and west neighbors. This communication step is fast since interactions between neighboring model cells only require on-chip, or hardware sup-

ported nearest neighbor communication between physical processors in the CM-2.

Simulation of sinus–atrial network models on a SIMD computer is somewhat more involved. The difficulty stems from the fact that the atrial cell model includes membrane currents and state variables not present in the sinus cell model. In a SIMD computer, every processor must execute the same instruction at the same time. Therefore, if the mapping between model cells to virtual processors preserves nearest neighbor relationships, then those processors modeling atrial cells must sit idle while those processors modeling sinus cells update state. In our sino-atrial node models, the ratio of sinus to atrial cells can be as high as 1:20, raising the likelihood that enormous computational power will be wasted while a large fraction of the total number of processors in the machine sit idle.

To avoid this scenario, two virtual processor sets are defined; one consisting of all the atrial cells, and the other all the sinus cells in the network. Two steps are used to sequentially update the state of cells in these two sets. First, all components of the velocity field, except membrane potential, are computed in parallel for each virtual processor set. Inspection of eqs. (2), (3) and (5) shows that this step requires no interprocessor communication. Total inward current through membrane ion channels, pumps, and exchangers is also computed. As long as the number of cells of each type is sufficiently large to allocate a nonnull subgrid to each physical processor, processors do not sit idle. The second step is to update membrane potential by computing the current which flows into each cell through gap functions, and to add this to the previously computed value of total inward current through ion channels and membrane pumps/ exchangers to get the total inward membrane current in every cell of both virtual processor sets. This total inward current determines the time rate of change of membrane potential according to eq. (6). Only nearest neighbor communication is required to perform this step for homogeneous and non-homogeneous network models. However, in the heterogeneous case, each physical processor simulates properties of both sinus and atrial cells. Membrane potential data for neighboring model cells may now be physically distant within the connective topology of the Connection Machine, and purely local NEWS communication is no longer sufficient to exchange this data. Instead, a general communication step is taken in which neighboring model cells of different type communicate their membrane potentials by dynamic routing of messages through the hypercube network, after which cells of similar type exchange membrane potential by performing NEWS communication. The rate of change of membrane potential may then be computed in all cells.

References

[1] G.W. Beeler and H. Rheuter, J. Physiol. 268 (1977) 177.

[2] W.K. Bleeker, A.J.C. Mackaay, M. Masson-Pevet, L.N. Bouman and A.E. Becker, Circ. Res. 46(1) (180) 11.

[3] L.N. Bouman, J.J. Duivenvoorden, F.F. Bukauskas and H.J. Jongsma, J. Mol. Cell Cardiol. 21 (1989) 407.

[4] D.G. Bristow and J.W. Clark, Am. J. Physiol. 243 (1982) H207.

[5] H.F. Brown, J. Kimura, D. Noble, S. Noble and A. Taupignon, Proc. R. Soc. London. B 222 (1984) 329.

[6] J. Denyer and H. Brown, Jpn. J. Physiol. 37 (1987) 963.

[7] J. Denyer and H. Brown, J. Physiol., in press.

[8] D. DiFrancesco, A. Ferroni, M. Mazzanti and C. Tromba, J. Physiol. 377 (1986) 61.

[9] D. DiFrancesco and D. Noble, Phil. Trans. R. Soc. London B. 307 (1985) 353.

[10] Y.E. Earm and D. Noble, Proc. R. Soc. London B 240 (1990) 83.

[11] R.P. Gould and J. Powell, J. Physiol. 225 (1972) p 16.

[12] D.W. Hilgemann and D. Noble, Proc. R. Soc. London B 230 (1987) 163.

[13] W.D. Hillis, The Connection Machine (MIT Press, Cambridge, MA, 1984).

[14] A.L. Hodgkin and A.F. Huxley, J. Physiol. 117 (1952) 500.

[15] R. Horn and A. Marty, J. Gen. Physiol. 92 (1988) 145.

[16] H. Irisawa and T. Nakayama, Jikeikai. Med. J. 30 (1984) 65.

[17] J. Jalife, J. Physiol. 356 (1984) 221.

[18] H. Jongsma, I. ten Velde, B. de Jonge and D. Gros, presented at: Symposium on Background Conductance Mechanisms in Cardiac Muscle, 16–17 July 1991, University Laboratory of Physiology, Oxford.

[19] C.J.H.J. Kirchoff, PhD thesis, University of Maastricht (1990).

[20] I. Kodama and M. Boyett, Pflugers Arch. 404 (1985) 214.

[21] M. Kubicek and M. Marek, Computational Methods in Bifurcation Theory and Dissipative Structures (Springer, New York, 1983).

[22] M. Masson-Pevet, W.K. Bleeker, A.J.C. Mackaay and L.N. Bouman, J. Molec. Cell. Cardiol. 11 (1979) 555.

[23] D.C. Michaels, D.R. Chialvo, E.P. Matyas and J. Jalife, in: Mathematical Approaches to Cardiac Arrhythmias, ed. J. Jalife (New York Acad. Sci., New York, 1990).

[24] D.C. Michaels, E.P. Matyas and J. Jalife, Circ. Res. 58(5) 1986) 706.

[25] D.C. Michaels, E.P. Matyas and J. Jalife, Circ. Res. 61(5) (1987) 704.

[26] T. Nakayama, Y. Kurachi, A. Noma and H. Irisawa, Pflugers Archiv. 402 (1984) 248.

[27] D. Noble, OXSOFT Heart Program Manual, OXSOFT Ltd., Oxford (1990).

[28] D. Noble, J. Denyer, R. Winslow and A. Kimball, in: Proc. 18th Int. Symp. Cardiac Electrophysiol (Korean Acad. Med. Sci., 1991) pp. 69–86.

[29] D. Noble, J.C. Denyer, H.F. Brown and D. DiFrancesco, J. Physiol. submitted.

[30] D. Noble, D. DiFrancesco and J. Denyer, In: Neuronal and Cellular Oscillators, ed. J.W. Jacklet (Marcel Dekker, New York, 1989).

[31] D. Noble and S. Noble, Proc. R. Soc. London B 222 (1984) 295.

[32] D. Noble and T. Powell, Electrophysiology of Single Cardiac Cells (Academic Press, New York, 1987).

[33] A. Noma and N. Tsuboi, J. Physiol. 382 (1987) 193.

[34] M.B. Rook, H.J. Jongsma and A.C.G. van Ginneken, Am. J. Physiol. 255 (1988) H770.

[35] J. Tanaguchi, S. Kokubun, A. Noma and H. Irisawa, Jpn. J. Physiol. 31 (1981) 547.

[36] Thinking Machines Corporation, Connection Machine Model CM-2 Technical Summary, Cambridge, MA (1989)

[37] V. Torre, J. Theor. Biol. 61 (1976) 55.

[38] R.L. Winslow, A. Kimball, D. Noble and J. Denyer, Med. Biol. Eng. Comput. 29(2) (1991) 832.

[39] R.L. Winslow, A. Kimball, D. Noble and J. Denyer, J. Physiol. London 438 (1991) 180P.

[40] R.L. Winslow, A. Varghese, D. Noble, J.C. Denyer and A. Kimball, J. Physiol. London 446 (1992) 242P.

[41] K. Yanagihara, A. Noma and H. Irisawa, Jpn. J. Physiol. 30 (1980) 841.

IEEE TRANSACTIONS ON BIOMEDICAL ENGINEERING, VOL. 41, NO. 3, MARCH 1994   217

Effects of Gap Junction Conductance on Dynamics of Sinoatrial Node Cells: Two-Cell and Large-Scale Network Models

Dongming Cai, Raimond L. Winslow, and Denis Noble

Abstract—A computational model of single rabbit sinoatrial (SA) node cells has been revised to fit data on regional variation of rabbit SA node cell oscillation properties. The revised model simulates differences in oscillation frequency, maximum diastolic potential, overshoot potential, and peak upstroke velocity observed in cells from different regions of the node. Dynamic properties of electrically coupled cells, each with different intrinsic oscillation frequency, are studied as a function of coupling conductance. Simulation results demonstrate at least four distinct regimes of behavior as coupling conductance is varied: a) independent oscillation ($G_c < 1$ pS); b) complex oscillation ($1 \leq G_c < 220$ pS); c) frequency, but not waveform entrainment ($G_c \geq 220$ pS); and d) frequency and waveform entrainment ($G_c \geq 50$ nS). The conductance of single cardiac myocyte gap junction channels is about 50 pS. These simulations therefore show that very few gap junction channels between each cell are required for frequency entrainment. Analyses of large-scale SA node network models implemented on the Connection Machine CM-200 supercomputer indicate that frequency entrainment of large networks is also supported by a small number of gap junction channels between neighboring cells.

I. Introduction

THE sinoatrial (SA) node is a thin sheet of cardiac muscle fibers located near the junction of the superior vena cava and the right atrium. The node is composed of several hundred thousand cells, each of which is an electrical oscillator. Studies of cells isolated enzymatically from the SA node indicate that the intrinsic oscillation frequency of each cell is different. Despite these differences, a coherent oscillatory electrical wave known as the pacemaker potential is generated within the node. This wave is conducted throughout the heart, determining its rate of beating.

There has been considerable interest in determining the mechanisms by which cells within the SA node coordinate their oscillation frequency to a common value, a process known as frequency entrainment. Early morphological studies demonstrated the presence of gap junctions between neighboring SA node cells, suggesting the possibility that frequency entrainment occurs as a result of ionic current flow through these junctions. However, the gap junctions were observed to

Manuscript received September 1, 1992; revised October 20, 1993.
The authors are with the Army High Performance Computing Center, University of Minnesota, Minneapolis, MN 55455. R. L. Winslow is also with the Department of Biomedical Engineering, The Johns Hopkins University School of Medicine, Baltimore, MD 21205. D. Noble is also with the University Laboratory of Physiology, Oxford University, Oxford, UK.
IEEE Log Number 9215011.

be smaller in size and far less numerous than those found in other regions of the heart [6], [7]. Electrophysiological studies of frequency entrainment using embryonic heart cell aggregates demonstrated that when aggregates, each beating at its own intrinsic frequency, were brought into physical contact, synchronization to a common frequency followed within minutes [8]–[10]. The transition to electrical synchronization was accompanied by a decrease in resistance of the junctional area between the aggregates. These results suggested that synchronization behavior is related to the time-dependent formation of gap junction channels between apposed aggregates. Mechanisms of synchronization between SA node cells were studied further using the sucrose gap technique [11], [12]. These experiments demonstrated that electrical coupling results in mutual entrainment with both pacemakers beating at simple harmonic ($1 : 1, 1 : 2$) or more complex ratios ($3 : 2, 5 : 4$, etc.) depending on the extent of coupling and intrinsic frequency of the two pace-maker regions.

Synchronization behavior of SA node cells has also been investigated using theoretical and computational models. Torre [13] modeled properties of individual cells using the Bonhoeff–Van der Pol (BVP) equations. Cells were coupled by a linear resistor. The model was used to prove the existence of a structurally stable periodic solution of the coupled system at low values of coupling resistance, demonstrating that electrical coupling alone could account for frequency entrainment. Computer simulations based on the early Purkinje fiber equations [14] were also used to demonstrate that relatively little coupling between cells leads to frequency entrainment, a conclusion reached using much simpler models by Bleeker [6, appendix] and Noble [10, appendix]. In 1980, Yanagihara and his colleagues developed the first SA node cell model based on voltage-clamp data obtained from multicellular preparations [15]. Michaels *et al.* [16] used these models in studies of frequency entrainment of resistively coupled cell pairs, the objective being to understand the mechanisms underlying frequency entrainment. Analyses of phase resetting curves [see also [17]–[19]] showed that phase delays produced by application of a brief depolarizing current pulse within an appropriate region of the oscillation cycle were due to a transient decrease of the second inward current I_{si} and the fast inward current I_{na}; phase advances were due to transient increases in these same currents. Most importantly, these phase-dependent interactions were shown to underlie the

218   IEEE TRANSACTIONS ON BIOMEDICAL ENGINEERING, VOL. 41, NO. 3, MARCH 1994

synchronization of oscillation frequency in coupled model cell pairs.

Considered together, these morphological, electrophysiological, and modeling results provide convincing evidence that electrical coupling of neighboring cells by gap junctions underlies frequency entrainment within the SA node.

During the past decade, improvements in cell dissociation methods, whole-cell voltage-clamp recording, and optical techniques for measuring intracellular ion concentration have provided a wealth of detailed information on the biophysical mechanisms that give rise to oscillation of membrane potential in single SA node cells [20]–[27]. This has led to the formulation of a biophysically detailed model of these cells based directly on patch-clamp data [1] and [2]. Our objective has been to study properties of frequency entrainment in SA node cell networks using this new single SA node cell

In this paper, we describe the results of studies investigating the dynamics of coupled cell pairs in which one model cell is adjusted to have a high and the other a low oscillation frequency. The way in which the magnitude of gap junction coupling determines cell dynamics is investigated. Frequency entrainment properties of coupled pairs of cells are compared to those observed in large networks modeled on a massively parallel supercomputer, the Connection Machine CM-200 [28]–[31]. The preliminary results of these studies have been published in abstract form [32].

II. METHODS

A. Models of Single SA Node Cells

Descriptions of the SA node cell model used in these studies are given in [1], [2]. Membrane currents included in the model are a) I_f—a hyperpolarizing activated current carried by sodium (Na) and potassium (K) ions; b) I_K—a delayed K current; c) I_{K1}—a background (instantaneous) K current; d) I_{Na}—a fast inward Na current; e) $I_{Si,Ca}$ and $I_{Si,K}$—calcium (Ca) and K components of the second inward current; and f) $I_{b,Na}$ and $I_{b,Ca}$—linear background Na and Ca currents. The time and voltage dependent properties of membrane currents are modeled using Hodgkin–Huxley type equations. The model also includes quantitative descriptions of the sodium–potassium (Na-K) ATP-dependent pump, the sodium–calcium (Na-Ca) ion exchanger, changes in intracellular ion concentrations over time, and intracellular calcium (Ca) sequestering and release mechanisms. Detailed comparisons of model properties to experimental data are given in [1], [33]–[36].

B. Modeling Cell Pairs

In two cell simulations, individual cells are modeled as described above, with model parameters adjusted to yield different intrinsic oscillation frequencies in each cell (see Section IIIA for details). Cells are coupled by a conductor G_c, yielding the following expression for the time-rate of change of membrane potential in the two cells (denoted by superscript 1 and 2):

$$\dot{V}_t^1 = \left[-\sum_j I_j^1(V_t) - \sum_j I_j^{1,pe} + (V_t^2 - V_t^1)G_c \right] \bigg/ C_m^1 \tag{1}$$

$$\dot{V}_t^2 = \left[-\sum_j I_j^2(V_t) - \sum_j I_j^{2,pe} + (V_t^1 - V_t^2)G_c \right] \bigg/ C_m^2 \cdot u \tag{2}$$

Variables are defined as follows: V_t—instantaneous transmembrane potential of a single cell; $I_j(V_t)$—outward membrane current through the jth ionic conductance; I_j^{pe}—outward current generated by the jth ion pump/exchanger; C_m—membrane capacitance of a single cell; G_c—coupling conductance between two cells.

C. Modeling Large Cell Networks

Large, two-dimensional networks of coupled cells are modeled as a rectangular $N \times N$ mesh of cells in which each interior cell has four nearest neighbors. Fixed linear conductances (G_c) model gap junction coupling between neighboring cells. Each cell is modeled as described in Section IIA above. State equations for the time-rate of change of membrane potential can be derived by applying Kirchoff's current law at each node of the mesh:

$$\dot{V}_t^{ij} = \left[-\sum_k I_k^{ij}(V_t) - \sum_k I_k^{ij,pe} + (S_{ij} - N_{ij}V_t^{ij})G_c \right] \bigg/ C_m^{ij}. \tag{3}$$

Variables are defined as: V_t^{ij}—membrane potential at time t in cell ij; $I_k^{ij}(V_t)$— kth outward membrane current in cell ij; $I_k^{ij,pe}$—kth pump/exchanger in cell ij; S_{ij}—sum of membrane potentials of the cells neighboring cell ij; N_{ij}—number of cells neighboring cell ij; C_m^{ij}—membrane capacitance of each cell.

Additional state equations are given by the time-rate of change of activation/inactivation variables in each cell described in Section IIA. This brings the total number of state variables N_{eq} to 14 per cell, and to $N^2 N_{eq}$ in an $N \times N$ mesh of cells. For the network simulations described in this paper, N was set to either 128 or 512. Simulation of network model properties therefore requires the integration of at least a quarter-million coupled nonlinear ordinary differential equations. This is a computationally demanding task requiring the use of high performance computing systems.

D. Integration Algorithm

Differential equations describing single- and two-cell models were integrated using an adaptive step fourth-order Runge–Kutta algorithm. In this algorithm, if the local error at the current step size is below a specified error tolerance (10^{-7}), the integration step is accepted; otherwise, a new smaller step is selected and the local error recomputed (for a detailed description of this algorithm, see [37]). Results were indistinguishable from those obtained using implicit Gear and Adams–Moulton methods (as implemented in the IMSL [38]), and explicit fixed step $(100 \, \mu s)$ second-order Runge–Kutta methods. Single cell results were verified by comparison to output of the OXSOFT Heart software [2].

Differential equations describing the networks were integrated on the Connection Machine CM-200 massively parallel supercomputer using either the adaptive step fourth-order Runge–Kutta algorithm described above, or a fixed step $(100 \, \mu s)$ second-order Runge–Kutta algorithm. Both gave identical results. However, the fixed step procedure ran about 6 times faster than the adaptive step method. All computations were performed using double precision arithmetic. Implementations have been described elsewhere [28]–[31]. Briefly, subgrids of model cells were mapped onto individual processors of a 32 768 processor Connection Machine. The state equations defining the properties of each subgrid were integrated concurrently over all the subgrids, thereby enabling very high throughput (in excess of 3.5 Gflops) in the computations.

E. Simulation Data Analysis

An objective of this study is to determine the ways in which the "steady-state" behaviors of coupled SA node cells change as a function of coupling conductance magnitude (steady state refers to behavior as time $t \rightarrow \infty$). A variety of procedures are used to analyze this behavior. These procedures include: a) phase portrait; b) power spectrum; c) attractor reconstruction; and d) Poincaré surface of section analyses [39].

The phase portrait of an N-dimensional system is a plot of system trajectories in an N-dimensional space, with axes defined by the N state variables. Such plots provide a qualitative method for studying features of the trajectories. It is clearly impractical to generate a complete phase portrait given the high dimension of our system ($N = 28$; 14 state equations per cell, two cells). We have therefore taken the approach of plotting values of voltage versus intracellular calcium concentration [the V-Ca phase plane; see [40]–[42]] for each of the two coupled cells. From a physiological viewpoint, these are the two most critical state variables, and the ones amenable to direct experimental measurement. All phase portraits were generated by imposing a set of initial conditions on the state equations. The system was then integrated until a steady-state behavior was achieved (typically after 30 s). Voltage and calcium values were then computed over an additional 20-s window at a sampling interval of 500 μs.

Power spectral analyses can be used to help distinguish between periodic, quasi-periodic, and chaotic regimes [39], [43]–[48]. In general, the power spectrum of a periodic solution with frequency f has energy at $\{f, 2f, 3f, \cdots\}$. A quasiperiodic solution with base frequencies $f_1, f_2, \cdots, f_k$ has energy at these base frequencies as well as all linear combinations of these base frequencies with integer coefficients. A chaotic solution typically has a broad-band spectrum similar to that of a noise signal. We have computed power spectra of simulated SA node cell oscillation patterns using time sequences with 4096 points sampled at 5-ms intervals, yielding a frequency resolution of 0.05 Hz.

As noted above, the trajectory of each cell is a set of points in an $N = 14$ dimensional space. Despite this high dimension, it is possible that the system state evolves towards an attractor of lower dimension. The technique of attractor reconstruction can be used to identify low dimensional attractors [46], [47], [49]–[51]. The idea of attractor reconstruction is to generate a multidimensional phase portrait from measurement of just a single state variable at successive time lags. For example, rather than plotting trajectories in a space defined by the $N = 14$ state variables, trajectories can be plotted in a k-dimensional space defined by time-lagged values of membrane potential $V(t), V(t - \tau), V(t - 2\tau), \cdots, V(t - (k - 1)\tau)$. The time lag τ cannot be too small or else the successive state variable values will be similar, and the resulting phase portrait will be stretched along the diagonal axis. Additionally, τ cannot be too large or else successive sample values may be uncorrelated. In our analyses, we have used a time lag of 25 ms (selected by trial and error). The value of k is determined by increasing it by one until new structures fail to appear in the phase portrait. As we will see, so far in our analyses k need be no larger than two.

The final technique we have employed is known as computation of the Poincaré surface section [39], [46], and [47]. A two-sided Poincaré surface section is formed by the points of intersection of the trajectories of an N-dimensional system with an $N - 1$ dimensional hyperplane transverse to the flow. Poincaré surface sections have particular forms depending on the nature of the flow that generates them. Poincaré surface sections consisting of a set of discrete points are generated by autonomous systems with periodic solutions. Poincaré surface sections consisting of sets of closed curves are generated by autonomous systems with quasi-periodic solutions. Poincaré surface sections for autonomous systems with chaotic solutions do not have a simple geometric form.

III. RESULTS

A. Modeling Oscillation Properties of SA Node Cells

Kodama and Boyett [3] have shown that the oscillation properties of rabbit SA node cells vary as a function of distance from the center of the node. Specifically, when moving from center to periphery, intrinsic oscillation frequency, overshoot potential (OV), and maximum upstroke velocity increase, and maximum diastolic potential (MDP) decreases. Increases in oscillation period due to increases in extracellular K concentration are also largest in peripheral cells.

Honjo and Boyett [4] have confirmed these regional difference in oscillation properties observed in multicellular preparations by recording from individual cells isolated from the rabbit SA node. Morphological data indicate that cells from the center of the SA node have smaller surface area than do cells from the periphery. Since cell capacitance is proportional to surface area, Honjo and Boyett used estimated cell capacitance to determine whether isolated cells were from the peripheral, transitional, or central region of the node. Cells categorized on this basis had oscillation properties similar to those in the previous study [3].

220

IEEE TRANSACTIONS ON BIOMEDICAL ENGINEERING, VOL. 41, NO. 3, MARCH 1994

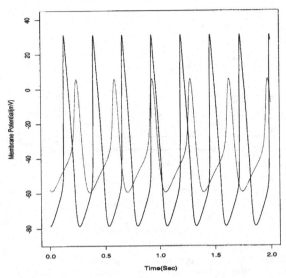

Fig. 1. Membrane potential (ordinate, in millivolts mV) versus time (abscissa, in seconds) for the model cell with low oscillation frequency (central cell; dotted line), and high oscillation frequency (peripheral cell; solid line).

TABLE I

	MDP	OV	Period	Upstroke Velocity
	(mV)	(mV)	(mSec)	(V/Sec)
Central Cell	−58.0 (−60.0)	4 (2.0)	347.0 (370.0)	1.2 (1.0–2.0)
Peripheral Cell	−78.0 (−70.0)	28.0 (18.0)	276.0 (300.0)	25.0 (30.0–40.0)

Experimentally measured properties [3] of peripheral and central cells are contrasted to those of the models in this table (the experimental values are listed within the parentheses).

4 mM extracellular K. The ordinate is membrane potential in millivolts, and the abscissa is time in seconds. Experimentally measured properties of peripheral and central cells are contrasted to those of the models in Table I (experimental values are listed within the parentheses). Model and experimentally measured oscillation properties are in reasonable agreement. Parameter values for model central and peripheral cells are summarized in the Appendix.

B. Dynamics of Coupled Cell Pairs:

Results described in the previous section provide a physiologically realistic way of modeling SA node cells with different intrinsic oscillation characteristics. In the next series of simulations, pairs of cells with differing oscillation frequency were electrically coupled, and their responses simulated over time. Membrane parameters of one member of the pair were adjusted to yield oscillation characteristics of a peripheral SA node cell (period 276 ms); the other was adjusted to model a central cell (period 347 ms). This was done since these two types of cells exhibit the greatest difference in oscillation frequency within the SA node. Simulations of these cells therefore constitute an extreme test of the ability of a given level of coupling conductance to support frequency entrainment.

Results of these simulations are shown in Fig. 2. Fig. 2(a)–(d) show representative oscillation patterns of the central (dashed line) and peripheral cell (solid line) at 0.1, 70, 1000, and 50 000 pS of coupling, respectively. Each plot starts at time $t = 30$ s (responses were simulated for an initial 30-s interval to assure that a steady state was reached following coupling). In Fig. 2(a) (0.1 ps coupling), the central and peripheral cells oscillate with the same intrinsic period (347 and 276 ms, respectively) and waveform as they do in isolation. In Fig. 2(b) (70-pS coupling), oscillation waveforms are more complex. These effects are present in the responses of both cells, but are seen most clearly in those of the central cell. Oscillation of the central cell appears to be modulated by a slowly varying envelope, reflected in the slow, repetitive variation of the MDP and OV amplitude. In addition, oscillation period varies in time, from a maximum of 399 ms during interval $T1$ to 387 ms during interval $T2$. Inspection of the oscillation waveform suggests that the responses of the peripheral and central cell are entrained at a complex ratio of 5 : 4 cycles. In Fig. 2(c), coupling conductance is increased to 1 nS. In this case, the oscillation period of both cells is 280 ms. However, cell waveforms are clearly different, being phase-shifted relative to one another. This phase shift

Motivated by these data, we have attempted to model the differences in ion channel and pump/exchanger current magnitudes required to account for observed differences in SA node cell oscillation properties. There have been relatively few direct experimental measurements of membrane current properties in cells isolated from different regions of the SA node. Our adjustment of SA node cell model parameters to fit these oscillation properties is therefore based largely on indirect reasoning. Kodama and Boyett [3] measured a maximum upstroke velocity of approximately 30–40 V/s in peripheral cells, and 1–2 V/s in central cells. Maximum upstroke velocity in the model is controlled primarily by the magnitude of the fast inward Na current. We have therefore increased this current in peripheral cells relative to that in central cells. Measured overshoot potential in peripheral cells is approximately 20 mV; that in central cells is about 0 mV. Model overshoot potential is determined primarily by properties of the inward Ca current. We have therefore increased the magnitude of the calcium current, $I_{Si,Ca}$ in peripheral cells. In addition, the oscillation period of central and peripheral cells differ by about 15%. Increasing the hyperpolarizing activated inward current I_f by increasing peak conductance or by shifting the activation range to a more depolarized level can produce moderate increases in cell oscillation frequency. We have therefore increased the peak conductance of the I_f current in peripheral cells, and have shifted the activation range +2 mV relative to that in central cells. Finally, increasing the magnitude of the instantaneous K current I_{K1} is necessary to account for differences in extracellular K sensitivity observed in peripheral versus central cells [3], and for the more negative maximum diastolic potential of peripheral cells.

Fig. 1 shows pacemaking activity of a model peripheral (solid line) and central (dashed line) cell in the presence of

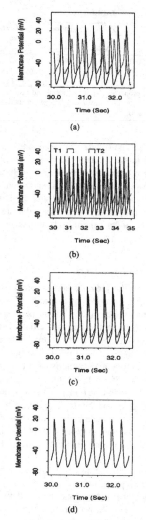

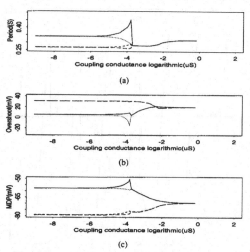

Fig. 3. Oscillation period (ms), overshoot potential (mV), and maximum diastolic potential (mV), as a function of coupling conductance. In each panel, solid and dotted lines show the maximum and minimum values (respectively) of the period, MDP, and OV for the central cell. The dashed and long dashed lines show corresponding values for the peripheral cell. The abscissa is the base 10 logarithm of the coupling conductance in μS (a) Oscillation period. (b) Overshoot potential. (c) Maximum diastolic potential.

Fig. 2. Oscillation waveforms of the coupled central (dotted line) and peripheral cell (solid line) as a function of coupling conductance. Various coupling conductance values, (a) 0.1 pS. (b) 70 pS. (c) 1 nS. (d) 50 nS.

results from the fact that both the inward Ca and Na current magnitudes are smaller in the central than the peripheral cell, giving the central cell a smaller maximum upstroke velocity. Despite this difference in oscillation waveforms, the cells are clearly frequency entrained. Fig. 2(d) shows oscillation waveforms when coupling is increased to 50 nS. In this case, both oscillation frequency and waveform are identical in the two cells.

The data of Fig. 2 suggest that there are at least four dynamic regimes as a function of coupling conductance: a) a regime of independent oscillation at low coupling values; b) a regime of complex dynamics at slightly higher coupling; c) a regime of frequency, but not waveform entrainment; and d) a regime of both frequency and waveform entrainment at high coupling values. This possibility is explored further in

Fig. 3, which shows values of the oscillation period (Fig. 3(a)), overshoot potential (Fig. 3(b)), and MDP (Fig. 3(c)) as a function of coupling conductance (abscissa; in units of the base 10 logarithm of conductance in microsiemens). Four curves are shown in each plot. Solid and dotted curves are derived from responses of the central cell. They show the maximum (solid line) and minimum (dotted line) values of the quantity displayed on the ordinate during the time course of the oscillation. In a similar fashion, short and long dashed lines plot maximum and minimum values of the same response measure for peripheral cells.

Inspection of the data in Fig. 3(a) shows that for coupling values less than approximately 10^{-6} μS (1 pS), changes in coupling conductance have little effect on the oscillation period of either the central or peripheral cell. Both beat at their intrinsic periods of approximately 347 ms (central cell) and 276 ms (peripheral cell). Similarly, at coupling values less than 1 pS, variation of coupling has little effect on either the overshoot (Fig. 3(b)) or maximum diastolic potentials (Fig. 3(c)). This indicates that independent oscillation of coupled cells occurs for coupling values less than approximately 1 pS.

Divergence of the solid and dotted lines at coupling values greater than 10^{-6} μS indicates that the central cell is beginning to exhibit complex oscillations of the form shown in Fig. 2(b). Specifically, both the oscillation period (Fig. 3(a)), overshoot potential (Fig. 3(b)), and maximum diastolic potential (Fig. 3(c)) are beginning to vary during the course of oscillation. Temporal variation of the peripheral cell period and maximum diastolic potential are also seen. Temporal variation of oscillation characteristics is largest at a coupling value slightly greater than approximately 10^{-4} μS, and cease to exist as coupling conductance approaches 220

222   IEEE TRANSACTIONS ON BIOMEDICAL ENGINEERING, VOL. 41, NO. 3, MARCH 1994

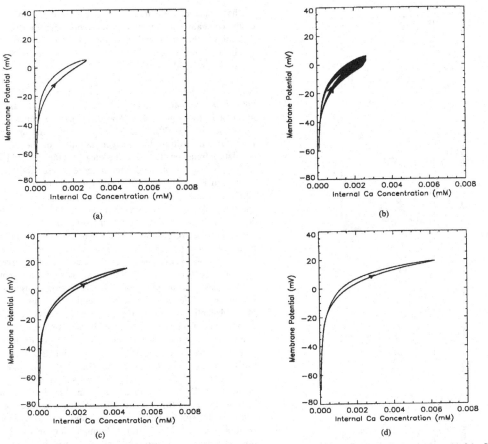

Fig. 4. V-Ca phase-space plots for the central cell as a function of coupling conductance. Various coupling conductance values, (a) 0.1 pS. (b) 70 pS. (c) 1 nS. (d) 50 nS.

pS. This is shown in each panel by the fact that the solid and dotted lines, and the short and long dashed lines, converge to a single value at a coupling conductance of about 220 pS. In particular, the oscillation period of the two cells takes on a common value of 280 ms (Fig. 3(a)) at this level of coupling. An abrupt change in system dynamics occurring with variation of an underlying model parameter is known as a bifurcation. In this case, the bifurcation occurs at a coupling value of approximately 220 pS, where the system makes a transition from complex oscillation to frequency entrainment. Note that the overshoot and maximum diastolic potentials remains different despite the fact that the two cells are frequency entrained. Thus, the cells oscillate with different waveforms. As coupling conductance is increased, the overshoot and maximum diastolic potential of the two cells gradually become closer, and are in approximate agreement for coupling values greater than about 50 nS. These data suggest a more precise partitioning of system dynamics as a function of coupling conductance G_c. Coupled peripheral and central cells exhibit: a) indepedent oscillation for $G_c < 1$ pS; b) complex oscillation for $1 \leq G_c < 220$ pS; c) frequency,

but not waveform entrainment for $G_c \geq 220$ pS; and d) both frequency and waveform entrainment for $G_c \geq 50$ nS.

The nature of the complex dynamics observed when G_c is in the range from $1 \leq G_c < 220$ pS remains to be quantified. As noted previously, inspection of the data in Fig. 2(b) suggests that the two cells are entrained at a complex ratio of 5 : 4. However, careful examination of the data from both cells at this level of coupling shows that successive values of oscillation period, overshoot, and maximum diastolic potential form what appear to be random sequences; that is, the oscillation waveforms of the two cells do not appear to be periodic, let alone frequency entrained at any ratio. This is shown in the V-Ca phase-space plots (see Methods, Section IIE) of Fig. 4. Data are shown for the central cell at the same coupling values used in computing the data of Fig. 2 (similar results have been obtained for the peripheral cell). In each case, values of membrane potential (mV, ordinate) are plotted versus internal Ca concentration (mM, abscissa) at successive instants of time (500-μs time steps over a 20-s interval). The arrow head on each curve shows the counterclockwise direction of

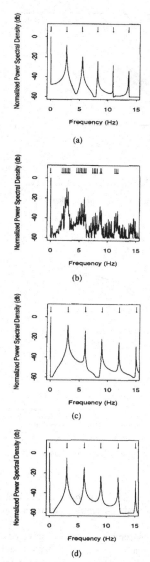

(a)

(b)

(c)

(d)

Fig. 5. Power spectra of the central cell as a function of coupling conductance. Various coupling conductance values, (a) 0.1 pS. (b) 70 pS. (c) 1 nS. (d) 50 nS. Arrows point to sum and difference integer combinations of the base frequencies 2.9 (central cell) and 3.60 (peripheral cell) Hz.

rotation during each action potential. Inspection of Figs. 2(a) (c), and (d) (corresponding to G_c values of 0.1, 1000, and 50 000 pS) shows that the V-Ca trajectories trace out a single closed curve in the phase space (a period oscillation). This is not true of the phase space plot in Fig. 4(b). Inspection of the plots shows that no particular trajectory is ever repeated during successive action potentials (or, if trajectories are repeated, the period is so long that it cannot be detected by observation over the 20 second window). Note that the magnitude of intracellular Ca^{+2} concentration can fluctuate by a factor of 2.

These data indicate that the two cells are not frequency entrained at any complex ratio, and provides evidence that cell dynamics within this coupling regime are either quasi-periodic or chaotic. Similar results are obtained based on response of the peripheral cell model.

Fig. 5 show normalized power spectra of the central cell, computed as described in Methods (Section IIE). The ordinate shows normalized power spectral density (dB), and the abscissa is frequency (Hz). Power spectra are computed at G_c values of 0.1, 70, 1000, and 50 000 pS. The arrows in Fig. 5(a) are positioned at integer multiples of the intrinsic oscillation frequency (2.9 Hz). It is clear that the spectral peaks in Fig. 5(a) occur at integer multiples of these frequencies. Arrows in Fig. 5(c) are positioned at integer multiples of the entrained frequency of 3.6 Hz, and account for each of the spectral peaks. The power spectrum shown in Fig. 5(b) is more complicated. Unlike the smooth spectra shown in the other panels, the spectrum shown in this figure is striated, exhibiting numerous local energy peaks. The arrows in Fig. 5(b) mark those spectral peaks which can be accounted for (given a spectral resolution of 0.05 Hz) as an integer combination of two base frequencies equal to the intrinsic oscillation frequencies of the central and peripheral cells. This accounts for a substantial fraction of the energy in each waveform, and is the defining property of a quasi-periodic signal.

Fig. 6(a) shows an attractor reconstruction using the method of Packard and Takens (see Methods, Section IIE; also [49], [50]). The plot was generated from the set of state variables $V(t_k), V(t_k - \tau), V(t_k - 2\tau)$, where τ is a time-lag ($\tau = 25$ ms in our studies). The attractor in Fig. 6(a) is clearly toroidal in structure. Fig. 6(b) is a two-sided Poincaré surface section generated by cutting the flow in Fig. 6(a) with a hyperplane positioned at $V(t_k) = -10$ mV. Points of intersection of the flow with the plane generate two closed curves, demonstrating that the dynamics of the coupled cell pair at this level of coupling evolve on the surface of a torus. We have performed similar Poincaré surface section analyses over the range of coupling values from $1 \leq G_c < 120$ pS in steps of 2 pS. Within this range of coupling values, the observed motions within were quasi-periodic or of long period (the two cannot be resolved numerically). Preliminary analyses in the regime from $120 \leq G_c < 220$ pS indicate a complex alteration between regimes of period-k and chaotic motion. These analyses will be described in a subsequent publication.

Comparison of Figs. 6(a) and 7(a) shows that as coupling conductance is increased from 70 to 220 pS, the toroidal attractor collapses, giving rise to a periodic attractor. Computation of the Poincaré surface section in Fig. 7(b) confirms the periodic nature of the attractor (the points of intersection of the flow and the hyperplane at $V(t_k) = -10$ mV are two discrete points).

C. Properties of Large-Scale SA Node Networks

Results of the cell-pair studies indicate that the conductance of as few as approximately 4 to 5 gap junction channels [50 pS per channel; [5]] is sufficient to support frequency entrainment of cell pairs. The question of whether or not this degree of coupling can support frequency entrainment in large

224   IEEE TRANSACTIONS ON BIOMEDICAL ENGINEERING, VOL. 41, NO. 3, MARCH 1994

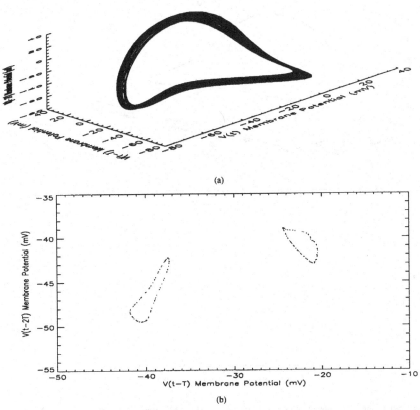

(a)

(b)

Fig. 6. (a) An attractor reconstruction using the set of variables $V(t_k), V(t_k - \tau), V(t_k - 2\tau)$, where τ is a time-lag equal to 25 ms, for the central cell when the coupling conductance is 70 pS. (b) The Poincaré section obtained by intersecting the flow of the attractor shown in Fig. 6(a) with a surface section at $V(t_k)$ equal to -10 mV.

networks remains unanswered. We have examined this issue by simulating properties of large networks of coupled SA node cells on the Connection Machine CM-200 [28]–[31]. Networks were 128×128 cells in size.

In order to investigate properties of frequency entrainment in large cell networks, parameters of SA node cells within the mesh were varied randomly to produce model networks in which each cell has a slightly different oscillation frequency and waveform. This was done by selecting a scaling factor S for each cell in the mesh from a uniform distribution on the interval [0,1]. This scaling factor was used to linearly interpolate between the parameter sets defining central and peripheral cells, with a scale factor of zero generating the central cell parameter set, and a scale factor of 1 generating the peripheral cell parameter set [for more details, see [31]]. The resulting assortment of cells have oscillation frequency in the range of 276 to 347 ms. A uniform coupling conductance and set of initial conditions was then imposed throughout the mesh, and the state equations defining the network were integrated on the Connection Machine.

Fig. 8 shows a perspective view of membrane potential in 128×128 SA node cell network models with random network parameters. The x- and y-axes in each plot specify the coordinates of cells within the mesh. Displacement along the z-axis represents depolarization of membrane potential. Figs. 8(a)–(d) show membrane potential throughout the mesh when coupling conductance equals 0, 220 pS, 1 nS, and 50 nS, respectively. In each panel, membrane potential is displayed at the same instant of time, after integrating network activity for 2.515 s. This particular time was selected because it is a time at which many cells in the network are undergoing the rapid upstroke phase of the cardiac cycle. Fig. 8(a) shows that with zero coupling between cells, activity is completely asynchronous, with some cells maximally depolarized and others maximally hyperpolarized at the same instant of time. The data in Fig. 8(b) was computed using a coupling conductance of 220 pS. Dispersion of cell membrane potential is reduced; network voltage rises and falls in a more synchronized fashion. At the third coupling value (1 nS; Fig. 8(c)), membrane potential oscillation exhibits even more synchrony. The data in Fig. 8(d) shows membrane potential in the grid with cell coupling set to 50 nS. In this case, membrane potential throughout the mesh is approximately uniform, even during the rapid upstroke phase of the pacemaker cycle.

While demonstrating that increased cell-to-cell coupling increases the synchrony of membrane potential oscillation in

The Selected Papers of Denis Noble CBE FRS

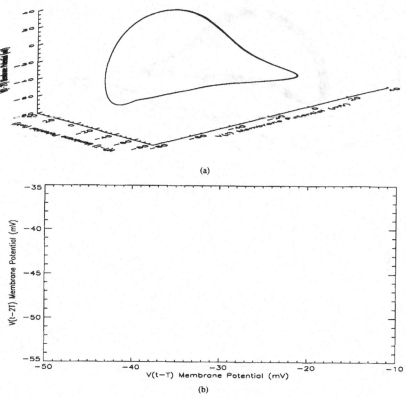

(a)

(b)

Fig. 7. (a) An attractor reconstruction using the set of variables $V(t_k), V(t_k - \tau), V(t_k - 2\tau)$, where τ is a time-lag equal to 25 ms, for the central cell when the coupling conductance is 220 pS. (b) The Poincaré section obtained by intersecting the flow of the attractor shown in Fig. 7(a) with a surface section at $V(t_k)$ equal to -10 mV.

network cells, these particular plots do not adequately demonstrate the extent of frequency entrainment in the network. Fig. 1 shows the reason for this. Cells with high oscillation frequency have a larger maximum upstroke velocity than do cells with low oscillation frequency. Thus, even if these two types of cells reach threshold at the same time, the difference in upstroke velocity means that the more rapidly oscillating cell fires an action potential prior to the slower cell. In the network model, these random differences in upstroke velocity phase. Thus, the extent of frequency entrainment is difficult to discern when examining the distribution of membrane potential throughout the mesh.

To overcome this problem we have developed a measure we refer to as instantaneous period. This is computed for each cell at time t by setting a threshold voltage equal to the membrane potential $V(t)$, and computing the time at which the last threshold crossing of the same sign occurred. Fig. 9 shows plots of instantaneous period at $t = 2.515$ s after onset of the simulation. Four plots are shown, each displaying instantaneous period at the same coupling values and time as in Fig. 8. With no coupling between cells (Fig. 9(a)), there is of course a wide dispersion of oscillation periods throughout the mesh. The mean and standard deviation of the oscillation period in the absence of cell coupling is 288.6 and 19.1 ms,

respectively. Fig. 9(b) shows instantaneous period throughout the mesh at time $t = 2.515$ s when cell-to-cell coupling is set to 220 pS. Clearly, very little coupling between cells is required to achieve a remarkable degree of frequency entrainment, even in large networks. The data shown in Fig. 9(b) have mean and standard deviation values of 282.3 and 2.7 ms. Thus, a cell-to-cell coupling value of 220 pS reduces standard deviation in oscillation period by a factor of 7. This trend continues as coupling is increased. In Figs. 9(c)–(d), periods are 280.3, and 278.5 ms, and standard deviations are 0.6 and 0.2 ms.

IV. DISCUSSION

One of the goals of this study was to develop models of the oscillation properties of cells from different regions of the rabbit SA node, since it is known that these cells exhibit differences in oscillation period, maximum upstroke velocity, maximum diastolic and overshoot potential, and sensitivity to external potassium. This was done in order to gain insight into the physiologically relevant membrane properties which should be varied in order to confer different oscillation frequencies and waveforms on model cells. Progressive increases of the magnitude of $I_{Na}, I_{Si,Ca}, I_f$, and I_{k1}, along with a shift of the I_f activation function in the depolarizing

226 IEEE TRANSACTIONS ON BIOMEDICAL ENGINEERING, VOL. 41, NO. 3, MARCH 1994

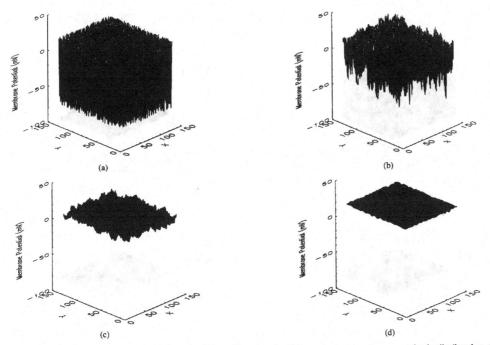

Fig. 8. Membrane potential (z-axis) in a 128×128 mesh of sinoatrial node cells. Cell membrane parameters are randomly distributed so that each cell has a different intrinsic oscillation frequency and waveform. Cell coupling is uniform throughout the mesh at various values. (a) 0 pS. (b) 220 pS. (c) 1000 pS. (d) 50 000 pS.

direction, enabled us to model qualitative properties of central, transitional, and peripheral cells. We use the term qualitative because our fits to particular oscillation properties of these cells are not in exact agreement with the data of Kodama and Boyett. Instead, we have focused on fitting the changes in cell oscillation properties observed with increasing distance from the SA node center. We have been reasonably successful in these efforts; differences in oscillation properties of model peripheral/central cells agree well with the corresponding differences measured by Kodama and Boyett [3].

It is essential that the validity of these models be examined experimentally by studying properties of membrane currents in single cells isolated from the SA node. The elegant approach of Honjo and Boyett [4] in which cell capacitance is used as a feature for determining the nodal location of a cell subsequent to isolation should make possible the correlation of membrane current data with cell location in the node. There is one specific test of the model which we think is of primary importance. One of the key features distinguishing cell oscillation properties in the SA node is maximum upstroke velocity. The recent studies of Honjo and Boyett [4] have shown that upstroke velocity in single cells isolated from the rabbit SA node is highest in the cells with the largest capacitance (surface-area), and is in the range of 30–40 V/s. The correlation of large capacitance and high upstroke velocity suggests that these cells are peripheral SA node cells, an interpretation which yields conclusions about cell oscillation properties which are consistent with their previous study [3]. The only way in

which the 30–40-V/s upstroke velocity could be achieved in the peripheral cell model was to increase the density of Na channels in the peripheral cell by a factor of 25. While extreme, this assumed difference in Na channel density is supported by the fact that the resulting peripheral SA node cell Na conductance is similar in magnitude to that in neighboring atrial cells [52]. Additionally, Nathan [53] has measured a TTX-sensitive fast inward current which is larger in Type II than Type I SA node cells. Type II cells have a large upstroke velocity and a high oscillation frequency, whereas Type I cells have a lower upstroke velocity and oscillation frequency. This suggests a correlation between the Type I and II cells described by Nathans and central and peripheral cells, respectively; a correlation which agrees with the assumptions in our model.

This study demonstrates that there are at least four distinct regimes of behavior for the coupled cell pair as a function of coupling conductance: a) independent oscillation ($G_c < 1$ pS); b) complex dynamics ($1 \leq G_c < 220$ pS); c) frequency but not waveform entrainment ($G_c \geq 220$ pS); and d) frequency and waveform entrainment ($G_c \geq 50$ nS). It is likely that other dynamic behaviors will be found within the complex regime. We are at present attempting to quantify the way in which the torus identified in this regime breaks down, giving rise to successive bands of period-k and chaotic motion, as coupling is increased.

We have investigated the effects of the relative phase of the central and peripheral cell at the time of coupling on cell pair dynamics by shifting the phase of the central cell

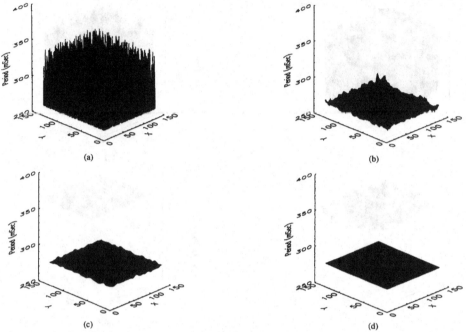

Fig. 9.   Instantaneous frequency of oscillation (z-axis) in a 128×128 mesh of sinoatrial node cells. Cell membrane parameters are randomly distributed so that each cell has a different intrinsic oscillation frequency and waveform. The parameter distribution is the same as that for the network described in Fig. 8. Cell coupling is uniform throughout the mesh at various values. (a) 0 pS. (b) 220 pS. (c) 1000 pS. (d) 50 000 pS.

in steps of 60°. These differences in initial conditions have no effect on the dynamics. We have also investigated dynamic properties of cell pairs with different intrinsic oscillation periods. The general results are identical to those reported here. However, the transition point from complex to periodic behavior is shifted to lower coupling conductance as the difference in intrinsic oscillation frequencies is decreased. In the course of developing these cell models, we examined the properties of systems based on a wide range of parameter values. While the precise location of the four regimes varied slightly from model to model, the generic properties described in this paper were always observed. In every case examined, small numbers of gap junction channels could support frequency entrainment. The unitary conductance of gap junction channels is known to be about 50 pS [5]; thus, our models predict that as few as about 4 to 5 gap junction channels can account for frequency entrainment in cell pairs. The finding that small coupling conductance can support frequency entrainment is consistent with recent work of Herve *et al.* [54] in which frequency entrainment has been studied in coupled pairs of neonatal rat heart cells. Their experimental data show that frequency entrainment occurs at a coupling conductance as low as 1 nS. At this value of coupling, waveforms were identical to the simulation data shown in Fig. 4(c). In addition, Jongsma *et al.* [55] have recently used immunocytochemical techniques to label the gap junction protein connexin-43 in rabbit SA node and atrium. Very heavy labeling was found within the atrium. However, labeling in the SA node was very sparse, indicating that relatively few gap junctions are present there (this assumes that the same gap junction protein is found in the atrium and SA node). Complex oscillation waveforms at intermediate coupling values, exhibiting $n : m$ patterns of entrainment (described as Wenkebach-like phenomena) have also been reported previously [16, Fig. 4]. These complex oscillations appear similar to the quasi-periodic dynamics described in our models. This agreement of a broad range of experimental and theoretical results suggest that the dynamics described in this study are a robust and general property, and that small numbers of gap junctions are sufficient to give rise to frequency entrainment of cell firing within the rabbit SA node.

It is interesting to compare the physical size of our network model to the rabbit SA node. A typical central cell is about 8 μm in diameter and 40 μm in length; a typical peripheral cell is about $10 \times 100 \, \mu m^2$ [25]. Therefore, a lattice of 128×128 cells will have physical dimensions in the range from 1.0×5.1 mm² to 1.3×12.8 mm². The numerical studies of 128×128 cell lattices have shown that 4–5 gap junctional channels are sufficient to achieve frequency entrainment. We have verified that this small number of gap junctional channels can also account for frequency entrainment in larger 512×512 network models. These larger lattice models correspond to physical dimensions in the range from 4×20.4 mm² to 5.2×51.2 mm². These dimensions are comparable to the size of SA node in species used in experimental studies [7] and [25].

The propagation velocity of the pacemaker potential is greatest in a direction which is parallel to the crista terminalis in the rabbit SA node; spread of excitation is slower in other directions [7]. Although the mechanism for this preferential conduction has not been completely clarified, it could result from fiber orientation or differences in gap junction conductance [7] and [56]. The development of an SA node model that accurately reproduces regional differences in conduction velocity is certainly important. Our study on large SA node network models sets the lower bound on the magnitude of gap junction coupling required to account for frequency entrainment, regardless of the anisotropy which may exist within the SA node.

The number of channels required for frequency entrainment could be increased if their stochastic gating properties are considered. Rook *et al.* [5] have shown that the ratio of mean channel open to closed time in rat neonatal heart cells at a transjunctional potential of 50 mV is about 5.5. This corresponds to an open channel probability of 0.85. The ratio is in the range of 0.3 to 0.4 at a tranjunctional potential of 100 mV (open channel probability 0.24). Thus, while the channel tends to close at extreme junctional potentials, the high channel open probability at lower junctional potentials suggests that the effect of stochastic channel opening and closing on the conclusions of our model may be minimal.

The gating properties of gap junction channels are modulated by factors such as pH and Ca concentration [57], [58]. Noma and Tsuboi [57] have shown that elevation of intracellular cytosolic Ca concentration in coupled pairs of cardiac myocytes decreases gap junction conductance. Specifically, they have demonstrated an order of magnitude reduction of gap junction conductance upon increasing cytosolic Ca levels from about 40 to 400 nM [57]. Elevated intracellular Ca levels are predicted to occur under certain pathological conditions. Noble *et al.* [1] have used computational models of isolated SA node cells to show that inhibition of the Na-K pump can lead to marked increases in intracellular Ca levels. Such changes occur over a long time course of several minutes following pump inhibition. These results suggest that under certain pathological conditions, slow increases in intracellular Ca concentration could produce a corresponding slow decrease in gap junction conductance. A particularly intriguing possibility is that such changes may occur during ischemia. While the cellular consequences of ischemia remain controversial, there is preliminary evidence that when whole rat hearts are made ischemic, there is a rise in both intracellular inorganic phosphate (a known Na-K pump inhibitor) and Na levels (G. Radda and K. Clarke, personal communication). This is consistent with the hypothesis that ischemic conditions lead to Na-K pump inhibition and potentially elevated intracellular Ca levels. Our model would predict that if coupling conductance is reduced as a consequence of slow increases in intracellular Ca levels to below the threshold value required to account for frequency entrainment, the oscillation pattern of coupled cells will make a transition from periodic, frequency entrained dynamics to more complex oscillation motion. Reductions of gap junction conductance may therefore be one possible mechanism underlying the generation of arrhythmias during ischemia.

APPENDIX

THE NOBLE–DIFRANCESCO–DENYER EQUATIONS FOR SINGLE SINOATRIAL CELL[1]

The following refer to central and peripheral sinoatrial node cell models.

A. Vector Field Equations

$$\dot{V} = \frac{-1}{C_m}(I_{fK} + I_{fNa} + I_K + I_{K1} + I_{bNa} + I_{NaK}$$
$$+ I_{NaCa} + I_{Na} + I_{SiCa} + I_{SiK} + I_{bCa}) \tag{4}$$

$$\dot{y} = 0.028e^{-(V-\varsigma)/16}(1-y) - 19.5e^{(V-\varsigma)/19}y \tag{5}$$

$$\dot{x} = \eta 2.1e^{V/28}(1-x) - \eta 0.96e^{-V/24}x \tag{6}$$

$$\dot{m} = \frac{200(V+41)}{1-e^{-0.1(V+41)}}(1-m)$$
$$- 8000e^{-0.056(V+66)}m \tag{7}$$

$$\dot{h} = 20e^{-0.125(V+75)}(1-h)$$
$$- \frac{2000}{1+320e^{-0.1(V+75)}}h \tag{8}$$

$$\dot{d} = \frac{30(V+19)}{1-e^{-(V+19)/4}}(1-d) - \frac{12(V+19)}{e^{(V+19)/10}-1}d \tag{9}$$

$$\dot{f} = \frac{6.25(V+34)}{e^{(V+34)/4}-1}(1-f) - \frac{50}{1+e^{-(V+34)/4}}f \tag{10}$$

$$\dot{f_2} = 10(1-f_2) - \frac{10}{k_{mf2}}[Ca^{++}]_i f_2 \tag{11}$$

$$\dot{p} = \frac{0.625(V+64)}{e^{(V+64)/4}-1}(1-p)$$
$$- \frac{5}{1+e^{-(V+64)/4}}p \tag{12}$$

$$[Na^+]_i = \frac{-1}{V_i F}$$
$$(I_{fNa} + I_{bNa} + 3I_{NaK} + 3I_{NaCa} + I_{Na}) \tag{13}$$

$$[K^+]_i = \frac{-1}{V_i F}(I_{fK} + I_K + I_{K1} - 2I_{NaK} + I_{siK}) \tag{14}$$

$$[Ca^{++}]_i = \frac{-1}{2V_i F}$$
$$(I_{bCa} - 2I_{NaCa} + I_{SiCa} + I_{up} - I_{rel}) \tag{15}$$

$$[Ca^{++}]_{up} = \frac{1}{2V_{up}F}(I_{up} - I_{tr}) \tag{16}$$

$$[Ca^{++}]_{rel} = \frac{1}{2V_{rel}F}(I_{tr} - I_{rel}) \tag{17}$$

B. Current Equations

The membrane component of this model comprises 10 channel currents:

I_{fK} hyperpolarization-activated potassium current:

$$I_{fK} = \frac{[K^+]_o}{[K^+]_o + k_{mf}}G_{fK}(V - E_K)y^2. \tag{18}$$

[1] Produced from *OXSOFT Heart 3.6*.

TABLE II

C_m	$60(10)^{-6}$ μF	F	96485 Coulombs/mole
k_{mf}	45 mM	R	8314.41 mJoules/(mole °K)
k_{mk1}	10 mM	T	310°K
k_{mk}	1 mM	radius	$10\,\mu$m
k_{mNa}	40 mM	length	$28\,\mu$m
k_{mf2}	$5(10)^{-4}$ mM	V_{ecs}	0.1
k_{mca}	$2(10)^{-3}$ mM	V_e	$V_{ecs}10^{-9}\pi$ radius^{2}length μL
k_{naca}	0.02 pA	V_i ^{2}length μL	$(1 - V_{ecs}10^{-9}\pi$ radius^{2}length μL
d_{naca}	10^{-4}	V_{up}	$0.05\,V_i\mu$L
γ	$\frac{1}{2}$	V_{rel}	$0.02\,V_i\mu$L
G_{k1}	22.5 nS	$\overline{[Ca^{++}]_{up}}$	5 mM
G_{bca}	0.1 nS	τ_{up}	$5(10)^{-3}$ s
I_{NaKmax}	0.45 nA	τ_{rel}	$(10)^{-2}$ s
P_{cak}	0.01	τ_{rep}	0.2 s
P_{cana}	0.01	α_{up}	$\dfrac{2V_iF}{\tau_{up}\overline{[Ca^{++}]_{up}}}$ nA/mM^2
$[Ca^{++}]_o$	2 mM	α_{tr}	$\dfrac{2V_{rel}F}{\tau_{rep}}$ nA/mM
$[K^+]_o$	4 mM	α_{rel}	$\dfrac{2V_{rel}F}{\tau_{rel}}$ nA/mM
$[Na^+]_o$	140 mM		

I_{fNa} hyperpolarization-activated sodium current:

$$I_{fNa} = \frac{[K^+]_o}{[K^+]_o + k_{mf}}G_{fNa}(V - E_{Na})y^2. \tag{19}$$

I_K time-dependent (delayed) potassium current:

$$I_K = \frac{I_{Kmax}}{140}([K^+]_i - [K^+]_o e^{-V/(RT/F)})x. \tag{20}$$

I_{K1} time-dependent (background) potassium current:

$$I_{K1} = G_{K1}\frac{[K^+]_o}{[K^+]_o + k_{mk1}}\left(\frac{V - E_K}{1 + e^{(V-E_K+10)/(RT/2F)}}\right). \tag{21}$$

I_{bNa} background sodium current:

$$I_{bNa} = G_{bNa}(V - E_{Na}). \tag{22}$$

I_{Na} fast sodium current:

$$I_{Na} = G_{Na}(V - E_{mh})m^3 h. \tag{23}$$

I_{bCa} background calcium current:

$$I_{bCa} = G_{bCa}(V - E_{Ca}). \tag{24}$$

I_{SiCa} slow inward calcium current:

$$I_{SiCa} = 4P_{Ca}df f_2 \frac{\dfrac{V-50}{RT/F}}{1 - e^{-(V-50)/(RT/2F)}}$$
$$\cdot [[Ca^{++}]_i e^{50/(RT/2F)}$$
$$- [Ca^{++}]_o e^{-(V-50)/(RT/2F)}]. \tag{25}$$

I_{SiK} slow inward potassium current:

$$I_{SiK} = P_{CaK}P_{Ca}df f_2 \frac{\dfrac{V-50}{RT/F}}{1 - e^{-(V-50)/(RT/F)}}$$
$$\cdot [[K^+]_i e^{50/(RT/F)} - [K^+]_o e^{-(V-50)/(RT/F)}]. \tag{26}$$

There are also two types of ion exchangers in the cell membrane:

I_{NaK} sodium-potassium exchange pump current:

$$I_{NaK} = I_{NaKmax}\frac{[K^+]_o}{[K^+]_o + k_{mk}}\frac{[Na^+]_i}{[Na^+]_i + k_{mNa}}. \tag{27}$$

I_{NaCa} sodium-calcium exchanger current (28), [shown at the bottom of the page].

The other currents are ones between the cytosolic Ca^{++} pool and the SR Ca^{++} compartments:

I_{up} calcium uptake from cytosol to SR uptake store:

$$I_{up} = \alpha_{up}[Ca^{++}]_i(\overline{[Ca^{++}]_{up}} - [Ca^{++}]_{up}). \tag{29}$$

I_{tr} calcium transfer from SR uptake store to release store:

$$I_{tr} = \alpha_{tr}p([Ca^{++}]_{up} - [Ca^{++}]_{rel}). \tag{30}$$

I_{rel} calcium release from SR release store to cytosol:

$$I_{rel} = \alpha_{rel}[Ca^{++}]_{rel}\frac{[Ca^{++}]_i^2}{[Ca^{++}]_i^2 + k_{mCa}^2}. \tag{31}$$

$$I_{NaCa} = k_{NaCa}\frac{e^{\gamma(V)/(RT/F)}[Na^+]_i^3[Ca^{++}]_o - e^{-(1-\gamma)[(V)/(RT/F)]}[Na^+]_o^3[Ca^{++}]_i}{1 + d_{NaCa}([Ca^{++}]_i[Na^+]_o^3 + [Ca^{++}]_o[Na^+]_i^3)}. \tag{28}$$

230

IEEE TRANSACTIONS ON BIOMEDICAL ENGINEERING, VOL. 41, NO. 3, MARCH 1994

TABLE III

Parameter	Central cells	Peripheral cells
G_{na}	12.5 nS	313 nS
P_{ca}	0.06 nA/mM	0.30 nA/mM
I_{Kmax}	1.2 nA	1.76 nA
G_{bna}	0.9 nS	0.7 nS
G_{fk}	60 nS	120 nS
G_{fna}	60 nS	120 nS
η	0.4	0.15
ζ	−6 mV	−8 mV

Note: ζ is used in (5)
and η in (6).

C. Reversal Potentials

E_{Na} reversal potential for I_{fNa} and I_{bNa}:

$$E_{Na} = \frac{RT}{F} \ln \left(\frac{[Na^+]_o}{[Na^+]_i} \right). \tag{32}$$

E_K potassium reversal potential:

$$E_K = \frac{RT}{F} \ln \left(\frac{[K^+]_o}{[K^+]_i} \right). \tag{33}$$

E_{Ca} calcium reversal potential:

$$E_{Ca} = \frac{RT}{2F} \ln \left(\frac{[Ca^{++}]_o}{[Ca^{++}]_i} \right). \tag{34}$$

E_{mh} reversal potential for I_{Na}:

$$E_{mh} = \frac{RT}{F} \ln \left(\frac{[Na^+]_o + 0.12[K^+]_o}{[Na^+]_i + 0.12[K^+]_i} \right). \tag{35}$$

D. Parameters

Units: M = moles/liter; S = siemens; A = amperes; s = seconds; K = Kelvin. The parameters given in Table II are applicable to all sinoatrial node cells.

Those parameters that vary in space (between the *central* and *peripheral* regions of the sinoatrial node) are listed in Table III. Each cell will have a parameter value within the bounds specified by the pairs of numbers given in Table III.

ACKNOWLEDGMENT

The Army High Performance Computing Research Center of The University of Minnesota provided support critical to all phases of this research. Dr. J. Denyer and Dr. H. Brown provided advice on modification of the single SA node cell models. Dr. J. Rinzel and Dr. G. Sell provided advice on analytical procedures. T. Kimball of Thinking Machines Corporation contributed to the development of CM-200 software.

REFERENCES

[1] D. Noble, D. DiFrancesco, and J. Denyer, "Ionic mechanisms in normal and abnormal cardiac pacemaker activity," in J. W. Jacklet, Ed., *Neuronal and Cellular Oscillators*, New York: Marcel Dekker, 1989, pp. 59–85.
[2] D. Noble, *OXSOFT Heart Program Manual*. OXSOFT Ltd., Oxford, England, 1990.
[3] I. Kodama and M. Boyett, "Regional differences in the electrical activity of the rabbit sinus node," *Pflugers Arch*, vol. 404, pp. 214–226, 1985.
[4] H. Honjo and M. R. Boyett, "Correlation between action potential parameters and the size of single sino-atrial node cells isolated from the rabbit heart," *J. Physiol.* (London), vol. 452, p. 128P, 1992.
[5] M. B. Rook, H. J. Jongsma, and A. C. G. van Ginneken, "Properties of single gap junctional channels between isolated neonatal rat heart cells," *Amer. J. Physiol.*, vol. 255, pp. H770–H782, 1988.
[6] M. Masson-Pevet, W. K. Bleeker, A. J. C. Mackaay, and L. N. Bouman, "Sinus node and atrium cells from the rabbit heart: A quantitative electron microscopic description after electrophysiological localisation," *J. Molec. Cell. Cardiol.*, vol. 11, pp. 555–568, 1979.
[7] W. K. Bleeker, A. J. C. Mackaay, M. Masson-Pevet, L. N. Bouman, and A. E. Becker, "Functional and morphological organization of the rabbit sinus node," *Circ. Res.*, vol. 46, no. 1, pp. 11–22, 1980.
[8] D. L. Ypey, D. E. Clapham, and R. L. DeHaan, "Development of electrical coupling and action potential synchrony between paired aggregates of embryonic heart cells," *Membrane Biol.*, vol. 51, pp. 75–96, 1979.
[9] R. L. DeHaan, D. L. Y. E. H. Williams, and D. E. Clapham, "Intercellular coupling of embryonic heart cells," in T. Pexieder, Ed., *Mechanisms of Cardiac Morphogenesis and Teratogenesis*, New York: Raven Press, 1981, pp. 299–316.
[10] R. L. DeHaan, "In vitro models of entrainment of cardiac cells," in L. N. Bouman and H. J. Jongsma, Eds., *Cardiac Rate and Rhythm*, The Hague: Martinus Nijhoff, 1982, pp. 323–361.
[11] J. Jalife, "Mutual entrainment and electrical coupling as mechanisms for synchronous firing of rabbit sino-atrial pace-maker cells," *J. Physiol.*, vol. 356, pp. 221–243, 1984.
[12] M. Delmar, J. Jalife, and D. C. Michaels, "Effects of changes in excitability and intercellular coupling on synchronization in the rabbit sino-atrial node," *J. Physiol.*, vol. 370, pp. 127–150, 1986.
[13] V. Torre, "A theory of synchronization of heart pace-maker cells," *J. Theor. Biol.*, vol. 61, pp. 55–71, 1976.
[14] D. Noble, "A modification of the Hodgkin-Huxley equations applicable to Purkenje fibre action and pacemaker potential," *J. Physiol.* (London), vol. 160, pp. 317–352, 1962.
[15] K. Yanagihara, A. Noma, and H. Irisawa, "Reconstruction of sinoatrial node pacemaker potential based on the voltage-clamp experiments," *Japan. J. Physiol.*, vol. 30, pp. 841–857, 1980.
[16] D. C. Michaels, E. P. Matyas, and J. Jalife, "Dynamic interactions and mutual synchronization of sinoatrial node pacemaker cells," *Circ. Res.*, vol. 58, no. 5, pp. 706–720, 1986.
[17] R. E. McAllister, D. Noble, and R. W. Tsien, "Reconstruction of the electrical activity of cardiac purkinje fibres," *J. Physiol.* (London), vol. 251, pp. 1–59, 1975.
[18] D. G. Bristow and J. W. Clark, "A mathematical model of primary pacemaking cell in sa node of the heart," *Amer. J. Physiol.*, vol. 243, pp. H207–H217, 1982.
[19] T. R. Chay and J. Rinzel, "Bursting, beating, and chaos in an excitable membrane model," *Biophys. J.*, vol. 47, pp. 357–366, 1985.
[20] R. P. Gould and T. Powell, "Intact isolated muscle cells from the adult rat heart," *J. Physiol.*, vol. 225, pp. 16–19P, 1972.
[21] J. Tanaguchi, S. Kokubun, A. Noma, and H. Irisawa, "Spontaneously active cells isolated from the sino- atrial and atrio-ventricular nodes of the rabbit heart," *Japan. J. Physiol.*, vol. 31, pp. 547–558, 1981.
[22] H. Irisawa and T. Nakayama, "Isolation of a single pacemaker cell from rabbit s-a node," *Jikeikai. Med. J.*, vol. 30, pp. 65–70, 1984.
[23] T. Nakayama, Y. Kurachi, A. Noma, and H. Irisawa, "Action potentials and membrane currents of single pacemaker cells of the rabbit heart," *Pflugers Archiv.*, vol. 402, pp. 248–257, 1984.
[24] D. DiFrancesco, A. Feroni, M. Mazzanti, and C. Tromba, "Properties of the hyperpolarizing-activated current if in cells isolated from the rabbit sino-atrial node," *J. Physiol.*, (London), vol. 377, pp. 61–88, 1986.
[25] J. Denyer and H. Brown, "A method for isolating rabbit sino-atrial node cells which maintains their natural shape," *Japan. J. Physiol.*, vol. 37, pp. 963–965, 1987.
[26] J. Denyer and H. Brown, "Rabbit sino-atrial node cells: Isolation and electrophysiological properties," *J. Physiol.*, vol. 428, pp. 405–424, 1990.
[27] H. Irisawa, H. F. Brown, and W. R. Giles, "Cardiac pacemaking in the sinoatrial node," *Physiol. Rev.*, vol. 73, pp. 197–227, 1992.
[28] R. L. Winslow, A. L. Kimball, D. Noble, and J. C. Denyer, "Computational model of the mammalian cardiac sinus node implemented on a connection machine cm-2," *Med. Biol. Eng. Comput.*, vol. 29, supplement 2, p. 832, 1991.
[29] R. L. Winslow, A. Kimball, D. Noble, and J. Denyer, "Simulation of large-scale sinus node and atrial cell network models on the connection machine cm-2," *J. Physiol.* (London), vol. 438, p. 180P, 1991.
[30] R. L. Winslow, A. Varghese, D. Noble, J. C. Denyer, and A. Kimball, "Modelling large sa node - atrial cell networks on a massively parallel processor," *J. Physiol.* (London), vol. 446, p. 242P, 1992.

[31] R. L. Winslow, A. Kimball, A. Varghese, and D. Noble, "Simulating cardiac sinus and atrial network dynamics on the connection machine," *Physica D: Nonlinear Phenomena*, vol. 64, pp. 281–298, 1993.

[32] D. Cai, R. L. Winslow, and D. Noble, "Effects of gap junction conductance on oscillation properties of coupled sinoatrial node cells," *Computer in Cardiology Proc. 1992*, IEEE Computer Soc. Press, Oct. 11–14, 1992, Durham, NC, 1992, pp. 579–582.

[33] D. Noble and S. Noble, "A model of sino-atrial node electrical activity based on a modification of the difrancesco-noble (1984) equations," *Proc. R. Soc. B*, vol. 222, pp. 295–304, 1984.

[34] H. F. Brown, J. Kimura, D. Noble, S. Noble, and A. Taupignon, "The ionic currents underlying pacemaker activity in rabbit sino-atrial node: Experimental results and computer simulation," *Proc. R. Soc. London B*, vol. 222, pp. 329–347, 1984.

[35] D. Noble, J. C. Denyer, H. F. Brown, and D. DiFrancesco, "Reciprocal role of the inward current ib,na and if in controlling and stabilizing pacemaker frequency of rabbit sino-atrial node cells," *Proc. R. Soc. London B*, vol. 250, pp. 199–207, 1992.

[36] D. Noble, J. Denyer, H. F. Brown, R. L. Winslow, and A. Kimball, "Cardiac pacemaker activity: from single cells to modeling the heart," in J. G. Taylor and C. L. T. Mannion, Eds., *Coupled Oscillators*, Heidelberg: Springer-Verlag, 1992, pp. 132–145.

[37] M. Kubicek and M. Marek, *Computational Methods in Bifurcation and Dissipative Structures*. New York: Springer-Verlag, 1983.

[38] IMSL Problem-Solving Software System, *IMSL User's Manual: MATH/LIBRAY FORTRAN Subroutine for Mathematical Applications Version 1.0*, 1987.

[39] T. S. Parker and L. O. Chua, *Practical Numerical Algorithms for Chaotic Systems*. New York: Springer-Verlag, 1989.

[40] T. R. Chay, "Abnormal discharges and chaos in a neuronal model system," *Biol. Cyber*, vol. 50, pp. 301–311, 1984.

[41] T. R. Chay and Y. S. Lee, "Impulse responses of automaticity in the purkinje fiber," *Biophys. J*, vol. 45, pp. 841–849, 1984.

[42] J. R. Clay, M. R. Guevara, and A. Shrier, "Phase resetting of the rhythmic activity of embryonic heart cell aggregates," *Biophys. J.*, vol. 45, pp. 699–714, 1984.

[43] J. S. Turner, J. C. Roux, W. D. McCormick, and H. L. Swinney, "Alternating periodic and chaotic regimes in a chemical reaction-experiment and theory," *Phys. Lett.*, vol. 85A, pp. 9–12, 1981.

[44] J. C. Roux, S. B. A. Rossi, and C. Vidal, "Experimental observations of complex dynamics in a chemical reaction," in (A. R. Bishop, D. K. Campbell, and B. Nicolaenki, Eds.,) *Nonlinear Problems: Present and Future*, Amsterdam: North Holland, 1982.

[45] J. C. Roux, R. H. Simoyi, and H. L. Swinney, "Observation of a strange attractor," *Physica D: Nonlinear Phenomena*, vol. 7, pp. 3–15, 1983.

[46] H. L. Swinney, "Observations of order and chaos in nonlinear systems," *Physica D: Nonlinear Phenomena*, vol. 7, pp. 3–15, 1983.

[47] J. P. Eckmann and D. Ruelle, "Ergodic theory of chaos and strange attractors," *Rev. Mod. Phys.*, vol. 57, pp. 617–656, 1985.

[48] M. R. Bassett and J. L. Hudson, "Experimental evidence of period doubling of tori during an electrochemical reaction," *Physica D: Nonlinear Phenomena*, vol. 35, pp. 289–298, 1989.

[49] N. H. Packard, J. D. F. J. P. Crutchfield, and R. S. Shaw, "Geometry from a time series," *Phys. Rev. Lett.*, vol. 45, pp. 712–716, 1980.

[50] F. Takens, "Detecting strange attractors in turbulence," in (D. A. Rand and L. S. Young, (Eds.), *Lecture Notes in Mathematics 898*, Berlin: Springer, 1980.

[51] P. Grassberger and I. Procaccia, "Measuring the strangeness of strange attractors," *Physica D: Nonlinear Phenomena*, vol. 9, pp. 189–208, 1983.

[52] Y. E. Earm and D. Noble, "A model of the single atrial cell: relation betwen calcium current and calcium release," *Proc. R. Soc. London B*, vol. 240, pp. 83–96, 1990.

[53] R. D. Nathan, "Action potentials and membrane currents of single pacemaker cells of the rabbit heart," *Amer. J. Physiol.*, vol. 250, pp. H325–H329, 1986.

[54] J. C. Herve, D. Noble, B. Bastide, L. Cronier, and J. Deleze, "Nexus channel requirements for action potential entrainment in entricular cardiac cells: Simulation and experimental approaches," *J. Physiol. (London)*, vol. 446, p. 343P, 1992.

[55] H. J. Jongsma, I. ten Velde, B. de Jonge, and D. Gros, "Distribution of gap junctions in and around the guinea-pig sinoatrial node in vitro," *J. Physiol. (London) Proc. Oxford Meeting*, July 18–20, 1991, p. 20P.

[56] L. N. Bouman, J. J. Duivenvoorden, F. F. Bukauskas, and H. J. Jongsma, "Anisotropy of electrotonus in the sinoatrial node of the rabbit heart," *J. Molec. Cell Cardiol.*, vol. 21, pp. 407–418, 1989.

[57] A. Noma and N. Tsuboi, "Dependence of junctional conductance on proton, calcium, and magnesium ions in cardiac paired cells of guinea-pig," *J. Physiol.*, vol. 382, pp. 193–211, 1987.

[58] D. C. Spray and J. M. Burt, "Structure-activity relations of the cardiac gap junction channel," *Amer. J. Physiol.*, vol. 258, pp. C195–C205, 1990.

Dongming Cai received the B.S. and M.S. degrees in biomedical engineering from Zhejiang University in 1984 and 1989, respectively. He is a Ph.D. candidate in the Biomedical Engineering Program and Research Assistant at the Army High Performance Computing Research Center of the University of Minnesota.

He currently is working on his Ph.D. dissertation in the Biomedical Engineering Department at Johns Hopkins University. His research interests are in the areas of theoretical modeling, high performance computing, and nonlinear dynamical analyses related to cardiac and vascular system.

Raimond L. Winslow received the B.S.E.E. degree from Worcester Polytechnic Institute, Worcester, MA, in 1978 and the Ph.D. degree in biomedical engineering from The Johns Hopkins University School of Medicine in 1985.

Following completion of his doctoral degree, he served as a Research Associate at The Institute for Biomedical Computing, Washington University, St. Louis, MO. He is currently Assistant Professor of Biomedical Engineering at The Johns Hopkins University School of Medicine, with a joint appointment in the Department of Computer Science of the Whiting School of Engineering, The Johns Hopkins University. He is Director of the Systems Neurobiology Program within the Department of Biomedical Engineering. Research interests are in the areas of computational biology, nonlinear dynamical systems theory, and high-performance computing.

Denis Noble received the B.Sc. and Ph.D. degrees from University College, London.

He is presently the Burdon Sanderson Professor of Cardiovascular Physiology at Oxford University. His interests are in both experimental and theoretical cardiac physiology. He has published extensively in these fields, authoring over 200 scientific papers and 4 books.

Dr. Noble is a Fellow of the Royal Society and an Honorary Member of the Royal College of Physicians.

Modeling the Heart—From Genes to Cells to the Whole Organ

Denis Noble

Successful physiological analysis requires an understanding of the functional interactions between the key components of cells, organs, and systems, as well as how these interactions change in disease states. This information resides neither in the genome nor even in the individual proteins that genes code for. It lies at the level of protein interactions within the context of subcellular, cellular, tissue, organ, and system structures. There is therefore no alternative to copying nature and computing these interactions to determine the logic of healthy and diseased states. The rapid growth in biological databases; models of cells, tissues, and organs; and the development of powerful computing hardware and algorithms have made it possible to explore functionality in a quantitative manner all the way from the level of genes to the physiological function of whole organs and regulatory systems. This review illustrates this development in the case of the heart. Systems physiology of the 21st century is set to become highly quantitative and, therefore, one of the most computer-intensive disciplines.

The amount of biological data generated over the past decade by new technologies has completely overwhelmed our ability to understand it. Genomics has provided us with a massive "parts catalog" for the human body; proteomics seeks to define these individual "parts" and the structures they form in detail. But there is as yet no "user's guide" describing how these parts are put together to allow those interactions that sustain life or cause disease. In many cases, the cellular, organ, and system functions of genes and proteins are unknown, although clues often come from similarity in the gene sequences. Moreover, even when we understand at the protein level, successful intervention, for example, in drug therapy, depends on knowing how a protein behaves in context, as it interacts with the rest of the relevant cellular machinery to generate function at a higher level. Without this integrative knowledge, we may not even know in which disease states a receptor, enzyme, or transporter is relevant, and we will certainly encounter side effects that are unpredictable from molecular information alone.

Inspecting genome databases alone will not get us very far in addressing these problems. The reason is simple. Genes code for protein sequences. They do not explicitly code for the interactions between proteins and other cell molecules and organelles that generate function. Nor do they indicate which proteins are on the critical path for supporting cell and organelle function in health and disease. Much of the logic of the interactions in living systems is implicit. Wherever possible, nature leaves that to the chemical properties of the molecules themselves and to the exceedingly complex way in which these properties have been exploited during evolution. It is as though the function of the genetic code, viewed as a program, is to build the components of a computer, which then self-assembles to run programs about which the genetic code knows nothing, although proteomics can show us some aspects of the grouping and interaction of proteins (*1*). Sydney Brenner (*2*) expressed this very effectively when he wrote: "Genes can only specify the properties of the proteins they code for, and any integrative properties of the system must be 'computed' by their interactions." Brenner meant not only that biological systems themselves "compute" these interactions but also that, in order to understand them, we need to compute them, and he concluded, "this provides a framework for analysis by simulation." In this review, I describe how far we have advanced in using simulation to understand these interactions in the case of the heart.

Cellular Models of the Heart Models of heart cells have become highly sophisticated and have benefited from four decades of iterative interaction between experimental and simulation work (*3*). Models of all the main types of cardiac myocyte exist (in many cases there are multiple models of the same cell type), and we are now able to represent the variations in the expression of particular genes, for example, across the ventricular wall (*4*), between the center and periphery of the sinoatrial node (*5*), and within the atrium (*6*). These variations are of fundamental importance in understanding global phenomena such as the electrocardiogram (ECG, see Fig. 1), and for analyzing the way in which cardiac rhythm is generated. They are also fundamental to understanding disease states, some of which, like heart failure (*7*), can be

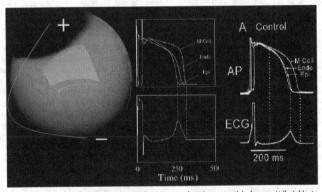

Fig. 1. Reconstruction of cardiac ventricular transmural action potential shapes attributable to variations in gene expression levels and the insertion of these cellular models into a 3D model of the ventricular wall capable of reproducing the T wave of the ECG. Left, supercomputer reconstruction of electrical field (color coded) when ventricular wall wedge is inserted into a conducting medium. Middle, in silico models of endocardial, mid-myocardial (M cell) and epicardial cells together with the reconstructed ECG obtained from the wedge model. Right, Experimental recordings of dog ventricle (*34, 35*). The in silico records (left and middle) are from the CardioPrism cardiac safety assessment program of Physiome Sciences, Inc. (*36*).

University Laboratory of Physiology, Parks Road, Oxford OX1 3PT, UK. E-mail: denis.noble@physiol.ox.ac.uk characterized by alterations in gene expression profiles.

Linking to Genetics An important strength of models based on reconstructing the functional properties of proteins is that it is possible for the models to reach down to the genetic level, for example, by reconstructing the effects of particular mutations when these are characterized by changes in protein function. An example of this approach is the use of multistate (Markov) models of the cardiac sodium channel (8) in which models of the wild-type and of a mutant sodium channel were formulated and validated. The simulated mutation was the ΔKPQ mutation, a three–amino acid (lysine-proline-glutamine) deletion that affects the channel inactivation and is associated with a congenital form of the long-QT syndrome, LQT3. The simulations showed that mutant channel reopenings from the inactivated state and channel bursting due to a transient failure of inactivation generate a persistent inward sodium current during the action potential plateau in the mutant cell. This causes major prolongation of repolarization and the development of arrhythmogenic early afterdepolarizations at slow pacing rates, a behavior that is consistent with the clinical presentation of bradycardia-related arrhythmogenic episodes during sleep or relaxation in LQT3 patients.

Another sodium-channel mutation that has been, at least partially, reconstructed is a missense mutation that affects the voltage dependence of sodium-channel inactivation; it is responsible for one form of idiopathic ventricular fibrillation [the Brugada syndrome (9)]. In this case, small shifts of the voltage dependence of inactivation generate early afterdepolarizations that may underlie fatal arrhythmia (10).

Early afterdepolarizations are also responsible for the arrhythmias of congestive heart failure. This process can be modeled on the basis of experimentally determined changes in gene expression for several of the transporter proteins involved (7).

These examples highlight the ability of cellular models to reconstruct the arrhythmogenic consequences of genetic and ion-channel abnormalities either of behavior or of expression levels. Given the present explosion of genetic information, such studies will continue to be at the forefront of modeling efforts in the next decade. Connecting the genome to physiology is one of the exciting prospects for computational biology.

Counterintuitive Predictions Characteristically, the results of modeling complex systems are frequently counterintuitive. This occurs because, beyond a certain degree of complexity, armchair (qualitative) thinking is not only inadequate for understanding such systems, it can even be misleading. A good example of this comes from the extension of

cellular models to include some of the biochemical changes that occur during ischemia (11). This work succeeds in reconstructing arrhythmias attributable to delayed afterdepolarizations that arise as a consequence of intracellular calcium oscillations in conditions in which intracellular concentrations of sodium and calcium become excessive. These oscillations generate an inward current carried by the sodium-calcium exchanger that can lead to premature excitation of the cell. This work has led to counterintuitive predictions concerning up- and down-regulation of sodium-calcium exchange in disease states involving metabolic damage, such as cardiac ischemia (12). This transporter is currently a focus of antiarrhythmia drug therapy. Simulation is playing an important role in clarifying and assessing the mechanism of action of such drugs, by unraveling the complex changes that occur is a consequence of the change in transporter activity.

Another area in which modeling has been rich in counterintuitive results is that of mechano-electric feedback, in which the contraction of the heart influences its electrical properties. This feedback mechanism has been unraveled in elegant experimental and computational work (13). Some of the results, particularly on the actions of changes in cell volume (which are important in many disease states) are unexpected and have been responsible for determining the next stage in experimental work. Indeed, it is hard to see how such unraveling of complex physiological processes can occur without the iterative interaction between experiment and simulation.

Assessing and Predicting Drug Actions Drugs act on proteins such as receptors, channels, transporters, and enzymes. Models that simulate effects of perturbing protein structure and function are therefore highly relevant to assessing and predicting drug actions. Simulations have already been used in assessing drug action by the U.S. Food and Drug Administration, and we can expect use of such biological models to increase greatly as their complexity and power grows (14, 15). One obvious use in the case of the heart is in assessing the cardiac safety of drugs. It should be noted that half the drug withdrawals that have occurred since 1998 in the USA when drugs have come on the market have been attributable to cardiac side effects, often in the form of effects on the ECG and consequent arrhythmias. This is a large and very expensive form of attrition. Because virtually all the ion transporters involved in cardiac repolarization are now modeled and because very realistic simulations of the T wave of the ECG can be obtained when these models are incorporated into three-dimensional (3D) cardiac tissue models, it is clearly becoming possible to use in silico screens for drug development. One of the reasons that this is

necessary is that the ECG is, unfortunately, an unreliable indicator of potential arrhythmogenicity. Similar changes in form of the QT interval and T waveform can be induced by very different molecular and cellular effects, some benign, others dangerous. We need to understand and predict the mechanisms all the way from individual channel properties through to the ECG. This goal is within reach, particularly as we acquire more experience of the incorporation of accurate cellular models into anatomically detailed organ models (see below).

Another use of simulation in drug discovery is screening drugs for multiple actions. Very few drugs that act on the heart bind to just one receptor. It is much more common for two, three, or even more receptors or channels to be affected. This is particularly true for drugs that act on the sodium-calcium exchanger (16). An important point to realize here is that multisite action may actually be beneficial. Many multireceptor drug actions are expected to be beneficial. I predict that this will be one of the ways in which more rational discovery of antiarrhythmic drugs may occur. In regulating cardiac function, nature has developed many multiple-action processes, particularly those regulated by G protein–coupled receptors. In seeking more "natural" ways of intervening in disease states, we should also be seeking to play the orchestra of proteins in more subtle ways. We need simulation to guide us through the complexity and to understand multiple action functionality.

Incorporation of Cellular Models into Whole-Organ Models There has been considerable debate over the best strategy for biological simulation, whether it should be "bottom-up," "top-down" or some combination of the two [see discussions in (17, 18)]. The consensus is that it should be "middle-out," meaning that we start modeling at the level(s) at which there are rich biological data and then reach up and down to other levels. In the case of the heart, we have benefited from the fact that, in addition to the data-rich cellular level, there has also been data-rich modeling of the 3D geometry of the whole organ (19, 20). Connecting these two levels has been an exciting venture (21, 22). Anatomically detailed models of the ventricles, including fiber orientations and sheet structure, have been used to incorporate the cellular models in an attempt to reconstruct the electrical and mechanical behavior of the whole organ.

Still pictures from a simulation in which the spread of the activation wavefront is reconstructed are shown in Fig. 2. This is heavily influenced by cardiac ultrastructure, with preferential conduction along the fiber-sheet axes, and the result corresponds well with that obtained from multielectrode recording from dog hearts in situ. I referred earlier (Fig. 1) to work that reconstructs the later phases of the ECG using detailed reconstruction of the dispersion of repolarization. Accurate reconstruction of the depolarization wavefront promises to provide reconstruction of the ECG during the early phases of ventricular excitation, i.e., the QRS complex, and as the sinus node, atrium, and conducting system are incorporated into this whole heart, we can look forward to the first example of reconstruction of a complete physiological process from the level of protein function right up to routine clinical observation. The whole ventricular model has already been incorporated into a virtual torso (23), including the electrical conducting properties of the different tissues, to extend the external field computations to reconstruction of multiple-lead chest and limb recording.

Blood Flow and the Coronary Circulation Blood flow within the chambers of the heart, including the movement of valves, has been elegantly modeled by McQueen and Peskin (24) and this has been extended to the study of diastolic mechanical function (25). Blood flow within the coronary circulation has also been modeled (26).

Ischemic heart disease is a major cause of serious incapacity and mortality. It is also a good example of the multifactorial character of most disease states. Very few diseases are attributable to a single gene or protein malfunction. As noted above, cellular reconstructions of the metabolic and electrophysiological processes that occur following deprivation of the energy supply to cardiac cells have already advanced to the point at which some arrhythmic mechanisms can be reproduced.

The initiating process in such energy deprivation is restriction or block of coronary arteries. This is another example where modeling at different data-rich levels is holding out the prospect of very exciting integration of function. Some of the spectacular modeling of the coronary circulation are shown in Fig. 3 (26). These are stills from a simulation in which the blood flow through an anatomically detailed model of the coronary circulation is computed while the ventricles are beating. The simulation, therefore, also included the deformation that occurs as mechanical events influence blood flow.

This model has already been used to investigate the changes in blood flow that occur following constriction or blockage of one of the main arterial branches, and work is in progress to connect this to the modeling of ischemia at the cell and tissue level (see Fig. 4). If we can also connect the cellular mechanisms of arrhythmia to the processes by which regular excitation breaks down into the multiple wavelets of ventricular fibrillation (27) then yet another "grand challenge" for integrative physiological computation will come within range: the full-scale reconstruction of a coronary heart attack.

The term "grand challenge" is chosen deliberately. This kind of work requires massive computer power. The whole organ simulations described here require many hours of computation using supercomputers. (By contrast, the single-cell models can be run faster than in real time on a PC or laptop!) Future progress will be determined partly by the availability of computing capacity. It is significant therefore that attempts to break

Moore's law (computing power doubles every 18 months) are in progress, notably that of IBM's blue gene project (28).

The Future: From Genome to Proteome to Physiome The computer modeling of biological systems is an important technique for organizing and integrating vast amounts of biological information. Although this review has focused on modeling of the heart, it is important to note that biological simulation is now being done for a wide range of pathways, cells, and systems (29). The role of in silico biology in medical and pharmaceutical research is likely to become increasingly prominent as we seek to exploit the data generated through rapid gene sequencing and proteomic mapping (1) through to creating the physiome (30, 31).

However, progress will be significantly enhanced by enabling ever greater numbers of researchers to use and verify models in the course of their everyday experimental work [for simulation and experiment must go together (3)]. It has been extremely difficult to transfer models between research centers, or to extend existing models so that more complex models can be constructed in an object-oriented or modular fashion. This process will be enhanced by the development of uniform standards for representing and communicating the content of models, and by the wide distribution of software tools that permit even nonmodelers to access, execute, and improve existing models. Increasingly, publication of models is accompanied by their availability on Web sites. Also, the process of establishing standards of communication and languages is developing (32, 33).

Fig. 2. Spread of the electrical activation wavefront in an anatomically detailed cardiac model (21). Earliest activation occurs at the left ventricular endocardial surface near the apex (left). Activation then spreads in endocardial-to-epicardial direction (outward) and from apex

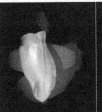

towards the base of the heart (upward, middle frames). The activation sequence is strongly influenced by the fibrous-sheet architecture of the myocardium, as illustrated by the nonuniform transmission of excitation. Red, activation wavefront; blue, endocardial surface.

Fig. 3. Flow calculations coupled to the deforming myocardium. The color coding represents transmural pressure acting on the coronary vessels from the myocardial stress (dark blue, zero pressure, red, peak pressure). The deformation states are (from left to right) zero pressure,

end-diastole, early systole, and late systole (26).

Fig. 4. Left, the coronary circulation model shown in Fig. 3 has been subjected to a constriction of one of the main branches leading to blocked blood flow in the regions colored blue. Right, simulation of ectopic beats in a Purkinje fiber model in conditions of calcium overload of the kind that occurs in ischemic tissue. Oscillatory calcium changes (bottom) induce inward sodium-calcium exchange current (middle) leading to initiation of action potentials (above). Linking these two levels of modeling to create a complete model of coronary heart attack is one of the grand challenges requiring massive computer power. [Left panel kindly provided by N. Smith. Right panel specially prepared for this review using the DiFrancesco-Noble 1985 Purkinje fiber model (37) as follows. To simulate sodium/calcium overload, [Na]$_i$ was increased from 8 to 12 mM. The first action potential is evoked by a current pulse. The second two are initiated by calcium oscillations. Note that the rise in [Ca]$_i$ and the flow of inward Na-Ca exchange current occur before the depolarization.]

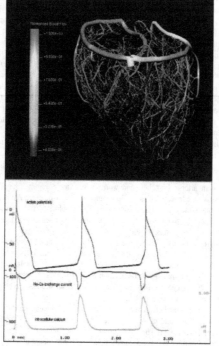

Once this is achieved, we can confidently predict an explosion in the development of integrated model cells, organs, and systems. In a few years' time we shall all wonder how we ever managed to do without them in biological research. For drug development, there will certainly be a major change as these tools come on line and rapidly increase in their power. This will grow in a nonlinear way with the degree of biological detail that is incorporated. The number of interactions modeled increases much faster than the number of components. Biology is set to become highly quantitative in the 21st century. It will become a computer-intensive discipline.

References and Notes

1. M. Gerstein, N. Lan, R. Jansen, *Science* **295**, 284 (2002).
2. S. Brenner, in *The Limits of Reductionism in Biology,* G. R. Bock and J. A. Goode, Eds. (Novartis Found. Symp. 213, John Wiley, London, 1998), pp. 106–XXX.
3. D. Noble, Y Rudy, [*PHILOS. Trans. R. Soc. London Ser. A Math. Phys. Sci. Math. Phys. Sci.* **359**, 1127 (2001).
4. C. Antzelevich et al., *Philos. Trans. R. Soc. London Ser. A Math. Phys. Sci.* **359**, 1201 (2001).
5. M. R. Boyett *ET AL.*, *Philos. Trans. R. Soc. London Ser. A Math. Phys. Sci.* **359**, 1091 (2001).
6. A. Nygren, L. J. Leon, W. R. Giles, [*PHILOS. Trans. R. Soc. London Ser. A Math. Phys. Sci.* **359**, 1111 (2001).
7. R. L. Winslow *ET AL.*, *Philos. Trans. R. Soc. London Ser. A Math. Phys. Sci.* **359**, 1187 (2001).
8. C. E. Clancy, Y. Rudy, *Nature* **400**, 566 (1999).
9. Q. Chen et al., *Nature* **392**, 293 (1998).
10. D. Noble, P. J. Noble, *J. Physiol. (London)* **518**, 2P (1999).
11. F. T. Ch'en et al., *Prog. Biophys. Mol. Biol.* **69**, 515 (1998).
12. D. Noble, *Ann. N.Y. Acad. Sci.*, in press.
13. P. Kohl, F. Sachs, *Philos. Trans. R. Soc. London Ser. A Math. Phys. Sci.* **359**, 1173 (2001).
14. D. Noble, J. Levin, W. Scott, *Drug Discov. Today* **4**, 10 (1999).
15. D. Noble, T. J. Colatsky, *Emerg. Ther. Targets* **4**, 39 (2000).
16. Y. Watanabe, J. Kimura, *Ann. N.Y. Acad. Sci.*, in press.
17. G. R. Bock, J. A. Goode, [EDS., *Complexity in Biological Information Processing* (Novartis Foundation Symposium **239**, Wiley, London, 2001).
18. G. R. Bock, J. A. Goode, Eds., *PROPOSED TITLE?* (Novartis Foundation Symposium **247**, Wiley, London, in press).
19. I. LeGrice et al *Philos. Trans. R. Soc. London Ser. A Math. Phys. Sci.* **359**, 1217 (2001).
20. K. D. Costa, J. W. Holmes, A. D. McCulloch, *Philos. Trans. R. Soc. London Ser. A Math. Phys. Sci.* **359**, 1233 (2001).
21. P. Kohl et al *Philos. Trans. R. Soc. London Ser. A Math. Phys. Sci.* **358**, 579 (2000).
22. D. P. Nickerson, N. P. Smith, P. J. Hunter, *Philos. Trans. R. Soc. London Ser. A Math. Phys. Sci.* **359**, 1159 (2001).
23. C. P. Bradley, A. J. Pullan, P. J. Hunter, *Annu. Biomed. Eng.* **25**, 96 (1997).
24. D. M. McQueen, C. S. Peskin, *Comput. Graph.* **34**, 56 (2000).
25. S . Kovacs, D. M. McQueen, C. S. Peskin, *Philos. Trans. R. Soc. London Ser. A Math. Phys. Sci.* **359**, 1299 (2001).
26. N. P. Smith, G. S. Kassab, *Philos. Trans. R. Soc. London Ser. A Math. Phys. Sci.* **359**, 1251 (2001).
27. A. Panfilov, A. Pertsov, *Philos. Trans. R. Soc. London Ser. A Math. Phys. Sci.* **359**, 1315 (2001).
28. See www.research.ibm.com/bluegene/index.html
29. See, for example, www.afcs.org
30. See www.physiome.org
31. See www.physiome.org.nz [I couldn't link to this.]
32. See www.cds.caltech.edu/erato/sbml/docs/index.html
33. See www.cellml.org
34. X. X. Yan, C. Antzelevich, *Circulation* **98**, 1928 (1998).
35. A. L. Muzikant, R. C. Penland, *Curr. Opin. Drug Discov. Dev.* **5**(1), 127 (2002)
36. See www.physiome.com
37. D. DiFrancesco, D. Noble, *Philos. Trans. R. Soc. London Ser. B Biol. Sci.* **307**, 353 (1985).
38. Work in the author's laboratory is supported by the British Heart Foundation, Medical Research Council, Wellcome Trust, and Physiome Sciences.

Article is 627 picas

Exp Physiol 94.5 pp 597–605   597

Experimental Physiology – *Review Article*

The Cardiac Physiome: perspectives for the future

James Bassingthwaighte[1], Peter Hunter[2] and Denis Noble[3]

[1] Bioengineering, N210G Foege Building, 1705 NE Pacific Street, University of Washington, Box 35-5061 Seattle, WA 98195-5061, USA
[2] University of Auckland, Auckland Bioengineering Institute (ABI), 70 Symonds Street, Auckland, New Zealand
[3] Department of Physiology, Anatomy & Genetics, Parks Road, Oxford, OX1 3PT, UK

The Physiome Project, exemplified by the Cardiac Physiome, is now 10 years old. In this article, we review past progress and future challenges in developing a quantitative framework for understanding human physiology that incorporates both genetic inheritance and environmental influence. Despite the enormity of the challenge, which is certainly greater than that facing the pioneers of the human genome project 20 years ago, there is reason for optimism that real and accelerating progress is being made.

(Received 8 October 2008; accepted after revision 12 December 2008; first published online 19 December 2008)
Corresponding author D. Noble: Department of Physiology, Anatomy & Genetics, Parks Road, Oxford OX1 3PT, UK.
Email: denis.noble@dpag.ox.ac.uk

The Physiome Project was formally launched at a satellite symposium of the International Union of Physiological Sciences (IUPS) Congress in St Petersburg in 1997. Just over a decade later, where are we? Are the aims and principles outlined at that time being fulfilled? In this article, we address these questions by discussing multiscale analysis and modularity in biological systems, the various approaches to mathematical analysis in biology, and the framework being established by the IUPS Physiome Project to help with the understanding of complex physiological systems through the use of biophysically based mathematical models that link genes to organisms.

Multiscale analysis

One of the central principles is that complex systems like the heart are inevitably multiscalar, composed of elements of diverse nature, constructed spatially in a hierarchical fashion. This requires linking together different types of modelling at the various levels. It is neither possible nor explanatory to attempt to model at the organ and system levels in the same way as at the molecular and cellular levels. To represent the folding, within microseconds, of a single protein using quantum mechanical calculations requires months of computation on the fastest existing parallel computers (such as IBM's Blue Gene). It would require unbelievably large numbers of such computers (one estimate is 10^{27}; Noble, 2006) to analyse just a single cell in this degree of detail. Even if we could do it, we would still need to abstract from the mountain of

computation some explanatory principles of function at the cellular level. Furthermore, we would be completely lost within that mountain of data if we did not include the constraints that the cell as a whole exerts on the behaviour of its molecules. This is the fundamental reason for employing the middle-out approach. In multiscalar systems with feedback and feedforward loops between the scale levels, there may be no privileged level of causation (Noble, 2008a).

The impressive developments in epigenetics over the last decade (Bird, 2007) have reinforced this conclusion by revealing the nature and extent of some of the molecular mechanisms by which the higher level constraints are exerted. In addition to regulation by transcription factors, the genome is extensively marked by methylation and binding to histone tails. It is partly through this marking that a heart cell achieves, with precisely the same genome, the distinctive pattern of gene expression that makes it a heart cell rather than, for example, a bone cell or a liver cell. The marking is transmitted down the cell lines as they divide to form more cells of the same kind. The feedbacks between physiological function and gene expression that must be responsible are still to be discovered. Since fine gradations of expression underlie important regional characteristics of cardiac cells, making a pacemaker cell different from a ventricular cell, and making different ventricular cells have different repolarization times, this must be one of the important targets of future work on the Cardiac Physiome. We need to advance beyond annotating those gradients of expression to understanding how they arise during development and how they are maintained in the adult. This is one of the ways in which quantitative physiological analysis will be connected to theories of development and of evolution. The logic of these interactions in the adult derives from what made them important in the process of natural selection. Such goals of the Physiome Project may lie far in the future, but they will ultimately be important in deriving comprehensive theories of the 'logic of life'.

A second reason why multiscale analysis is essential is that a goal of systems analysis must be to discover at which level each function is integrated. Thus, pacemaker activity is integrated at the cell level; single sinus node cells show all the necessary feedback loops that are involved. Below this level, it does not even make sense to speak of cardiac rhythm. At another level, understanding fibrillation requires analysis at least at the level of large volumes of tissue and even of the whole organ. Likewise, understanding the function of the heart as a mechanical pump is, in the end, an organ-level property. Another way of expressing this point is to say that high-level functions are emergent properties that require integrative analysis and a systems approach. The word 'emergent' is itself problematic. These properties do not 'emerge' blindly from the molecular events; they were originally guided by natural selection and have become hard-wired into the system. Perhaps 'system properties' would be a better description. They exist as a property of the system, not just of its components.

A third reason why multiscale analysis is necessary is that there is no other way to circumvent the 'genetic differential effect problem' (Noble, 2008*b*). This problem arises because most interventions at the level of genes, such as gene knockouts and mutations, do not have phenotypic effects. The system as a whole is very effective in buffering genetic manipulations at the level of DNA, through a variety of back-up systems. This is one of the bases of the robustness of biological systems. Moreover, when we manipulate a gene, e.g. through a mutation, even when phenotypic effects do result they reveal simply the consequences of the difference at the genetic level; they do not reveal all the effects of that gene that are common to both the wild and mutated gene. This is the reason for calling this the 'genetic differential effect problem'. Reverse engineering through modelling at a high level that takes account of all the relevant lower level mechanisms enables us to assign quantitatively the relative roles of the various genes/proteins involved. Thus, a model of pacemaker activity allows absolute quantitative assignment of contributions of different protein transporters to the electric current flows involved in generating the rhythm. Only a few models within the Cardiac Physiome Project are already detailed enough to allow this kind of reverse engineering that succeeds in connecting down to the genetic level, but it must be a goal to achieve this at all levels. This is the reason why top-

down analysis, on its own, is insufficient, and is therefore another justification for the middle-out approach.

Modularity in biological systems

Another major principle is that of modularity. A module represents a component of a system that can be relatively cleanly separated from other components. An example is a model of a time- and voltage-dependent ion channel, where the model represents kinetically the behaviour of a large number of identical channel proteins opening more or less synchronously in the same conditions. A model for a cellular action potential would be composed of an assemblage of such modules, each providing the current flow through a different channel type for different ions. Each module is linked to the same environment, but the modules interact with that environment each in their own way. The key to the separability of the modules is that they should be relatively independent of one another, though dependent on their common environment though the effects of each module's behaviour on the environment itself. The separation of modular elements at the same level in the hierarchy works best when the changes in the extramodular environments (concentrations, temperature, pH) do not change too rapidly, that is, more slowly than do the individual channel conductances. The reason is that, when the environmental conditions also change rapidly, the computational 'isolation' of a module becomes less realistic; the kinetic processes represented must extend beyond the module. Choosing the boundaries of modules is important, since a major advantage of modularization is that a limited number of variables are needed to define the interface between models relative to the number required to capture function within the module.

At another level, one might consider the heart, the liver and the lung, etc., as individual modules within a functioning organism, while their common environment (body temperature, blood composition and blood pressure) is relatively stable (homeostasis in Claude Bernard's terms; Bernard, 1865, 1984). At an intermediate level, a module might be composed to represent a part of an organ with a different functional state than other parts, for example, an ischaemic region of myocardium having compromised metabolism and contractile function. Such a module, in an acute phase of coronary flow reduction, might be parametrically identical to the other, normal regions, but have a reduced production of ATP. At a later stage, the regional properties might change, stiffening with increasing collagen deposition, and requiring a different set of equations, so that there would be a substitution for the original module.

In the normal state, a module for any particular region within an organ is inevitably a multiscale

Exp Physiol 94.5 pp 597–605
The Cardiac Physiome: perspectives for the future   599

model, containing elements at the protein (channel or enzyme) level, of subcellular regions, of interacting cell groups such as endothelial–smooth muscle–cardiomyocyte arrangements for blood–tissue exchange of nutrients, and with the intracellular responses of each cell type. This level of complexity invites 'model reduction', to save computational time when one has to account for regional heterogeneity within the organ in order to characterize its overall behaviour. No organ yet studied has been found to be homogeneous in all of its functions. Livers, hearts, lungs and brains all exhibit internal heterogeneities; for example, blood flows vary with standard deviations of about 25% of the mean at any particular point in time within an intact healthy organ, and are undoubtedly associated with similar variation in other aspects of their function.

Modules can be envisaged as computational units. Having such units well defined provides for security in archiving, in model sharing and for ease of reproducibility, and for selection in model construction. It also renders those units more accessible and independently modifiable. A given module, e.g. for force generation by muscle contraction, might be cast in several different forms that represent different degrees of fidelity, robustness and biophysical detail. Some versions might be grossly simplified compared with a detailed and thermodynamically correct biophysical/biochemical reference model; such simplified versions could then be used effectively within a multiscale cardiac model for particular physiological states, for example with the onset of cardiac dyssynchrony with left bundle branch block, a situation in which local cardiac contractile work and cardiac glucose metabolism diminish dramatically in the early-activated septal region and increase greatly in the late-activated left ventricular free wall. In this case, the parameters of the metabolic or contractile modules change, but the modules are not necessarily replaced. A principle of modularity is that modules should also be replaceable to allow an appropriate choice for a particular purpose, e.g. when infarction and replacement by scar render the tissue incapable of contracting so that it acts simply as passive elastic material.

Multiscale models are inherently hierarchical; an organ-level module comprises a set of tissue-level modules, and a tissue-level module is composed of a larger set of cell and structural modules. The modules higher in the hierarchy (organ, tissue) are necessarily representing more complex biological functions, so are usually simplified for computation. The result is a loss of the robustness, which lies in the adaptability in cell signalling, protein transcription rates, ATP generation rates, vasomotion, etc. Let us define robustness as the ratio of a perturbing force or demand to the degree of disturbance of the system; an example of strong robustness would be the large change in cardiac output demanded by the body in going from

rest to exercise divided by the small change in cytosolic ATP levels in a normal heart. A reduced form module, lacking the cell's metabolic regulatory system, would not be able to respond by increasing its substrate uptake, metabolic reaction rates and ATP production in a finely tuned, automated way.

Technically, module-to-module compatibility requires some standardization in design. In addition to having a name compatible with an ontology, each needs to be identified as to domain and to the inputs and outputs that are needed to communicate with the environment. For an ion channel, the inputs would be the concentrations of solutes on either side of the membrane and the membrane potential. The output would be the current flux as a function of time. The equations for the environmental state (inside and outside the cell) would take the flux and calculate the concentrations and transmembrane potential. The parameters governing the channel conductance can remain hidden from the environment; they are used in computing the conductance as a function of time, but if their values are not needed outside the module, they need not be conveyed, so the information flow is minimized.

Module reduction is presumed desirable; if a governing set of parameters can be held constant then the behaviour of the module (its current flux) is all that needs representation, and the simplest algorithm that does this is adequate. If this is true for a channel, then the same statement can be made for the next level composite module, representing the whole cell excitation–contraction coupling or the whole region of tissue. Successive reductions, each capturing the physiological behaviour of the particular level, can then be made progressively simpler. We might end up, for example, with just a varying elastance representation of each of a set of regions in the heart. While this works in an unchanging physiological state (Bassingthwaighte, 2008) and is useful for limited purposes, there is both a risk and a clear deficiency in the approach. The risk is that the resulting reduced version may be correct over such a small range of physiological variation that the model is incorrect a good fraction of the time, like a stopped clock being correct twice a day. The deficiency is that the model cannot adapt to a change in environment or demand for changes in rate or cardiac output. This difficulty can be partly offset by not taking the reduction quite so far, retaining some links to the subcellular level where adaptations in metabolism, force generation and signal sensing occur; one can even develop sets of alternative, partly reduced model forms, substituting in the version that is most appropriate to the occasion. This begs the next question, how does one automate or use artificial intelligence to make these substitutions or to choose to return to the unreduced, fully detailed model form, as the model is being run. Such automation is critical to the use of models in diagnostic or clinical monitoring situations.

The criteria for modularity we have outlined above are essentially descriptive criteria, i.e. the criteria to be taken into account when designing modules within computational models. There is a separate question, which is whether the modules we find necessary or convenient in computational models correspond to any modularity displayed by nature. In fact there is no guarantee that nature is organized in modules that correspond to our choices in dividing up the task of simulation. As an example, in the fruit fly, the same gene (the period gene, *per*) may be involved in circadian rhythm, in embryonic development and in modulated wing-beat frequencies used in communication. Many, perhaps most, genes have such multiple functionality, sometimes surprising in their range, like pieces of a child's construction kit that can be re-used to build many different models. Thus, while it may be necessary for us to divide function up according to what we need mathematically and computationally at the higher levels, we should remember that natural reality at the lower levels may more closely resemble tangled forest undergrowth rather than a neatly laid out park.

This may be one of the reasons for the extensive genetic buffering to which we have already referred. The 'mapping' of lower level interactions may not easily correspond to that at higher levels. Many different lower level networks must be capable of subserving a higher level function. Yet connecting low-level genetic and protein network processes to high-level organ function, and the reverse engineering required to use high-level simulation to assess the relative contributions of different genes to overall function (thus solving the 'genetic differential effect problem'; see '*Multiscale analysis*' above) is necessary. It also poses major challenges that have yet to be resolved in the Physiome Project, since the mathematics required at the different levels is usually very different.

A genetic program?

The problem of modularity is related to another deep question in simulating organisms. The discovery of the structure of DNA and of the triplet genetic code almost inevitably led to analogies between organisms and computers. After all, the code itself can be represented digitally, and Monod and Jacob, when they introduced the concept of a genetic program (Monod & Jacob, 1961), specifically noted the analogy with early valve-based digital computers. The DNA corresponded to the tape of instructions and data fed into the machine, while the egg cell corresponded to the machine itself. This analogy also fuelled the concept of organisms as Turing machines. If the genetic code really was a complete 'program of life', readable like the tape of a Turing machine, then it might

follow the Church–Turing computability thesis that every effective computation can be carried out by a Turing machine (Church, 1936; Turing, 1936).

An organism, however, breaks many of the restrictive requirements for a Turing machine. First, information in biological systems is not only digital, it is also analog. Even though the CGAT code within DNA strings can be represented as a digital string, expression levels of individual genes are continuously variable. As we noted earlier, it is the continuous variation in patterns of gene expression that accounts for a heart cell being what it is despite having the same genome as a bone cell. Some computer scientists have argued that analog processing is precisely what is required to go beyond the Turing limit (Siegelmann, 1995). Moreover, gene expression is a stochastic process, displaying large variations (not just experimental noise) even within cells from the same tissue. Such stochasticity is incompatible with determinate Turing-type programming (Kupiec, 2008, 2009).

Second, DNA is not the sole information required to 'program' the organism. Cellular, tissue, organ and system architectures are also involved, including in particular what Cavalier-Smith calls the membranome (Cavalier-Smith, 2000, 2004). Some of these structures (mitochondria, chloroplasts) are now thought to have been incorporated independently of nuclear DNA during evolution by the process known as endosymbiosis (Margulis, 1998). The DNA is therefore not the only 'book of life'. This organelle and other structural information can only be said to be digital in the sense in which we can represent any image (of whatever dimension) digitally within a certain degree of resolution. But, of course, the organism does not use such representation. Finally, there is continuous interaction between genomes and their environment. This interaction can even include environmental and behavioural influences on epigenetic marking of DNA (Weaver *et al.* 2004, 2007). Organisms are therefore interaction machines, not Turing machines (Neuman, 2008).

Nevertheless, the 'genetic program' metaphor has had a powerful effect historically on the way in which we think about modelling life. The idea that we could represent organisms in a fully bottom-up manner is seductive. We suggest that it also underlies the general approach used by many systems biologists, which is to neglect the higher level structural and organizational features. The Physiome Project, in contrast, by including structural and organizational features, provides a mathematical framework for incorporating both genetic and environmental influences on physiological function. In fact, imaging data is central to many of its successes, starting with fully anatomical models of cardiac structure.

While it would be a mistake to reduce organisms to algorithms (in the sense used in the Church–Turing thesis), there is an important role for mathematical

Exp Physiol 94.5 pp 597–605

analysis. Brute force computation, however impressive in reproducing biological function and however much computing capacity it may use, is not in itself an explanation. Computation needs to be complemented by mathematical analysis, involving simplifying assumptions to reduce highly complex models to a tractable form to which mathematical analysis might be applied. If we are to unravel the 'logic of life' via the Physiome Project, then such mathematical insights will play an important role.

Mathematical modelling in biology

Before we describe the multiscale modelling framework being developed by the Physiome Project, it is instructive to review the areas in which mathematical modelling is currently being applied to biology and the techniques being used across the huge range of spatial and temporal scales required to understand integrative physiological function from genes to organisms. In order of historical development and using commonly accepted terminology, the major areas of application of mathematical modelling to biology can be summarized as follows.

(1) *Evolutionary biology and genetics* is a long-established and sophisticated area of quantitative (i.e. mathematical and statistical, if not model based) analysis that deals with mutation and selection, genotype–phenotype maps, morphogenesis, etc. This field dates from the early 1900s (Pearson, 1898) and is ripe for connection to modern computational physiology as the connection between gene regulatory networks and engineering models of whole organ phenotype is established. In fact, the Physiome Project could be one of the ways in which physiology can reconnect with evolutionary and developmental biology, as advocated, for example, by Jared Diamond (1993).

(2) *Biophysics and electrophysiology* comprise a historically distinct modelling area that deals with membrane-bound ion channels, transporters, exchangers and the membranes themselves. This field dates from the 1940s work on neural action potentials that culminated in the Hodgkin–Huxley model (Hodgkin & Huxley, 1952) and for cardiac cells from the 1960s, with the series of papers that began with the first model of a cardiac pacemaker (Noble, 1962). Biological applications that began with giant squid now include many mammalian species. Models are based on combinations of ordinary differential equations, representing membrane capacitative effects and the voltage-dependent gating behaviour of ion channels, and algebraic equations representing biophysical constraints such as charge conservation and the mass conservation of ion species, cast in ordinary and partial differential equations (ODEs and PDEs) and differential algebraic equations (DAEs), the latter being more problematic to solve (Fábián *et al.* 2001).

Recent developments include more sophisticated Markov state models of ion channel gating. This is an advanced field, thanks to 50 years of close interaction between modelling and experiments, especially in the neural and cardiac fields but now also in relation to endothelial cells (Vargas *et al.* 1994), epithelial cells (Thomas *et al.* 2006), pancreatic β-cells (e.g. Sherman & Rinzel, 1991) and links with the more general areas of facilitated membrane transport and cellular metabolism, e.g. in endothelial cells (Bassingthwaighte *et al.* 1989) and hepatic cells (Goresky *et al.* 1973). It is a field that will benefit greatly from the development of coarse-grained approximations of molecular dynamics/quantum mechanics simulations of ion channel proteins, as the three-dimensional structure for these becomes available from X-ray diffraction or nuclear magnetic resonance imaging or, in the future, predictable from sequence.

(3) *Mathematical biology* is a more general field dating from the 1960s, based primarily on ODEs, DAEs and PDEs (primarily reaction–diffusion systems). Areas of study include cancer modelling, the cell cycle and pattern formation in embryogenesis. The link with experiments is less developed than in the electrophysiology field and the link to biophysical mechanisms has also not been a priority, possibly because the practitioners are mathematicians working in isolation from physiologists, although this is now beginning to change. There have been some attempts to link this field to clinical issues, such as the work on diagnosing ECG patterns with low-dimensional state space dynamics (Peng *et al.* 1993).

(4) *Computational physiology* is the study of structure–function relations at the levels of cells, tissues and organs using anatomically and biophysically based models. The field, which has traditionally had a strong clinical focus, dates from the 1960s, when it was confined to tissues and organs and was dominated by engineers studying stress analyses in bone and the influence of fluid mechanics on arteriosclerosis. Biomechanics is now a mature field, particularly in its application to musculo-skeletal mechanics and to arterial flow mechanics. The multiscale version of computational physiology is now encompassed in the Physiome Project and deals both with the interaction of different types of biophysical equation at the tissue level (reaction–diffusion equations, large deformation solid mechanics, fluid mechanics, etc.) and the connection with models of subcellular function (ion channels, myofilament mechanics, signal transduction, metabolic pathways, gene regulation, etc.). The mathematical techniques derive from engineering physics (conservation laws, continuum mechanics, finite element methods, etc., coupled to DAEs at the cell level) and are heavily dependent on high-performance computation (Hunter *et al.* 2003). It is a field that is also increasingly dependent on computer science (markup languages, ontologies, the Semantic Web, etc.) in order to cope with the explosion of complexity that

occurs when tissue modelling is extended down to protein pathways and associated bioinformatic databases (Hunter *et al.* 2008). An important direction for this field is the use of anatomically based models at the cell level. One example of how bioengineering approaches are contributing to our understanding of structure–function relations at the cell level is the cellular biomechanics area, which has successfully incorporated molecular mechanisms into a continuum mechanics framework (Kamm & Mofrad, 2006).

(5) *Computational chemistry* (as applied to biology) is the analysis of protein–protein and protein–ligand interactions using three-dimensional atomic structure. Quantum mechanics is used to a limited extent, but the dominant method is molecular dynamics (MD), which applies $F = ma$ to the atomic masses, where F is the combination of covalent bonds, van der Waal forces and electrostatic interactions, m is mass and a is acceleration. A major challenge for this area is that biological models require hundreds of thousands of atoms, so it is not computationally feasible to run them for more than a few nanoseconds of real time (equivalent to months of high-performance computer time). This is enough to compute the permeability of an aquaporin water channel but is not enough to predict the gating behaviour of an ion channel. There is some (not enough) work on coarse-grained approximations of MD. This is needed in order to derive the ion channel gating, possibly via the parameters of Markov gating models, from the atomic structure of ion channels. Hopefully, at some point this field will also allow us to predict protein folding from residue sequence.

(6) *Network systems biology* is the analysis of networks, primarily gene regulatory networks, signal transduction pathways and metabolic networks. The field has been strongly promoted by Hiroaki Kitano (Kitano, 2002) and Leroy Hood (Facciotti *et al.* 2004) over the last 10 years and has benefitted greatly from the development of the Systems Biology Mark-up Language (SBML) standard and a rich set of software tools coupled with the SBML-encoded BioModels Database (www.ebi.ac.uk/biomodels-main). Yeast is the common biological species studied by systems biologists, because it is one of the simplest eukaryotic organisms. The equations are DAEs, and the network properties of interest are, for example, modularity, transient/oscillatory responses, state space attractors and robustness to stochastic perturbation. The equations include biophysics only to the extent of the mass balance of chemical species (and usually including Michaelis–Menten enzyme-catalysed reaction kinetics). Mathematical analysis is based on state space techniques (e.g. low dimensional dynamics, bifurcation theory, etc.), non-linear control theory and Bayesian statistical methods, as well as a whole range of classical linear algebra techniques such as singular value decomposition, dynamic programming, etc.

(7) *Systems physiology* is the study of integrated physiological function at the organ system level using DAE models. The field was initiated with the blood pressure control work of Guyton in the 1960s (Guyton *et al.* 1972) but has stalled. It is ripe for a renaissance based on modern computational techniques and the use of the CellML and FieldML encoding standards. For example, one-dimensional bioengineering models of the cardiovascular system and the lymphatic system could be used with cellular models of exocrine signalling and lumped parameter versions of organs that are linked through multiscale techniques to computational physiology models. Similarly, we need to connect the field of clinical musculo-skeletal biomechanics to computational physiology models of the three-dimensional structure of bones, muscles, tendons, ligaments and cartilage.

Physiome Project infrastructure

To address the challenges of multiphysics and multiscale modelling in computational physiology, the Physiome Project is developing modelling standards, model repositories and modelling tools. The key elements of this modelling infrastructure are CellML and its related model repositories and tools, and FieldML and its related model repositories and tools.

CellML (www.cellml.org) is an XML markup language developed to encode models based on systems of ODEs and DAEs. CellML deals with the structure of a model and its mathematical expression (using the MathML standard) and also contains additional information about the model in the form of metadata; such things as: (1) bibliographic information about the journal publication in which the model is described; (2) annotation of model components in order to link them to biological terms and concepts defined by bio-ontologies such as GO (the gene ontology project); (3) simulation metadata to encode parameters for use in the numerical solution of the equations; (4) graphing metadata to specify how the output of the model should be described; and (5) information about the curation status of the model (Cooling *et al.* 2008). Once a model is encoded in CellML, the mathematical equations can be automatically rendered in presentation MathML or can be converted into a low-level computer language such as C, C++, Fortran, Java or Matlab. A number of simulation tools are available to run CellML models, for example, PCEnv (www.cellml.org/tools/pcenv/), COR (cor.physiol.ox.ac.uk/), JSim (www.physiome.org/jsim/) and Virtual Cell (www.vcell.org/).

Note that another similar markup language, SBML (www.sbml.org), has also been developed for models of gene regulation, protein signalling pathways and

Exp Physiol 94.5 pp 597–605

metabolic networks. This language has widespread acceptance in the systems biology community, and many systems biology analysis tools have been developed that are compatible with SBML. CellML has a broader but complementary scope in that it deals with biophysically based models, whereas SBML is focused on biochemical networks. The CellML and SBML developers frequently exchange ideas, and models can fairly readily be converted between the two formats. At some point it may be advantageous to merge the two languages.

FieldML (www.fieldml.org) is an XML markup language being developed to work in conjunction with CellML for encoding spatially varying and time-varying fields within regions of an organ or tissue. The language is designed to support the definition and sharing of models of biological processes by including information about model structure (how the parts of a model are organizationally related to one another), mathematics (such as 2-dimensional and 3-dimensional partial differential equations describing concentrations or other variables over the fields) and metadata (additional information about the model). FieldML will describe spatially varying quantities such as the geometric co-ordinates and structure of an anatomical object, or the variation of a dependent variable field such as temperature or oxygen concentration over that anatomical region.

These standards are now being applied across a very wide range of physiological function. The model repositories for CellML (www.cellml.org/models), Bio-Models (www.biomodels.org), the National Simulation Resource Physiome site (www.physiome.org/Models), and JWSmodels (http://jjj.biochem.sun.ac.za) cover most aspects of cellular function and many areas of organ system physiology, albeit to varying degrees, across the body's twelve organ systems. The recently established European Network of Excellence for the Virtual Physiological Human (VPH; www.vph-noe.eu/index.php) is providing a major boost to the development of the standards, model repositories and tools and in particular their clinical applications.

Another integrative effort was initiated within the US federal science support system. The National Institute of General Medical Sciences began issuing requests for applications in support of integrative biology and modelling in 1998. In April of 2003, an Interagency Modeling and Analysis Group (IMAG) was formed, starting from a working group comprised of program staff from nine Institutes of the National Institutes of Health (NIH) and three directorates of the National Science Foundation (NSF). The IMAG now represents 17 NIH components, four NSF directorates, two Department of Energy (DOE) components, five Department of Defense (DOD) components, the National Aeronautics and Space Administration (NASA), the United States Department of Agriculture (USDA) and the United States

Department of Veterans Administration (USDVA) (see www.nibib.nih.gov/Research/MultiScaleModeling/IMAG). Since its creation, this group has convened with monthly meetings at various locations of the IMAG agency participants, and less frequent meetings of the investigators from the 30 funded projects, often as phone or Web presentation/discussions, and annually as workshops. The ten IMAG investigator-led Working Groups develop collaborations in technologies and in science, and share models and technologies.

Clinical applications

The intention of the Physiome Project is to span medical science and its applications 'From Genes to Health', the title of a 1997 Coldspring Harbor symposium and the central theme of the 1997 Physiome IUPS Satellite meeting near St Petersburg. Clinical applications include clinical image interpretation using positron emission tomography (PET), magnetic resonance imaging (MRI) and ultrasound. The interpretation of cardiac PET image sequences, for example, requires models for blood–tissue exchange and metabolic processes using PDEs and ODEs. The result is the production of 'functional images' displaying cardiac three-dimensional maps of flow, metabolic rates of utilization of oxygen or glucose or other substrates (e.g. thymidine in tumours), and the regional densities of receptors in people with arrhythmia or cardiac failure (Caldwell *et al.* 1990, 1998; Wilke *et al.* 1995). Current work in progress concerns the development of automated algorithms for image capture, segmentation and region-of-interest selection, estimation of the particular function through optimization of model fits to the data and construction of the three-dimensional functional image and report.

Where to now?

Over the last 10 years, the IUPS Physiome Project has focused more on computational physiology, while the NSR Physiome group has worked on clinical and research applications and teaching models. These groups and others have been developing computational infrastructure for combining models at the cell, tissue, organ and organ system levels. The Physiome Project is now well underway, with the markup languages, model repositories and modelling tools advancing rapidly. But this is only the very beginning, and there are many challenges ahead. One challenge is the connection to networks systems biology, which should be relatively straightforward because the field has already widely adopted the SBML standard that is closely related to the Physiome standard, CellML. Another challenge is to link the Physiome models to clinical data in a broader way than as described under '*Clinical applications*' and to link the models to the standard image formats, such as DICOM (http://medical.nema.org/), that are widely used in clinical imaging devices such as MRI and computed tomography (CT) scanners. The use of FieldML should help greatly with this, and discussions are now underway to include FieldML files within the public or private header tags of a DICOM file in order to include the fitted parametric model with the clinical images used in its creation. A related goal is to use models for bedside assessment of clinical status, requiring real-time simulation and data analysis to provide (or suppress) alarms and to guide therapy (Neal & Bassingthwaighte, 2007). A particular computational problem is to link three-dimensional finite-element and finite-volume models to one- or two-dimensional models that can provide the boundary conditions; this technology is needed in order to create the setting for the proper physiological behaviour of the three-dimensional models, since they are far too computationally intensive to be used for a whole system at once. Another challenge is to bring the computational physiology models of the Physiome Project to bear on the fascinating genotype–phenotype questions that have occupied the minds of evolutionary biologists for over 100 years. As we noted earlier in this article (see '*Multiscale analysis*'), high-level phenotype modelling can in principle solve a major problem in genotype–phenotype relations, i.e. what we have called the 'genetic differential effect problem'.

To date, most of the insights gained into physiological processes from mathematical models have been derived from models that deal with one or more physical processes but at only one spatial scale. Examples are models of mechanical processes in the heart, gas transport in the lungs, arterial blood flow dynamics and lipid uptake into endothelial cells, and stress–strain analysis to assist with prosthetic implant design in musculo-skeletal joints. There are a few examples of multiscale analysis, for example, models of electrical activation waves in the heart that are linked to ion channel kinetics. There is a pressing need to be doing much more multiscale analysis. For example, the most important aspect of joint implant design is how to avoid the bone remodelling that leads to implant loosening, and this problem can only be tackled by linking tissue-level stress analysis to protein-level cell signalling pathways. A similar requirement exists in the cardiac mechanics field, where the processes underlying heart failure are governed by a combination of tissue-level stress from raised blood pressure (for example) and gene regulatory processes that alter the protein composition of the tissue.

Probably the biggest challenge facing the Physiome Project now, and one that is crucial to the computational feasibility of multiscale analysis, is that of model reduction. Automated methods are needed to analyse a complex model defined at a particular spatial and temporal scale in order to compute the parameters of a simpler model that

captures the model behaviour relevant to the scales above. For example, if the three-dimensional atomic structure of an ion channel is known, a molecular dynamics model can be formed to compute channel conductance, but one would like to compute the parameters of a much simpler (Hodgkin–Huxley or Markov state) model of the current–voltage channel phenotype appropriate for understanding its behaviour at the cell level. Similarly, it would be highly desirable to be able to derive a model of multichannel cell-level action potential phenotype that accounted for current load from surrounding cells and could be used efficiently in larger scale models of myocardial activation patterns in the whole heart.

The mathematical challenges of deriving automated model reduction methods are greatly facilitated by the model encoding standards that have been put in place over the last 10 years.

Conclusion

We are optimistic that within the next 10 years multiscale analysis based on automated model reduction will be a well-honed tool in the hands of physiologists and bioengineers. Over that time scale, we can expect a significant number of important applications of the Cardiac Physiome Project in the healthcare field.

What of the prospects of more fundamental contributions to the conceptual foundations of biology, i.e. the questions with which we began this article? It is inherently hazardous to predict the development of concepts. If we could, they wouldn't be predictions. But we hope that by drawing attention to these issues and indicating how the Physiome Project may also contribute to the conceptual foundations of biology, we will have encouraged adventurous physiologists, mathematicians, engineers and computer scientists to tackle those problems.

References

Bassingthwaighte JB (2008). Linking cellular energetics to local flow regulation in the heart. *Ann N Y Acad Sci* **1123**, 126–133.

Bassingthwaighte JB, Wang CY & Chan I (1989). Blood-tissue exchange via transport and transformation by endothelial cells. *Circ Res* **65**, 997–1020.

Bernard C (1865, 1984). *Introduction à l'étude de la médecine expérimentale*. Flammarion, Paris.

Bird A (2007). Perceptions of epigenetics. *Nature* **447**, 396–398.

Caldwell JH, Kroll K, Seymour KA, Link JM & Krohn KA (1998). Quantitation of presynaptic cardiac sympathetic function with carbon-11-meta-hydroxyephedrine. *J Nucl Med* **39**, 1327–1334.

Caldwell JH, Martin GV, Link JM, Gronka M, Krohn KA & Bassingthwaighte JB (1990). Iodophenylpentadecanoic acid–myocardial blood flow relationship during maximal exercise with coronary occlusion. *J Nucl Med* **30**, 99–105.

Exp Physiol 94.5 pp 597–605   The Cardiac Physiome: perspectives for the future   605

Cavalier-Smith T (2000). Membrane heredity and early choroplast evolution. *Trends Plant Sci* **5**, 174–182.

Cavalier-Smith T (2004). The membranome and membrane heredity in development and evolution. In *Organelles, Genomes and Eukaryite Phylogeny: An Evolutionary Synthesis in the Age of Genomics*, ed. Hirt RP & Horner DS. CRC Press, Boca Raton.

Church, A (1936). An unsolvable problem of elementary number theory. *Am J Math* **58**, 345–363.

Cooling MT, Hunter PJ & Crampin EJ (2008). Modeling biological modularity with CellML. *IET Syst Biol* **2**, 73–79.

Diamond JM (1993). Evolutionary physiology. In *The Logic of Life*, ed. Boyd CAR & Noble D. Oxford University Press, Oxford.

Fábián G, van Beek DA & Rooda JE (2001). Index reduction and discontinuity handling using substitute equations. *Math Comput Model Dynam Syst* **7**, 173–187.

Facciotti MT, Bonneau R, Hood L & Baliga NS (2004). Systems biology experimental design – considerations for building predictive gene regulatory network models for prokaryotic systems. *Curr Genomics* **5**, 527–544.

Goresky CA, Bach GG & Nadeau BE (1973). On the uptake of materials by the intact liver: the transport and net removal of galactose. *J Clin Invest* **52**, 991–1009.

Guyton AC, Coleman TG & Granger HJ (1972). Circulation: overall regulation. *Annu Rev Physiol* **34**, 13–46.

Hodgkin AL & Huxley AF (1952). A quantitative description of membrane current and its application to conduction and excitation in nerve. *J Physiol* **117**, 500–544.

Hunter PJ, Crampin EJ & Nielsen PMF (2008). Bioinformatics, multiscale modeling and the IUPS Physiome Project. *Brief Bioinform* **9**, 333–343.

Hunter PJ, Pullan AJ & Smaill BH (2003). Modelling total heart function. *Rev Biomed Eng* **5**, 147–177.

Kamm RD & Mofrad MK (2006). *Cytoskeletal Mechanics: Models and Measurements*. Cambridge University Press, Cambridge, UK.

Kitano H (2002). Systems biology: towards systems-level understanding of biological systems. In *Foundations of Systems Biology*, ed. Kitano H. MIT Press, Cambridge, MA, USA.

Kupiec J-J (2008). *L'origine des individus*. Fayard, Paris.

Kupiec J-J (2009). *The Origin of Individuals: A Darwinian Approach to Developmental Biology*. World Scientific Publishing Company, London.

Margulis L (1998). *Symbiotic Planet: A New Look at Evolution*. Basic Books, New York.

Monod J & Jacob F (1961). Teleonomic mechanisms in cellular metabolism, growth and differentiation. *Cold Spring Harb Symp Quant Biol* **26**, 389–401.

Neal ML & Bassingthwaighte JB (2007). Subject-specific model estimation of cardiac output and blood volume during hemorrhage. *Cardiovasc Eng* **7**, 97–120.

Neuman Y (2008). *Reviving the Living: Meaning Making in Living Systems*. Elsevier, Amsterdam and Oxford.

Noble D (1962). A modification of the Hodgkin–Huxley equations applicable to Purkinje fibre action and pacemaker potentials. *J Physiol* **160**, 317–352.

Noble D (2006). *The Music of Life*. OUP, Oxford.

Noble D (2008*a*). Claude Bernard, the first systems biologist, and the future of physiology. *Exp Physiol* **93**, 16–26.

Noble D (2008*b*). Genes and causation. *Philos Transact A Math Phys Eng Sci* **366**, 3001–3015.

Pearson K (1898). Mathematical contributions to the theory of evolution. On the law of ancestral heredity. *Proc R Soc B Biol Sci* **62**, 386–412.

Peng C-K, Mietus J, Hausdorff JM, Havlin S, Stanley HE & Goldberger AL (1993). Long-range anticorrelations and non-Gaussian behavior of the heartbeat. *Phys Rev Lett* **70**, 1343–1346.

Sherman A & Rinzel J (1991). Model for synchronization of pancreatic β-cells by gap junction coupling. *Biophys J* **59**, 547–559.

Siegelmann HT (1995). Computation beyond the Turing limit. *Science* **268**, 545–548.

Thomas SR, Layton AT, Layton HE & Moore LC (2006). Kidney modelling: status and perspectives. *Proc IEEE* **94**, 740–752.

Turing AM (1936). On computable numbers, with an application to the Entscheidungsproblem. *Proc Lond Math Soc 2* **42**, 230–265.

Vargas FF, Caviedes PF & Grant DS (1994). Electrophysiological characteristics of cultured human umbilical vein endothelial cells. *Microvasc Res* **47**, 153–165.

Weaver ICG, Cervoni N, Champagne FA, D'Alessio AC, Sharma S, Sekl JR, Dymov S, Szyf M & Meaney MJ (2004). Epigenetic programming by maternal behavior. *Nat Neurosci* **7**, 847–854.

Weaver ICG, D'Alessio AC, Brown SE, Hellstrom IC, Dymov S, Sharma S, Szyf M & Meaney MJ (2007). The transcription factor nerve growth factor-inducible protein A mediates epigenetic programming: altering epigenetic marks by immediate-early genes. *J Neurosci* **27**, 1756–1768.

Wilke N, Kroll K, Merkle H, Wang Y, Ishibashi Y, Xu Y, Zhang J, Jerosch-Herold M, Mühler A, Stillman AE, Bassingthwaighte JB, Bache R & Ugurbil K (1995). Regional myocardial blood volume and flow: first-pass MR imaging with polylysine-Gd-DTPA. *J Magn Res Imaging* **5**, 227–237.

Generation and propagation of ectopic beats induced by spatially localized Na–K pump inhibition in atrial network models

RAIMOND L. WINSLOW[1], ANTHONY VARGHESE[2], DENIS NOBLE[3], CHARU ADLAKHA[4] AND ADAM HOYTHYA[1]

[1] *Department of Biomedical Engineering, The Johns Hopkins University School of Medicine, Traylor Research Building, Baltimore, Maryland 21205, U.S.A.*
[2] *Army High Performance Computing Research Center, The University of Minnesota, Minneapolis, Minnesota 55455, U.S.A.*
[3] *University Laboratory of Physiology, Oxford University, Parks Road, Oxford OX1 3PJ, U.K.*
[4] *Department of Electrical Engineering, Massachusetts Institute of Technology, Cambridge, Massachusetts, U.S.A.*

SUMMARY

A biophysically detailed two-dimensional network model of the cardiac atrium has been implemented on the Thinking Machines massively parallel CM-5 supercomputer. The model is used to study the effects of spatially localized inhibition of the Na–K pump. Na overloading produced by pump inhibition can induce spontaneous, propagating ectopic beats within the network. At a cell-to-cell coupling value yielding a realistic plane wave conduction velocity of 0.6 m s^{-1}, pump inhibition in roughly 1000 cells can induce propagating ectopic beats in a 512×512 lattice of cells.

1. INTRODUCTION

It is now well established that inhibition of the energy-requiring Na–K pump present in cardiac cells is arrhythmogenic. This effect has been studied by Ferrier & Moe (1973) and Lederer & Tsien (1976), who inhibited pump activity in Purkinje cells by using strophanthidin. Damped oscillatory inward current was observed in voltage-clamped cells during repolarizing steps applied following depolarizing clamps (Lederer & Tsien 1976). This current was referred to as the transient inward current (TIC), and was in some cases sufficiently large to produce afterpotentials or triggered activity. Kass *et al.* (1978) proposed the idea that the TIC was generated by Ca efflux through the Na–Ca exchanger in response to oscillatory release of Ca ions from intracellular stores. Specifically, they suggested that inhibition of Na–K pump activity leads to an increase of internal Na concentration [Na]$_i$, reduced Ca extrusion by the Na–Ca exchanger, and elevated internal Ca concentration [Ca]$_i$. This in turn produced oscillatory Ca-induced release of Ca from sarcoplasmic reticulum (Fabiato 1983). This hypothesis was strengthened following direct measurement of oscillation of [Ca]$_i$ in voltage-clamped cardiac cells following Na–K pump inhibition (Allen *et al.* 1984).

Development of Purkinje (DiFrancesco & Noble 1985), sinus node (Noble *et al.* 1989), atrial (Hilgemann & Noble 1987; Earm & Noble 1990) and, most recently, ventricular cell models (Noble 1992; Luo 1991; Luo & Rudy 1991), which include voltage-dependent ionic currents, Na–K and Na–Ca pump–exchanger currents, time-dependent changes in internal ion concentration, Ca buffering, and sequestration-release of Ca from sarcoplasmic reticulum, has made possible the numerical simulation of the short and long term effects of Na–K pump blockade. The models have been used to reproduce a range of experimental observations on effects of Na–K pump inhibition including: (i) reconstruction of the damped, oscillatory TIC seen during voltage-clamp protocols (Brown *et al.* 1986; Noble *et al.* 1989); (ii) shortening of the action potential with increases in [Na]$_i$ (Noble 1992); (iii) an initial positive inotropic effect subsequent to pump blockade followed in the long term by a negative inotropic effect (Noble 1992); and (iv) production of afterdepolarizations and oscillation of [Ca]$_i$ during Na overload (Noble 1992; Luo 1991).

Although numerical integration of model equations has reproduced many features of the TIC and oscillations of [Ca]$_i$ following Na–K pump inhibition, the stability properties of these numerical solutions were unknown until recently. Varghese & Winslow (1993) used bifurcation analyses to investigate the properties of the system of differential equations describing the Ca subsystem in the DiFrancesco–Noble Purkinje fibre model. A particular interest was to investigate the origins of [Ca]$_i$ oscillations that emerge under voltage-clamp conditions. To do this, internal sodium concentration [Na]$_i$ was treated as a variable control parameter, a step justified because of the slow rate of change of internal Na concentration following pump inhibition. Stability properties of asymptotic solutions were then studied as a function of [Na]$_i$. It was shown that, at low [Na]$_i$ values, there was a single stable fixed point corresponding to a low internal Ca concen-

55

tration. As $[Na]_i$ was increased, a critical point was reached at which the fixed point became unstable, giving rise to a stable periodic oscillation of $[Ca]_i$ through a supercritical Hopf bifurcation. Stable periodic orbits existed for $[Na]_i$ levels in the range 14–22 mM.

The results described above demonstrate the explanatory power of the single-cell models with regard to both short- and long-term effects of Na–K pump inhibition. In this paper, we extend these analyses to networks of cells. We demonstrate that Na overloading induced in small subsets of network cells as a consequence of long-term Na–K pump inhibition can lead to the development of an ectopic focus, a region of normally quiescent cardiac tissue which generates spontaneous, repetitive, propagating electrical activity. We explore the relation between the magnitude of cell-to-cell coupling and the number of Na-overloaded cells required to generate propagating ectopic beats in two-dimensional cardiac network models.

2. METHODS

(a) Single cell analyses

The cardiac cell model used in these studies is the rabbit atrial cell model developed by Earm & Noble (1990), based on previous work by Hilgemann & Noble (1987). This model is included in the OXSOFT Heart Version 3.8 software distribution (Noble 1990). Complete equations for the model are given in Appendix 1. Briefly, the general form of the system of differential-algebraic equations defining the model is:

$$\dot{V}_t = -(1/C_m)\left[\sum_k I_k(V_t) + \sum_k P_k(C_i^m, C_i^n, C_o^m, C_o^n, V_t)\right], \quad (1)$$

$$I_k(V_t) = (V_t - E_k)\, G_k^{\max}\, m_k(V_t)^{p_k} h_k(V_t)^{q_k}, \quad (2)$$

$$\dot{m}_k(V_t) = \alpha_k^m(V_t)\,[1 - m_k(V_t)] - \beta_k^m(V_t)\, m_k(V_t), \quad (3)$$

$$\dot{h}_k(V_t) = \alpha_k^h(V_t)\,[1 - h_k(V_t)] - \beta_k^h(V_t)\, h_k(V_t), \quad (4)$$

$$\dot{C}_i^k = -\sum_k I_k^{\text{total}} / z_k F V_i(V_t). \quad (5)$$

Equation (1) gives the time rate of change of membrane potential, V_t. C_m is total cell membrane capacitance, $I_k(V_t)$ is the kth outward membrane current, $P_k(C_i^m, C_i^n, C_o^m, C_o^n, V_t)$ is an algebraic function specifying the kth outward pump-exchanger current, and C_i^m and C_o^m are internal and external concentrations of the mth ionic species. Equations (2–4) specify the general form of voltage-gated membrane currents. E_k is the reversal potential of the ion(s) to which the kth membrane conductance is permeable, $G_k^{\max}$ is the peak conductance, and $m_k(V_t)$ and $h_k(V_t)$ are activation and inactivation factors. Variables m_k and h_k are solutions of nonlinear ordinary differential equations (3) and (4). The functions $\alpha_k^{m,h}(V_t)$ and $\beta_k^{m,h}(V_t)$ are forward and backward voltage-dependent rate constants for the activation and inactivation variables m_k and h_k.

Equation (5) specifies the time rate of change of the intracellular concentration of the kth ionic species. Extracellular concentrations were assumed to remain constant. In equation (5), I_k^{total} is the total outward current associated with the kth ionic species, V_i the intracellular volume, F Faraday's constant, and z_k the valence of the kth ionic species. In the case of calcium, I_{ca}^{total} includes the sarcoplasmic reticulum (SR) uptake and release currents, as well as the effects of Ca buffering by troponin and calmodulin. Complete

equations specifying the time rate of change of internal Ca, Ca buffering equations, and SR uptake and release equations are given in Appendix 1.

(b) Cell network analyses

Atrial networks were modelled as two-dimensional 512×512 lattices of resistively coupled cells. The time rate of change of membrane potential of the ijth cell, V_t^{ij}, is derived by application of Kirchof's current law at node ij:

$$\dot{V}_t^{ij} = (-1/C_m)\{\sum_k I_k^{ij}(V_t^{ij}) + \sum_k P_k^{ij}[C_i^m(ij),$$
$$C_i^n(ij), C_o^m(ij), C_o^n(ij), V_t^{ij}] + (S_{ij} - N_{ij}\, V_t^{ij})\, G_c\}, \quad (6)$$

where S_{ij} is the sum of the membrane potentials of the cells neighbouring cell ij, N_{ij} is the number of such neighbours, and G_c is the coupling conductance between each cell. Coupling conductance was set to a uniform value throughout the mesh; no attempt was made to model anisotropic conduction within the atrium. The remaining state variables for each network cell are defined as in equations (2–5).

Network models were implemented on the Thinking Machines massively parallel CM-5 supercomputer in CMFORTRAN (Version 2.1 Beta 0.1). Details of implementation have been described previously (Winslow et al. 1993). Briefly, subgrids of cells were mapped onto separate vector processing units of the CM-5 and integrated by using an adaptive step Merson modified fourth-order Runge-Kutta algorithm (Kubicek & Marek 1983). Absolute error tolerance was set to 10^{-6}. For each cell in the array, computed values of selected state variables (V_t and $[Ca]_i$) spaced at 1 ms intervals were stored in the local memory of the CM-5. Blocks of data representing the state of equations (1–5) were written to the CM-5 datavault (a parallel disk array) at intervals of 100 ms. For runs used to measure atrial plane wave conduction velocity, data points were stored at 500 μs intervals.

3. RESULTS

Figure 1a shows a train of spontaneous action potentials (solid line) and Ca transients (broken line) computed by using the single atrial cell model. To induce Na overload, the peak Na–K pump current was inhibited by 90% at time $t = 0$. The cell was then stimulated with 1 nA, 2 ms duration inward current pulses at intervals of 2 s during the interval from 0–300 s. At the end of this interval, $[Na]_i$ had risen to 16.24 mM. Integration was continued for another 100 s after cessation of the current pulse stimuli. The cell continued to oscillate during this interval. The data shown in figure 1 are the final 3 s of oscillation, from $t = 400$ to $t = 403$ s. During this time, the Ca concentration oscillates between a minimum value of about 64 nM to a peak value of 962 nM. The oscillation period is 438 ms. Theoretical analyses of the DiFrancesco–Noble (DiFrancesco & Noble 1985) Purkinje fibre equations have shown that, at this level of internal Na, the Ca subsystem undergoes a stable periodic oscillation. This theoretical result, in conjunction with the greater than 100 s numerically evaluated duration of oscillation, suggests that the data of figure 1a represent a stable, sustained, periodic oscillation of $[Ca]_i$ in the atrial model. Figure 1b shows a higher resolution view of the sixth voltage and Ca

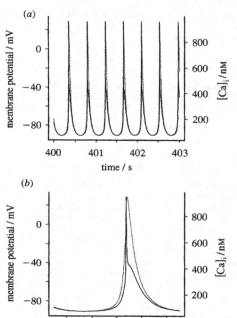

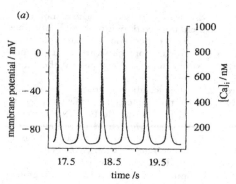

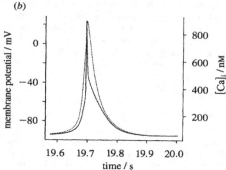

Figure 1. (*a*) Spontaneous action potentials evoked by long-term Na–K pump inhibition (90%) in a single atrial cell model (OXSOFT Heart V3.8). Membrane potential shown by the solid line and $[Ca]_i$ by the broken line. (*b*) Spike 6 of figure 1*a* shown on an expanded timescale.

Figure 2. (*a*) Spontaneous action potentials evoked by long-term Na–K pump inhibition (90%) in a single cell within an atrial network simulated on the CM-5. Network size was 512×512 cells. Region of Na overload was a circular region of about 1000 cells positioned in the centre of the array. Initial conditions for perturbed network cells are those at $t = 400$ ms for the single-cell model shown in figure 1. Network cell-to-cell coupling was set to a uniform value of $1 \mu S$, yielding a planar conduction velocity of 0.6 m s^{-1}. Membrane potential shown by the solid line and $[Ca]_i$ by the broken line. (*b*) Spike 6 of figure 2*a* shown on an expanded timescale.

spike shown in figure 1*a*. Note that the Ca spike leads the voltage spike by about 15 ms.

An important question is whether or not abnormal pacemaker potentials of the type shown in figure 1 can be generated and propagate in cell networks. Unfortunately, it is not feasible to reproduce within a network model a simulation with the same timecourse as that described for figure 1; required run times on the CM-5 are excessive. Instead, an alternative approach has been used to induce Na overload in a subset of network cells. The approach has been to simulate long-term effects of Na–K pump blockade in a single cell, and then to use asymptotic values of the state variables as initial conditions for a small subset of network cells in which Na overload is to be induced.

Figure 2*a* shows membrane potential (solid line) and internal Ca concentration (broken line) as a function of time for a single cell within a 512×512 lattice of atrial cells simulated on the CM-5. This particular cell was positioned at the centre of a circular region of cells (radius 26 cells) in which initial conditions corresponding to Na overload were imposed; the Na overloaded region was positioned in the centre of the 512×512 atrial mesh. The initial conditions were those obtained at $t = 400$ s for the single-cell simulation described previously. Initial conditions for the remaining, quiescent network cells were set to their steady-state values. Cell-to-cell coupling was uniform

at a value of $1 \mu S$. This particular coupling conductance was selected because it yields an atrial plane wave conduction velocity of 0.6 s^{-1}, a value similar to that observed within the atria (Kirchof 1990). Network activity was simulated for a total of 20 s.

These data show a sustained oscillation of the central cell in the Na overloaded region. Slight variation in amplitude of the action potentials and the Ca transients is an artefact resulting from storing data points at 1 ms intervals (Each second of simulation data is 2 Gbytes in size, so increasing temporal resolution in any substantial way is not practical). Calcium concentration ranges from 62 nM to 964 nM, and the period of oscillation is 485 ms. Subtle differences in oscillation waveform, and the longer oscillation period relative to that in figure 1, are due to current shunting by the surrounding, more hyperpolarized normal atrial cells (Winslow *et al.* 1993). A higher resolution view of the sixth spike in the train is shown in figure 2*b*. Note again that the Ca spike leads the voltage spike.

Figure 3 shows the distribution of voltage for each cell in the central 256×256 region of the 512×512

58   R. L. Winslow and others   *Ectopic beats in atrial network models*

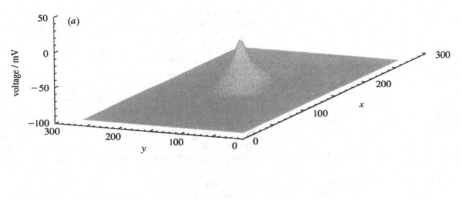

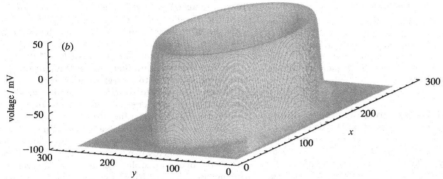

Figure 3. (*a*) Distribution of voltage in the central 256×256 region of the model described in figure 2*a* at time $t = 19.690$ s; *x*- and *y*-axes are cell coordinates, and the *z*-axis is membrane potential. (*b*) Distribution of voltage at time $t = 19.703$ s.

atrial lattice at two instants of time. Figure 3*a* shows initiation of the sixth spike shown in figure 2*a* at $t = 19.690$ s. At $t = 19.703$ s, this ectopic beat has spread beyond the confines of the central Na-overloaded region into the surrounding normal atrial cells. Thus, not only is the oscillation in the Na-overloaded region sustained, but the size of the region is sufficiently large that these cells are capable of activating a propagating wave that spreads throughout the atrial network.

For a Na-overloaded region of size N, there will be a range of small cell-to-cell coupling values at which cells within the region of Na overload oscillate, but do not drive the surrounding normal atrial cells. At this same N, there may be values of G_c which are sufficiently large so that the shunting of current from the Na overloaded region by the surrounding normal atrial cells prevents oscillation. As a consequence, for any N, there may be a range of intermediate coupling values that support the generation and propagation of ectopic beats. There may also be values of N too small to support propagating ectopic beats at any value of G_c.

This relation between N and G_c was explored numerically in a series of simulations on the CM-5. The results are summarized in figure 4. The abscissa of this figure plots the number of cells within a circular region in the centre of the 512×512 atrial mesh in which Na

overloading was imposed. The ordinate plots the coupling conductance used in the simulations. The filled triangles indicate $G_c - N$ pairs which did not support the generation and propagation of ectopic beats; the open squares mark pairs which did.

As predicted, at any N except for values large relative to the total number of cells in the network, there is an upper and lower bound on values of G_c that support propagating ectopic beats. Also note that cell-to-cell coupling values less than about 3 nS do not support the propagation of ectopic beats for any size region of Na overload (up to the maximum $N = 64516$ cells tested).

The data in figure 4 can be used to determine the minimum size region of Na overload that can support the generation and propagation of ectopic beats within the atrium given an estimate of atrial cell-to-cell coupling. Conduction velocity in the atrium is known to be about 0.6 m s^{-1} in the direction parallel to the long axis of each cell (Kirchof 1990), and is dependent on the magnitude of cell-to-cell coupling. We have therefore undertaken a series of simulations to determine the relation between planar wave-front conduction velocity and coupling conductance for the Earm–Noble atrial cell model. This was done by stimulating every cell on one edge of a 512×512 mesh of atrial cells with a 2 ms duration, 1 nA inward

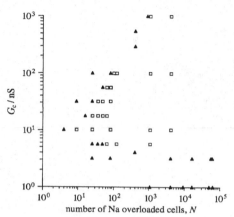

Figure 4. Results of varying the number of centrally positioned Na overloaded cells, N, and cell-to-cell coupling, G_c, in a 512×512 atrial network model. Filled triangles mark G_c–N pairs for which spontaneous, propagating beats were not obtained. Open squares mark G_c–N pairs producing spontaneous propagating beats.

current pulse, and then simulating the propagation of the wave across the mesh. Velocity in the direction of the long axis of the cells was computed by assuming the dimensions of each cell were 10×80 µm. An atrial conduction velocity of 0.6 m s^{-1} is achieved with a cell-to-cell coupling value of 1 µS. From figure 4, at this level of coupling, only about 1000 Na-overloaded cells are required to sustain a propagating ectopic beat within the network.

4. DISCUSSION

We have demonstrated that Na overloading in response to spatially localized Na–K pump inhibition can induce propagating ectopic beats in two-dimensional models of atrial networks. The size of the Na overload region generating propagating ectopic beats depends on the cell-to-cell coupling magnitude. At a realistic coupling value yielding an atrial conduction velocity of 0.6 m s^{-1}, a surprisingly small perturbed region (about 1000 cells) can become an ectopic focus.

There is substantial evidence showing that $[Na]_i$ measured in whole heart increases over a timecourse of tens of minutes during experimentally induced global ischemia (perfused ferret heart (Pike *et al.* 1990); rat heart (Malloy *et al.* 1990; Tani & Neely 1990)). Malloy *et al.* (1990) measured control levels of $[Na]_i$ using atomic spectroscopy (6.2 mM). They then measured $[Na]_i$ during global ischemia by using Na nuclear magnetic resonance (NMR) spectroscopy, and saw a linear increase of $[Na]_i$ at a rate of 0.43 mM min^{-1}. Na levels reached 9.5 mM after 10 min. Pike *et al.* (1990) also observed a linear increase in $[Na]_i$ during global ischemia, with $[Na]_i$ increasing by more than fivefold after 20 min. Na levels may therefore reach the range of values required to evoke spontaneous activity of the type seen in this model after 15–20 min of global ischemia. Large increases in $[Ca]_i$ are also observed

during global ischemia (rat heart (Steenbergen *et al.* 1990)). The increase in $[Ca]_i$ is somewhat delayed with respect to those of $[Na]_i$, but still occurs over a similar timecourse of tens of minutes (Steenbergen *et al.* 1990; Marban *et al.* 1987).

The mechanisms giving rise to Na accumulation during global ischemia are not yet clear. Three can be identified: (i) Na entry through Na-permeable membrane channels; (ii) Na entry via the Na–H exchanger; and (iii) reduction of Na efflux through reductions in Na–K pumping. Pike *et al.* (1990) noted that mechanism (i) is unlikely because membrane depolarization occurs rapidly following onset of global ischemia. This would inactivate Na channels, and reduce inward Na flux. They have argued that mechanisms (ii) and (iii) are more likely. During global ischemia, increases in $[Na]_i$ are accompanied by decreases in pH, suggesting that the Na–H exchanger is extruding H ions at the expense of increased intracellular Na levels. Finally, Pike *et al.* (1990) argued that, because of the small dissociation constant for the Na-K-ATPase, ATP depletion itself was unlikely to be the cause of pump inhibition. Rather, they argued that accumulation of inorganic phosphate during ischemia produces a decrease in the free energy of hydrolysis of ATP, thus inhibiting Na–K pump function.

Our model predicts that Na overloading in spatially localized regions of tissue can induce spontaneous ectopic beats. However, the model only captures properties of one possible mechanism of Na overloading during ischemia. Modelling the detailed temporal evolution of cardiac cell dynamics during ischemia will undoubtedly require incorporation of the Na–H exchanger into the single-cell and network models.

This work was supported by NSF grants DIR91-17874, NSF Research Associateship ASC-9211131, U.S. Army Research Office contract DAAL03-89-C-0038, the Wellcome Trust, and the BHF.

REFERENCES

Allen, D. G., Eisner, D. A. & Orchard, C. A. 1984 Characteristics of oscillations of intracellular calcium concentration in ferret ventricular muscle. *J. Physiol., Lond.* **352**, 113–128.

Brown, H. F., Noble, D., Noble, S. J. & Taupignon, A. I. 1986 Relationship between the transient inward current and slow inward currents in the sino-atrial node of the rabbit. *J. Physiol., Lond.* **370**, 299–315.

DiFrancesco, D. & Noble, D. 1985 A model of cardiac electrical activity incorporating ionic pumps and concentration changes. *Phil. Trans. R. Soc. Lond.* B **307**, 353–398.

Earm, Y. E. & Noble, D. 1990 A model of the single atrial cell: relation between calcium current and calcium release. *Proc. R. Soc. Lond.* B **240**, 83–96.

Fabiato, A. 1983 Calcium-induced release of calcium from the cardiac sarcoplastic reticulum. *Am. J. Physiol.* **245**, C1–C14.

Hilgemann, D. W. & Noble, D. 1987 Excitation-contraction coupling and extracellular calcium transients in rabbit atrium: reconstruction of basic cellular mechanisms. *Proc. R. Soc. Lond.* B **230**, 163–205.

Kass, R. S., Tsien, R. W. & Weingart, R. 1978 Ionic basic

60   R. L. Winslow and others   *Ectopic beats in atrial network models*

of transient inward current induced by strophanthidin in cardiac Purkinje fibres. *J. Physiol., Lond.* **281**, 209–226.

Kirchof, C. J. H. J. 1990 The sinus node and atrial fibrillation. Unpublished PhD thesis, University of Maastricht, The Netherlands.

Kubicek, M. & Marek, M. 1983 *Computational methods in bifurcation and dissipative structures.* New York: Springer-Verlag.

Lederer, W. J. & Tsien, R. W. 1976 Transient inward current underlying arrhythmogenic effects of cardiotonic steroids in Purkinje fibres. *J. Physiol., Lond.* **263**, 73–100.

Luo, C.-H. 1991 A dynamic model of the mammalian ventricular action potential: formulation and physiological simulations. Unpublished PhD thesis, Case Western Reserve University.

Luo, C. H. & Rudy, Y. 1991 A model of the ventricular action potential: Depolarization, repolarization, and their interaction. *Circulation Res.* **68**, 1501–1526.

Malloy, C., Buster, D., Magarida, M., Castro, C., Geraldes, C. & Jeffrey, M. 1990 Influence of global ischemia on intracellular sodium in the perfused rat heart. *Mag. Res. Med.* **15**, 33–44.

Marban, E., Kitakaze, M., Kusuoka, H., Porterfield, J. K., Yue, D. T. & Chacko, V. P. 1987 Intracellular free calcium concentration measured with ^{19}f nmr spectroscopy in intact ferret hearts. *Proc. natn. Acad. Sci. U.S.A.* **84**, 6005–6009.

Noble, D. 1990 OXSOFT *Heart program manual.* Oxford: OXSOFT Ltd.

Noble, D. 1992 Ionic mechanisms determining the timing of ventricular repolarization: Significance for cardiac arrhythmias. *Ann. N.Y. Acad. Sci.*, **644**, 1–22.

Noble, D., DiFrancesco, D. & Denyer, J. 1989 Ionic mechanisms in normal and abnormal cardiac pacemaker activity. In *Neuronal and cellular oscillators* (ed. J. W. Jacklet). New York: Marcel Dekker.

Pike, M. M., Kitakaze, M. & Marban, E. 1990 Na-NMR measurements of intracellular sodium in intact perfused ferret hearts during ischemia and reperfusion. *Am. J. Physiol.* **259**, H1767–H1773.

Steenbergen, C., Murphy, E., Levy, L. & London, R. E. 1990 Elevation in cytosolic free calcium concentration in myocardial ischemia in perfused rat heart. *Circulation Res.* **60**, 700–707.

Tani, M. & Neely, J. R. 1990 Na$^+$ accumulation increases Ca^{2+} overload and impairs function in anoxic rat heart. *J. molec. Cell. Cardiol.* **22**, 57–72.

Varghese, A. & Winslow, R. L. 1993 Dynamics of the Ca subsystem in cardiac Purkinje fibers. *Physica D: Nonlin. Phenom.* **66**, 101–115.

Winslow, R. L., Kimball, A., Varghese, A. & Noble, D. 1993 Simulating cardiac sinus and atrial network dynamics on the connection machine. *Physica D: Nonlin. Phenom.* **64**, 281–298.

Received 13 July 1993; accepted 30 July 1993

APPENDIX 1: THE EARM–HILGEMANN–NOBLE EQUATIONS FOR SINGLE ATRIAL CELLS

(a) Vector field equations

$$\dot{V} = (-1/C_m)\,(I_{K1} + I_{to} + I_{siK} + I_{bK} + I_{NaK} + I_{Na} + I_{bNa} + I_{siNa}$$
$$+ I_{siNa} + I_{NaCa} + I_{siCa} + I_{bCa}),$$

$$\dot{m} = [200(V+41)/(1-e^{-0.1)V+41)})]\,(1-m)$$
$$- 8000\,e^{-0.056(V+66)}m,$$

$$\dot{h} = 20\,e^{-0.125(V+75)}(1-h) - [2000/(1+320\,e^{-0.1(V+75)})]\,h,$$

$$\dot{d} = \frac{90(V+19)}{1-e^{-(V+19)/4}}(1-d) - \frac{36(V+19)}{e^{(V+19)/10}-1}d,$$

$$\dot{f} = \frac{12}{1+e^{-(V+34)/4}}\left(\frac{119[Ca^{2+}]_i}{k_{cachoff}+[Ca^{2+}]_i}+1\right)(1-f)$$
$$- \frac{6.25(V+34)}{e^{(V+34)/4}}f,$$

$$\dot{q} = 333[1/(1+e^{-(V+4)/5})-q],$$

$$\dot{r} = 0.033\,e^{-V/17}(1-r) - \frac{33}{1+e^{-(V+10)/8}}r,$$

$$[\dot{Na^+}]_i = (-1/V_i F)\,[I_{Na} + I_{bNa}([Na^+]_o/140) + 3I_{NaK}$$
$$+ 3I_{NaCa} + I_{siNa}],$$

$$[\dot{K^+}]_i = (-1/V_i F)\,(I_{K1} + I_{siK} + I_{bK} + I_{to} - 2I_{NaK}),$$

$$[\dot{Ca^{2+}}]_o = (1/2V_{cell}\,V_{ecs}\,F)\,(I_{siCa} + I_{bCa} - 2I_{NaCa})$$
$$- d_{NaCa}([Ca^{2+}]_o - [Ca^{2+}]_b),$$

$$[\dot{Ca^{2+}}]_i = (-1/2V_i F)\,(I_{siCa} + I_{bCa} - 2I_{NaCa}) - I_{up}$$
$$+ I_{rel}(V_{SRup}\,V_{rel}/V_i\,V_{up}) - [Ca^{2+}]_{calmod} - [Ca^{2+}]_{troponin},$$

$$[\dot{Ca^{2+}}]_{up} = (V_i/V_{SRup})\,I_{up} - I_{tr},$$

$$[\dot{Ca^{2+}}]_{rel} = (V_{up}/V_{rel})\,I_{tr} - I_{rel},$$

$$[\dot{Ca^{2+}}]_{calmod} =$$
$$10^5(M_{trop} - [Ca^{2+}]_{calmod})\,[Ca^{2+}]_i - 50[Ca^{2+}]_{calmod},$$

$$[\dot{Ca^{2+}}]_{troponin} =$$
$$10^5(C_{trop} - [Ca^{2+}]_{troponin})\,[Ca^{2+}]_i - 200[Ca^{2+}]_{troponin},$$

$$\dot{f}_{activator} =$$
$$(1-f_{activator}-f_{product})\,\{500-([Ca^{2+}]_i/[Ca_{2+}]_i + k_{mca}))^2 + 600$$
$$e^{(V-40)\,0.08}\} - f_{activator}\{500[(Ca_{2+}]_i/[Ca^{2+}]_i + k_{mca}]^2 + 60\},$$

$$\dot{f}_{product} = f_{activator}\{500\{(Ca^{2+}]_i/[Ca^{2+}]_i + k_{mca})]^2 + 60\} - f_{product}.$$

(b) Membrane current equations

$$I_{K1} = G_{K1}\frac{[K^+]_o}{[K^+]_o + k_{mk1}}\left(\frac{V-E_K}{1+e^{\frac{V-E_K+10}{RT/2F}}}\right),$$

$$I_{to} = G_{to}(V-E_K)\,qr,$$

$$I_{bK} = G_{bK}(V-E_K),$$

$$I_{Na} = G_{na}(V-E_{mh})\,m^3 h,$$

$$I_{bNa} = G_{bna}(V-E_{Na}),$$

$$I_{siCa} = 4P_{Ca}\,d(1-f)\frac{k_{cachoff}}{k_{cachoff}+[Ca^{2+}]_i}\frac{\frac{V-50}{RT/F}}{1-e^{\frac{-(V-50)}{RT/2F}}}$$
$$\times ([Ca^{2+}]_i\,e^{\frac{50}{RT/2F}} - [Ca^{2+}]_o\,e^{\frac{-(V-50)}{RT/2F}}),$$

$$I_{siK} = P_{CaK}\,P_{Ca}\,d(1-f)\frac{k_{cachoff}}{k_{cachoff}+[Ca^{2+}]_i}\frac{\frac{V-50}{RT/F}}{1-e^{\frac{-(V-50)}{RT/F}}}$$
$$\times ([K^+]_i\,e^{\frac{50}{RT/F}} - [K^+]_o\,e^{\frac{-(V-50)}{RT/F}}),$$

$$I_{\text{siNa}} = P_{\text{CaNa}} P_{\text{Ca}} d(1-f) \frac{k_{\text{cachoff}}}{k_{\text{cachoff}} + [\text{Ca}^{2+}]_i} \frac{\frac{V-50}{RT/F}}{1 - e^{\frac{-(V-50)}{RT/F}}}$$

$$\times \, ([\text{Na}^+]_i \, e^{\frac{50}{RT/F}} - [\text{Na}^+]_o \, e^{\frac{-(V-50)}{RT/F}}),$$

$$I_{b\text{Ca}} = G_{b\text{ca}}(V - E_{\text{Ca}}).$$

(*c*) *Pump–exchanger current equations*

$$I_{\text{NaK}} = I_{\text{NaKmax}} \frac{[\text{K}^+]_o}{[\text{K}^+]_o + k_{mk}} \frac{[\text{Na}^+]_i}{[\text{Na}^+]_i + k_{m\text{Na}}},$$

$$I_{\text{NaCa}} = k_{\text{NaCa}}$$

$$\frac{e^{\gamma \frac{V}{RT/F}}[\text{Na}^+]_i^3[\text{Ca}^{2+}]_o - e^{-(1-\gamma)\frac{V}{RT/F}}[\text{Na}^+]_o^3[\text{Ca}^{2+}]_i}{1 + d_{\text{NaCa}}([\text{Ca}^{2+}]_i[\text{Na}^+]_o^3 + [\text{Ca}^{2+}]_o[\text{Na}^+]_i^3)}.$$

(*d*) *Ca sequestration equations*

$$I_{\text{up}} = \frac{3[\text{Ca}^{2+}]_i - 0.23[\text{Ca}^{2+}]_{\text{up}}\frac{k_{\text{cyca}}k_{\text{xcs}}}{k_{\text{srca}}}}{[\text{Ca}^{2+}]_i + [\text{Ca}^{2+}]_{\text{up}}\frac{k_{\text{cyca}}k_{\text{xcs}}}{k_{\text{srca}}} + k_{\text{cyca}}k_{\text{xcs}} + k_{\text{cyca}}}.$$

$$I_{\text{tr}} = 50([\text{Ca}^{2+}]_{\text{up}} - [\text{Ca}^{2+}]_{\text{rel}}),$$

$$I_{\text{rel}} = [f_{\text{activator}}/(f_{\text{activator}} + 0.25)]^2 k_{\text{mca2}}[\text{Ca}^{2+}]_{\text{rel}}.$$

(*e*) *Reversal potentials*

$$E_{\text{Na}} = (RT/F)\ln([\text{Na}^+]_o/[\text{Na}^+]_i),$$

$$E_{\text{K}} = (RT/F)\ln([\text{K}^+]_o/[\text{K}^+]_i),$$

$$E_{\text{Ca}} = (RT/2F)\ln([\text{Ca}^{2+}]_o/[\text{Ca}^{2+}]_i),$$

$$E_{mh} = (RT/F)\ln(([\text{Na}^+]_o + 0.12[\text{K}^+]_o)/([\text{Na}^+]_i + 0.12[\text{K}^+]_o)).$$

(*f*) *Parameters*

parameter	value	parameter	value
C_m	$40(10)^{-6}$ μF	k_{naca}	10^{-4} nA
k_{cachoff}	0.001 mM	d_{naca}	10^{-4}
k_{mk1}	10 mM	γ	$\frac{1}{2}$
k_{mk}	1 mM	n_{NaCa}	3
$k_{m\text{Na}}$	40 mM	k_{cyca}	$3(10)^{-4}$ mM
		k_{xcs}	0.4 mM
k_{mca}	$5(10)^{-4}$ mM	k_{srca}	0.5 mM
		F	96 485 C mol^{-1}
I_{NaKmax}	0.14 nA	R	8314.41 mJ (mol K)$^{-1}$
G_{na}	0.5 μS	T	310 K
G_{to}	0.01 μS	V_{ecs}	0.4
$G_{b\text{K}}$	0.0017 μS	radius	10 μm
G_{k1}	0.017 μS	length	80 μm
$G_{b\text{na}}$	0.000 12 μS	V_{cell}	$10^{-9}\,\pi$ radius^2 length μl
$G_{b\text{ca}}$	0.000 05 μS	V_i	$(1 - V_{\text{ecs}} - V_{\text{up}} - V_{\text{rel}})\,V_{\text{cell}}$ μl
P_{ca}	0.05 nA mm^{-1}	V_{up}	0.01
P_{cak}	0.002	V_{rel}	0.1
P_{cana}	0.002	V_{SRup}	$V_{\text{cell}}\,V_{\text{up}}$ μl
$[\text{Ca}^{2+}]_b$	2 mM	k_{mca2}	250 nA mM
$[\text{K}^+]_o$	4 mM	M_{trop}	0.02 mM
$[\text{Na}^+]_o$	140 mM	C_{trop}	0.15 mM

Am J Physiol Heart Circ Physiol 286: H1573–H1589, 2004.
First published December 4, 2003; 10.1152/ajpheart.00794.2003.

A model for human ventricular tissue

K. H. W. J. ten Tusscher,[1] D. Noble,[2] P. J. Noble,[2] and A. V. Panfilov[1,3]

[1]*Department of Theoretical Biology, Utrecht University, 3584 CH Utrecht, The Netherlands;*
and [2]*University Laboratory of Physiology, University of Oxford, Oxford OX1 3PT; and*
[3]*Division of Mathematics, University of Dundee, Dundee DD1 4HN, United Kingdom*

Submitted 9 August 2003; accepted in final form 2 December 2003

Ten Tusscher, K. H. W. J., D. Noble, P. J. Noble, and A. V. Panfilov. A model for human ventricular tissue. *Am J Physiol Heart Circ Physiol* 286: H1573–H1589, 2004. First published December 4, 2003; 10.1152/ajpheart.00794.2003.—The experimental and clinical possibilities for studying cardiac arrhythmias in human ventricular myocardium are very limited. Therefore, the use of alternative methods such as computer simulations is of great importance. In this article we introduce a mathematical model of the action potential of human ventricular cells that, while including a high level of electrophysiological detail, is computationally cost-effective enough to be applied in large-scale spatial simulations for the study of reentrant arrhythmias. The model is based on recent experimental data on most of the major ionic currents: the fast sodium, L-type calcium, transient outward, rapid and slow delayed rectifier, and inward rectifier currents. The model includes a basic calcium dynamics, allowing for the realistic modeling of calcium transients, calcium current inactivation, and the contraction staircase. We are able to reproduce human epicardial, endocardial, and M cell action potentials and show that differences can be explained by differences in the transient outward and slow delayed rectifier currents. Our model reproduces the experimentally observed data on action potential duration restitution, which is an important characteristic for reentrant arrhythmias. The conduction velocity restitution of our model is broader than in other models and agrees better with available data. Finally, we model the dynamics of spiral wave rotation in a two-dimensional sheet of human ventricular tissue and show that the spiral wave follows a complex meandering pattern and has a period of 265 ms. We conclude that the proposed model reproduces a variety of electrophysiological behaviors and provides a basis for studies of reentrant arrhythmias in human ventricular tissue.

reentrant arrhythmias; human ventricular myocytes; restitution properties; spiral waves; computer simulation

CARDIAC ARRHYTHMIAS and sudden cardiac death are among the most common causes of death in the industrialized world. Despite decades of research their causes are still poorly understood. Theoretical studies into the mechanisms of cardiac arrhythmias form a well-established area of research. One of the most important applications of these theoretical studies is the simulation of the human heart, which is important for a number of reasons. First, the possibilities for doing experimental and clinical studies involving human hearts are very limited. Second, animal hearts used for experimental studies may differ significantly from human hearts [heart size, heart rate, action potential (AP) shape, duration, and restitution, vulnerability to arrhythmias, etc.]. Finally, cardiac arrhythmias, especially those occurring in the ventricles, are three-dimensional phenomena whereas experimental observations are still largely constrained to surface recordings. Computer simulations of arrhythmias in the human heart can overcome some of these problems.

To perform simulation studies of reentrant arrhythmias in human ventricles we need a mathematical model that on the one hand reproduces detailed properties of single human ventricular cells, such as the major ionic currents, calcium transients, and AP duration (APD) restitution (APDR), and important properties of wave propagation in human ventricular tissue, such as conduction velocity (CV) restitution (CVR). On the other hand, it should be computationally efficient enough to be applied in the large-scale spatial simulations needed to study reentrant arrhythmias.

Currently, the only existing model for human ventricular cells is the Priebe-Beuckelman (PB) model and the reduced version of this model constructed by Bernus et al. (3, 51). The PB model is largely based on the phase 2 Luo-Rudy (LR) model for guinea pig ventricular cells (38). Although the model incorporates some data on human cardiac cells and successfully reproduces basic properties of APs of normal and failing human ventricular cells, it has several limitations. First, several major ionic currents are still largely based on animal data, and second, the APD is 360 ms, which is much longer than the values typically recorded in tissue experiments (~270 ms; Ref. 36). The aim of this work was to formulate a new model for human ventricular cells that is based on recent experimental data and that is efficient for large-scale spatial simulations of reentrant phenomena.

We formulated a model in which most major ionic currents [fast Na^+ current (I_{Na}), L-type Ca^{2+} current (I_{CaL}), transient outward current (I_{to}), rapid delayed rectifier current (I_{Kr}), slow delayed rectifier current (I_{Ks}), and inward rectifier K^+ current (I_{K1})] are based on recent experimental data. The model includes a simple calcium dynamics that reproduces realistic calcium transients and a positive human contraction staircase and allows us to realistically model calcium-dominated I_{CaL} inactivation, while at the same time maintaining a low computational load.

The model fits experimentally measured APDR properties of human myocardium (42). In addition, the CVR properties of our model agree better with experimental data—which are currently only available for dog and guinea pig (6, 19)—than those of existing ionic models. Both APD and CV restitution are very important properties for the occurrence and stability of reentrant arrhythmias (6, 17, 31, 52, 67). Our model is able to reproduce the different AP shapes corresponding to the endo-, epi-, and midmyocardial regions of the ventricles and their

Address for reprint requests and other correspondence: K. H. W. J. ten Tusscher, Utrecht Univ., Dept. of Theoretical Biology, Padualaan 8, 3584 CH Utrecht, The Netherlands (E-mail: khwjtuss@hotmail.com).

different rate dependencies (9, 10, 36). Finally, we model spiral wave dynamics in a two-dimensional (2D) sheet of human ventricular tissue and study the dynamics of its rotation, the ECG manifestation of the spiral wave, and membrane potential recordings during spiral wave rotation. In conclusion, we propose a new model for human ventricular tissue that is feasible for large-scale spatial computations of reentrant sources of cardiac arrhythmias.

MATERIALS AND METHODS

General

The cell membrane is modeled as a capacitor connected in parallel with variable resistances and batteries representing the different ionic currents and pumps. The electrophysiological behavior of a single cell can hence be described with the following differential equation (23)

$$\frac{dV}{dt} = -\frac{I_{ion} + I_{stim}}{C_m} \quad (1)$$

where V is voltage, t is time, I_{ion} is the sum of all transmembrane ionic currents, I_{stim} is the externally applied stimulus current, and C_m is cell capacitance per unit surface area.

Similarly, ignoring the discrete character of microscopic cardiac cell structure, a 2D sheet of cardiac cells can be modeled as a continuous system with the following partial differential equation (23)

$$\frac{\partial V}{\partial t} = -\frac{I_{ion} + I_{stim}}{C_m} + \frac{1}{\rho_x S_x C_m}\frac{\partial^2 V}{\partial x^2} + \frac{1}{\rho_y S_y C_m}\frac{\partial^2 V}{\partial y^2} \quad (2)$$

where ρ_x and ρ_y are the cellular resistivity in the x and y directions, S_x and S_y are the surface-to-volume ratio in the x and y directions, and I_{ion} is the sum of all transmembrane ionic currents given by the following equation

$$I_{ion} = I_{Na} + I_{K1} + I_{to} + I_{Kr} + I_{Ks} + I_{CaL} + I_{NaCa} + I_{NaK}$$
$$+ I_{pCa} + I_{pK} + I_{bCa} + I_{bNa} \quad (3)$$

where I_{NaCa} is Na^+/Ca^{2+} exchanger current, I_{NaK} is Na^+/K^+ pump current, I_{pCa} and I_{pK} are plateau Ca^{2+} and K^+ currents, and I_{bCa} and I_{bK} are background Ca^{2+} and K^+ currents.

Physical units used in our model are as follows: time (t) in milliseconds, voltage (V) in millivolts, current densities (I_x) in picoamperes per picofarad, conductances (G_x) in nanosiemens per picofarad, and intracellular and extracellular ionic concentrations (X_i, X_o) in millimoles per liter. The equations for the ionic currents are specified in *Membrane Currents*.

For one-dimensional (1D) computations cell capacitance per unit surface area is taken as $C_m = 2.0\ \mu F/cm^2$ and surface-to-volume ratio is set to $S = 0.2\ \mu m^{-1}$, following Bernus et al. (3). To obtain a maximum planar conduction velocity (CV) of 70 cm/s, the velocity found for conductance along the fiber direction in human myocardium by Taggart et al. (61), a cellular resistivity $\rho = 162\ \Omega cm$ was required. This is comparable with the $\rho = 180\ \Omega cm$ used by Bernus et al. (3) and the $\rho = 181\ \Omega cm$ used by Jongsma and Wilders (29), and it results in a "diffusion" coefficient $D = 1/(\rho S C_m)$ of 0.00154 cm²/ms. Because in 2D we did not intend to study the effects of anisotropy, we use the same values for ρ_x and ρ_y and for S_x and S_y. Parameters of the model are given in Table 1.

For 1D and 2D computations, the forward Euler method was used to integrate *Eq. 1*. A space step of $\Delta x = 0.1$–0.2 mm and a time step of $\Delta t = 0.01$–0.02 ms were used. To integrate the Hodgkin-Huxley-type equations for the gating variables of the various time-dependent currents (m, h, and j for I_{Na}, r and s for I_{to}, x_{r1} and x_{r2} for I_{Kr}, x_s for I_{Ks}, d, f, and f_{Ca} for I_{CaL}, and g for I_{rel}) the Rush and Larsen scheme (54) was used.

Table 1. *Model parameters*

Parameter	Definition	Value
R	Gas constant	8.3143 J·K⁻¹·mol⁻¹
T	Temperature	310 K
F	Faraday constant	96.4867 C/mmol
C_m	Cell capacitance per unit surface area	2 μF/cm²
S	Surface-to-volume ratio	0.2 μm⁻¹
ρ	Cellular resistivity	162 Ω·cm
V_C	Cytoplasmic volume	16,404 μm³
V_{SR}	Sarcoplasmic reticulum volume	1,094 μm³
K_O	Extracellular K^+ concentration	5.4 mM
Na_O	Extracellular Na^+ concentration	140 mM
Ca_O	Extracellular Ca^{2+} concentration	2 mM
G_{Na}	Maximal I_{Na} conductance	14.838 nS/pF
G_{K1}	Maximal I_{K1} conductance	5.405 nS/pF
G_{to}, epi, M	Maximal epicardial I_{to} conductance	0.294 nS/pF
G_{to}, endo	Maximal endocardial I_{to} conductance	0.073 nS/pF
G_{Kr}	Maximal I_{Kr} conductance	0.096 nS/pF
G_{Ks}, epi, endo	Maximal epi- and endocardial I_{Ks} conductance	0.245 nS/pF
G_{Ks}, M	Maximal M cell I_{Ks} conductance	0.062 nS/pF
p_{KNa}	Relative I_{Ks} permeability to Na^+	0.03
G_{CaL}	Maximal I_{CaL} conductance	1.75^{-4} cm³·μF⁻¹·s⁻¹
k_{NaCa}	Maximal I_{NaCa}	1,000 pA/pF
γ	Voltage dependence parameter of I_{NaCa}	0.35
K_{mCa}	Ca_i half-saturation constant for I_{NaCa}	1.38 mM
K_{mNai}	Na_i half-saturation constant for I_{NaCa}	87.5 mM
k_{sat}	Saturation factor for I_{NaCa}	0.1
α	Factor enhancing outward nature of I_{NaCa}	2.5
P_{NaK}	Maximal I_{NaK}	1.362 pA/pF
K_{mK}	K_O half-saturation constant of I_{NaK}	1 mM
K_{mNa}	Na_i half-saturation constant of I_{NaK}	40 mM
G_{pk}	Maximal I_{pK} conductance	0.0146 nS/pF
G_{pCa}	Maximal I_{pCa} conductance	0.025 nS/pF
K_{pCa}	Ca_i half-saturation constant of I_{pCa}	0.0005 mM
G_{bNa}	Maximal I_{bNa} conductance	0.00029 nS/pF
G_{bCa}	Maximal I_{bCa} conductance	0.000592 nS/pF
V_{maxup}	Maximal I_{up}	0.000425 mM/ms
K_{up}	Half-saturation constant of I_{up}	0.00025 mM
a_{rel}	Maximal Ca_{SR}-dependent I_{rel}	16.464 mM/s
b_{rel}	Ca_{SR} half-saturation constant of I_{rel}	0.25 mM
c_{rel}	Maximal Ca_{SR}-independent I_{rel}	8.232 mM/s
V_{leak}	Maximal I_{leak}	0.00008 ms⁻¹
Buf_c	Total cytoplasmic buffer concentration	0.15 mM
K_{bufc}	Ca_i half-saturation constant for cytoplasmic buffer	0.001 mM
Buf_{sr}	Total sarcoplasmic buffer concentration	10 mM
K_{bufsr}	Ca_{SR} half-saturation constant for sarcoplasmic buffer	0.3 mM

We test the accuracy of our numerical simulations in a cable of cells by varying both the time and space steps of integration. The results of these tests are shown in Table 2. From Table 2 it follows that, with a $\Delta x = 0.2$ mm, decreasing Δt from 0.02 to 0.0025 ms leads to a 3.7% increase in CV. Similarly, with $\Delta t = 0.02$ ms, decreasing Δx from 0.2 to 0.1 mm leads to a an increase in CV of 4.6%. The changes in CV occurring for changes in space and time integration steps are similar to those occurring in other models (see, for example, Ref. 52). The time and space steps used in most computations are $\Delta t = 0.02$ ms and $\Delta x = 0.2$ mm, similar to values used in other studies (3, 6, 52, 69). Major conclusions of our model were tested for smaller space and time steps; the results were only slightly different.

Action potential duration (APD) is defined as action potential duration at 90% repolarization (APD_{90}). Two different protocols were used to determine APD restitution (APDR). The S1-S2 restitution protocol, typically used in experiments, consists of 10 S1 stimuli

Table 2. *Numerical accuracy of conduction velocity for different Δt and Δx*

	Conduction Velocity, cm/s			
Δx, cm	$\Delta t = 0.0025$ ms	$\Delta t = 0.005$ ms	$\Delta t = 0.01$ ms	$\Delta t = 0.02$ ms
0.010	75.4	75.0	74.2	72.5
0.015	74.4	73.8	73.0	71.5
0.020	71.9	71.5	70.8	69.3
0.030	67.8	67.4	66.8	65.7
0.040	63.2	63.0	62.6	61.7

applied at a frequency of 1 Hz and a strength of two times the threshold value, followed by a S2 extrastimulus delivered at some diastolic interval (DI) after the AP generated by the last S1 stimulus. The APDR curve is generated by decreasing DI and plotting APD generated by the S2 stimulus against DI. The second restitution protocol is called the dynamic restitution protocol. It was first proposed by Koller et al. (32) as being a more relevant determinant of spiral wave stability than S1-S2 restitution. The protocol consists of a series of stimuli at a certain cycle length until a steady-state APD is reached; after that, cycle length is decreased. The APDR curve is obtained by plotting steady-state APDs against steady-state DIs. CV restitution (CVR) was simulated in a linear strand of 400 cells by pacing it at one end at various frequencies and measuring CV in the middle of the cable.

Spiral waves were initiated in 2D sheets of ventricular tissue with the S1-S2 protocol. We first applied a single S1 stimulus along the entire length of one side of the tissue, producing a planar wave front propagating in one direction. When the refractory tail of this wave crossed the middle of the medium, a second S2 stimulus was applied in the middle of the medium, parallel to the S1 wave front but only over three-quarters of the length of the medium. This produces a second wave front with a free end around which it curls, thus producing a spiral wave. Stimulus currents lasted for 2 (S1) and 5 (S2) ms and were two times diastolic threshold. The trajectory of the spiral tip was traced with an algorithm suggested by Fenton and Karma (16). It is based on the idea that the spiral tip is defined as the point where excitation wave front and repolarization wave back meet. This point can be found as the intersection point of an isopotential line (in our case, -35 mV) and the $dV/dt = 0$ line.

Electrograms of spiral wave activity were simulated in 2D by calculating the dipole source density of the membrane potential V in each element, assuming an infinite volume conductor (50). The electrogram was recorded with a single electrode located 10 cm above the center of the sheet of tissue.

All simulations were written in C^{++}. Single-cell and cable simulations were run on a PC Intel Pentium III 800-MHz CPU; 2D simulations were run on 32 500-MHz processors of a SGI Origin 3800 shared-memory machine, using OpenMP for parallellization (Source code available at http://www-binf.bio.uu.nl/khwjtuss/HVM).

A description of the membrane currents of the model and the experimental data on which they are based is given in *Membrane Currents*. For most currents a comparison is made with the formulations used in existing models for human ventricular myocytes by Priebe and Beuckelmann (51)—for the rest of the text referred to as the PB model—and for human atrial myocytes by Courtemanche and coworkers (8)—for the rest of the text referred to as the CRN model. For the fast Na^{2+} current a comparison is made to the widely used I_{Na} formulation first used in phase 1 of the Luo-Rudy (LR) model (37) that is used in both the PB and CRN model. The LR I_{Na} formulation is largely based on the I_{Na} formulation by Ebihara and Johnson (11), which is fitted to data from embryonic chicken heart cells to which a slow inactivation gate j, as first proposed by Beeler and Reuter (1), was added. A detailed listing of all equations can be found in the APPENDIX.

Membrane Currents

Fast Na$^+$ current: I_{Na}. We use the three gates formulation of I_{Na} first introduced by Beeler and Reuter (1)

$$I_{Na} = G_{Na}m^3hj(V - E_{Na}) \qquad (4)$$

where m is an activation gate, h is a fast inactivation gate, and j is a slow inactivation gate. Each of these gates is governed by Hodgkin-Huxley-type equations for gating variables and characterized by a steady-state value and a time constant for reaching this steady-state value, both of which are functions of membrane potential (see APPENDIX).

The steady-state activation curve (m_{∞}^3) is fitted to data on steady-state activation of wild-type human Na^{2+} channels expressed in HEK-293 cells from Nagatomo et al. (44). Experimental data were extrapolated to 37°. Because there is no equivalent to the Q$_{10}$ values used to extrapolate time constants to different temperatures, a linear extrapolation was used based on a comparison of values obtained at 23° and 33°. Note that similar Na$^+$ channel activation data were obtained by others (64, 40, 55). Figure 1A shows the steady-state activation curve used in our model. For comparison, temperature-corrected experimental data are added.

The steady-state curve for inactivation ($h_{\infty} \times j_{\infty}$) is fitted to steady-state inactivation data from Nagatomo et al. (44). Again, data were extrapolated to 37°. Similar inactivation data were obtained by others (55, 64). Figure 1B shows the steady-state inactivation curve used in our model together with temperature-corrected experimental data. Note that for resting membrane potentials the h and j gates are partially inactivated.

The time constants τ_h and τ_j are derived from current decay (typically V greater than -50 mV) and current recovery experiments (typically V less than -80 mV) (40, 44, 55, 58, 63–65). In both types of experiments a double-exponential fit is made to the data, allowing interpretation of the fast and slow inactivation time constants as τ_h and τ_j, respectively. To convert all data to 37° a Q$_{10} = 2.79$ (derived from a comparison of fast inactivation time constants obtained at 23° and 33 by Nagatomo et al.) was used. Figure 1D shows our fit for the fast inactivation time constants, and Fig. 1E shows our fit for the slow inactivation time constants of the model. Temperature-adjusted experimental data points are added for comparison.

Activation time constants are derived from time to peak data from Nagatomo et al., converted as discussed above to 37°. τ_m can be calculated from the peak time (where $dI_{Na}/dt = 0$), assuming that j is constant and knowing m_{∞}, h_{∞} and τ_h. Figure 1C shows our fit of τ_m together with experimentally derived, temperature-corrected time constants.

In Fig. 1F the time course of recovery from inactivation for our I_{Na} is shown. Recovery from inactivation was established by applying a double-pulse protocol: from the holding potential, a 1-s duration pulse to -20 mV was applied to fully inactivate I_{Na}, the voltage was then stepped back to the holding potential to allow I_{Na} to recover for variable durations, and finally a second 30-ms pulse to -20 mV was applied. The I_{Na} elicited during the second pulse is normalized relative to the I_{Na} elicited during the first pulse to establish the amount of recovery. Figure 1F shows normalized I_{Na} as a function of the duration of the recovery interval between the two pulses for various values of the holding potential. Similar to experimental observations by Viswanathan et al. (63), Nagatomo et al. (44), Schneider et al. (58), and Makita et al. (40), recovery is slower for higher recovery potentials and has a sigmoid shape when plotted on a logarithmic scale. Note that in experiments τ_{Na} recovery is often slower than observed in our model, because our model is at physiological temperature whereas most experiments are performed at room temperature.

Figure 1G displays the rate dependence of the I_{Na} current. Rate dependence was tested by applying 500-ms pulses to -10 mV from a holding potential of -100 mV with different interpulse intervals. Steady-state current was normalized to the current elicited by the first pulse. The graph shows that for increasing frequency (decreasing interpulse interval) I_{Na} decreases and that this decrease is considerably faster for 21° than for 37°. Experiments performed by Wang et al. (65) at 32° with an interpulse interval of 20 ms (1.92 Hz) show a reduction to 0.17 of the original current level, which lies between the reduction to 0.5 we measure at 37° and the reduction to 0.13 we measure at 21°.

For comparison purposes we also added LR m_∞, h_∞, τ_m, τ_h, and τ_j curves to Fig. 1, *A–E*. The following observations can be made. Our steady-state activation curve lies 8 mV to more negative potentials (Fig. 1*A*). Our steady-state inactivation curve lies 12 mV to more negative potentials (Fig. 1*B*). Activation time constants are in the same range of values (Fig. 1*C*). Our τ_h curve has a similar shape, but for voltages smaller than −40 mV time constants are a factor of 1.6 larger, resulting in slower recovery dynamics (Fig. 1*D*). Similarly, our τ_j is a factor of 3–5 larger for voltages smaller than −30 mV, leading to a considerably slower recovery from inactivation (Fig. 1*E*). G_{Na}

was fitted to reproduce a $\dot{V}_{max}$ = 260 mV/ms, which is in the range of experimental data found by Drouin et al. (10).

L-type Ca^{2+} current: I_{CaL}. The L-type calcium current is described by the following equation

$$I_{CaL} = G_{CaL} d f f_{Ca} 4 \frac{VF^2}{RT} \frac{Ca_i e^{2VF/RT} - 0.341 Ca_o}{e^{2VF/RT} - 1} \quad (5)$$

where d is a voltage-dependent activation gate, f is a voltage-dependent inactivation gate, f_{Ca} is an intracellular calcium-dependent inactivation gate, and driving force is modeled with a Goldmann-Hodgkin-Katz equation.

The steady-state activation d_∞ and steady-state voltage inactivation curve f_∞ are fitted to I_{CaL} steady-state data from human ventricular myocytes reported by Benitah et al. (2), Mewes and Ravens (41), Pelzmann et al. (46),and Magyar et al. (39). Figure 2*A* shows the steady-state activation, and Fig. 2*B* shows the steady-state inactivation curve of our model together with experimental data from Pelzmann et al. (46).

Experimental data show that calcium-mediated inactivation is rapid, increases with calcium concentration, but is never complete (20, 60). More quantitative data about the precise dependence of amount and speed of inactivation on calcium concentration are unavailable and hard to obtain because intracellular calcium cannot be clamped to a constant value. As shown in Fig. 2*C* our $f_{ca\infty}$ curve has a switchlike shape, switching from no inactivation to considerable but incomplete inactivation if calcium concentration exceeds a certain threshold. For suprathreshold concentrations the amount of inactivation depends mildly on calcium concentration.

There are hardly any experimental data on activation times of I_{CaL} in human myocytes. Therefore, as was done in the CRN model, we used the curve from the phase-2 LR model. Limited data on I_{CaL} activation times from Pelzmann et al. (46) were used to adjust the shape of the τ_d curve of the LR model. Figure 2*D* displays the voltage-dependent activation time constant of our model.

The time constant τ_{fca} is derived from experiments performed by Sun et al. (60). They show that during current decay experiments a fast and a slow time constant can be distinguished, with the fast time constant being independent of voltage and depending only on calcium, allowing interpretation as τ_{fca}. Sun et al. (60) find a time constant of ~12 ms at 23°; no data on the concentration dependence of the time

Fig. 1. Steady-state and time constant curves describing the gating of the fast sodium current. Model curves of our model (cur model) and the Luo-Rudy (LR) model are shown together with experimental data. *A*: steady-state activation curves. Experimental data are from Nagatomo et al. (44) and are extrapolated to 37°. *B*: steady-state inactivation curves. Experimental data are from Nagatomo et al. *C*: activation time constants. Experimental activation time constants are derived from time to peak data from Nagatomo et al. Experimental data are converted to 37° (Q_{10} = 2.79). *D*: fast inactivation constants. *E*: slow inactivation constants. Fast and slow experimental inactivation time constants are taken from Nagatomo et al. (44), Schneider et al. (58), Sakakibara et al. (55), Makita et al. (40), Wan et al. (64), Viswanathan et al. (63), and Wang et al. (65). Experimental time constants are converted to 37° (Q_{10} = 2.79). *F*: recovery from inactivation for different recovery potentials. The time course of recovery from inactivation was established using a double-pulse protocol: from the holding potential a first 1-s duration pulse to −20 mV was applied to fully inactivate I_{Na}, then voltage was stepped back to the holding potential to allow recovery for periods ranging from 1 ms to 5s, and finally a second 30-ms pulse to −20 mV was applied to measure the amount of recovery. I_{Na} elicited by the second pulse was normalized to the I_{Na} elicited by the first pulse and is shown as a function of recovery time. *G*: frequency dependence of I_{Na}. From a holding potential of −100 mV, 500-ms pulses to −10 mV are given with interpulse intervals of 10, 20, 50, 100, 250, and 500 ms, corresponding to stimulus frequencies of 1.96, 1.92, 1.82, 1.67, 1.3, and 1 Hz. The steady-state I_{Na} obtained for the different frequencies is normalized to the I_{Na} elicited by the first pulse. Frequency dependence was determined for 37° and 21°. For the 21° experiments gate dynamics were adapted with a Q_{10} = 2.79.

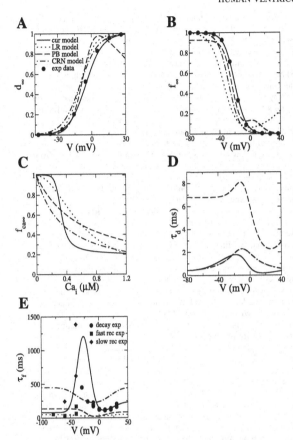

Fig. 2. Steady-state and time constant curves describing the gating of the L-type calcium current. Model curves of our model (cur model), the LR model, the PB model, and the CRN model are shown together with experimental data. *A*: steady-state voltage activation curves. Experimental steady-state activation data are from Pelzmann et al. (46). *B*: steady-state voltage inactivation curves. Experimental steady-state inactivation data are from Pelzmann et al. *C*: steady-state calcium inactivation curves. *D*: voltage activation time constants. Note that the LR and CRN models have the same time constant formulation: curves are overlapping. *E*: voltage inactivation time constants. Experimental inactivation time constants are taken from Beuckelmann et al. (4), Benitah et al. (2), Mewes et al. (41), Sun et al. (60), Li et al. (35), Pelzmann et al. (46), and Magyar et al. (39). Experimental data are converted to 37° degrees (Q_{10} = 2.1). In addition, time constants derived from current decay experiments (V less than −30 mV) are also corrected for the presence of extracellular calcium (correction factor 2.2). For an explanation of the latter, see text.

constant is available. We assumed a single time constant of 2 ms to be reasonable at 37°, comparable to the immediate inactivation used in the PB model and the 2-ms time constant used in the CRN model.

The time constants τ_f are derived from experiments on calcium current decay and recovery in human ventricular and atrial myocytes by Beuckelmann et al. (4), Benitah et al. (2), Mewes et al. (41), Sun et al. (60), Li and Nattel (35), Pelzmann et al. (46), and Magyar et al. (39). Sun et al. (60) show that during current decay experiments the slow inactivation time constant depends both on voltage and extracellular calcium. After removal of extracellular calcium, an even slower, purely voltage-dependent inactivation time constant arises. It is this time constant that should be interpreted as τ_f. Therefore, slow inactivation time constants found in current decay experiments were not only converted to 37° with a Q_{10} = 2.1 (based on a comparison of time constants obtained by Li et al. and

Pelzmann et al. at physiological temperatures and data obtained by Benitah et al. at 21°) but were also corrected for the presence of extracellular Ca^{2+} with a slowing-down correction factor of 2.2 (based on a comparison of time constants obtained under normal conditions and under conditions in which extracellular Ca^{2+} was replaced by Sr^{2+} performed by Sun et al.).

In current recovery experiments also, two time constants are derived. However, assuming that recovery from calcium inactivation is fast and given the clear voltage dependence of both fast and slow recovery time constants, both can be considered as voltage-dependent time constants. Because our formulation of I_{CaL} incorporates only a single voltage-dependent inactivation gate, our τ_f is constructed to form an intermediate between these fast and slow recovery time constants (−40 < V < −80 mV). Figure 2*E* displays the voltage-dependent inactivation time constant of our model. For comparison, experimentally found inactivation time constants are added.

For comparison purposes we also added LR phase-2, PB, and CRN model curves for d_∞, f_∞, $f_{ca\infty}$, τ_d, and τ_f to Fig. 2, *A–E*. From this the following observations can be made. Our steady-state activation curve is similar to the curves used in the LR and CRN model. The PB model has, for unknown reasons, a curve with a somewhat different shape (Fig. 2*A*). Our steady-state inactivation curve inactivates completely, similar to the curve used in the CRN model, whereas the LR and PB model use incompletely inactivating curves (Fig. 2*B*). In experiments, inactivation is more complete if prepulse duration is longer or temperature is higher (35, 60), implying that inactivation is slow rather than incomplete.

Our $f_{ca\infty}$ curve has a switch shape, with a high level of inactivation beyond the threshold, whereas the LR, PB, and CRN curves are gradually declining functions of calcium (Fig. 2*C*). The high level of calcium inactivation, together with slow voltage inactivation, causes calcium to be the dominant mechanism of I_{CaL} inactivation in our model. This agrees well with experimental data. Incorporating calcium-dominated inactivation in models without local control may easily result in AP instability: in a local control model (21, 53)—in which individual I_{CaL} and calcium-induced calcium release (CICR) channels interact in subspaces—a smaller I_{CaL} current implies fewer open channels and hence fewer local calcium transients (sparks); the individual sparks still have the same effectiveness in closing nearby calcium channels. However, in a nonlocal control model, a smaller I_{CaL} generates a smaller global calcium transient that might be less effective in inactivating the I_{CaL} current. By using a switch shape for $f_{ca\infty}$, we ensure effective calcium inactivation for a wide range of systolic calcium levels in our model.

The τ_d curve of our model has a similar shape as that used in the PB model, but with a factor of 2–3 shorter time constants, and is a minor adaptation of the curves used in the LR and CRN model (Fig. 2*D*). Our τ_f curve, which is fitted to the experimental data, has a maximum between −50 and 0 mV, whereas the curves used in the LR, PB, and CRN models have a minimum in this voltage region, in strong contrast with the experimental data (Fig. 2*E*). The f gate inactivation in our model is much slower than in the LR and PB model; f gate inactivation in the CRN model is even slower. The slow f gate inactivation in our model and the CRN model is important for making calcium the dominant mechanism of I_{CaL} inactivation. The f gate recovery in our model is similar to that in the LR and PB model and a factor of 4 faster than in the CRN model. It plays an important role in determining APDR, which in our model agrees well with experimental data (see Fig. 8). P_{CaL} is fitted to reproduce peak current values found by Li et al. (35) under physiological temperatures.

We modeled the driving force of the calcium current with the Goldmann-Hodgkin-Katz equation. For simplicity, we ignored the small permeability the channel also has for sodium and potassium ions. In the Luo-Rudy phase-2 model, a Goldmann-Hodgkin-Katz-like equation for calcium, sodium, and potassium is used. In the PB and CRN models, a constant-valued reversal potential is used.

Transient outward current: I_{to}. For I_{to} the following formulation is used

$$I_{\text{to}} = G_{\text{to}}rs(V - E_{\text{K}}) \qquad (6)$$

where r is a voltage-dependent activation gate and s is a voltage-dependent inactivation gate.

The steady-state activation curve (r_∞) is fitted to data on steady-state activation of I_{to} current in human ventricular myocytes of epicardial and endocardial origin at 35° from Nabauer et al. (43). Because no significant difference between activation in epicardial and endocardial cells was found, a single formulation was used. Figure 3A shows the steady-state activation curve used in the model together with experimental data. A 10-mV positive shift was performed on the experimental data to account for the use of Cd^{2+} to block I_{CaL} current, similar to the approach taken by Courtemanche et al. (8). The greater steepness of the model curve relative to the experimental curve was necessary to make sure that no significant reactivation of I_{to} occurs on repolarization.

The steady-state inactivation curve (s_∞) is fitted to data on steady-state inactivation from Nabauer et al. Because of significant differ-

ences between curves obtained for epicardial and endocardial cells, two separate model formulations were used. Figure 3B shows the steady-state inactivation curves used in the model together with Cd^{2+}-corrected experimental data.

Inactivation time constants are fitted to data from Nabauer et al. (43) and Wettwer et al. (68). Current decay experiments show similar time constants for epicardial and endocardial I_{to}, whereas current recovery experiments show much slower recovery from inactivation for endocardial than epicardial I_{to}, thus making two separate formulations for τ_s necessary. Figure 3D shows our fits for epicardial τ_s, and Fig. 3E shows our fit for endocardial τ_s. For comparison, experimental inactivation time constants are added.

Activation time constants are derived from time to peak data from expression of hKv4.3–2— encoding an epicardial type transient outward channel—in mouse fibroblast cells by Greenstein et al. (22) in a manner similar to the derivation of τ_m. Figure 3C shows our τ_r curve together with experimentally derived time constants.

For comparison purposes we also added PB and CRN model curves for r_∞, s_∞, τ_r, and τ_s to Fig. 3, A–D. From this the following observations can be made. The steady-state activation of our model almost coincides with that of the PB model and has a steeper slope and more positive half-activation point than that of the CRN model (Fig. 3A). In our model we distinguish epicardial and endocardial steady-state inactivation, whereas the PB and CRN model only have a single steady-state inactivation curve. The epicardial steady-state inactivation of our model lies 15 and 25 mV to more positive potentials than the curves used in the PB and CRN models, respectively. For the endocardial steady-state inactivation of our model these numbers are 8 and 18 mV, respectively (Fig. 3B).

The activation time constant of our model results in faster inactivation and slower recovery than that of the PB model, whereas it is a factor of 2 slower than the CRN model time constant (Fig. 3C). However, in the CRN model three activation gates are used (r^3), causing net activation to be slower and net deactivation to be faster than that of a single gate and thus complicating the comparison. In our model we distinguish epicardial and endocardial inactivation time constants, whereas the PB and CRN model only have a single inactivation time constant. The inactivation time constant of the PB model resembles our epicardial inactivation time constant in magnitude but has hardly any voltage dependence. The inactivation time constant of the CRN model is similar to our epicardial inactivation time constant (Fig. 3D).

G_{to} is fitted to experimental data on current density from Wettwer et al. (68) and Nabauer et al. (43). Both show large differences in I_{to} size between epicardial and endocardial cells. We use $G_{\text{to}} = 0.294$ nS/pF for epicardial and $G_{\text{to}} = 0.073$ nS/pF for endocardial cells (25% of the value for epicardial cells). Figure 4 shows the current voltage (I-V) relationships for epicardial (Fig. 4A) and endocardial (Fig. 4B) I_{to} together with experimental data from Nabauer et al. (43).

We assume that I_{to} is specific for potassium ions and used the reversal potential E_{K}. A similar approach is taken in the CRN model, whereas in the PB model it is assumed that the channel is also permeable to sodium ions.

Slow delayed rectifier current: I_{Ks}. For the slow delayed rectifier current the following formulation is used

$$I_{\text{Ks}} = G_{\text{Ks}}x_s^2(V - E_{\text{Ks}}) \qquad (7)$$

where x_s is an activation gate and E_{Ks} is a reversal potential determined by a large permeability to potassium and a small permeability to sodium ions (see APPENDIX).

The steady-state activation curve ($x_{s\infty}^2$) is fitted to I_{Kr} activation data obtained for human ventricular myocytes at 36° from Li et al. (34). In Fig. 5A the steady-state activation curve used in the model is shown together with the experimental data.

Activation time constants are based on data from Virag et al. (62) and Wang et al. (66). Both sets of data were obtained in human ventricular myocytes at physiological temperatures. Figure 5B shows

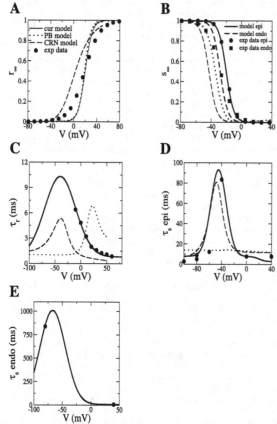

Fig. 3. Steady-state and time constant curves describing the gating of the transient outward current. Model curves of our model (cur model), the PB, and the CRN model are shown together with experimental data. *A*: steady-state activation curves. Experimental steady-state data are from Nabauer et al. (43). A 10-mV positive shift was performed on the experimental curve to account for the use of Cd^{2+}. *B*: steady-state inactivation curves. Cd^{2+}-corrected experimental data are from Nabauer et al. (43). *C*: activation time constants. Experimental activation time constants are derived from time to peak data by Greenstein et al. (22). *D*: epicardial inactivation time constants. Experimental inactivation time constants are from Nabauer et al. and are obtained at physiological temperatures. *E*: endocardial inactivation time constant. Experimental data are from Nabauer et al.

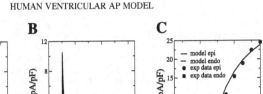
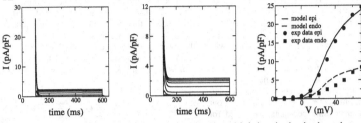

Fig. 4. *A* and *B*: epicardial (*A*) and endocardial (*B*) I_{to} current traces generated by the model during simulated voltage-clamp experiments. Currents are recorded during 500-ms voltage steps to potentials ranging from −30 to 80 mV from a holding potential of −80 mV and an intermittent 100-ms prepulse to −60 mV. Note that during simulation both the I_{to} and I_{pK} currents were switched on. The sustained part of the current traces is generated by I_{pK}. *C*: epicardial and endocardial I_{to} current-voltage (*I-V*) curves obtained from the series of experiments shown in *A* and *B*, respectively. Peak current minus maintained current was measured. Experimental *I-V* curves from Nabauer et al. (43), obtained with a similar voltage-clamp protocol, are added for comparison. The experimental curves are rescaled by a factor of 1.4. We choose I_{to} density to be somewhat larger than experimentally found to obtain a realistic notch in epicardial action potential morphology.

our fit of τ_{xs}. Experimentally obtained activation time constants are added for comparison.

Fitting G_{Ks} to experimentally obtained current densities would result in a small I_{Ks} that has little effect on APD: simulating M cells by a 75% reduction in I_{Ks} density (the principal difference with epicardial cells) would result in M cell APD being only 10 ms longer than epicardial APD, in strong contrast with the 100-ms difference in APD found experimentally (9, 36). Thus there is a discrepancy between current density measured in voltage-clamp experiments and the apparent contribution of the current to APD. This discrepancy is probably due to sensitivity of I_{Ks} channels to the cell isolation procedures used for voltage-clamp experiments (70), resulting in considerable degradation of I_{Ks} channels before current density measurements. Therefore, instead of fitting G_{Ks} to voltage-clamp data, we based it on APD measurements: by using a $G_{Ks} = 0.327$ nS/pF for epicardial cells and a $G_{Ks} = 0.082$ nS/pF in M cells, we get an epicardial APD at 1 Hz of 276 ms and an M cell APD of 336 ms, resulting in an APD difference of 60 ms, which is in the range of experimental values (9, 36). This approach of basing a conductance value on electrophysiological properties rather than measured current density is also used in the development of other models; e.g., in the CRN model, G_{Na} is fitted to get the right $\dot{V}_{max}$, G_{to} to get the right AP morphology, and G_{Kr} and G_{Ks} to get the right APD (8) and in a later version of the LR model in which I_K was replaced by I_{Kr} and I_{Ks}, G_{Ks} was fitted to get the right APD prolongation if I_{Ks} current is blocked

(71). In addition, the values used for G_{Ks} and G_{Kr} in the CRN model and the LR model are similar to the values we use in our model. In Fig. 5*C* the *I-V* relationship of I_{Ks} is shown together with rescaled experimental data from Li et al.

We use a sodium-to-potassium permeability ratio of $p_{KNa} = 0.03$, resulting in a reversal potential E_{Ks} that forms a compromise between reversal potentials found experimentally (34, 62). In the PB model a permeability ratio of 0.018 is used, whereas in the CRN model it is assumed that I_{Ks} is permeable to potassium ions only. Our steady-state activation curve lies 5 and 15 mV to more positive potentials than the curves used in the CRN and PB models, respectively. Compared with the τ_{xs} formulations used by both of these models, our I_{Ks} displays slower activation and more rapid deactivation.

Rapid delayed rectifier current: I_{Kr}. The rapid delayed rectifier current is described by the following equation

$$I_{Kr} = G_{Kr}\sqrt{\frac{K_o}{5.4}}\, x_{r1} x_{r2} (V - E_K) \qquad (8)$$

where x_{r1} is an activation gate and x_{r2} is an inactivation gate. $\sqrt{K_o/5.4}$ represents the K_o dependence of the current. Note that because no data are available on the K_o dependence of human ventricular I_{Kr}, a similar dependence as measured in and implemented for animal myocytes is assumed (38).

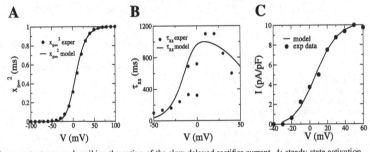

Fig. 5. Steady-state and time constant curves describing the gating of the slow delayed rectifier current. *A*: steady-state activation curve. Experimental steady-state data are from Li et al. (34). *B*: activation time constants. Experimental activation time constants are from Virag et al. (62) (for voltages ranging from −50 to 0 mV) and Wang et al. (66) (for voltages ranging from −10 to 40 mV) and are obtained at physiological temperatures. *C*: *I-V* relationship for I_{Ks}. Tail currents are measured during 5,000-ms voltage steps to potentials ranging from −10 to 50 mV from a holding potential of −40 mV. Experimental *I-V* curves obtained with a similar protocol by Li et al. (34) were included for comparison. Experimental data are rescaled by a factor of 12.5. See text for an explanation of why our model I_{Ks} is much larger than the experimentally measured I_{Ks}.

The steady-state activation curve (x_{r1}^∞) is fitted to activation data on the expression of HERG channels in HEK 293 cells by Zhou et al. (72), in Chinese hamster ovary cells by Johnson et al. (28), and in *Xenopus* oocytes by Smith and Yellen (59). Steady-state inactivation (x_{r2}^∞) is fitted to data from Johnson et al. (28) and Smith and Yellen (59). Figure 6A shows steady-state curves of the model together with experimental data.

Activation time constants (τ_{xr1}) are fitted to data from Zhou et al. (72) obtained at physiological temperatures. τ_{xr12} is fitted to inactivation time constants obtained at physiological temperatures by Johnson et al. (28). Figure 6B shows our fit of τ_{xr1}, and Fig. 6C shows our fit of τ_{xr2}. Experimentally obtained time constants are added for comparison.

Fitting G_{Kr} to experimentally obtained current densities would result in an I_{Kr} that has a lower contribution to APD than suggested by experiments in which I_{Kr} is blocked (34). Therefore, similar to our approach for G_{Ks}, we used such a value for G_{Kr} (0.128 nS/pF) that a complete blocking of I_{Kr} leads to 44 ms of APD prolongation, which is in the range of values found experimentally by Li et al. (34). The *I-V* relationship of I_{Kr} is shown in Fig. 6D. For comparison, rescaled experimental data from Iost et al. (25) and Li et al. (34) are added. Note that our model curve is shifted toward more negative potentials relative to the experimental curves. This difference may be due to the fact that we fitted I_{Kr} to data from expression experiments in hamster and *Xenopus* cells, whereas the experimental *I-V* curves are from experiments on human cardiac cells.

The steady-state activation curve of our I_{Kr} lies 10 and 20 mV to more negative potentials than the curves used in the PB and CRN

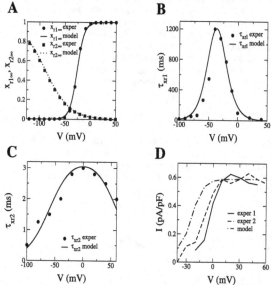

Fig. 6. Steady-state and time constant curves describing the gating of the rapid delayed rectifier current. *A*: steady-state activation and inactivation curves. Experimental data on steady-state activation are from Zhou et al. (72), experimental data on inactivation are from Johnson et al. (28) and Smith and Yellen (59). *B*: activation time constants. Experimental activation time constants are from Zhou et al. (72). *C*: inactivation time constants. Experimental inactivation time constants are from Johnson et al. (28). Experimental time constants are obtained at physiological temperatures. *D*: *I-V* relationship for I_{Kr}. Tail currents are measured during 3,000-ms voltage steps to potentials ranging from −40 to 60 mV from a holding potential of −60 mV. For comparison experimental *I-V* curves obtained using a similar protocol by Iost et al. (25) (exper 1) and Li et al. (34) (exper 2) are added. Experimental curves are rescaled by factors of 2.4 and 1.4, respectively. See text for an explanation of why our model I_{Kr} is larger than the experimentally measured I_{Kr}.

models, respectively. τ_{xr1} has a size and a shape similar to those used in these two models. We modeled inward rectification as a time-dependent inactivation gate, whereas the PB and the CRN model implemented this inactivation as being instantaneous. Our steady-state inactivation curve lies 50 and 60 mV to more negative potentials than the inward rectification curves used in the PB and CRN models, which seem to have no clear experimental basis.

Inward rectifier K^+ current: I_{K1}. For I_{K1} the following formulation is used

$$I_{K1} = G_{K1}\sqrt{\frac{K_o}{5.4}}\,x_{K1\infty}(V - E_K) \tag{9}$$

where $x_{K1\infty}$ is a time-independent inward rectification factor that is a function of voltage. $\sqrt{K_o/5.4}$ represents the K_o dependence of the current. As for I_{Kr}, because of a lack of data on K_o dependence of human I_{K1} we assumed a dependence similar to that in animal myocytes.

Experimental data on I_{K1} current size are highly variable, as previously discussed by Courtemanche et al. (8). We used the formulation for I_{K1} used in the PB model but increased G_{K1} by a factor of 2 to account for the larger current densities found by Koumi et al. (33) in the I_{K1}-relevant voltage range (−90 mV to −40 mV). This results in a value for G_{K1} that is approximately five times larger than in the CRN model, which agrees with data from Koumi et al. (33) showing that I_{K1} is a factor of 5.6 higher in human ventricular than in atrial myocytes.

Na^+/Ca^{2+} exchanger current, Na^+/K^+ pump current, and plateau and background currents. For I_{NaCa} the following equation is used

$$I_{NaCa} = k_{NaCa}\frac{e^{\gamma VF/RT}Na_i^3Ca_o - e^{(\gamma-1)VF/RT}Na_o^3Ca_i\alpha}{(K_{mNai}^3 + Na_o^3)(K_{mCa} + Ca_o)(1 + k_{sat}e^{(\gamma-1)VF/RT})} \tag{10}$$

This formulation is similar to the equation used in the LR model, except for the extra factor α (= 2.5) that accounts for the higher concentration of calcium in the subspace close to the sarcolemmal membrane where the Na^+/Ca^{2+} exchanger is actually operating. Our approach is similar to that used in the Noble et al. model with diadic subspace (45), where I_{NaCa} was made dependent on diadic calcium rather than bulk cytoplasmic calcium. Otherwise, during the AP plateau phase the increase in bulk cytoplasmic calcium is not enough to counteract the increase in voltage and I_{NaCa} current becomes outward oriented during a long period of the plateau phase, which is unrealistic (27).

For I_{NaK} we use the following formulation

$$I_{NaK} = R_{NaK}\frac{K_oNa_i}{(K_o + K_{mK})(Na_i + K_{mNa})(1 + 0.1245e^{-0.1VF/RT} + 0.0353e^{-VF/RT})} \tag{11}$$

This formulation is similar to the formulations used in the LR, PB, and CRN models.

For I_{pCa} the following commonly applied equation is used

$$I_{pCa} = G_{pCa}\frac{Ca_i}{K_{pCa} + Ca_i} \tag{12}$$

For I_{pK} the following equation is used

$$I_{pK} = G_{pK}\frac{V - E_K}{1 + e^{(25-V)/5.98}} \tag{13}$$

This is similar to the equation used by Luo and Rudy.

The background sodium and calcium leakage currents are given by the following equations

$$I_{bNa} = G_{bNa}(V - E_{Na}) \tag{14}$$

$$I_{bCa} = G_{bCa}(V - E_{Ca}) \qquad (15)$$

For P_{NaCa}, P_{NaK}, G_{pCa}, G_{pK}, G_{bNa}, and G_{bCa}, values were chosen such that a frequency change results in Na_i, K_i, and Ca_i transients with a time scale of ~10 min, similar to experimental recordings (5) and which result in equilibrium concentrations—for different frequencies—that lie in the range of experimental observations (49). The values used lie in the range of values used in the PB and CRN models (for parameter values see Table 1).

Intracellular ion dynamics. The calcium dynamics of our model are described using the following set of equations

$$I_{leak} = V_{leak}(Ca_{sr} - Ca_i) \qquad (16)$$

$$I_{up} = \frac{V_{maxup}}{1 + K_{up}^2/Ca_i^2} \qquad (17)$$

$$I_{rel} = \left(a_{rel} \frac{Ca_{sr}^2}{b_{rel}^2 + Ca_{sr}^2} + c_{rel} \right) dg \qquad (18)$$

$$Ca_{ibufc} = \frac{Ca_i \times Buf_c}{Ca_i + K_{bufc}} \qquad (19)$$

$$\frac{dCa_{itotal}}{dt} = -\frac{I_{CaL} + I_{bCa} + I_{pCa} - 2I_{NaCa}}{2V_C F} + I_{leak} - I_{up} + I_{rel} \qquad (20)$$

$$Ca_{srbufsr} = \frac{Ca_{sr} \times Buf_{sr}}{Ca_{sr} + K_{bufsr}} \qquad (21)$$

$$\frac{dCa_{srtotal}}{dt} = \frac{V_C}{V_{SR}}(-I_{leak} + I_{up} - I_{rel}) \qquad (22)$$

where I_{leak} is a leakage current from the sarcoplasmic reticulum to the cytoplasm, I_{up} is a pump current taking up calcium in the SR, I_{rel} is the calcium-induced calcium release (CICR) current, d is the activation gate of I_{CaL}, here reused as the activation gate of I_{rel}, following a similar approach as in Chudin et al. (7), and g is the calcium-dependent inactivation gate of I_{rel}. Ca_{itotal} is the total calcium in the cytoplasm, it consists of Ca_{ibufc}, the buffered calcium in the cytoplasm, and Ca_i, the free calcium in the cytoplasm. Similarly, $Ca_{srtotal}$ is the total calcium in the SR, it consists of $Ca_{srbufsr}$, the buffered calcium in the SR, and Ca_{sr}, the free calcium in the SR. Ratios between free and buffered calcium are analytically computed assuming a steady-state for the buffering reaction (*Eqs. 19* and *21*), following the same approach as first used by Zeng et al. (71) (for a description and values of the parameters see Table 1). Our model for calcium dynamics has a complexity similar to that of most of the current models that are used to study the dynamics of wave propagation in cardiac tissue (LR, CRN, and PB models). Recently, complex models for intracellular calcium handling have been developed that model individual L-type calcium and ryanodine channels, discrete calcium release subunits, and sparks (53, 21). Because of their huge computational demands these models for calcium dynamics have not yet been incorporated in models for cardiac wave propagation.

The changes in the intracellular sodium (Na_i) and potassium (K_i) concentrations are governed by the following equations

$$\frac{dNa_i}{dt} = -\frac{I_{Na} + I_{bNa} + 3I_{NaK} + 3I_{NaCa}}{V_C F} \qquad (23)$$

$$\frac{dK_i}{dt} = -\frac{I_{K1} + I_{to} + I_{Kr} + I_{Ks} - 2I_{NaK} + I_{pK} + I_{stim} - I_{ax}}{V_C F} \qquad (24)$$

To avoid the model being overdetermined, as is the case for a lot of second-generation electrophysiological models (13, 14, 24), we followed the approach suggested by Hund et al. (24) and accounted for the external stimulus current (I_{stim}) and the axial current flow (I_{ax}) in the equation for K_i dynamics. As mentioned above, conductances of background leakage, plateau, and pump currents were chosen such

that transient time scales and equilibrium concentrations lie in the range of experimental observations.

RESULTS

Single Cell

Figure 7 shows an AP, a calcium transient, and the major ionic currents generated by the model under 1-Hz pacing for a parameter setting corresponding to a human epicardial ventricular cell. The AP shows the characteristic spike notch dome

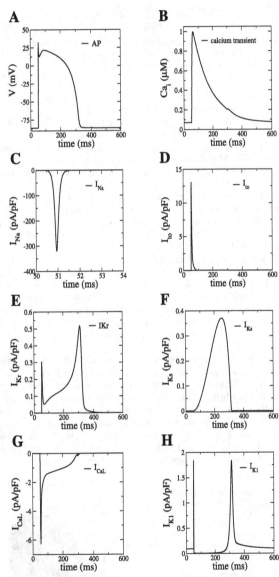

Fig. 7. Steady-state action potential, calcium transient, and major ionic currents under 1-Hz pacing. *A*: action potential. *B*: calcium transient. *C*: fast sodium current (note the difference in time scale used here). *D*: transient outward current. *E*: rapid delayed rectifier current. *F*: slow delayed rectifier current. *G*: L-type calcium current. *H*: inward rectifier current.

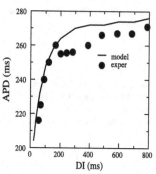

Fig. 8. Action potential duration (APD) restitution curve obtained by using the S1-S2 protocol with a basic cycle length of 1,000 ms. For comparison, experimental data from Morgan et al. (42) are included. DI, diastolic interval.

architecture found for epicardial cells. Resting potential is −87.3 mV, maximum plateau potential is 21.7 mV, and $\dot{V}_{max}$ = 288 mV/ms, all in agreement with experimental data (10, 36). The calcium transient shows the characteristic rounded-off triangular shape found in experiments. Diastolic calcium level is 0.07 μM and maximum systolic calcium level under 1-Hz pacing is 1.0 μM, comparable to values in the PB model and experimentally obtained values by Beuckelmann et al. (26). The ionic currents presented in Fig. 7 have shapes and values similar to those recorded experimentally. Note that the initial spikelike increase of I_{Kr} in our model is absent in the PB and CRN models. This initial increase is also observed in experiments (18, 72) and is achieved by modeling the inward rectification as a time-dependent process.

Figure 8 shows the APDR curve for a single epicardial cell obtained with the S1-S2 restitution protocol (see MATERIALS AND METHODS) with a basic cycle length (BCL) of 1,000 ms. For comparison, experimental data found by Morgan and coworkers (42) are added. It can be seen that the APDR curve of our model in a wide range closely matches the experimentally measured curve.

Figure 9A shows the change in diastolic and systolic calcium levels when pacing frequency is increased in a stepwise fashion from 0.25 to 0.5 to 1 to 1.5 to 2 to 2.5 to 3 Hz. From the figure it can be seen that systolic calcium level first increases substantially up to a frequency of 2 Hz, saturates, and than starts to decrease. In Fig. 9B the corresponding increase in intracellular sodium levels with increasing pacing frequency are shown. It can be seen that sodium keeps increasing but the speed of increase decreases for higher frequencies. Sodium

concentrations for different frequencies are in the range of values measured by Pieske et al. (49). Figure 9C shows the normalized systolic calcium level as a function of pacing frequency. Assuming that generated force is linearly dependent on systolic calcium, the calcium frequency staircase of our model is similar to the force-frequency relationship obtained experimentally for human myocardial cells by Pieske et al. (48, 49) and Schmidt et al. (57).

Different Cell Types

The parameter setting in *Single Cell* reproduces the AP of an epicardial cell. By changing a few parameters, our model is capable of reproducing the AP shapes of the two other ventricular cell types: endocardial and M cells.

In our model endocardial cells differ from epicardial cells in their lower I_{to} density (G_{to} =0.073 nS/pF instead of 0.294 nS/pF, factor 4 difference) and in their slower recovery from inactivation (different τ_s, see APPENDIX). These differences are based on data from Nabauer et al. (43) and Wettwer et al. (68). According to Pereon et al. (47), I_{Ks} of epicardial and endocardial cells are similar. In our model, M cells differ from epicardial and endocardial cells by having an I_{Ks} density of 0.062 instead of 0.245 nS/pF (factor 4 difference). This is based on data from Pereon et al. (47). M cells have I_{to} density and dynamics similar to those of epicardial cells (36).

Figure 10 shows APs recorded under steady-state conditions at BCLs of 1,000, 2,000 and 5,000 ms for epicardial (Fig. 10A), endocardial (Fig. 10B), and M (Fig. 10C) cells. From a comparison of Fig. 10, A and B, it follows that the smaller I_{to} of endocardial cells results in the virtual absence of the notch that is clearly present in the APs of epicardial and M cells. From a comparison of Fig. 10, A and C, it follows that the smaller I_{Ks} of M cells results in a longer APD relative to epicardial and endocardial cells (336 ms in M cells vs. 276 ms in epicardial and 282 ms in endocardial cells for BCL = 1,000 ms) and in a stronger rate dependence of M cell APD.

Note that simulated APD and rate dependence differences between M cells and epi- and endocardial cells are a bit smaller than the experimentally observed differences (9, 36). This is probably due to the fact that M cells differ from the other cell types not only in their I_{Ks} density but also with respect to other current densities. Currently, differences in the late component of I_{Na} have been described for guinea pig (56) and canine myocardium (73), with guinea pig M cells having a smaller and canine M cells having a larger $I_{Na late}$ than the other cell types, and differences in the density of I_{NaCa} have been described for canine myocardium (74), with M cells having a larger ex-

Fig. 9. Changes in calcium dynamics under increasing pacing frequencies. Pacing frequency is varied from 0.25 to 3 Hz; each frequency is maintained for 10 min. *A*: positive contraction staircase: increase in calcium transient amplitude when pacing frequency is increased from 0.25 to 0.5 to 1 to 1.5 to 2 to 2.5 Hz. For the transition from 2.5 to 3 Hz, a mild decrease of systolic calcium level occurs. *B*: concurrent increase in intracellular sodium when pacing frequency is increased. *C*: normalized systolic calcium as a function of pacing frequency. Normalization relative to level at 0.25 Hz.

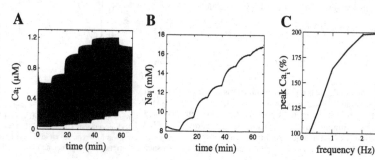

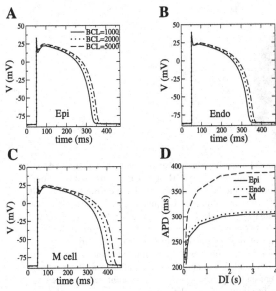

Fig. 10. Different behavior of epicardial, endocardial, and midmyocardial cell types. Steady-state action potentials for BCL of 1,000, 2,000, and 5,000 ms for epicardial (*A*), endocardial (*B*), and M (*C*) cells. *D*: APD restitution curves of the 3 cell types obtained with the dynamic restitution protocol.

changer current. Because of the partially contradictory nature of the data and the lack of data for human myocardium, we decided not to incorporate any of these other differences in our M cell description in the current version of our model.

Figure 10*D* shows restitution curves for epicardial, endocardial, and M cells obtained with the dynamic restitution protocol. Again, the longer APD and stronger rate dependence of M cells can be observed. In addition, we can see that the APD of endocardial cells is slightly longer than that of epicardial cells, because of the lower I_{to} density, and their APD rate dependence is slightly different, because of the slower I_{to} recovery dynamics. Results are very similar to experimentally obtained restitution curves by Drouin et al. (10).

1D Propagation

Figure 11 shows CV restitution curves for a cable of 400 cells. We plotted results for our model and for the PB model, which uses the LR formulation of I_{Na} dynamics. For comparison, we also added experimental CV data. Because no experimental data on human CVR are available, guinea pig CV data measured by Giruoard et al. (19) were used. It can be seen that the CVR of our model agrees much better with experimental data; it declines less steeply and over a much broader range of diastolic intervals than is the case with the LR I_{Na} formulation.

Essential for the shape of the CVR curve is the recovery of I_{Na}, which is mainly determined by the recovery time constant of the slow inactivation gate *j*. Slowing down the dynamics of the *j* gate in the LR I_{Na} formulation makes CVR less steep and similar to that of our model (6). Note, however, that our CVR curve was not obtained by rescaling of a time constant but is based on our τ_j formulation, which fits experimental data on human I_{Na}.

Spiral Waves

The 2D simulations were performed on a 600×600 square lattice of epicardial ventricular cells with $\Delta x = 200$ μm. Spiral waves were initiated with the S1-S2 stimulation protocol described in MATERIALS AND METHODS. The results of these computations are shown in Fig. 12. Figure 12*A* shows a typical spiral wave pattern after an initial period of spiral formation and stabilization (at 1.38 s after the S2 stimulus). The average period of the spiral wave is 264.71 ± 10.49 ms, with an average APD of 217.36 ± 9.19 ms and an average diastolic interval of 47.21 ± 6.13 ms. The spiral wave meanders with a typical tip trajectory shown in Fig. 12*B*; the size of the core is ~3 cm; the rotation type is similar to the "Z" core (see Fig. 12*C*) described by Fast et al. (15) and Efimov et al. (12), which combines regions of fast rotation of the tip (see Fig. 12*C*, A-B-C and D-E-F), typical of a circular core, with regions of laminar motion (see Fig. 12*C*, C-D) typical of a linear core.

Figure 13*A* shows an ECG recorded during spiral wave rotation. The ECG is similar to ECGs recorded during ventricular tachycardia. Figure 13, *B* and *C*, show recordings of membrane voltage in a point far away from the spiral core (star in Fig. 12*A*) and close to the spiral core (circle in Fig. 12*B*), respectively. Note the regular AP pattern in the point far from the core and the irregular pattern recorded close to the spiral core.

Similar results were obtained for simulations of sheets of endocardial and M cells. Wave patterns, tip trajectories, and ECG and membrane potential recordings were very similar (data not shown), the only real difference being the period of spiral wave rotation, which is 264.23 ± 10.37 ms for endocardial and 285.60 ± 6.70 ms for M cell tissue.

DISCUSSION

In this paper we propose a model for human ventricular tissue. An important feature of our model is that all major ionic currents are fitted to recent data on human ventricular myocytes and expression experiments of human cardiac channels. This results in several important differences between our and previous models, the most important of which are the follow-

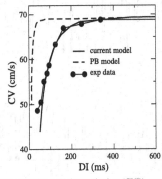

Fig. 11. Model conduction velocity restitution (CVR) curves for both the current model and the PB model obtained by pacing 1 end of a cable of a length of 400 cells with the dynamic restitution protocol. For comparison, experimental guinea pig CVR data measured by Girouard et al. (19) are added. The experimental data are rescaled by a factor of 0.92 to get the same maximum CV level as measured in human tissue.

A   **B**   **C**

Fig. 12. *A*: spiral wave pattern. The star denotes a measurement electrode at position (100,100), and the circle denotes a measurement electrode at position (200,375). *B*: spiral tip trajectory during 1 s of rotation 8 s after the S2 stimulus. *C*: tip trajectory for a single full rotation showing "Z"-type core.

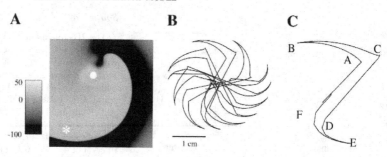

ing: slower recovery dynamics of the fast sodium current, leading to a more gradual CVR that agrees better with available experimental data; differentiated formulations for epicardial and endocardial transient outward current, allowing the modeling of these different cell types; and an L-type calcium current with a fast, dominant, and stable calcium inactivation and slow voltage inactivation dynamics. In Table 3 the most important differences between the major ionic currents in our model and the LR, PB, and CRN models are summarized together with their electrophysiological consequences. In addition, conductance parameters used for these currents in our model and the other models are compared.

Our model reproduces three different cell types: endocardial, epicardial, and M cells with different characteristic action potential morphologies and rate dependencies. The APDR of

our model closely matches experimentally obtained APDR curves. In addition, the CVR of our model resembles experimentally obtained CVR curves, which are currently only available for animal cardiac tissue, much closer than CVR curves generated with models using the Luo-Rudy I_{Na} formulation. Both APDR and CVR are very important determinants for the stability of reentrant arrhythmias. Our calcium dynamics formulation is of a complexity comparable to that of the LR, PB, and CRN models and allows us to simulate a realistic calcium transient and a typical positive human contraction staircase.

The feasibility of spatial simulations is demonstrated with a simulation of a reentrant spiral wave in a 2D sheet of epicardial tissue. The period of spiral wave rotation is 254 ms, vs. 304 ms in the reduced PB model (3), because of the shorter APD in our model. The spiral wave tip trajectory is somewhat different from the linear tip trajectory found by Bernus et al. (3), because of the larger horizontal parts of the Z-type core. Another difference is that in the model by Bernus et al. (3) the core pattern has a cross section of ~5 cm, whereas in our model the core pattern has a cross section of ~3 cm, again because of the shorter APD in our model.

Limitations

A limitation of our model is that differences in APD and rate dependence between M cells and epicardial and endocardial cells in our model are smaller than the experimentally observed differences. This is because of the limited current knowledge of basic electrophysiological differences between M cells and the other cell types causing these differences. If these differences are further characterized, they can be easily incorporated in our model in terms of different current densities and/or dynamics for M cells, similar to what already has been done for I_{Ks} density differences.

We put considerable effort in obtaining, evaluating, rescaling to physiological conditions, and fitting of experimental data to get a model closely resembling true human ventricular cells. However, limitations are unavoidable because of the limited availability of data, the extensive variability among experimental data, the considerable variation in experimental conditions, and the potentially deleterious effects of cell isolation procedures used in voltage-clamp experiments.

There were no data available on I_{CaL} activation time constants, so formulations from another model based on animal experiments had to be used. The precise nature of calcium-mediated inactivation of the L-type calcium current is unknown. We used a simple description with a constant-valued time constant and a dependence on intracellular calcium only,

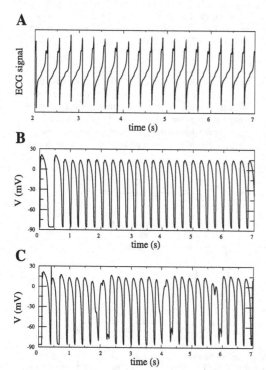

Fig. 13. *A*: ECG recorded during the spiral wave activity shown in Fig. 12*A*. *B* and *C*: membrane potential (*V*) recorded at point (100,100) (*B*) and point (200,375) (*C*) during 7 s after the application of S1. Point (100,100) corresponds to the star, and point (200,375) corresponds to the circle in Fig. 12*A*.

Table 3. *Summary of most important differences in major currents between our model and original LR phase 2, PB, and CRN models and electrophysiological properties for which they are relevant*

Current	Our Model	LR Model	PB Model	CRN Model
I_{Na}	Slow τ_j recovery from inactivation More gradual CV restitution	Faster τ_j recovery from inactivation Steep CV restitution	Same as in LR model	Same as in LR model
G_{Na}	14.838 nS/pF	16 nS/pF	16 nS/pF	7.8 nS/pF
I_{CaL}	Typical τ_f shape Slow f gate inactivation Switchlike f_{Ca}, large calcium inactivation Calcium-dominated inactivation GHK current model	Atypical τ_f shape Fast f gate inactivation f_{Ca} gradually declining Voltage-dominated inactivation GHK current model	Similar to LR model Linear current model	Atypical τ_f shape Slow f gate inactivation f_{Ca} gradually declining Calcium-dominated inactivation Linear current model
G_{CaL}	1.75^{-4} cm$^3\cdot\mu$F$^{-1}\cdot$s^{-1}	5.4^{-4} cm$^3\cdot\mu$F$^{-1}\cdot$s^{-1}	0.064 nS/pF	0.1238 nS/pF
I^{to}	Different G_{to} and recovery from inactivation dynamics for epicardial and endocardial cells	No I_{to} current	Only epicardial I_{to}	Only epicardial I_{to}
G_{to}	0.294 nS/pF (epi), 0.073 nS/pF (endo)		0.3 nS/pF	0.1652 nS/pF
I_{Ks}	Slower activation and faster inactivation	No I_{Ks} current	Faster activation and slower inactivation	Similar to PB model
G_{Ks}	0.245 nS/pF (epi/endo), 0.062 nS/pF (M cell)		0.02 nS/pF	0.129 nS/pF
I_{Kr}	Time-dependent inactivation gate Initial peak of current during AP	No I_{Kr} current	Immediate inactivation gate No initial peak	Similar to PB model
G_{Kr}	0.096 nS/pF		0.015 nS/pF	0.0294 nS/pF

LR, Luo-Rudy; PB, Priebe-Beuckelman; CRN, Courtemanche and coworkers; GHK, Goldmann-Hodgkin-Katz.

although data suggest that extracellular calcium also plays a role in inactivation (60). Finally, no data were available on the nature of calcium dynamics under high frequencies. We assumed calcium dynamics to stay stable under high frequencies. In addition, we assumed CICR to depend in a saturating manner on sarcoplasmic reticulum calcium content.

There is no agreement of experimental data on steady-state inactivation of I_{Na}. Some researchers give a value of -85 mV for the voltage of half-inactivation (44, 55, 64), whereas others give a value of around -65 mV (30, 40, 58). Rather than making a compromise, we decided to base our steady-state inactivation curves on data from Nagatomo et al. (44) because we also used their data for steady-state activation and time constants and wanted to maintain consistency.

Experiments are often performed at different temperatures. This is especially true for I_{Na} voltage-clamp experiments, where the temperature is usually below the physiological temperature to slow dynamics and limit current size. We derived a single Q_{10} factor and used this to rescale all I_{Na} time constants to 37°. Because it is not clear how steady-state curves change with temperature, we performed a linear extrapolation of the half-(in)activation voltages and slopes.

Experiments are also performed in the presence of different combinations of pharmacological agents used to suppress other currents. It is known that some of these chemicals have an impact on the dynamics of the measured current, e.g., Cd^{2+} used to block L-type calcium current is known to influence steady-state curves of I_{to} current. We corrected experimental steady-state curves of I_{to} for the presence of Cd^{2+} by shifting them 10 mV in the positive direction.

To perform single-cell voltage-clamp experiments, cells must be isolated. These isolation procedures can have profound effects on the density of particular currents. A clear example of this are the I_{Ks} and I_{Kr} currents. I_{Ks} and I_{Kr} current densities found during voltage-clamp experiments are in sharp contrast with the clear relevance of these currents for action potential duration and restitution and the different behavior of M cells. We therefore decided to apply current densities for I_{Ks} and I_{Kr}

that are substantially larger than experimentally measured densities and that allow us to realistically simulate the consequences of I_{Kr} block and the behavior of M cells.

In conclusion, we propose a new model for human ventricular epicardial, endocardial, and M cells based on ionic currents measured in human ventricular cells or measured in expression experiments using human cardiac channels. The model reproduces a number of experimental observations, ranging from voltage-clamp current traces and I-V curves, to AP morphology, as well as APD and CV restitution curves and the contraction staircase. Because of its relative computational simplicity, the model is suitable for application in large-scale spatial simulations, which are necessary for investigating the dynamics of reentrant arrhythmias. The latter is illustrated with the simulation of a reentrant spiral wave in a 2D sheet of epicardial ventricular tissue.

APPENDIX

Reversal Potentials

$$E_X = \frac{RT}{zF} \log \frac{X_o}{X_i} \quad \text{for} \quad X = Na^+, K^+, Ca^{2+} \quad (25)$$

$$E_{Ks} = \frac{RT}{F} \log \frac{K_o + p_{KNa}Na_o}{K_i + p_{KNa}Na_i} \quad (26)$$

Fast Na$^+$ Current

$$I_{Na} = G_{Na}m^3hj(V - E_{Na}) \quad (27)$$

$$m_\infty = \frac{1}{[1 + e^{(-56.86 - V)/9.03}]^2} \quad (28)$$

$$\alpha_m = \frac{1}{1 + e^{(-60 - V)/5}} \quad (29)$$

$$\beta_m = \frac{0.1}{1 + e^{(V+35)/5}} + \frac{0.1}{1 + e^{(V-50)/200}} \quad (30)$$

$$\tau_m = \alpha_m\beta_m \quad (31)$$

$$h_\infty = \frac{1}{[1 + e^{(V+71.55)/7.43}]^2} \tag{32}$$

$$\alpha_h = 0 \quad \text{if} \quad V \geq -40 \tag{33}$$

$$\alpha_h = 0.057 e^{-(V+80)/6.8} \qquad \text{otherwise}$$

$$\beta_h = \frac{0.77}{0.13[1 + e^{-(V+10.66)/11.1}]} \quad \text{if} \quad V \geq -40 \tag{34}$$

$$\beta_h = 2.7 e^{0.079V} + 3.1 \times 10^5 e^{0.3485V} \qquad \text{otherwise}$$

$$\tau_h = \frac{1}{\alpha_h + \beta_h} \tag{35}$$

$$j_\infty = \frac{1}{[1 + e^{(V+71.55)/7.43}]^2} \tag{36}$$

$$\alpha_j = 0 \quad \text{if} \quad V \geq -40 \tag{37}$$

$$\alpha_j = \frac{\left(\begin{array}{c}-2.5428 \times 10^4 e^{0.2444V} - 6.948 \\ \times 10^{-6} e^{-0.04391V}\end{array}\right)(V + 37.78)}{1 + e^{0.311(V+79.23)}} \quad \text{otherwise}$$

$$\beta_j = \frac{0.6 e^{0.057V}}{1 + e^{-0.1(V+32)}} \quad \text{if} \quad V \geq -40 \tag{38}$$

$$\beta_j = \frac{0.02424 e^{-0.01052V}}{1 + e^{-0.1378(V+40.14)}} \qquad \text{otherwise}$$

$$\tau_j = \frac{1}{\alpha_j + \beta_j} \tag{39}$$

L-type Ca²⁺ Current

$$I_{\text{CaL}} = G_{\text{CaL}} d f f_{\text{Ca}} 4 \frac{VF^2}{RT} \frac{\text{Ca}_i e^{2VF/RT} - 0.341 \text{Ca}_o}{e^{2VF/RT} - 1} \tag{40}$$

$$d_\infty = \frac{1}{1 + e^{(-5-V)/7.5}} \tag{41}$$

$$\alpha_d = \frac{1.4}{1 + e^{(-35-V)/13}} + 0.25 \tag{42}$$

$$\beta_d = \frac{1.4}{1 + e^{(V+5)/5}} \tag{43}$$

$$\gamma_d = \frac{1}{1 + e^{(50-V)/20}} \tag{44}$$

$$\tau_d = \alpha_d \beta_d + \gamma_d \tag{45}$$

$$f_\infty = \frac{1}{1 + e^{(V+20)/7}} \tag{46}$$

$$\tau_f = 1125 e^{-(V+27)^2/240} + \frac{165}{1 + e^{(25-V)/10}} + 80 \tag{47}$$

$$\alpha_{f\text{ca}} = \frac{1}{1 + (\text{Ca}_i/0.000325)^8} \tag{48}$$

$$\beta_{f\text{ca}} = \frac{0.1}{1 + e^{(\text{Ca}_i - 0.0005)/0.0001}} \tag{49}$$

$$\gamma_{f\text{ca}} = \frac{0.2}{1 + e^{(\text{Ca}_i - 0.00075)/0.0008}} \tag{50}$$

$$f_{\text{ca}\infty} = \frac{\alpha_{f\text{ca}} + \beta_{f\text{ca}} + \gamma_{f\text{ca}} + 0.23}{1.46} \tag{51}$$

$$\tau_{f\text{ca}} = 2 \text{ ms} \tag{52}$$

$$\frac{df_{\text{ca}}}{dt} = k \frac{f_{\text{ca}\infty} - f_{\text{ca}}}{\tau_{f\text{ca}}} \tag{53}$$

$$k = 0 \quad \text{if} \quad f_{\text{ca}\infty} > f_{\text{ca}} \quad \text{and} \quad V > -60 \text{ mV}$$
$$k = 1 \qquad \text{otherwise} \tag{54}$$

Transient Outward Current

$$I_{\text{to}} = G_{\text{to}} rs(V - E_K) \tag{55}$$

For all cell types

$$r_\infty = \frac{1}{1 + e^{(20-V)/6}} \tag{56}$$

$$\tau_r = 9.5 e^{-(V+40)^2/1800} + 0.8 \tag{57}$$

For epicardial and M cells

$$s_\infty = \frac{1}{1 + e^{(V+20)/5}} \tag{58}$$

$$\tau_s = 85 e^{-(V+45)^2/320} + \frac{5}{1 + e^{(V-20)/5}} + 3 \tag{59}$$

For endocardial cells

$$s_\infty = \frac{1}{1 + e^{(V+28)/5}} \tag{60}$$

$$\tau_s = 1{,}000 e^{-(V+67)^2/1{,}000} + 8 \tag{61}$$

Slow Delayed Rectifier Current

$$I_{\text{Ks}} = G_{\text{Ks}} x_s^2 (V - E_{\text{Ks}}) \tag{62}$$

$$x_{s\infty} = \frac{1}{1 + e^{(-5-V)/14}} \tag{63}$$

$$\alpha_{xs} = \frac{1{,}100}{\sqrt{1 + e^{(-10-V)/6}}} \tag{64}$$

$$\beta_{xs} = \frac{1}{1 + e^{(V-60)/20}} \tag{65}$$

$$\tau_{xs} = \alpha_{xs} \beta_{xs} \tag{66}$$

Rapid Delayed Rectifier Current

$$I_{\text{Kr}} = G_{\text{Kr}} \sqrt{\frac{K_o}{5.4}} x_{r1} x_{r2}(V - E_K) \tag{67}$$

$$x_{r1\infty} = \frac{1}{1 + e^{(-26-V)/7}} \tag{68}$$

$$\alpha_{xr1} = \frac{450}{1 + e^{(-45-V)/10}} \tag{69}$$

$$\beta_{xr1} = \frac{6}{1 + e^{(V+30)/11.5}} \tag{70}$$

$$\tau_{xr1} = \alpha_{xr1} \beta_{xr1} \tag{71}$$

$$x_{r2\infty} = \frac{1}{1 + e^{(V+88)/24}} \tag{72}$$

$$\alpha_{xr2} = \frac{3}{1 + e^{(-60-V)/20}} \tag{73}$$

$$\beta_{xr2} = \frac{1.12}{1 + e^{(V-60)/20}} \tag{74}$$

$$\tau_{xr2} = \alpha_{xr2}\beta_{xr2} \tag{75}$$

Inward Rectifier K^+ Current

$$I_{K1} = G_{K1}\sqrt{\frac{K_o}{5.4}}x_{K1\infty}(V - E_K) \tag{76}$$

$$\alpha_{K1} = \frac{0.1}{1 + e^{0.06(V - E_K - 200)}} \tag{77}$$

$$\beta_{K1} = \frac{3e^{0.0002(V - E_K + 100)} + e^{0.1(V - E_K - 10)}}{1 + e^{-0.5(V - E_K)}} \tag{78}$$

$$x_{K1\infty} = \frac{\alpha_{K1}}{\alpha_{K1} + \beta_{K1}} \tag{79}$$

Na^+/Ca^{2+} Exchanger Current

$$I_{NaCa} = k_{NaCa}\frac{e^{\gamma VF/RT}Na_i^3Ca_o - e^{(\gamma-1)VF/RT}Na_o^3Ca_i\alpha}{(K_{mNai}^3 + Na_o^3)(K_{mCa} + Ca_o)(1 + k_{sat}e^{(\gamma-1)VF/RT})} \tag{80}$$

Na^+/K^+ Pump Current

$$I_{NaK} =$$

$$P_{NaK}\frac{K_oNa_i}{(K_o + K_{mK})(Na_i + K_{mNa})(1 + 0.1245e^{-0.1VF/RT} + 0.0353e^{-VF/RT})} \tag{81}$$

I_{pCa}

$$I_{pCa} = G_{pCa}\frac{Ca_i}{K_{pCa} + Ca_i} \tag{82}$$

I_{pK}

$$I_{pK} = G_{pK}\frac{V - E_K}{1 + e^{(25 - V)/5.98}} \tag{83}$$

Background Currents

$$I_{bNa} = G_{bNa}(V - E_{Na}) \tag{84}$$

$$I_{bCa} = G_{bCa}(V - E_{Ca}) \tag{85}$$

Calcium Dynamics

$$I_{leak} = V_{leak}(Ca_{sr} - Ca_i) \tag{86}$$

$$I_{up} = \frac{V_{maxup}}{1 + K_{up}^2/Ca_i^2} \tag{87}$$

$$I_{rel} = \left(a_{rel}\frac{Ca_{sr}^2}{b_{rel}^2 + Ca_{sr}^2} + c_{rel}\right)dg \tag{88}$$

$$g_\infty = \frac{1}{1 + Ca_i^6/0.00035^6} \quad \text{if} \quad Ca_i \leq 0.00035$$
$$g_\infty = \frac{1}{1 + Ca_i^{16}/0.00035^{16}} \quad \text{otherwise} \tag{89}$$

$$\tau_g = 2\,\text{ms} \tag{90}$$

$$\frac{dg}{dt} = k\frac{g_\infty - g}{\tau_g} \tag{91}$$

$$k = 0 \quad \text{if} \quad g_\infty > g \quad \text{and} \quad V > -60\,\text{mV}$$
$$k = 1 \quad \text{otherwise} \tag{92}$$

$$Ca_{ibufc} = \frac{Ca_i \times Buf_c}{Ca_i + K_{bufc}} \tag{93}$$

$$\frac{dCa_{itotal}}{dt} = -\frac{I_{CaL} + I_{bCa} + I_{pCa} - 2I_{NaCa}}{2V_CF} + I_{leak} - I_{up} + I_{rel} \tag{94}$$

$$Ca_{srbufsr} = \frac{Ca_{sr} \times Buf_{sr}}{Ca_{sr} + K_{bufsr}} \tag{95}$$

$$\frac{dCa_{srtotal}}{dt} = \frac{V_C}{V_{SR}}(-I_{leak} + I_{up} - I_{rel}) \tag{96}$$

Sodium and Potassium Dynamics

$$\frac{dNa_i}{dt} = -\frac{I_{Na} + I_{bNa} + 3I_{NaK} + 3I_{NaCa}}{V_CF} \tag{97}$$

$$\frac{dK_i}{dt} = -\frac{I_{K1} + I_{to} + I_{Kr} + I_{Ks} - 2I_{NaK} + I_{pK} + I_{stim} - I_{ax}}{V_CF} \tag{98}$$

ACKNOWLEDGMENTS

We are thankful to Dr. O. Bernus and Dr. R. Wilders for valuable discussions.

This work was initiated during a visit of A. V. Panfilov at the Isaac Newton Institute for Mathematical Sciences in 2001.

GRANTS

This work was funded by the Netherlands Organization for Scientific Research (NWO) through Grant 620061351 of the Research Council for Physical Sciences (EW) (to K. H. W. J. ten Tusscher), the Netherlands National Computer Facilities Foundation (NCF) through Grant SG-095, the British Heart Foundation, the British Medical Research Council, the Welcome Trust, and Physiome Sciences (to D. Noble and P. J. Noble)

REFERENCES

1. **Beeler GW and Reuter H.** Reconstruction of the action potential of ventricular myocardial fibers. *J Physiol* 268: 177–210, 1977.
2. **Benitah J, Bailly P, D'Agrosa M, Da Ponte J, Delgado C, and Lorente P.** Slow inward current in single cells isolated from adult human ventricles. *Pflügers Arch* 421: 176–187, 1992.
3. **Bernus O, Wilders R, Zemlin CW, Verschelde H, and Panfilov AV.** A computationally efficient electrophysiological model of human ventricular cells. *Am J Physiol Heart Circ Physiol* 282: H2296–H2308, 2002.
4. **Beuckelmann DJ, Nabauer M, and Erdmann E.** Characteristics of calcium-current in isolated human ventricular myocytes from patients with terminal heart failure. *J Mol Cell Cardiol* 23: 929–937, 1991.
5. **Boyett MR and Fedida D.** A computer simulation of the effect of heart rate on ion concentrations in the heart. *J Theor Biol* 132: 15–27, 1988.
6. **Cao J, Qu Z, Kim Y, Wu T, Garfinkel A, Weiss JN, Karagueuzian HS, and Chen P.** Spatiotemporal heterogeneity in the induction of ventricular fibrillation by rapid pacing, importance of cardiac restitution properties. *Circ Res* 84: 1318–1331, 1999.
7. **Chudin E, Goldhaber J, Garfinkel J, Weiss A, and Kogan B.** Intracellular Ca^{2+} dynamics and the stability of ventricular tachycardia. *Biophys J* 77: 2930–2941, 1999.
8. **Courtemanche M, Ramirez RJ, and Nattel S.** Ionic mechanisms underlying human atrial action potential properties: insights from a mathematical model. *Am J Physiol Heart Circ Physiol* 275: H301–H321, 1998.
9. **Drouin E, Charpentier F, Gauthier C, Laurent K, and Le Marec H.** Electrophysiologic characteristics of cells spanning the left ventricular wall of human heart: evidence for the presence of M cells. *J Am Coll Cardiol* 26: 185–192, 1995.
10. **Drouin E, Lande G, and Charpentier F.** Amiodarone reduces transmural heterogeneity of repolarization in the human heart. *J Am Coll Cardiol* 32: 1063–1067, 1998.
11. **Ebihara L and Johnson EA.** Fast sodium current in cardiac muscle, a quantitative description. *Biophys J* 32: 779–790, 1980.
12. **Efimov IR, Krinsky VI, and Jalife J.** Dynamics of rotating vortices in the Beeler-Reuter model of cardiac tissue. *Chaos Solitons Fractals* 5: 513–526, 1995.
13. **Endresen LP, Hall K, Hoye JS, and Myrheim J.** A theory for the membrane potential of living cells. *Eur Biophys J* 29: 90–103, 2000.
14. **Endresen LP and Skarland N.** Limit cycle oscillations in pacemaker cells. *IEEE Trans Biomed Eng* 47: 1134–1137, 2000.

15. **Fast VG, Efimov IR, and Krinsky VI.** Transition from circular to linear cores in excitable media. *Phys Lett A* 151: 157–161, 1990.

16. **Fenton F and Karma A.** Vortex dynamics in three-dimensional continuous myocardium with fiber rotation: filament instability and fibrillation. *Chaos* 8: 20–47, 1998.

17. **Garfinkel A, Kim YH, Voroshilovsky O, Qu Z, Kil JR, Lee MH, Karagueuzian HS, Weiss JN, and Chen PS.** Preventing ventricular fibrillation by flattening cardiac restitution. *Proc Natl Acad Sci USA* 97: 6061–6066, 2000.

18. **Gintant GA.** Characterization and functional consequences of delayed rectifier current transient in ventricular repolarization. *Am J Physiol Heart Circ Physiol* 278: H806–H817, 2000.

19. **Girouard SD, Pastore JM, Laurita KR, Greogry KW, and Rosenbaum DS.** Optical mapping in a new guinea pig model of ventricular tachycardia reveals mechanisms for multiple wavelengths in a single reentrant circuit. *Circulation* 93: 603–613, 1996.

20. **Grantham CJ and Cannell MB.** Ca^{2+} influx during the cardiac action potential in guinea pig ventricular myocytes. *Circ Res* 79: 194–200, 1996.

21. **Greenstein JL and Winslow RL.** An integrative model of the cardiac ventricular myocyte incorporating local control of Ca^{2+} release. *Biophys J* 83: 2918–2945, 2002.

22. **Greenstein JL, Wu R, Po S, Tomaselli GF, and Winslow RL.** Role of the calcium-independent transient outward current I_{to1} in shaping action potential morphology and duration. *Circ Res* 87: 1026–1033, 2000.

23. **Hodgkin AL and Huxley AF.** A quantitative description of membrane current and its application to conduction and excitation in nerve. *J Physiol* 117: 500–544, 1952.

24. **Hund TJ, Kucera JP, Otani NF, and Rudy Y.** Ionic charge conservation and long-term steady state in the Luo-Rudy dynamic cell model. *Biophys J* 81: 3324–3331, 2001.

25. **Iost N, Virag L, Opincariu M, Szecsi J, Varro A, and Papp JG.** Delayed rectifier potassium current in undiseased human ventricular myocytes. *Cardiovasc Res* 40: 508–515, 1998.

26. **Beuckelmann DJ, Nabauer M, and Erdmann E.** Intracellular calcium handling in isolated ventricular myocytes from patients with terminal heart failure. *Circulation* 94: 992–1002, 1992.

27. **Janvier NC and Boyett MR.** The role of the Na-Ca exchange current in the cardiac action potential. *Cardiovasc Res* 32: 69–84, 1996.

28. **Johnson JP, Mullins FM, and Bennet PB.** Human Ether-a-go-go-related gene K^+ channel gating probed with extracellular Ca^{2+}, evidence for two distinct voltage sensors. *J Gen Physiol* 113: 565–580, 1999.

29. **Jongsma HJ and Wilders R.** Gap junctions in cardiovascular disease. *Circ Res* 86: 1193–1197, 2000.

30. **Kambouris NG, Nuss HB, Johns DC, Marban E, Tomaselli GF, and Balser JR.** A revised view of cardiac sodium channel "blockade" in the long-QT syndrome. *J Clin Invest* 105: 1133–1140, 2000.

31. **Karma A.** Electrical alternans and spiral wave breakup in cardiac tissue. *Chaos* 4: 461–472, 1994.

32. **Koller ML, Riccio ML, and Gilmour RF Jr.** Dynamic restitution of action potential duration during electrical alternans and ventricular fibrillation. *Am J Physiol Heart Circ Physiol* 275: H1635–H1642, 1998.

33. **Koumi S, Backer CL, and Arentzen CE.** Characterization of inwardly rectifying K^+ channel in human cardiac myocytes. *Circulation* 92: 164–174, 1995.

34. **Li G, Feng J, Yue L, Carrier M, and Nattel S.** Evidence for two components of delayed rectifier K^+ current in human ventricular myocytes. *Circ Res* 78: 689–696, 1996.

35. **Li G and Nattel S.** Properties of human atrial I_{Ca} at physiological temperatures and relevance to action potential. *Am J Physiol Heart Circ Physiol* 272: H227–H235, 1997.

36. **Li GR, Feng J, Yue L, and Carrier M.** Transmural heterogeneity of action potentials and I_{to1} in myocytes isolated from the human right ventricle. *Am J Physiol Heart Circ Physiol* 275: H369–H377, 1998.

37. **Luo C and Rudy Y.** A model of the ventricular cardiac action potential, depolarization, repolarization, and their interaction. *Circ Res* 68: 1501–1526, 1991.

38. **Luo C and Rudy Y.** A dynamic model of the cardiac ventricular action potential. I. simulations of ionic currents and concentration changes. *Circ Res* 74: 1071–1096, 1994.

39. **Magyar J, Szentandrassy N, Banyasz T, Fulop L, Varro A, and Nanasi PP.** Effects of thymol on calcium and potassium currents in canine and human ventricular cardiomyocytes. *Br J Pharmacol* 136: 330–338, 2002.

40. **Makita N, Shirai N, Wang DW, Sasaki K, George AL, Kanno M, and Kitabatake A.** Cardiac Na^+ channel dysfunction in Brugada syndrome is aggravated by β_1-subunit. *Circulation* 101: 54–60, 2000.

41. **Mewes T and Ravens U.** L-type calcium currents of human myocytes from ventricle of non-failing and failing hearts and atrium. *J Mol Cell Cardiol* 26: 1307–1320, 1994.

42. **Morgan JM, Cunningham D, and Rowland E.** Dispersion of monophasic action potential duration: demonstrable in humans after premature ventricular extrastimulation but not in steady state. *J Am Coll Cardiol* 19: 1244–1253, 1992.

43. **Nabauer M, Beuckelmann DJ, Uberfuhr P, and Steinbeck G.** Regional differences in current density and rate-dependent properties of the transient outward current in subepicardial and subendocardial myocytes of human left ventricle. *Circulation* 93: 168–177, 1996.

44. **Nagatomo T, Fan Z, Ye B, Tonkovich GS, January CT, Kyle JW, and Makielski JC.** Temperature dependence of early and late currents in human cardiac wild-type and long Q-T ΔKPQ Na^+ channels. *Am J Physiol Heart Circ Physiol* 275: H2016–H2024, 1998.

45. **Noble D, Varghese A, Kohl P, and Noble P.** Improved guinea-pig ventricular cell model incorporating a diadic space, I_{Kr} and I_{Ks}, and length- and tension-dependent processes. *Can J Cardiol* 14: 123–134, 1998.

46. **Pelzmann B, Schaffer P, Bernhart E, Lang P, Machler H, Rigler B, and Koidl B.** L-type calcium current in human ventricular myocytes at a physiological temperature from children with tetralogy of Fallot. *Cardiovasc Res* 38: 424–432, 1998.

47. **Pereon Y, Demolombe S, Baro I, Drouin E, Charpentier F, and Escande D.** Differential expression of KvLQT1 isoforms across the human ventricular wall. *Am J Physiol Heart Circ Physiol* 278: H1908–H1915, 2000.

48. **Pieske B, Maier LS, Bers DM, and Hasenfuss G.** Ca^{2+} handling and sarcoplasmic reticulum Ca^{2+} content in isolated failing and non-failing human myocardium. *Circ Res* 85: 38–46, 1999.

49. **Pieske B, Maier LS, Piacentino V, Weisser J, Hasenfuss G, and Houser S.** Rate dependence of $(Na^+)_i$ and contractility in nonfailing and failing human myocardium. *Circ Res* 106: 447–453, 2002.

50. **Plonsey R and Barr RC.** *Bioelectricity*. New York: Plenum, 1989.

51. **Priebe L and Beuckelmann DJ.** Simulation study of cellular electric properties in heart failure. *Circ Res* 82: 1206–1223, 1998.

52. **Qu Z, Weiss JN, and Garfinkel A.** Cardiac electrical restitution properties and stability of reentrant spiral waves: a simulation study. *Am J Physiol Heart Circ Physiol* 276: H269–H283, 1999.

53. **Rice JJ, Jafri MS, and Winslow RL.** Modeling gain and gradedness of Ca^{2+} release in the functional unit of the cardiac diadic space. *Biophys J* 77: 1871–1884, 1999.

54. **Rush S and Larsen H.** A practical algorithm for solving dynamic membrane equations. *IEEE Trans Biomed Eng* 25: 389–392, 1978.

55. **Sakakibara Y, Wasserstrom JA, Furukawa T, Jia H, Arentzen CE, Hartz RS, and Singe DH.** Characterization of the sodium current in single human atrial myocytes. *Circ Res* 71: 535–546, 1992.

56. **Sakmann BFAS, Spindler AJ, Bryant SM, Linz KW, and Noble D.** Distribution of a persistent sodium current across the ventricular wall in guinea pig. *Circ Res* 87: 910–914, 2000.

57. **Schmidt U, Hajjar RJ, Helm PA, Kim CS, Doye AA, and Gwathmey JK.** Contribution of abnormal sarcoplasmic reticulum ATPase activity to systolic and diastolic dysfunction in human heart failure. *J Mol Cell Cardiol* 30: 1929–1937, 1998.

58. **Schneider M, Proebstle T, Hombach V, Hannekum A, and Rudel R.** Characterization of the sodium currents in isolated human cardiocytes. *Pflügers Arch* 428: 84–90, 1994.

59. **Smith PL and Yellen G.** Fast and slow voltage sensor movements in HERG potassium channels. *J Gen Physiol* 119: 275–293, 2002.

60. **Sun H, Leblanc N, and Nattel S.** Mechanisms of inactivation of L-type calcium channels in human atrial myocytes. *Am J Physiol Heart Circ Physiol* 272: H1625–H1635, 1997.

61. **Taggart P, Sutton PMI, Opthof T, Coronel R, Trimlett R, Pugsley W, and Kallis P.** Inhomogeneous transmural conduction during early ischemia in patients with coronary artery disease. *J Mol Cell Cardiol* 32: 621–639, 2000.

62. **Virag L, Iost N, Opincariu M, Szolnoky J, Szecsi J, Bogats G, Szenohradszky P, Varro A, and Papp JP.** The slow component of the delayed rectifier potassium current in undiseased human ventricular myocytes. *Cardiovasc Res* 49: 790–797, 2001.

63. **Viswanathan PC, Bezzina CR, George AL, Roden DM, Wilde AAM, and Balser JR.** Gating-dependent mechanism for flecainide action in SCN5A-linked arrhythmia syndromes. *Circulation* 104: 1200–1205, 2001.

64. **Wan X, Chen S, Sadeghpour A, Wang Q, and Kirsch GE.** Accelerated inactivation in a mutant Na$^+$ channel associated with idiopathic ventricular fibrillation. *Am J Physiol Heart Circ Physiol* 280: H354–H360, 2001.

65. **Wang DW, Makita N, Kirabatake A, Balser JR, and George AL.** Enhanced Na$^+$ channel intermediate inactivation in Brugada syndrome. *Circ Res* 87: e37–e43, 2000.

66. **Wang Z, Fermini B, and Nattel S.** Rapid and slow components of delayed rectifier current in human atrial myocytes. *Cardiovasc Res* 28: 1540–1546, 1994.

67. **Watanabe MA, Fenton FH, Evans SJ, Hastings HM, and Karma A.** Mechanisms for discordant alternans. *J Cardiovasc Electrophysiol* 12: 196–206, 2001.

68. **Wettwer E, Amos GJ, Posival H, and Ravens U.** Transient outward current in human ventricular myocytes of subepicardial and subendocardial origin. *Circ Res* 75: 473–482, 1994.

69. **Xie F, Qu Z, Garfinkel A, and Weiss JN.** Effects of simulated ischemia on spiral wave stability. *Am J Physiol Heart Circ Physiol* 280: H1667–H1673, 2001.

70. **Yue L, Feng J, Li GR, and Nattel S.** Transient outward and delayed rectifier currents in canine atrium: properties and role of isolation methods. *Am J Physiol Heart Circ Physiol* 270: H2157–H2168, 1996.

71. **Zeng J, Laurita KR, Rosenbaum DS, and Rudy Y.** Two components of the delayed rectifier K$^+$ current in ventricular myocytes of the guinea pig type. Theoretical formulation and their role in repolarization. *Circ Res* 77: 140–152, 1995.

72. **Zhou Z, Gong Q, Ye B, Fan Z, Makielski JC, Robertson GA, and January CT.** Properties of HERG channels stably expressed in HEK 293 cells studied at physiological temperatures. *Biophys J* 74: 230–241, 1973.

73. **Zygmunt AC, Eddlestone GT, Thomas GP, Nesterenko VV, and Antzelevitch C.** Larger late sodium conductance in M cells contributes to electrical heterogeneity in canine ventricle. *Am J Physiol Heart Circ Physiol* 281: H689–H697, 2001.

74. **Zygmunt AC, Goodrow RJ, and Antzelevitch C.** I_{NaCa} contributes to electrical heterogeneity within the canine ventricle. *Am J Physiol Heart Circ Physiol* 278: H1671–H1678, 2000.

J Physiol 580.1 (2007) pp 15–22

Hodgkin–Huxley–Katz Prize Lecture

From the Hodgkin–Huxley axon to the virtual heart

Denis Noble

Department of Physiology, Anatomy and Genetics, Parks Road, Oxford OX1 3PT, UK

Experimentally based models of the heart have been developed since 1960, starting with the discovery and modelling of potassium channels. The early models were based on extensions of the Hodgkin–Huxley nerve impulse equations. The first models including calcium balance and signalling were made in the 1980s and have now reached a high degree of physiological detail. During the 1990s these cell models have been incorporated into anatomically detailed tissue and organ models to create the first virtual organ, the Virtual Heart. With over 40 years of interaction between simulation and experiment, the models are now sufficiently refined to begin to be of use in drug development.

(Received 16 August 2006; accepted 27 September 2006; first published online 28 September 2006)

Corresponding author D. Noble: Department of Physiology, Anatomy and Genetics, Parks Road, Oxford, OX1 3PT, UK.   Email: denis.noble@physiol.ox.ac.uk

Extending the Hodgkin–Huxley equations

Modelling excitable cells took a giant leap forward when Alan Hodgkin & Andrew Huxley (1952) published their equations for nerve conduction in the giant axon of the squid, an achievement for which they received the Nobel Prize for Physiology and Medicine in 1963. It was the first model to use mathematical reconstruction of experimentally determined kinetics of ion channel transport and gating, rather than abstract equations (e.g. Van der Pol & Van der Mark, 1928), and it correctly predicted the shape of the action potential, the impedance changes and the conduction velocity. As a brilliant demonstration of the need for biological modelling to respect the details of experimental results to achieve accurate predictions, I was fascinated by this work as a young medical student at University College London in the 1950s. As soon as Otto Hutter and I (Hutter & Noble, 1960; Hall *et al.* 1963) had demonstrated the existence of two kinds of potassium channels in Purkinje fibres of the heart, the inward rectifier i_{K1} and the delayed rectifier i_K, I was keen to see whether the Hodgkin–Huxley model could be adapted to apply to the cardiac action potential and pacemaker rhythm.

Fitting equations to the data we had on the K^+ channels was the first stage. For the delayed rectifier I used slowed-down versions of the Hodgkin–Huxley (HH) K^+ current equations, representing the fact that a process that takes a few milliseconds in nerve requires hundreds

of milliseconds in heart. For the inward rectifier, I fitted my own equations, including notably and successfully, the dependence of this current on external potassium (Noble, 1965). For the sodium current, I used the Hodgkin–Huxley equations since Weidmann (1956) had shown that the sodium inactivation process was very similar in the heart.

The next problem was how to solve the resulting differential equations. It took Andrew Huxley months with a Brunsviga mechanical calculator to solve for just a few milliseconds of nerve activity. At that rate, solving for hundreds of milliseconds of cardiac activity would have taken years. It was therefore necessary to use one of those huge valve machines, an early electronic computer. Fortunately, there was one in London University, though getting time on it was not easy, nor was the programming. Computer languages had hardly emerged from the endless gibberish of machine code (Noble, 2006*b*).

The first results, however, were disappointing. Despite the fact that the HH equations predict a steady plateau sodium current, later discovered experimentally (Attwell *et al.* 1979), it wasn't large enough during strong depolarizations. In retrospect, we can see that this calculation predicted either that the sodium current in heart differs from that in nerve or that other inward current channels were present. Both predictions are correct, but the demonstration of calcium channels had to wait until Reuter's pioneering work (Reuter, 1967). So, I opted in 1960 for modifying the sodium activation equations to generate current over a wider range of potentials. The inward current equations in the 1960 model are therefore seriously incomplete, not only by omitting calcium channels but also because we now know that

This article is based on the Hodgkin–Huxley–Katz Prize Lecture delivered to the Physiological Society in October 2004.

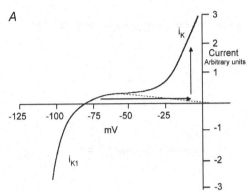

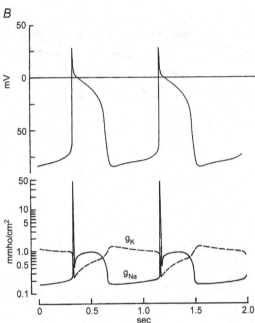

Figure 1. The first cardiac model
A, the first analysis of potassium channel currents in the heart and their incorporation into a heart cell model. Redrawn from Hall *et al.* (1963). The continuous line shows the total membrane current recorded in a Purkinje fibre in a sodium-depleted solution. The inward-rectifying current was identified as i_{K1}, which is extrapolated here as nearly zero at positive potentials. The outward-rectifying current, i_K, is now known to be mostly formed by the component i_{Kr}. The horizontal arrow indicates the trajectory at the beginning of the action potential, while the vertical arrow indicates the time-dependent activation of i_K which initiates repolarization. *B*, sodium and potassium conductance changes computed from the first biophysically detailed model of cardiac cells (Noble, 1962). Two cycles of activity are shown. The conductances are plotted on a logarithmic scale to accommodate the large changes in sodium conductance. Note the persistent level of sodium conductance during the plateau of the action potential, which is about 2% of the peak conductance. Note also the rapid fall in potassium conductance at the beginning of the action potential. This is attributable to the properties of the inward rectifier i_{K1}, and it helps to maintain the long duration of the action potential, and in energy conservation by greatly reducing the ionic exchanges involved. The

there are late components of sodium current that are not predicted by the HH equations (Kiyosue & Arita, 1989; Maltsev *et al.* 1998; Sakmann *et al.* 2000; Zygmunt *et al.* 2001) and that these can play an important role in drug action (Bottino *et al.* 2006; Noble & Noble, 2006).

Nevertheless, the broad outline of the time course of potassium current was successfully reconstructed (Fig. 1). Moreover, the slow time course of decay of i_K readily allowed the equations to account for pacemaker rhythm at about the right frequency. As a first working model, it also illustrated an important property of the electro-physiology of repetitive activity: that pacemaker activity is an integrative characteristic of the system as a whole; there is no 'molecular driver'. This kind of analysis is now one of the major features of the systems biology approach (Noble, 2006a, 2006b). I think I can claim therefore to have been doing the 'new' subject of systems biology for around 45 years!

The 1962 model also revealed the nature of the balance evolution had struck in developing long action potentials in the heart. The inward rectifier, i_{K1}, is energy saving. By greatly reducing potassium ion flow during depolarization, it enables much smaller inward sodium and calcium currents to maintain the depolarization and so requires much less energy expenditure by ion pumps to restore the transmembrane gradients. This is important since the ATP consumption by ionic pumps is significant.

These insights have stood the test of time. They are also features of all subsequent cardiac cell models.

Developments of the cell models

We now know of course that there are many more ion channels and other transporters than were represented in that early model. I will review the key developments according to the decades in which they appeared.

The 1960s revealed three major changes. The first was Reuter's discovery of calcium current in the heart (Reuter, 1967). The second was the discovery with Dick Tsien that there are multiple components of slow changes in potassium currents (Noble & Tsien, 1968, 1969). These two developments formed the basis of the McAllister–Noble–Tsien Purkinje fibre model (McAllister *et al.* 1975), which was later developed by Beeler & Reuter (1977) into the first model of a ventricular cell, which in turn was a precursor to the Luo–Rudy models (Luo & Rudy, 1994, 1994). There was also a third important discovery in the 1960s, which was that of the sodium–calcium exchange

current i_K is then responsible for repolarization. The arrows correspond to those shown in the top panel. These insights into the main potassium current changes have been incorporated into all subsequent models of cardiac cells.

J Physiol 580.1   From axon to virtual heart

in cardiac muscle (Reuter & Seitz, 1969). This was not, however, incorporated into electrophysiological models at that time since the early results suggested that the exchange was electrically neutral (two sodium ions for each calcium ion).

Almost as soon as these 1970s models had been formulated, problems developed. I have reviewed these problems in considerable detail in a previous review lecture (Noble, 1984) and in a recent essay (Noble, 2002b). Here, I will briefly refer to the key developments that became important for the development of later models. The first was DiFrancesco's (1981) discovery that the slow ionic current change in the pacemaker range of potentials in Purkinje fibres was not a pure potassium current. He showed that the channel responsible carries both sodium and potassium ions and is activated by hyperpolarization, not by depolarization. In fact, it is the same mechanism, i_f, as that already discovered in sinus node cells (see reviews by Irisawa *et al.* 1993; DiFrancesco, 1993, 2006). The second was the progressive realization that not all of the 'slow inward current' was attributable to ionic current flow through calcium channels. Some of it was found to be carried by sodium–calcium exchange, for which the stoichiometry was revised upwards to 3 : 1 instead of 2 : 1, which made it electrogenic. These developments were used to develop the DiFrancesco–Noble model (DiFrancesco & Noble, 1985). This was the first model to take into account changes in ionic concentrations during electrical activity and to formulate a model of intracellular calcium signalling. These extensions were necessary not only to account fully for DiFrancesco's experimental findings on i_f but also to explain how the new model could account for previous experimental work on slow current changes at negative potentials.

Although DiFrancesco and I had detailed experimental evidence for most of the major developments in this model, there was one very significant gap. We needed to incorporate an electrogenic sodium–calcium exchanger but we did not know its voltage dependence. We chose to develop some equations that had been developed theoretically by Mullins (1981), and in doing this we had a remarkable stroke of luck. Figure 2B shows the voltage dependence given by these equations at various external sodium concentrations. Figure 2A shows the relations determined experimentally later by Kimura *et al.* (1986, 1987). It is easy to imagine the pleasure DiFrancesco and I experienced when these experimental results appeared.

It was the prediction from this model that there must be inward sodium–calcium exchange current flowing during the action potential that led to the development of the Hilgemann–Noble model (Hilgemann & Noble, 1987) and its single cell version (Earm & Noble, 1990). This model opened up the field to simulation of calcium handling by cardiac cells, an area that has now

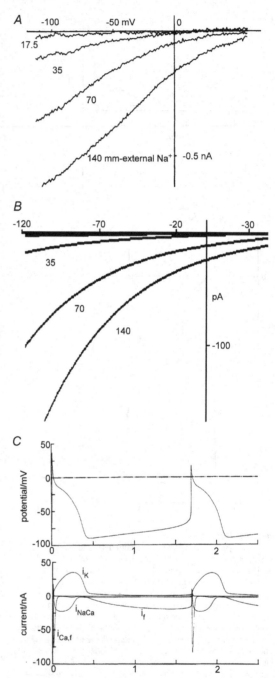

Figure 2. The DiFrancesco-Noble model
A, experimental results obtained by Kimura *et al.* (1987).
B, current–voltage relations given by the equations for sodium–calcium exchange used in the DiFrancesco-Noble model. The curves show the relations at various external sodium concentrations. *C*, action and pacemaker potentials computed from the DiFrancesco–Noble modelling highlighting the roles played by activation of i_f during the pacemaker depolarization and of i_{NaCa} during the action potential.

18   D. Noble   *J Physiol* 580.1

expanded into intense activity in a variety of laboratories (Eisner *et al.* 2000; Bers, 2001; Hinch, 2004; Hinch *et al.* 2006; Sobie *et al.* 2006). The DiFrancesco–Noble and Hilgemann–Noble models became the generic models from which all the modern models of excitation–contraction coupling derive (Luo & Rudy, 1994, 1994; Jafri *et al.* 1998; Noble *et al.* 1998; Winslow *et al.* 1999, 2000).

The Hilgemann–Noble model addressed a number of important questions concerning calcium balance, as follows.

(1) How quickly is calcium balance achieved? Net calcium efflux is established as soon as 20 ms after the beginning of the action potential (Hilgemann, 1986), which was considered to be surprisingly soon. In the model this was achieved by calcium activation of efflux via the Na^+–Ca^{2+} exchanger, thus revealing the time course of one of the important functions of this transporter in the heart. These results apply to the relatively short action potential of the atrium. The details vary according to the location of cells in the heart and which species we are dealing with.

(2) Since the exchanger is electrogenic, where was the current that this would generate and did it correspond to the quantity of calcium that the exchanger needed to pump? Mitchell *et al.* (1984) provided the first experimental evidence that the action potential plateau is maintained by sodium–calcium exchange current. The Hilgemann–Noble model showed that this is precisely what one would expect, both qualitatively and quantitatively. Subsequent experimental and modelling work has fully confirmed this conclusion (Egan *et al.* 1989; LeGuennec & Noble, 1994; Eisner *et al.* 2000; Bers, 2001). In particular, Earm *et al.* (1990) performed the necessary experiments to show this phase of exchanger current during the late plateau of atrial cells (see Fig. 3).

(3) Could a model of the SR that reproduced the major features of Fabiato's (1983) experiments be incorporated? The model followed as much of the Fabiato data as possible, but while broadly consistent with the Fabiato work it could not be based on that alone. Fabiato's experiments were heroic but they were done on skinned fibres, which removes many of the relevant membrane-based mechanisms. It is an important function of simulation to reveal when experimental data needs extending.

(4) Were the quantities of calcium, free and bound, at each stage of the cycle consistent with the properties of the

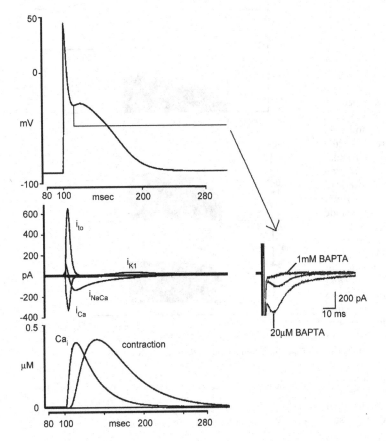

Figure 3. Predictions and experiments concerning the role of sodium–calcium exchange current during the late plateau of atrial cells
A, the single atrial cell model (Earm & Noble, 1990) developed from the Hilgemann–Noble model showing the action potential (top), some of the ionic currents, including sodium–calcium exchange, and the calcium transient and contraction (bottom). *B*, experimental recordings of the sodium–calcium exchange current during voltage clamp at a potential in the range of the late plateau (Earm *et al.* 1990). The current trace is similar to that predicted by the model and, as a calcium chelator is diffused into the cell, the current change is eliminated.

J Physiol 580.1   From axon to virtual heart

cytosol buffers? The great majority of the cytosol calcium is bound so that, although large calcium fluxes are involved, the free calcium transients are much smaller, as they are experimentally.

The major deficiency of this model was that it could not account for graded release of calcium from the SR. Much more complex models, incorporating finer detail of the excitation–contraction process, including the communication between L-type calcium channels and the SR calcium release channels, are required to achieve this (Stern, 1992; Jafri *et al.* 1998; Rice *et al.* 1999; Winslow *et al.* 2000; Greenstein & Winslow, 2002).

Modelling of cardiac cells is now a highly active field. There are more than 30 curated cell models on the CellML website (www.cellml.org) that can be run with compatible software such as COR (http://cor.physiol.ox.ac.uk/). The days when such biological computation was the preserve of just a few with the relevant maths and computing expertise are over. We can look forward to simulation being as central to work in physiology as it is in physics and engineering, used by experimentalists as well as by theoreticians. It will be the iterative interaction between the two that will be important in refining our understanding and predictive ability.

Linking levels: building the virtual heart

I have been privileged to collaborate with several of the key people involved in extending cardiac modelling to levels higher than the cell. The earliest work was with Raimond Winslow who had access to the Connection Machine at Minnesota, a huge parallel computer with 64 000 processors. We were able to construct models in which up to this number of cell models were connected together to form 2D or 3D blocks of atrial or ventricular tissue. This enabled us to study the factors determining whether ectopic beats generated during sodium overload would propagate across the tissue (Winslow *et al.* 1993) and to study the possible interactions between sinus node and atrial cells (Noble *et al.* 1995).

At about the same time, I stayed for a period as a visiting professor at the University of Auckland where Peter Hunter, Bruce Smaill and their colleagues in bioengineering and physiology were constructing the first anatomically detailed models of a whole ventricle. These models include fibre orientations and sheet structure (Hooks *et al.* 2002; Crampin *et al.* 2004), and have been used to incorporate the cellular models in an attempt to reconstruct the electrical and mechanical behaviour of the whole organ.

This work includes simulation of the activation wave front (Smith *et al.* 2001; Noble, 2002*a*). This is heavily influenced by cardiac ultra-structure, with preferential conduction along the fibre–sheet axes, and the result

corresponds well with that obtained from multielectrode recording from dog hearts *in situ*. Accurate reconstruction of the depolarization wavefront promises to provide reconstruction of the early phases of the ECG to complement work already done on the late phases (Antzelevitch *et al.* 2001) and as the sinus node, atrium and conducting system are incorporated into the whole heart model we can look forward to the first example of reconstruction of a complete physiological process from the level of protein function right up to routine clinical observation. The whole ventricular model has already been incorporated into a virtual torso (Bradley *et al.* 1997), including the electrical conducting properties of the different tissues, to extend the external field computations to reconstruction of multiple-lead chest and limb recording. Incorporation of biophysically detailed cell models into whole organ models (Noble, 2002*c*, 2002*a*; Trayanova 2006; Crampin *et al.* 2004) is still at an early

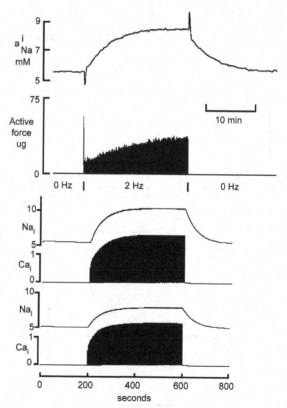

Figure 4. Sodium loading during repetitive activity
A, experimental recording of the rise in intracellular sodium during repetitive activity in a Purkinje fibre (from Boyett *et al.* 1987). *B*, computation of the same effect in a model ventricular cell in which the persistent sodium current has been incorporated (Sakmann *et al.* 2000). *C*, results of the same computation after block of the persistent sodium current. Sodium leading is reduced by 48%.

stage of development, but it is essential to attempts to understand heart arrhythmias. So also is the extension of modelling to human cells (Nygren *et al.* 1998; Ten Tusscher *et al.* 2003).

Applications of the models

Understanding the electrophysiology of the heart is critical to resolving a major problem for the drug industry. This is that a large number of compounds target the proteins involved in cardiac repolarization, so causing arrhythmic side-effects that can be fatal. The cardiac models are now being used in drug development and in attempting to understand these problems (Noble & Colatsky, 2000; Bottino *et al.* 2006). Since I began with a problem in modelling the sodium current, it is appropriate therefore to finish this review lecture with an example of application to drug development that depends on understanding the role of sodium channels. There are compounds that have been developed recently that reduce or block the persistent component of sodium current and we have recently helped to understand the mechanism of action of one of these, Ranolazine (Belardinelli *et al.* 2006; Noble & Noble, 2006; Undrovinas *et al.* 2006). In addition to helping to prevent repolarization failure, block of persistent sodium current underlies the main area of application of this drug since it reduces sodium loading of cardiac cells during ischaemia and heart failure (Undrovinas *et al.* 2006). Figure 4 shows the basis of this action. Figure 4*A* is from the work of Boyett *et al.* (1987) showing the slow increase in intracellular sodium loading during repetitive activity. Figure 4*B* is a model simulation of the same effect using a ventricular cell model incorporating persistent sodium current (Sakmann *et al.* 2000). The predicted rise in sodium concentration is similar to that seen experimentally. Figure 4*C* shows that block of persistent sodium current would be expected to reduce sodium loading by nearly 50%. We have yet to repeat these computations in conditions corresponding to ischaemia, but such an effect may easily be large enough to prevent internal sodium rising to the point at which arrhythmias are triggered (Noble & Varghese, 1998).

Concluding remarks

I started work on modelling cardiac cells just over 45 years ago. If we follow Moore's law on computing speed and power doubling every 18 months, that corresponds to 30 doubling periods, which is an increase in computing power of 2^{30} or roughly 1 billion fold. The calculations that required 2 h of time on a huge valve computer in 1960 can now be done in microseconds, and much more complex cell models can be run faster than real time even on a small laptop. This brings closer the prospect of physiology becoming a quantitative mathematical science.

Remember too that even the grandfather of experimental physiology, Claude Bernard, looked forward to such a day. In his seminal book, *Introduction a l'étude de la médecine expérimentale*, he wrote 'Cette application des mathématiques aux phénomènes naturels est le but de toute science' (This application of mathematics to natural phenomena is the aim of all science) (Bernard, 1865, 1984). But he also cautioned that, in 1865, it was too early to achieve that in physiology and medicine. Perhaps, 140 years later, we are progressively getting there.

References

Antzelevitch C, Nesterenko VV, Muzikant AL, Rice JJ, Chien G & Colatsky T (2001). Influence of transmural gradients on the electrophysiology and pharmacology of ventricular myocardium. Cellular basis for the Brugada and long-QT syndromes. *Philos Trans R Soc Lond A* **359**, 1201–1216.

Attwell D, Cohen I, Eisner D, Ohba M & Ojeda C (1979). The steady state TTX sensitive ('window') sodium current in cardiac Purkinje fibres. *Pflugers Arch* **379**, 137–142.

Beeler GW & Reuter H (1977). Reconstruction of the action potential of ventricular myocardial fibres. *J Physiol* **268**, 177–210.

Belardinelli L, Shryock J & Fraser H (2006). Inhibition of the late sodium current as a potential cardioprotective principle: effects of the selective late sodium current inhibitor, Ranolazine. *Heart* **92**, iv6–iv12.

Bernard C (1865, 1984). *Introduction a l'étude de la Médecine Expérimentale*. Flammarion, Paris.

Bers D (2001). *Excitation-Contraction Coupling and Cardiac Contractile Force*. Kluwer, Dordrecht.

Bottino D, Penland RC, Stamps A, Traebert M, Dumotier B, Georgieva A, Helmlinger G & Lett GS (2006). Preclinical cardiac safety assessment of pharmaceutical compounds using an integrated systems-based computer model of the heart. *Prog Biophysics Mol Biol* **90**, 414–443.

Boyett MR, Hart G, Levi AJ & Roberts A (1987). Effects of repetitive activity on developed force and intracellular sodium in isolated sheep and dog Purkinje fibres. *J Physiol* **388**, 295–322.

Bradley CP, Pullan AJ & Hunter PJ (1997). Geometric modeling of the human torso using cubic Hermite elements. *Ann Biomed Eng* **25**, 96–111.

Crampin EJ, Halstead M, Hunter PJ, Nielsen P, Noble D, Smith N & Tawhai M (2004). Computational physiology and the physiome project. *Exp Physiol* **89**, 1–26.

DiFrancesco D (1981). A new interpretation of the pace-maker current, i_{K2}, in Purkinje fibres. *J Physiol* **314**, 359–376.

DiFrancesco D (1993). Pacemaker mechanisms in cardiac tissue. *Annu Rev Physiol* **55**, 455–472.

DiFrancesco D (2006). Serious workings of the funny current. *Prog Biophysics Mol Biol* **90**, 13–25.

DiFrancesco D & Noble D (1985). A model of cardiac electrical activity incorporating ionic pumps and concentration changes. *Philos Trans R Soc B* **307**, 353–398.

Earm YE, Ho WK & So IS (1990). Inward current generated by Na-Ca exchange during the action potential in single atrial cells of the rabbit. *Proc R Soc Lond B Biol Sci* **240**, 61–81.

J Physiol 580.1 From axon to virtual heart

Earm YE & Noble D (1990). A model of the single atrial cell: relation between calcium current and calcium release. *Proc R Soc Lond B Biol Sci* **240**, 83–96.

Egan T, Noble D, Noble SJ, Powell T, Spindler AJ & Twist VW (1989). Sodium-calcium exchange during the action potential in guinea-pig ventricular cells. *J Physiol* **411**, 639–661.

Eisner DA, Choi HS, Diaz ME & O'Neill SC (2000). Integrative analysis of calcium cycling in cardiac muscle. *Circ Res* **87**, 1087–1094.

Fabiato A (1983). Calcium induced release of calcium from the sarcoplasmic reticulum. *Am J Physiol Cell Physiol* **245**, C1–C14.

Greenstein JL & Winslow RL (2002). An integrative model of the cardiac ventricular myocyte incorporating local control of Ca^{2+} release. *Biophys J* **83**, 2918–2945.

Hall AE, Hutter OF & Noble D (1963). Current-voltage relations of Purkinje fibres in sodium-deficient solutions. *J Physiol* **166**, 225–240.

Hilgemann DW (1986). Extracellular calcium transients and action potential configuration changes related to post-stimulatory potentiation in rabbit atrium. *J Gen Physiol* **87**, 675–706.

Hilgemann DW & Noble D (1987). Excitation-contraction coupling and extracellular calcium transients in rabbit atrium: Reconstruction of basic cellular mechanisms. *Proc R Soc Lond B Biol Sci* **230**, 163–205.

Hinch R (2004). A mathematical analysis of the generation and termination of calcium sparks. *Biophys J* **86**, 1293–1307.

Hinch R, Greenstein JL & Winslow RL (2006). Multi-scale modelling of local control of calcium induced calcium release. *Prog Biophysics Mol Biol* **90**, 136–150.

Hodgkin AL & Huxley AF (1952). A quantitative description of membrane current and its application to conduction and excitation in nerve. *J Physiol* **117**, 500–544.

Hooks DA, Tomlinson KA, Marsden SG, LeGrice IJ, Smaill BH, Pullan AJ & Hunter PJ (2002). Cardiac microstructure: Implications for electrical propagation and defibrillation in the heart. *Circ Res* **91**, 331–338.

Hutter OF & Noble D (1960). Rectifying properties of heart muscle. *Nature* **188**, 495.

Irisawa H, Brown HF & Giles WR (1993). Cardiac pacemaking in the sinoatrial node. *Physiol Rev* **73**, 197–227.

Jafri S, Rice JJ & Winslow RL (1998). Cardiac Ca^{2+} dynamics: the roles of ryanodine recptor adaptation and sarcoplasmic reticulum load. *Biophys J* **74**, 1149–1168.

Kimura J, Miyamae S & Noma A (1987). Identification of sodium-calcium exchange current in single ventricular cells of guinea-pig. *J Physiol* **384**, 199–222.

Kimura J, Noma A & Irisawa H (1986). Na-Ca exchange current in mammalian heart cells. *Nature* **319**, 596–597.

Kiyosue T & Arita M (1989). Late sodium current and its contribution to action potential configuration in guinea pig ventricular myocytes. *Circ Res* **64**, 389–397.

LeGuennec JY & Noble D (1994). The effects of rapid perturbation of external sodium concentration at different moments of the action potential in guinea-pig ventricular myocytes. *J Physiol* **478**, 493–504.

Luo C & Rudy Y (1994). A Dynamic model of the cardiac ventricular action potential – simulations of ionic currents and concentration changes. *Circ Res* **74**, 1071–1097.

Luo C & Rudy Y (1994). A dynamic model of the cardiac ventricular action potential. II. Afterdepolarizations, triggered activity and potentiation. *Circ Res* **74**, 1097–1113.

Maltsev VA, Sabbah HN, Higgins RSD, Silverman N, Lesch M & Undrovinas AI (1998). Novel, ultraslow inactivating sodium current in human ventricular myocytes. *Circulation* **98**, 2545–2552.

McAllister RE, Noble D & Tsien RW (1975). Reconstruction of the electrical activity of cardiac Purkinje fibres. *J Physiol* **251**, 1–59.

Mitchell MR, Powell T, Terrar DA & Twist VA (1984). The effects of ryanodine, EGTA and low-sodium on action potentials in rat and guinea-pig ventricular myocytes: evidence for two inward currents during the plateau. *Br J Pharmacol* **81**, 543–550.

Mullins L (1981). *Ion Transport in the Heart*. Raven Press, New York.

Noble D (1962). A modification of the Hodgkin-Huxley equations applicable to Purkinje fibre action and pacemaker potentials. *J Physiol* **160**, 317–352.

Noble D (1965). Electrical properties of cardiac muscle attributable to inward-going (anomalous) rectification. *J Cell Comp Physiol* **66** (Suppl. 2), 127–136.

Noble D (1984). The surprising heart: a review of recent progress in cardiac electrophysiology. *J Physiol* **353**, 1–50.

Noble D (2002a). Modelling the heart: from genes to cells to the whole organ. *Science* **295**, 1678–1682.

Noble D (2002b). Modelling the heart: insights, failures and progress. *Bioessays* **24**, 1155–1163.

Noble D (2002c). The rise of computational biology. *Nat Rev Mol Cell Biol* **3**, 460–463.

Noble D (2006a). Multilevel modelling in systems biology: from cells to whole organs. In *Systems Modelling in Cellular Biology* eds. Szallasi Z, Stelling J & Periwal V. MIT Press, Cambridge, MA, USA.

Noble D (2006b). *The Music of Life*. Oxford University Press, Oxford.

Noble D, Brown HF & Winslow R (1995). Propagation of pacemaker activity: interaction between pacemaker cells and atrial tissue. In *Pacemaker Activity and Intercellular Communication*, ed.Huizinga JD, pp. 73–92. CRC Press, Boca Raton, FL, USA.

Noble D & Colatsky TJ (2000). A return to rational drug discovery: computer-based models of cells, organs and systems in drug target identification. *Expert Opin Ther Targets* **4**, 39–49.

Noble D & Noble PJ (2006). Late sodium current in the pathophysiology of cardiovascular disease: consequences of sodium-calcium overload. *Heart* **92**, iv1–iv5.

Noble D & Tsien RW (1968). The kinetics and rectifier properties of the slow potassium current in cardiac Purkinje fibres. *J Physiol* **195**, 185–214.

Noble D & Tsien RW (1969). Outward membrane currents activated in the plateau range of potentials in cardiac Purkinje fibres. *J Physiol* **200**, 205–231.

Noble D & Varghese A (1998). Modeling of sodium-calcium overload arrhythmias and their suppression. *Can J Cardiol* **14**, 97–100.

Noble D, Varghese A, Kohl P & Noble PJ (1998). Improved guinea-pig ventricular cell model incorporating a diadic space, iKr and iKs, and length- and tension-dependent processes. *Can J Cardiol* **14**, 123–134.

Nygren A, Fiset C, Firek L, Clark JW, Lindblad DS, Clark RB & Giles WR (1998). A mathematical model of an adult human atrial cell: the role of K$^+$ currents in repolarization. *Circulation Res* **82**, 63–81.

Reuter H (1967). The dependence of slow inward current in Purkinje fibres on the extracellular calcium concentration. *J Physiol* **192**, 479–492.

Reuter H & Seitz N (1969). The dependence of calcium efflux from cardiac muscle on temperature and external ion composition. *J Physiol* **195**, 451–470.

Rice JJ, Jafri MS & Winslow RL (1999). Modeling gain and gradedness of Ca^{2+} release in the functional unit of the cardiac diadic space. *Biophys J* **77**, 1871–1884.

Sakmann BFAS, Spindler AJ, Bryant SM, Linz KW & Noble D (2000). Distribution of a persistent sodium current across the ventricular wall in guinea pigs. *Circ Res* **87**, 910–914.

Smith NP, Mulquiney PJ, Nash MP, Bradley CP, Nickerson DP & Hunter PJ (2001). Mathematical modelling of the heart: cell to organ. *Chaos Solitons Fractals* **13**, 1613–1621.

Sobie EA, Guatimosim S, Gomez-Viquez L, Song L-S, Hartmann H, Saleet JM & Lederer WJ (2006). The Ca^{2+} leak paradox and 'rogue ryanodine receptors': SR Ca^{2+} efflux theory and practice. *Prog Biophysics Mol Biol* **90**, 172–185.

Stern MD (1992). Theory of excitation-contraction coupling in cardiac muscle. *Biophys J* **63**, 497–517.

Ten Tusscher KHWJ, Noble D, Noble PJ & Panfilov AV (2003). A model of the human ventricular myocyte. *Am J Physiol Heart Circ Physiol* **286**, H1573–1589.

Trayanova N (2006). Defibrillation of the heart: Insights into mechanisms from modelling studies. *Exp Physiol* **91**, 323–337.

Undrovinas AI, Belardinelli L, Nidas A, Undrovinas RN & Sabbah HN (2006). Ranolazine improves abnormal repolarization and contraction in left ventricular myocytes of dogs with heart failure by inhibiting late sodium current. *J Cardiovasc Electrophysiol* **17**, S169–S177.

Van der Pol B & Van der Mark J (1928). The heartbeat considered as a relaxation oscillation and an electrical model of the heart. *Phil Mag* (Suppl) **6**, 763–775.

Weidmann S (1956). *Elektrophysiologie der Herzmuskelfaser*. Huber, Bern.

Winslow RL, Greenstein JL, Tomaselli GF & O'Rourke B (1999). Computational models of the failing myocyte: relating altered gene expression to cellular function. *Philos Trans R Soc Lond A* **359**, 1187–1200.

Winslow RL, Scollan DF, Holmes A, Yung CK, Zhang J & Jafri MS (2000). Electrophysiological modeling of cardiac ventricular function: from cell to organ. *Annu Rev Biomed Eng* **2**, 119–155.

Winslow R, Varghese A, Noble D, Adlakha C & Hoythya A (1993). Generation and propagation of triggered activity induced by spatially localised Na-K pump inhibition in atrial network models. *Proc Biol Sci* **254**, 55–61.

Zygmunt AC, Eddlestone GT, Thomas GP, Nesterenko VV & Antzelevitch C (2001). Larger late sodium conductance in M cells contributes to electrical heterogeneity in canine ventricle. *Am J Physiol Heart Circ Physiol* **281**, H689–H697.

Acknowledgements

Research in the author's laboratory is funded by the EU BioSim Consortium, the BBSRC and The Wellcome Trust.

Exp Physiol 93.1 pp 16–26

16

| **Experimental Physiology – *Paton Lecture*** |

Claude Bernard, the first systems biologist, and the future of physiology

Denis Noble

Department of Physiology, Anatomy and Genetics, Parks Road, Oxford OX1 3PT, UK

The first systems analysis of the functioning of an organism was Claude Bernard's concept of the constancy of the internal environment (*le milieu intérieur*), since it implied the existence of control processes to achieve this. He can be regarded, therefore, as the first systems biologist. The new vogue for systems biology today is an important development, since it is time to complement reductionist molecular biology by integrative approaches. Claude Bernard foresaw that this would require the application of mathematics to biology. This aspect of Claude Bernard's work has been neglected by physiologists, which is why we are not as ready to contribute to the development of systems biology as we should be. In this paper, I outline some general principles that could form the basis of systems biology as a truly multilevel approach from a physiologist's standpoint. We need the insights obtained from higher-level analysis in order to succeed even at the lower levels. The reason is that higher levels in biological systems impose boundary conditions on the lower levels. Without understanding those conditions and their effects, we will be seriously restricted in understanding the logic of living systems. The principles outlined are illustrated with examples from various aspects of physiology and biochemistry. Applying and developing these principles should form a major part of the future of physiology.

(Received 4 August 2007; accepted after revision 3 October 2007; first published online 26 October 2007)
Corresponding author D. Noble: Department of Physiology, Anatomy and Genetics, Parks Road, Oxford OX1 3PT, UK. denis.noble@dpag.ox.ac.uk

Historical introduction

Claude Bernard was Sir William Paton's great physiological hero. When the Physiological Society celebrated its centenary in 1976, Bill contributed a paper to the historical part of the meeting concerning one of Bernard's experiments on curare and drawing attention to the important role his ideas played in the foundation of the Society in 1876 (Paton, 1976). The reasons for his admiration of Claude Bernard are not hard to find. Bernard was a superb experimentalist, as the history of his work on digestion shows (Holmes, 1974). He also displayed his skills in many other areas of physiology and he laid out the principles of his science in his highly influential *Introduction à l'étude de la Médecine Expérimentale* (Bernard, 1865, 1984), in which he revealed himself to be a great thinker as well as a great experimentalist. The theoretical problem he addressed is one that is very relevant

both to my claim that he was the first systems biologist and to the challenge that physiology faces today.

What was Claude Bernard's problem? It was that the chemists had created 'organic' molecules. This was a major development, since people had thought since Lémery's *Cours de Chymie* (published in 1675) that there were three completely separate classes of compounds: mineral, vegetable and animal. The first break in this idea came from the work of Lavoisier (1784), who showed that all compounds from vegetable and animal sources always contained at least carbon and hydrogen, and frequently nitrogen and phosphorus. This work bridged the vegetable–animal chemical boundary, but it left intact the boundary between the living and non-living. In fact, Berzelius (1815) even proposed that organic compounds were produced by laws different from inorganic compounds; the idea that there was a specific vital force that could not operate outside living systems. In 1828, however, Wöhler succeeded in creating urea from ammonium cyanate. The distinction between organic and non-organic origins was further weakened by Kolbe who, in 1845, synthesized acetic acid from its elements. Many

---

This article is based on the Paton Lecture delivered with the same title to the Life Sciences 2007 meeting in Glasgow in July 2007.

Exp Physiol 93.1 pp 16–26   Systems biology and the future of physiology   17

other discoveries of this kind (Finar, 1964) led to the idea that life itself could be reduced to chemistry and physics.

This was the challenge that physiologists such as Claude Bernard faced. His answer was precise. Neither vitalism nor chemical reductionism characterized living organisms. To the challenge that 'There are ... chemists and physicists who ... try to absorb physiology and reduce it to simple physico-chemical phenomena', Bernard responded, 'Organic individual compounds, though well defined in their properties, are still not active elements in physiological phenomena. They are only passive elements in the organism.' The reason, he explained, is that 'The living organism does not really exist in the *milieu extérieur* but in the liquid *milieu intérieur* .... a complex organism should be looked upon as an assemblage of simple organisms .... that live in the liquid *milieu intérieur*.'

His response to vitalism was equally robust: 'Many physicians .... assume a vital force in opposition to physico-chemical forces. I propose therefore to prove that the science of vital phenomena must have the same foundations as the science of the phenomena of inorganic bodies, and that there is no difference between the principles of biological science and those of physico-chemical science.'

By 'principles' here Bernard meant the laws governing the behaviour of the components. The control of the *milieu intérieur* meant not that the individual molecules did anything different from what they would do in non-living systems, but rather that the *ensemble* behaves in a controlled way, the controls being those that maintain the constancy of the internal environment. How could that be formalized? Could there be a theoretical physiology? Physical scientists had long since used mathematics to formalize their theories. Could that also be done in physiology? Bernard's answer to this question was 'yes, but not yet.' He cautioned, 'The most useful path for physiology and medicine to follow now is to seek to discover new facts instead of trying to reduce to equations the facts which science already possesses.' I believe that this view has been in part responsible for the broadly antitheoretical stance of British and American Physiology. It is important, therefore, to recognize that it represents only half of Bernard's views on the matter. For the emphasis in that statement should be on the word *now*. He also wrote that it was necessary to 'fix numerically the relations' between the components. He continued: 'This application of mathematics to natural phenomena is the aim of all science, because the expression of the laws of phenomena should always be mathematical.' His caution, therefore, was purely practical and temporal. In 1865 he saw, correctly of course, that physiology simply did not have enough data to make much mathematical application worthwhile *at that time*. But he clearly foresaw that the day would come when there would be sufficient data and that mathematical analysis would then become necessary.

The problem physiology faces today both resembles that faced by Bernard and differs from it. We face a new form of reductionism: that of genetic determinism, exemplified by the idea that there is a genetic program, what Jacob and Monod called '*le programme génétique*' (Monod & Jacob, 1961; Jacob, 1970). This challenge strongly resembles that of 'reducing life to physics and chemistry', the chemical being DNA. The major difference from Bernard's day is that we now have more facts than we can handle. There is a data explosion at all levels of biology. The situation is almost the reverse of that in Bernard's time. I have no doubt, therefore, that if he were alive today he would be championing his 'application of mathematics to natural phenomena.' I will illustrate why this is necessary and how it can be achieved by outlining some principles of systems biology from a physiologist's viewpoint. The principles are derived from my book on systems biology, *The Music of Life* (Noble, 2006), but their arrangement as a set of 10 was first presented by Noble (2007).

The principles of systems biology

First principle: biological functionality is multilevel. I start with this principle because it is obviously true, all the other principles can be shown to follow from it, and it is therefore the basis on which a physiological understanding of the phenomenon of life must be based. It is also a more general statement of the insight contained in Claude Bernard's idea of the constancy of the internal environment. That functionality is attributable to the organism as a whole and it controls all the other levels. This is the main reason why I describe Bernard as the first systems biologist. It is hard to think of a more important overall systems property than the one Bernard first identified.

Yet, the language of modern reductionist biology often seems to deny this obvious truth. The enticing metaphor of the 'book of life' made the genome into the modern equivalent of the 'embryo-homunculus', the old idea that each fertilized egg contains within it a complete organism in miniature (Mayr, 1982; p. 106). That the miniature is conceived as a digital 'map' or 'genetic program' does not avoid the error to which I am drawing attention, which is the idea that the living organism is simply the unfolding of an already-existing program, fine-tuned by its interaction with its environment, to be sure, but in all essentials, already there in principle as a kind of zipped-up organism. In its strongest form, this view of life leads to gene-selectionism and to gene-determinism: 'They [genes] created us body and mind' (Dawkins, 1976).

Dawkins himself does not really believe that. In a more recent book, he entitles one chapter 'Genes aren't us' (Dawkins, 2003) and, even in *The Selfish Gene*, the bold, simple message of the early chapters is qualified at the

18   D. Noble   *Exp Physiol* 93.1 pp 16–26

end. My reservations, however, go much further than his. For, in truth, the stretches of DNA that we now call genes do nothing on their own. They are simply databases used by the organism as a whole. This is the reason for replacing the metaphor of the 'selfish' gene by genes as 'prisoners' (Noble, 2006; chapter 1). As Maynard Smith & Szathmáry (1999) express it, 'Co-ordinated replication prevents competition between genes within a compartment, and forces co-operation on them. They are all in the same boat.' From the viewpoint of the organism, genes as DNA molecules are therefore captured entities, no longer having a life of their own independent of the organism.

Second principle: transmission of information is not one way. The central dogma of molecular biology (Crick, 1970) is that information flows from DNA to RNA, from RNA to proteins, which can then form protein networks, and so on up through the biological levels to that of the whole organism. Information does not flow the other way. This is the dogma that is thought to safeguard modern neo-Darwinian theory from the spectre of 'Lamarckism', the inheritance of acquired characteristics. Applied to all the levels, this view is illustrated in Fig. 1. It encourages the bottom-up view of systems biology, the idea that if we knew enough about genes and proteins we could reconstruct all the other levels. Bioinformatics alone would be sufficient.

There are two respects in which the dogma is at least incomplete. The first is that it defines the relevant information uniquely in terms of the DNA code, the sequence of C, G, A, T bases. But the most that this information can tell us is *which* protein will be made. It does not tell us *how much* of each protein will be made. Yet, this is one of the most important characteristics of any living cell. Consider the speed of conduction of a nerve or muscle impulse, which depends on the density of rapidly activated sodium channels: the larger the density, the greater the ionic current and the faster the conduction. But this relationship applies only up to a certain optimum density, since the channel gating also contributes to the cell capacitance, which itself slows conduction, so there is a point beyond which adding more channel proteins is counter-productive (Hodgkin, 1975; Jack *et al.* 1975; p. 432). A feedback mechanism must therefore operate between the electrical properties of the nerve and the expression levels of the sodium channel protein. We now refer to such feedback mechanisms in the nervous system, which take many forms, as electro-transcription coupling (e.g. Deisseroth *et al.* 2003).

Similar processes must occur in the heart (e.g. Bers & Guo, 2005) and all the other organs. One of the lessons I have learnt from many attempts to model cardiac electrophysiology (Noble, 2002) is that, during the slow phases of repolarization and pacemaker activity, the ionic currents are so finely balanced that it is inconceivable that

nature arrives at the correct expression and activity levels without some kind of feedback control. We don't yet know what that control might be, but we can say that it must exist. Nature cannot be as fragile as our computer models are! Robustness is an essential feature of successful biological systems.

There is nothing new in the idea that such feedback control of gene expression must exist. It is, after all, the basis of cell differentiation. All nucleated cells in the body contain exactly the same genome (with the exception of course of the germ cells, with only half the DNA). Yet the expression pattern of a cardiac cell is completely different from, say, a hepatic or bone cell. Moreover, whatever is determining those expression levels is accurately inherited during cell division. This cellular inheritance process is robust; it depends on some form of gene marking. It is this information on relative gene expression levels that is critical in determining each cell type.

By what principle could we possibly say that this is not relevant information? In the processes of differentiation and growth it is just as relevant as the raw DNA sequences. Yet, it is clear that this information *does* travel 'the other way'. The genes are told by the cells and tissues what to do, how frequently they should be transcribed and when to stop. There is 'downward causation' (Noble, 2006; chapter 4) from those higher levels that determines how the genome is 'played' in each cell (Fig. 2). Moreover, the possible number of combinations that could arise from so many gene components is so large (Feytmans *et al.* 2005) that there wouldn't be enough material in the whole universe for nature to have tried more than a small fraction

The reductionist causal chain

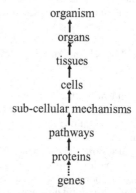

Figure 1. The reductionist 'bottom-up' causal chain (reproduced with permission from Noble, 2006)
This begins with the central dogma that information flows from DNA to proteins (bottom dotted arrow), never the other way, and extends the same concept through all the higher levels.

Exp Physiol 93.1 pp 16–26   Systems biology and the future of physiology   19

of the possible combinations even over the billions of years of evolution (Noble, 2006; chapter 2).

So the dogma is at least incomplete. But I also think it is incorrect in several important ways. Sure, protein sequences are not back-translated to form DNA sequences. In this limited original form, as formulated by Crick (1970), the central dogma is correct. But there is growing evidence from work on plants and microbes that environmental factors *do* change the genome, particularly by gene transfer (Goldenfeld & Woese, 2007). We cannot, therefore, use the original central dogma to exclude information transfer *into* the genome, determined by the organism and its environment.

Moreover, the DNA code itself is marked by the organism. This is the focus of the rapidly growing field of epigenetics (Qiu, 2006). At least two such mechanisms are now known at the molecular level: methylation of cytosine bases and control by interaction with the tails of histones around which the DNA is wound. Both of these processes modulate gene expression. The terminological question then arises: do we regard this as a form of code-modification? Is a cytosine, the C of the code, a kind of C* when it is methylated? That is a matter of definition of code, and one which I will deal with in the next section, but what is certain is that it is relevant information determining levels of gene expression, and that this information does flow against the direction of the central dogma. In fact, a form of inheritance of acquired characteristics (those of specific cell types) is rampant within all multicellular organisms with very different specialized cell types (Noble,

Downward causation

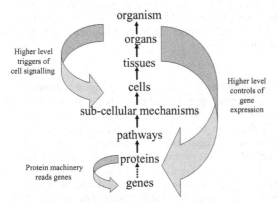

Figure 2. Figure 1 has been completed by adding the downward forms of causation, such as higher levels triggering cell signalling and gene expression
Note the downward-pointing arrow connecting from proteins to genes to indicate that it is protein machinery that reads and interprets gene coding. Loops of interacting downward and upward causation can be built between all levels of biological organization. Reproduced with permission from Noble (2006).

2006; chapter 7). At the least we have to say that, during the lifetime of the individual organism, transmission of information is far from being one way.

Third principle: DNA is not the sole transmitter of inheritance. The defenders of the original version of the central dogma would argue that, while my conclusions regarding the second principle are correct, what happens when information is transmitted to the next generation through the germ-line nevertheless involves wiping the slate clean of epigenetic effects. Methylation of cytosine bases and other forms of genome marking are removed. The genome is reset so that 'Lamarckism' is impossible.

But this is to put the matter the wrong way round. We need to explain *why* the genome (usually) reverts to an unmarked state. We don't explain that by appealing to the central dogma, for that dogma is simply a restatement of the same idea. We are in danger of circular logic here. Later, I will suggest a plausible reason why, at least most of the time, the resetting is complete, or nearly so. In order to do that, we first need to analyse the idea that genetics, as originally understood, is just about DNA.

This is not the original biological meaning of 'gene'. The concept of a gene has changed (Kitcher, 1982; Mayr, 1982; Dupré, 1993; Pichot, 1999). Its original biological meaning was an inheritable phenotype characteristic, such as eye/hair/skin colour, body shape and weight, number of legs/arms, to which we could perhaps add more complex traits like intelligence, personality, sexuality, etc. Genes, as originally conceived, are not just the same as stretches of DNA unless we subscribe to the view that the inheritance of all such characteristics is attributable entirely to DNA sequences. That is clearly false, since the egg cell is also inherited, together with any epigenetic characteristics transmitted by sperm (Anway *et al.* 2005), perhaps via RNA in addition to its DNA, and all the epigenetic influences of the mother and environment. Of course, the latter (environment) begins to be about 'nurture' rather than 'nature', but one of my points is that this distinction is fuzzy. The proteins that initiate gene transcription in the egg cell and impose an expression pattern on the genome are initially from the mother, and other such influences continue throughout development in the womb. Where we draw the line between nature and nurture is not at all obvious. There is an almost seamless transition from one to the other. 'Lamarckism', the inheritance of acquired characteristics, lurks in this fuzzy crack to a degree yet to be defined (Jablonka & Lamb, 1995, 2005). As the evolutionary geneticist Maynard Smith says, 'It [Lamarckism] is not so obviously false as is sometimes made out' (Maynard Smith, 1998).

Inheritance of the egg cell is important for two reasons. First, it is the egg cell DNA-reading machinery (a set of around 100 proteins and the associated cellular ribosome architecture) that enables the DNA to be used as a template to make more proteins. Second, the set of other cellular elements, mitochondria, endoplasmic reticulum, microtubules, nuclear and other membranes, and a host of chemicals arranged specifically in cellular compartments, is also inherited. Most of this is not coded for by DNA sequences. Lipids certainly are not so coded. But they are absolutely essential to all the cell architecture. There would be no cells, nuclei, mitochondria, endoplasmic reticulum, ribosomes and all the other cellular machinery and compartments without the lipids. The specific details of all this cellular machinery matter. We can't make any old DNA do its thing in any old egg cell. Most attempts at interspecies cloning simply don't work. Invariably, a block occurs at an early stage in development. The only successful case so far is that of a wild ox (*Bos javanicus*) cloned in a domestic cow egg. The chances are that it will work only in very closely related species. The egg cell information is therefore also species specific.

Could epigenetic inheritance and its exclusion from the germ cell line be a requirement of multicellular harmony? The exact number of cell types in a human is debatable. It is partly a question of definition. A project that seeks to model all the cell types in the body, the Human Physiome Project (Crampin *et al.* 2004), estimates that there are around 200, all with completely different gene expression patterns. There would be even more if one took account of finer variations, such as those that occur in various regions of the heart and which are thought to protect the heart against fatal arrhythmias.

The precise number is not too important. The important fact is that it is large and that the range of patterns of gene expression is therefore also large and varied. Their patterns must also be harmonious in the context of the organism as a whole. They are all in the same boat; they sink or swim together. Disturbing their harmony would have serious consequences. It was arrived at after more than 2 billion years of experimentation.

Each cell type is so complex that the great majority of genes are expressed in many cell types. So it makes sense that all the cells in the body have the same gene complement, and that the coding for cell type is transmitted by gene marking, rather than by gene complement. I think that this gives the clue to the purpose of re-setting in germ-line inheritance. Consider what would happen if germ-line inheritance reflected adaptive changes in individual cell types. Given that all cell types derive ultimately from the fused germ-line cells, what would the effect be? Clearly, it would be to alter the patterns of expression in nearly all the cell types. There would be no way to transmit an improvement in, say, heart function to the next generation via gene marking of the germ cells without *also* influencing the gene expression patterns in many other types of cell in the body. And of course there is no guarantee that what is beneficial for a heart cell will be so in, say, a bone cell or a liver cell. On the contrary, the

chances are that an adaptation beneficial in one cell type would be likely to be deleterious in another.

Much better, therefore, to let the genetic influences of natural selection be exerted on undifferentiated cells, leaving the process of differentiation to deal with the fine-tuning required to code for the pattern of gene expression appropriate to each type of cell. If this explanation is correct, we would not necessarily expect it to be 100% effective. It is conceivable that some germ-line changes in gene expression patterns might be so beneficial for the organism as a whole, despite deleterious effects on a few cell lines, that the result would favour selection. This could explain the few cases where germ-line 'Lamarckian' inheritance seems to have occurred. It also motivates the search for other cases. The prediction would be that it will occur in multicellular species only when beneficial to overall intercellular harmony. It might be more likely to occur in simpler species. That makes sense in terms of the few examples that we have so far found (Maynard Smith, 1998). Notice that, in contrast to the central dogma, this explanation is a systems level explanation.

Finally, in this section, I will comment on the concept of code. Applied to DNA, this is clearly metaphorical. It is also a useful metaphor, but we should beware of its limitations. One of these is to imply that only information that is coded is important, as in talk of the genome as the 'book of life'. The rest of cellular inheritance is not so coded; in fact, it is not even digital. The reason is very simple. The rest of the cellular machinery doesn't need to 'code for' or get 'translated into' anything else for the simple reason that it 'represents' itself; cells divide to form more cells, to form more cells, and so on. In this sense, germ-line cells are just as 'immortal' as DNA but a lot of this information is transmitted directly without having to be encoded. We should beware of thinking that only digitally 'coded' information is what matters in genetic inheritance.

Fourth principle: the theory of biological relativity; there is no privileged level of causality. A fundamental property of systems involving multiple levels between which there are feedback control mechanisms is that there is no privileged level of causality. Consider, as an example, the cardiac pacemaker mechanism. This depends on ionic current generated by a number of protein channels carrying sodium, calcium, potassium and other ions. The activation, de-activation and inactivation of these channels proceed in a rhythmic fashion in synchrony with the pacemaker frequency. We might therefore be tempted to say that their oscillations generate that of the overall cell electrical potential, i.e. the higher-level functionality. But this is not the case. The kinetics of these channels varies with the electrical potential. There is therefore feedback between the higher-level property, the cell potential, and

Exp Physiol 93.1 pp 16–26   Systems biology and the future of physiology

the lower level property, the channel kinetics (Noble, 2006; chapter 5). This form of feedback was originally identified by Alan Hodgkin working on the nerve impulse, so it is sometimes called the Hodgkin cycle. If we remove the feedback, e.g. by holding the potential constant, as in a voltage clamp experiment, the channels no longer oscillate (Fig. 3). The oscillation is therefore a property of the system as a whole, not of the individual channels or even of a set of channels unless they are arranged in a particular way in the right kind of cell.

Nor can we establish any priority in causality by asking which comes first, the channel kinetics or the cell potential. This fact is also evident in the differential equations we use to model such a process. The physical laws represented in the equations themselves, and the initial and boundary conditions, operate *at the same time* (i.e. during every integration step, however infinitesimal), not sequentially.

It is simply a prejudice that inclines us to give some causal priority to lower-level, molecular events. The concept of level in biology is itself metaphorical. There is no literal sense in which genes and proteins lie *underneath* cells, tissues and organs. It is a convenient form of biological classification to refer to different levels, and we would find it very hard to do without the concept (Fig. 4). But we should not be fooled by the metaphor into thinking that 'high' and 'low' here have their normal meanings. From the metaphor itself, we can derive no justification for referring to one level of causality as privileged over others. That would be a misuse of the metaphor of level.

One of the aims of my book, *The Music of Life* (Noble, 2006), is to explore the limitations of biological metaphors. This is a form of linguistic analysis that is rarely applied in science, though a notable exception is Steven J. Gould's monumental work on the theory of evolution

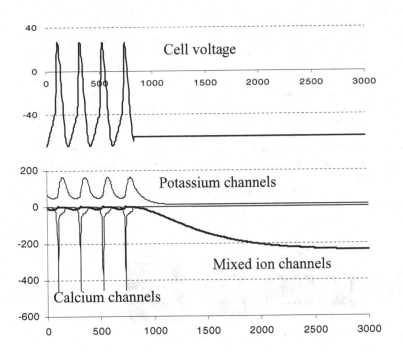

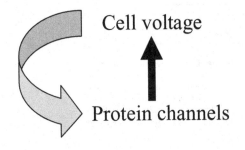

Figure 3. Computer model of pacemaker rhythm in the heart (reproduced with permission from Noble & Noble, 1984)
For the first four beats, the model is allowed to run normally and generates rhythm closely similar to a real heart. Then the feedback from cell voltage to protein channels is interrupted. All the protein channel oscillations then cease. They slowly change to steady, constant values. The diagram shows the causal loop involved. Protein channels carry current that changes the cell voltage (upward arrow), while the cell voltage changes the protein channels (downward arrow). In the simulation, this downward arrow was broken at 800 ms.

The Selected Papers of Denis Noble CBE FRS

22

D. Noble

Exp Physiol 93.1 pp 16–26

(Gould, 2002), in which he analyses the arguments for the multiplicity of levels at which natural selection operates.

These points can be generalized to any biological function. The only sense in which a particular level might be said to be privileged is that, in the case of each function, there is a level at which the function is integrated, and it is one of our jobs as biological scientists to determine what that level may be.

The idea that there is no privileged level of causality has a much wider range of applications than purely biological ones (Dupré, 1993; Cartwright, 1999; Keller, 2002), though the idea is rarely expressed in this bold, relativistic form. I use the word 'relativity' in formulating the principle because it shares certain features with theories of scale relativity proposed by some theoretical physicists, in particular the idea that there is no privileged scale, which is at the foundation of the theory of scale relativity (Nottale, 1993). There is an obvious correlation between scale and level, since lower and higher levels in any system operate at different scales. For this reason, some have proposed the application of the scale relativity theory framework and its associated mathematical tools to tackle the challenge of multiscale integration in systems biology (Nottale, 2000; Auffray & Nottale, 2008; Nottale & Auffray, 2008). But it is too early to judge whether this can provide a firm basis to a fully fledged theory of systems biology. Although the theory of scale relativity has already delivered a number of predictions in the realm of astrophysics which have been validated by subsequent observations, it still has to establish fully its position within theoretical physics. Nor is it possible yet to decide which principles are specific to systems biology and which are of general importance beyond the boundaries of biology.

Fifth principle: gene ontology will fail without higher-level insight. Genes, as defined by molecular genetics to be the coding regions of DNA, code for proteins. Biological function then arises as a consequence of multiple interactions between different proteins in the context of the rest of the cell machinery. Each function therefore depends on many genes, while many genes play roles in multiple functions. What then does it mean to give genes names in terms of functions? The only unambiguous labelling of genes is in terms of the proteins for which they code. Thus, the gene for the sodium–calcium exchange protein is usually referred to as *ncx*. Ion channel genes are also often labelled in this way, as in the case of sodium channel genes being labelled *scn*.

This approach, however, naturally appears unsatisfactory from the viewpoint of a geneticist, since the original question in genetics was not which proteins are coded for by which stretches of DNA [in fact, early ideas on where the genetic information might be found (Schrödinger, 1944) favoured the proteins], but rather what is responsible for higher-level phenotype characteristics. There is no one-to-one correspondence between genes or proteins and higher-level biological functions. Thus, there is no 'pacemaker' gene. Cardiac rhythm depends on many proteins interacting within the context of feedback from the cell electrical potential.

Let's do a thought experiment. Suppose we could knock out the gene responsible for L-type calcium channels and still have a living organism (perhaps because a secondary pacemaker takes over and keeps the organism viable – and something else would have to kick-in to enable excitation–contraction coupling, and so on throughout the body because L-type calcium channels are ubiquitous!). Since

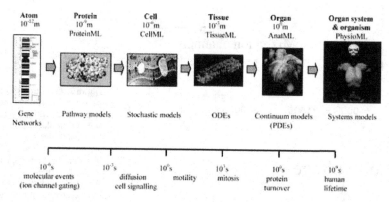

Figure 4. Spatial (top) and temporal (bottom) scales encompassed by the Human Physiome Project
The types of mathematical model appropriate to each spatial scale are also indicated. The last two images on the right in this figure, and all subsequent anatomical images, are from anatomically based models developed by the Auckland Bioengineering group. The tissue image is a three-dimensional confocal microscopy reconstruction of a transmural segment of rat heart by the Auckland group led by Peter Hunter (Hunter *et al.* 2002). Abbreviations: ML, markup language; ODE, ordinary differential equations; PDE, partial differential equations. Reproduced with Permission from Hunter *et al.* (2002).

Exp Physiol 93.1 pp 16–26   Systems biology and the future of physiology   23

L-type calcium current is necessary for the upstroke of the action potential in the SA node of most species, we would find that we had abolished normal pacemaker rhythm. Do we then call the gene for L-type calcium channels the 'pacemaker' gene? The reason why this is unsatisfactory, even misleading, to a systems-level biologist is obvious. Yet it is the process by which we label many genes with high-level functions. The steadily growing list of 'cancer genes' have been identified in this way, by determining which mutations (including deletions) change the probability of cancer occurring. We can be fairly sure though that this characteristic is not why they were selected during the evolutionary process. In this sense, there are no 'cancer genes'. As the Gene Ontology (GO) Consortium (http://geneontology.org/) puts it, 'oncogenesis is not a valid GO term because causing cancer is not the normal function of any gene'.

Another good example of this approach is the discovery of what are called clock genes, involved in circadian rhythm. Mutations in a single gene (now called the *period* gene) are sufficient to abolish the circadian period of fruit flies (Konopka & Benzer, 1971). This discovery of the first 'clock gene' was a landmark, since it was the first time that a single gene had been identified as playing such a key role in a high-level biological rhythm. The expression levels of this gene are clearly part of the rhythm generator. They vary (in a daily cycle) in advance of the variations in the protein for which they code. The reason is that the protein is involved in a negative feedback loop with the gene that codes for it (Hardin *et al.* 1990). The idea is very simple. The protein levels build up in the cell as the *period* gene is read to produce more protein. The protein then diffuses into the nucleus, where it inhibits further production of itself by binding to the promoter part of the gene sequence. With a time delay, the protein production falls off and the inhibition is removed so that the whole cycle can start again. So, we not only have a single gene capable of regulating the biological clockwork that generates circadian rhythm, it is itself a key component in the feedback loop that forms the rhythm generator.

However, such rhythmic mechanisms do not work in isolation. There has to be some connection with light-sensitive receptors (including the eyes). Only then will the mechanism lock on to a proper 24 h cycle rather than free-running at say 23 or 25 h. In the mouse, for example, many other factors play a role. Moreover, the clock gene itself is involved in other functions. That is why Foster and Kreitzman have written 'What we call a clock gene may have an important function within the system, but it could be involved in other systems as well. Without a complete picture of all the components and their interactions, it is impossible to tell what is part of an oscillator generating rhythmicity, what is part of an input, and what is part of an output. In a phrase, it ain't that simple!' (Foster & Kreitzman, 2004).

Indeed not. The *period* gene has also been found to be implicated in embryonic development as the adult fly is formed over several days, and it is deeply involved in coding for the male love songs generated by wing-beat oscillations which are specific to each of around 5000 species of fruit fly and ensure that courtship is with the right species. Perhaps it should be renamed the 'fruit fly love gene'!

The point is obvious. We should not be misled by gene ontology. The first function a gene is found to be involved in is rarely, if ever, the only one and may not even be the most important one. Gene ontology will require higher-level insight to be successful in its mission. Moreover, current methods of relating genotype to phenotype suffer from a major methodological limitation: by determining the effects of *changes* (mutations) in the genome, we can say little *a priori* on the direct causal relations between wild-type genes and the phenotype. They reveal simply the *differences* produced as a result of the *change* in genotype. All the causal effects *common* to both the wild-type and the mutated gene are hidden. What is observed may be just the tip of the iceberg.

Gene ontology in its fullest sense, as originally conceived by geneticists to relate genes to high-level features, is therefore very difficult and subject to many traps for the unwary. This would explain why projects such as the GO Consortium are more limited in their scope. Thus, GO assigns three categories to a gene, namely molecular function, biological process and cellular component, which are not intended to deal with higher-level function. It specifically excludes protein domains or structural features, protein–protein interactions, anatomical or histological features above the level of cellular components, including cell types, and it excludes the environment, evolution and expression. In other words, it excludes virtually all of what we classically understand by physiology and most aspects of evolutionary biology.

Sixth principle: there is no genetic program. No genetic programs? Surely, they are all over the place! They are the crown jewels of the molecular genetic revolution, invented by none other than the famous French Nobel Prize winners, Monod and Jacob (Monod & Jacob, 1961; Jacob, 1970). Their enticing idea was born during the early days of electronic computing, when computers were fed with paper tape or punched cards coded with sequences of instructions. Those instructions were clearly separate from the machine itself that performed the operations. They dictated those operations. Moreover, the coding is digital. The analogy with the digital code of DNA is obvious. So, are the DNA sequences comparable to the instructions of a computer program?

An important feature of such computer programs is that the program is separate from the activities of the machine that it controls. Originally, the separation was physically complete, with the program on the tape or cards only loaded temporarily into the machine. Nowadays, the programs are stored within the memory of the machine, and the strict distinction between the program, the data and the processes controlled may be breaking down. Perhaps computers are becoming more like living systems, but in any case the concept of a genetic program was born in the days when programs were separate, identifiable sets of instructions.

So, what do we find when we look for genetic programs in an organism? We find no genetic programs! There are no sequences of instructions in the genome that could possibly play a role similar to that of a computer program. The reason is very simple. A database, used by the system as a whole, is not a program. To find anything comparable to a program we have to extend our search well beyond the genome itself. Thus, as we have seen above, the sequence of events that generates circadian rhythm includes the *period* gene, but it necessarily also includes the protein for which it codes, the cell in which its concentration changes and the nuclear membrane across which it is transported with the correct speed to effect its inhibition of transcription. This is a gene–protein–lipid–cell network, not simply a gene network. The nomenclature matters. Calling it a gene network fuels the misconception of genetic determinism. In the generation of a 24 h rhythm, none of these events in the feedback loop is privileged over any other. Remove any of them, not just the gene, and you no longer have circadian rhythm.

Moreover, it would be strange to call this network of interactions a program. The network of interactions *is itself the circadian rhythm process.* As Enrico Coen, the distinguished plant geneticist, put it, 'Organisms are not simply manufactured according to a set of instructions. There is no easy way to separate instructions from the process of carrying them out, to distinguish plan from execution' (Coen, 1999). In short, the concept of a program here is completely redundant. It adds nothing to what a systems approach to such processes can reveal.

Seventh principle: there are no programs at any other level. I have introduced the analogy of the genome as a database and the metaphor of 'genes as prisoners' in order to provoke the change in mindset that is necessary for a fully systems approach to biology to be appreciated. The higher levels of the organism 'use the database' and 'play the genome' to produce functionality. If the genome can be likened to a huge pipe organ (Noble, 2006; chapter 2), then it seems correct to ask who is the player, who was the composer? If we can't find the program of life at the level of the genome, at what level do we find it? The answer is 'nowhere'!

We should view all such metaphors simply as ladders of understanding. Once we have used them we can, as it were, throw them away. This way of thinking can seem strange to some scientists for whom there must be just one correct answer to any scientific question. I explore this important issue in *The Music of Life* by analysing the 'selfish gene' and 'prisoner gene' metaphors linguistically to reveal that no conceivable experiment could decide which is correct (Noble, 2006; chapter 1). They highlight totally different aspects of the properties of genes. This philosophy is applied throughout the book as it answers questions like 'where is the program of life?' The conclusion is simply that there are no such programs at any level. At all levels, the concept of a program is redundant since, as with the circadian rhythm network, the networks of events that might be interpreted as programs are themselves the functions we are seeking to understand. Thus, there is no program for the heart's pacemaker separate from the pacemaker network itself.

While causality operates within and between all levels of biological systems, there are certain levels at which so many functions are integrated that we can refer to them as important levels of abstraction. Sydney Brenner wrote, 'I believe very strongly that the fundamental unit, the correct level of abstraction, is the cell and not the genome' (unpublished Lecture, Columbia University, 2003). He is correct, since the development of the eukaryotic cell was a fundamental stage in evolutionary development, doubtless requiring at least a billion years to be achieved. To systems physiologists though there are other important levels of abstraction, including whole organs and systems.

Eighth principle: there are no programs in the brain. In his book *The Astonishing Hypothesis*, Francis Crick proclaimed, 'You, your joys and your sorrows, your memories and your ambitions, your sense of personal identity and free will, are in fact no more than the behaviour of a vast assembly of nerve cells and their associated molecules' (Crick, 1994). This is a variation of the idea that in some sense or other, the mind is just a function of the brain. The pancreas secretes insulin, endocrine glands secrete hormones ... and the brain 'secretes' consciousness! All that's left is to find out how and where in the brain that happens. In one of his last statements, Crick has even hinted at where that may be: 'I think the secret of consciousness lies in the claustrum' (Francis Crick, 2004, quoted by V. S. Ramachanran, in *The Astonishing Francis Crick*, Edge, 18 October, 2004, http://www.edge.org/3rd culture/crick04/crick04 index. html). This structure is a thin layer of nerve cells in the brain. It is very small and it has many connections to other parts of the brain, but the details are of no importance to the argument. The choice of brain location for the 'secret of consciousness' varies greatly according to the author. Descartes even thought that it was in the pineal gland. The mistake is always the same, which is to think that in some way or other the brain is a kind of performance space in which the world of perceptions is reconstructed

Exp Physiol 93.1 pp 16–26   Systems biology and the future of physiology   25

inside our heads and presented to us as a kind of Cartesian theatre. But that way of looking at the brain leaves open the question: where is the 'I', the conscious self that sees these reconstructions? Must that be another part of the brain that views these representations of the outside world?

We are faced here with a mistake similar to that of imagining that there must be programs in the genomes, cells, tissues and organs of the body. There are no such programs, even in the brain. The activity of the brain and of the rest of the body simply *is* the activity of the person, the self. Once again, the concept of a program is superfluous. When a guitarist plays the strings of his guitar at an automatic speed that comes from frequent practice, there is no separate program that is making him carry out this activity. The patterns and processes in his nervous system and the associated activities of the rest of his body simply *are* him playing the guitar. Similarly, when we deliberate intentionally, there is no nervous network 'forcing' us to a particular deliberation. The nervous networks, the chemistry of our bodies, together with all their interactions within the social context in which any intentional deliberation makes sense, *are* us acting intentionally. Looking for something in addition to those processes is a mistake.

Ninth principle: the self is not an object. In brief, the mind is not a separate object competing for activity and influence with the molecules of the body. Thinking in that way was originally the mistake of the dualists, such as Sherrington and Eccles, led by the philosophy of Descartes. Modern biologists have abandoned the separate substance idea, but many still cling to a materialist version of the same mistake (Bennett & Hacker, 2003), based on the idea that somewhere in the brain the self is to be found as some neuronal process. The reason why that level of integration is too low is that the brain, and the rest of our bodies which are essential for attributes such as consciousness to make sense (Noble, 2006; chapter 9), are tools (back to the database idea again) in an integrative process that occurs at a higher level involving social interactions. We cannot attribute the concept of self-ness to ourselves without also doing so to others (Strawson, 1959). Contrary to Crick's view, therefore, our selves are indeed much 'more than the behaviour of a vast assembly of nerve cells and their associated molecules' precisely because the social interactions are essential even to understanding what something like an intention might be. I analyse an example of this point in much more detail in chapter 9 of *The Music of Life*. This philosophical point is easier to understand when we take a systems view of biology, since it is in many ways an extension of that view to the highest level of integration in the organism.

Conclusions

Tenth principle: there are many more to be discovered; a genuine 'theory of biology' does not yet exist. Well, of course, choosing just 10 principles was too limiting. This last one points the way to many others of whose existence we have only vague ideas. We do not yet have a genuine theory of biology. The Theory of Evolution is not a theory in the sense in which I am using the term. It is more an historical account, itself standing in need of explanation. We don't even know yet whether it consists of events that are difficult, if not impossible, to analyse fully from a scientific perspective, or whether it was a process that would have homed in to the organisms we have, regardless of the conditions. My own suspicion is that it is most unlikely that, if we could turn the clock right back and let the process run again, we would end up with anything like the range of species we have today on earth (Gould, 2002).

But, whichever side of this particular debate you may prefer, the search for general principles that could form the basis of a genuine theory of biology is an important aim of systems biology. Can we identify the logic by which the organisms we find today have succeeded in the competition for survival? In searching for that logic, we should not restrict ourselves to the lower levels. Much of the logic of living systems is to be found at the higher levels, since these are often the levels at which selection has operated (Keller, 1999; Gould, 2002) and determined whether organisms live or die. This is the level at which physiology works. Physiology therefore has a major contribution to make to systems biology.

In conclusion, I return to the theme with which this article began. Claude Bernard's concept of the constancy of the internal environment was the first example of multilevel functionality. It was critical in defining physiology as a subject distinct from the applications of physics and chemistry. The challenge we face today resembles that faced by Bernard in the mid-nineteenth century, but the chemistry involved is that of the molecule DNA. The answer though should be much the same. Higher-level control cannot be reduced to lower-level databases like the genome. A major part of the future of physiology surely lies in returning to our roots. Higher-level systems biology is, I suggest, classical physiology by another name.

References

Anway MD, Cupp AS, Uzumcu M & Skinner MK (2005). Epigenetic transgenerational actions of endocrine disruptors and male fertility. *Science* **308**, 1466–1469.

Auffray C & Nottale L (2008). Scale relativity theory and integrative systems biology 1. Founding principles and scale laws. *Prog Biophys Mol Biol*, in press.

26   D. Noble   *Exp Physiol* 93.1 pp 16–26

Bennett MR & Hacker PMS (2003). *Philosophical Foundations of Neuroscience*. Blackwell Publishing, Oxford.

Bernard C (1865, 1984). *Introduction a L'étude de la Médecine Expérimentale*. Flammarion, Paris.

Bers DM & Guo T (2005). Calcium signaling in cardiac ventricular myocytes. *Ann New York Acad Sci* **1047**, 86–98.

Berzelius (1815). *Afhandlingar I Fysik, Kemi och Mineralogi*, Stockholm: **4**, 307.

Cartwright N (1999). *The Dappled World. A Study of the Boundaries of Science*. Cambridge University Press, Cambridge.

Coen E (1999). *The Art of Genes*. Oxford University Press, Oxford.

Crampin EJ, Halstead M, Hunter PJ, Nielsen P, Noble D, Smith N & Tawhai M (2004). Computational physiology and the physiome project. *Exp Physiol* **89**, 1–26.

Crick FHC (1970). Central dogma of molecular biology. *Nature* **227**, 561–563.

Crick FHC (1994). *The Astonishing Hypothesis: the Scientific Search for the Soul*. Simon and Schuster, London.

Dawkins R (1976). *The Selfish Gene*. Oxford University Press, Oxford.

Dawkins R (2003). *A Devil's Chaplain*. Weidenfeld and Nicolson, London.

Deisseroth K, Mermelstein PG, Xia H & Tsien RW (2003). Signaling from synapse to nucleus: the logic behind the mechanisms. *Curr Opin Neurobiol* **13**, 354–365.

Dupré J (1993). *The Disorder of Things*. Harvard, Cambridge, MA, USA.

Feytmans E, Noble D & Peitsch M (2005). Genome size and numbers of biological functions. *Trans Comput Systems Biol* **1**, 44–49.

Finar IL (1964). *Organic Chemistry*. Longmans, London.

Foster R & Kreitzman L (2004). *Rhythms of Life*. Profile Books, London.

Frankland E & Kolbe H (1845). Upon the chemical constitution of metacetonic acid, and some other bodies related to it. *Mem. Proc. Chem. Soc.* **1865**, 386–391.

Goldenfeld N & Woese C (2007). Biology's next revolution. *Nature* **445**, 369.

Gould SJ (2002). *The Structure of Evolutionary Theory*. Harvard, Cambridge, MA, USA.

Hardin PE, Hall JC & Rosbash M (1990). Feedback of the *Drosophila* period gene product on circadian cycling of its messenger RNA levels. *Nature* **343**, 536–540.

Hodgkin AL (1975). The optimum density of sodium channels in an unmyelinated nerve. *Proc Royal Soc Lond B Biol Sci* **270**, 297–300.

Holmes FL (1974). *Claude Bernard and Animal Chemistry. The Emergence of a Scientist*. Harvard, Cambridge, MA, USA.

Hunter PJ, Robbins P & Noble D (2002). The IUPS human physiome project. *Pflugers Arch* **445**, 1–9.

Jablonka E & Lamb M (1995). *Epigenetic Inheritance and Evolution. The Lamarckian Dimension*. Oxford University Press, Oxford.

Jablonka E & Lamb M (2005). *Evolution in Four Dimensions*. MIT Press, Boston, MA, USA.

Jack JJB, Noble D & Tsien RW (1975). *Electric Current Flow in Excitable Cells*. Oxford University Press, Oxford.

Jacob F (1970). *La Logique Du Vivant, une Histoire de L'hérédité*. Gallimard, Paris.

Keller EF (2002). *Making Sense of Life. Explaining Biological Development with Models, Metaphors and Machines*. Harvard, Cambridge, MA, USA.

Keller L (1999). *Levels of Selection in Evolution*. Princeton University Press, Princeton, NJ, USA.

Kitcher P (1982). Genes. *Br J Philosophy Sci* **33**, 337–359.

Konopka RJ & Benzer S (1971). Clock mutants of *Drosophila melanogaster*. *Proc Natl Acad Sci U S A* **68**, 2112–2116.

Lémery N (1675). *Cours de Chymie*. Paris: Michallet.

Lavoisier A (1784). *Traité élémentaire de chimie, présenté dans un ordre nouveau et d'après les découvertes modernes*, 2 vols. Paris: Chez Cuchet.

Maynard Smith J (1998). *Evolutionary Genetics*. Oxford University Press, New York.

Maynard Smith J & Szathmáry E (1999). *The Origins of Life*. Oxford University Press, New York.

Mayr E (1982). *The Growth of Biological Thought*. Harvard, Cambridge, MA, USA.

Monod J & Jacob F (1961). Teleonomic mechanisms in cellular metabolism, growth and differentiation. *Cold Spring Harb Symp Quant Biol* **26**, 389–401.

Noble D (2002). Modelling the heart: insights, failures and progress. *Bioessays* **24**, 1155–1163.

Noble D (2006). *The Music of Life*. Oxford University Press, Oxford.

Noble D (2007). Mind over molecule: activating biological demons. *Annals N Y Acad Sci*, in press.

Noble D & Noble SJ (1984). A model of sino-atrial node electrical activity using a modification of the DiFrancesco-Noble (1984) equations. *Proc Royal Soc Lond B Biol Sci* **222**, 295–304.

Nottale L (1993). *Fractal Space-Time and Microphysics: Towards a Theory of Scale Relativity*. World Scientific, Singapore.

Nottale L (2000). *La Relativité Dans Tous Ses Etats. Du Mouvements Aux Changements D'échelle*. Hachette, Paris.

Nottale L & Auffray C (2008). Scale relativity and integrative systems biology 2. Macroscopic quantum-type mechanics. *Prog Biophys Mol Biol*, in press.

Paton WDM (1976). An experiment of Claude Bernard on curare: the origins of the Physiological Society. *J Physiol* **263**, 26P–29P.

Pichot A (1999). *Histoire de la Notion de Gène*. Flammarion, Paris.

Qiu J (2006). Unfinished symphony. *Nature* **441**, 143–145.

Schrödinger E (1944). *What Is Life?* Cambridge University Press, Cambridge, UK.

Strawson PF (1959). *Individuals*. Routledge, London.

Wöhler F (1828). Ueber künstliche Bildung des Harnstoffs. *Ann. Chim. Phys.* **37**, 330.

Phil. Trans. R. Soc. A (2008) **366**, 3001–3015
doi:10.1098/rsta.2008.0086
Published online 17 June 2008

REVIEW

Genes and causation

By Denis Noble*

*Department of Physiology, Anatomy and Genetics, University of Oxford,
Parks Road, Oxford OX1 3PT, UK*

Relating genotypes to phenotypes is problematic not only owing to the extreme complexity of the interactions between genes, proteins and high-level physiological functions but also because the paradigms for genetic causality in biological systems are seriously confused. This paper examines some of the misconceptions, starting with the changing definitions of a gene, from the cause of phenotype characters to the stretches of DNA. I then assess whether the 'digital' nature of DNA sequences guarantees primacy in causation compared to non-DNA inheritance, whether it is meaningful or useful to refer to genetic programs, and the role of high-level (downward) causation. The metaphors that served us well during the molecular biological phase of recent decades have limited or even misleading impacts in the multilevel world of systems biology. New paradigms are needed if we are to succeed in unravelling multifactorial genetic causation at higher levels of physiological function and so to explain the phenomena that genetics was originally about. Because it can solve the 'genetic differential effect problem', modelling of biological function has an essential role to play in unravelling genetic causation.

Keywords: genes; genetic causation; genetic program; digital coding;
analogue representation; cell inheritance

1. Introduction: what is a gene?

At first sight, the question raised by this paper seems simple. Genes transmit inherited characteristics; so in each individual they must be the cause of those characteristics. And so it was when the idea of a gene was first mooted. The word itself was coined by Johannsen (1909), but the concept already existed and was based on 'the silent assumption [that] was made almost universally that there is a 1:1 relation between genetic factor (gene) and character' (Mayr 1982).

Since then, the concept of a gene has changed fundamentally (Kitcher 1982; Mayr 1982; Dupré 1993; Pichot 1999; Keller 2000a,b), and this is a major source of confusion when it comes to the question of causation. Its original biological meaning referred to the cause of an inheritable phenotype characteristic, such as

*denis.noble@physiol.ox.ac.uk

One contribution of 12 to a Theme Issue 'The virtual physiological human: building a framework for computational biomedicine I'.

eye/hair/skin colour, body shape and weight, number of legs/arms/wings, to which we could perhaps add more complex traits such as intelligence, personality and sexuality.

The molecular biological definition of a gene is very different. Following the discovery that DNA codes for proteins, the definition shifted to locatable regions of DNA sequences with identifiable beginnings and endings. Complexity was added through the discovery of regulatory elements, but the basic cause of phenotype characteristics was still the DNA sequence since that determined which protein was made, which in turn interacted with the rest of the organism to produce the phenotype.

But unless we subscribe to the view that the inheritance of all phenotype characteristics is attributable entirely to DNA sequences (which I will show is just false) then genes, as originally conceived, are not the same as the stretches of DNA. According to the original view, genes were necessarily the cause of inheritable phenotypes since that is how they were defined. The issue of causation is now open precisely because the modern definition identifies them instead with DNA sequences.

This is not a point that is restricted to the vexed question of the balance of nature versus nurture. Even if we could separate those out and arrive at percentages attributable to one or the other (which I believe is misconceived in a system of nonlinear interactions and in which either on its own is equal to zero), we would still be faced with the fact that not all the 'nature' characteristics are attributable to DNA alone. Indeed, as we will see as we come to the conclusion of this paper, strictly speaking no genetic characteristics as originally defined by geneticists in terms of the phenotype could possibly be attributable to DNA alone.

My first point therefore is that the original concept of a gene has been taken over and significantly changed by molecular biology. This has undoubtedly led to a great clarification of *molecular* mechanisms, surely one of the greatest triumphs of twentieth-century biology, and widely acknowledged as such. But the more philosophical consequences of this change for higher level biology are profound and they are much less widely understood. They include the question of causation by genes. This is also what leads us to questions such as 'how many genes are there in the human genome?', and to the search to identify 'genes' in the DNA sequences.

2. Where does the genetic code lie?

Of course, it is an important question to ask which stretches of DNA code for proteins, and that is a perfectly good *molecular biological* question. It also leads us to wonder what the other stretches of DNA are used for, a question to which we are now beginning to find answers (Pearson 2006). But genetics, as originally conceived, is not just about what codes for each protein. Indeed, had it turned out (as in very simple organisms) that each coding stretch of DNA translates into just one protein, then it would have been as valid to say that the genetic code lies in the protein sequences, as was originally thought (Schrödinger 1944). We are then still left with the question 'how do these sequences, whether DNA or protein, generate the phenotypic characteristics that we wish to explain?' Looked at from this viewpoint, modern molecular biology, starting with Watson and

Crick's work, has succeeded brilliantly in mapping sequences of DNA to those of amino acids in proteins, but not in explaining phenotype inheritance. Whether we start from DNA or protein sequences, the question is still there. It lies in the complexity of the way in which the DNA and proteins are used by the organism to generate the phenotype. Life is not a soup of proteins.

The existence of multiple splice variants and genetic 'dark matter' (only 1–2% of the human genome actually codes for proteins, but much of the rest codes for non-protein coding RNA; Bickel & Morris 2006; Pearson 2006) has made this question more complicated in higher organisms, while epigenetics (gene marking) makes it even more so (Qiu 2006; Bird 2007), but the fundamental point remains true even for higher organisms. In a more complicated way, the 'code' could still be seen to reside in the proteins. Some (e.g. Scherrer & Jost 2007) have even suggested that we should redefine genes to be the completed mRNA before translation into a polypeptide sequence (see also Noble 2008, in press). In that case, there would be as many as 500 000 genes rather than 25 000. The more complex genome structure (of multiple exons and introns and the way in which the DNA is folded in chromosomes) could then be viewed as an efficient way of preserving and transmitting the 'real' causes of biological activity, the proteins. It is still true that, if we identify genes as just the stretches of DNA and identify them by the proteins they code for, we are already failing to address the important issues in relation to genetic determinism of the phenotype. By accepting the molecular biological redefinition of 'gene', we foreclose some of the questions I want to ask. For, having redefined what we mean by a gene, many people have automatically taken over the concept of necessary causation that was correctly associated with the original idea of a gene, but which I will argue is incorrectly associated with the new definition, except in the limited case of generating proteins from DNA. This redefinition is not therefore just an arcane matter of scientific history. It is part of the mindset that needs to change if we are to understand the full nature of the challenge we face.

3. Digital versus analogue genetic determinism

The main reason why it is just false to say that all nature characteristics are attributable to DNA sequences is that, by itself, DNA does nothing at all. We also inherit the complete egg cell, together with any epigenetic characteristics transmitted by sperm (in addition to its DNA), and all the epigenetic influences of the mother and environment. Of course, the latter begins to be about 'nurture' rather than nature, but one of my points in this paper is that this distinction is fuzzy. The proteins that initiate gene transcription in the egg cell and impose an expression pattern on the genome are initially from the mother, and other such influences continue throughout development in the womb and have influences well into later life (Gluckman & Hanson 2004). Where we draw the line between nature and nurture is not at all obvious. There is an almost seamless transition from one to the other. 'Lamarckism', the inheritance of acquired characteristics, lurks in this fuzzy crack to a degree yet to be defined (Jablonka & Lamb 1995, 2005).

This inheritance of the egg cell machinery is important for two reasons. First, it is the egg cell gene reading machinery (a set of approx. 100 proteins and the associated cellular ribosome architecture) that enables the DNA to be used to make more proteins. Second, the complete set of other cellular elements, mitochondria, endoplasmic reticulum, microtubules, nuclear and other membranes and a host (billions) of chemicals arranged specifically in cellular compartments, is also inherited. Much of this is not coded for by DNA sequences since they code only for RNA and proteins. Lipids certainly are not so coded. But they are absolutely essential to all the cell architecture. The nature of the lipids also determines how proteins behave. There is intricate two-way interaction between proteins and lipids (see Roux *et al.* 2008).

One way to look at this situation therefore is to say that there are two components to molecular inheritance: the genome DNA, which can be viewed as digital information, and the cellular machinery, which can, perhaps by contrast, be viewed as analogue information. I will refer to both of these as 'molecular inheritance' to emphasize that the distinction at this point in my argument is not between genetic molecular inheritance and higher-level causes. The egg cell machinery is just as molecular as the DNA. We will come to higher-level causation later.

The difference lies elsewhere. Both are used to enable the organism to capture and build the new molecules that enable it to develop, but the process involves a coding step in the case of DNA and proteins, while no such step is involved in the rest of the molecular inheritance. *This is the essential difference.*

The coding step in the case of the relationship between DNA and proteins is what leads us to regard the information as digital. This is what enables us to give a precise number to the base pairs (3 billion in the case of the human genome). Moreover, the CGAT code could be completely represented by binary code of the kind we use in computers. (Note that the code here is metaphorical in a biological context—no one has determined that this should be a code in the usual sense. For that reason, some people have suggested that the word 'cipher' would be better.)

By contrast, we cannot put similar precise numbers to the information content of the rest of the molecular inheritance. The numbers of molecules involved (trillions) would be largely irrelevant since many are exactly the same, though their organization and compartmentalization also need to be represented. We could therefore ask how much digital information would be required to 'represent' the non-DNA inheritance but, as with encoding of images, that depends on the resolution with which we seek to represent the information digitally. So, there is no simple answer to the question of a quantitative comparison of the DNA and non-DNA molecular inheritance. But given the sheer complexity of the egg cell—it took evolution at least 1 or 2 billion years to get to the eukaryotic cellular stage—we can say that it must be false to regard the genome as a 'vast' database while regarding the rest of the cell as somehow 'small' by comparison. At fine enough resolution, the egg cell must contain even more information than the genome. If it needed to be coded digitally to enable us to 'store' all the information necessary to recreate life in, say, some distant extra-solar system by sending it out in an 'Earth-life' information capsule, I strongly suspect that most of that information would be non-genomic. In fact, it would be almost useless to send just DNA information in such a capsule. The chances of any recipients anywhere in the Universe having egg cells and a womb capable of permitting the DNA of life on Earth to 'come alive' may be close to zero. We might as well pack the capsule with the bar codes of a supermarket shelf!

4. Is digital information privileged?

Of course, quantity of information is not the only criterion we could choose. Whatever its proportion would be in my imagined Earth-life capsule, some information may be more important than others. So, which is privileged in inheritance? Would it be the cell or the DNA? 'How central is the genome?' as Werner puts the question (Werner 2007). On the basis of our present scientific knowledge, there are several ways in which many people would seek to give primacy to the DNA.

The first is the fact that, since it can be viewed as digital information, in our computer-oriented age, that can appear to give it more security, to ensure that it is more reliable, much as the music recorded on a CD is said to be 'clearer' and less 'noisy' than that on a vinyl disc. Digital information is discrete and fixed, whereas analogue information is fuzzy and imprecise. But I wonder whether that is entirely correct. Large genomes actually require correcting machinery to ensure their preciseness. Nevertheless, with such machinery, it clearly is secure enough to act as reliably inheritable material. By contrast, it could be said that attempting to reduce analogue information, such as image data, to digital form is always fuzzy since it involves a compromise over questions such as resolution. But this criterion already biases us towards the DNA. We need to ask the fundamental question 'why do we need to prioritize digital information?' After all, DNA needs a digital code simply and precisely because it does not code only for itself. It codes for another type of molecule, the proteins. The rest of the cellular machinery does not need a code, or to be reduced to digital information, *precisely because it represents itself.* To Dawkins' famous description of DNA as the eternal replicator (Dawkins 1976, ch. 2), we should add that egg cells, and sperm, also form an eternal line, just as do all unicellular organisms. DNA cannot form an eternal line on its own.

So, although we might characterize the cell information as analogue, that is only to contrast it with being digital. But it is *not* an analogue *representation.* It itself *is* the self-sustaining structure that we inherit and it reproduces itself directly. Cells make more cells, which make more cells (and use DNA to do so), ..., etc. The inheritance is robust: liver cells make liver cells for many generations of liver cells, at each stage marking their genomes to make that possible. So do all the other 200 or so cell types in the body (Noble 2006, ch. 7). Yet, the genome is the same throughout. That common 'digital' code is made to dance to the totally different instructions of the specific cell types. Those instructions are 'analogue', in the form of continuous variations in imposed patterns of gene expression. The mistake in thinking of gene expression as digital lies in focusing entirely on the CGAT codes, not on the continuously variable degree of expression. It is surely artificial to emphasize one or the other. When it comes to the pattern of expression levels, the information is analogue.

So, I do not think we get much leverage on the question of privileged causality (DNA or non-DNA) through the digital–analogue comparison route. We might even see the digital coding itself as the really hazardous step—and indeed it does require complex machinery to check for errors in large genomes (Maynard Smith & Szathmáry 1995; Maynard Smith 1998). Having lipid membranes that automatically 'accept' certain lipids to integrate into their structure and so to grow, enable cells to divide and so on seems also to be chemically reliable. The lipid membranes are also good chemical replicators. That process was probably 'discovered' and 'refined' by evolution long before cells 'captured' genes and started the process towards the full development of cells as we now know them. I suspect that primitive cells, probably not much more than lipid envelopes with a few RNA enzymes (Maynard Smith & Szathmáry 1995, 1999), 'knew' how to divide and have progeny long before they acquired DNA genomes.

5. An impossible experiment

Could we get a hold on the question by a more direct (but currently and probably always impossible; Keller 2000*a*,*b*) biological experiment? Would the complete DNA sequence be sufficient to 'resurrect' an extinct species? Could dinosaur DNA (let us forget about all the technical problems here), for example, be inserted into, say, a bird egg cell. Would it generate a dinosaur, a bird, or some extraordinary hybrids?

At first sight, this experiment seems to settle the question. If we get a dinosaur, then DNA is the primary, privileged information. The non-DNA is secondary. I suspect that this is what most 'genetic determinists' would expect. If we get a bird, then the reverse is true (this is *highly* unlikely in my or anyone else's view). If we get a hybrid, or nothing (I suspect that this would be the most likely outcome), we could maintain a view of DNA primacy by simply saying that there is, from the DNA's point of view, a fault in the egg cell machinery. But note the phrase 'DNA's point of view' in that sentence. It already gives the DNA primacy and so begs the question.

The questions involved in such experiments are important. Cross-species clones are of practical importance as a possible source of stem cells. They could also reveal the extent to which egg cells are species specific. This is an old question. Many early theories of what was called 'cytoplasm inheritance' were eventually proved wrong (Mayr 1982), though Mayr notes that 'The old belief that the cytoplasm is important in inheritance ... is not dead, although it has been enormously modified.' I suspect that the failure of most cross-species clones to develop to the adult stage is revealing precisely the extent to which 'the elaborate architecture of the cytoplasm plays a greater role than is now realized' (Mayr 1982). Since we cannot have the equivalent of mutations in the case of the non-DNA inheritance, using different species may be our only route to answering the question.

Interspecies cloning has already been attempted, though not with extinct animals. About a decade ago, J. B. Cibelli of Michigan State University tried to insert his own DNA into a cow egg cell and even patented the technique. The experiment was a failure and ethically highly controversial. Cibelli has since failed to clone monkey genes in cow's eggs. The only successful case is of a wild ox (a banteng *Bos javanicus*) cloned in domestic cow's eggs. The chances are that the technique will work only on very closely related species. At first sight, a banteng looks very much like a cow and some have been domesticated in the same way. More usually, interspecies clones fail to develop much beyond the early embryo.

But however interesting these experiments are, they are misconceived as complete answers to the question I am raising. Genomes and cells have evolved together (Maynard Smith & Szathmáry 1995). Neither can do anything without the other. If we got a dinosaur from the imagined experiment, we would have to conclude that dinosaur and bird egg cells are sufficiently similar to make that possible. The difference (between birds and dinosaurs) would then lie in the DNA not in the rest of the egg cell. Remember that eukaryotic cells evolved aeons before dinosaurs and birds and so all cells necessarily have much of their machinery in common. But that difference does not give us grounds for privileging one set of information over the other. If I play a PAL video tape on a PAL reading machine, surely, I get a result that depends specifically on the information on the tape, and that would work equally well on another PAL reader, but I would get nothing at all on a machine that does not read PAL coding. The egg cell in our experiment still ensures that we get an organism at all, if indeed we do get one, and that it would have many of the characteristics that are common between dinosaurs and birds. The egg cell inheritance is not limited merely to the *differences* we find. It is essential for the *totality* of what we find. Each and every high-level function depends on effects attributable to *both* the DNA and the rest of the cell. 'Studying biological systems means more than breaking the system down into its components and focusing on the digital information encapsulated in each cell' (Neuman 2007).

6. The 'genetic differential effect problem'

This is a version of a more general argument relating to genes (defined here as DNA sequences) and their effects. Assignment of functions to genes depends on observing differences in phenotype consequent upon *changes* (mutations, knockouts, etc.) in genotype. Dawkins made this point very effectively when he wrote 'It is a fundamental truth, though it is not always realized, that *whenever* a geneticist studies a gene 'for' any phenotypic character, he is always referring to a *difference* between two alleles' (Dawkins 1982).

But differences cannot reveal the totality of functions that a gene may be involved in, since they cannot reveal all the effects that are common to the wild and mutated types. We may be looking at the tip of an iceberg. And we may even be looking at the wrong tip since we may be identifying a gene through the pathological effects of just one of its mutations rather than by what it does for which it must have been selected. This must be true of most so-called oncogenes, since causing cancer is unlikely to be a function for which the genes were selected. This is why the Gene Ontology (GO) Consortium (http://geneontology.org/) excludes oncogenesis: 'oncogenesis is not a valid GO term because causing cancer is not the normal function of any gene'. Actually, causing cancer could be a function if the gene concerned has other overwhelming beneficial effects. This is a version of the 'sickle cell' paradigm (Jones 1993, p. 219) and is the reason why I do not think oncogenesis could *never* be a function of a gene: nature plays with balances of positive and negative effects of genes (see 'Faustian pacts with the devil'; Noble 2006, p. 109).

Identifying genes by *differences* in phenotype correlated with *those* in genotype is therefore hazardous. Many, probably most, genetic modifications are buffered. Organisms are robust. They have to be to have succeeded in the evolutionary process. Even when the function of the gene is known to be significant, a knockout or mutation may not reveal that significance. I will refer to this problem as the genetic differential effect problem. My contention is that it is a very severe limitation in unravelling the causal effects of genes. I will propose a solution to the problem later in this paper.

It is also important to remember that large numbers (hundreds or more) of genes are involved in each and every high-level function and that, at that level, individual genes are involved in many functions. We cannot assume that the first phenotype–genotype correlation we found for a given gene is its only or even its main function.

7. Problems with the central dogma

The video reader is a good analogy so far as it goes in emphasizing that the reading machinery must be compatible with the coding material, but it is also seriously limited in the present context. It is best seen as an analogy for the situation seen by those who take an extension of the central dogma of biology as correct: information passes from the coded material to the rest of the system but not the other way. What we now know of epigenetics requires us to modify that view. The cell machinery does not just read the genome. It imposes extensive patterns of marking and expression on the genome (Qiu 2006). This is what makes the precise result of our imagined experiment so uncertain. According to the central dogma, if the egg cell is compatible, we will automatically get a dinosaur, because the DNA dictates everything. If epigenetic marking is important, then the egg cell also plays a determining, not a purely passive, role. There are therefore two kinds of influence that the egg cell exerts. The first is that it is totally necessary for any kind of organism at all to be produced. It is therefore a primary 'genetic cause' in the sense that it is essential to the production of the phenotype and is passed on between the generations. The second is that it exerts an influence on what kind of organism we find. It must be an empirical question to determine how large the second role is. At present, we are frustrated in trying to answer that question by the fact that virtually all cross-species clones do not develop into adults. As I have already noted, that result itself suggests that the second role is important.

It would also be an interesting empirical question to determine the range of species across which the egg cell machinery is sufficiently similar to enable different genomes to work, but that tells us about similarities of the match of different genomes with the egg cells of different species, and their mutual compatibility in enabling development, not about the primacy or otherwise of DNA or non-DNA inheritance. In all cases, the egg cell machinery is as necessary as the DNA. And, remember, as 'information' it is also vast.

Note also that what is transferred in cross-species cloning experiments is not just the DNA. Invariably, the whole nucleus is inserted, with all its machinery (Tian *et al.* 2003). If one takes the contribution of the egg cell seriously, that is a very serious limitation. The nucleus also has a complex architecture in addition to containing the DNA, and it must be full of transcription factors and other molecules that influence epigenetic marking. Strictly speaking, we should be looking at the results of inserting the raw DNA into a genome-free nucleus of an egg cell, not at inserting a whole nucleus, or even just the chromosomes, into an enucleated egg cell. No one has yet done that. And would we have to include the histones that mediate many epigenetic effects? This is one of the reasons, though by no means the only one, why the dinosaur cloning experiment may be impossible.

To conclude this section, if by genetic causation we mean the totality of the inherited causes of the phenotype, then it is plainly incorrect to exclude the non-DNA inheritance from this role, and it probably does not make much sense to ask which is more important, since only an interaction between DNA and non-DNA inheritance produces anything at all. Only when we focus more narrowly on changes in phenotype attributable to differences in genotype (which is how functionality of genes is currently assessed) could we plausibly argue that it is all down to the DNA, and even that conclusion is uncertain until we have carried out experiments that may reveal the extent to which egg cells are species specific, since nuclear DNA marking may well be very important.

8. Genetic programs?

Another analogy that has come from comparison between biological systems and computers is the idea of the DNA code being a kind of program. This idea was originally introduced by Monod & Jacob (1961) and a whole panoply of metaphors has now grown up around their idea. We talk of gene networks, master genes and gene switches. These metaphors have also fuelled the idea of genetic (DNA) determinism.

But there are no purely gene networks! Even the simplest example of such a network—that discovered to underlie circadian rhythm—is not a gene network, nor is there a gene for circadian rhythm. Or, if there is, then there are also proteins, lipids and other cellular machinery for circadian rhythm.

The circadian rhythm network involves at least three other types of molecular structures in addition to the DNA code. The stretch of DNA called the period gene (*per*) codes for a protein (PER) that builds up in the cell cytoplasm as the cellular ribosome machinery makes it. PER then diffuses slowly through the nuclear (lipid and protein) membrane to act as an inhibitor of *per* expression (Hardin *et al.* 1990). The cytoplasmic concentration of PER then falls, and the inhibition is slowly removed. Under suitable conditions, this process takes approximately 24 hours. It is the whole network that has this 24 hour rhythm, not the gene (Foster & Kreitzman 2004). However else this network can be described, it is clearly not a gene network. At the least, it is a gene–protein–lipid–cell network. It does not really make sense to view the gene as operating without the rest of the cellular machinery. So, if this network is part of a 'genetic program', then the genetic program is not a DNA program. It does not lie within the DNA coding. Moreover, as Foster & Kreitzman emphasized, there are many layers of interactions overlaid onto the basic mechanism—so much so that it is possible to knock out the *CLOCK* gene in mice and retain circadian rhythm (Debruyne *et al.* 2006). I prefer therefore to regard the DNA as a database rather than as a program (Atlan & Koppel 1990; Noble 2006). What we might describe as a program uses that database, but is not controlled by it.

The plant geneticist Coen (1999) goes even further. I will use my way of expressing his point, but I would like to acknowledge his ideas and experiments as a big influence on my thinking about this kind of question. In the early days of computing, during the period in which Monod & Jacob (1961) developed their idea of *le programme génétique*, a program was a set of instructions separate from the functionality it serves. The program was a complete piece of logic, a set of instructions, usually stored on cards or tapes, that required data to work on and outputs to produce. Pushing this idea in relation to the DNA/non-DNA issue, we arrive at the idea that there is a program in the DNA, while the data and output is the rest: the cell and its environment. Jacob was quite specific about the analogy: 'The programme is a model borrowed from electronic computers. It equates the genetic material with the magnetic tape of a computer' (Jacob 1982). That analogy is what leads people to talk of the DNA 'controlling' the rest of the organism.

Coen's point is that there is no such distinction in biological systems. As we have seen, even the simplest of the so-called gene networks are not 'gene programs' at all. The process is the functionality itself. There is no separate program. I see similar conclusions in relation to my own field of heart rhythm. There is no heart rhythm program (Noble 2008, in press), and certainly not a heart rhythm genetic program, separate from the phenomenon of heart rhythm itself. Surely, we can refer to the functioning networks of interactions involving genes, proteins, organelles, cells, etc. as programs if we really wish to. They can also be represented as carrying out a kind of computation (Brenner 1998), in the original von Neumann sense introduced in his theory of self-reproducing machines. But if we take this line, we must still recognize that this computation does not tell something else to carry out the function. It is itself the function.

Some will object that computers are no longer organized in the way they were in the 1960s. Indeed not, and the concept of a program has developed to the point at which distinctions between data and instructions, and even the idea of a separate logic from the machine itself, may have become outdated. Inasmuch as this has happened, it seems to me that such computers are getting a little closer to the organization of living systems.

Not only is the *period* gene not the determinant of circadian rhythm, either alone or as a part of a pure gene network, but also it could be argued that it is incorrect to call it a 'circadian rhythm' gene. Or, if it is, then it is also a development gene, for it is used in the development of the fly embryo. And it is a courtship gene! It is used in enabling male fruitflies to sing (via their wing-beat frequencies) to females of the correct species of fruitfly (more than 3000 such species are known). Genes in the sense of the stretches of DNA are therefore like pieces of re-usable Lego. That is, in principle, why there are very few genes compared with the vast complexity of biological functions. Needless to say, human courtship uses other genes! And all of those will be used in many other functions. My own preference would be to cease using high-level functionality for naming genes (meaning here DNA sequences), but I realize that this is now a lost cause. The best we can do is to poke fun at such naming, which is why I like the Fruit Fly Troubadour Gene story (Noble 2006, p. 72).

9. Higher-level causation

I have deliberately couched the arguments so far in molecular terms because I wish to emphasize that the opposition to simplistic gene determinism, gene networks and genetic programs is not based only on the distinction between higher- and lower-level causation, but also there are additional factors to be taken into account as a consequence of multilevel interactions.

The concept of level is itself problematic. It is a metaphor, and a very useful one in biology. Thus, there is a sense in which a cell, for example, and an organ or an immune system, is much more than its molecular components. In each of these cases, the molecules are constrained to cooperate in the functionality of the whole. Constrained by what? A physicist or an engineer would say that the constraints do not lie in the laws governing the behaviour of the individual components—the same quantum mechanical laws will be found in biological molecules as in molecules not forming part of a biological system. The constraints lie in the boundary and initial conditions: 'organisation becomes cause in the matter' (Strohman 2000; Neuman 2006). These conditions, in turn, are constrained by what? Well, ultimately by billions of years of evolution. That is why I have used the metaphor of evolution as the composer (Noble 2006, ch. 8). But that metaphor is itself limited. There may have been no direction to evolution (but for arguments against this strict view, see Jablonka & Lamb 2005). We are talking of a set of historical events, even of historical accidents. The information that is passed on through downward causation is precisely this set of initial and boundary conditions without which we could not even begin to integrate the equations representing molecular causality.

To spell this out in the case of the circadian rhythm process, this is what determines the cytoplasm volume in which the concentration of the protein changes, the speed with which it crosses the nuclear membrane, the speed with which ribosomes make new protein and so on. And those characteristics will have been selected by the evolutionary process to give a roughly 24 hour rhythm. Surely, each molecule in this process does not 'know' or represent such information, but the ensemble of molecules does. It behaves differently from the way in which it would behave if the conditions were different or if they did not exist at all. This is the sense in which molecular events are different as a consequence of the life process. Moreover, the boundary and initial conditions are essentially global properties, identifiable at the level at which they can be said to exist.

What is metaphorical here is the notion of 'up and down' (Noble 2006, ch. 10) —it would be perfectly possible to turn everything conceptually upside down so that we would speak of upward causation instead of downward causation. The choice is arbitrary, but important precisely because the principle of reductionism is always to look for 'lower-level' causes. That is the reductionist prejudice and it seems to me that it needs justification; it is another way in which we impose our view on the world.

Although the concept of level is metaphorical, it is nevertheless an essential basis for the idea of multilevel causation. The example I often give is that of pacemaker rhythm, which depends on another global property of cells, i.e. the electrical potential, influencing the behaviour of the individual proteins, the ionic channels, which in turn determine the potential. There is a multilevel feedback network here: channels→ionic current→electrical potential→channel opening or closing→ionic current and so on. This cycle is sometimes called the Hodgkin cycle, since it was Alan Hodgkin who originally identified it in the case of nerve excitation (Hodgkin & Huxley 1952).

Similarly, we can construct feedback networks of causation for many other biological functions. I see the identification of the level at which such networks are integrated, i.e. the highest level involved in the network, as being a primary aim of systems biology. This will also be the lowest level at which natural selection can operate since it is high-level functionality that determines whether organisms live or die. We must shift our focus away from the gene as the unit of selection to that of the whole organism (Tautz 1992).

But I also have hesitations about such language using the concepts of levels and causation. My book, in its last chapter, recommends throwing all the metaphors away once we have used them to gain insight (Noble 2006, ch. 10). In the case of the cycles involving downward causation, my hesitation is because such language can appear to make the causation involved be sequential in time. I do not see this as being the case. In fact, the cell potential influences the protein kinetics at exactly the same time as they influence the cell potential. Neither is primary or privileged as causal agency either in time or in space. This fact is evident in the differential equations we use. The physical laws represented in the equations themselves, and the initial and boundary conditions, operate *at the same time* (i.e. during every integration step, however infinitesimal), not sequentially.

This kind of conceptual problem (causality is one of our ways of making sense of the world, not the world's gift to us) underlies some knotty problems in thinking about such high-level properties as intentionality. As I show in *The music of life* (Noble 2006, ch. 9), looking for neural or, even worse, genetic 'causes' of an intention is such a will-of-the-wisp. I believe that this is the reason why the concept of downward causation may play a fundamental role in the philosophy of action (intentionality, free will, etc.).

I am also conscious of the fact that causality in any particular form does not need to be a feature of all successful scientific explanations. General relativity theory, for example, changes the nature of causality through replacing movement in space by geodesics in the structure of space–time. At the least, that example shows that a process that requires one form of causality (gravity acting at a distance between bodies) in one theoretical viewpoint can be seen from another viewpoint to be unnecessary. Moreover, there are different forms of causality, ranging from proximal causes (one billiard ball hitting another) to ultimate causes of the kind that evolutionary biologists seek in accounting for the survival value of biological functions and features. Genetic causality is a particularly vexed question partly not only because the concept of a gene has become problematic, as we have seen in this paper, but also because it is not usually a proximal cause. Genes, as we now define them in molecular biological terms, lie a long way from their phenotypic effects, which are exerted through many levels of biological organization and subject to many influences from both those levels and the environment. We do not know what theories are going to emerge in the future to cope with the phenomenon of life. But we can be aware that our ways of viewing life are almost certainly not the only ones. It may require a fundamental change in the mindset to provoke us to formulate new theories. I hope that this paper will contribute to that change in the mindset.

10. Unravelling genetic causation: the solution to the genetic differential effect problem

Earlier in this paper, I referred to this problem and promised a solution. The problem arises as an inherent difficulty in the 'forward' (reductionist) mode of explanation. The consequences of manipulations of the lowest end of the causal chain, the genes, can be hidden by the sheer cleverness of organisms to hide genetic mistakes and problems through what modern geneticists call genetic buffering and what earlier biologists would call redundancy or back-up mechanisms that kick in to save the functionality. The solution is not to rely solely on the forward mode of explanation. The backward mode is sometimes referred to as reverse engineering. The principle is that we start the explanation at the higher, functional level, using a model that incorporates the forward mode knowledge but, crucially, also incorporates higher level insights into functionality. For example, if we can successfully model the interactions between all the proteins involved in cardiac rhythm, we can then use the model to assess qualitatively and quantitatively the contribution that each gene product makes to the overall function. That is the strength of reverse engineering. We are no longer dealing just with differences. If the model is good, we are dealing with the totality of the gene function within the process we have modelled. We can even quantify the contribution of a gene product whose effect may be largely or even totally buffered when the gene is manipulated (see Noble 2006, p. 108). This is the reason why higher level modelling of biological function is an essential part of unravelling the functions of genes: 'Ultimately, *in silico* artificial genomes and *in vivo* natural genomes will translate into each other, providing both the possibility of forward and reverse engineering of natural genomes' (Werner 2005).

11. Conclusions

The original notion of a gene was closely linked to the causes of particular phenotype characteristics, so the question of causal relationships between genes and phenotype were circular and so hardly had much sense. The question of causality has become acute because genes are now identified more narrowly with particular sequences of DNA. The problem is that these sequences are uninterpretable outside the cellular context in which they can be read and so generate functionality. But that means that the cell is also an essential part of the inheritance and therefore was, implicitly at least, a part of the original definition of a gene. Depending on how we quantify the comparison between the contributions, it may even be the larger part. Genetic information is not confined to the digital information found in the genome. It also includes the analogue information in the fertilized egg cell. If we were ever to send out through space in an Earth-life capsule the information necessary to reconstruct life on Earth on some distant planet, we would have to include both forms of information. Now that we can sequence whole genomes, the difficult part would be encoding information on the cell. As Sydney Brenner has said, 'I believe very strongly that the fundamental unit, the correct level of abstraction, is the cell and not the genome' (Lecture to Columbia University in 2003). This fundamental insight has yet to be adopted by the biological science community in a way that will ensure

3014   *D. Noble*

success in unravelling the complexity of interactions between genes and their environment. In particular, the power of reverse engineering using mathematical models of biological function to unravel gene function needs to be appreciated. Multilevel systems biology requires a more sophisticated language when addressing the relationships between genomes and organisms.

Work in the author's laboratory is supported by EU FP6 BioSim network, EU FP7 PreDiCT project, BBSRC and EPSRC. I would like to acknowledge valuable discussions with Jonathan Bard, John Mulvey, James Schwaber, Eric Werner and the critical comments of the referees.

References

Atlan, H. & Koppel, M. 1990 The cellular computer DNA: program or data. *Bull. Math. Biol.* **52**, 335–348. (doi:10.1007/BF02458575)

Bickel, K. S. & Morris, D. R. 2006 Silencing the transcriptome's dark matter: mechanisms for suppressing translation of intergenic transcripts. *Mol. Cell* **22**, 309–316. (doi:10.1016/j.molcel. 2006.04.010)

Bird, A. 2007 Perceptions of epigenetics. *Nature* **447**, 396–398. (doi:10.1038/nature05913)

Brenner, S. 1998 Biological computation. In *The limits of reductionism in biology* (eds G. R. Bock & J. A. Goode). Novartis Foundation Symposium, no. 213, pp. 106–116. London, UK: Wiley.

Coen, E. 1999 *The art of genes.* Oxford, UK: Oxford University Press.

Dawkins, R. 1976 *The selfish gene.* Oxford, UK: Oxford University Press.

Dawkins, R. 1982 *The extended phenotype.* London, UK: Freeman.

Debruyne, J. P., Noton, E., Lambert, C. M., Maywood, E. S., Weaver, D. R. & Reppert, S. M. 2006 A clock shock: mouse CLOCK is not required for circadian oscillator function. *Neuron* **50**, 465–477. (doi:10.1016/j.neuron.2006.03.041)

Dupré, J. 1993 *The disorder of things.* Cambridge, MA: Harvard University Press.

Foster, R. & Kreitzman, L. 2004 *Rhythms of life.* London, UK: Profile Books.

Gluckman, P. & Hanson, M. 2004 *The fetal matrix. Evolution, development and disease.* Cambridge, UK: Cambridge University Press.

Hardin, P. E., Hall, J. C. & Rosbash, M. 1990 Feedback of the *Drosophila* period gene product on circadian cycling of its messenger RNA levels. *Nature* **343**, 536–540. (doi:10.1038/343536a0)

Hodgkin, A. L. & Huxley, A. F. 1952 A quantitative description of membrane current and its application to conduction and excitation in nerve. *J. Physiol.* **117**, 500–544.

Jablonka, E. & Lamb, M. 1995 *Epigenetic inheritance and evolution. The Lamarckian dimension.* Oxford, UK: Oxford University Press.

Jablonka, E. & Lamb, M. 2005 *Evolution in four dimensions.* Boston, MA: MIT Press.

Jacob, F. 1982 *The possible and the actual.* New York, NY: Pantheon Books.

Johannsen, W. 1909 *Elemente der exakten Erblichkeitslehre.* Jena, Germany: Gustav Fischer.

Jones, S. 1993 *The language of the genes.* London, UK: HarperCollins.

Keller, E. F. 2000*a The century of the gene.* Cambridge, MA: Harvard University Press.

Keller, E. F. 2000*b* Is there an organism in this text? In *Controlling our destinies. Historical, philosophical, ethical and theological perspectives on the human genome project* (ed. P. R. Sloan), pp. 273–288. Notre Dame, IN: University of Notre Dame Press.

Kitcher, P. 1982 Genes. *Br. J. Philos. Sci.* **33**, 337–359. (doi:10.1093/bjps/33.4.337)

Maynard Smith, J. 1998 *Evolutionary genetics.* New York, NY: Oxford University Press.

Maynard Smith, J. & Szathmáry, E. 1995 *The major transitions in evolution.* Oxford, UK: Oxford University Press.

Maynard Smith, J. & Szathmáry, E. 1999 *The origins of life.* New York, NY: Oxford University Press.

Mayr, E. 1982 *The growth of biological thought.* Cambridge, MA: Harvard University Press.

Monod, J. & Jacob, F. 1961 Teleonomic mechanisms in cellular metabolism, growth and differentiation. *Cold Spring Harb. Symp. Quant. Biol.* **26**, 389–401.

Neuman, Y. 2006 Cryptobiosis: a new theoretical perspective. *Progr. Biophys. Mol. Biol.* **92**, 258–267. (doi:10.1016/j.pbiomolbio.2005.11.001)

Neuman, Y. 2007 The rest is silence. *Perspect. Biol. Med.* **50**, 625–628. (doi:10.1353/pbm.2007.0053)

Noble, D. 2006 *The music of life.* Oxford, UK: Oxford University Press.

Noble, D. 2008 Claude Bernard, the first systems biologist, and the future of physiology. *Exp. Physiol.* **93**, 16–26. (doi:10.1113/expphysiol.2007.038695)

Noble, D. In press. Commentary on Scherrer & Jost (2007) Gene and genon concept: coding versus regulation. *Theory Biosci.* **127**.

Pearson, H. 2006 Genetics: what is a gene? *Nature* **441**, 398–401. (doi:10.1038/441398a)

Pichot, A. 1999 *Histoire de la notion de gène.* Paris, France: Flammarion.

Qiu, J. 2006 Epigenetics: unfinished symphony. *Nature* **441**, 143–145. (doi:10.1038/441143a)

Roux, A., Cuvelier, D., Bassereau, P. & Goud, B. 2008 Intracellular transport. From physics to biology. *Ann. NY Acad. Sci.* **1123**, 119–125. (doi:10.1196/annals.1420.014)

Scherrer, K. & Jost, J. 2007 Gene and genon concept: coding versus regulation. *Theory Biosci.* **126**, 65–113. (doi:10.1007/s12064-007-0012-x)

Schrödinger, E. 1944 *What is life?* Cambridge, UK: Cambridge University Press.

Strohman, R. C. 2000 Organisation becomes cause in the matter. *Nat. Biotechnol.* **18**, 575–576. (doi:10.1038/76317)

Tautz, D. 1992 Redundancies, development and the flow of information. *Bioessays* **14**, 263–266. (doi:10.1002/bies.950140410)

Tian, X. C., Kubota, C., Enright, B. & Yang, X. 2003 Cloning animals by somatic cell nuclear transfer—biological factors. *Reprod. Biol. Endocrinol.* **1**, 98–105. (doi:10.1186/1477-7827-1-98)

Werner, E. 2005 Genome semantics. *In silico* multicellular systems and the central dogma. *FEBS Lett.* **579**, 1779–1782. (doi:10.1016/j.febslet.2005.02.011)

Werner, E. 2007 How central is the genome? *Science* **317**, 753–754. (doi:10.1126/science.1141807)

Phil. Trans. R. Soc. A (2010) **368**, 1125–1139
doi:10.1098/rsta.2009.0245

REVIEW

Biophysics and systems biology

BY DENIS NOBLE*

*Department of Physiology, Anatomy and Genetics, University of Oxford,
Parks Road, Oxford OX1 3PT, UK*

Biophysics at the systems level, as distinct from molecular biophysics, acquired its most famous paradigm in the work of Hodgkin and Huxley, who integrated their equations for the nerve impulse in 1952. Their approach has since been extended to other organs of the body, notably including the heart. The modern field of computational biology has expanded rapidly during the first decade of the twenty-first century and, through its contribution to what is now called systems biology, it is set to revise many of the fundamental principles of biology, including the relations between genotypes and phenotypes. Evolutionary theory, in particular, will require re-assessment. To succeed in this, computational and systems biology will need to develop the theoretical framework required to deal with multilevel interactions. While computational power is necessary, and is forthcoming, it is not sufficient. We will also require mathematical insight, perhaps of a nature we have not yet identified. This article is therefore also a challenge to mathematicians to develop such insights.

Keywords: cell biophysics; systems biology; computational biology; mathematical biology

1. Introduction: the origins of biophysics and systems biology

As a young PhD student at University College London, I witnessed the celebrations of the 300th anniversary of the Royal Society in 1960. As the magnificent procession of red-gowned Fellows of the Royal Society (FRS) paraded into the Royal Albert Hall, two black gowns suddenly appeared. They were worn by Alan Hodgkin and Andrew Huxley. The founders of the field of cellular biophysics, with their ground-breaking mathematical reconstruction of the nerve impulse (Hodgkin & Huxley 1952), were simply Mr Hodgkin and Mr Huxley—neither had submitted a thesis for a PhD. With 'FRS' to their names, they hardly needed to! A year later, Alan Hodgkin examined my PhD thesis, which applied their ideas to reconstructing the electrical functioning of the heart (Noble 1960, 1962), and 3 years later we were celebrating their Nobel Prize.

It is highly appropriate to recall these events in a volume to celebrate the 350th anniversary, but they also remind us that the field that is now called systems biology has important historical roots. Hodgkin and Huxley themselves were not

*denis.noble@dpag.ox.ac.uk

One contribution of 17 to a Theme Issue 'Personal perspectives in the physical sciences for the Royal Society's 350th anniversary'.

1126   *D. Noble*

the first. I would nominate Claude Bernard as the first systems biologist (Noble 2008*a*), since in the middle of the nineteenth century he formulated the systems principle of control of the internal environment (Bernard 1865). This is well known and is widely recognized as the homeostatic basis of modern physiological science. It is much less well known that Bernard also presaged the development of mathematical biology when he wrote 'this application of mathematics to natural phenomena is the aim of all science, because the expression of the laws of phenomena should always be mathematical.'[1] Other historical roots can be found in the work of Harvey (Auffray & Noble 2009) and Mendel (Auffray 2005). Despite these strong historical roots, however, the field did not flourish in the second half of the twentieth century. Soon after Hodgkin and Huxley's achievement it was to be swept aside as molecular biology took the centre stage.

2. The achievements and problems of molecular biology

Physicists and mathematicians contributed greatly to the spectacular growth of molecular biology. The double-helical structure of DNA was discovered in the Cavendish laboratory in Cambridge (Watson & Crick 1953*a,b*) and in the biophysics laboratory at King's College London (Franklin & Gosling 1953*a,b*; Wilkins *et al.* 1953), while some of the seminal ideas of molecular biology were first developed by Schrödinger (1944). In addition to correctly predicting that the genetic material would be found to be an aperiodic crystal, his book, *What is Life?,* followed a proposal by Max Delbrück (see Dronamrajua 1999) that was to prove fundamental in the twentieth century interpretation of molecular biology. This was that physics and biology are essentially different disciplines in that while physics is about the emergence of order from disorder, such as the ordered global behaviour of a gas from the disordered Brownian motion of the individual molecules, biology dealt with order even at the molecular level. The paradigm for this view was the effects of mutations of the genetic material. Even a single switch from one nucleotide to another, corresponding to a single amino acid change in the protein for which the DNA sequence acts as a template, can have dramatic effects on the phenotype at higher levels. A good example in the case of the heart is that of the various sodium channel mutations that can cause arrhythmia (Clancy & Rudy 1999), and there are excellent examples in the processes of embryonic development (Davidson 2006).

The attribution of control to the DNA was strongly reinforced by Monod and Jacob (Jacob *et al.* 1960), who interpreted their work as evidence for the existence of a 'genetic program', an analogy explicitly based on comparison with an electronic computer: 'The programme is a model borrowed from electronic computers. It equates the genetic material with the magnetic tape of a computer' (Jacob 1982), while the rest of the organism, particularly the fertilized egg cell, could be compared with the computer itself. Specific instructions at the level of DNA could then be seen to 'program' or control the development and behaviour of the organism. These ideas married well with the gene-centred theories of evolution and the metaphor of 'selfish' genes (Dawkins 1976, 1982, 2006), which relegated the organism to the role of a disposable transient carrier of its DNA.

---

[1]Cette application des mathématiques aux phénomènes naturels est le but de toute science, parce que l'expression de la loi des phénomènes doit toujours être mathématique.

It is not surprising therefore that the peak of the achievement of molecular biology, the sequencing of the complete human genome, was widely signalled as finally reading the 'book of life'. However, the main architects of that project are much more circumspect: 'One of the most profound discoveries I have made in all my research is that you cannot define a human life or any life based on DNA alone...'. Why? Because 'An organism's environment is ultimately as unique as its genetic code' (Venter 2007). Sulston is also cautious: 'The complexity of control, overlaid by the unique experience of each individual, means that we must continue to treat every human as unique and special, and not imagine that we can predict the course of a human life other than in broad terms' (Sulston & Ferry 2002). So also is Sydney Brenner, whose work has contributed so much to the field: 'I believe very strongly that the fundamental unit, the correct level of abstraction, is the cell and not the genome' (lecture at Columbia University 2003).

I have briefly summarized some of these aspects of the development of molecular biology because, in fulfilling my brief to look into the crystal ball and give my own perspective on where my subject is heading in the next 50 years, I am going to turn some of the concepts derived from the successes of molecular biology upside down. I suggest that the next stage in the development of biological science will be revolutionary in its conceptual foundations (Shapiro 2005; see also Saks *et al.* 2009) and strongly mathematical in its methods. I also see this as the fulfilment of Claude Bernard's dream of the role of mathematics in his discipline, a dream that certainly could not be achieved in his lifetime.

3. Digital, analogue and stochastic genetic causes

Since the C, G, A, T sequences can be represented digitally (two bits are sufficient to represent four different entities, so the three billion base pairs could be represented by six billion bits), the idea of a determinate genetic program in the DNA, controlling the development and functioning of the organism, rather like the digital code of a computer program, was seductive, but for it to be correct, three conditions need to be satisfied. The first is that the relevant program logic should actually be found in the DNA sequences. The second is that this should control the production of proteins. The third is that this should be a determinate process. It is now known that none of these conditions are fulfilled. Molecular biology itself has revealed these deficiencies in at least six different ways.

(i) The C, G, A, T sequences of nucleotides in the genome do not themselves form a program as normally understood, with complete logic (i.e. one that could be subjected to syntactic analysis) of a kind that could separately run a computer. We cannot therefore predict life using these sequences alone. Instead, the sequences form a large set of templates that the cell uses to make specific proteins, and a smaller bank of switches, the regulatory genes, forming about 10 per cent of human genes, and the regulatory sites on which the regulatory proteins and other molecules act. Impressive switching circuits can be drawn to represent these (Levine & Davidson 2005). But they require much more than the DNA sequences themselves to operate since those switches depend on input from the rest of the organism, and from the environment. Organisms are interaction machines, not Turing machines (Shapiro 2005; Neuman 2008; Noble 2008c). There is therefore no computer into which we could insert the DNA sequences to generate life, other than life itself. Far from being just a transient vehicle, the organism itself contains the key to interpreting its DNA, and so to give it meaning. I will return later to this question (see §7).

(ii) In higher organisms, the sequences are broken into sometimes widely dispersed fragments, the exons, which can be combined in different ways to form templates for many different proteins. Something else must then determine which combination is used, which protein is formed and at which time. The DNA sequences therefore better resemble a database on which the system draws rather than a logical program of instructions (Atlan & Koppel 1990; Shapiro 2005; Noble 2006). For that we must look elsewhere, if indeed it exists at all. The dispersed nature of the exons and the combinatorial way in which they are used also challenges the concept of genes as discrete DNA sequences (Keller 2000a; Pearson 2006; Scherrer & Jost 2007).

(iii) What determines which proteins are made and in what quantity is not the DNA alone. Different cells and tissues use precisely the same DNA to produce widely different patterns of gene expression. This is what makes a heart cell different from, say, a bone cell or a pancreatic cell. These instructions come from the cells and tissues themselves, in the form of varying levels of transcription factors and epigenetic marks (Bird 2007) that are specific to the different types of cell. These processes are robust and inherited. Differentiated heart cells always form new heart cells as the heart develops, not new bone cells. They would need to be 'de-differentiated' to form multipotent stem cells in order to give rise to a different differentiated cell. This should not surprise us. Some kinds of cellular inheritance, perhaps starting with the ability of a lipid membrane-enclosed globule to divide, almost certainly predated genome inheritance (Maynard Smith & Szathmáry 1995).

(iv) The resulting patterns of gene expression are not only widely variable from one tissue to another, they themselves are not digital. The expression levels vary continuously in a way that is better described as an analogue. Since we must include these analogue levels in any description of how the process works, any 'program' we might identify is not based on digital coding alone. It is significant therefore that the inclusion of analogue processing is seen by some computer scientists as an important way in which a system can perform beyond the Turing limits (Siegelmann 1995, 1998, 1999). Organisms are, at the least, 'super-Turing' machines in this sense.

(v) Gene expression is a stochastic process (Kaern *et al.* 2005). Even within the same tissue, there are large variations in gene expression levels in different cells. Such stochasticity is incompatible with the operation of a determinate Turing machine (Kupiec 2008; Neuman 2008).

(vi) Finally, there is continuous interaction between DNA and its environment. As Barbara McClintock put it in her Nobel prize lecture (1983) for her work on 'jumping genes', the genome is better viewed as 'a highly sensitive organ of the cell' that can be reorganized in response to challenges (Keller 1983). We now also understand the extent to which organisms can swap DNA between each other, particularly in the world of micro-organisms (Goldenfeld & Woese 2007).

Another way to express the significance of these developments in molecular biology is to say that not much is left of the so-called 'central dogma of biology' (see Shapiro (2009) for more details) other than that part of Crick's original statement of it that is correct, which is that while DNA is a template for amino acid sequences in proteins, proteins do not form a template from which DNA can be produced by a reverse version of the DNA→protein transcription process. But in the extended sense in which it is frequently used in a neo-Darwinist context, as forbidding the passage of information from the organism and environment to DNA, the 'dogma' is seriously incorrect. Information is continually flowing in the opposite direction. I will return later to the significance of this fact for neo-Darwinism itself.

To these facts we must add a few more before we reassess the comparison between physics and biology.

(vii) Many genetic changes, either knockouts or mutations, appear not to have significant phenotypic effects; or rather they have effects that are subtle, often revealed only when the organism is under stress. For example, complete deletion of genes in yeast has no obvious phenotypic effect in 80 per cent of cases. Yet, 97 per cent have an effect on growth during stress (Hillenmeyer *et al.* 2008). The reason is that changes at the level of the genome are frequently buffered, i.e. alternative processes kick in at lower levels (such as gene–protein networks) to ensure continued functionality at higher levels (such as cells, tissues and organs). And even when a phenotype change does occur there is no guarantee that its magnitude reveals the full quantitative contribution of that particular gene since the magnitude of the effect may also be buffered. This is a problem I have recently referred to as the 'genetic differential effect problem' (Noble 2008*c*) and it has of course been known for many years. There is nothing new about the existence of the problem. What is new is that gene knockouts have revealed how extensive the problem is. Moreover, there is a possible solution to the problem to which I will return later.

(viii) The existence of stochastic gene expression allows some form of selection operating at the level of tissues and organs (Laforge *et al.* 2004; Kaern *et al.* 2005; Kupiec 2008, 2009). In fact, such selection may be a prerequisite of successful living systems which can use only those variations that are fit for purpose. As Kupiec has noted, Darwinian selection could also be very effective within the individual organism, as well as between organisms.

(ix) Not only is gene expression stochastic, the products of gene expression, the proteins, each have many interactions (at least dozens) with other elements in the organism. Proteins are not as highly specific as was once anticipated. Bray (Bray & Lay 1994; Bray 2009) has highlighted the role of multiple interactions in comparing the evolution of protein networks with that of neural networks.

4. The multifactorial nature of biological functions

So, while it is true to say that changes at the molecular level can *sometimes* have large effects at the higher phenotype levels, these effects are frequently buffered. Even the sodium channel mutations I referred to earlier do not, by themselves,

1130   *D. Noble*

trigger cardiac arrhythmia. The picture that emerges is that of a multifactorial system. Biology, it turns out, must also create order from stochastic processes at the lower level (Auffray *et al.* 2003). Physics and biology do not after all differ in quite the way that Schrödinger thought. This is a point that has been forcibly argued recently by Kupiec (2008, 2009). There is absolutely no way in which biological systems could be immune from the stochasticity that is inherent in Brownian motion itself. It is essential therefore that biological theory, like physical theory, should take this into account.

The systems approach has already pointed the way to achieve this. The massively combinatorial nature of biological interactions could have evolved precisely to overcome stochastic effects at the molecular level (Shapiro 2009). As Bray (2009) notes, protein networks have many features in common with the neural networks developed by artificial intelligence researchers. They can 'evolve' effective behaviour strategies from networks initialized with purely random connections, and once they have 'evolved' they show a high degree of tolerance when individual components are 'knocked out'. There is then what Bray calls 'graceful degradation', which can take various forms (not necessarily requiring random connectivity). This provides an insight into the nature of the robustness of biological systems. Far from stochasticity being a problem, it is actually an advantage as the system evolves. 'Graceful degradation' is also a good description of what happens in knockout organisms. All may appear to be well when the organism is well-fed and protected. The deficiency may reveal itself only when the conditions are hostile.

I suspect that more relevant insights will come from analysis of such artificial networks and even more so from the modelling of real biological networks. Note that such networks do not require a separate 'program' to operate. The learning process in the case of artificial networks, and evolutionary interaction with the environment in the case of biological networks, is the 'programming' of the system. So, if we still wish to use the program metaphor, it is important to recognize that the program is the system itself (Noble 2008*c*). The plant geneticist Enrico Coen expressed this point well when he wrote 'Organisms are not simply manufactured according to a set of instructions. There is no easy way to separate instructions from the process of carrying them out, to distinguish plan from execution' (Coen 1999). This is another version of the points made earlier about the limitations of regarding the DNA sequences as a program.

5. The multilevel nature of biological functions

This takes me to the question of multilevel analysis. Organisms are not simply protein soups. Biological functions are integrated at many different levels. Thus, pacemaker rhythm in the heart is integrated at the level of the cell. There is no oscillator at the biochemical level of subcellular protein networks (Noble 2006). Tempting though it may be to think so, there is therefore no 'gene for' pacemaker rhythm. A set of genes, or more correctly the proteins formed from their templates, is involved, together with the cellular architecture—and which set we choose to represent depends on the nature of the questions we are asking. But that does not prevent us from building computer programs that mimic pacemaker rhythm. Simulation of cardiac activity has been developed over a period of nearly five decades and is now sufficiently highly developed that it can be used in the pharmaceutical industry to clarify the actions of drugs (Noble 2008*b*).

Does not the fact that we can succeed in doing this prove that, after all, there are genetic programs? Well no, for two reasons. First the logic represented by such computer simulation programs is certainly not to be found simply in the DNA sequences. The programs are representations of the processes involved at *all* the relevant biological levels, right up to and including the intricate architecture of the cell itself. And when even higher levels are modelled, the structural biology included is that of tissues or the entire organ (Hunter *et al.* 2003; Garny *et al.* 2005). In the case of the heart, the three-dimensional imaging technology to achieve this has now advanced to paracellular or even subcellular levels (Plank *et al.* 2009).

Second, reflecting Coen's point above, the processes represented in our modelling programs *are* the functionality itself. To the extent that the program succeeds in reproducing the behaviour of the biological system it reveals the *processes* involved, not a separate set of *instructions*.

Multilevel simulation will be a major development in biology as the project known as the Human Physiome Project develops. Recent issues of this journal have been devoted to one of its components, the Virtual Physiological Human (VPH) project (Clapworthy *et al.* 2008; Fenner *et al.* 2008) and some of the achievements and future challenges of the Physiome Project (Bassingthwaighte *et al.* 2009) and its relation to systems biology (Kohl & Noble 2009) have recently been reviewed.

6. A theory of biological relativity?

One of the major theoretical outcomes of multilevel modelling is that causation in biological systems runs in both directions: upwards from the genome and downwards from all other levels.[2] There are feedforward and feedback loops between the different levels. Developing the mathematical and computational tools to deal with these multiple causation loops is itself a major challenge. The mathematics that naturally suits one level may be very different from that for another level. Connecting levels is not therefore trivial. Nor are the problems simply mathematical and computational. They also require biological insight to determine how much detail at one level is relevant to functionality at other levels. These problems are now exercising the minds of interdisciplinary teams of researchers involved in the Physiome Project and they offer great opportunities for physical and mathematical scientists in the future. They have also led some physicists and biologists to develop what might be called theories of biological relativity. My own version of this idea is that, in multilevel systems, there is no privileged level of causation (Noble 2008*a,c*). Others have also pointed out that such a principle need not be restricted to biological systems. It could become a

---

[2]'Upwards' and 'downwards' in this context are metaphorical. A more neutral terminology would refer to different (larger and smaller) scales. But the concept of level is strongly entrenched in biological science so I have continued to use it here. There is also possible confusion with 'scale' as used in scale relativity, though I believe that one of the key questions for the future is that of relating the ideas of scale relativity to multilevel systems biology.

1132   *D. Noble*

general theory of relativity of levels. Such a theory, called scale relativity (Nottale 1993, 2000), already exists in physics and its possible applications to biological systems have been the subject of major recent reviews (Auffray & Nottale 2008; Nottale & Auffray 2008).

I will not review these theories in detail here. I wish rather to draw attention to a related general question. Is multilevel analysis simply a matter of including downward causation (Noble 2006)? And what exactly do we mean by that term?

In my own field the paradigm example originated with Alan Hodgkin. The proteins that form ion channels in excitable cells generate electric current that charges or discharges the cell capacitance. That can be seen as upward causation. But the electrical potential of the cell also controls the gating of the ion channel proteins. This downward causation closes the loop of the 'Hodgkin cycle'.

Is downward causation always discrete feedback or feedforward? The answer is no and the basis for that answer is profound, forming one of the reasons why I think that systems biology is revolutionary. A feedback loop can be closed. Feedback loops could exist between the levels of an organism, while the organism itself could still be modelled as a closed system. Yet, we know that organisms are not closed systems. Firstly they exchange energy and matter with the environment, including particularly other organisms whose existence forms a major part of the selection pressure. That is well recognized as a reason for regarding organisms as open systems. But there are other reasons also. I think that the best way to explain that is mathematical.

We model many biological processes as systems of differential equations. These equations describe the rates at which those processes occur. The number of such equations depends on the kind of question we are asking. At a cellular or subcellular (protein network) level, there may be a few dozen equations for the protein and other chemical entities involved. When we include structural details at the tissue or organ level, we may be dealing with millions of equations. Whatever the number, there is an inescapable requirement before we can begin to solve the equations. We must know or make plausible guesses for the initial and boundary conditions. They are not set by the differential equations themselves. These conditions restrain the solutions that are possible. In fact, beyond a certain level of complexity, the more interesting question becomes the explanation of that restraining set of conditions, not just the behaviour of the system, since the restraints may completely change the behaviour of the system. A restraint, therefore, is not necessarily a feedback. Restraints can be simply the background set of conditions within which the system operates, i.e. its environment. Through these interactions organisms can adapt to many different conditions. Their robustness in doing so distinguishes them from complex nonlinear systems that are highly sensitive to initial conditions or which end up unable to escape attractors.

7. 'Genetic programs'

This is a suitable point at which to return to the question of 'genetic programs'. As we have seen, DNA sequences act as templates for proteins and as switches for turning genes on and off when they are in an organism, starting with the fertilized egg cell and maternal environment in the case of higher animals. A possible objection to my conclusion that the DNA sequences are better viewed as a database rather than as a program is that all programs require a computer to implement them. It was part of Monod and Jacob's idea that, if DNA is the program, the organism is equivalent to the computer. Programs also do nothing outside the context of a computer. Could we somehow update this approach to save the 'program' metaphor? It is so ingrained into modern thought, among laypeople as well as most scientists, that it may now be difficult to convince people to abandon it. It is therefore worth spelling out, once again, what the difficulties are.

DNA sequences alone are not capable of being parsed as the complete logic of a program. Whenever we talk of a genetic program we must also include steps that involve the rest of the organism (e.g. my discussion of the 'circadian rhythm' program in Noble (2006, pp. 69–73), and this is certainly true for the analysis of cardiac rhythm (Noble 2006, pp. 56–65)). Much of the logic of living systems lies beyond DNA. To save the program metaphor therefore we would have to say that the 'program' is distributed between the tape and the machine. This would, incidentally, explain an important fact. Virtually all attempts at cross-species cloning fail to develop to the adult (Chung *et al.* 2009). A possible explanation is that the egg cell information is too specific (Chen *et al.* 2006). In fact, in the only case so far, that of a carp nucleus and goldfish egg, the egg cytoplasm clearly influences the phenotype (Sun *et al.* 2005). Strathmann (1993) also refers to the influence of the egg cytoplasm on gene expression during early development as one of the impediments to hybridization in an evolutionary context. There is no good reason why cells themselves should have ceased to evolve once genomes arose. But if we need a specific (special purpose) 'computer' for each 'program', the program concept loses much of its attraction.

The way to save the genetic program idea would therefore be to abandon the identification of genes with specific sequences of DNA alone and return to the original idea of genes as the causes of particular phenotypes (Kitcher 1982; Mayr 1982; Dupré 1993; Pichot 1999; Keller 2000*b*; Noble 2008*c*) by including other relevant processes in the organism. The problem with this approach is that the closer we get to characterizing the 'program' for a particular phenotype, the more it looks like the functionality itself. Thus, the process of cardiac rhythm can be represented as such a 'program' (indeed, modellers write computer programs to reproduce the process), but it is not a sequence of instructions separate from the functionality itself. This is another way to understand the quotation from Coen referred to earlier. The clear distinction between the replicator and the vehicle disappears and, with it, a fundamental aspect of the 'selfish gene' view.

If we do wish to retain the idea of a program, for example in talking about embryonic development where the concept of a 'developmental program' has its best applications (Keller 2000*a*), it might be better to think in the same terms in which we talk of neural nets being programmed. They are programmed by the initial setting up of their connections and then by the learning process, the set of restraints that allows them to 'home in' to a particular functionality. Those open-ended restraints are as much a part of the 'program' as the initial setting up of the system. The analogy with organisms as interaction machines is obvious. I am not proposing that organisms function as neural nets; only that the example of neural nets expands our concept of the word 'program' in a relevant way. The program is a distributed one (Siegelmann 1998) involving much more than DNA sequences, and is therefore far removed from Monod and Jacob's original concept of a genetic program.

8. Systems biology and evolution

Where do the restraints come from in biological systems? Clearly, the immediate environment of the system is one source of restraint. Proteins are restrained by the cellular architecture (where they are found in or between the membrane and filament systems), cells are restrained by the tissues and organs they find themselves in (by the structure of the tissues and organs and by the intercellular signalling) and all levels are restrained by the external environment. Even these restraints though would not exhaust the list. Organisms are also a product of their evolutionary history, i.e. the interactions with past environments. These restraints are stored in two forms of inheritance—DNA and cellular. The DNA sequences restrict which amino acid sequences can be present in proteins, while the inherited cellular architecture restricts their locations, movements and reactions.

This is one of the reasons why systems biology cannot be restricted to the analysis of protein and gene circuits. The structural information is also crucial. Much of its evolution may have been independent of the cell's own DNA since the early evolution of the eukaryotic cell involved many forms of symbiosis. The best known example is the mitochondria, which are now accepted to have originally been invading (or should we say 'captured'?) bacteria, as were chloroplasts (Cavalier-Smith 2000, 2004). They even retain some of the original DNA, though some also migrated to the nucleus. There are other examples of symbiosis (Margulis 1981; Margulis & Sagan 2002; Williamson 2003, 2006; Williamson & Vickers 2007). Cooperativity may have been quite as important as competition in evolution (see also Goldenfeld & Woese 2007).

Cavalier-Smith has described some of these inherited features of animal and plant cells as the 'membranome', an important concept since lipids are not formed from DNA templates. An organism needs to inherit the membranome, which it does of course—it comes complete with the fertilized egg cell—yet another reason why it does not make sense to describe the organism as merely a vehicle for DNA. As I have argued elsewhere (Noble 2008c), the relative contributions of DNA and non-DNA inheritance are difficult to estimate (one is largely digital and so easy to calculate, whereas the other is analogue and hard to calculate), but the non-DNA inheritance is very substantial. It also contains many historical restraints of evolution.

This is the point at which I should attempt to explain the neo-Darwinian model and the modern synthesis and what is wrong with them from a systems viewpoint.

Neo-Darwinism brings together natural selection and nineteenth century genetics, while the modern synthesis (Huxley 1942) fuses Darwinism with twentieth century genetics. 'Neo-Darwinism' is the term often used for both of these syntheses. Darwin knew nothing of Mendel's work on genetics. Moreover, he also accepted the idea of the inheritance of acquired characteristics, as did Lamarck (Lamarck 1809; Corsi 2001), who is incorrectly represented in many texts as inventing the idea. Darwin's disagreements with Lamarck were not over the mechanisms of inheritance. Both were ignorant of those mechanisms. Their disagreement was more over the question of whether evolution had a direction or whether variation was random. Historically, we would do better to recognize Lamarck as the inventor of the term 'biology' as a separate science, and as championing the idea that species change (transformationism). Darwin can then be seen as discovering one of the mechanisms in his theory of natural selection, involved not only in transformations but also in the origin of species.

The problem with both revisions of Darwinism is that they involve a version of genetics that we need to revise. This version was one in which the central dogma of biology was taken to mean that the genetic material is never modified by the rest of the organism and the environment. Francis Crick's original statements of the 'central dogma of molecular biology' (Crick 1958, 1970) do not in fact make such a strong claim. He stated a more limited chemical fact: that DNA sequences are used as templates to make proteins, but proteins are not used as reverse templates to make DNA. So, even if its proteins were to become modified during the lifetime of an individual, that modification cannot be inherited. The 'dogma' was then interpreted by many biologists to mean that information flows only one way. As we have seen, it does not. The *quantities* of proteins synthesized count as relevant information just as much as their amino acid sequences. But those quantities are most certainly dependent on signals from the rest of the system through the levels of transcription factors (including proteins and RNA) and the epigenetic marking of DNA itself and of the histone tails. All of this is open to the rest of the organism and to the environment to degrees we have yet to fully determine.

I will give just one example here to illustrate the potential significance of this openness. More examples can be found elsewhere (Jablonka & Lamb 1995, 2005). Neuroscientists have recently studied the epigenetic factors involved in maternal grooming behaviour in colonies of rats. Grooming depends on the environment. Colonies that are safe groom their young a lot. Colonies that are fighting off predators do not. This behaviour is inherited. The mechanisms are a fascinating example of epigenetic effects. The genome in the hippocampal region of the brain is epigenetically marked by the grooming behaviour and this predisposes the young to show that behaviour (Weaver *et al.* 2004, 2007). This is an important development, but as Weaver himself points out (Weaver 2009) it is currently restricted to one gene and one region of the brain. That underlines the importance of further research in this area. The implications of this form of epigenetic influence, however, are profound since it can transmit patterns of epigenetic marking through the generations even though they are not transmitted via the germline. This constitutes another form of inheritance of acquired characteristics to add to those reviewed by Jablonka and Lamb.

There is a tendency to dismiss such challenges to extensions of the central dogma as merely examples of cultural evolution. They seem to show rather that the boundaries between the different evolutionary processes are fuzzy. Once such interactions between behaviour and epigenetics are established and transmitted through the generations they can favour genetic combinations that lock them into the genome (Jablonka & Lamb 2005, pp. 260–270). This mechanism was originally

1136   *D. Noble*

described by Waddington (1942, 1957, 1959; Bard 2008), who demonstrated that, in fruitflies, just 14 generations of induced phenotype change could be assimilated into the genome. Mutations and genetic recombinations themselves are not random (Shapiro 2005). Moreover, they do not occur in a random context. They occur in the context of all the restraints exerted on the organism, including those of the environment. In such a process, it is the phenotype, not individual genes, that are the targets of selection (Keller 1999). Central building blocks of the neo-Darwinian synthesis are now known to be incompatible with the most recent discoveries in molecular biology.

9. Reverse engineering in systems biology

I referred earlier to the 'genetic differential effect problem'. In a previous article in this journal I have proposed that computational systems biology could provide a solution (Noble 2008*c*). The idea is basically simple. If our understanding and simulations are good enough they should include the robustness of biological systems, including their resistance to damage from mutations and knockouts. Moreover, if the models include representations of specific gene products (i.e. they extend down to the protein level) then it should be possible to reverse engineer to arrive at *quantitative* estimates of the contribution of each gene product to the functionality represented. That may be possible even if the system completely buffers the mutation or knockout so that no effect is observed in the phenotype. I give an example of this in the previous article from work on the heart (Noble 2008*c*). However, I would readily agree that, in its present state of development, computational systems biology is a long way from being able to do this in general. But it is worth bearing this in mind as an important long-term goal.

Work in the author's laboratory is funded by the EU (Framework 6 and Framework 7), The British Heart Foundation, EPSRC and BBSRC. I acknowledge valuable criticisms from Charles Auffray, Jonathan Bard, Evelyn Fox Keller, Peter Kohl, Jean-Jacques Kupiec, Lynn Margulis, Laurent Nottale, James Shapiro, Hava Siegelmann, Eric Werner and Michael Yudkin.

References

Atlan, H. & Koppel, M. 1990 The cellular computer DNA: program or data? *Bull. Math. Biol.* **52**, 335–348.

Auffray, C. 2005 Aux sources de la biologie des systèmes et de la génétique: la pertinence des expérimentations de Gregor Mendel sur le développement des plantes hybrides (2e volet). *L'Observatoire de la génétique* **21**.

Auffray, C. & Noble, D. 2009 Conceptual and experimental origins of integrative systems biology in William Harvey's masterpiece on the movement of the heart and the blood in animals. *Int. J. Mol. Sci.* **10**, 1658–1669. (doi:10.3390/ijms10041658)

Auffray, C. & Nottale, L. 2008 Scale relativity theory and integrative systems biology. I. Founding principles and scale laws. *Prog. Biophys. Mol. Biol.* **97**, 79–114. (doi:10.1016/j.pbiomolbio.2007.09.002)

Auffray, C., Imbeaud, S., Roux-Rouquie, M. & Hood, L. 2003 Self-organized living systems: conjunction of a stable organization with chaotic fluctuations in biological space-time. *Phil. Trans. R. Soc. Lond. A* **361**, 1125–1139. (doi:10.1098/rsta.2003.1188)

Bard, J. B. L. 2008 Waddington's legacy to developmental and theoretical biology. *Biol. Theory* **3**, 188–197. (doi:10.1162/biot.2008.3.3.188)

Bassingthwaighte, J. B., Hunter, P. J. & Noble, D. 2009 The cardiac physiome: perspectives for the future. *Exp. Physiol.* **94**, 597–605. (doi:10.1113/expphysiol.2008.044099)

Bernard, C. 1865 *Introduction à l'étude de la médecine expérimentale*. Paris, France: J. B. Baillière. (Reprinted by Flammarion 1984.)

Bird, A. 2007 Perceptions of epigenetics. *Nature* **447**, 396–398. (doi:10.1038/nature05913)

Bray, D. 2009 *Wetware. A computer in every cell*. New Haven, CT: Yale University Press.

Bray, D. & Lay, S. 1994 Computer simulated evolution of a network of cell-signalling molecules. *Biophys. J.* **66**, 972–977. (doi:10.1016/S0006-3495(94)80878-1)

Cavalier-Smith, T. 2000 Membrane heredity and early chloroplast evolution. *Trends Plant Sci.* **5**, 174–182. (doi:10.1016/S1360-1385(00)01598-3)

Cavalier-Smith, T. 2004 The membranome and membrane heredity in development and evolution. In *Organelles, genomes and eukaryote phylogeny: an evolutionary synthesis in the age of genomics* (eds R. P. Hirt & D. S. Horner), pp. 335–351. Boca Raton, FL: CRC Press.

Chen, T., Zhang, Y.-L., Jiang, Y., Liu, J.-H., Schatten, H., Chen, D.-Y. & Sun, Q.-Y. 2006 Interspecies nuclear transfer reveals that demethylation of specific repetitive sequences is determined by recipient ooplasm but not by donor intrinsic property in cloned embryos. *Mol. Reprod. Dev.* **73**, 313–317. (doi:10.1002/mrd.20421)

Chung, Y. *et al.* 2009 Reprogramming of human somatic cells using human and animal oocytes. *Cloning Stem Cells* **11**, 1–11. (doi:10.1089/clo.2009.0004)

Clancy, C. E. & Rudy, Y. 1999 Linking a genetic defect to its cellular phenotype in a cardiac arrhythmia. *Nature* **400**, 566–569. (doi:10.1038/23034)

Clapworthy, G., Viceconti, M., Coveney, P. & Kohl, P. (eds) 2008 Editorial. *Phil. Trans. R. Soc. A* **366**, 2975–2978. (doi:10.1098/rsta.2008.0103)

Coen, E. 1999 *The art of genes*. Oxford, UK: Oxford University Press.

Corsi, P. 2001 *Lamarck. Genèse et enjeux du transformisme*. Paris, France: CNRS Editions.

Crick, F. H. C. 1958 On protein synthesis. *Symp. Soc. Exp. Biol.* **XII**, 138–163.

Crick, F. H. C. 1970 Central dogma of molecular biology. *Nature* **227**, 561–563. (doi:10.1038/227561a0)

Davidson, E. H. 2006 *The regulatory genome: gene regulatory networks in development and evolution*. New York, NY: Academic Press.

Dawkins, R. 1976, 2006 *The selfish gene*. Oxford, UK: Oxford University Press.

Dawkins, R. 1982 *The extended phenotype*. London, UK: Freeman.

Dawkins, R. 2006 *The selfish gene* (revised edn). Oxford, UK: Oxford University Press.

Dronamrajua, K. R. 1999 Erwin Schrödinger and the origins of molecular biology. *Genetics* **153**, 1071–1076.

Dupré, J. 1993 *The disorder of things*. Cambridge, MA: Harvard University Press.

Fenner, J. W. *et al.* 2008 The EuroPhysiome, STEP and a roadmap for the virtual physiological human. *Phil. Trans. R. Soc. A* **366**, 2979–2999. (doi:10.1098/rsta.2008.0089)

Franklin, R. E. & Gosling, R. G. 1953*a* Evidence for 2-chain helix in crystalline structure of sodium deoxyribonucleate. *Nature* **172**, 156–157. (doi:10.1038/172156a0)

Franklin, R. E. & Gosling, R. G. 1953*b* Molecular configuration in sodium thymonucleate. *Nature* **171**, 740–741. (doi:10.1038/171740a0)

Garny, A., Noble, D. & Kohl, P. 2005 Dimensionality in cardiac modelling. *Prog. Biophys. Mol. Biol.* **87**, 47–66. (doi:10.1016/j.pbiomolbio.2004.06.006)

Goldenfeld, N. & Woese, C. 2007 Biology's next revolution. *Nature* **445**, 369. (doi:10.1038/445369a)

Hillenmeyer, M. E. *et al.* 2008 The chemical genomic portrait of yeast: uncovering a phenotype for all genes. *Science* **320**, 362–365. (doi:10.1126/science.1150021)

Hodgkin, A. L. & Huxley, A. F. 1952 A quantitative description of membrane current and its application to conduction and excitation in nerve. *J. Physiol.* **117**, 500–544.

Hunter, P. J., Pullan, A. J. & Smaill, B. H. 2003 Modelling total heart function. *Rev. Biomed. Eng.* **5**, 147–177. (doi:10.1146/annurev.bioeng.5.040202.121537)

Huxley, J. S. 1942 *Evolution: the modern synthesis*. London, UK: Allen & Unwin.

Jablonka, E. & Lamb, M. 1995 *Epigenetic inheritance and evolution. The Lamarckian dimension*. Oxford, UK: Oxford University Press.

1138   *D. Noble*

Jablonka, E. & Lamb, M. 2005 *Evolution in four dimensions.* Boston, MA: MIT Press.

Jacob, F. 1982 *The possible and the actual.* New York, NY: Pantheon Books.

Jacob, F., Perrin, D., Sanchez, C., Monod, J. & Edelstein, S. 1960 The operon: a group of genes with expression coordinated by an operator. *C. R. Acad. Sci. Paris* **250**, 1727–1729.

Kaern, M., Elston, T. C., Blake, W. J. & Collins, J. J. 2005 Stochasticity in gene expression: from theories to phenotypes. *Nat. Rev. Genet.* **6**, 451–464. (doi:10.1038/nrg1615)

Keller, E. F. 1983 *A feeling for the organism: the life and work of Barbara McClintock.* New York, NY: W.H. Freeman.

Keller, E. F. 2000*a The century of the gene.* Cambridge, MA: Harvard University Press.

Keller, E. F. 2000*b* Is there an organism in this text? In *Controlling our destinies. historical, philosophical, ethical and theological perspectives on the human genome project* (ed. P. R. Sloan), pp. 273–288. Notre Dame, IN: University of Notre Dame Press.

Keller, L. 1999 *Levels of selection in evolution.* Princeton, NJ: Princeton University Press.

Kitcher, P. 1982 Genes. *Br. J. Phil. Sci.* **33**, 337–359. (doi:10.1093/bjps/33.4.337)

Kohl, P. & Noble, D. 2009 Systems biology and the virtual physiological human. *Mol. Syst. Biol.* **5**. (doi:10.1038/msb.2009.51)

Kupiec, J.-J. 2008 *L'origine des individus.* Paris, France: Fayard.

Kupiec, J.-J. 2009 *The origin of individuals: a Darwinian approach to developmental biology.* London, UK: World Scientific Publishing Company.

Laforge, B., Guez, D., Martinez, M. & Kupiec, J.-J. 2004 Modeling embryogenesis and cancer: an approach based on an equilibrium between the autostabilization of stochastic gene expression and the interdependence of cells for proliferation. *Prog. Biophys. Mol. Biol.* **89**, 93–120. (doi:10.1016/j.pbiomolbio.2004.11.004)

Lamarck, J.-B. 1809 *Philosophie Zoologique.* Paris, France: Dentu. (Reprinted by Flammarion 1994 as original edition with introduction by André Pichot.)

Levine, M. & Davidson, E. H. 2005 Gene regulatory networks for development. *Proc. Natl Acad. Sci. USA* **102**, 4936–4942. (doi:10.1073/pnas.0408031102)

Margulis, L. 1981 *Symbiosis in cell evolution.* London, UK: W.H. Freeman Co.

Margulis, L. & Sagan, D. 2002 *Acquiring genomes.* New York, NY: Basic Books.

Maynard Smith, J. & Szathmáry, E. 1995 *The major transitions in evolution.* Oxford, UK: Oxford University Press.

Mayr, E. 1982 *The growth of biological thought.* Cambridge, MA: Harvard University Press.

Neuman, Y. 2008 *Reviving the living: meaning making in living systems.* Amsterdam, The Netherlands: Elsevier.

Noble, D. 1960 Cardiac action and pacemaker potentials based on the Hodgkin-Huxley equations. *Nature* **188**, 495–497. (doi:10.1038/188495b0)

Noble, D. 1962 A modification of the Hodgkin-Huxley equations applicable to Purkinje fibre action and pacemaker potentials. *J. Physiol.* **160**, 317–352.

Noble, D. 2006 *The music of life.* Oxford, UK: Oxford University Press.

Noble, D. 2008*a* Claude Bernard, the first systems biologist, and the future of physiology. *Exp. Physiol.* **93**, 16–26. (doi:10.1113/expphysiol.2007.038695)

Noble, D. 2008*b* Computational models of the heart and their use in assessing the actions of drugs. *J. Pharmacol. Sci.* **107**, 107–117. (doi:10.1254/jphs.CR0070042)

Noble, D. 2008*c* Genes and causation. *Phil. Trans. R. Soc. A* **366**, 3001–3015. (doi:10.1098/rsta.2008.0086)

Nottale, L. 1993 *Fractal space-time and microphysics: towards a theory of scale relativity.* Singapore: World Scientific.

Nottale, L. 2000 *La relativité dans tous ses états. Du mouvements aux changements d'échelle.* Paris, France: Hachette.

Nottale, L. & Auffray, C. 2008 Scale relativity and integrative systems biology. II. Macroscopic quantum-type mechanics. *Prog. Biophys. Mol. Biol.* **97**, 115–157. (doi:10.1016/j.pbiomolbio.2007.09.001)

Pearson, H. 2006 What is a gene? *Nature* **441**, 399–401. (doi:10.1038/441398a)

Pichot, A. 1999 *Histoire de la notion de gène.* Paris, France: Flammarion.

Plank, G. *et al.* 2009 Generation of histo-anatomically representative models of the individual heart: tools and application. *Phil. Trans. R. Soc. A* **367**, 2257–2292. (doi:10.1098/rsta.2009.0056)

Saks, V., Monge, C. & Guzun, R. 2009 Philosophical basis and some historical aspects of systems biology: from Hegel to Noble—applications for bioenergetic research. *Int. J. Mol. Sci.* **10**, 1161–1192. (doi:10.3390/ijms10031161)

Scherrer, K. & Jost, J. 2007 Gene and genome concept. Coding versus regulation. *Theory Biosci.* **126**, 65–113. (doi:10.1007/s12064-007-0012-x)

Schrödinger, E. 1944 *What is life?* Cambridge, UK: Cambridge University Press.

Shapiro, J. A. 2005 A 21st century view of evolution: genome system architecture, repetitive DNA, and natural genetic engineering. *Gene* **345**, 91–100. (doi:10.1016/j.gene.2004.11.020)

Shapiro, J. A. 2009 Revisiting the central dogma in the 21st century. *Ann. N Y Acad. Sci.* **1178**, 6–28. (doi:10.1111/j.1749-6632.2009.04990.x)

Siegelmann, H. T. 1995 Computation beyond the Turing Limit. *Science* **268**, 545–548. (doi:10.1126/science.268.5210.545)

Siegelmann, H. T. 1998 *Neural networks and analog computation: beyond the Turing limit.* Boston, MA: Birkhauser.

Siegelmann, H. T. 1999 Stochastic analog networks and computational complexity. *J. Complexity* **15**, 451–475. (doi:10.1006/jcom.1999.0505)

Strathmann, R. R. 1993 Larvae and evolution: towards a new zoology. *Q. Rev. Biol.* **68**, 280–282. (doi:10.1086/418103)

Sulston, J. & Ferry, G. 2002 *The common thread.* London, UK: Bantam Press.

Sun, Y. H., Chen, S. P., Wang, Y. P., Hu, W. & Zhu, Z. Y. 2005 Cytoplasmic impact on cross-genus cloned fish derived from transgenic common carp (*Cyprinus carpio*) nuclei and goldfish (*Carassius auratus*) enucleated eggs. *Biol. Reprod.* **72**, 510–515. (doi:10.1095/biolreprod.104.031302)

Venter, C. 2007 *A life decoded.* London, UK: Allen Lane.

Waddington, C. H. 1942 Canalization of development and the inheritance of acquired characteristics. *Nature* **150**, 563–565. (doi:10.1038/150563a0)

Waddington, C. H. 1957 *The strategy of the genes.* London, UK: Allen and Unwin.

Waddington, C. H. 1959 Canalization of development and genetic assimilation of acquired characteristics. *Nature* **183**, 1654–1655. (doi:10.1038/1831654a0)

Watson, J. D. & Crick, F. H. C. 1953*a* Genetical implications of the structure of deoxyribonucleic acid. *Nature* **171**, 964–967. (doi:10.1038/171964b0)

Watson, J. D. & Crick, F. H. C. 1953*b* Molecular structure of nucleic acids. A structure for deoxyribose nucleic acid. *Nature* **171**, 737–738. (doi:10.1038/171737a0)

Weaver, I. C. G. 2009 Life at the interface between a dynamic environment and a fixed genome. In *Mammalian brain development* (ed. D. Janigro), pp. 17–40. Totowa, NJ: Humana Press.

Weaver, I. C. G., Cervoni, N., Champagne, F. A., D'Alessio, A. C., Sharma, S., Sekl, J. R., Dymov, S., Szyf, M. & Meaney, M. J. 2004 Epigenetic programming by maternal behavior. *Nat. Neurosci.* **7**, 847–854. (doi:10.1038/nn1276)

Weaver, I. C. G., D'Alessio, A. C., Brown, S. E., Hellstrom, I. C., Dymov, S., Sharma, S., Szyf, M. & Meaney, M. J. 2007 The transcription factor nerve growth factor-inducible protein a mediates epigenetic programming: altering epigenetic marks by immediate-early genes. *J. Neurosci.* **27**, 1756–1768. (doi:10.1523/JNEUROSCI.4164-06.2007)

Wilkins, M. H. F., Stokes, A. R. & Wilson, H. R. 1953 Molecular structure of deoxypentose nucleic acids. *Nature* **171**, 738–740. (doi:10.1038/171738a0)

Williamson, D. I. 2003 *The origins of larvae.* Dordrecht, The Netherlands: Kluwer Academic Publishers.

Williamson, D. I. 2006 Hybridization in the evolution of animal form and life cycle. *Zool. J. Linn. Soc.* **148**, 585–602. (doi:10.1111/j.1096-3642.2006.00236.x)

Williamson, D. I. & Vickers, S. E. 2007 The origins of larvae. *Am. Sci.* **95**, 509–517.

nature publishing group

STATE OF THE ART

Systems Biology: An Approach

P Kohl[1], EJ Crampin[2], TA Quinn[1] and D Noble[1]

In just over a decade, Systems Biology has moved from being an idea, or rather a disparate set of ideas, to a mainstream feature of research and funding priorities. Institutes, departments, and centers of various flavors of Systems Biology have sprung up all over the world. An Internet search now produces more than 2 million hits. Of the 2,800 entries in PubMed with "Systems Biology" in either the title or the abstract, only two papers were published before 2000, and >90% were published in the past five years. In this article, we interpret Systems Biology as an approach rather than as a field or a destination of research. We illustrate that this approach is productive for the exploration of systems behavior, or "phenotypes," at all levels of structural and functional complexity, explicitly including the supracellular domain, and suggest how this may be related conceptually to genomes and biochemical networks. We discuss the role of models in Systems Biology and conclude with a consideration of their utility in biomedical research and development.

SYSTEMS BIOLOGY AS AN APPROACH

Origins

The use of Systems Biology approaches in analyzing biochemical networks is well established,[1] and it is now also gaining ground in explorations of higher levels of physiological function, as exemplified by the Physiome[2] and Virtual Physiological Human[3,4] projects. However, the use of the term "system" in the field of biology long predates "Systems Biology."

Throughout its existence as a discipline, physiology has concerned itself with the systems of the body (circulatory, nervous, immune, and so on). Back in 1542, Jean Fernel wrote, "So, if the parts of a complete Medicine are set in order, physiology will be the first of all; it concerns itself with the nature of the wholly healthy human being, all the powers and functions."[5] Claude Bernard is widely credited with introducing one of the key biological concepts—control of the internal environment—and he may therefore be viewed as the first "systems biologist,"[6] although good claims can also be made for William Harvey,[7] Gregor Mendel,[8] and others.

Essence

In order to explore the essence of Systems Biology—a notion that, in spite of its broad appeal, is still lacking a definition—it may be helpful to start by considering the meaning of each of the two words. "Biology" is easy to define: it is the science (Greek λόφ ; "reason[ed] account") that is concerned with living matter (Greek βίọ ; "life").

Although perhaps less well appreciated in the biological field, the term "system" is equally well defined, as "an entity that maintains its existence through the mutual interaction of its parts."[9] Systems research, therefore, necessarily involves the combined application of "reductionist" and "integrationist" research techniques, to allow identification and detailed characterization of the parts, investigation of their interaction with one another and with their wider environment, and elucidation of how parts and interactions give rise to maintenance of the entity[10] (Figure 1).

Systems Biology, therefore, can be seen to stand for an *approach* to bioresearch, rather than a field or a destination.

This approach consciously combines reduction and integration from the outset of research and development activities, and it necessarily involves going across spatial scales of structural and functional integration (i.e., between the parts and the entity). There is no inherent restriction on the level at which "the system" may be defined. In fact, there is no such thing as *the* system because structures that are parts of one system (say, a mitochondrion in a cell) may form systems in their own right at a different level of integration (for example, in the contexts of electron transport chains and ATP synthesis). The focus of Systems Biology can be, but is not required to be, at the single-cell level (a predominant target so far). As an approach, Systems Biology is equally applicable to small or large biological entities.

In addition to addressing the relationship between structure and function from the nano- to the macroscale, Systems Biology interprets biological phenomena as dynamic processes whose inherent time resolution depends on the behavior studied. This range extends from submicroseconds for molecular-level

[1]Department of Physiology, Anatomy and Genetics, University of Oxford, Oxford, UK; [2]Auckland Bioengineering Institute, The University of Auckland, Auckland, New Zealand. Correspondence: P Kohl (peter.kohl@dpag.ox.ac.uk)

Received 25 March 2010; accepted 20 April 2010; advance online publication 9 June 2010. doi:10.1038/clpt.2010.92

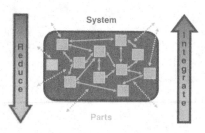

Figure 1 A system as an "entity that maintains its existence through the mutual interaction of its parts."[9] Systems research must combine (i) the identification and (ii) the detailed characterization of parts (orange boxes; as opposed to "lookalikes" (pale blue box), which need to be identified and excluded), with the exploration of their interactions (iii) with each other (orange arrows) and (iv) with the environment (pale blue dashed arrows) affecting parts either directly or indirectly, via modulation of internal interactions, to develop (v) a systemic understanding of the entity. An important, but often overlooked, aspect is that the system itself not only enables but also restricts the type and extent of functions and interactions that may occur (dark blue box). Systems research therefore requires a combination of tools and techniques for reduction and integration. Reprinted from ref. 10.

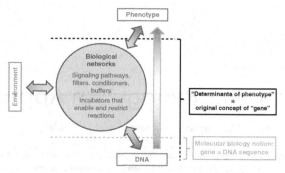

Figure 2 General relationships between genes, environment, and phenotype characters according to current physiological and biochemical understanding. The division of the conceptual entities—environment, phenotype, DNA, and biological networks—is neither strict nor mutually exclusive (and it does not specifically address the presence of any epigenomic information processing). Depending on the point of view, DNA, for example, is part of biological network activity (when you look "down" from the phenotype level), whereas biological networks are part of the environment (if you look "up" from DNA). It is hoped that this scheme will help to emphasize the complexity of interactions mediated by biological networks, which perform a whole host of key functions, such as enabling, filtering, conditioning, and buffering of the interplay between environment, phenotype, and DNA sequences. As shown on the right, the "determinants of a phenotype" (the original concept of genes) include much more than DNA sequences (the currently prevailing concept).

interactions to days, months, and years, e.g., for the development of a disease in humans.

Thus, Systems Biology explores how parts of biological entities function and interact to give rise to the behavior of the system as a whole. It is important to realize that "the entity," for example a cell, enables and restricts the range of components and interactions that are conceivable (e.g., a saline-based solute environment affects lipid bilayers in ways that are principally different from those of an alcohol-based solvent system, prescribing functional properties that need not be "encoded" other than in the basic biochemical and biophysical properties of the matter involved). However, the interrelation between genomic code and phenotypic representation deserves consideration in this context.

THE CONNECTION BETWEEN GENOMES AND PHENOTYPES
In order to understand biological systems, it is necessary to understand the relationship between the genome and the phenotype. When the concept of a gene was first introduced more than a century ago (see p.124 in Johannsen, 1909, where the term was derived from Greek γίνομαι; "to become"),[11] the relationship was thought to be simple. For each inheritable character, there was postulated to be a "gene" transmitting that character through the generations. This seemed to be the best interpretation of Mendel's experiments, implying discrete genetic elements that were responsible for phenotype characters. Later, even after this broad concept of genes was replaced by one focusing on DNA sequences as an equivalent information carrier, this idea persisted in the "one gene, one protein" hypothesis, even though proteins themselves are not the same as phenotype characters of complex organisms. Incidentally, this hypothesis is generally, but falsely, attributed to a 1941 PNAS paper by George W Beadle and Edward L Tatum.[12] In that paper, the authors show an example in fungi of "one gene, one enzyme" control of a step in vitamin B6 synthesis, but they highlight in the introduction "...it would

appear that there must exist orders of directness of gene control ranging from simple one-to-one relations to relations of great complexity." The "one gene, one protein" hypothesis was developed over the following decade, and earned Beadle and Tatum the Nobel Prize in 1958, 5 years after the structural description of DNA by James D Watson and Francis Crick.

We now know that the relationships between "genotype" and "phenotype" are even more complex. Protein-coding DNA is assumed to form only 1% of metazoan genomes. It is controlled through multiple mechanisms involving DNA that is stably transcribed (i.e., functional) yet not protein-coding. The proportion of functional, non-protein-coding DNA is understood to be an order of magnitude larger than that of protein-coding DNA; however total functional DNA represents only ~10% of overall DNA content.[13] Many questions regarding the spatio-temporal organization of the regulatory genome remain to be resolved.[14] Also, whether the other 90% of DNA really has no function at all is an interesting question, particularly if one allows the notion of functionality to extend beyond its use as an RNA template (such as for scaffolding). Complete removal of the "junk DNA" is experimentally difficult (it does not form a coherent set of large segments). Interestingly, one study that removed two very large blocks of non-coding DNA (2.3 Mb) in mice found no significant changes in phenotype.[15] However, this is equivalent to just under 0.1% of the mouse genome (which would make it feasible, at least, to assume that structural effects of such deletion would have been minor or absent). It should also be recalled that many deletions, even of protein-coding regions, do not necessarily manifest themselves as a phenotypic change, unless the system is stressed.[16] Further complexity arises from the fact that multiple

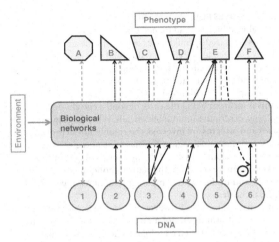

Figure 3 Simplified examples of interrelation between genes, environment, and phenotype characters according to current physiological and biochemical understanding. Interactions between particular DNA sequences and particular phenotype characters are mediated by biological networks. There is therefore no reason to assume direct causal relations between particular DNA sequences and particular phenotype characters in complex biological systems. To emphasize this, we have drawn each arrow of causation between a DNA sequence and a character as changing (from continuous to dotted) as it is transmitted through, and modified by, the biological interaction networks. Strictly speaking, not only do the causal arrows change, they interact within the network. The dotted arrows should therefore not be seen as mere continuations of the solid-line arrows. Green arrows highlight the fact that environmental influences (whether "external" or "internal" to the biological networks in this scheme) affect DNA sequences, their expression, and the shaping of phenotypic traits. Any diagram of these complex relationships is limited in what it can show. For details, see the text.

splice variants, even of the same DNA sequence, can give rise to alternative proteins. These effects are open to influences by the environment (here broadly defined as what is external to the system in question), and actual "DNA sequences" may not be as compact or uniquely defined as was initially assumed.[17]

There is therefore a (at least) three-way interaction between DNA, the environment, and the phenotype. **Figure 2** is an attempt to represent this interaction in a simplified scheme. Interactions are mediated through the networks within and between cells, including subcellular components such as proteins and organelles. These networks not only provide signaling pathways but also filter and condition the transmission of signals between environment, DNA, and phenotype. This is the basic explanation for the finding that interventions at the level of functional DNA (knockouts, insertions, and mutations) do not necessarily show a phenotypic effect. They are buffered by the networks, so that, even when changes at the level of proteins occur, there may be alternative (and normally redundant or quiescent) ways to ensure the retention of phenotype characters.

The influences of the phenotype and the environment on DNA are mediated by various mechanisms. DNA itself is chemically marked, e.g., by methylation of cytosines,[18,19] and control of expression is affected by interactions with histones (the histone

code[20]). Together, these form part of the epigenome (http://www.epigenome.org) that constitutes a cellular memory, which can be transmitted to the subsequent generation(s). Longer-term effects include many forms of modification of the DNA itself through environment-induced genome rearrangement, nonrandom mutations, and gene transfer.[21] These have played a major role in the evolution of eukaryotic cells,[22] as have "gene" and "genome" duplication.[23] Similar mechanisms also play a major role in the immune system, in which targeted hypermutation in B cells can generate changes in the genome that are as much as 10^6 times greater than the normal mutation rates in the genome as a whole. This effectively extends the already huge range of antibodies that can be produced to an infinite one. Whereas the exact mechanism by which the recognition of a foreign antigen triggers or selects such DNA changes is not known, the existence of the process is well established.[24] This behavior is entirely somatic (restricted to the cells of the immune system) and is therefore not transmitted through the germline. It was originally thought that epigenetic marking was also restricted to somatic processes. There is, however, increasing evidence to show that some epigenetic marking can be transmitted via the germline[25] or via behavioral re-marking in each generation.[26]

The existence of these mechanisms makes the definition of a gene even more problematic. The horizontal lines in **Figure 2** indicate the difference between the original concept of genes and the modern definition. The original notion of a gene as the sufficient determinant of a phenotype includes everything below the black dashed line in **Figure 2** (although those who introduced the concept, such as Johannsen,[11] would not have known that). A "gene" in this sense is now understood to be a distributed cause, all of which is inherited (i.e., inheritance includes both DNA and other cellular components; here, conceptually separated—although they are, of course, usually combined). The modern molecular-biology definition of a gene is DNA alone (below the gray broken line in **Figure 2**) and is therefore very different from the original meaning, also from a causal viewpoint. This confusion in terminology lies at the heart of many arguments over the role of genes in physiological function, with an extremely simplified variant represented by the vertical arrow on the right in **Figure 2**. Genes, defined as DNA sequences, may form necessary but not sufficient causes of phenotype characters.

Figure 3 elaborates on this by depicting the relationships between individual DNA sequences and phenotype characters. To simplify what would otherwise be an illegible tangle of connections, we show just six DNA sequences and six phenotype characters and indicate only some of the connections that could exist between these 12 elements.

DNA sequence 1 does not contribute to any of the given phenotype characters, and its modification may give rise to irrelevant data and interpretations. Similarly (but unrelatedly), phenotype A is not affected by any of the given DNA sequences, and therefore assessment of causal relationships between the six DNA sequences shown and "A" may lead to false-negative conclusions (as DNA sequences outside the given range may be relevant). These two will be the most frequently encountered "causal" relations.

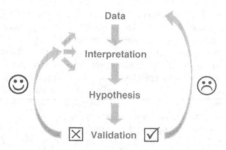

Figure 4 Schematic illustration of the scientific process and the role of validation. Emphasis is placed on the fact that, contrary to the common perception, the intellectual benefit of hypothesis rejection (left) may exceed that of confirmation (right). The value of successful hypothesis validation lies in increasing the level of confidence in a particular conceptual approach. Rejection highlights shortcomings in the approach and can be productive in guiding improved data acquisition, interpretation, and hypothesis formation.

DNA sequence 2 directly, and solely, contributes to phenotype characteristic B. This is the "ideal" scenario, which was once thought to be generally applicable. It is, in fact, either extremely rare or simply does not occur, except at the level of proteins in lower organisms such as prokaryotes.

DNA sequence 3 contributes to multiple phenotype characters (C, D, and E), whereas phenotype character E depends on DNA sequences 3–5. Such multiple connections are now known to be universal. The DNA–phenotype effects are, therefore, conditional. For example, a change in sequence 3 may not be translated into character E unless sequences 4 and 5 are knocked out as well; this again may contribute to potentially false-negative findings.

In addition, DNA–phenotype effects may affect other links, such as the one depicted by the dashed-line black arrow from phenotype characteristic E to DNA sequence 6 and, consequently, to characteristic F (this is merely one example and does not even begin to address the complexity of feedback from phenotype characteristics to underlying genetic determinants); this type of interaction may give rise to false-positive interpretations of data.

Each phenotype character also depends on cellular inheritance and on the influence of the environment via epigenetic and/or acute effects (see green arrows in Figure 3). All these influences are mediated by networks within cells and tissues. The traditional, "differential" view of genetics avoids acknowledging this mediation by focusing on a single change (usually a mutation, addition, or deletion) in a DNA sequence and the observed net change in phenotype. It then defines this as "the gene for" that characteristic (or, more precisely, the observed "difference" in characteristics). Clearly, this ignores the great majority of the components that, in combination, give rise to a phenotype character.

The logic of these conditional effects may be very complex, with various combinations forming a sufficient set of parameters that may give rise to similar or identical phenotypes. The major goal of a Systems Biology approach to genome–phenotype relations is to work out this logic. An "integral" view of genetics, which takes these complexities into account, is therefore essential to the success of Systems Biology.[10,27,28]

ROLE OF MODELS FOR SYSTEMS RESEARCH

Conceivably, if biology had turned out to be as simple as early geneticists envisaged, it could have continued to be an essentially descriptive subject. Identifying functions and their genetic causes could have been viewed as simply linking the two together, bit by bit, a function or a gene at a time. The complexity represented (albeit only partially and simplistically) in Figures 2 and 3 shows that this is far from being the case. Beyond a certain degree of complexity, descriptive intuition often fails. When large numbers of genes and proteins are involved, the combinatorial problems become seriously challenging.[29] This is one of the reasons for another major characteristic of the Systems Biology approach: it makes extensive use of mathematical modeling in order to represent and understand complex interactions of parts and biological entities.

Mathematical models, however, need to be used with care. They are aids to thought, not a replacement for it. The only serious difference between a biologist who uses mathematical modeling and one who does not is that the former explores the consequences of his ideas quantitatively, including implementation of computational experiments to assess the plausibility of those ideas. The potential benefits of doing so are obvious because quantitatively plausible predictions improve subsequent hypothesis-driven experimental research. William Harvey[30] used this approach in his convincing arguments for the circulation of blood, when he calculated how quickly the blood in the body would run out if it did not recirculate (see also ref. 7). Using mathematics for quantitative prediction, Harvey arrived at an assessment of the plausibility of a certain hypothesis (or lack thereof, as the case may be).

Modeling of the electrophysiology of the heart, in particular, has repeatedly been used to direct new experimental approaches. In this process, the "failures" (predictions that were shown wrong in subsequent experimental assessment) have been as important as the "successes,"[31] as Figure 4 illustrates. Let us assume, for a moment, that we all agree that proper scientific process is based on review of the available data and knowledge, followed by interpretation to form a falsifiable hypothesis, which is then subjected to validation.[32] Falsifiability of a theory as a virtue has been highlighted before, for example, by leading philosopher of science, Sir Karl Popper, who stated: "A theory which is not refutable by any conceivable event is non-scientific. Irrefutability is not a virtue of a theory (as people often think) but a vice."[32]

This view holds for the exploration of biological behavior. For the purpose of this argument, it does not matter whether this process is aided by formalized theoretical models (e.g., computer simulations) or is based entirely on conceptualization by an individual or group. If the validation shows agreement with the hypothesis, all it does is reconfirm what has been anticipated. Thus, arguably, no new insight is generated, although the data that emerge from the validation can be fed back into the scientific process (see Figure 4, right), and the same models (or concepts) will be applied in the future with a higher degree of confidence. Compare that to rejection of a hypothesis (Figure 4, left). Often seen as a less desirable outcome, it is when we show our best-conceived predictions to be wrong that we

learn something about shortcomings in input data, their interpretation (including any formalisms applied to aid this process), and/or the ensuing hypothesis (assuming that the approach to validation was suitable and applied correctly). This is the stage of the scientific process in which new insight is generated and the seeds for further progress are laid.[33]

Therefore, experimental information is the key to proper model development and validation, suggesting that "dry" computational modeling should not be pursued in isolation from "wet" lab or clinical studies. Incidentally, the reverse statement is prudent, too. Studies involving biological samples benefit from theoretical assessment of most likely outcomes, helping in the selection of promising approaches, supporting experimental design, and avoiding ill-conceived studies.[34] In other words, the cycle of "wet" data generation, "dry" interpretation and hypothesis formation, "wet" validation, and so on, should be seen as a continuous chain. Theoretical and practical research approaches do not thrive in isolation from each other.

The main limitations of mathematical modeling in biology arise from the very complexity that makes such modeling necessary.[35] By definition (model = simplified representation of reality), all models are partial descriptions of the original, whether they are conceptual (to think is to model!), mathematical/computational, or experimental/clinical. Of note, even an individual human would not be a perfect model system for the entire species, calling for patient-specific tools (including models) for prevention, diagnosis, and treatment.

Of course, a full representation of all aspects of a given reality in a "model" would render it a copy (or a clone). This would suffer exactly the same shortcomings with regard to the insight generated, ranging from complexity-related difficulty in identifying causal interrelations to ethico-legal boundaries on permissible interventions and data-gathering approaches. By the very definition of the term, an "all-inclusive" model would cease to be a model. The attempt to make such a model strip it of all its advantages. It would be overburdened by what stands in need of simplification or explanation and offer no advantages for targeted assessment of hypotheses.

Like tools in a toolbox, each model has its inherent limitations and its specific utility. As an illustration, let us consider models of a train. Depending on purpose (toddler's toy, collector's replica, miniature railway), emphasis may be on simplicity, mechanical sturdiness, and color; on "to-scale" representation of appearance; or on mechanical function and ride comfort. An "all-inclusive model" of a train that captures every aspect, however, would be another train (and, as in patients, there are no two truly identical ones either). The copy train would not be suitable for application to the aforementioned model purposes, whether for the toddler, for the collector's display cabinet, or for your local landscaped gardens. Therefore, models can be good or bad only with respect to a particular purpose (in fact, well-suited or ill-suited would be more appropriate categories), but modeling *per se*—the utilization of simplified representations of reality—is neither: it is simply necessary. We all do it, in one way or another.

The difficulty in the case of complex biological systems (as opposed to man-made items) is that, on the basis of our present level of understanding, models remain very partial indeed. Therefore, for some time to come, there will be a place for both negative and positive validation to drive model improvement and to calibrate confidence. A problem to be wary of, not only in the context of formalized (mathematical) modeling, is what we can call the plausibility trap—just because a model reproduces observed behavior does not mean that implicated mechanisms are major contributors or even that they are involved at all. All that such models can do is to illustrate quantitative plausibility (which, in its own right, is certainly a major achievement). Even established theoretical models, therefore, require continual validation of predictions against the above described outcome-dependent consequences.

SYSTEMS BIOLOGY APPLICATION

If Systems Biology is accepted as an approach to biomedical research and development that, from the outset, consciously combines reduction and integration across a wide range of spatio-temporal scales, then one can explore different starting points for this systematic exploration of biological function.

Bottom–up

This is the classic molecular biology approach and can also be termed the "forward approach." It starts with "bottom" elements of the organism—genes and proteins—and represents these by equations that describe their known interactions. "Bottom" here is, of course, metaphorical. Genes and proteins are everywhere, in all cells of the body. It is a conceptual convenience to place them at the bottom of any multiscale representation, that is, with structures of low spatial dimensionality. From these components and their interactions, the modeler aims to reconstruct the system, including multiple feed-forward properties. It is conceivable that this might work in the case of the simplest organisms, such as prokaryotes, which can be represented as a relatively formless set of molecules with their networks surrounded by a lipid cell membrane. In the case of eukaryotes, many of the interactions between the components are restricted by the complex cell structure, including organelles. The forward approach would necessarily include these structures, in which case it is no longer purely bottom–up because, as we have already noted, many of these structural features are inherited independently of DNA sequences. Levels higher than DNA and proteins would be necessary for successful modeling. This does not imply that a bottom–up approach is of no value. It simply means that this approach, and the vast databanks that are being developed through genomics, proteomics, and bioinformatics, need to be complemented by other approaches. This need is underlined by studies showing that the great majority of DNA knockouts do not afford any insight into normal physiological function (for an example, see ref. 16).

Top–down

This may be regarded as the classic physiology approach, somewhat akin to reverse engineering. First, study the system at a high level, then burrow down to lower levels in an attempt to arrive at an inverse solution. In this case, we start with the system and try to infer its parts and their functionality. This

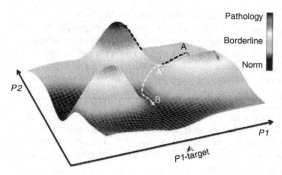

Figure 5 Schematic illustration of the landscape concept in parameter space. The value of a hypothetical biological function (color-coded, z axis) varies as a function of multiple parameters, including *P1* and *P2*. Assume a patient whose biological profile places him in position *A*, where the desired action (or a "side effect" associated with another treatment) is a reduction in the *P1* value toward a new target level. Direct reduction in *P1* (black trajectory) leads to severe negative consequences. Covariation in both *P1* and *P2* (white trajectory) allows transition toward the desired *P1* levels without detrimental changes. An isolated reduction in *P2* to the same extent (gray trajectory) would also be detrimental, showing that the combined action (passage from *A* to *B*) would not have been an intuitively obvious path to take.

approach has succeeded in some cases. The study of circulatory and respiratory physiology started off with the successful identification and characterization of a system (closed circulatory loop, pump function of the heart, gas exchange in lungs and tissues), leading eventually to identification of cells (red blood cells) and molecules (such as hemoglobin) that transport oxygen, and so on. It must be admitted, of course, that this approach has had its failures. High in the list of these failures is the classic view of genetics. Burrowing down to the level of DNA using differences in the phenotype to infer the existence of particular genes and then identifying individual properties from these DNA sequences can be seen as one of the great success stories of twentieth-century biology. Unfortunately, however, it works in only a small proportion of cases. The reasons are explained in Figure 2. There is no basis for supposing that we can always correctly infer the existence of particular DNA sequences from observations based on the phenotype because the relations between genotypes and phenotypes are massively multifactorial (Figure 3). In cross-species cloning, for example, cytoplasmic networks can even influence phenotypes (such as numbers of vertebrae), contradicting the expected genome influence.[36] In this case, the "gene" (in the classic sense of the term) is in the egg cytoplasm networks!

Middle–out

The limitations of the bottom–up and top–down approaches used in isolation have led to the adoption of the middle–out approach in a major proportion of work in Systems Biology at higher levels.[37] It can be represented as locally combining the bottom–up and top–down approaches, but that is only part of the story. Its success in the Physiome Project was possible precisely because it is pragmatic. Modeling begins at any level of the

organization at which there are sufficient reliable data to build a model. This is the starting point of the middle–out approach. It involves exploration of parameter spaces at the chosen level. The next step is to reach toward both higher and lower levels of structural complexity (the "out" part of the metaphor). A good example of this approach is the modeling of the heart, which started at the level of the cell by representing processes and components that contribute to electrical, mechanical, or metabolic functions (see refs. 38, 39). It then reached upward to tissue and organ levels by incorporating the cell models into detailed models of higher-level tissue and organ structure (see refs. 40, 41) and downward to the genome by representing the effects of known genetic changes on the proteins represented in the model (see refs. 42, 43).

Whichever approach is adopted, successful models span different levels of organization. Causes of particular phenotype characteristics are unraveled as multidimensional interactions—the networks depicted in Figure 2. This leads us to a discussion of a very important conceptual tool: the multidimensionality of the many complex interactions in biological systems can be represented by what can be termed "landscape diagrams."

The landscape concept

Appreciation of the complexity and multidimensionality of the relationships between the components of organisms is not new. The idea of representing these relationships in the form of landscapes was introduced by Wright[44] and Waddington[45,46] (for a review, see ref. 47). When Waddington introduced his landscape metaphor, he used it to depict the rearrangements of genes in the gene pool that trigger the expression of different combinations of pre-existing alleles in response to environmental stress, a process he called epigenetics (note that the modern definition of epigenetics is different—it usually refers to chemical marking of the DNA). However, the landscape concept can usefully be applied much more broadly, relating the function of the biological system (or phenotype) to properties that we may seek to vary clinically (such as by pharmacological or device-based interventions) in order to manipulate the system toward a state of stability, safety, or health. Because of its focus on interactions, the landscape approach is already being used in Systems Biology.[48]

The underlying concept is that networks of interactions in a biological system can be represented as a multidimensional space in which variations in any of the parameters can be seen to correspond to perturbations in one (or more) of the dimensions. These effects find representation as changes, either in the landscape itself, as a translocation of functional states from one point to another within a given landscape, or a combination of both. Figure 5 illustrates a conceptual example of state translocation to show how covariation of two parameters (P1, P2) may give rise to principally different effects on systems behavior (see the color scale) than one would have predicted from changing either of these parameters in isolation.

The importance of parameter interaction in complex systems has long been appreciated by engineers, and, correspondingly, mathematical theories to deal with this issue have been developed. In one such approach, parameter interactions can be explored using "response surface methodology,"[49] a subset of "design of experiments" theory.[50] This collection of statistical techniques is tailored for parameter space exploration, with the aim of identifying maximally effective target combinations with the minimal number of experiments. Initially applied to optimization of production processes in various industries, the potential of these techniques for parameter optimization in drug- and device-based diagnosis and therapy has begun to be explored.[51,52]

The landscape approach aims to proceed beyond parameter optimization, to identify trajectories for dynamic parameter variation while keeping responses within a certain range. In Figure 5, for example, a straight connection from A to B would involve transition via a response range that, depending on dynamics (e.g., dwell times along parts of the trajectory), could be detrimental. This is avoided by moving through the intermediate target A′. Trajectory identification can be conducted in multiple ways. One option is to acquire a thorough knowledge of the entire landscape. This can be done using brute-force multidimensional parameter space exploration or with the guidance of coarse (or even adaptive) grid-point characterization, followed by detailed mapping of regions of interest (e.g., areas of steep changes in biological function or regions near known sites of desirable/undesirable functional behavior). Alternatively, one can conduct neighborhood mapping from (multiple) known source or target locations and try to interrelate identified fragments.

This is not a mere conceptual pastime; it is relevant to the development of therapeutic interventions. Early forays include the mid-nineteenth-century studies of Fraser, who noted the "hyperantagonistic" effect of two drugs: the herbal poison "physostigma" (a cholinesterase inhibitor) and "atropia" (atropine, a competitive antagonist for the muscarinic acetylcholine receptor that can act as a therapeutic antidote, unless given in excess).[53] Today, multi-drug combinations are common in medical treatments, and the effects of drugs can be additive, synergistic, antagonistic, or give rise to qualitatively different side effects (for example, via changes in compound metabolism). A good practical example is the evolution of knowledge concerning the actions of ranolazine (CV Therapeutics, now Gilead, Palo Alto, CA). This compound blocks the hERG channel (human Ether-à-go-go Related Gene, underlying the rapid delayed rectifying potassium current, I_{Kr}) and thereby prolongs the action potential in cardiac muscle cells. This type of response can be associated with an increased likelihood of heart rhythm disturbances. This is not the case here, however, because ranolazine also partially blocks the persistent sodium current ($i_{Na,p}$).[54] This combined action has two beneficial effects: it suppresses the development of so-called "early after-depolarizations" (which can cause acute initiation of heart rhythm disturbances), and it reduces sodium loading of the cell (which is a risk factor in the longer-term development of arrhythmias[55,56]). The blocking of $i_{Na,p}$ in isolation can also have negative side effects, in that this channel subtype is important for the initiation and conduction of the heart's electrical activation. Therefore, similar to what is shown in Figure 5, the combination of two wrongs can actually make a right. To date, ranolazine has been given US Food and Drug Administration approval for use in chest pain of cardiac origin (*angina pectoris*); further studies evaluate whether it is also an effective antiarrhythmic drug.

Similarly, the landscape concept can be productive in the development and application of medical devices. An example comes from the study of biventricular pacing optimization. Initial multiparameter pacing studies relied largely on varying one pacing parameter at a time, neglecting possible parameter interdependence that may give rise to nonlinear or cumulative effects. The advantage of exploring multiple variables simultaneously has been demonstrated in studies of simultaneous optimization of left ventricular pacing site and interventricular[57,58] or atrioventricular[59,60] pacing delay. Here, independent variation of single parameters may cause hemodynamic deterioration, whereas covariation improves patient status. The best trajectory of parameter variation for biventricular pacing optimization, for example, has been identified using a gradient method for targeted neighborhood mapping to guide the user through optimal parameter combinations.[61]

There are also many physiological examples of similar relationships in the heart. For example, hyperkalemia on its own can be fatal, as can be an excess of adrenaline. But when the two increase together, such as in exercise, the result is "safe."[62] The covariation of parameters can also go in opposite directions. For instance, when the background sodium current $i_{Na,b}$ is progressively reduced in a sinus-node pacemaker model, the hyperpolarization-activated "funny" current, i_f, automatically increases. The net result of this is a minimal change in beating rate.[63] This kind of reciprocal variation must be a basis for the robustness that biological systems display in response to interventions such as gene knockouts, many of which appear to have no phenotypic effect. Hillenmeyer *et al.*[16] studied this phenomenon in yeast and found that 80% of knockouts had no effect on the phenotype, as measured by cell growth and division, in a normal physiological environment. But when the organisms were metabolically stressed, 97% of the same knockouts did affect growth. In this example, the phenotypic expression of any given gene was therefore conditional on what the metabolic networks were experiencing. When backup networks are called into play because a particular metabolite is in short supply, the deficiency at the level of DNA may be revealed.

In mathematical models, robustness—that is, lack of significant changes in systems behavior despite significant parameter variation (for an example, see ref. 64)—is also referred to as "parameter sloppiness."[65] Determining safe areas in a functional landscape (Figure 5) is therefore equivalent to identifying regions of sloppiness. This is done by systematically exploring the range of parameter changes to which critical behavior of the system is insensitive. Such "insensitivity analysis" can be conducted either locally or in global parameter space. Estimates of global parameter sensitivity are typically based on sampling local sensitivities over multiple regions of a landscape (for example, by using the Morris method, see ref. 66). This requires close iteration between experimental data input and theoretical modeling

and is somewhat akin to the daunting task of drawing a map of a city by taking underground train transportation and characterizing the landscapes that present themselves at each overground exit without knowing the precise spatial interrelation among the stations.

What helps is that "sloppiness" is thought to be a universal property of Systems Biology models (much as "robustness" is common among biological systems). If this is true, it will be of great importance, for both the development of mathematical models and their practical application. Knowledge of critical parameter ranges is essential for producing reliable and predictive models, while insight into "uncritical" aspects will allow parameter reduction and model simplification. In the ideal scenario, models will be as complex as necessary yet as simple as possible to address a given problem.[67]

CONCLUSIONS

Systems Biology is an approach to biomedical research that consciously combines reduction and integration of information across multiple spatial scales to identify and characterize parts and explore the ways in which their interaction with one another and with the environment results in the maintenance of the entire system. In this effort, it faces the difficult task of connecting genomes and phenotypes, which are linked in a bidirectional manner and through complex networks of interaction, including modulation by the environment of the system itself. This process would be impossible without the use of advanced computational modeling techniques to explore the landscapes that are constituted by mutually interacting and highly dynamic parameters. The challenge for Systems Biology is to use multiparameter perturbations to identify the safe areas, in which covariation of multiple processes supports the maintenance of stability. Valleys in the landscape interconnect such areas, and their topography can guide the selection of patient-specific and safe treatment options.

This approach can be of use to the pharmaceutical industry in three ways. First, we may identify multitarget drug profiles that would be beneficial for a given purpose or condition. In fact, there may well be multiple solutions to the same problem, thereby expanding the range of available options for individual patients. Second, we should be able to predict tectonic changes, which involve the landscape itself being altered in such a way that the system shifts to a principally different, perhaps unstable, state outside the normal physiological range. Characterizing the factors that determine a switch from normal, or even disturbed, cardiac rhythms with a regular pattern (e.g., bradycardias or tachycardias) to chaotic behavior (e.g., fibrillation) is a good example. Achieving this, and then relating it to known properties of drug compounds, would greatly help the pharmaceutical discovery process (see ref. 68 for a comprehensive account of why this shift toward virtual R&D strategies will be vital for the industry as a whole). Third, if we have identified one (or several) safe combination(s) of background activity and intervention profiles, we may be able to map out isolines that demarcate the safe from the unsafe directions ("map out the valleys"). Patient-specific insensitivity analysis in particular could hold the key to

identifying and eliminating the main obstacle to many otherwise efficient pharmacological treatments—drug side effects.

ACKNOWLEDGMENTS
Our work was supported by the European FP6 grants BioSim and normaCOR; by FP7 grants VPH NoE, preDiCT, and euHeart; and by the UK Biotechnology and Biological Sciences Research Council, the UK Medical Research Council, and the Wellcome Trust. P.K. is a senior fellow of the British Heart Foundation. E.J.C. acknowledges support from KAUST through a visiting fellowship held at the Oxford Centre for Collaborative Applied Mathematics. T.A.Q. is a postdoctoral research fellow of the UK Engineering and Physical Sciences Research Council.

CONFLICT OF INTEREST
The authors declared no conflict of interest.

1. Kitano, H. *Systems biology: towards systems-level understanding of biological systems.* In *Foundations of Systems Biology* (ed. Kitano, H.) (MIT Press, Cambridge, MA, 2002).
2. Bassingthwaighte, J., Hunter, P. & Noble, D. The Cardiac Physiome: perspectives for the future. *Exp. Physiol.* **94**, 597–605 (2009).
3. Fenner, J.W. *et al.* The EuroPhysiome, STEP and a roadmap for the virtual physiological human. *Philos. Transact. A. Math. Phys. Eng. Sci.* **366**, 2979–2999 (2008).
4. Hunter, P. *et al.* A vision and strategy for the VPH in 2010 and beyond. *Philos. Trans. R. Soc. A* **368**, 2595–2614 (2010).
5. Fernel, J. *Physiologia* (1542). Translated and annotated by Forrester, J.M. *Trans. Am. Philos. Soc.* **93/1**: 636 pp. (2003).
6. Noble, D. Claude Bernard, the first systems biologist, and the future of physiology. *Exp. Physiol.* **93**, 16–26 (2008).
7. Auffray, C. & Noble, D. Conceptual and experimental origins of integrative systems biology in William Harvey's masterpiece on the movement of the heart and the blood in animals. *Int. J. Mol. Sci.* **10**, 1658–1669 (2009).
8. Auffray, C. & Nottale, L. Scale relativity theory and integrative systems biology: I founding principles and scale laws. *Prog. Biophys. Mol. Biol.* **97**, 79–114 (2008).
9. von Bertalanffy, L. *General System Theory* (George Braziller, Inc, New York, 1968).
10. Kohl, P. & Noble, D. Systems biology and the virtual physiological human. *Mol. Syst. Biol.* **5**, 292 (2009).
11. Johannsen, W. *Elemente der Exakten Erblichkeitslehre* (Gustav Fischer, Jena, Germany, 1909).
12. Beadle, G.W. & Tatum, E.L. Genetic control of biochemical reactions in neurospora. *Proc. Natl. Acad. Sci. U.S.A.* **27**, 499–506 (1941).
13. Ponting, C.P. The functional repertoires of metazoan genomes. *Nat. Rev. Genet.* **9**, 689–698 (2008).
14. Alonso, M.E., Pernaute, B., Crespo, M., Gómez-Skarmeta, J.L. & Manzanares, M. Understanding the regulatory genome. *Int. J. Dev. Biol.* **53**, 1367–1378 (2009).
15. Nóbrega, M.A., Zhu, Y., Plajzer-Frick, I., Afzal, V. & Rubin, E.M. Megabase deletions of gene deserts result in viable mice. *Nature* **431**, 988–993 (2004).
16. Hillenmeyer, M.E. *et al.* The chemical genomic portrait of yeast: uncovering a phenotype for all genes. *Science* **320**, 362–365 (2008).
17. Pennisi, E. Genomics. DNA study forces rethink of what it means to be a gene. *Science* **316**, 1556–1557 (2007).
18. Bird, A. DNA methylation patterns and epigenetic memory. *Genes Dev.* **16**, 6–21 (2002).
19. Bird, A. Perceptions of epigenetics. *Nature* **447**, 396–398 (2007).
20. Turner, B.M. Cellular memory and the histone code. *Cell* **111**, 285–291 (2002).
21. Shapiro, J.A. A 21st century view of evolution: genome system architecture, repetitive DNA, and natural genetic engineering. *Gene* **345**, 91–100 (2005).
22. Embley, T.M. & Martin, W. Eukaryotic evolution, changes and challenges. *Nature* **440**, 623–630 (2006).
23. Veron, A.S., Kaufmann, K. & Bornberg-Bauer, E. Evidence of interaction network evolution by whole-genome duplications: a case study in MADS-box proteins. *Mol. Biol. Evol.* **24**, 670–678 (2007).
24. Li, Z., Woo, C.J., Iglesias-Ussel, M.D., Ronai, D. & Scharff, M.D. The generation of antibody diversity through somatic hypermutation and class switch recombination. *Genes Dev.* **18**, 1–11 (2004).
25. Anway, M.D., Memon, M.A., Uzumcu, M. & Skinner, M.K. Transgenerational effect of the endocrine disruptor vinclozolin on male spermatogenesis. *J. Androl.* **27**, 868–879 (2006).

26. Weaver, I.C.G. *Life at the interface between a dynamic environment and a fixed genome.* In *Mammalian Brain Development* (ed. Janigrom, D.) 17–40 (Humana Press, Springer, New York, 2009).
27. Noble, D. Genes and causation. *Philos. Transact. A. Math. Phys. Eng. Sci.* **366**, 3001–3015 (2008).
28. Noble, D. Biophysics and systems biology. *Philos. Transact. A. Math. Phys. Eng. Sci.* **368**, 1125–1139 (2010).
29. Feytmans, E., Noble, D. & Peitsch, M. Genome size and numbers of biological functions. *Trans. Comput. Syst. Biol.* **1**, 44–49 (2005).
30. Harvey, W. *An Anatomical Disputation Concerning the Movement of the Heart and Blood in Living Creatures.* (Blackwell, Oxford, UK, 1627).
31. Noble, D. Modelling the heart: insights, failures and progress. *Bioessays* **24**, 1155–1163 (2002).
32. Popper, K. *Conjectures and Refutations* 33–39 (Routledge and Keagan Paul, London, 1963).
33. Kohl, P., Noble, D., Winslow, R. & Hunter, P.J. Computational modelling of biological systems: tools and visions. *Philos. Trans. R. Soc. A* **358**, 579–610 (2000).
34. Fink, M., Noble, P.J. & Noble, D. Mathematical models in cardiac electrophysiology research can help the 3Rs. NC3R <http://www.nc3rs.org.uk/news.asp?id=1162> (2008).
35. Hunter, P.J., Kohl, P. & Noble, D. Integrative models of the heart: achievements and limitations. *Philos. Trans. R. Soc. A* **359**, 1049–1054 (2001).
36. Sun, Y.H., Chen, S.P., Wang, Y.P., Hu, W. & Zhu, Z.Y. Cytoplasmic impact on cross-genus cloned fish derived from transgenic common carp (Cyprinus carpio) nuclei and goldfish (Carassius auratus) enucleated eggs. *Biol. Reprod.* **72**, 510–515 (2005).
37. Brenner, S. *et al.* Understanding complex systems: top-down, bottom-up or middle-out? In *Novartis Foundation Symposium: Complexity in Biological Information Processing*, Vol. 239, 150–159 (Wiley, Chichester, UK, 2001).
38. Luo, C.H. & Rudy, Y. A dynamic model of the cardiac ventricular action potential. I. Simulations of ionic currents and concentration changes. *Circ. Res.* **74**, 1071–1096 (1994).
39. Ten Tusscher, K.H.W.J., Noble, D., Noble, P.J. & Panfilov, A.V. A model of the human ventricular myocyte. *Am. J. Physiol.* **286**, H1573–H1589 (2004).
40. Plank, G. *et al.* Generation of histo-anatomically representative models of the individual heart: tools and application. *Philos. Transact. A. Math. Phys. Eng. Sci.* **367**, 2257–2292 (2009).
41. Vetter, F.J. & McCulloch, A.D. Three-dimensional analysis of regional cardiac function: a model of rabbit ventricular anatomy. *Prog. Biophys. Mol. Biol.* **69**, 157–183 (1998).
42. Noble, D., Sarai, N., Noble, P.J., Kobayashi, T., Matsuoka, S. & Noma, A. Resistance of cardiac cells to NCX knockout: a model study. *Ann. NY Acad. Sci.* **1099**, 306–309 (2007).
43. Sung, R.J., Wu, S.N., Wu, J.S., Chang, H.D. & Luo, C.H. Electrophysiological mechanisms of ventricular arrhythmias in relation to Andersen-Tawil syndrome under conditions of reduced IK1: a simulation study. *Am. J. Physiol. Heart Circ. Physiol.* **291**, H2597–H2605 (2006).
44. Wright, S. The roles of mutation, inbreeding, crossbreeding and selection in evolution. *Proc. 6th Int. Congr. Genet.* **1**, 356–366 (1932).
45. Waddington, C.H. Canalization of development and the inheritance of acquired characteristics. *Nature* **150**, 563–565 (1942).
46. Waddington, C.H. Canalization of development and genetic assimilation of acquired characters. *Nature* **183**, 1654–1655 (1959).
47. Bard, J.B.L. Waddington's legacy to developmental and theoretical biology. *Biol. Theory* **3**, 188–197 (2008).
48. Ao, P. Global view of bionetwork dynamics: adaptive landscape. *J. Genet. Genomics* **36**, 63–73 (2009).
49. Myers, R.H. & Montgomery, D.C. *Response Surface Methodology* (Wiley, New York, 2002).
50. Montgomery, D.C. *Design and Analysis of Experiments* (Wiley, New York, 1984).
51. Carter, W.H. Jr & Wampler, G.L. Review of the application of response surface methodology in the combination therapy of cancer. *Cancer Treat. Rep.* **70**, 133–140 (1986).
52. Tirand, L. *et al.* Response surface methodology: an extensive potential to optimize *in vivo* photodynamic therapy conditions. *Int. J. Radiat. Oncol. Biol. Phys.* **75**, 244–252 (2009).
53. Fraser, T.R. The antagonism between the actions of active substances. *Br Med J* **2**, 485–487 (1871).
54. Sakmann, B.F., Spindler, A.J., Bryant, S.M., Linz, K.W. & Noble, D. Distribution of a persistent sodium current across the ventricular wall in guinea pigs. *Circ. Res.* **87**, 910–914 (2000).
55. Noble, D. Computational models of the heart and their use in assessing the actions of drugs. *J. Pharmacol. Sci.* **107**, 107–117 (2008).
56. Noble, D. & Noble, P.J. Late sodium current in the pathophysiology of cardiovascular disease: consequences of sodium-calcium overload. *Heart* **92** (suppl. 4), iv1–iv5 (2006).
57. Berberian, G., Cabreriza, S.E., Quinn, T.A., Garofalo, C.A. & Spotnitz, H.M. Left ventricular pacing site-timing optimization during biventricular pacing using a multi-electrode patch. *Ann. Thorac. Surg.* **82**, 2292–2294 (2006).
58. Quinn, T.A., Cabreriza, S.E., Richmond, M.E., Weinberg, A.D., Holmes, J.W. & Spotnitz, H.M. Simultaneous variation of ventricular pacing site and timing with biventricular pacing in acute ventricular failure improves function by interventricular assist. *Am. J. Physiol. Heart Circ. Physiol.* **297**, H2220–H2226 (2009).
59. Whinnett, Z.I. *et al.* Haemodynamic effects of changes in atrioventricular and interventricular delay in cardiac resynchronisation therapy show a consistent pattern: analysis of shape, magnitude and relative importance of atrioventricular and interventricular delay. *Heart* **92**, 1628–1634 (2006).
60. Zuber, M., Toggweiler, S., Roos, M., Kobza, R., Jamshidi, P. & Erne, P. Comparison of different approaches for optimization of atrioventricular and interventricular delay in biventricular pacing. *Europace* **10**, 367–373 (2008).
61. Quinn, T.A. Optimization of biventricular pacing for the treatment of acute ventricular dysfunction. PhD thesis, Columbia University (2008).
62. Sears, C.E., Noble, P., Noble, D. & Paterson, D.J. Vagal control of heart rate is modulated by extracellular potassium. *J. Auton. Nerv. Syst.* **77**, 164–171 (1999).
63. Noble, D., Denyer, J.C., Brown, H.F. & DiFrancesco, D. Reciprocal role of the inward currents ib,Na and if in controlling and stabilizing pacemaker frequency of rabbit sino-atrial node cells. *Proc. R. Soc. B* **250**, 199–207 (1992).
64. Dassow, G.v., Meir, E., Munro, E.M. & Odell, G.M. The segment polarity network is a robust developmental module. *Nature* **406**, 188–192 (2000).
65. Gutenkunst, R.N., Waterfall, J.J., Casey, F.P., Brown, K.S., Myers, C.R. & Sethna, J.P. Universally sloppy parameter sensitivities in systems biology models. *PLoS Comput. Biol.* **3**, 1871–1878 (2007).
66. Cooling, M., Hunter, P. & Crampin, E.J. Modeling hypertrophic IP3 transients in the cardiac myocyte. *Biophys. J.* **93**, 3421–3433 (2007).
67. Garny, A., Noble, D. & Kohl, P. Dimensionality in cardiac modelling. *Prog. Biophys. Mol. Biol.* **87**, 47–66 (2005).
68. PricewaterhouseCoopers. Pharma 2020: Virtual R&D—which path will you take? <http://www.pwc.com/gx/en/pharma-life-sciences/pharma-2020/pharma2020-virtual-rd-which-path-will-you-take.jhtml> (2008).